DEVELOPMENTS IN HEAT TRANSFER

Edited by
Marco Aurélio dos Santos Bernardes

INTECHOPEN.COM

Developments in Heat Transfer

Edited by Marco Aurélio dos Santos Bernardes

CBS, Edition **2017**

Published by InTech

Janeza Trdine 9, 51000 Rijeka, Croatia

Publishing Process Manager Ivana Lorković

Technical Editor InTech DTP team

Cover Designer InTech Design Team

Image Copyright slava_vn, 2010. Used under license from Shutterstock.com

Additional hard copies can be obtained from orders@intechopen.com

Developments in Heat Transfer, Edited by Marco Aurélio dos Santos Bernardes
p. cm.
ISBN 978-953-307-569-3

INTECH

open science | open minds

Contents

Contents

Contents

Contents

Preface

Recent 40 years have witnessed considerable advances in experimental, theoretical and calculation approaches as well as in industrial utilizations. In this way, this book aims to bring together current progresses in some special heat transfer fields contributing to the dissemination of high-quality research efforts. Therefore, issues comprising heat transfer fundamental concepts and modes (specifically conduction, convection and radiation), bioheat, entransy theory development, micro heat transfer, high temperature applications, turbulent shear flows, mass transfer, heat pipes, design optimization, medical therapies, fiber-optics, heat transfer of in surfactant solutions, landmine detection, heat exchangers, radiant floor, packed bed thermal storage systems, inverse space marching method, heat transfer in short slot ducts, freezing an drying mechanisms, variable property effects in heat transfer, heat transfer in electronics and process industries, fission-track thermochronology, combustion, heat transfer in liquid metal flows, human comfort in underground mining, heat transfer on electrical discharge machining, mixing convection are included in this book aiming. The experimental and theoretical investigations, assessment and enhancement techniques illustrated here aspire to be useful for many researchers, scientists, engineers and graduate students.

Marco Aurélio dos Santos Bernardes
CRP Henri Tudor, CRTE
Esch-sur-Alzette
Luxembourg

1

Thermal Effects in Optical Fibres

Paulo André, Ana Rocha, Fátima Domingues and Margarida Facão
Instituto de Telecomunicações and Departamento de Física, Universidade de Aveiro
Portugal

1. Introduction

Optical fibres are essential components in the modern telecommunication scenario. From the first works dealing with the optimization of optical fibres transmission characteristics to accommodate long distance data transmission, realized by Charles Kao (Nobel Prize of Physics in 2009), until the actual optical fibre communication networks, a long way was paved.

The developments introduced in the optical communication systems have been focused in 3 main objectives: increase of the propagation distance, increase of the transmission capacity (bitrate) and reduction of the deployment and operation costs. The achievement of these objectives was only possible due to several technological breakthroughs, such as the development of optical amplifiers and the introduction of wavelength multiplexing techniques. However, the consequence of those developments was the increase of the total optical power propagating along the fibres.

Moreover, in the last years, the evolution of the optical networks has been toward the objective of deploying the fibre link end directly to the subscribers home (FTTH - fibre to the home).

Thus, the conjugation of high power propagation and tight bending, resulting from the actual FTTH infrastructures, is responsible for fibre lifetime reduction, mainly caused by the local increase of the coating temperature. This effect can lead to the rupture of the fibre or to the fibre fuse effect ignition with the consequent destruction of the optical fibre along kilometres.

In this work, we analyze the thermal effects occurring in optical fibres, such as the coating heating due to high power propagation in bent fibres and the fibre fuse effect. We describe the actual state of the art of these phenomena and our contribution to the subject, which consists on both experimental and numerical simulation results.

2. Literature review

The fibre fuse effect, named due to the similarity with a burning fuse, was first observed in 1987 (Kashyap, 1987; Kashyap et al., 1988). At that time, the effect was observed on a single mode silica fibre illuminated by an optical signal with an average power density higher than 5MW/cm^2. Like a burning fuse, after the optical fibre fuse ignition, the fuse zone propagates towards the light source while a visible white light is emitted. After the fuse zone propagation, the fibre core shows a string of voids, being permanently damaged. The phenomenon was always associated with a thermal effect and although there are not yet very accurate experimental data for the actual temperature achieved in the fibre core, it is believe that the peak temperature is up above the silica vaporization point, around 3300 K. Some authors also refer that the white light emission characteristic of this effect may indicate temperatures that would allow plasma like fuse zone (10^4) (Hand et al., 1988[b;] Dianov et al., 2006; Shuto, 2010).

The first explanation for the effect related it to a thermal self-focusing mechanism (Kashyap et al., 1988). Afterwards, the fuse zone was identified as a soliton-like thermal shock-wave which would occur by strong thermal dependent absorption due to the creation of Ge-related defects in Ge doped core fibre. To sustain this hypothesis, the a Ge doped fibre was heated up to 1000°C in the absence of any propagating optical signal and the same kind of periodic damaged pattern was produced (Driscoll et al., 1991), however this result has not been reproduced by any others research groups.

At the time of these first observations the fuse effect did not represent a practical problem, since the total power injected in the network optical fibres was well below the power densities used in the experiments. However, the rise of optical communications demand and the consequent increase of the injected power have promoted the fuse effect to one of the fundamental issues which should be considered while developing and maintaining optical networks. Hence, for several years the phenomenon was referred as the origin of the optical fibres damage, but only in the presence of high powers. It was only a few years ago that the scientific community turned to this effect in order to explain it better but also to design devices able to detect and halt this catastrophic effect.

Nowadays, the most accepted explanation for the fuse effect describes it as an absorption enhanced temperature rise that propagates toward the light source by thermal conduction and driven by the optical power itself. The first numerical simulation of the fuse propagation used an explicit finite-difference method where it was assumed that the electrical conductivity and consequently the absorption of the core increase rapidly above a given temperature, *Tc*. Using this thermally induced optical absorption, Tc of 1100 °C and an optical power of 1 W, the core temperature was shown to reach 100000 °C (Shuto et al., 2003), which is well above the temperature of the fuse zone measured by (Dianov et al., 2006).

Also, the trigger to ignite this effect was studied. The trigger is a high loss local point in the fibre network, usually in damaged or dirty connectors or in tight fibre bends that, combined with high power signals, generate a heating point (Andre et al., 2010[b]; Seo et al., 2003; Martins et al., 2009; Andre et al., 2010[a]). The specific mechanism associated with the fuse effect generation in optical connectors was also studied and correlated with the absorption of the dust particles in the connector end face (Shuto et al., 2004[c]).

Another important issue is the power density threshold to initiate and maintain the fibre fuse propagation. The investigation so far indicates that the power density threshold is ~1-5 MW/cm^2, depending on the type of fibre and on the signal wavelength (Davis et al., 1997; Seo et al., 2003). Note that the first experiments using microstructured fibres have shown that the optical power density threshold value to ignite the phenomena is 10 times higher in these fibres than in traditional step index silica fibres (Dianov et al., 2004[b]).

The increase of absorption that is believed to take place during fuse propagation was related with Ge′ defects, as mentioned above, but also with Si E′ defects in the Germanium doped silica core optical fibres. These defects are induced at high temperatures, like the temperatures present in the fibre drawing process (Hanafusa et al., 1985). The E′ defects are associated with oxygen vacancies ≡Ge–Si≡ and are stable at temperatures above 870 K. The conjugation of this temperature dependent absorption mechanism with the absorption of the SiO specimen, produced by the thermal decomposition reaction of the Silica glass at high temperatures, occurring for temperatures above 3000 °C, was considered by Shuto et al to numerically simulate the fuse effect ignition and propagation. He reported estimate for the Silica absorption coefficient was 10^7 m^{-1} at 6000 K for a wavelength of 1064 nm (Shuto, 2010). *Dianov et al* has experimentally demonstrated that the radiation spectrum for the optical discharge, propagating through the silica fibre, is close to that of the blackbody with plasma temperature values of 10^4 K. The observed optical discharge velocities were up to 10 m/s on step index single mode fibre (Dianov et al., 2006) and 30 m/s for Erbium doped fibre(Davis et al., 1997).

Atkins et al propose a model for the bubble and voids tracks based on the Rayleigh instability due to the capillary effects in the molten silica that surrounds the vaporised fibre core(Atkins et al., 2003). The void formation and other dynamics of the fibre fuse propagation were exhaustively studied, leading to models for the voids and bubbles shape (Todoroki, 2005[b;] Todoroki, 2005[c;] Yakovlenko, 2006[a]), and profile models for the optical discharge (Todoroki, 2005[a]). Todoroki has also shown that is possible to have optical discharge without the formation of voids, along short distances, being this responsible by the irregular patterns on the voids trail (Todoroki, 2005[d]).

Other authors have also observed and studied the fibre fuse effect in special fibres like hole assisted fibre (Hanzawa et al., 2010), high numerical aperture fibres (Wang et al., 2008), polarization maintaining fibres(Lee et al., 2006) or in dispersion shift and non zero dispersion shift fibres (Rocha et al., 2010; Andre et al., 2010[a]).

Recently, more accurate simulation models for fuse propagation have been proposed (Yakovlenko, 2006[b]), or even alternative models based on ordinary differential equations that represent time saving in the numerical integration (Facao et al., 2011).

The concern with the effects for the network structure caused by the triggering of the fuse effect imposes the development of devices with the capacity to stop the fuse zone propagation. An early solution proposed in 1989 was the use of single mode tapers (Hand et al., 1989). The decrease of the fibre cladding led to expansion of the optical discharge plasma and to decrease of the power density, this results in the termination of the fuse propagation (Dianov et al., 2004[a]). Others proposed solutions to detect the fuse effect that are based in the analysis of the electric spectrum of the back reflected optical signal (Abedin et al., 2009), or in the fast temperature increase in the fibre outer surface (Rocha et al., 2011).

The deployment of FTTH networks imposes a new challenge, the dissemination of the optical fibre infrastructure in the access networks, where the fibre installation conditions are not always the more adequate. In these conditions, the deployed fibre is subject to tight bending, which impose an additional attenuation for the network power budget. The additional attenuation of waveguides subject to tight bending is a well know phenomenon, studied in 1976 by Marcuse(Marcuse, 1976). Marcuse associated the additional losses in bent waveguides with the optical signal radiated to the cladding region, this model was later improved by other authors (Harris et al., 1986; Valiente et al., 1989; Schermer et al., 2007).

Besides the new attenuation limits imposed by the bending, other constrain was observed. For high propagation power signals, the optical modes irradiated to the cladding, are absorbed in the primary coating, resulting in a temperature increase. This local heating point can induce the fuse effect already referred above or can reduce the fibre lifetime (Percival et al., 2000; Glaesemann et al., 2006; Sikora et al., 2007).

The first works on this subject have shown that the temperature achieved by the fibre coating was linearly correlated to the propagated optical power (Logunov et al., 2003). However, this linear model fails to describe the coating temperature for high power propagation (> 1 W), and recently an improved model that considers a nonlinear absorption coefficient for the coating was proposed (Andre et al., 2010[b]).

This topic has attracted the focus of the scientific community and many new achievements have been reported in the last years technical conferences. Namely, the correlation of temperature and fibre time failure (Davis et al., 2005), the definition of the safety bending limits (Andre et al., 2009; Rocha et al., 2009[a]). Recently, this topic was also studied in the new bend insensitive fibres (G.657), showing that the maximum power that can be injected safely in these fibres without coating risk is > 3 W (Bigot-Astruc et al., 2008).

3. Fibre fuse effect

As described in the previous section, the fibre fuse effect is a phenomenon that can occur in optical fibres in the presence of high optical powers and that may lead to the destruction of the optical fibre, along several kilometres, and also reach the optical emitter equipment, resulting in a permanent damage of the network active components.

However, the presence of high optical powers is not enough to ignite the fibre fuse but a trigger consisting of a initial heating point is also required. During the fuse effect ignition, this initial heating point causes a strong light absorption, due to the thermal induced absorption increase, which in turn leads to a catastrophic temperature increase, up to values that are high enough to vaporize the optical fibre core. This fuse zone propagates towards the light source melting and vaporizing the fibre core while a visible white light is emitted, as schematically illustrated in Fig 1. The propagation of the fuse zone only stops if the input power is reduced below the threshold value or even shut down. After the fuse zone propagation, the fibre core shows a string of voids, being permanently damaged.

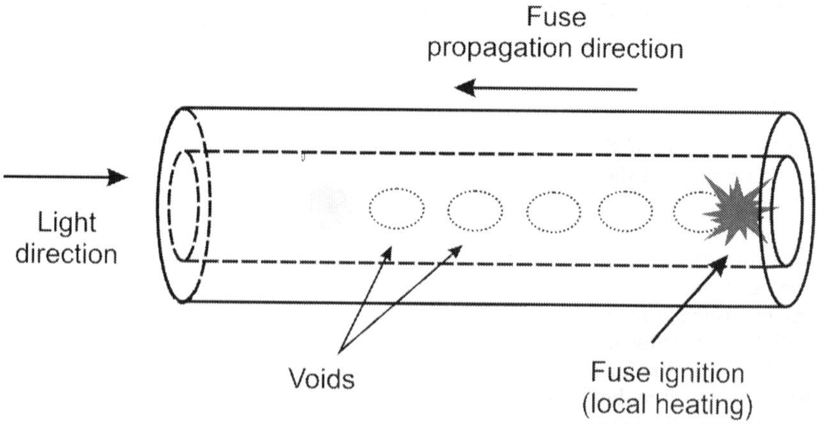

Fig. 1. Schematic representation of the fuse effect ignition and propagation in an optical fibre

3.1 Experimental characterization of the fuse effect

As referred above, the fibre fuse effect is initiated in a local heating point, whenever the optical signal have powers above a certain threshold value. Fig. 2 presents a controllable experimental setup for the fibre fuse ignition.

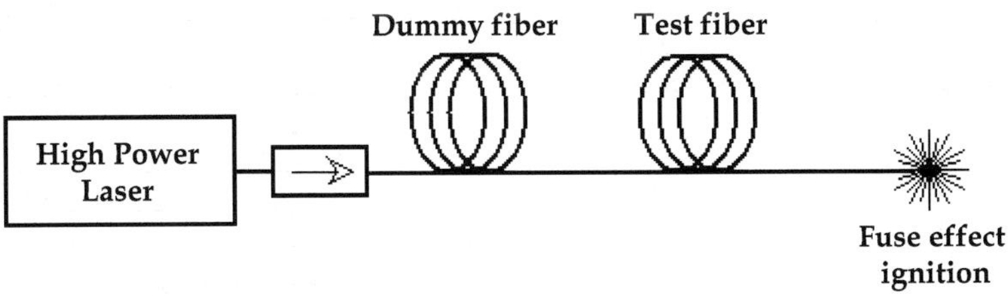

Fig. 2. Experimental setup implemented to study the fibre fuse effect (Rocha et al., 2011)

This setup consists in a short length of fibre (~3m) connected to a high power laser. The other end of the fibre is placed in contact with a metallic foil in order to produce a local heating and promote the fuse effect ignition. In order to protect the optical source, an optical isolator and a dummy fibre with 20 m are used between the test fibre and the laser.

Fig. 3 shows three frames from a movie, displaying the fuse propagation. In this movie, the white light emitted (optical discharge) from the fuse zone is clearly seen. The fuse discharge propagates at constant velocity towards the light source.

Fig. 3. Sequence of frames of the fibre fuse propagation in a SMF fibre, the time difference between pictures is 0.1s

As mentioned above, if the optical power is reduced below a threshold value, the fuse propagation stops and the optical discharge extinguishes. For standard single mode fibre (SMF-28, manufactured by Corning) and a laser signal with a wavelength of 1480 nm, the optical discharge extinguishes for an optical power of 1.39 W.

The fibre fuse propagation velocity increases with the optical power density, and could reach values high as 10 m/s (Dianov et al., 2006; Rocha et al., 2011). Fig. 4 presents the experimental velocities for the fuse effect propagation, ignited with a laser signal at 1480 nm in a SMF-28 fibre. These experimental results indicate that, for this limited range of optical power values, the fibre fuse propagation velocity is linearly dependent on the optical power launched into the fibre, however, if we consider higher optical power values, the velocity will be no longer a linear function function of the optical power (Dianov et al., 2006; Facao et al., 2011).

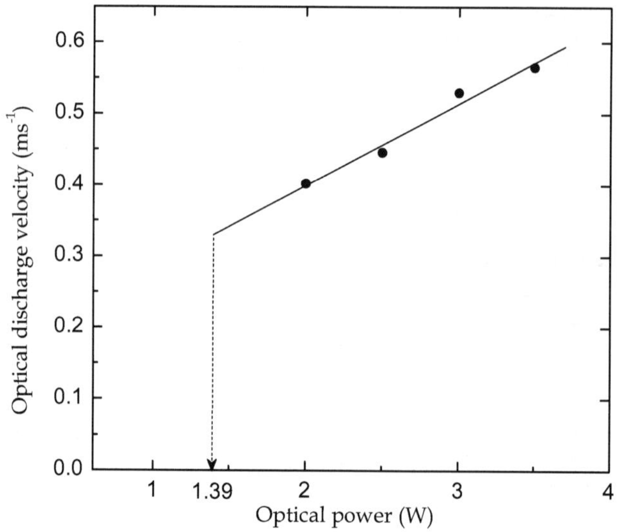

Fig. 4. Fuse discharge velocity as function of the injected optical power. The arrow represents the power threshold and the line correspond to the data linear fit (slope=0.110±0.002 m s^{-1} W^{-1}, intercept= 0.190±0.004 m s^{-1}, correlation coefficient > 0.993)

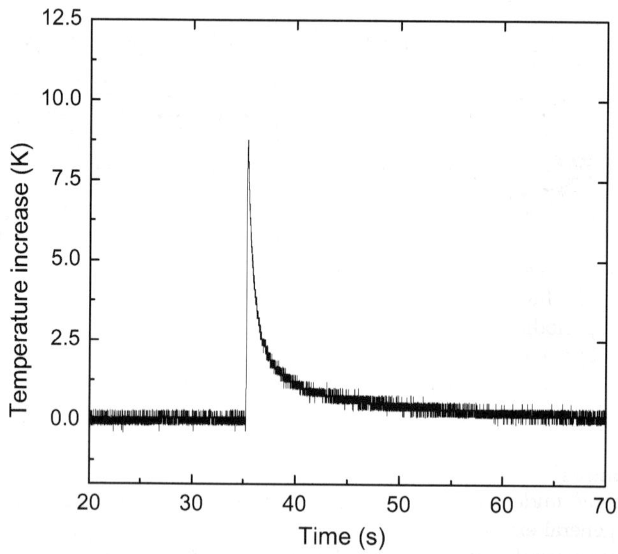

Fig. 5. Fibre surface temperature increase as a function of time, recorded by the FBG sensor during the optical discharge for an optical power of 3 W

The propagation velocity of the fuse zone was measured using a setup based on FBG (Fibre Bragg Grating) temperature sensors that measure the fibre outer interface temperature (Andre et al., 2010[a]; Rocha et al., 2011). Two fibres Bragg gratings were placed in contact with the optical fibre outer interface, in two positions separated by 2 m.

The optical discharge leads to a temperature increase in the outer fibre surface, which is monitored by the FBG sensors. The time difference between the temperature peaks, recorded at each FBG, is then used to obtain the velocity of the optical discharge. Fig. 5 displays the temperature increase in the fibre surface measured by one FBG. This graph presents an abrupt temperature increase, followed by an exponential decrease. The temperature peak corresponds to the optical discharge passing through the FBG location. Although, the fiber core is believed to achieve temperatures around 10^4 K during the optical discharge, the fiber surface temperature increases just a few degrees above the environmental temperature, as results of the heat transfer mechanisms (conduction, radiation and convection) that dissipate the thermal energy along the optical fiber and to the surrounding environment.

After the optical discharge propagation, the fibre presents a chain of voids in the core region that can be observed with an optical microscope. Fig 6 displays the optical microscopic images of the SMF fibre, obtained after the optical discharge propagation.

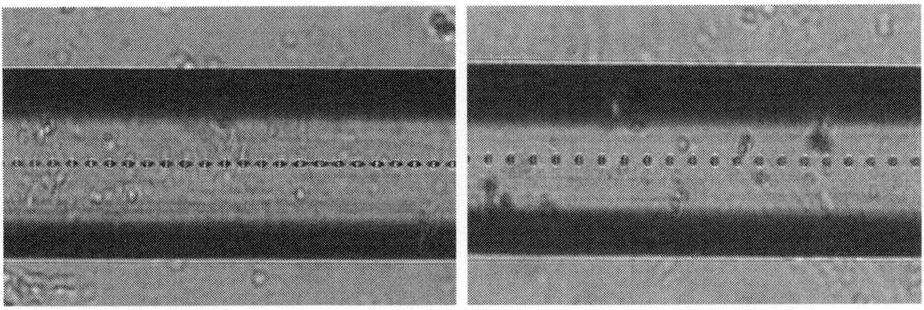

Fig. 6. Microscopic images of the optical fibre after the optical discharge propagation for optical powers of 2.5 W (right) and 4.0 W (left) (pictures obtained using an optical magnification of ×50)

These pictures were taken after the removal of the fibre coating. In these pictures, the damage caused by the fuse is clearly visible, the voids are created in the melted/vaporized core region with a periodic spatial distribution. The size and the spatial interval of the voids vary with the input power and the type of fibre (Andre et al., 2010[a]). Fig 7 shows the relation between the void period and the optical signal power. For this limited range of optical powers, the void period is linearly dependent on the optical power level.

3.2 Theoretical model

Even though many underlining phenomena that sustain the fuse effect are still not understood, the general explanation says that the initial high temperature zone, that ignite the effect, increases strongly the light absorption that, in turn, is responsible for the increase of the fibre temperature around 10^4 K (Dianov et al., 2006) well above the silica vaporization temperature. The localized high temperature zone spreads to neighbouring regions, due to heat conduction, and propagates into the laser direction, where the optical power signal is present to drive the spike up of the temperature. The process repeats causing the propagation of the optical discharge.

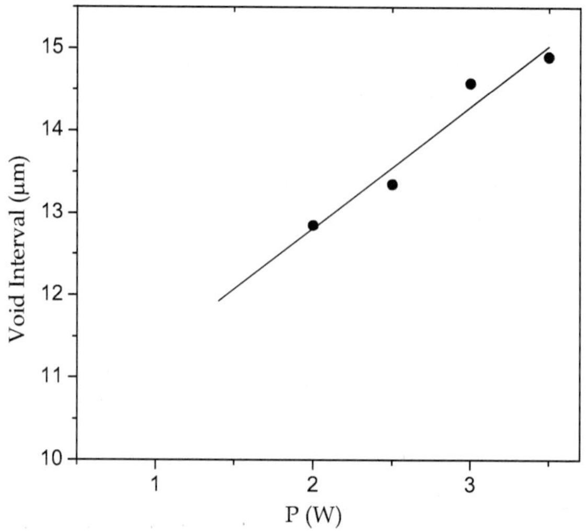

Fig. 7. Void period as function of the injected optical power at 1480 nm in a SMF-28 fibre. Points are experimental data and the line corresponds to the data linear fit (slope=1.38±0.06 μm W^{-1}, intercept=10.1±0.2 μm, correlation coefficient > 0.944)

To summarise, we assume that the main process taking place in the fibre during the fuse effect is a positive feedback heating process induced by temperature enhanced light absorption.

In the recent years, there has been substantial interest in the development of theoretical models for the fibre fuse phenomenon. Several hypotheses have been put forward to explain the strong absorption, but as we mentioned previously a lot of mechanisms are still to be understood, especially because it has been hard to measure the optical absorption at such high temperatures or even to chemically analyse the contents of the voids and their surrounding on a fuse damaged fibre. Nevertheless, most of these works propose a propagation model based on a heat conduction equation with a heat source term that corresponds to the optical signal absorption which itself is enhanced by the temperature rise. This equation is coupled to an ordinary differential equation (ODE) for the spatial evolution of the optical signal power (Shuto et al., 2003; Shuto et al., 2004a; Facao et al., 2011). Hence, let us model the fuse effect by a one-space-dimensional heat conduction equation coupled to an equation for the optical power evolution along the fibre length, namely:

$$\begin{aligned} \rho c_p \frac{\partial T}{\partial t} &= k\frac{\partial^2 T}{\partial z^2} + \frac{\alpha P}{\pi R^2} - \frac{2\sigma_s \varepsilon}{R}(T^4 - T_r^4), \\ \frac{dP}{dz} &= -\alpha P, \end{aligned} \tag{1}$$

where $T(t,z)$ is the fibre temperature, P is the optical power, t is time and z is the longitudinal coordinate along the fibre distance, ρ, C_p and k are the density, the specific heat, and the thermal conductivity of the fibre, respectively and R is the optical signal mode field radius.

The second term of the heat conduction equation is the heat source, caused by light absorption, where $\alpha(T)$ is the local absorption coefficient and the last term represents the loss by radiation which is written in terms of the Stefan-Boltzmann constant, σ_s, the surface emissivity, ε, and the environment temperature, T_r.

The increase of the optical signal absorption coefficient, α, with temperature plays the most important role in the generation of the fibre fuse. It was reported that the absorption coefficient is temperature dependent and rapidly increases above a critical temperature (1000 °C), moreover it achieves a very large value for temperatures above 2000°C (Hand et al., 1988[a]; Hand et al., 1988[b]; Shuto et al., 2004[b]). In 1988, Hand and Russell suggested that the absorption increase is closely related with Ge defects that are supposedly created in the core of the fiber once the temperature rises. In their model the absorption dependence with temperature is described by an Arrhenius law (Hand et al., 1988[a]). Shuto *et al* (Shuto et al., 2004[b]) have also proposed that the formation of Ge related defects could increase the number of free electrons and the subsequent electrical conductivity of the fibre core then enhances the absorption. In their opinion, this mechanism would explain the absorption values reported by Kashyap (Kashyap et al., 1997). This latter model also results in an Arrhenius law. But since the fuse effect also occurs in fibres without germanium, other models of absorption increase that do not rely on the presence of germanium should be put forward. One of them, also proposed by *Shuto et al* (Shuto et al., 2004[b]) relates the absorption increase with the thermochemically formation of SiO at higher temperatures, this model would allow absorption values as large as 10^4 m^{-1} for 2293 K. They also propose that, for lower temperatures, the Ge-defects should be the main absorption mechanism but for higher temperatures the presence of SiO should be more relevant. Moreover, an Arrhenius law was also proposed to model this absorption of SiO. Therefore, even if more than one process promotes the temperature rise in fibers, whether they are doped with germanium or not, the experimental data that have been collected up to now seem to manifest that all of them are thermally induced, so they are the kind of processes that are frequently modeled by an Arrhenius law.

For the reasons presented above and since here we are mainly concerned with common fibres used in telecommunications, with Ge doped cores, we use the following Arrhenius law, proposed by Hand and Russell, to model the temperature dependent optical absorption coefficient (Hand et al., 1988[a]).

$$\alpha = \alpha_0 \exp\left(-\frac{E_f}{k_b T}\right) \tag{2}$$

where E_f = 2.5 eV (Shuto et al., 2004[a]) is the formation energy of the Ge defects, k_B is Boltzmann's constant and α_0 is a constant dependent on the light wavelength and on the optical fibre type.

As already stated, the fibre fuse phenomenon is initiated only in a local heating point, thus in our model we assume an initial hot zone with the temperature above the critical value T_c.

3.2.1 Simulation of the fibre fuse effect for a single mode fibre (SMF)

The system of equations (1) and (2) was integrated using a numerical routine from the NAG toolbox, *d03pp*, that integrates nonlinear parabolic differential equations with automatic adaptive spatial remeshing.

In this calculation it was assumed an optical signal source at 1480 nm, a mode field radius of 5.025 μm, that is the mode field radius for the optical fibre used in the experimental characterization (SMF-28) at this wavelength and the same thermal proprieties of the silica glass, since the Ge concentration is very low. In this study the variation of the thermal proprieties and energy loss during changes of phase or even chemical reactions that should occur for such high temperatures were neglected, therefore, we have used the constant thermal coefficients of the silica glass listed in table 1. The equations were solved for a 3 cm long fiber with a initial hot zone of 0.2 mm centered within the fiber, and considering T_c=2900K and an environment temperature of 300 K.

ρ	2200 kg m^{-1}
k	2 Wm^{-1}K^{-1}
C_p	1430 JKg^{-1}K^{-1}

Table 1. Thermal coefficients of silica glass (Sergeev et al., 1982; Facao et al., 2011)

Since the absorption dependence with temperature is not exactly known and, particularly, the parameter of the Arrhenius law here proposed were not yet determined for this wavelength neither for such high temperatures, the parameter α_0 was adjusted with the model described in (Facao et al., 2011) which gave $\alpha_0 = 4.56 \times 10^6$ m^{-1}.

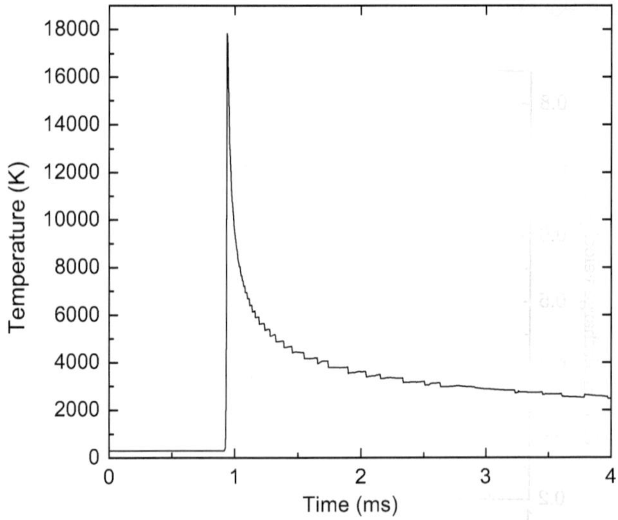

Fig. 8. Fibre core temperature at a fixed point as function of time for an optical power of 3 W

Fig. 8 displays the temperature peak that corresponds to the hot zone passing through a fixed point in the fiber. It shows an abrupt change in temperature followed by an exponential decrease. This temperature pulse is similar to the one obtained in the experimental characterization but achieves a much higher temperature. Recall that, in the experimental characterization, the measured temperature was done in the outer surface of the fiber and the temperature obtained numerically is estimated in the fiber core. Then, the temperature pulses moves in the negative-z direction (direction of the light source) with constant velocity (see Fig 9 (a)).

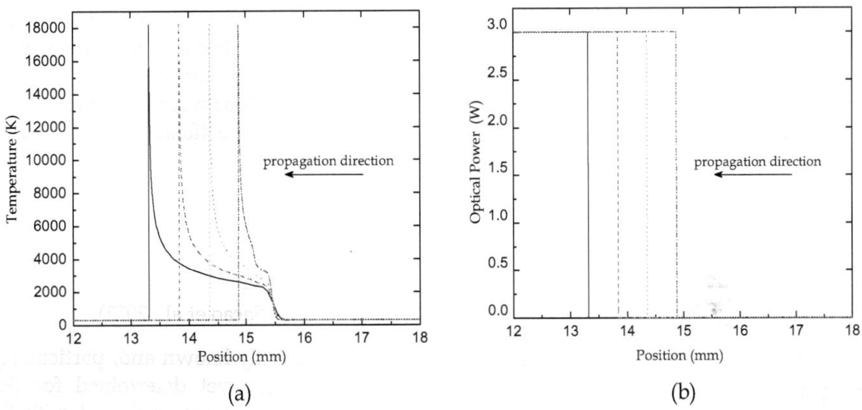

Fig. 9. Temperature (a) and power (b) distribution profiles along the propagation axis at several temporal moments spaced by 1 ms, for an optical power of 3W (the profile timing increases from the right to the left)

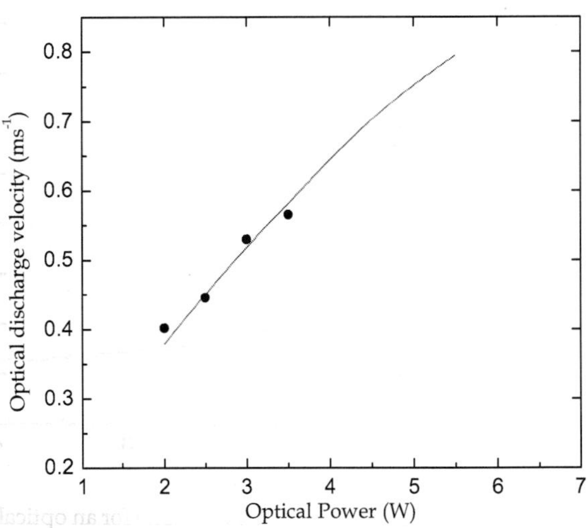

Fig. 10. Experimental (points) and numerical values (line) for the optical discharge propagation velocity as function of the injected optical power

Fig. 9 shows the travelling pulses of the optical discharge and the travelling fronts of optical power in several temporal moments. In the fuse zone all the incident optical signal is absorbed, scattered or reflected and the transmitted optical power decreases abruptly down to a null value.
The velocities obtained in this calculation are compared with the experimental values in Fig. 10, showing a good agreement with a relative error smaller than 6%.
As mentioned in section 3.2, for a limited range of relatively low optical power values, the fiber fuse propagation velocity is linearly dependent on the optical power, yet Fig 10 shows that the relation of the velocity with power density is not linear if a wider range is considered.

4. Optical fiber coating temperature increase

The increase of the temperature in an optical fibre is one of the responsible causes for its degradation. This increase in the temperature can be due to the high optical powers propagating in regions with small bending diameters. In these regions, part of the optical signal transfers from the core of the fibre to the surrounding cladding and coating (Bigot-Astruc et al., 2006; Giraldi et al., 2009; Rocha et al., 2009[a]; Rocha et al., 2009[b]). This effect can lead to the degradation of the protection layers of the fibre, with the consequent rupture, or even to the fibre fuse effect ignition.
Some recent works have studied the optical fiber resistance to high optical power (Logunov et al., 2003). The maximum attained temperature value has been studied, based on a thermal model for low power propagation signals (<1W) (Percival et al., 2000).
Here, we report our study about coating temperatures of an optical fiber, when subjected to low bending and high power optical signals. The coating temperature and the optical power loss were measured for different bend diameters. For that purpose, we have implemented an experimental setup consisting in a circular loop whose radius is controllable, as sketched and imaged in Fig. 11. The setup was designed in order to assure that the fiber suffer a bend of 360° without any change of radius. The bend diameters under study comprised values between 2.95 mm and 20.14 mm.

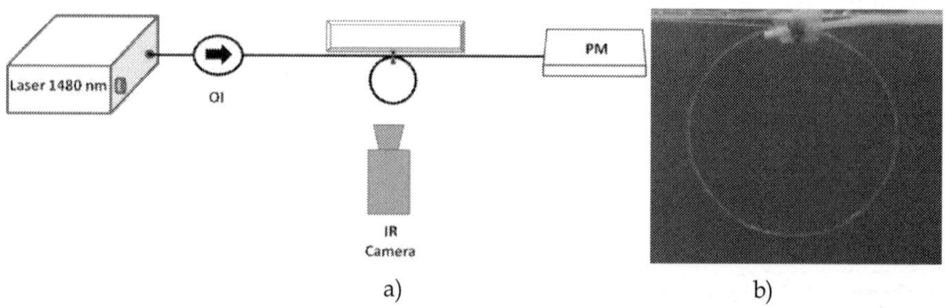

Fig. 11. a) Diagram of the experimental setup, and b) photograph of the circular loop with a diameter of 12.50mm

The temperature in the bent region was measured with an infrared thermal camera (ThermaCAM™ Flir i40). The optical signal source was a Raman laser (IPG - RLR-10-1480), emitting at the wavelength of 1480 nm, with a maximum optical power of 2W. To determine

the total attenuation in the curved section of the fiber, the optical signal output was analyzed with an optical power meter (EXFO FPM-600).
The most common type of fiber used in optical networks is the SMF28.G652.D, thus this fiber was the one studied on the work presented here. The fiber, produced by Corning, has an outer diameter of 125 μm, a core diameter of 10 μm and a primary acrylate coating with an external diameter of 250 μm. The environmental temperature at which the tests took place was 23°C.
Examples of thermal images captured during the experiments are presented in Fig. 12. These two images were taken for two different bend diameters, after one minute of exposure to high optical powers (1.75 W). It is perceptible that the temperature in the curved section is rising considerably with the decrease of the bend diameter.

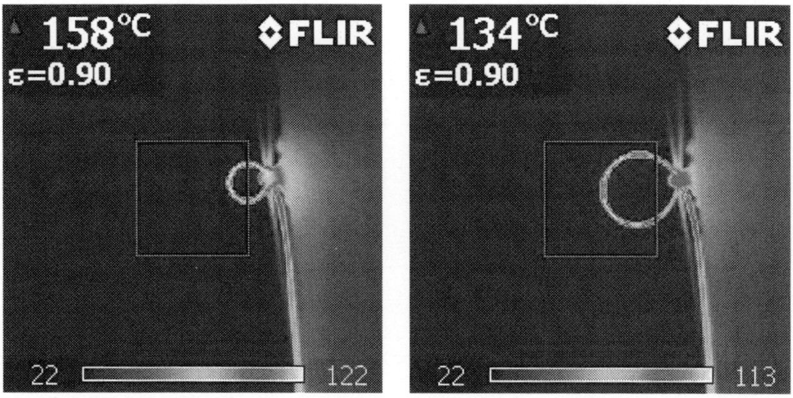

a) b)

Fig. 12. Thermal image of the fiber bending section for an injected optical power of 1.75W and bending diameters of a) 4.86 mm and b) 9.95 mm

Through the analysis of the thermal images, it is evident that, after an initial increase, the temperature stabilizes with time. In Fig. 13, the fiber temperature values obtained for several bending diameters are represented as a function of time. The input optical power was 0.5W for this test.
Considering the above results, all the subsequent measures were made in the stationary regime, i.e., 60 s after the optical power has been turned on. The critical bend diameter is 20 mm, since for higher diameters the rise of the injected optical power as no significant impact in the temperature increase (Andre et al., 2009).
Fig. 14 shows the maximum temperature increase values as function of the injected optical power for different bending diameters.
The observed behavior confirms the relation previously described, between the maximum temperature value obtained and the fiber bend diameter (Andre et al., 2010[b]; Andre et al., 2010[c]).
After the exposure to high optical powers, the physical condition of the fiber bent section was observed using an optical microscope (Olympus SZH-ILLD). Fig. 15 displays two obtained images, showing a fiber that has been submitted to an optical power of 1.5W, and a bend diameter of 2.9 mm during 60 s.

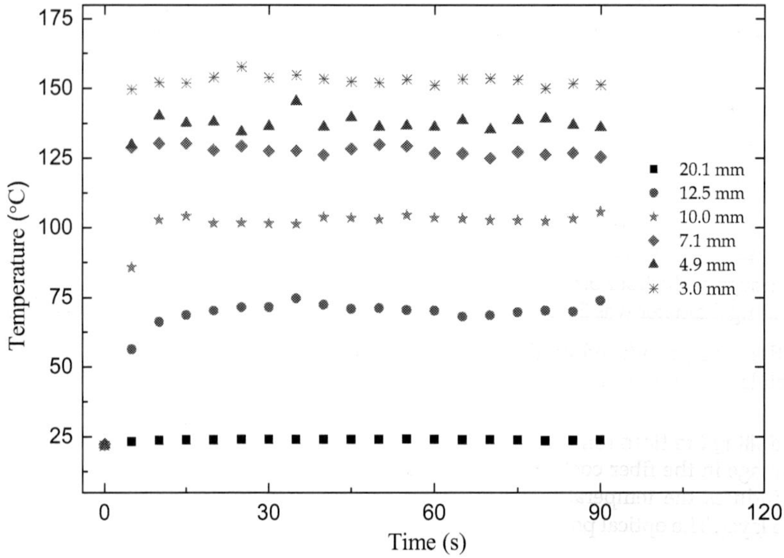

Fig. 13. Temperature values in the bent region along time

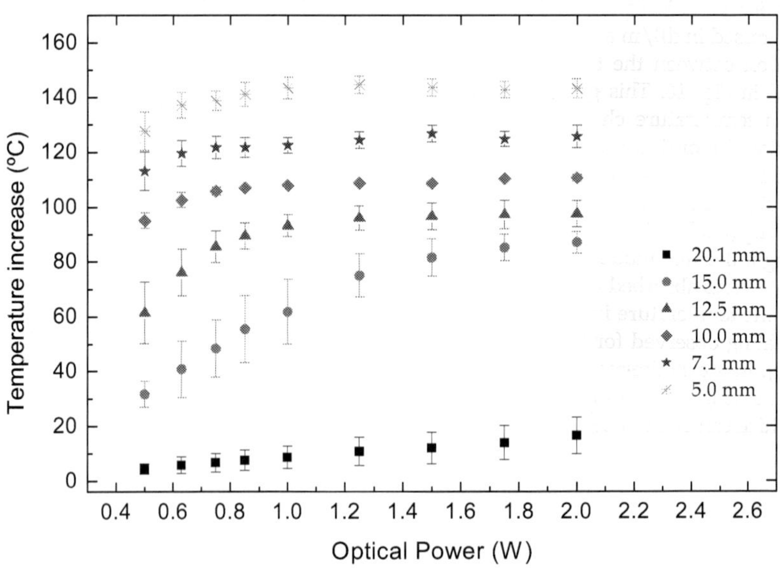

Fig. 14. Maximum temperature increase achieved as function of the optical power injected for several values of bending diameters

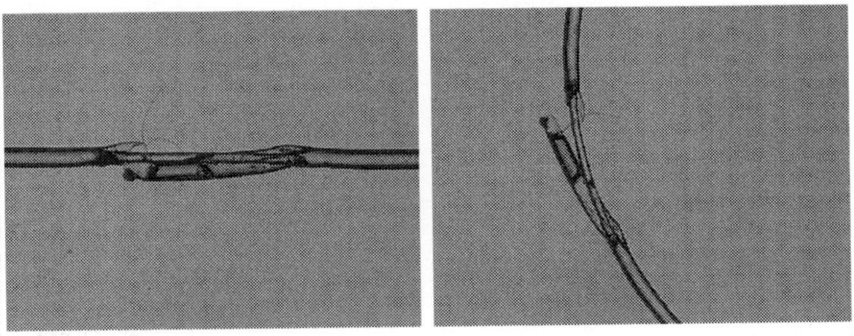

Fig. 15. Image of the fiber bend section, after being submitted to an optical power of 1.5W. The bending diameter was 2.9mm and the exposure time was 60 s (magnification of X5)

There, the damage induced by the optical power on the fiber coating in the bent section is well visible. There is degradation of the acrylate and its physical detachment from the silica.

4.1 Modelling the fibre temperature

The damage in the fiber coating is inflicted by the optical power loss in the bent section , which induces the temperature increase of the coating, resulting in an oxidation of the acrylate layer. The optical power irradiated from the optical fiber core is given by:

$$\Delta P = P_{in}\left(1-10\left(-\frac{\alpha_{fiber}L}{10}\right)\right) \tag{3}$$

where P_{in} is the optical power injected in the fiber, α_{fiber} is the power loss associated to the bend, expressed in dB/m and L is the length of bent fiber section.

The relation between the temperature increase and the power loss in the bent region is presented in Fig. 16. This graph shows that, independently of the bending diameter, the maximum temperature changes nonlinearly with the optical power loss, showing some stabilization for optical power losses around 0.3 W (Andre et al., 2010[b]). The differences observed in the maximum temperature for the several bending diameter can be explained by the fact that some of the energy that is lost from the core guided mode can be guided in the cladding (Andre et al., 2010[b]).

Assuming a thermal model that considers that the optical signal energy transferred to non guided modes is absorbed by the acrylate layer and then converted into heat, it is possible to estimate the temperature in the bent region. Nevertheless, due to the absorption saturation of the coating observed for higher optical powers not all the energy is absorbed (Andre et al., 2010[b]). In the stationary regime, the heat source (P_{in}) given by the radiation absorbed in the acrylate is converted in stored heat in the fiber (Q_{stored}) and in heat loss to the exterior (Q_{out}), which can be modeled considering a heat transfer coefficient (h) (Andre et al., 2010[b]).

$$Q_{stored} = \rho c_p A l \frac{dT}{dt} \tag{4}$$

$$Q_{out} = \pi d l h (T - T_0) \tag{5}$$

where d and l are the diameter and length of the fiber bent section, respectively. T_0 is the environment temperature, A is the optical fiber area and ρ and c_p are the density and specific heat of the fiber (considering an average value for the silica and acrylate), respectively (Andre et al., 2010[b])

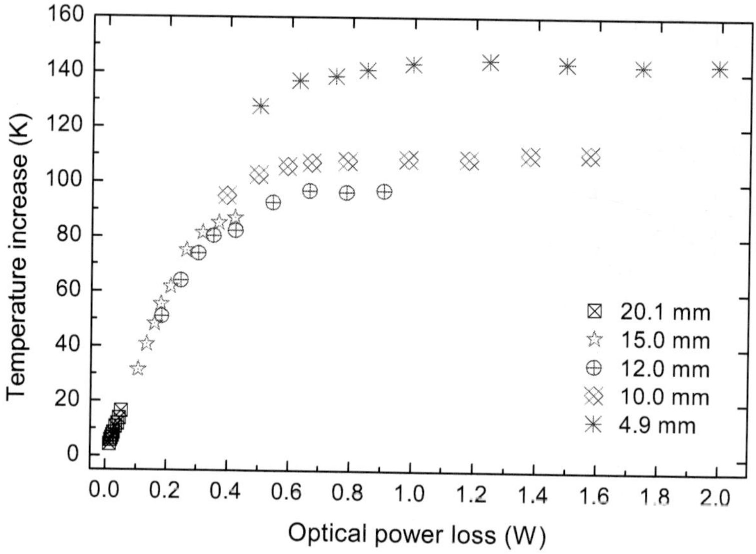

Fig. 16. Temperature as function of the optical power loss for different bending diameters for a propagation signal at 1480 nm

As stated before, in the presence of high optical powers, the absorption in the acrylate layer exhibits a non-linear behavior and thus the power absorbed by the acrylate saturates in a maximum value given by:

$$P_{in} = P_{\max}\left(1 - \exp(-\beta \Delta P)\right) \tag{6}$$

So, in equilibrium, the maximum temperature rise can be expressed as (Andre et al., 2010[b]):

$$\Delta T = \frac{P_{\max}\left(1 - \exp(-\beta \Delta P)\right)}{\pi d l h} \tag{7}$$

Where ΔT is the maximum temperature increase relatively to the environment equilibrium temperature, P_{max} is the maximum optical power absorbed by the coating and β is the activation constant (Andre et al., 2010[b]). The experimental results referred above for maximum temperature increase versus injected optical power for different bending diameters were fitted to the theoretical curve given by equation (7). The results are presented in Fig. 17. The parameters for this theoretical fit were P_{max}/h and β and the reduced chi-square value obtained was of 0.4542.

These results show that is possible to correlate the bend diameter, the injected optical power and the temperature increase in the optical fiber coating. This model can be useful for the design of future systems, being an approach to limit the bending diameter to values that depend on the injected power, in order to maintain the operational conditions below a safety limit.

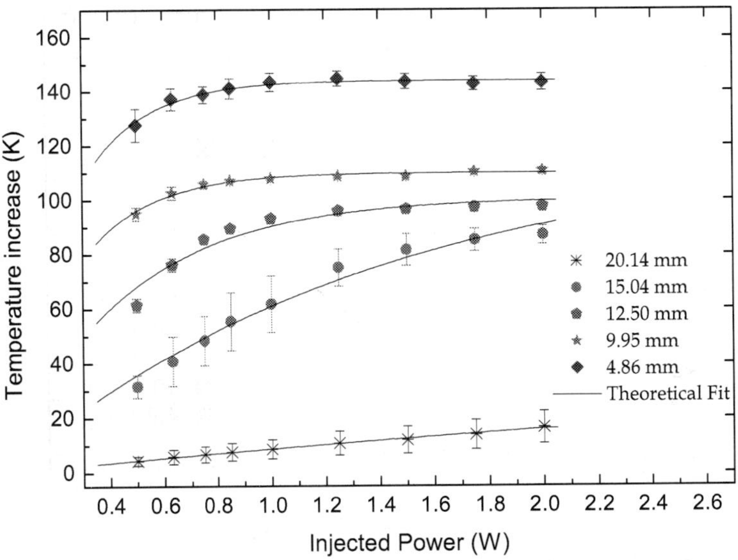

Fig. 17. Optical fiber maximum temperature increase as function of injected optical power for different values of bending diameter. Dots are experimental results and the lines are the fitting results, with $P_{max}/h=0.00245\pm0.00010$ K·m^2 and $\beta=5.03\pm0.68$ W^{-1}

5. Conclusions

In this work, we have analyzed the thermal effects occurring in an optical fiber. We have shown that the maximum power required to extinguish the optical discharge propagation is 1.39 W. It was also shown that the critical bending diameter for SMF fibers with high power signal is 20 mm.
These results describe the main thermal affects occurring in an optical fibre used in telecommunications.

6. References

Abedin, K. S. and T. Morioka (2009). Remote Detection of Fiber Fuse Propagating in Optical Fibers, *Proceedings of Conference on Optical Fiber Communication, collocated National Fiber Optic Engineers Conference* Vol. 1-5, pp. OThD5, San Diego, CA, USA.

Andre, P. S., M. Facao, A. M. Rocha, P. Antunes and A. Martins (2010[a]). Evaluation of the fuse effect propagation in networks infrastructures with different types of fibers, *Proceedings of Conference on Optical Fiber Communication, collocated National Fiber Optic Engineers Conference* pp. JWA10, San Diego, CA, USA.

Andre, P. S., A. M. Rocha, F. Domingues and A. Martins (2010[b]). Improved thermal model for optical fibre coating owing to small bending diameter and high power signals, *Electronics Letters*, Vol. 46, No. 10, pp. 695-696, ISSN: 0013-5194.

Andre, P. S., A. M. Rocha, F. Domingues and A. Martins (2010[c]). Thermal model for optical fiber coating under tight bending diameters, *Proceedings of 15th OptoeElectronics and Communications Conference* pp. 318-319, Sapporo, Japan.

Andre, P. S., A. M. Rocha, B. Neto, A. Martins, M. Facao, J. L. Pinto, A. L. J. Teixeira, R. N. Nogueira, M. J. Lima, G. Incerti, D. Forin and G. T. Beleffi (2009). Optical Fiber Bending Limits for Optical Fiber Infraestructures, *Proceedings of IEEE Africon*, pp. 882-884, Nairobi, Kenya.

Atkins, R. M., P. G. Simpkins and A. D. Yablon (2003). Track of a fiber fuse: a Rayleigh instability in optical waveguides, *Optics Letters*, Vol. 28, No. 12, pp. 974-976, ISSN: 0146-9592.

Bigot-Astruc, M., L. A. de Montmorillon and P. Sillard (2008). High-Power Resistance of Bend-Optimized Single-Mode Fibers, *Proceedings of Conference on Optical Fiber communication / National Fiber Optic Engineers Conference*, pp. JWA2, San Diego, CA, USA.

Bigot-Astruc, M., P. Sillard, S. Gauchard, P. Le Roux and E. Brandon (2006). Analysis of coating temperature increase in fibers under high power and tight bending, *Proceedings of Optical Fiber Communication Conference/National Fiber Optic Engineers Conference*, pp. OFK4, Anaheim, CA, USA.

Davis, D. D., S. C. Mettler and D. J. DiGiovanni (1997). A comparative evaluation of fiber fuse models, *Proceedings of SPIE*, Vol. 2966, pp. 592-606.

Davis, I. M., G. S. Glaesemann, S. Ten and M. J. Winningham (2005). Optical fibres resilient to failure in bending under high power, *Proceedings of 31st European Conference on Optical Communication*, Vol. 3, pp. We3.4.5, ISBN 0537-9989.

Dianov, E. M., I. A. Bufetov and A. A. Frolov (2004[a]). Destruction of silica fiber cladding by the fuse effect, *Optics Letters*, Vol. 29, No. 16, pp. 1852-1854, ISSN: 0146-9592.

Dianov, E. M., I. A. Bufetov, A. A. Frolov, Y. K. Chamorovsky, G. A. Ivanov and I. L. Vorobjev (2004[b]). Fiber fuse effect in microstructured fibers, *Ieee Photonics Technology Letters*, Vol. 16, No. 1, pp. 180-181, ISSN: 1041-1135.

Dianov, E. M., V. E. Fortov, I. A. Bufetov, V. P. Efremov, A. E. Rakitin, M. A. Melkumov, M. I. Kulish and A. A. Frolov (2006). High-speed photography, spectra, and temperature of optical discharge in silica-based fibers, *Ieee Photonics Technology Letters*, Vol. 18, No. 5-8, pp. 752-754, ISSN: 1041-1135.

Driscoll, T. J., J. M. Calo and N. M. Lawandy (1991). Explaining the Optical Fuse, *Optics Letters*, Vol. 16, No. 13, pp. 1046-1048, ISSN: 0146-9592.

Facao, M., A. M. Rocha and P. S. Andre (2011). Traveling Solutions of the Fuse Effect in Optical Fibers, *Journal of Lightwave Technology*, Vol. 29, No. 1, pp. 109-114, ISSN: 0733-8724.

Giraldi, M. T. M. R., A. M. Rocha, B. Neto, C. Correia, M. E. V. Segatto, M. J. Pontes, A. P. L. Barbero, J. C. W. Costa, M. A. G. Martinez, O. Frazao, J. M. Baptista, H. M. Salgado, M. B. Marques, A. L. J. Teixeira and P. S. Andre (2009). Brillouin Effects in Distributed Raman Amplifiers under Saturated Conditions, *Proceedings of Ieee Mtt-S International Microwave and Optoelectronics Conference* pp. 831-835

Glaesemann, G. S., M. J. Winningham, D. A. Clark, J. Coon, S. E. DeMartino, S. L. Logunov and C. K. Chien (2006). Mechanical failure of bent optical fiber subjected to high power, *Journal of the American Ceramic Society*, Vol. 89, No. 1, pp. 50-56, ISSN: 0002-7820.

Hanafusa, H., Y. Hibino and F. Yamamoto (1985). Formation Mechanism of Drawing-Induced E' Centers in Silica Optical Fibers, *Journal of Applied Physics*, Vol. 58, No. 3, pp. 1356-1361, ISSN: 0021-8979.

Hand, D. P. and T. A. Birks (1989). Single-Mode Tapers as Fiber Fuse Damage Circuit-Breakers, *Electronics Letters*, Vol. 25, No. 1, pp. 33-34, ISSN: 0013-5194.

Hand, D. P. and P. S. Russell (1988[a]). Solitary thermal shock waves and optical damage in optical fibers: the fiber fuse, *Optics Letters*, Vol. 13, No. 9, pp. 767-769, ISSN: 0146-9592

Hand, D. P. and P. S. J. Russell (1988[b]). Soliton-like thermal shock-waves in optical fibres: origin of periodic damage tracks, *Proceedings of Fourteenth European Conference on Optical Communication*, pp. 111-114 vol.111.

Hanzawa, N., K. Kurokawa, K. Tsujikawa, K. Takenaga, S. Tanigawa, S. Matsuo and S. Tomita (2010). Observation of a propagation mode of a fiber fuse with a long-period damage track in hole-assisted fiber, *Optics Letters*, Vol. 35, No. 12, pp. 2004-2006, ISSN: 0146-9592.

Harris, A. J. and P. F. Castle (1986). Bend Loss Measurements on High Numerical Aperture Single-Mode Fibers as a Function of Wavelength and Bend Radius, *Journal of Lightwave Technology*, Vol. 4, No. 1, pp. 34-40, ISSN: 0733-8724.

Kashyap, R. (1987). Self-propelled self-focusing damage in optical fibers, *Proceedings of X International Conference on Lasers*, Lake Tahoe.

Kashyap, R. and K. J. Blow (1988). Observation of Catastrophic Self-Propelled Self-Focusing in Optical Fibers, *Electronics Letters*, Vol. 24, No. 1, pp. 47-49, ISSN: 0013-5194.

Kashyap, R., A. H. Sayles and G. F. Cornwell (1997). Heat-flow modeling and visualization of catastrophic self-propagating damage in single-mode optical fibers at low powers, *Proceedings of SPIE*, Vol. 2966, pp. 586-591.

Lee, M. M., J. M. Roth, T. G. Ulmer and C. V. Cryan (2006). The fiber fuse phenomenon in polarization-maintaining fibers at 1.55 μm, *Proceedings of Lasers and Electro-Optics and Quantum Electronics and Laser Science Conference*, pp. JWB66, Long Beach, CA, USA.

Logunov, S. L. and M. E. DeRosa (2003). Effect of coating heating by high power in optical fibres at small bend diameters, *Electronics Letters*, Vol. 39, No. 12, pp. 897-898, ISSN: 0013-5194.

Marcuse, D. (1976). Curvature Loss Formula for Optical Fibers, *Journal of the Optical Society of America*, Vol. 66, No. 3, pp. 216-220, ISSN: 0030-3941.

Martins, A., A. M. Rocha and P. S. Andre (2009). High power effects on fiber optic connectors, *Proceedings of International Microwave and Optoelectronics Conference (IMOC)*, ISBN 1679-4389, Belem, Brazil.

Percival, R. M., E. S. R. Sikora and R. Wyatt (2000). Catastrophic damage and accelerated ageing in bent fibres caused by high optical powers, *Electronics Letters*, Vol. 36, No. 5, pp. 414-416, ISSN: 0013-5194.

Rocha, A. M., P. Antunes, F. Domingues, M. Facao and P. S. Andre (2011) Detection of Fiber Fuse Effect Using FBG Sensors. *IEEE Sensors Journal*, DOI: 10.1109/JSEN.2010.2094183

Rocha, A. M., M. Facão and P. S. André (2010). Study of fiber fuse effect on different types of single mode optical fibers, *Proceedings of Network and Optical Communications*, pp. 75-79, Faro, Portugal.

Rocha, A. M., A. Martins, M. Facao and P. S. Andre (2009[a]). Effect of bending in SMF fibers under high power, *Proceedings of 11th International Conference on Transparent Optical Networks*, pp. Th.B2.1, Azores, Portugal.

Rocha, A. M., B. Neto, A. Martins, G. Incerti, D. Forin, G. Bellefi, M. Facao, J. L. Pinto, A. L. J. Teixeira, R. N. Nogueira, M. J. Lima and P. S. Andre (2009[b]). The effect of high power propagation in bended fibers, *Proceedings of 10th International Conference onTelecommunications.*, pp. 303-304.

Schermer, R. T. and J. H. Cole (2007). Improved bend loss formula verified for optical fiber by simulation and experiment, *Ieee Journal of Quantum Electronics*, Vol. 43, No. 9-10, pp. 899-909, ISSN: 0018-9197.

Seo, K., N. Nishimura, M. Shiino, R. Yuguchi and H. Sasaki (2003). Evaluation of High-power Endurance in Optical Fiber Links, *Furukawa Review*, Vol. 24, pp. 17-22,

Sergeev, O. A., A. G. Shashkov and A. S. Umanskii (1982). Thermophysical properties of quartz glass, *Journal of Engineering Physics and Thermophysics*, Vol. 43, No. 6, pp. 1375-1383, ISSN: 1062-0125.

Shuto, Y. (2010). Evaluation of High-Temperature Absorption Coefficients of Ionized Gas Plasmas in Optical Fibers, *Ieee Photonics Technology Letters*, Vol. 22, No. 3, pp. 134-136, ISSN: 1041-1135.

Shuto, Y., S. Yanagi, S. Asakawa, M. Kobayashi and R. Nagase (2004[a]). Fiber fuse phenomenon in step-index single-mode optical fibers, *Ieee Journal of Quantum Electronics*, Vol. 40, No. 8, pp. 1113-1121, ISSN: 0018-9197.

Shuto, Y., S. Yanagi, S. Asakawa, M. Kobayashi and R. Nagase (2004[b]). Evaluation of high-temperature absorption coefficients of optical fibers, *Ieee Photonics Technology Letters*, Vol. 16, No. 4, pp. 1008-1010, ISSN: 1041-1135.

Shuto, Y., S. Yanagi, S. Asakawa, M. Kobayashi and R. Nagase (2004[c]). Fiber fuse generation in single-mode fiber-optic connectors, *Ieee Photonics Technology Letters*, Vol. 16, No. 1, pp. 174-176, ISSN: 1041-1135.

Shuto, Y., S. Yanagi, S. Asakawa and R. Nagase (2003). Generation mechanism on fiber fuse phenomenon in single-mode optical fibers, *Electronics and Communications in Japan Part Ii-Electronics*, Vol. 86, No. 11, pp. 11-20, ISSN: 8756-663X.

Sikora, E. S. R., D. J. McCartney and J. V. Wright (2007). Impact of coating ageing on susceptibility to high-power damage at fibre bends, *Electronics Letters*, Vol. 43, No. 4, pp. 208-210, ISSN: 0013-5194.

Todoroki, S. (2005[a]). In-situ observation of fiber-fuse propagation, *Japanese Journal of Applied Physics Part 1-Regular Papers Short Notes & Review Papers*, Vol. 44, No. 6A, pp. 4022-4024, ISSN: 0021-4922.

Todoroki, S. (2005[b]). Origin of periodic void formation during fiber fuse, *Optics Express*, Vol. 13, No. 17, pp. 6381-6389, ISSN: 1094-4087.

Todoroki, S. (2005[c]). Animation of fiber fuse damage, demonstrating periodic void formation, *Optics Letters*, Vol. 30, No. 19, pp. 2551-2553, ISSN: 0146-9592.

Todoroki, S. (2005[d]). Transient propagation mode of fiber fuse leaving no voids, *Optics Express*, Vol. 13, No. 23, pp. 9248-9256, ISSN: 1094-4087.

Valiente, I. and C. Vassallo (1989). New Formalism for Bending Losses in Coated Single-Mode Optical Fibers, *Electronics Letters*, Vol. 25, No. 22, pp. 1544-1545, ISSN: 0013-5194.

Wang, J., S. Gray, D. Walton and L. Zantano (2008). Fiber Fuse in High power optical fiber, *Proceedings of SPIE*, Vol. 7134, pp. 2E-1-9.

Yakovlenko, S. I. (2006[a]). Mechanism for the void formation in the bright spot of a fiber fuse, *Laser Physics*, Vol. 16, No. 3, pp. 474-476, ISSN: 1054-660X.

Yakovlenko, S. I. (2006[b]). Physical processes upon the optical discharge propagation in optical fiber, *Laser Physics*, Vol. 16, No. 9, pp. 1273-1290, ISSN: 1054-660X.

2

Heat Transfer for NDE: Landmine Detection

Fernando Pardo, Paula López and Diego Cabello
Universidad de Santiago de Compostela
Spain

1. Introduction

Although land mine problems existed in many regions, Bosnia (1995) and Afghanistan (2001) gave the land mine issue a particular sense of urgency. Intended for warfare, these mines remain buried after the end of the conflict. These mines are triggered by civilians causing around 15,000-20,000 victims per year in 90 countries, ICBL (2006). The U.S. State Department estimates that there are around 40-50 million of buried mines that need to be cleared. According to Horowitz (1996) 100,000 mines are found and destroyed per year; thus 450 years will be necessary to clean all mines. However, each year, 1.9 million of new mines are buried. In addition, the presence of mines also causes economic decline being one of the major limitations to agricultural work on these regions, Cameron & Lawson (1998). Thus, it is necessary to develop new techniques which allow to detect mines quickly and with high precision. The *Ottawa treaty*, Ottawa (1997), banning the production and use of AP mines was signed by 158 countries in 2007 however the most important AP manufacturers, China, Russia, India and EE.UU, have not yet signed it.
Nowadays more than 350 types of mines exist, Vines & Thompson (1999); but they can be broadly divided into two main categories:

- Antipersonnel (AP) mines.
- Antitank (AT) mines.

AT mines are relative big and heavy (2-5 Kg) and are usually laid on the ground forming regular patterns and shallowly buried. AT mines have enough explosive to destroy a tank or a truck, as well as to kill people in or around the vehicle; they also require more pressure to be detonated than AP mines. On the contrary AP mines contain less explosive and are lighter than AT mines. AP mines can be buried anywhere, they may lie on the surface or be shallowly buried. Sometimes they are placed in a regular pattern to protect AT mines, however in most cases they are placed randomly. Moreover, as AP mines are light and small, wind or rain can easily move them making their location, even with the original pattern, more difficult. AP mines are designed to damage foot soldiers avoiding their penetration into an specific area. These mines can kill or disable their victims and are activated by pressure, tripwire or remote detonation. These characteristics make AT mine detection and clearance easier than AP mine detection.
Detection and clearance of buried mines is a big problem with lots of humanitarian, environmental and economic implications. Current techniques for non-destructive evaluation

of soils fail to address the detection of small plastic landmines, which are the most difficult to detect. There is no universal technique capable of detecting buried landmines in all situations. The most widespread techniques in mine detection are metal detector and magnetometers. Magnetometers are used to detect ferromagnetic objects and they measure the disturbance of the earth's natural electromagnetic field. Many modern mines have almost no metal parts except for the small striker pin. Although metal detectors can be tuned to be sensitive enough to detect these small items (current detectors can track a tenth of a gram of metal at a depth of 10 cm), such sensitivity detects more metal debris and increases considerably the rate of false alarms. Increasing the sensitivity of metal detectors, therefore, does not solve the problem of non-metal mines satisfactorily. Taking in mind this limitation new techniques are appearing to detect the plastic landmines. Several techniques have been proposed for mine detection, such as acoustic techniques, Sabatier & Xiang (2001); X-rays techniques, Lockwood et al. (1997); biosensors, Larsson & Abrahamsson (1993); ion mobility spectrometers, Jankowski et al. (1992); nuclear quadrupole resonance, Englelbeen (1998); neutron analysis, Bach et al. (1996); ground penetrating radar (GPR) and infrared thermography (IRT), López et al. (2009); Thanh et al. (2009; 2007; 2008). Each technique has its advantages and disadvantages, for a more detailed description of theses techniques see Furuta & Ishikawa (2009); Gros & Bruschini (1998); López (2003); Robledoa et al. (2009); Siegel (2002).

Infrared thermography is an attractive technique for some mine detection tasks because it can be used from a considerable standoff distance, it provides information on several mine properties, and it can rapidly survey large areas. Its ability to detect mines has been recognized since the 1950s, Maksymonko & Le (1999). IRT sensors respond to electromagnetic radiation in a sensor-specific wavelength range. The source of the received signal may be either natural (i.e., thermal emission from the target or scattering of sunlight) or artificial (e.g., an infrared illuminator), which leads to both passive and active sensor concepts. Each material shows a characteristic thermal response to a given stimulus, also known as the *thermal signature*. Thus, the cooling or heating process affects buried objects and the surrounding soil in a different way. This difference is due to the fact that the mines are a better insulators than the soil. Thus, in the case of passive thermography, the soil layer over the mine tends to accumulate thermal energy during the day because the mine blocks the transport of thermal energy. As a result of this process the soil over the mine tends to be warmer than the surrounding soil. In the evening, the soil over the mine gives up thermal energy faster than the surrounding soil and results cooler. Around midday the soil and the mine reach the thermal equilibrium, which makes it impossible to perform the detection in this temporal gap. The main limitation of this technique is the fact that temperature differences strongly depend on the atmospheric conditions. On the other hand, active thermography uses artificial energy to heat the soil under study, avoiding the dependence on the atmospheric conditions. In spite of their long history, there is little compelling performance data available for infrared detection of antipersonnel mines.

The use of IRT as non-destructive evaluation (NDE) for landmine detection consists of subjecting the area under inspection to a source of natural or artificial heating/cooling process and studying the soil's response by means of the analysis of its thermal evolution given by a temporal sequence of infrared images. In this sense the study of the basic phenomenology lead to the development of mathematical models of the soil, England et al. (1992); England (1990); Kahle (1977); Liou & England (1998); Pregowski et al. (2000). The idea underneath is to characterize, and therefore predict, the thermal behavior of the unperturbed soil under given conditions, i.e., its *thermal signature*. The presence of buried mines or other objects will alter this signature, which if manifested as a thermal contrast on the surface. From the analysis of this thermal contrast and, in particular of its dynamic evolution, it is possible to extract relevant information about the nature of the objects. A deep analysis of the evolution of the thermal contrast over a diurnal cycle can be found in Khanafer & Vafai (2002), where the conditions for detection are studied, and sunrise and sunset are established as the periods of the day in which the presence of buried mines induces a greater thermal contrast on the surface (around 4-6°C).

An efficient way of extracting information from this data regarding the presence of landmines using a 3D thermal model of the soil based on the solution of the heat equation was presented in López et al. (2009; 2004). The process is divided in two steps. On the first one, the forward problem, the soil is subjected to a heating process and a comparison between temperatures measured at the soil surface (through IR imaging) and those obtained by simulation using the model under the assumption of homogeneous soil and mine absence is made. The differences between measured and simulated data put into evidence the presence of unexpected objects on the soil. The second step is an inverse engineering problem where the thermal model must be run for multiple soil configurations representing different types of possible targets (mine, stone, ...) and depths of burial. The nearest configuration to the measured data give us the estimated nature and location of the targets. This approach and, particularly, the inverse engineering process, makes an intensive use of the 3D thermal model that needs to be solved iteratively involving complex, coupled sets of partial differential equations. The extensive computing power required makes impractical its software implementation in personal computers . An alternative solution is the hardware implementation of the Finite Difference (FD) representation of the thermal model. In fact, different hardware implementations of FD solutions in the electromagnetics domain can be found in the literature, Durbano et al. (2004); Placidi et al. (2002); Schneider et al. (2002). In previous works, Pardo et al. (2009; 2010), we have presented an FPGA implementation of a FD Heat Equation solver which speeds the computation up by a factor of 10 compared to the purely software solution. The bottleneck of such an implementation is the access to memory and the amount of available memory, which dramatically reduces the performance of the system. However, this solution is hardware dependent an its cost is quite high. In recent years, Graphic Processing Units (GPUs) have been proved to be a valuable hardware platform to solve problems with a high degree of parallelism, Hwu et al. (2008). The total speedup of the system in a typical simulation setup by a factor 40 compared to a Core2Duo 2.8 GHz implementation in C++, using a NVIDIA GTS250 GPU, but even higher speedups could be achieved with more advanced GPUs.

The chapter is outlined as follows. In Section 2 the thermal model and the detection algorithm are introduced. Section 3 addresses the architecture of the hardware implementation and its GPU projection details. In Section 4 the main results are shown and, finally, the conclusions are summarized.

2. Infrared thermography for NDE

Infrared thermography sensors, which respond to electromagnetic radiation in a sensor-specific wavelength range, constitute an attractive NDE technique because they can

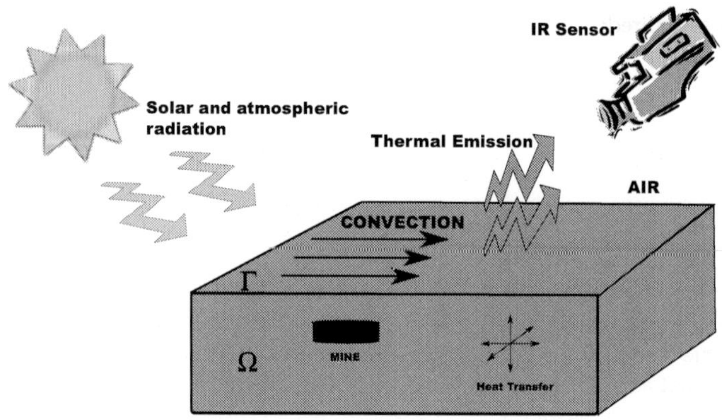

Fig. 1. Different heat transfer processes taking place in the soil volume and at the air-soil interface.

be used from a considerable standoff distance to rapidly survey large areas while providing information on several properties. Different materials show a characteristic response to a given thermal stimulus over time, a property that can be successfully used for the identification of buried targets. We will present a general procedure for the inspection of soils based on the solution of the heat equation applied, but not limited, to the detection of plastic antipersonnel mines.

2.1 Thermal model of the soil

The main processes considered are summarized in Fig. 1. We consider a soil volume, Ω, subjected to a known thermal stimulus through the soil-air interface, Γ, where both the soil and the targets are modeled as isotropic objects. We also assume that the thermal diffusivity is constant and that the temporal variation of the moisture content and the mass transference during the time of analysis are negligible, which are fair assumptions as long as the duration of the experiment does not exceed a couple of hours and the depth of inspection is limited to 10-15cm. A typical example is antipersonnel mine detection, given that mines are usually either laid on the surface or only shallowly buried. Moreover, empirical evidence demonstrates that it is possible to reduce the interval of interest to roughly one hour around sunrise or sunset as it is at these times when their different thermal evolution is more clearly manifested. Under these assumptions the overall process is described by the time-dependent single-phase 3D heat equation:

$$\frac{\partial T(\overrightarrow{r},t)}{\partial t} - div\left(\alpha(\overrightarrow{r})\ grad\ T(\overrightarrow{r},t)\right) = 0\ ,\alpha = \frac{k}{\rho c_p} \tag{1}$$

where $\overrightarrow{r} = (x,y,z)$ with $r \in \Omega$, ρ [kg/m^3] is the density, c_p[J/kg K] is the specific heat, k[W/m K] is the thermal conductivity, α [m^2/s] is the thermal diffusivity and T [K] is the distribution of temperatures inside the soil. To solve this equation the initial and boundary conditions are required:

$$k\frac{\partial T(\overrightarrow{r},t)}{\partial n} = q_{net}(t) \quad for\ \tau \times \Gamma \tag{2}$$

$$\frac{\partial T(\overrightarrow{r},t)}{\partial n} = 0 \quad for\ \tau \times \partial\Omega\backslash\Gamma \tag{3}$$

$$T(x,y,z \to \infty,t) = T_\infty \tag{4}$$

$$T(\overrightarrow{r},t = t_0) = T_0(\overrightarrow{r}) \quad in\ \Omega \tag{5}$$

where τ is the time interval of analysis, q_{net} is the net heat that flows through the soil-air interface Γ, n is the normal to the surface under consideration $\partial\Omega$. Eq. (2) gives the boundary condition at the air-soil interface; Eq. (3) shows the boundary conditions applied to the sides of the volume not accessible for measurements, imposing a vanishing heat flux across them. Eq. (4) is the deep-ground condition and it establishes that the temperature at a large enough depth remains constant. Finally, Eq. (5) gives the initial conditions for the system. The net heat flux at the soil-air interface, q_{net}, in Eq. (2) can be written as:

$$q_{net}(t) = q_{sun}(t) + q_{rad}(t) + q_{conv}(t) \tag{6}$$

where q_{sun} is the short-wave radiation emitted by the sun and absorbed by the soil; q_{conv} represents the convection term at the soil-air interface; and q_{rad} is the heat flux exchange due to radiation. The first term of this expression, q_{sun}, can be easily measured with the help of low-cost equipment. The second term is given by $q_{\text{rad}}(t) = q_{\text{sky}}(t) - q_{\text{soil}}(t)$. The term $q_{\text{sky}}(t)$ is the longwave radiation from the atmosphere given by Stefan's law. Finally, the third term in Eq. (6) can be approximated by, $q_{\text{conv}}(t) = h(T_{\text{air}}(t) - T_{\text{soil}}(t))$, where h [W/m^{2}K] is the convective heat transfer coefficient that is known to depend strongly on the wind speed.

2.2 Algebraic equations

Eq. (1) can be solved applying a FD approach, using either implicit or explicit methods. Using the explicit method a set of equations that can be solved sequentially is obtained, its main advantages being the simplicity of the formulation and its direct hardware translation. This method, however, is only conditionally stable, Bejan (1993). On the other hand, the implicit schemes are unconditionally stable but require the solution of the whole set of equations simultaneously. The ADI method, Wang & Chen (2002), can overcome the drawbacks of these two methods using a mixed approach, where only a subset of equations must be solved simultaneously while retaining the unconditional stability property. Thus, high temporal discretization steps can be used, effectively reducing the computing time when, for instance, a system evolving toward a steady state is studied. However, in our case, the boundary conditions change during the simulation time making necessary the use of small temporal discretization steps and, therefore, the benefit of using an ADI method is lost. Due to this fact and to the simplicity of the equations obtained, which allows to easily perform a hardware implementation, we have chosen an explicit scheme to solve the heat equation. Under this approach the derivatives can be approximated as:

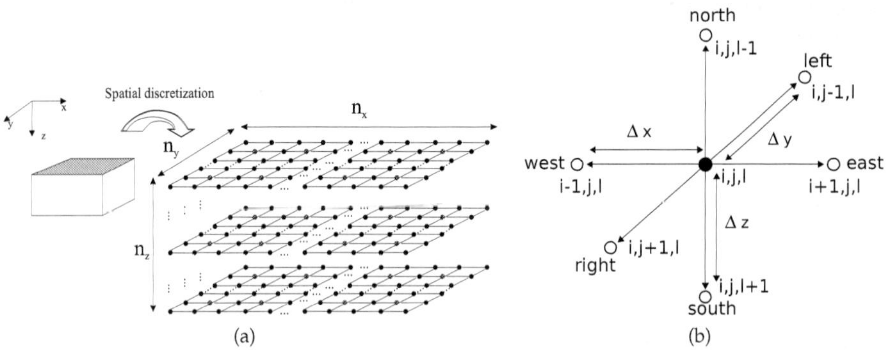

Fig. 2. (a) Spatial discretization scheme of the continuous soil into a set of discrete nodes. (b) Considered neighbors to perform the updating process.

$$\frac{\partial^2 T(\mathbf{r},t)}{\partial x^2} \approx \frac{T_{i+1,j,l} - 2T_{i,j,l} + T_{i-1,j,l}}{(\Delta x)^2} \tag{7}$$

$$\frac{\partial T(\mathbf{r},t)}{\partial t} \approx \frac{T^{m+1} - T^{m}}{\Delta t} \tag{8}$$

where Δx and Δt are, respectively, the spatial and temporal discretization steps; i, j, l are the discretization indexes in the x, y and z directions respectively and m is the time discretization index. The spatial discretization scheme of the continuous soil into a mesh of discrete nodes can be seen in Fig. 2(a), where n_x, n_y and n_z represent the number of nodes in the x, y and z directions respectively ($i = 0, 1, ...n_x - 1$, $j = 0, 1, ...n_y - 1$ and $l = 0, 1, ...n_z - 1$). Applying this discretization scheme to Eq. (1), we obtain the following equations for a surface ($T_{i,j,0}$) and an internal node ($T_{i,j,l}$, with $l > 0$), assuming, without loss of generality, that the spatial discretization steps are the same in all directions, that is, $\Delta x = \Delta y = \Delta z$,

$$\begin{aligned} T^{m+1}_{i,j,0} = T^{m}_{i,j,0} + F_0[&\sum_{neighbors\ surface} (T^{m}_{neighbor} - T^{m}_{i,j,0}) + 2(T^{m}_{i,j,1} - T^{m}_{i,j,0})] \\ &+ 2\alpha_{sun} F_0 S q^{m}_{sun} + (2F_0 H + 8F_0 R T^3_{air})(T_{air} - T^{m}_{i,j,0}), \quad \forall i,j \end{aligned} \tag{9}$$

$$T^{m+1}_{i,j,l} = T^{m}_{i,j,l} + F_0 \sum_{neighbors} (T^{m}_{neighbor} - T^{m}_{i,j,l}), \quad \forall i,j,l > 0 \tag{10}$$

where $F_0 = F_0(i,j,l)$, R, S and H are defined as,

$$F_0 = \frac{\alpha \Delta t}{(\Delta x)^2} \qquad R = \frac{\sigma\epsilon}{k}\Delta x \qquad S = \frac{\Delta x}{k} \qquad H = hS. \tag{11}$$

As seen, every new temperature, T^{m+1}, is calculated as a function of the temperatures in the previous time step, T^{m}, and only the nearest neighbors are considered in order to facilitate the hardware implementation, see Fig. 2(b). In the case of the surface layer, only the neighbors in the same layer ($z = 0$) and in the layer beneath ($z = 1$) are considered. Using the explicit FD method Eq. (10) can be written as,

$$T_{i,j,l}^{m+1} = T_{i,j,l}^{m}(1 - 6F_0) + F_0(T_{i+1,j,l}^{m} + T_{i-1,j,l}^{m} + T_{i,j+1,l}^{m} + T_{i,j-1,l}^{m} + T_{i,j,l+1}^{m} + T_{i,j,l-1}^{m}) \tag{12}$$

Thus, for an internal node the instability is avoided if the term $(1 - 6F_0)$ of Eq. (12) is non negative, (Bejan, 1993; Incropera & DeWitt, 2004), which is translated into a small enough Δt such that,

$$F_0 = \frac{\alpha \Delta t}{(\Delta x)^2} \leq \frac{1}{6} \tag{13}$$

As can be noted, the stability criteria is a trade-off between the spatial and temporal discretization steps. The small size of the mines constraints the value of Δx, which must be small enough to have several nodes representing the mines. Thus, a small value of Δx must be chosen which also implies to choose a small value of Δt. Meanwhile for a surface node Eq. (9) the stability criteria can be written as:

$$\begin{aligned} T_{i,j,0}^{m+1} &= T_{i,j,0}^{m}(1 - 6F_0 - 2F_0 H - 8F_0 R T_{air}^3) + \\ &F_0(T_{i+1,j,0}^{m} + T_{i-1,j,0}^{m} + T_{i,j+1,0}^{m} + T_{i,j-1,0}^{m} + 2\,T_{i,j,1}^{m}) + \\ &(2F_0 H + 8F_0 R T_{air}^3) T_{air} + 2\alpha_{sun} F_0 S q_{sun}^{m} \end{aligned} \tag{14}$$

Resulting in the following stability condition for a surface node:

$$F_0 \leq \frac{1}{6 + 2H + 8RT_{air}^3}. \tag{15}$$

As shown, the maximum Δt for a given Δx is limited by the condition on F_0.
In summary, applying the FD method, the thermal model of the soil is discretized into a set of nodes whose thermal behavior is given by algebraic equations involving additions, subtractions and multiplications, making it suitable for a hardware implementation.

2.3 IRT-based soil inspection for the detection of buried objects

Many physical systems are characterized by the solution of a differential equation or system of equations subject to known boundary conditions. This is the called *forward problem*. The representation of a non-linear forward operator can take the form of a functional equation involving a map F, which represents the connection between the model and the data:

$$y = F[p] \tag{16}$$

where F is a non-linear operator between Hilberts space Y and P, y is the measured data and p the original distribution of parameters that gives rise to y under the application of operator F. An *inverse problem* of this will be the reconstruction of the original distribution of parameters based on the measurements of the resulting data.
Solving an inverse problem implies approximating the best solution $p^{\dagger} = F^{-1}[y]$ of Eq. (16). In general, $y \in Y$ is never known exactly but up to an error of $\delta \neq 0$. Therefore, we assume

that we know $\delta > 0$ and $y^\delta \in Y$ with $\|y - y^\delta\| \leq \delta$. Thus, y^δ is the *noisy* data and δ is the noise level. Under these situations is, generally, impossible to compute numerically a solution of the problem unless making use of the regularization techniques, which makes it possible to restore stability and existence/uniqueness of the solution and develop efficient numerical algorithms, Kirsch (1996).

An example of such a technique is the non-linear Landweber iteration method which is defined by the following recursive procedure, Engl et al. (1996):

$$p_k^\delta = p_{k-1}^\delta + F'[p_{k-1}^\delta]^T (y^\delta - F[p_{k-1}^\delta]) \quad k \in \mathbb{N} \tag{17}$$

As initial guess $p_{k=0} = p_0$ is set, incorporating a *priori* knowledge of an exact solution $p^\dagger$. For non-linear problems, additional conditions about the stopping rule have to be imposed to guarantee convergence rates, Engl et al. (1996). The inequality known as the *discrepancy principle* will be used to define the stopping index $k(\delta, y^\delta)$ with

$$\|y^\delta - F[p_{k(\delta,y^\delta)}]\| \leq \mu\delta < \|y^\delta - F[p_k^\delta]\| \quad 0 \leq k \leq k(\delta, y^\delta) \tag{18}$$

if the parameter μ is chosen subject to the constraint

$$\mu > 2\frac{1+\eta}{1-2\eta} \tag{19}$$

being $\eta < 1/2$. The proof that this stopping rule regularizes the Landweber iteration method can be found in Engl et al. (1996).

Once the convergence criteria and the stopping rule have been established, the iterative procedure is applied to update the estimate of the solution in practical situations. It can be seen that the update term in the Landweber iterative method, see Eq. (17), is the negative gradient at $p = p_{k-1}^\delta$ of the functional

$$J = \|F[p] - y^\delta\| \tag{20}$$

that is the misfit between the model and the measured data. Solving the inverse problem is hence equivalent to minimize this functional.

The solution of the forward problem, F, is then the solution of the system of Eqs. (1)-(4) given the boundary conditions in the surface of the soil, Γ, and a distribution of soil parameters p within the volume Ω. According to Eq. (1), p corresponds to the value of the thermal diffusivity at every point within the soil volume, $\alpha = \alpha(x, y, z)$. Given the 3D nature of the model, the application of the operator F to the set of parameters, p, produces a 3D solution where the distribution of temperatures both on the surface and within the soil volume is calculated. For the sake of clarity we denote as $F[p]$ the distribution of the temperatures on the surface of the soil, which is compared with the temperature distribution acquired by the IR camera, our noisy data y^d. The reconstruction of the internal composition of the soil can be viewed as an inverse problem of this,that can be formulated as: *given the boundary conditions of the system and a distribution of measurements y on Γ, derive the soil parameter distribution, p, within Ω, that is, $p = F^{-1}[y]$.*

Now we will introduce the procedure for the non-destructive inspection of soils for the identification and classification of mines and mine-like targets on infrared images based on the solution of the heat equation and the use of inverse problems techniques, López (2003); López et al. (2009; 2004). The process starts with the acquisition of a sequence of infrared images of the surface of the soil under known heating and atmospheric conditions. As explained before, sunrise and sunset are the preferred times for detection. We will also assume that a pre-processing stage is run on a conventional PC in order to align the images and map grayscale colors to temperature values on the surface. Next, the soil inspection procedure itself starts. First, we run a detection procedure, as will be explained in the following section, to obtain the mask of potential targets. Then, a quasi-inverse process operator is used to identify the presence of antipersonnel mines among the potential targets. For those targets that failed to be classified as mines (and are therefore labeled as *unknown*), a full inverse procedure to extract their thermal diffusivity will be run in order to gain information about their nature. The overall detection process is summarized in Fig. 3, where the processes that require the use of the 3D thermal model are indicated with an ellipse. The detection, quasi-inverse and full-inverse procedures are based on the solution of the heat equation for different soil configurations. As explained, this is a very time consuming task that makes the whole algorithm inefficient for real on-field applications.

2.3.1 Target detection

The use of IR cameras taking images of the soil under inspection gives us the exact distribution of temperatures on the surface. On the other hand, the thermal model described previously and extensively validated with experimental data permits us to predict the thermal signature of the soil under given conditions, López (2003); López et al. (2004). The detection of the presence of potential targets on the soil is then made by comparing the measured IR images with the expected thermal behavior of the soil given by the solution of the forward problem under the assumption of absence of mines on the field, mathematically,

$$\alpha(x,y,z) = \alpha_{soil}, \qquad \forall x,y,z. \tag{21}$$

For this set of soil parameters, p, the application of the functional in Eq. (20) determines the surface positions (x,y) where the behavior is different from that expected under the assumption of mine absence, therefore revealing the presence of unexpected objects on the soil. These positions will be classified as potential targets, whereas the rest of the pixels (those that follow the expected pure-soil behavior) will be automatically classified as *soil*. This process is not trivial. The most straightforward approach, the thresholded detection, has the drawback of setting the threshold, which will vary not only for different image sequences, but it is also likely to depend on the particular frame of the sequence, and on the characteristics of the measured data such as lighting conditions and the nature and duration of the heating. For this reason, the use of a reconfigurable structure, capable of adapting to varying experimental conditions was proposed on López (2003); López et al. (2004). In this work they demonstrated that it is possible to reduce the time frame of analysis to roughly one hour around sunrise as it is at this time when the maximum thermal contrast at the surface is expected. This phenomena can be better appreciated in Fig. 4, where a sequence of IR images of a mine field taken between 07:40 am and 08:40 am is shown. Taking into account the short time interval we can consider that the properties of the soil remain unaltered and that there is no mass transference process during the simulation. The output of the detection stage is a black and white image with the mask of the potential targets.

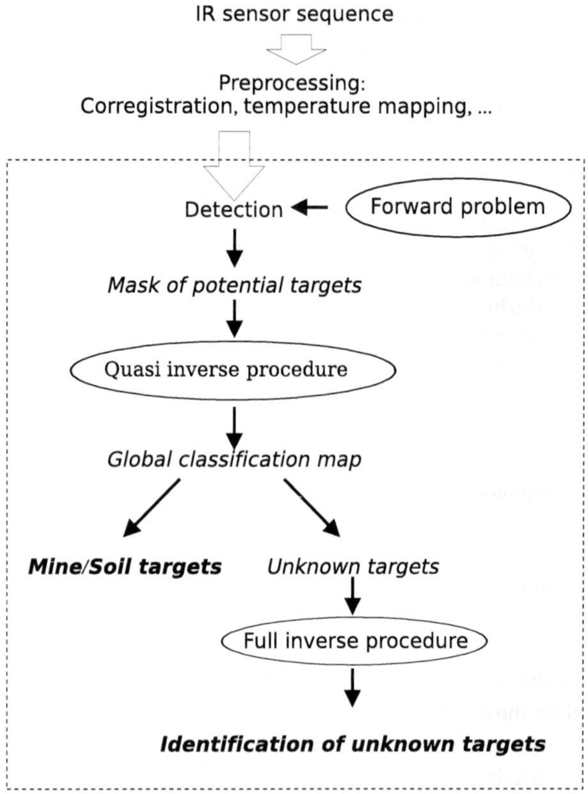

Fig. 3. Structure of the approach used to detect buried landmines using infrared thermography.

(a) 7:40 am (b) 08:00 am (c) 08:30 am (d) 09:00 am

Fig. 4. Measured IR images of a minefield at sunrise.

2.3.2 Quasi-inverse operator for the classification of the detected targets

In the previous section we dealt with the identification of the (x, y) position of the potential targets on the soil. In this section we will propose an operator for their classification into either *mine* or *soil* categories; any target that fails to fit into these categories will be classified as *unknown* (a procedure for the retrieval of further information about the nature of the *unknown* targets will be explained in the next section). For the *mine* category, the depth of burial will be also estimated. In general, this reconstruction is not possible unless additional information on the solution is incorporated in the model by means of the so-called *regularization techniques* Engl et al. (1996); Kirsch (1996). It is, however, possible to solve the inverse problem without the explicit use of a regularization strategy under proper initialization conditions and the use of iteration methods.

The iterative procedure is based on evaluating Eq. (20), which expresses the deviation between the observed IR data, y^{δ}, and the one given by the solution of the forward problem using known parameter distributions, $F[p]$. Therefore the heat equation needs to be solved for each of these distributions during the time of analysis (usually one hour around sunrise). In the case of mine targets, we will assume that their thermal evolution is driven by the thermal properties of the explosive used, which is commonly TNT composition B-3 or, less frequently, Tetryl. Our initial guess will be to assume that, *(i)* all the targets detected in the detection step are mines, that is,

$$\alpha_{\text{target}} = \alpha_{\text{mine}}, \tag{22}$$

and *(ii)* the possible depths of burial constitute a discrete set $z \in \tilde{Z}$ being,

$$\tilde{Z} = \{k\,\Delta z, \quad \forall k = 0, 1, ..., d\}, \tag{23}$$

with Δz the discretization step and $d\,\Delta z$ the a depth of burial at which is satisfied the deep-ground condition, see Eq. (4). The situation $k = 0$ corresponds to surface-laid mines. These two assumptions imply a reduction of the search space, therefore the *quasi*-inverse nature of the classification effort that will either confirm or reject them. Let,

- $\{y_s^{\delta}\}$, $s = 1, ..., S$, be the acquired IR image sequence, being S the total number of frames.
- $F[p]_{s,k}$, the modeled temperature distribution on the soil surface at time s. $F[p]_{s,k}$ is estimated by considering that all the detected targets are landmines buried at the depth given by index k in Eq. (23).

Note that, in the following, we will concentrate only on those areas of the image that were marked as possible targets in the detection phase. The classification map for the detected targets is obtained through the definition of a classification operator which includes the following computations:

1. For each time instant $s = 1, ..., S$ and burial depth $k = 0, ..., d$, an error map, $J_{s,k} = \|F[p]_{s,k} - y_s^{\delta}\|$, is estimated by evaluating Eq. (20) for each pixel position (x, y);
2. For each time instant s, a global error map (J_s) and a global classification map (Y_s) are estimated iteratively by comparing the error maps $J_{s,k}$, $k = 0, ..., d$, as follows:
   - *Initialization step:* For each pixel (x, y), we set $J_s(x, y) = \varepsilon$ (where ε is a predefined threshold error value); and $Y_s(x, y) = soil$.
   - *Iterative update step:* For each depth of burial, k, with $k = 0, ..., d$, $J_s(x, y) = min(J_{s,k}(x, y), J_s(x, y))$ and $Y_s(x, y) = argmin_k(J_{s,k}(x, y), J_s(x, y))$ (the category for which the error is smaller, i.e. the depth of burial). If $J_s(x, y) > \varepsilon$ then $Y_s(x, y)$ is set to $Unknown$.
3. Once J_s and Y_s have been obtained, we combine all these partial maps (J_s, resp. Y_s) into single ones (**J**, resp. **Y**) in the following way:

- **Y**: Pixels classified as mines at any processing step are kept in the final classification map. For the others, we keep the category that appears more times.
- $J(x,y) = max_s(J_s(x,y))$. This is a very conservative approach aiming at reducing the number of false negatives (failure to detect a buried mine) even at the cost of increasing the false alarm rate of the system.
- To find a trade-off between the accuracy of the classification and the number of false alarms, we define a *cutoff error*, e_{max}. If the entry on the error map, J for a pixel exceeds e_{max}, the pixel will be automatically assigned to the category of *Unknown*. e_{max} is estimated empirically, however it could be estimated taking into account the pixels classified as non-mine based their temperature variance using bootstrap techniques, Zoubir & Iskander (2004).

2.3.3 Full-inverse procedure for the classification of non-mine targets

In this case, no assumption about the nature of the targets found in the detection phase is made, although the set of possible depths at which the targets can be placed is still bounded by Eq. (23). Under these assumptions, Eq. (22) does not hold and α_{target} is unknown and could take any value depending on the nature of the object. For this reason, it is necessary in this case to use a systematic approach for the minimization of the functional J, which implies the calculation of the gradient $\partial J/\partial p$.

Let us consider the existence of a buried target in a 3D soil volume, Ω, with an unknown $\alpha = \alpha(r)$, $r = (x,y,z) \in \Omega$. The thermal experiment is the following: at time $t = t_0$, the solid is subject to a prescribed flux, $q_{\text{net}}(r',t)$, on its surface Γ, being Γ the portion of the surface $\partial\Omega$ accessible for measurements. We then measure the temperature response $\theta(r',t)$ at the boundary $r' \in \Gamma$, during the time interval $[t_0, t_f]$. We rewrite our 3D forward problem in Eq. (1) as,

$$-\text{div}\{\alpha(r)\,\text{grad}\theta\} + \frac{\partial\theta}{\partial t} = 0, \quad r \in \Omega \tag{24a}$$

$$\theta(r, t = t_0) = \theta_0, \quad r \in \Omega \tag{24b}$$

$$\frac{\partial}{\partial n}\theta(r',t) = q_{\text{net}}(r',t) \quad r' \in \Gamma,\ t \in [t_0, t_f]. \tag{24c}$$

We look at the reconstruction of $\alpha(r)$ from the knowledge of the surface response of temperature, $y^\delta = \theta(r',t)$, to prescribed flux applied on the boundary $q_{\text{net}}(r',t)$. We call *data* the pair $(\theta(r',t), q_{\text{net}}(r',t))$. As mentioned before, this is an ill-posed problem. It is intuitive that the data parameters (r',t) belong to a 3D subset, because $r' \in \Gamma$ and $t \in [t_0, t_f]$. This is sufficient enough for the reconstruction of the function $\alpha(r)$, defined in a 3D volume. Let us now introduce the *model* problem as an initial guess p, such that $p(r') = \alpha(r')$ (known data on the boundary), with the following governing equations and boundary conditions,

$$-\text{div}\{p(r)\,\text{grad}u\} + \frac{\partial u}{\partial t} = 0 \quad r \in \Omega \tag{25a}$$

$$u(r, t = t_0) = u_0, \quad r \in \Omega \tag{25b}$$

$$\frac{\partial}{\partial n}u(r,t) = q_{\text{net}}(r',t) \quad r' \in \Gamma, t \in [t_0, t_f]. \tag{25c}$$

The solution of Eq. (25) is a well-posed problem, as opposed to Eq. (24), and will be denoted by $u(r,t;p)$. Our aim will be to control p in such a way that the difference between the model and the observed data tends to zero. This goal is quantified by an objective function J to be minimized. The functional to be minimized is the L^2 norm of the misfit between the model and the observation given by,

$$J(u(p)) \equiv \frac{1}{2}\int_{t_0}^{t_f}\int_{\Gamma}\|u(r',t;p)-\theta(r',t)\|^2 dS\,dt. \tag{26}$$

This is a classic optimization problem which implies the calculation of the gradient of the functional J. To this aim we will make use of the variational method. If we introduce the notation,

$$\begin{aligned} <u,v>_\Omega &= \textstyle\int_\Omega u(r)\,v(r)\,d\Omega \\ <q_{\text{net}},v>_\Gamma &= \textstyle\int_\Gamma q_{\text{net}}(r')\,v(r')\,dS \\ a_p<u,v> &= \textstyle\int_\Omega p\,\text{grad}u\,\text{grad}v\,d\Omega, \end{aligned}$$

then the model problem, Eq. (25), is equivalent to the variational problem,

$$\int_{t_0}^{t_f} <\frac{\partial u}{\partial t},v>_\Omega\,dt + \int_{t_0}^{t_f}(a_p<u,v>)\,dt - \int_{t_0}^{t_f}<q_{\text{net}},v>_\Gamma\,dt = 0,\ \forall v \tag{27}$$

Eq. (27) can be considered as the constraints in the minimization problem, see Eq. (26). Therefore, we can introduce the Lagrange multiplier $\lambda(r,t)$ and define the Lagrangian L as,

$$L(u,p,\lambda) \equiv J(u) + \int_{t_0}^{t_f}\{<\frac{\partial u}{\partial t},\lambda>_\Omega + a_p<u,\lambda> - <q_{\text{net}},\lambda>_\Gamma\}\,dt. \tag{28}$$

Note that $L = J$ if u is the solution of the model problem, Eq. (25), since Eq.(27) holds for any λ. Thus, the minimum of J under the constraints in Eq. (27) is the stationary point of the Lagrangian L. Conversely, if $\delta L = 0$ for arbitrary $\delta\lambda$, u and p being held fixed, it follows necessarily that Eq. (27) holds. We consider,

$$\delta L = \frac{\partial L}{\partial u}\delta u + \frac{\partial L}{\partial p}\delta p, \tag{29}$$

where,

$$\tfrac{\partial L}{\partial u}\delta u \equiv \textstyle\int_{t_0}^{t_f}(u-\theta,\partial u)_\Gamma\,dt + \int_{t_0}^{t_f}\{(\delta\frac{\partial u}{\partial t},\lambda)_\Omega + a_p(\delta u,\lambda)\}\,dt \tag{30a}$$

$$\tfrac{\partial L}{\partial p}\delta p \equiv \textstyle\int_{t_0}^{t_f}\int_\Omega \delta p\,\text{grad}u\,\text{grad}\lambda\,d\Omega\,\delta\mu. \tag{30b}$$

We can restrict the choice of λ such that

$$\frac{\partial L}{\partial u}\delta u = 0. \tag{31}$$

This condition can be written as,

$$\int_{t_0}^{t_f}<u-\theta,\delta u>_\Gamma\,dt + \int_{t_0}^{t_f}\{-<\delta u,\frac{\partial\lambda}{\partial t}>_\Omega + a_p<\delta u,\lambda>\}\,dt + <\delta u,\lambda>_\Omega|_{t_0}^{t_f} = 0, \tag{32}$$

where $\delta u(x,0) = 0$. The last term of (32) vanishes if we impose,

$$\lambda(r, t \geq t_f) = 0. \tag{33}$$

By doing so we obtain the equation for the adjoint field λ,

$$-\text{div}\{p\,\text{grad}\lambda\} - \tfrac{\partial \lambda}{\partial t} = 0, \quad r \in \Omega \tag{34a}$$

$$\lambda(r, t \geq t_f) = 0, \quad r \in \Omega \tag{34b}$$

$$\tfrac{\partial}{\partial n}\lambda(r', t) = \theta - u, \quad r \in \Gamma. \tag{34c}$$

This is the so called *back diffusion* equation for the adjoint field, and it is also a well posed problem. With this choice of the adjoint field $\lambda(r,t)$, the variation ∂J becomes

$$\partial J = \int_{t_0}^{t_f} \int_\Omega \partial p\, \text{grad}u\, \text{grad}\lambda\, d\Omega\, dt. \tag{35}$$

It results from Eq. (35) that the derivative of J in the $p(r)$ direction is known explicitly by solving two problems, the direct problem for the field u and the adjoint problem for the field λ. That is, the Z-integral,

$$\frac{\partial J}{\partial p} \equiv Z = \int_{t_0}^{t_f} \text{grad}u\, \text{grad}\lambda\, dt. \tag{36}$$

Solving Eq. (25) and Eq. (34), both of them well-posed forward problems, and using Eq. (36), the expression of the update of Eq. (17)) can be calculated in a straightforward manner. With respect to the number of iterations of the Landweber method, the selection of the stopping criteria of the algorithm must be made according to the discrepancy principle in Eq. (18)). The bigger the η, the lower the number of iterations is, and the higher the error is. The selection of η for a particular application must then be a trade-off between computational time and accuracy of the solution.

2.4 Estimation of the computational cost

The algorithm described above is based on iterative procedures involving multiple solutions of the heat equation for different soil configurations. This constitutes a time consuming process not feasible for its use on the field as the computational complexity of the FD method, if $N = n_x \cdot n_y \cdot n_z$ is the total number of grid nodes, is $O(N \cdot IT)$, where IT is the number of iterations. As an example, we consider the analysis of a piece of soil ($\alpha_{soil} = 6.4 \cdot 10^{-7}$ m/s^2) of moderate dimensions of 1m$\times$1m with a shallowly buried mine ($\alpha_{mine} = 2.64 \cdot 10^{-7}$ m/s^2). Even if the depth resolution of IRT is barely 10-15 cm, the depth of analysis must be set to at least 40-50 cm in order to apply the boundary condition in Eq. (4). Using a uniform spatial discretization of $\Delta x = \Delta y = \Delta z = 0.8$ cm and assuming a temporal discretization step of $\Delta t = 6.25$ s ($F_0 = 0.06$), for a typical example the simulation of the behavior of the soil during one hour using C++ (optimized for speed using O2 flag from Microsoft Visual C++ compiler) on a Intel Core2Duo 2.8GHz takes 30 seconds if single precision arithmetic is used to represent the temperatures. Taking into account that the proposed inverse procedure requires the solution of the model for multiple soil configurations, the total computing time assuming that only 100 iterations are needed (a soft approach) will add up to 50 minutes. As this jeopardizes its use for field experiments we have developed a hardware implementation of a

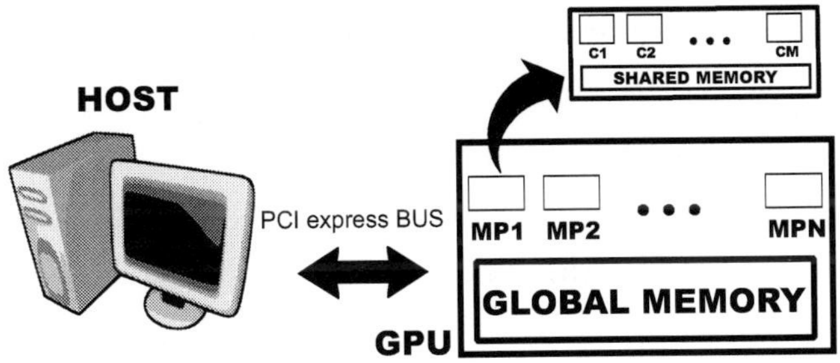

Fig. 5. GPU internal structure and memory hierarchy.

heat equation solver. In Pardo et al. (2009; 2010) we presented an FPGA-based implementation of such a solver. However, the main drawback of an FPGA implementation is the requirement of the system in terms of memory. The FPGA has a little amount of distributed memory and the FPGA's logic blocks can also be configured to behave like memory, however this is an inefficient way of FPGA using. Some vendors offer cards where external memory and FPGA are integrated on the same board, allowing to use the FPGA to deal with processing issues. However, these are expensive solutions. GPUs offers a structure which perfectly fits with the proposed problem and they have the advantage of being cheaper than FPGAs. GPUs are present in all computers and therefore we avoid the necessity of having a dedicated and expensive hardware to deal with our problem. Moreover, the GPU implementation is hardware independent, in the sense that it can be used on GPUs from NVIDIA with none or little changes, depending on GPU's computing capabilities.

3. GPU thermal model implementation

The system that solves the thermal model using the explicit FD method was implemented using CUDA language, NVIDIA (2010), and projected in a GPU from NVIDIA. The computing structure of GPUs makes them a suitable candidate to implement algorithms requiring high computing power. First we will introduce GPU characteristics and some basics abouts its programming mode. Then, we will present the proposed GPU implementation that simulates the thermal behavior of the soil and that speeds the computations up compared to a personnel computer.

3.1 GPU structure

GPUs are made up of several multiprocessors that can perform parallel processing data, which makes them suitable for processing in systems where it can be split up in independent portions and processed independently. The structure of such a GPU can be seen in Fig. 5. The GPU is made up of several multiprocessors, labeled as MP1 ... MPN in Fig. 5. Moreover, inside each multiprocessor there are several cores, labeled as C1 ... CM in Fig. 5. One important issue of GPU programming concerns to the use of the different memories available in the GPU, see Fig. 5. The *Global Memory* is available to all multiprocessors and cores, whereas the *Shared Memory* inside each multiprocessor is only available to the corresponding multiprocessor's cores. Additionally, each core has its own and private memory space. One key aspect of a GPU-based system is the memory data organization and access, as they can impose a

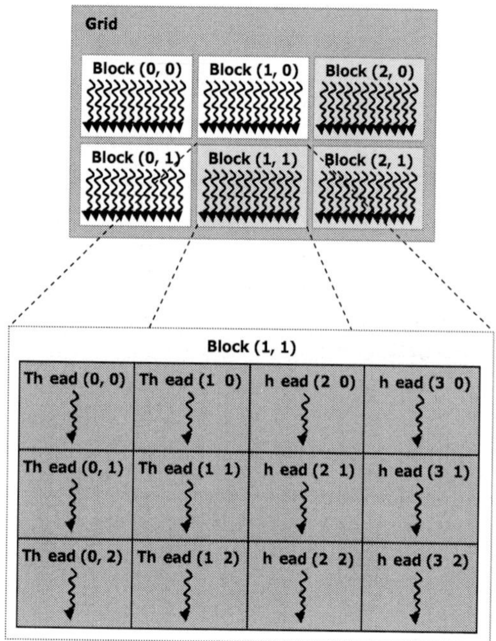

Fig. 6. Structure of threads hierarchy in a GPU (reprinted from NVIDIA (2010)).

bottleneck in the system performance. The global memory has an access latency two orders of magnitude higher than the access to the shared memory. Thus, it is important to minimize the use of global memory and maximize, as far as possible, the use of shared memory because this will increase the performance of the system.

Once the structure of the GPU has been briefly described we will introduce the basic aspects of GPU programming required to understand the structure of the proposed system . Functions in CUDA are called *kernels* and each kernel can be executed in parallel by several threads[1], as contrary to ordinary C/C++ functions that can only be executed by one processor. A kernel is not executed as a single thread, but it is executed as a block of threads, each of them processing the same function on different data, following a single-program multiple data (SPMD) computing model. Each thread inside the block has a 1D, 2D or 3D identifier (ID), depending on the applications, which distinguishes the concrete thread, to compute elements from a vector, matrix or volume of data. All the threads of a block are executed on the same multiprocessor and therefore they must fit within the available resources. This sets a limit on the maximum threads per block, which is limited to 512 in current GPUs. To avoid this limitation a kernel can be executed in several blocks of threads, which are organized as 1D or 2D groups of threads. The only requirement concerning the block of threads is that they must

[1] The thread is the basic element of processing

CUDA cores	128
CUDA Multiprocessors	8
Graphics Clock	738 MHz
Processor Clock	1836 MHz
Global Memory	512 MB
Memory Clock	1100 MHz
Memory Bandwidth	70.4 GB/s

Table 1. GTS 250 NVIDIA GPU main characteristics.

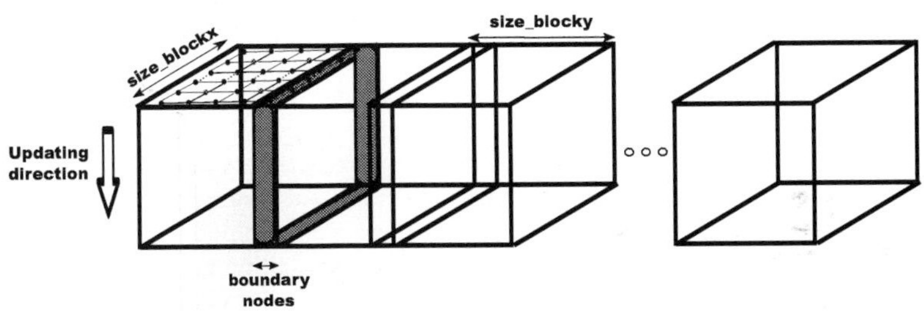

Fig. 7. Temperatures updating scheme on the GPU.

be independent from each other. Fig. 6 shows threads' hierarchy and its organization in the GPU.

3.2 GPU implementation of heat equation solver

GPU's structure fits perfectly our problem, where the full data can be split up in independent blocks that can be process the data in parallel. Each multiprocessor can work with a portion of grid's nodes increasing the performance of the system. The GPU used in this work was a GTS-250 from NVIDIA (cost around 250€- 300 $), whose characteristics are summarized in Table 1.

As was pointed, one of the main important aspects in an efficient CUDA-based system is the correct management of the memory to reduce the access to the global memory. To this aim the full grid of points, see Fig. 2(a), was divided into volume slices of size *size_blockx* × *size_blocky*, where the nodes' temperature of each slice is computed in a block of threads, see Fig. 7. Each thread of the block is responsible for updating the temperature of the nodes with the same (x,y) coordinates within the considered piece of soil. The threads advance as a wavefront, updating the nodes' temperature starting from the superficial layers to the inside of the soil, see Fig 7. It can be noted that there are overlapping areas between different blocks of threads, labeled as *boundary nodes* and indicated in grey in Fig. 7, which must be taking into account to compute only once the new temperature value.

Concerning the memory usage, the initial temperatures are stored in the global memory, and they have been transferred from the HOST memory to GPU global memory prior to the computation of the new temperatures. The temperatures are duplicated in the memory, as during one iteration we need to use one location to read temperatures and the other to write

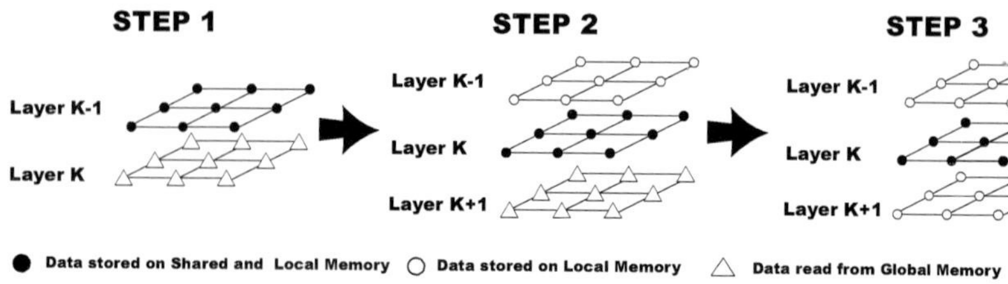

Fig. 8. Data memory transferences during the updating process.

the updated values and in the following iteration the roles are interchanged. The remainder constant values needed in the computations, such as F_0 and values related to the boundary conditions, see Eq. (14), are also stored in the global memory. The access to the global memory should be minimized to increase the speed of the computations, because the global memory has a high latency access. Thus, we use, during the updating process, the shared memory of the multiprocessors to accelerate the access to the data. The memory operations are shown in Fig. 8 where we can see the data transferences between the different memories of the GPU. In Fig. 8 we will consider the temperature updating process from nodes in *Layer K*. In *STEP 1* the temperatures of *Layer K-1* nodes are stored in both the multiprocessor's shared memory and in the cores' local memory. Moreover, nodes' temperatures from *Layer k* are read from the global memory and stored in the local cores' memory . During *STEP 2* the nodes' temperatures from *Layer K* replace those from *Layer K-1* in the shared memory, at the same time, the nodes' temperatures from *Layer K+1* are read from global memory and stored in cores' local memory. In *STEP 3* all data required to perform *Layer K* nodes' temperature updating is available on the local memory and shared memory. The same temperature of a *Layer K* is required to update the temperature of several nodes (the node itself and its north, south, west and east neighbors). If all nodes had to access global memory to read these values the process would be slowed, however once they are read from cores' local memory they are transferred to the shared memory, where they are available to all threads of the block, thus reducing the time access to the data. Once a thread has updated the temperature of a node, it uploads to the main memory the updated value and it continues computing the following temperature node updating. There is a synchronization process when a thread finishes one node's temperature updating because we must ensure that prior to continue with a node of the following layer all threads have finished the temperatures updating of the current layer.

4. Results

In this section we will introduce the results of the complete system. We divide this section into two main topics. On the one hand the results of the detection algorithm are shown for a scenario from the TNO Physics and Electronics Laboratory, Jong et al. (1999). On the other hand, we will show the performance of the GPU implementation and how it improves the usability of the detection system reducing the processing time.

4.1 Landmine detection algorithm

Next, we will show the result of the previously described detection algorithm to images acquired in a real test field. The scenario considered corresponds to the sand lane of the

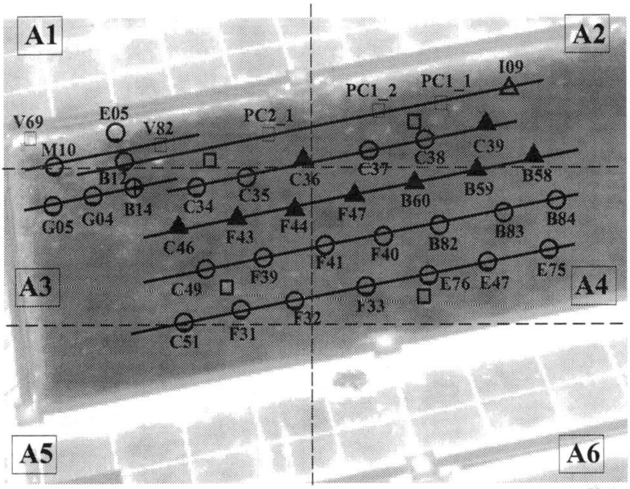

Fig. 9. Ground truth of the test field used during the experiment corresponding to a sand lane with different types of surrogated mines and non-mine targets.

Symbol	Category	Total
●	Surface mine	14
○	Mine at 1 cm	9
▲	Mine at 6 cm	9
⊕	Mine at 10cm	1
△	Mine at 15cm	1
■	Undefined test object	5
□	Shell marker	4

Table 2. Symbols used to represent the different categories of targets present in the test field in Fig. 9.

test facilities of the TNO Physics and Electronics Laboratory. For the experimental setup considered, the measured thermal diffusivity of the soil was $\alpha_{\text{sand}} = 6 \times 10^{-7}$ m^2/s. With respect to the test mine targets present, they are surrogated mines and most of them have been built at TNO-FEL. In all test mines the same substitute for the explosive has been used, RTV, having the same relevant properties as the real explosive and particularly $\alpha_{\text{RTV}} = 1.13 \times 10^{-7}$ m^2/s. Fig. 9 shows a sample image of the sand lane acquired with the IR sensor and the position of the different targets considered. Table 2 summarizes the symbols used for the different categories of targets present. The total number of targets is 43, 34 of which correspond to landmines. The remaining nine targets are five undefined test objects and four shells used as markers. We will concentrate on the results of the quasi-inverse and full-inverse procedures for the classification of mines and non-mine targets respectively.

Location	Detected and classified	Total
Surface	12	12
Buried at 1 cm	6	6
Buried at 6 cm	5	6
Buried at 10 cm	1	1
Buried at 15 cm	0	1

Table 3. Summary of the mines correctly detected and classified after the application of the quasi-inverse operator.

4.1.1 Results of the quasi-inverse operator for the classification of mine targets

The quasi-inverse operator classifies the detected targets as *Mine* or *Unknown*. Moreover, for the *Mine* class, sub-categories corresponding to their depth of burial are produced. For the experimental setup in Fig. 9, the results of the application of the quasi-inverse operator are summarized in Table 3, showing the distribution of mine targets correctly detected and classified according to their depth of burial with $e_{max} = 2.6$. As can be seen, all the mine targets on the surface or at 1cm depth were correctly detected and classified, and so were five out of six of the mines buried at 6 cm. The results for depths of 10 and 15 cm are not conclusive since only one of each is present, but a degraded performance of the quasi-inverse operator with depth is to be expected.

The performance of the classification operator is evaluated by making use of two properties, sensitivity and specificity, Hanley & McNeil (1982),

$$\text{Sensitivity} = \frac{TP}{TP + FN} \qquad \text{Specificity} = \frac{TN}{TN + FP} \tag{37}$$

being TP the number of true positives, TN the number of true negatives, FP the number of false positives and FN the number of false negatives. The performance depends strongly on the election of e_{max}, being a trade-off between sensitivity and specificity, i.e. , between the number of mine targets correctly classified and the number of false alarms. In humanitarian operations, the stress is put on the correct location of mines, while reducing the number of false alarms, although highly desirable, is a secondary goal. With respect to the global performance, it is clearly a function of the particular value of e_{max}. For $e_{max} = 2.6$ we find that 24 mines were correctly detected and classified over a total of 26.At the same time, the number of false positives is 13 compared to the 27 after the application of the detection stage alone.

4.1.2 Results of the full-inverse approach for the classification of non-mine targets

Now, we will illustrate the process of estimating the thermal parameters of non-mine targets making use of the full inverse process previously described. To this aim, we will consider the test object V82 present on the minefield (see Fig. 9). This target was classified as *unknown* by the quasi-inverse operator and we now aim to infer what type of object it is by estimating its thermal properties. If we estimate the measurement error to be $\delta = 0.3°C$ and setting $\eta = 0.2$, we have,

$$\mu > 2\frac{1+\eta}{1-2\eta} = 4. \tag{38}$$

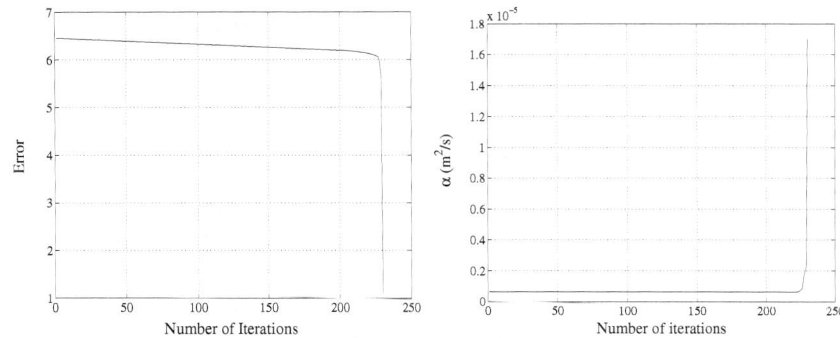

(a) (b)

Fig. 10. Test target V82: (a) Evolution of the error in the estimation of α; (b) Evolution of the value of α during the inverse problem procedure.

For $\mu = 4.1$, the discrepancy principle determines the stopping rule as,

$$\|y^\delta - F[p^\delta_{k(\delta,y^\delta)}]\| \leq 1.23 < \|y^\delta - F[p^\delta_k]\|, \quad 0 \leq k \leq k(\delta, y^\delta) \tag{39}$$

The evolution of the error for the Landweber iteration method is shown in figure 10(a). The stopping criteria in (39) corresponds to a number of iterations of the algorithm $N_{\text{it}} = 230$. Figure 10(b) shows the evolution of the estimation of the α parameter in this case. The final result obtained for the non-mine target V82 is,

$$\alpha_{\text{V82}} = 170 \times 10^{-7} \text{m}^2/\text{s}.$$

This value is two orders of magnitude bigger than that of the sand ($\alpha_{\text{sand}} \approx 6 \times 10^{-7} \text{m}^2/\text{s}$) which coincides with typical values of the thermal diffusivity of metallic solids.

4.2 GPU heat equation solver

We will now introduce the results obtained with the GPU implementation in terms of system throughput. As was pointed in the introduction a NVIDIA GTS 250 GPU, a low-cost GPU, was used to perform the comparison between a purely CPU implementation (Core2Duo 2.8 GHz implementation in C++) of the heat equation solver and a GPU implementation. One of the first issues is to think about the blocks threads' distribution and partitioning. The full volume of nodes which form the grid of points must be divided into blocks of threads, each of which is responsible for the nodes' temperature updating. The idea can be seen in Fig. 7, where the volume has been divided into blocks of threads of size *size_blockx*×*size_blocky* which are sent to the MP of th GPU, in this case a 1D array of blocks is shown for the shake of clarity. Fig. 11(a) shows the performance of the GPU for various block sizes, where we have chosen *size_blockx* = *size_blocky*, and a volume of 800×800×50 nodes. The results from these simulations can be seen in Table 4, where the speedup is compared to a purely CPU simulation of the full volume. It can be noted that the throughput of the system raises up as the size of the block is increased. This is due to the fact that when small blocks are used there are a lot

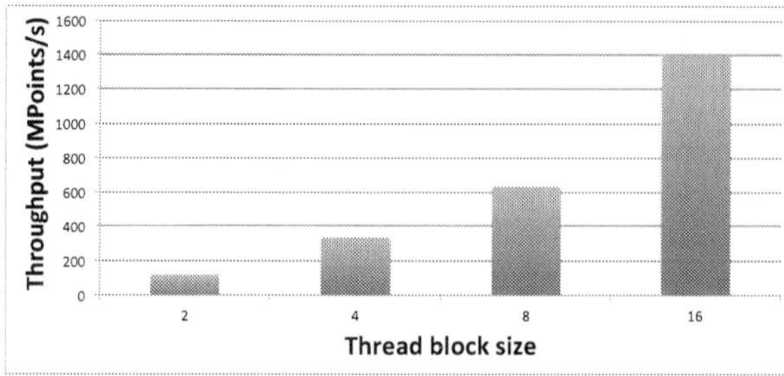

(a) Throughput of GPU heat equation solver for different blocks' size.

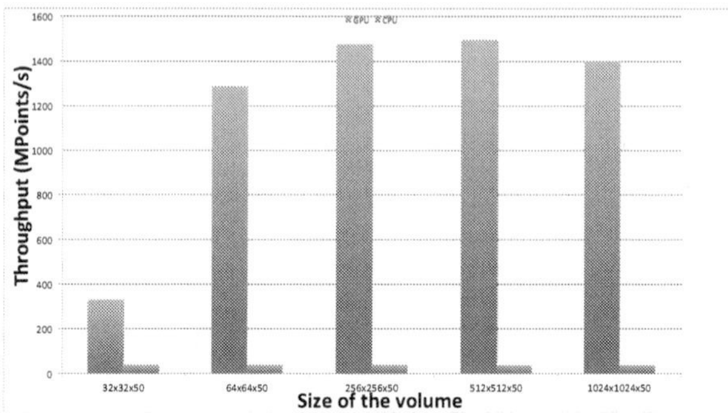

(b) Throughput of GPU and CPU heat equation solvers implementations for different volume of simulated points. In the GPU implementation the block grids' size was set to 16×16.

Fig. 11. Performance results comparing GPU and CPU throughputs for different setups.

of such small blocks spread over all MP, and therefore there will be a long cue of pending blocks to be processed. On the contrary, if the size of the blocks is increased we will have less blocks and the cue of pending blocks to be processed by the MP will be reduced. There is a limit, imposed by GPU's structure, given by the maximum number of threads that a MP can process (512). It can be seen from Fig. 11(a) and Table 4 that the throughput of the system is increased one order of magnitude when we go from 2×2 to 16×16 blocks of threads. Thus in the following simulations we will used this block's size for GPU simulations. Fig. 11(b) shows the GPU and CPU throughput for different nodes volumes, the data can be seen in Table 5. It can be noted how the performance of the GPU grows up one order of magnitude when the size of the volume is increased. This is due to the fact that for small volume of nodes not all GPU's resources are being used, whereas for big enough size volumes the inherent parallelism of the GPU increases the throughput of the system. It is obvious that for very big volumes the throughput will be low because there will be a cue of pending blocks to be processed which degrades the throughput of the system (note the reduction of the throughput for the $1024\times1024\times50$ volume).

block_dimx × block_dimy	GPU throughput (Mpoints/s)	GPU time (s)	Speedup
2×2	120.0	25.6	3.1
4×4	336.5	9.1	8.7
8×8	634.3	4.8	16.7
16×16	1400.0	2	40

Table 4. GPU throughput for a given volume of points and varying block thread's dimensions.

Volume	GPU throughput (Mpoints/s)	CPU throughput (Mpoint/s)	Speedup
32×32×50	332	39.3	8.4
64×64×50	1289.33	39.3	32.6
256×256×50	1476.67	39.3	37.0
512×512×50	1496.02	39.3	39.3
1024×1024×50	1402.8	39.3	36.7

Table 5. Throughput and CPU and GPU solvers of the heat equation solver for different size volumes.

5. Conclusions

In this chapter, two inverse procedures for the inspection of soils by non-invasive means with application in antipersonnel mines detection have been presented. The first quasi-inverse procedure aims at the detection of surface-laid and shallowly buried mines, giving an estimation of their depth of burial that will be of critical importance during the removal stage. In the second approach, a full inverse procedure for the identification of the thermal properties of other objects present on the soil was presented. Both procedures need the recursive solution of the heat equation problem for different soil configurations, which constitutes a very time consuming task on a conventional computer. The efficient solution of the aforementioned procedures is successfully solved using a heat equation solver accelerator based on the use of GPUs, obtaining speed-up factors over 40. The speedup obtained with the proposed system with respect to nowadays computers, together with its low-cost and portability justifies the implementation as it permits its use on the field during demining operations.

6. References

Bach, P., Toumeur, P. L., Poumarkde, B. & Bretteand, M. (1996). Neutron activation and analysis, *EUREL International Conference Detection of Abandoned Landmines*, Vol. 431, pp. 58–61.

Bejan, A. (1993). *Heat Transfer*, John Wiley & Sons, Inc.

Cameron, M. & Lawson, R. (1998). *To Walk Without Fear: The Global Movement to Ban Landmines*, Toronto: Oxford University Press.

Durbano, J., Ortiz, F., Humphrey, J. R., Curt, P. & Prather, D. (2004). Fpga-based acceleration of the 3d finite-difference time-domain method, *Proceedings of the 12th annual IEEE symposium on Field-Programmable Custom Computing Machines*, pp. 156–163.

Engl, H. W., Hanke, M. & Neubauer, A. (1996). *Regularization of Inverse problems*, Kluwer Academic Publishers.

England, A., Galantowiz, J. & Schretter, M. (1992). The radiobrightness thermal inertia measure of soil moisture, *IEEE Transactions on Geoscience and Remote Sensing* 30(1): 132–139.

England, A. W. (1990). Radiobrigthness of diurnally heated, freezing soil, *IEEE Transactions on Geoscience and Remote Sensing* 28(4): 464–476.

Englelbeen, A. (1998). Nuclear quadrupole resonance mine detection, *CLAWAR'98*, pp. 249–253.

Furuta, K. & Ishikawa, J. (eds) (2009). *Anti-personnel landmine detection for humanitarian demining: the current situation and future direction for Japanese research and development*, Springer-Verlag.

Gros, B. & Bruschini, C. (1998). A survey on sensor technology for landmine detection, *Journal of Humanitarian Demining* pp. 172–187.

Hanley, J. & McNeil, B. (1982). The meaning and use of the area under a receiver operating characteristic (ROC) curve, *Radiology* 143: 29–36.

Horowitz, P. (1996). New technological approaches to humanitarian demining, *Technical Report JSR-96-115*, JASON MITRE.

Hwu, W. W., Keutzer, K. & Mattson, T. (2008). The concurrency challenge, *IEEE Design & Test of Computers* 25(4): 312 – 320.

ICBL (2006). *Landmine Monitor Report 2006*, International campaign to can landmines (ICBL).

Incropera, F. & DeWitt, D. (2004). *Introduction to Heat Transfer*, 4th edn, John Wiley & Sons.

Jankowski, P., Mercado, A. & Hallowell, S. (1992). FAA explosive vapor/particle detection technology, *Applications of Signal and Image Processing in Explosives Detection Systems*, Vol. 1824, pp. 13–27.

Jong, W., Lensen, H. & Janssen, H. (1999). Sophisticated test facilities to detect land mines, *Detection and Remediation Technologies for Mines and Minelike Targets IV*, Vol. 3710 of *Proceedings of the SPIE*, pp. 1409–1418.

Kahle, A. B. (1977). A simple thermal model of the earth's surface for geologic mapping by remote sensing, *Journal of Geophysical Research* 82: 1673–1680.

Khanafer, K. & Vafai, K. (2002). Thermal analysis of buried land mines over a diurnal cycle, *IEEE Transactions on Geoscience and Remote Sensing* 40(2): 461–473.

Kirsch, A. (1996). *An introduction to the Mathematical Theory of Inverse problems*, Vol. 120 of *Applied mathematical sciences*, Springer-Verlag, New York.

Larsson, C. & Abrahamsson, S. (1993). Radar, multispectral and biosensor techniques for mine detection, *Symposium on Anti-Personnel Mines*, pp. 179–202.

Liou, Y. & England, A. (1998). A land surface process/radiobrightness model with couple heat and moisture transport for freezing soils, *IEEE Transactions on Geoscience and Remote Sensing* 36(2): 669–677.

Lockwood, G., Shope, S., Bishop, L., Selph, M. & Jojola, J. (1997). Mine detection using backscatered x-ray imaging of antitank and antipersonnel mines, *Detection and Remediation Technologies for Mines and Minelike Targets II* 3079: 408–417.

López, P. (2003). *Detection of Landmines from Measured Infrared Images using Thermal Modeling of the Soil*, PhD thesis, Universidad de Santiago de Compostela.

López, P., Pardo, F., Sahli, H. & Cabello, D. (2009). Non-destructive soil inspection using an efficient 3d softwareâĂŞhardware heat equation solver, *Inverse Problems in Science and Engineering* 6(17): 755–775.

López, P., van Kempen, L., Sahli, H. & Cabello, D. (2004). Improved thermal analysis of buried landmines, *IEEE Transactions Geoscience and Remote Sensing* 42(9): 1955–1964.

Maksymonko, G. B. & Le, N. (1999). Performance comparison of standoff minefield detection algorithms using thermal IR image data, Vol. 3710, SPIE, pp. 852–863. URL: *http://link.aip.org/link/?PSI/3710/852/1*

NVIDIA (2010). *NVIDIA CUDA C Programming Guide 3.1*, NVIDIA Corporation Technical Staff.

Ottawa (1997). Convention on the prohibition of the use, stockpiling, production and transfer of anti-personnel mines and on their destruction.

Pardo, F., López, P., Cabello, D. & Balsi, M. (2009). Effcient software-hardware 3d heat equation solver with applications on the non-destructive evaluation of minefields, *Computers & Geoscience* 35: 2239–2249.

Pardo, F., López, P., Cabello, D. & Balsi, M. (2010). Fpga computation of the 3d heat equation, *Computational Geoscience* 14: 649–664.

Placidi, P., Verducci, L., Matrella, G., Roselli, L. & Ciampiolini, P. (2002). A custom VLSI architecture for the solution of FDTD equations, *IEICE Transactions on Electronics* E85-C: 572–577.

Pregowski, P., Walczack, W. & Lamorski, K. (2000). Buried mine and soil temperature prediction by numerical model, *Proceedings of the SPIE, Detection and Remediation Technologies for Mines and Minelike Targets V*, Vol. 4038, pp. 1392–1403.

Robledoa, L., Carrascoa, M. & Merya, D. (2009). A survey of land mine detection technology, *International Journal of Remote Sensing* 30(9): 2399–2410.

Sabatier, J. & Xiang, N. (2001). An investigation of a system that uses acoustic seismic coupling to detect buried anti-tank mines, *IEEE Transactions on Geoscience and Remote Sensing* 39(6): 1146–1154.

Schneider, R., Turner, L. & Okoniewski, M. (2002). Application of FPGA technology to accelerate the Finite-Difference Time-Domain (FD-TD) method, *Proceedings of the 10th ACM/SIGDA International Symposium on Field-Programmable Gate Arrays*, pp. 97–105.

Siegel, R. (2002). Land mine detection, *IEEE Instrumentation & Measurement Magazine* pp. 22–28.

Thanh, N., Hao, D. N. & Sahli, H. (2009). *Augmented Vision Perception in Infrared Algorithms and Applied Systems*, Advances in Pattern Recognitiob, SpringerLink, chapter Infrared Thermography for Land Mine Detection, p. 471.

Thanh, N., Sahli, H. & Hao, D. (2007). Finite-Difference methods and validity of a thermal model for landmine detection with soil propierty estimation, *IEEE Transactions on Geoscience and Remote Sensing* (4): 656–674.

Thanh, N., Sahli, H. & Hao, D. (2008). Infrared thermography for buried landmine detection: inverse problem settin, *IEEE Transactions on Geoscience and Remote Sensing* (12): 3987–4004.

Vines, A. & Thompson, H. (1999). Thompson, beyond the landmine ban: Eradicating a lethal legacy, *Technical report*, Research Institute for the Study of Conflict and Terrorism.

Wang, T. & Chen, C. (2002). 3-D thermal-ADI: A linear-time chip level transient thermal simulator, *IEEE Transactions on Computer-Aided Design of Integrated Circuits and Systems* 21(12): 1434–1445.

Zoubir, A. M. & Iskander, R. (2004). *Bootstrap Techniques for Signal Processing*, Cambridge University Press.

3

The Heat Transfer Enhancement Analysis and Experimental Investigation of Non-Uniform Cross-Section Channel SEMOS Heat Pipe

Shang Fu-Min[1], Liu Jian-Hong[1] and Liu Deng-Ying[2]
[1]*Changchun Institute of Technology, Changchun 130012, Jilin,*
[2]*Institute of Engineering Thermo physics, Chinese Academy of Sciences, Beijing 100080, China*

1. Introduction

Along with the development of modern high-technology, the heat generation of many equipments that were heated in unit area becomes more and more, usually it is over 100W/cm^2, even 10^3~10^4W/cm^2, such as super power integrate circuit, high temperature battery for electricity generation, nuclear reactor, high power mini-machinery, mini-sized steam turbine, superconductive electricity generation and distribution devices and airship. Due to the trend of high rate heat generation, a new cooling technology is needed to solve the problem between design flexibility, needed space and heat-dissipating performance.

A new period of heat pipe began with Pulsation Heat Pipe, invented by Japanese Doctor H. Akachi in 1994, later is called Self-Exciting Mode Oscillating-Flow Heat Pipe[1, 2] (SEMOS Heat Pipe). Although it is just known less than sixteen years, the SEMOS Heat Pipe is drawing worldwide attention because of its immeasurable application potency. A lot of research about the SEMOS Heat Pipe is doing in Japan, U.S.A, Germany, Russia, Ukraine and China[3-5].

Notwithstanding all the research done is to understand the working principles of SEMOS Heat Pipe because it is still at its initial stage, this new high effective heat transfer device is catching attention on heat transfer due to phase transition all over the world. Thus the contradiction between cooling device downsizing and heat generation increasing is still the major technical problem to improve and bring the heat-dissipating performance to a higher lever for all of the researchers along with the development of new high-technology.

2. Heat transfer characteristics of SEMOS Heat Pipe

According to the working principles of SEMOS Heat Pipe, there are two basic methods to improve the heat transfer performance, one is to enhance the thermal conduction between the interface of the pipe and the working gas-liquid medium in the pipe, the other method is to increase the surge frequency and circulating power. Actually the thermal conduction of the SEMOS Heat Pipe between the pipe surface and the working liquid in the pipe is the non-steady phase transition heat transfer between the biphasic fluid and the interface of pipe. The first method to enhance the thermal conductive performance is to improve the liquid evaporation, the clotted phases transforming frequency, and intensity of phase transforming. Another method is to increase the heat transfer rate between the working liquid and the working surface. What needed to do to enhance the surge frequency and the reliable circulating power is to increase the difference in temperatures between the hot and cold liquid via enhancing the pulsing process inside the pipe. Obviously the two enhancing methods discussed above could supplement each other.

On the principle of field coordination heat transfer enhancement[6, 7], which was put forward by Guo Zengyuan academician, the heat transfer rate increases as field coordination coefficient between velocity vector field and temperature grads field increases. Although the convection heat transfer theory of single phase has been demonstrated, there still exists the problem about the heat transfer during the phases transforming between two phases, especially in a limited heat-dissipating space. Thus it needs further study on that if the coordination theory could be used universally.

This experiment to enhance the heat transfer rate of SEMOS Heat Pipe is to validate the application of field cooperation theory on the heat transfer field with phases transforming. The SEMOS Heat Pipe with non-uniform cross-section is used for this experiment so as to improve surging frequency and circulating power.

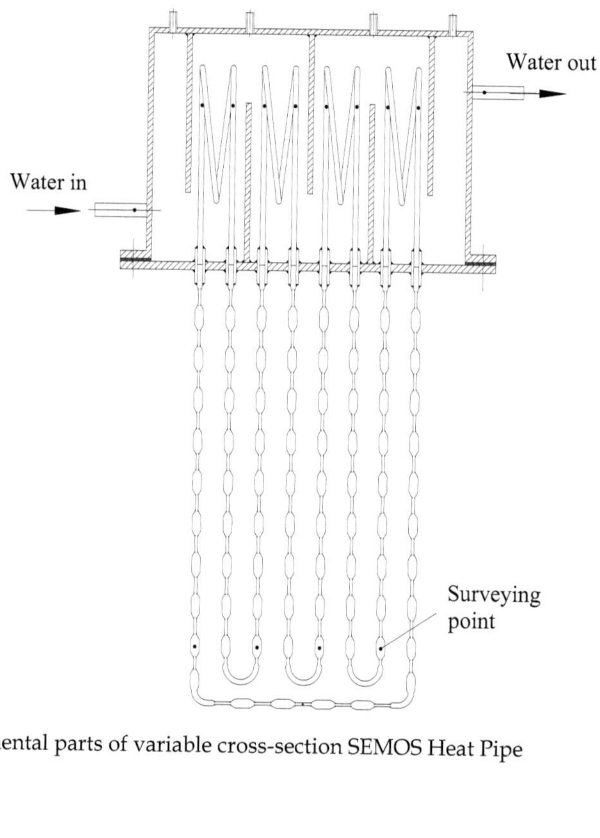

Fig. 1. Experimental parts of variable cross-section SEMOS Heat Pipe

3. Experiment set up and method

The object for this experiment is SEMOS Heat Pipes with closed loop, one is a SEMOS Heat Pipe with a uniform cross-section of 3mm inner diameter, the other heat pipe based on the uniform one with elliptic non-uniform cross-section is consisted of vertically intervened heating section and insulating section of the pipes. As shown in figure 1 the material of the heat pipe is brass, the working fluid in the pipe is distilled water with high purity, the amount of working is φ =42%, the obliquity of the heat pipe is θ =55^0, the pressure in the heat pipe is p =1.8×10^{-3}Pa.
Figure 2 shows the experiment set up for the thermal performance measurement consisting of the main test apparatus, a laser supply and its cooler and its power supply system, a data acquisition system combined with personal computer to show the data collected. As shown in figure 2, the laser supply is consisted of eight passages Quantum Well Laser Diode Arrays, the maximum output power of every single channel is 50W, the heating electrical current range is 5~40A, the wave length of the laser is 94nm. The heat input comes from continual laser heating while every single could work individually or together as heater, and 20 K-type thermocouples were mounted on the surface at diameter of 1mm totaling 20 in number. Accordingly, the data acquisition frequency is 1/s based on the 20 data collecting channels and the data acquisition precision is at ms.

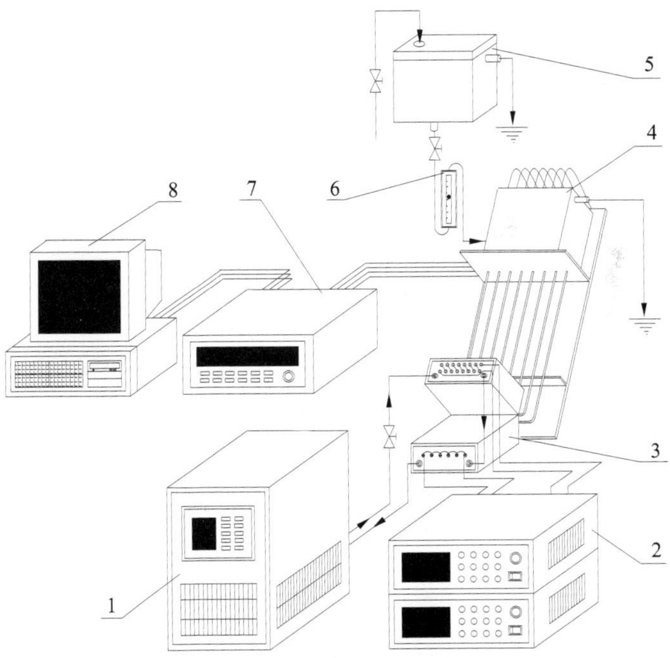

1—Unit of refrigeration cycle, 2—Power supply, 3—Laser supply, 4—Experimental table of SEMOS Heat Pipe, 5—Water tank, 6—Flow meter, 7—Data acquisition system, 8—Personal computer

Fig. 2. Experimental system of SEMOS Heat Pipe heat transfer enhancement

4. Experiment result and discussions

4.1 Compare and contrast heat pipes uniform with non-uniform cross-section

Judging from the experiments above mentioned to the heat pipes with uniform cross-section and non-uniform cross-section, figure 3 shows the transfer efficiency under different heating power (heat electrical current *I*). The transfer efficiency is given by the following equation.

$$P_o = Gc_p(T_2 - T_1) \tag{1}$$

Where, G denotes the amount of the cooling water, kg/s; c_p is the specific heat of water under constant pressure, J/(kg·K). T_2 and T_1 denote the output and input temperature of cooling water, K.

As can be seen from the figure, the transfer rate of the heat pipe with non-uniform cross-section is lower than that of the heat pipe with uniform cross-section when the heating electrical current is relatively low, while the rate of heat pipe with non-uniform cross-section would exceed that of the heat pipe with uniform cross-section as the heating electrical current of laser supply increases. It is suggested from the trend of the graph of transfer rate P_o~I that the more heat input the more of the difference of transfer rate between the heat pipe with non-uniform cross-section and the heat pipe with uniform cross-section becomes. The transfer rate of heat pipe with non-uniform cross-section is 13.6% higher than that of the heat pipe with uniform cross-section at the maximum heat input in this experiment, the heating electrical current of every channel is 23A that is about 25.5W.

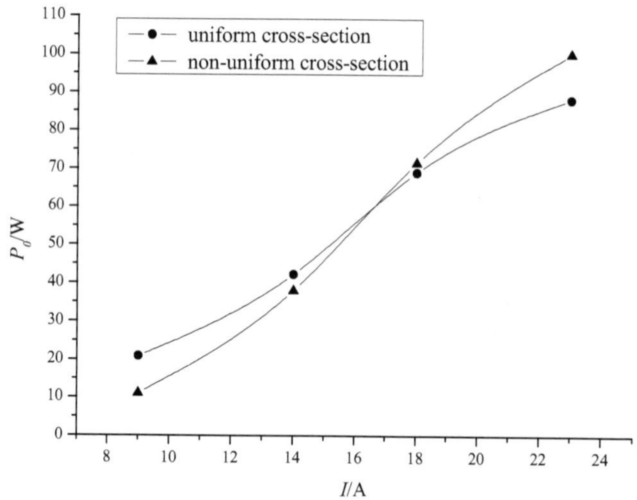

Fig. 3. Compare transferred power uniform with non-uniform cross-section heat pipes

4.2 Compare uniform and non-uniform cross-section with pure conductor equivalent thermal conductivity

Figure 4 shows the relationship between heat input and equivalent weight heat transfer coefficient under three different conditions, SEMOS Heat Pipes with non-uniform cross-section and uniform cross-section and the heat pipe without working fluid that means pure conductor. According to the different output and the difference in temperature of corresponding hot and cold end of the heat pipe, the equation of heat transfer coefficient is shown as below, while

$$\lambda_d = \frac{P_o l}{\left(\overline{T_h} - \overline{T_c}\right)F} \tag{2}$$

Where, P_o denotes transfer rate, W. l is the distance between the hot and cold end, m. $\overline{T_h}$ and $\overline{T_c}$ Mean temperature of hot end and cold end, K. F Total heat transfer area, m^2, total circulate cross section area for heat transfer, total cross section area for pure conductor.

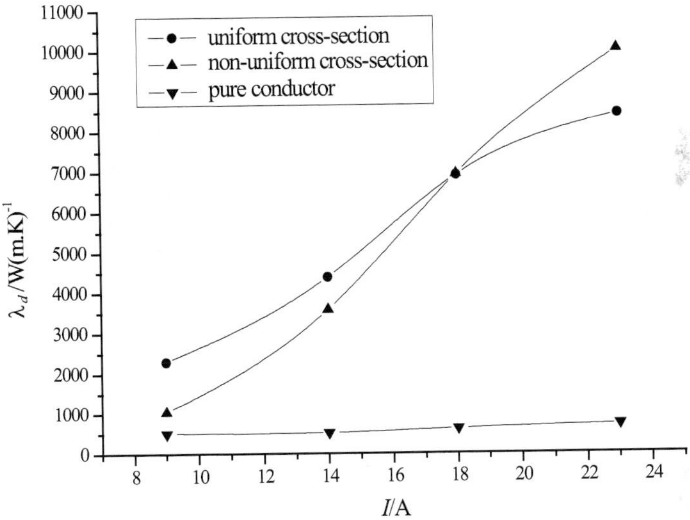

Fig. 4. Compare equivalent thermal conductivity in varying cases

It can be seen from the figure that the equivalent weight heat transfer coefficients under all the three conditions increase as the heat input increases, while the coefficient of the non-uniform heat pipe increased the most and that of the pure conductor heat pipe the lest. When the heat input is low, the heat transfer coefficient of uniform cross-section heat pipe is higher than that of the non-uniform one. As the heat input increases, the heat transfer coefficient of the non-uniform cross-section heat pipe becomes higher that that of the uniform one and the difference between them becomes bigger and bigger, so it is suggested that under certain condition the heat transfer rate could be improved by changing the form of the cross-section of the SEMOS Heat Pipe. The figure also shows that the function on heat transferring of heat pipes especially SEMOS Heat Pipes is better than the function on heat conducting of brass pipe. Under that heat input range, the equivalent weight heat transfer coefficient of heat pipe is as 4~15 times as that of the pure conductor one, and this difference becomes bigger as the heat input increases. At maximum heat input (23A for every single channel) in this experiment, the equivalent weight heat transfer rate of non-uniform cross-section heat pipe is 19% higher than that of uniform cross-section one, 14 times higher than that of the pure conductor one.

4.3 Heat transfer performance of the heat pipe with non-uniform cross-section

Theoretically, there are two reasons to answer that why non-uniform heat pipe consisted of the heating section and insulating section could improve heat transfer rate. One is that the portrait eddy formed for the Non-uniform Cross-Section of cross and adds parts of the velocity on the pipe surface vertically, that means the temperature grads field and velocity vector field co-operate together, then the heat transfer rate could be improved. Meanwhile the resistance on the working fluid increases because of the non-uniformity of the cross-section, certain amount of the heat input is needed for the SEMOS Heat Pipe with non-uniform cross-section to improve heat transfer performance in word and deed. Because the advantage of the SEMOS Heat Pipe with non-uniform cross-section on improving heat transferring could be demonstrated easily when the portrait eddy becomes stronger and circulating power becomes declining, the SEMOS Heat Pipe with non-uniform cross-section is more suitable for the high density working fluid.

On the other side, as shown in figure 5 it could be found by monitoring the temperature of the cold and hot end surface that the hot end surface temperature surging amplitude of the non-uniform heat pipe is lower than that of the uniform one, while the surging frequency is higher. In SEMOS Heat Pipe with non-uniform cross-section the circulating power increases because of the alternation of liquid evaporation and gas clot. The accelerated motion inside the pipe because of unsteady expanding and contracting process along with the vertical velocity on the interface could finally improve the heat transfer performance.

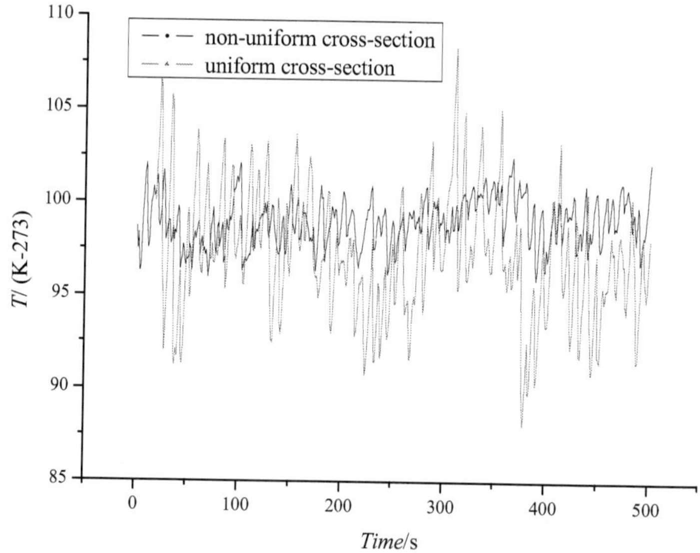

Fig. 5. Compare wave with hot end surface temperature

5. Conclusions

1. The precondition to improve heat transfer performance with non-uniform SEMOS Heat Pipe is that heat input must be high enough to overcome the resistance brought by the non-uniform cross-section.
2. The heat transfer rate of SEMOS Heat Pipes with non-uniform cross-section and uniform cross-section is much higher than that of the pure conductor under same experimental condition. The SEMOS Heat Pipe with non-uniform cross-section is more suitable for high density heat current because it especially shows its advantage when the heat input is high.
3. The surging frequency of SEMOS heat pipe with non-uniform cross-section becomes higher than that of the SEMOS Heat Pipe with uniform cross-section as the heat input increase.
4. It is confirmed in this experiment that field coordination heat transfer enhancement theory suit for heat transfer due to phase transition of the SEMOS Heat Pipe, which is an exemplification that this law could be used in the heat transfer domain of heterogeneous fluid. Which is the base of the future research and development for new SEMOS Heat Pipe.

6. Acknowledgement

This work was Supported by the Key Project of Chinese Ministry of Education (No: 210050).

7. Nomenclature

P heating power, W
G rate of mass flow, $kg\cdot s^{-1}$
T temperature, K
$\bar{T}$ mean temperature, K
I heating electrical current, A
F Total heat transfer area, m^2
c specific heat at constant pressure, $J\cdot kg^{-1}\cdot K^{-1}$
l distance between the hot and cold end, m
λ heat transfer coefficient, $W\cdot m^{-1}\cdot K^{-1}$

Subscripts

p pressure
o out
d equivalent
h hot end
c cold end

8. References

[1] H. Akachi. Looped Capillary Tube Heat Pipe. Proceedings of 71th General Meeting Conference of JSME, 1994, 3(9): 1-10
[2] H. Akachi, F. Polasek, P. Stulc. Pulsating Heat Pipe. Proceedings of the 5th International Heat Pipe Symposium, Australia, 1996: 17-20

[3] Qu Wei, Ma Tongze. Experimental Investigation on Flow and Heat Transfer of Pulsating Heat Pipe. Journal of Engineering Thermo physics, 2002, 23(5): 596-598
[4] Cui Xiaoyu, Weng Jianhua, M. Groll. Experimental Investigation of Heat Performance for The Copper/Water Pulsating Heat Pipe. Journal of Engineering Thermo physics, 2003, 24(5): 864-866
[5] Yasushi KATO, Takao NAGASAKI and Yutaka ITO. Study on Looped Heat Pipe with Non-uniform Cross section. Proceedings of the 40th Heat Transfer Conference. Japan, 2003, 5: 313-314
[6] Guo Zengyuan. Physical Mechanism and Control of Heat Convection: Speed and Temperature Field Synergy. Chinese Science Bulletin, 2000, 45(19): 2118-2122
[7] Meng Jian, Chen Zejing, Li Zhixin and Guo Zengyuan. Field Coordination Analysis and Convection Heat Transfer Enhancement in Duct. Journal of Engineering Thermo physics, 2003, 24(4): 652-654

4

Magneto Hydro-Dynamics and Heat Transfer in Liquid Metal Flows

J. S. Rao and Hari Sankar
Altair Engineering India Pvt. Ltd., Bengaluru, 560103
India

1. Introduction

Liquid metals are considered to be the most promising coolants for high temperature applications like nuclear fusion reactors because of the inherent high thermal diffusivity, thermal conductivity and hence excellent heat transfer characteristics. The coolant used in nuclear reactor should have high heat extraction rate. The high melting point and boiling point which eliminates the possibility of local boiling makes liquid metals more attractive to high temperature applications. The thermal entrance length of liquid metals are relatively high leading to flow never reaching fully developed condition which is always advantageous for heat transfer applications as the Nusselt number value is higher in a developing flow than a fully developed flow. The molecular properties of liquid metals are such that the thermal diffusion is faster than momentum diffusion having Prandtl number << 1. The thermal boundary layer for liquid metal flow is not only confined to the near wall region but also extends to the turbulent core region which makes the turbulent structures important in transfer of heat. As turbulence plays an important role in transfer of heat from the viscous sub-layer to the core flow, it is necessary to maintain high turbulence to achieve high convective heat transfer rates. It may also be noted that surrounding magnetic fields reduce the turbulence and flow becomes more streamlined.

Lithium is the lightest of all metals and has the highest specific heat per unit mass. Lithium is characterized by large thermal conductivity and thermal diffusivity, low viscosity, low vapor pressure as shown in table 1. Lithium is the most promising coolant for thermonuclear power installations. Tritium is a component of fusionable fuel. Tritium does not occur in nature in large amounts and is unstable with its half life 12 years. Tritium can be obtained from lithium with nuclear reactions in fission nuclear reactors or blankets of fusion reactors. Thus, lithium provides raw nuclear fuel to implement fusion reaction.

Density, kg/m^3	Dynamic Viscosity, kg/m-s	Specific Heat, KJ/kg-K	Thermal Conductivity, W/m-K
497	0.0004	4167	50

Table 1. Physical properties of liquid lithium at 650 °C, Davidson (1968)

The plasma generated in the nuclear fusion reactors is confined using an intense magnetic field created by a series of toroidal and poloidal magnets generating magnetic field intensity values of the order 10 Tesla, Kirillov (1994) and is very high compared to the magnetic field intensity of the earth which is of the order 10^{-4} Tesla, Roberts (1967). The electrical conductivity of liquid metals is of the order 10^6 to 10^7 $1/\Omega$-m, Davidson (1968). This introduces the term magneto-hydrodynamics relevant to design of liquid metal flows especially in the application as coolant and breeding material in Tritium breeders of nuclear fusion reactors.

Magneto-hydro-dynamics involves study of magnetic fields and fluid flows and their interaction. Alfven (1942) was the first to introduce the term Magneto-hydrodynamics, though Hartmann and Lazarus (1937) earlier performed both theoretical and experimental studies in Magneto-hydrodynamic flows in ducts. Moffatt (1967) showed analytically that the turbulent velocity fluctuation in a flow is suppressed by application of a uniform magnetic field. This chapter is dedicated to the study of how magnetic field affects the convective heat transfer characteristics in liquid metal flows by affecting the following mechanisms.

1. Mean velocity distribution in the flow
2. Velocity fluctuations in time and space

2. The governing equations of magneto-hydrodynamic flows

The hydrodynamic governing equations are the mass (continuity), momentum and energy equations. They are

Continuity equation

$$\frac{\partial \rho}{\partial t} + \nabla \cdot \rho V = 0 \qquad (1)$$

where ρ is the density of the fluid and V is the velocity represented by $V = u\mathrm{i} + v\mathrm{j} + z\mathrm{k}$. The first term on the left is the accumulation/unsteady term which can be neglected for steady state condition and second term on the left is mass flux term.

Momentum equation

$$\frac{\partial V}{\partial t} + (V \cdot \nabla)V = -\frac{1}{\rho}\nabla p + \nu\nabla^2 V + S \qquad (2)$$

where ν is the kinematic viscosity which is a molecular property of the fluid, p is the pressure. The first term on the left is the unsteady term and the second term on the left is the convection term. The first term on the right is the pressure force term; the second term on the right is the diffusion term which is due to the viscous effects. The third term on the right is the source term which includes other forces such as gravitational forces, electromagnetic forces etc.

Energy equation

$$\rho C_p\left[\frac{\partial T}{\partial t} + (V \cdot \nabla)T\right] = k\nabla^2 T + q''' \qquad (3)$$

where k is the thermal conductivity which is a molecular property of the fluid, T is the temperature and C_p is the specific heat of the fluid. The first term on the left is the unsteady term and the second term on the left is the convection term. The first term on the right is the diffusion term. The third term on the right is the source term which includes volumetric heat sources.

The equations given above are generic equations and what differentiates the equation when used for laminar and turbulent flow regimes is the replacement of molecular viscosity, ν and thermal conductivity, k by effective viscosity, ν_{eff} and effective thermal conductivity, k_{eff} in the momentum and energy equation when the flow is turbulent.

$$\nu_{\text{eff}} = \nu_{\text{m}} + \nu_{\tau} \tag{4}$$

$$k_{\text{eff}} = k_{\text{m}} + k_{\tau} \tag{5}$$

where ν_{τ} is the turbulent viscosity which is the increment in momentum transfer due to diffusion created by turbulence and k_{τ} is the turbulent thermal conductivity which is the increment in heat transfer due to diffusion created by turbulence.
The effect of magnetic field will be included by addition of Lorentz force term $j \times B$ and force due to electric charge $\rho_c E$ as a source term in the momentum equation and inclusion of Joule dissipation term $\frac{j^2}{\sigma}$ as a volumetric heat generation term in the energy equation. The effect of magnetic field on turbulence can be included by altering the turbulent viscosity and turbulent thermal conductivity in the momentum and energy equation respectively.
The modified momentum equation including the magnetic effects is given by,

$$\frac{\partial V}{\partial t} + (V \cdot \nabla)V = -\frac{1}{\rho}\nabla p + \nu\nabla^2 V + \frac{1}{\rho}(j \times B) + \frac{1}{\rho}(\rho_c E) + S \tag{6}$$

The modified energy equation including the Joule dissipation is given by,

$$\rho C_{\text{p}}\left(\frac{\partial T}{\partial t} + (V \cdot \nabla)T\right) = k\nabla^2 T + \frac{j^2}{\sigma} + q''' \tag{7}$$

The inclusion of new variables j, E and B in the hydrodynamic equations requires closure equations given using Maxwell's equations

$$\nabla \cdot D = \rho_c \tag{8}$$

$$\frac{\partial B}{\partial t} = -\nabla \times E \tag{9}$$

$$\nabla \cdot B = 0 \tag{10}$$

$$\nabla \times H = j + \frac{\partial D}{\partial t} \tag{11}$$

and Ohms law

$$j = \sigma(E + V \times B) \tag{12}$$

where D is the electric displacement.
The above equations (1), (6)-(12) can be used for a range of problems, but the equations are complex and may not be fully modeled to capture the physics of liquid metal flows and heat transfer.

The Magneto-hydro-dynamic approximation

In applications where the flow velocity is low as compared to the speed of light, the force due to electric charge $\rho_c E$ can be neglected when $U << c$ based on

$$\frac{|\rho_c E|}{|j \times B|} \approx \frac{U^2}{c^2} \tag{13}$$

where c is the speed of light. This approximation will eliminate the inclusion of the term $\partial D/\partial t$ from equation (11) and equation (8) can be ignored when $U << c$.
Using $B = \mu^* H$ and combining (11) and (12)

$$\frac{1}{\sigma\mu^*}(\nabla \times B) = (E + V \times B) \tag{14}$$

Considering $\nabla \times$ of equation (14) assuming $\eta^* = \frac{1}{\sigma\mu^*}$ is constant, we have

$$\eta^*\left\{\nabla(\nabla \cdot B) - \nabla^2 B\right\} = \nabla \times E + \nabla \times (V \times B) \tag{15}$$

Using (9) and (10), (15) becomes

$$\frac{\partial B}{\partial t} = \nabla \times (V \times B) + \eta^* \nabla^2 B \tag{16}$$

Equation (16) is called the magnetic induction equation; the first term on the right is the induction term and the second term the diffusive term. The induction term describes the interaction of the field with flow. If $Re_m << 1$ then the diffusion term is more significant than the induction term and hence the induction equation, for $Re_m << 1$ becomes

$$\frac{\partial B}{\partial t} = \eta^* \nabla^2 B \tag{17}$$

$$j = \frac{1}{\mu^*}(\nabla \times B) \tag{18}$$

A summary of equations used for liquid metal flows involving heat transfer in presence of magnetic field is given below namely, mass, momentum, energy and magnetic induction.

$$\frac{\partial \rho}{\partial t} + \nabla \cdot \rho V = 0 \tag{19}$$

$$\frac{\partial V}{\partial t} + (V \cdot \nabla)V = -\frac{1}{\rho}\nabla p + \nu\nabla^2 V + \frac{1}{\rho}\left[\frac{1}{\mu^*}B \cdot \nabla B - \nabla\left(\frac{B^2}{2\mu^*}\right)\right] + S \tag{20}$$

$$\rho C_p\left(\frac{\partial T}{\partial t} + (V \cdot \nabla)T\right) = k\nabla^2 T + \frac{\left(\frac{1}{\mu^*}(\nabla \times B)\right)}{\sigma} + q''' \tag{21}$$

$$\frac{\partial B}{\partial t} = \eta * \nabla^2 B \tag{22}$$

The total number of unknowns is 9 including three scalars of velocity, three scalars of magnetic field, pressure, density and temperature and the number of equations is 8 including continuity, 3 momentum equations, energy and 3 induction equations. We need a closure equation for the set of equations. If the flow is incompressible then we can treat density as constant and hence removing the need of an extra equation. If the flow is compressible then we can use an equation for pressure as a function of temperature and density i.e., $p = f(T, \rho)$ for example like ideal gas law for gases.

3. Non-dimensional numbers

Reynolds Number: The non-dimensional number that gives the ratio of the inertial forces to the viscous forces is Reynolds number defined by

$$\mathrm{Re} = \frac{\text{Inertial Forces}}{\text{Viscous Forces}} = \frac{\rho U_0 L}{\mu}$$

where ρ is the density of the fluid (kg/m³), L is the characteristic length (m), U_0 is the velocity (m/s) and μ is the dynamic viscosity of the fluid (kg/m-s). For liquid metal flows in fusion reactors, the typical values encountered is of the order 10^4 to 10^5. The transition from laminar to turbulent flows occurs at Re ≈ 2300.
Hartmann Number: The non-dimensional number that gives the ratio of the electromagnetic forces to the viscous forces is Hartmann number defined by

$$\mathrm{Ha} = \left(\frac{\text{Electromagnetic Forces}}{\text{Viscous Forces}}\right)^{\frac{1}{2}} = B_0 L \sqrt{\frac{\sigma}{\nu \rho}}$$

where, B_0 is the magnetic field intensity (Tesla), σ is the electrical conductivity of the fluid (1/Ωm) and ν is the kinematic viscosity of the fluid (m^2/s^2). For liquid metal flows in fusion reactors, the typical values encountered for Ha are of the order 2×10^4 and the effect of viscosity and turbulence on the flow is negligible in those situations.
Stuart Number: The non-dimensional number that gives the ratio of the electromagnetic forces to the inertia forces is Stuart number (also called Interaction Parameter N).

$$\mathrm{N} \equiv \mathrm{St} = \frac{\mathrm{Ha}^2}{\mathrm{Re}} = \frac{\text{Electromagnetic Forces}}{\text{Intertial Forces}} = \frac{\sigma B_0^2 L}{\rho U_0}$$

Stuart number can be used to characterize the effect of magnetic field on turbulence as it represents the ratio of large-eddy turnover time τ to the Joule time τ_m. Stuart number shows the ability of the magnetic field to change the isotropic turbulent structures to anisotropic two-dimensional turbulence. When N is small, the anisotropy induced by Joule dissipation is negligible and when it is larger than a critical value N_c, Lorentz force will drive turbulence to two-dimensional situation. The significance of St is found in recent days as the transition of turbulent flow will happen in the range 10^1 to 10^2 as reported by Uda *et al.* (2001) and a sudden increase of turbulence is noted to happen at Stuart number equal to 10. Another interpretation of Stuart number can be that when N is small, the electromagnetic forces will be weak and will not be able to affect the flow.

Magnetic Reynolds Number: The non-dimensional number that gives the ratio of the momentum diffusivity to the electromagnetic diffusivity defined by

$$\mathrm{Re_m} = \frac{\text{Momentum Advection}}{\text{Magnetic Diffusion}} = \frac{U_0 L}{\dfrac{1}{\mu^* \sigma}}$$

where, μ^* is the fluid magnetic permeability (h/m). It characterizes the effect of induced magnetic field on the resultant magnetic field due to the imposed and induced magnetic fields. At $\mathrm{Re}_m << 1$ the imposed magnetic field will guide the flow and will be independent of the induced magnetic field as it will be very small compared to the imposed magnetic field. The Lorentz force will be a linear function of velocity when $\mathrm{Re}_m << 1$ and quasi-static approximation can be applied. At $\mathrm{Re}_m >> 1$ the dependence between the magnetic field and velocity is non-linear as the imposed magnetic field is a function of velocity.

Another interpretation of magnetic Reynolds number is based on the ratio of characteristic time scale of diffusion of magnetic field to the time scale of turbulence defined by

$$\mathrm{Re_m} = \frac{\text{Time scale of diffusion of magnetic field}}{\text{Time scale of turbulence}} = \frac{vL}{\dfrac{1}{\mu^* \sigma}}$$

where v is the rms fluctuating velocity represented as

$$v = \sqrt{\frac{R_{\mathrm{ij}}}{3}} \text{ where } R_{\mathrm{ij}} = \overline{u_{\mathrm{i}} u_{\mathrm{j}}}$$

and u_i is the fluctuating velocity. At $\mathrm{Re_m} << 1$ the distortion of magnetic field lines by the fluid turbulence is small and the induced magnetic fluctuations with respect to the mean magnetic field is also small.

Batchelor number (Magnetic Prandtl Number): The non-dimensional number that gives the ratio of the momentum diffusivity to the electromagnetic diffusivity is Batchelor number defined by

$$\mathrm{Bt} \equiv \mathrm{Pr_m} = \frac{(\text{Viscous diffusion length})^2}{(\text{Elecro-magnetic diffusion length})^2} = \frac{\nu}{\nu_\mathrm{m}} = \mu^* \sigma \nu$$

It characterizes the ratio of thickness of hydrodynamic boundary layer to magneto-hydrodynamic layers. For liquid metal flows typical values encountered are of the order of 10^{-5} to 10^{-7}. When $\mathrm{Pr_m}$ is small, the effect of an external magnetic field is known to be stabilizing and weakly dependent on $\mathrm{Pr_m}$ in a variety of flow configurations.

4. Effect of MHD on mean velocity distribution in duct flows

Many analytical studies have been conducted in fully developed duct flows starting from 1930 onwards with some significant contributions by Hartmann (1937), Shercliff (1953) and Hunt (1965). Most of the analytical studies are performed on rectangular cross-section because of the ease of assigning the direction of the magnetic field and also due to the ease of non-dimensionalizing the MHD equations over the channel height and width. An axial magnetic field does not have much significance in redistribution of mean velocities in duct flow whereas a transverse magnetic field significantly influences the mean velocity distribution. This is because the value of the Lorentz force acting on the body is not significant for a flow subjected to an axial magnetic field Kirillov (1994) as shown in figure 1, because the net Lorentz force is zero (the cross product of two vectors in the same direction is zero). Studies have shown that a transverse magnetic field affects the mean velocity distribution near the walls based on the distribution of induced currents in the duct cross-section depending on the electrical conductivity of the duct walls. The direction of the Lorentz force will act against the flow direction as per right hand thumb rule.

Classical analytical studies with transverse magnetic fields showed that for Ha >> 1 there exist exact solutions for the governing equations or solutions in the form of infinite series as shown in Hartmann (1937), Shercliff (1953) and Hunt (1965). It was found that there exists a core region in the flow where the pressure forces are balanced by the Lorentz force and velocity is uniform in the region. Special boundary layers exist depending on the orientation of the wall with respect to the magnetic field.

Hartmann Layer: Hartmann layer refers to the thin boundary layers formed near the walls perpendicular to the magnetic field and hence has a normal magnetic field. The thickness of the Hartmann layer is of the order O (Ha^{-1}) and the viscous forces balance with the magnetic forces in the Hartmann layer. Hartmann layers have a major influence on the core region when the electrical conductivity of the Hartmann walls is low, Fink and Beaty (1999).

Shercliff/Shear/Side Layers: The boundary layers formed on the walls parallel to the magnetic field (side walls) can have different characteristics based on the electrical conductivity of the walls. The layers can either stay near the walls or exist some distance away from the wall in the form of shear layers. The thickness of the boundary layers formed on the side wall is of the order O ($Ha^{-1/2}$) and is higher than that of Hartmann layers. The boundary layers formed near the side walls influence the core region significantly only if the Hartmann walls are electrically conducting and the side walls are insulating, Fink and Beaty (1999).

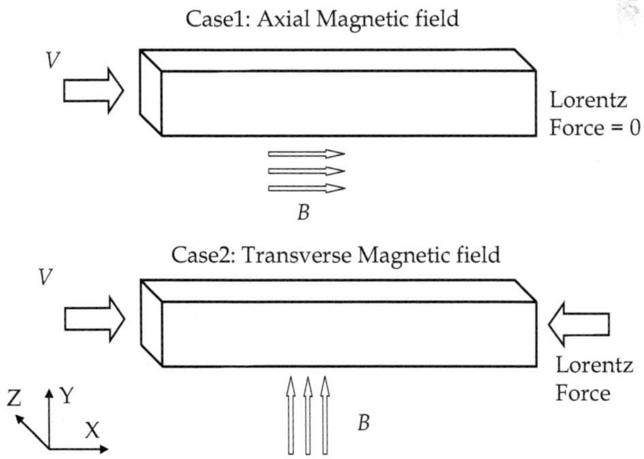

Fig. 1. Schematic showing the direction of Lorentz Force

The schematic of the cross-section of a rectangular duct used for the analytical studies is shown in figure 2. The direction of flow is along the z coordinate and the external magnetic field B_0 is applied in the y direction. The walls BB are the Hartmann walls as they are perpendicular to the magnetic field and walls AA are the side walls as they are parallel to the magnetic field. The dimensions of the duct in x and y directions are $2a$ and $2b$ respectively. The dimensions of the duct are non-dimensionalized as $\eta = y/a$ and $\xi = x/b$.

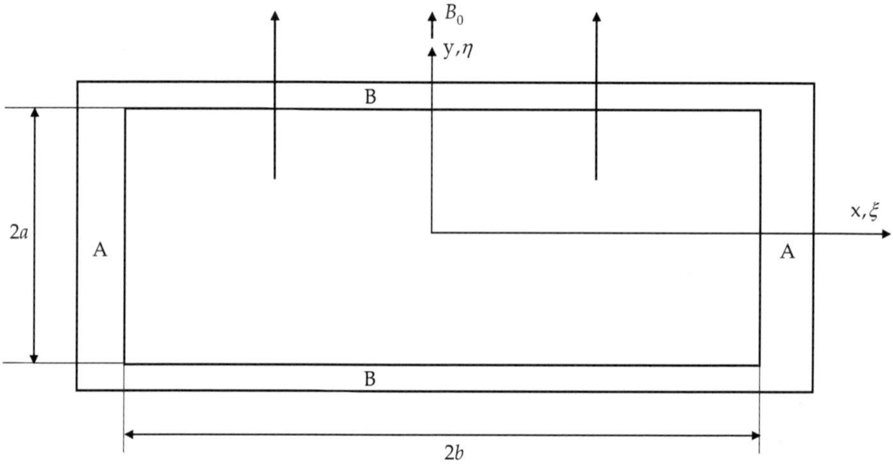

Fig. 2. Schematic of rectangular duct with magnetic field applied in y direction

4.1 Mean velocity distribution parallel to magnetic field (Hartmann layers)

The significance of Hartmann layer on the flow depends on the electrical conductivity of the walls. If the electrical conductivity of the Hartmann walls is low (insulating), then the currents induced in the core regions close in the Hartmann layers and hence making it important in deciding the resistance offered by magnetic filed on the flow. If the Hartmann walls are electrically conducting, then the electric current lines close in the walls and hence the significance of the Hartmann layer on the flow reduces. The mean velocity distribution perpendicular to the Hartmann wall is given using the relation given by Hartmann (1937). The thickness of Hartmann layer is proportional to Ha^{-1}. The normalized axial velocity perpendicular to the Hartmann wall is solved analytically by Hartmann (1937) for a duct having insulating Hartmann walls.

$$\frac{U}{U_c} = \frac{\mathrm{Ha}}{\mathrm{Ha\text{-}tanh}(\mathrm{Ha})}\left[1 - \frac{\cosh(\mathrm{Ha}\eta)}{\cosh(\mathrm{Ha})}\right] \tag{23}$$

If Ha grows, the velocity profile becomes more and more flattened as shown in figure 3. This effect is known as the *Hartmann effect*. The thin layer near the wall where the flow velocity changes from zero to U_c is called the *Hartmann layer*. The Hartmann effect is caused by the Lorentz force, which accelerates the fluid in the Hartmann layers and slows it down in the bulk. The case of Hartmann walls having high electrical conductivity is given by Hunt and

Stewartson (1965) and is similar to that of the solution shown in figure 3. This states that the Hartmann layer velocity profile is independent of the electrical conductivity of the Hartmann walls but its control on the core flow will be high if the electrical conductivity of the Hartmann walls is high because the electric field will be stronger.

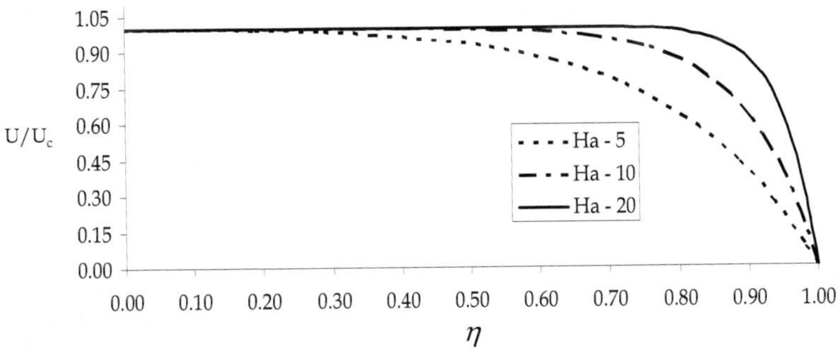

Fig. 3. Normalized axial velocity profiles in the Hartmann layer (centre of duct)

4.2 Mean velocity distribution perpendicular to magnetic field (side layers)

The mean velocity distribution depends on the distribution of electric current lines in the duct which depends on the electrical conductivity of the duct walls. The side layer thickness varies as a function of $Ha^{-1/2}$. Different cases based on the electrical conductivity of walls AA and BB, denoted by d_A and d_B respectively is given in sections 4.2.1 to 4.2.4.

4.2.1 Case 1: Rectangular duct with $d_B = 0$, $d_A = 0$

The mean velocity distribution perpendicular to the side wall for a rectangular duct with all walls electrically insulating is given using the relation Shercliff (1953) and then developed by Muller and Buhler (2001) as given by equation (24).

$$U \approx \sum_{i=1,3,5}^{\infty} u_i(y)\cos(\lambda_i x) \tag{24}$$

$$u_i(y) = \frac{k_i}{\lambda_i^2}\left\{1 - \frac{f_i(y)}{f_i(1)}\right\}$$

$$f_i(y) = \alpha_{i2}\cosh(p_{i1}y) - \alpha_{i1}\cosh(p_{i2}y)$$

$$\lambda_i = \frac{i\pi}{2d}$$

$$k_i \equiv \frac{2\sin(\lambda_i d)}{\lambda_i d}$$

$$\gamma \equiv \sqrt{Ha^2 + 4\lambda_i^2}$$

$$p_{i\,1,2} = \frac{1}{2}(\mathrm{Ha} \mp \gamma)$$

$$\alpha_{i\,1,2} = \sinh\left(p_{i\,1,2}\right)$$

The normalized axial velocity with respect to the normalized distance ξ from the side walls is plotted for various Hartmann numbers as shown in figure 4.

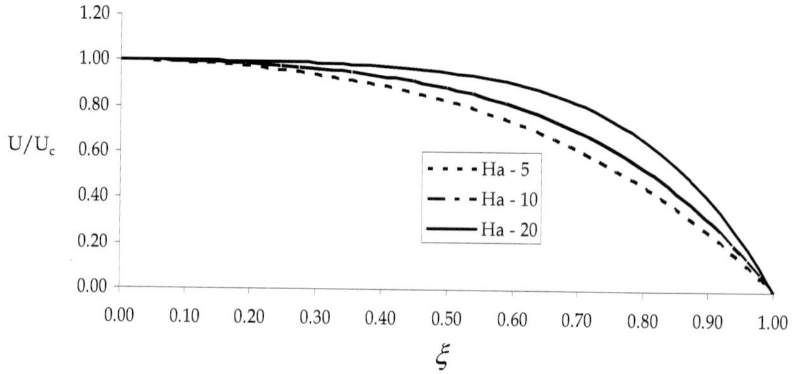

Fig. 4. Normalized axial velocity profiles in the side wall boundary layer

4.2.2 Case 2: Rectangular duct with $d_B = 0$, $d_A = \infty$

The mean velocity distribution perpendicular to the side wall for a rectangular duct with Hartmann walls electrically insulating and electrically conducting side walls is given by Hunt and Stewartson (1965). The axial velocity normalized with respect to the core velocity plotted against the $\mathrm{Ha}^{1/2}(b-\xi)$ is shown in figure 5. It can be seen that the velocity profile is similar to case 1 i.e. Rectangular duct with d_B=0, d_A=0. This states that when the Hartmann walls are electrically insulating then the velocity profile near the side walls is independent of the electrical conductivity of the side walls.

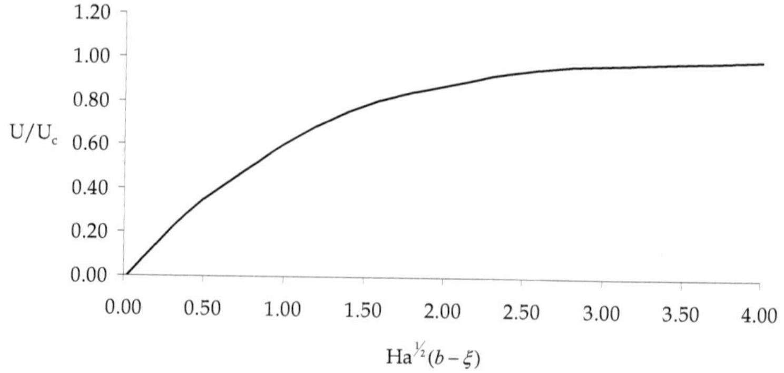

Fig. 5. Normalized axial velocity profiles in the side wall boundary layer

4.2.3 Case 3: Rectangular duct with $d_B = \infty$, $d_A = 0$

The mean velocity distribution perpendicular to the side wall for a rectangular duct with all walls electrically conducting is given by Hunt (1965). The mean velocity distribution perpendicular to the side wall is given using the relation Hunt (1965) as $\mathrm{Ha} \to \infty$

$$U \approx \sum_{j=0}^{\infty} \frac{2(-1)^j \cos\alpha_j \eta}{\mathrm{Ha}^2 \alpha_j}\left[1-\exp\left(-\lambda_j \xi\right)\left\{\cos\left(\lambda_j \xi\right)-\frac{\mathrm{Ha}}{\alpha_j}\sin\left(\lambda_j \xi\right)\right\}\right] \tag{25}$$

$$\alpha_j \approx \left(j+\frac{1}{2}\right)\pi$$

$$\lambda_j \approx \left(\frac{1}{2}\alpha_j \mathrm{Ha}\right)^{\frac{1}{2}}$$

The axial velocity normalized with respect to the normalized distance from the side walls is plotted for various Hartmann numbers as shown in figure 6. The maximum velocity in the side layers is of the order O(Ha)V_c and the maximum velocity on the side wall jet tends to 0.25HaV_c as $\mathrm{Ha} \to \infty$, while the minimum velocity becomes locally negative for Ha > 89 and tends to - 0.011 HaV_c The high velocity gradients near the side walls generate instabilities at high Hartmann numbers. The high velocity jets formed near the side walls results in velocity deficit in the core region and in the Hartmann layers.

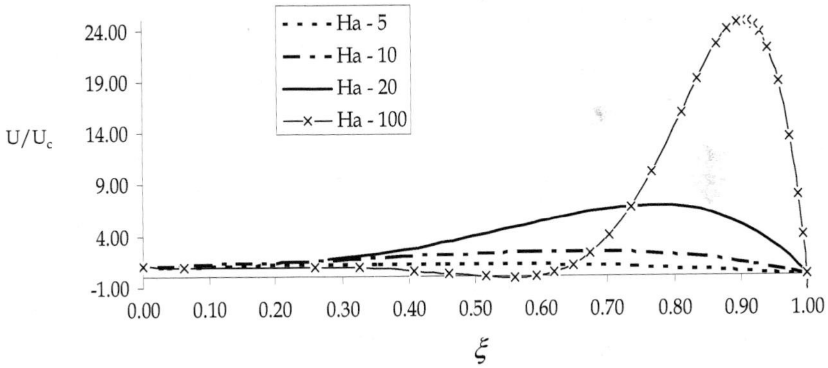

Fig. 6. Normalized axial velocity profiles in the side wall boundary layer

4.2.4 Case 4: Rectangular duct with $d_B = \infty$, $d_A = \infty$

The mean velocity distribution perpendicular to the side wall is given using the relation given by Hunt (1965). The maximum velocities in the boundary layer are higher than the core velocity but are of the same order unlike the case with $d_A = 0$

$$U \approx \sum_{j=0}^{\infty} \frac{2(-1)^j \cos\alpha_j \eta}{\mathrm{Ha}^2 \alpha_j}\left[1-\exp\left(-\lambda_j \xi\right)\left\{\cos\left(\lambda_j \xi\right)-\sin\left(\lambda_j \xi\right)\right\}\right] \tag{26}$$

The axial velocity normalized with respect to the normalized distance from the side walls is plotted for various Hartmann numbers as shown in figure 7. The velocity near the side wall increases with increase in magnetic field, but the maximum value of velocity jets approaches $1.25 V_c$ when $\text{Ha} \to \infty$.

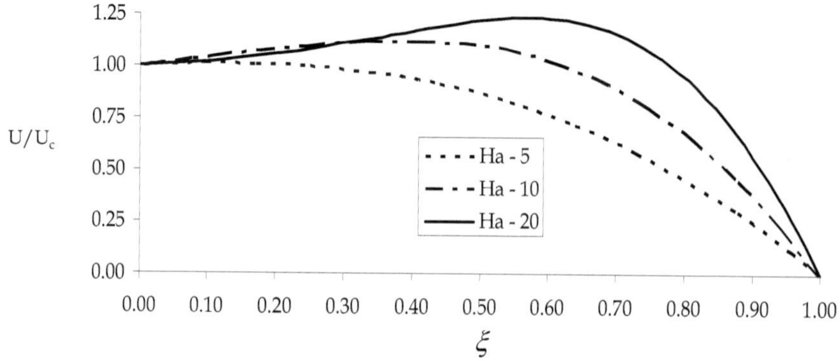

Fig. 7. Normalized axial velocity profiles in the side wall boundary layer

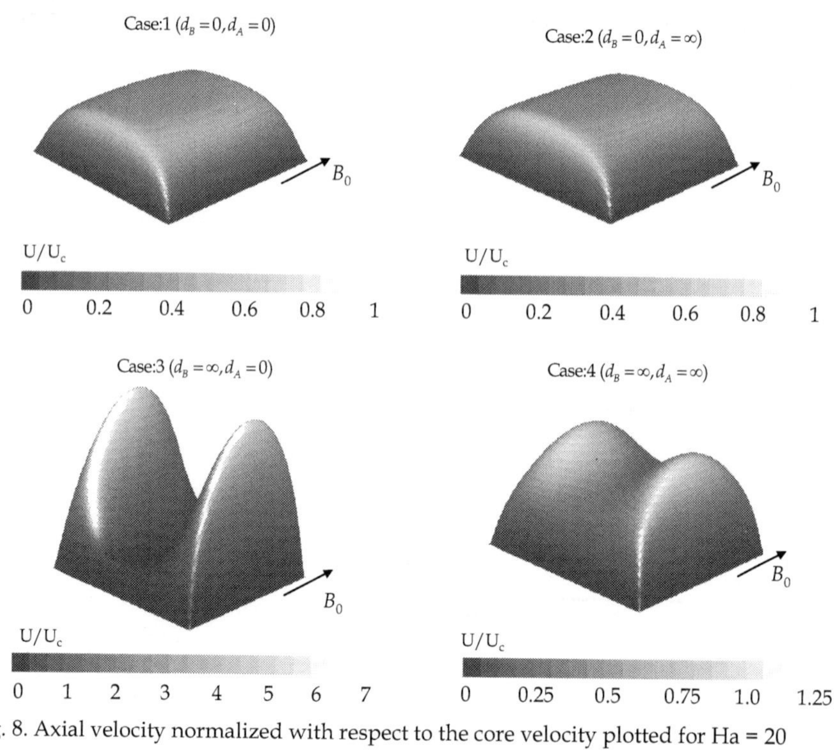

Fig. 8. Axial velocity normalized with respect to the core velocity plotted for Ha = 20

4.3 Comparison on mean axial velocity profiles

The analytical solutions of the mean velocity distribution presented in the previous sections are supported using numerical simulations performed for laminar ($Re \approx 10$) steady state flow of an electrically conducting fluid in a square duct using FLUENT (version 6.3.26), see figure 8. The velocity profile near the Hartmann walls is independent of the electrical conductivity of the walls as stated in section 4.1 and the velocity gradients and thickness seems to be similar. The velocity profiles near the side walls shows different characteristics depending on the electrical conductivity of the walls. The velocity profiles for case 1 in section 4.2.1 and case 2 in section 4.2.2 are similar as the induced field has low strength due to the high resistance offered by the Hartmann walls.

In the cases where $d_B = 0$ the net electrical current in the circuit will be low and current lines will not be strong enough to create a redistribution in the mean velocity field as shown in figure 8 and the imbalance between the Lorentz forces and pressure will not be significantly high.

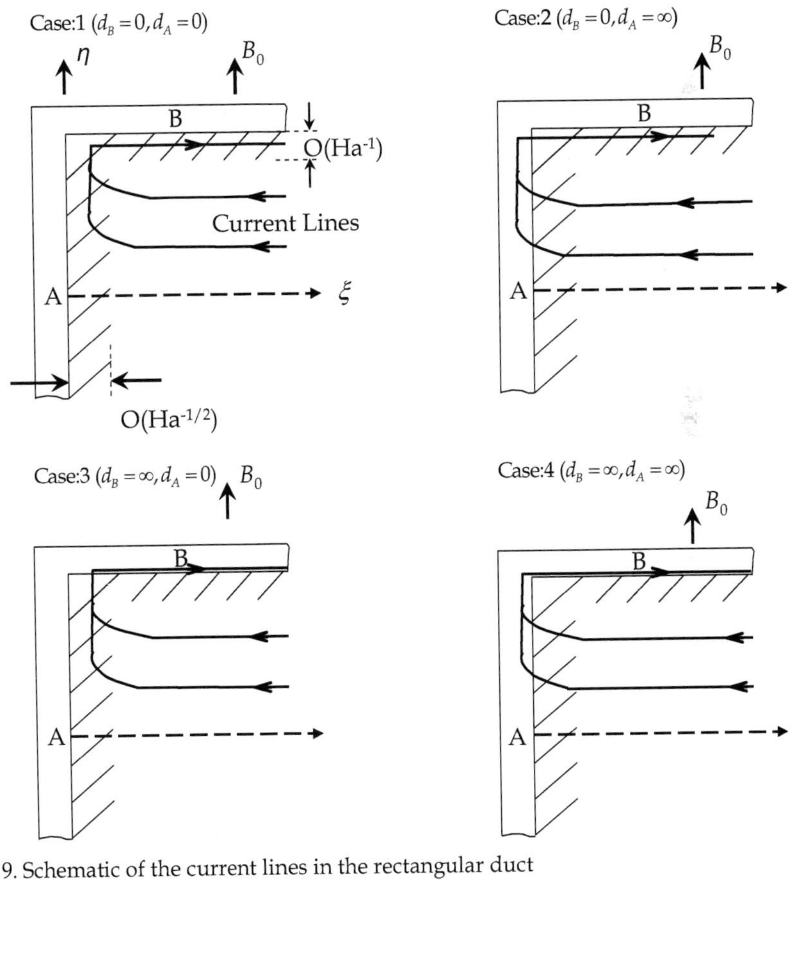

Fig. 9. Schematic of the current lines in the rectangular duct

The distribution of induced current lines in the cross section for different cases is given in figure 9. The formation of shear layers in the case 3 given in section 4.2.3 where $d_B = \infty$ and $d_A = 0$ is due to the imbalance in the Lorentz forces and pressure forces. If $d_A \mathrm{Ha}^{1/2} << 1$, then the current lines return to walls BB through boundary layers on wall AA and hence the Lorentz force term drops near the boundary layer as compared to other regions making the velocity near the side walls to shoot up as shown in figure 9.

If $d_A \mathrm{Ha}^{1/2} >> 1$ then the current lines returns to wall BB through walls AA which makes the induced magnetic field and hence the Lorentz Force ($j \times B$) term uniform in the duct cross-section which happens in case 4 of section 4.2.4 shown in figure 8 and figure 9. Higher the value of electrical conductivity of side walls, lower the value of the maximum velocity of the jet for case 3 and case 4.

5. Effect of MHD on turbulence suppression

An external magnetic field suppresses the turbulent fluctuations in liquid metals (Low magnetic Prandtl number fluid) independent of the direction of magnetic field with respect to the flow and independent of the mean velocity distribution in the flow. The flow regimes in magneto-hydrodynamic flows of liquid metals, i.e. $R_m << 1$, see Fink and Beaty (1989) is shown in figure 10. It can be seen that the flow regime above line 'A' is the laminar regime.

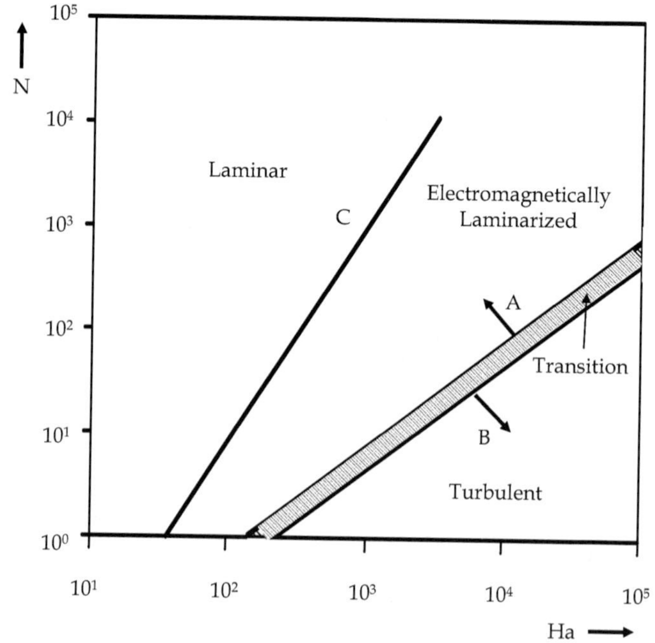

Fig. 10. Turbulent flow regimes in low magnetic Reynolds number flows, Fink and Beaty (1989)

The flow regime below line 'B' is turbulent regime with low values of Stuart number N, in which the influence of the magnetic field on the flow is less. The shaded region between line 'A' and line 'B' is the transition region in which the turbulence levels increase and will be greater than the flow without magnetic field as stated in Uda *et al.* (2001 and 2002). The line 'C' in figure 10 represents the critical Reynolds number for laminar to turbulent transition equal to 2000. The region on the left side of line 'C' is the classical laminar flow regime. The region on the right side of line 'C' and above line 'A' is the region of electromagnetically laminarized flow in which the turbulent flow is converted to so-called Stokes flow in which the inertial term is negligible as mentioned in Happel (1981).

Axial magnetic field suppresses turbulent fluctuations and also increases the critical Reynolds number for transition from laminar to turbulent regime as stated in Kirillov (1994). A transverse magnetic field also has the effect of turbulence reduction. Moffat (1967). The turbulent structures get aligned with the magnetic field lines. As shown in figure 11, when the magnetic field increases, the MHD effect suppresses the turbulence significantly and the large vortex structure breaks up into several small vortex structures. At the same time, the vortex structure was dampened and stretches along the flow direction (perpendicular to the magnetic field).

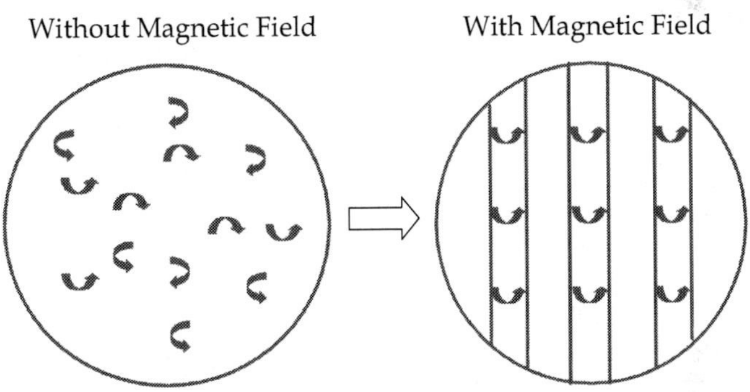

Fig. 11. Change of isotropic turbulent structures to anisotropic structures, Luo *et al.* (2003)

6. Effects of mean velocity distribution and turbulence suppression on convective heat transfer

The convective heat transfer characteristic is affected by magnetic field due to two main mechanisms. Evtushenko (1995)

1. Mean velocity distribution in the flow
2. Velocity fluctuations in time and space

The first mechanism, mean velocity distribution has to take into account two important factors: (1a) the increase of velocity and velocity gradients near side walls parallel to the magnetic field and (1b) Velocity (flow rate) redistribution in the magnetic field direction. The second mechanism, effect of MHD on turbulent fluctuations and hence heat transfer has to take into account important factors like (2a) Damping of turbulent fluctuations, (2b) transition to two-dimensional structure and (2c) the instabilities created in the high velocity jets near the side wall.

6.1 Effect of axial magnetic field on convective heat transfer

In case of flow of liquid metals in heated channels under the influence of a uniform axial magnetic field shows a decrease of convective heat transfer at low and moderate Hartmann numbers whereas the convective heat transfer and hence Nu increases at higher Hartmann numbers as shown in Miyazaki (1988). It was stated in section 4 that an axial magnetic field does not affect the mean velocity distribution so the modification of convective heat transfer is due the variation in the turbulent fluctuations in time and space.

Reynolds number	Hartmann Number	$Nu/Nu_{B=0}$
$(2.5-5)\cdot 10^3$	360	0.83
$(1-2)\cdot 10^4$	700	0.50
$(3-4)\cdot 10^4$	1400	0.30
$1\cdot 10^4$	3600	2.75

Table 2. Variation of Nusselt Number with Axial Magnetic Field, Miyazaki (1988)

The values of Nu for various value of Hartmann numbers is shown in table 2. It can be seen that the values of Nu decreases from its value, $Nu/Nu_{B=0} = 1$, at Ha = 0. The decrease in Nu value for small and moderate Ha is more when the Reynolds number is high because of the higher turbulence content in the flow. At lower values of Reynolds numbers, the flow will be inherently laminar and therefore the reduction in Nu due to suppression of turbulent fluctuations will be low.

At high values of Hartmann numbers, the Nusselt number was found to increase, violating the earlier theories and studies, see Miyazaki (1988). Miyazaki attributes the increase in the Nusselt number is due to the increase in turbulence levels in the flow as the effect of buoyancy can be ruled out because the flow direction upwards.

6.2 Effect of transverse magnetic field on convective heat transfer

The studies in the field of effect of magnetic field on convective heat transfer in ducts subjected to transverse magnetic field can be classified into two cases

1. Absence of high velocity jets near the side walls
2. Presence of high velocity jets near the side walls

6.2.1 Case 1: Absence of high velocity jets near the side walls

In case of ducts having Hartmann walls with zero conductivity, high velocity jets will not be formed near the side walls. Gardener and Lykoudis (1971b) performed experiments with flow of Mercury in horizontal electrically insulated pipe subjected to transverse magnetic field. It was found that the velocity profile near the Hartmann wall becomes flat with increase in magnetic field as discussed in section 4.1 and the velocity profile near side walls becomes round as discussed in sections 4.2.1 and 4.2.2. The mean velocity distribution is not much different with the increase in magnetic field, so the modification of turbulence phenomenon by the magnetic field will affect the convective heat transfer predominantly for this case. The Nusselt number distribution near the Hartmann and side walls for a range of Reynolds numbers and Hartmann numbers is shown in figure 12. The decrease of Nusselt number with increase in magnetic field is lesser at lower Reynolds number because the turbulence content in the flow at low Reynolds number will be lesser.

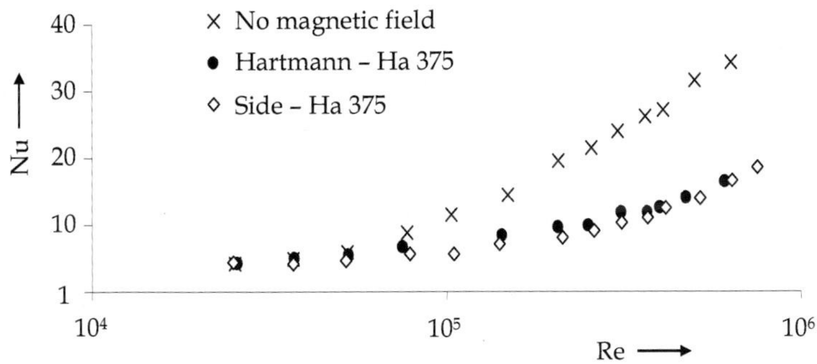

Fig. 12. Nusselt number with magnetic field intensity, Gardener and Lykoudis (1971b)

The reduction in Nusselt number with increase in magnetic field is because of the reduction in turbulence quantified using turbulence kinetic energy as shown in figure 13. It was found that the turbulent kinetic energy decreases both near the Hartmann and side walls with increase in magnetic field where r/R = 0 represents the centre of the duct and r/R = 1 represents the walls. The damping force within the Hartmann layer is much higher than at the side region due to the high local electric current density. The turbulence in core is suppressed initially and then the turbulence in the Hartmann layer followed by the turbulence near the side wall.

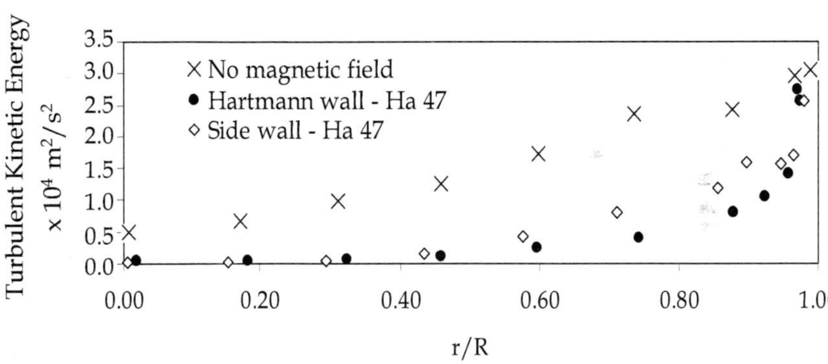

Fig. 13. Turbulent kinetic energy vs. r/R for Re = 50,000, Gardener and Lykoudis (1971a)

A correlation for Nusselt number values is created from various experimental results by Ji and Gardener (1997) and is given using the following relation as a function of Peclet number Pe and Hartmann number Ha

$$\mathrm{Nu} = 7 + \frac{0.00782\mathrm{Pe}^{0.811}}{\left(1 + 0.0004\mathrm{Ha}^{1.5} f(\mathrm{Pe})\right)} \tag{27}$$

$$f(\mathrm{Pe}) = \left(0.3 + 4.75 \times 10^{-5}\mathrm{Pe} - 2.10 \times 10^{-9}\mathrm{Pe}^2\right)$$

6.2.2 Case 2: Presence of high velocity jets near the side walls

In case of ducts having Hartmann walls with finite conductivity, high velocity jets will be formed near the side walls. The side layers with high velocity jets (M shaped profile, figure 8 case 3) carry high mass flux is the prime reason for increase of heat transfer near the side walls. Miyazaki et al. (1986) performed experiments to determine the heat transfer characteristics for liquid metal Lithium flow in annular duct with electrically conducting walls under the influence of transverse magnetic fields. The Nusselt number plotted with magnetic field is shown in figure 14. It can be seen that the Nusselt number increases near the side walls and decreases near the Hartmann walls. A singular rise of Nusselt number can be seen near both the walls.

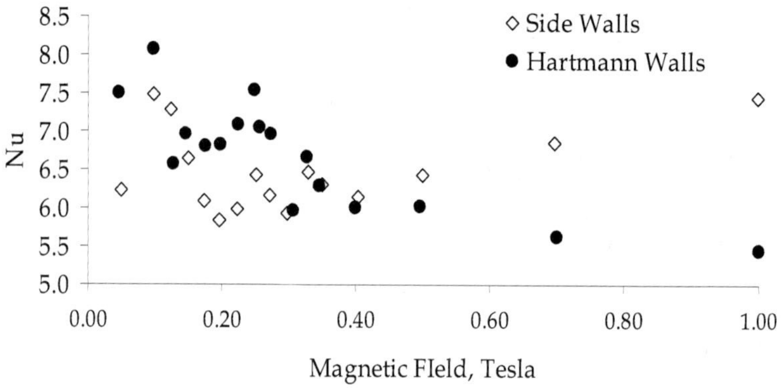

Fig. 14. Nusselt number plotted with magnetic field intensity, Miyazaki (1986)

This effect of heat transfer enhancement near the side walls is caused by the generation and development of large scale velocity fluctuations in the near wall area. The reduction in Nusselt number near the Hartmann walls is created due to the turbulence reduction as shown in figure 15.

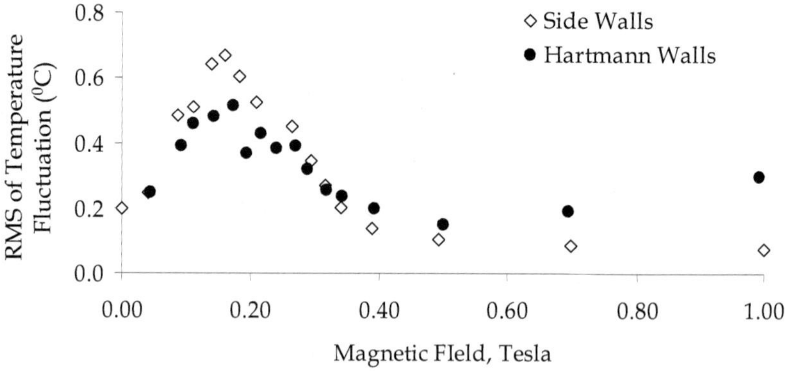

Fig. 15. RMS of temperature fluctuation with magnetic field intensity, Miyazaki (1986)

7. Application of numerical codes

A difficulty in experimental study of the flow of liquid metals arises as the visualization is not possible because of the opaque nature of liquid metals. Application of closed form analytical solutions is limited to simple cases where the equations are not very complex. This makes the application of numerical simulations useful for the study of liquid metal magneto-hydro-dynamic flows. An example of application of a numerical code to explain the mechanisms affecting heat transfer for flow subjected to transverse magnetic field is explained using a series of simulations given in Rao and Sankar (2010), see figure 16.

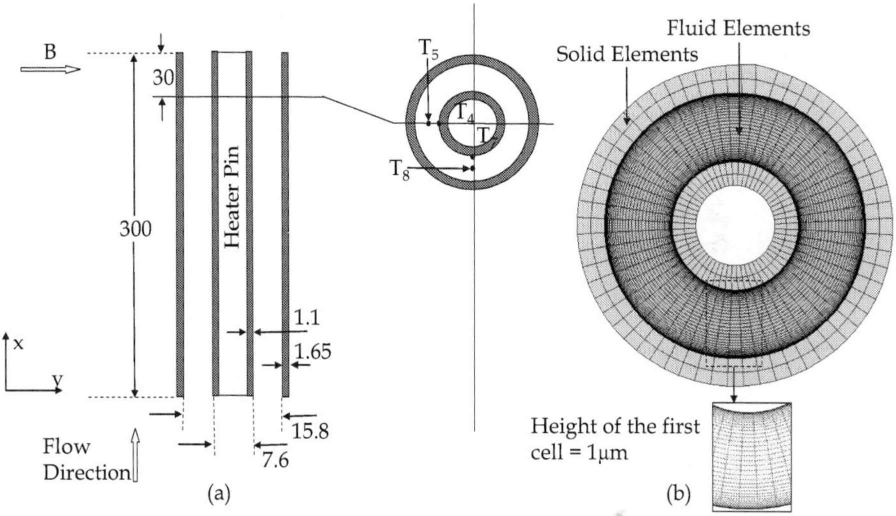

Fig. 16. (a) Schematic of model (b) Details of the computational mesh, Rao and Sankar (2010)

A numerical study is conducted in an annular duct formed by a SS316 circular tube with electrically conducting walls and a coaxial heater pin, with liquid Lithium as the working fluid for magnetic field ranging from 0 - 1 Tesla The Hartmann and Stuart number of the study ranges from 0 - 700 and 0 - 50 respectively. The Reynolds number of the study is 10^4. It was shown that the convective heat transfer and hence the Nusselt number decreases near the walls perpendicular to the magnetic field due to reduction in turbulent fluctuations with increase of magnetic field. It was observed that the Nusselt number value increases near the walls parallel to the magnetic field as the mean velocity increases near the walls. A singular rise was observed near both the walls near Stuart number ~ 10 which is due to the increase of turbulence levels in the process of changing from turbulent to electromagnetically laminarized flow, see figure 17.

When a very low Reynolds number ~ 300 is used, the reduction in Nusselt number near the Hartmann walls is less as shown in figure 18. This shows that the reduction in Nusselt number near the Hartmann walls for the high Reynolds number study is due to the reduction in turbulent fluctuations. The Nusselt number was found to increase near the side walls as the mean velocity increases near the walls. When an insulating duct is used the Nusselt number near the parallel walls did not increase for the case with insulating walls as in the case with conducting walls showing the contribution of the 'M' shaped velocity profile in the Nusselt number increase near the parallel walls. The Nusselt number near the perpendicular walls was found to decrease at a higher rate in case of insulating walls than that of the study with conducting walls as shown in figure 19.

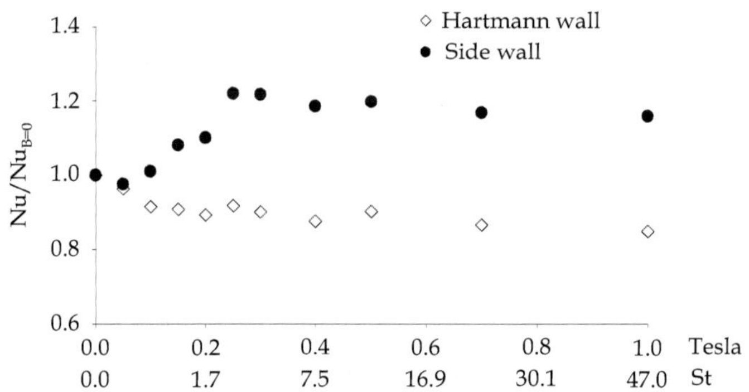

Fig. 17. High Reynolds number with conducting walls, Rao and Sankar (2010)

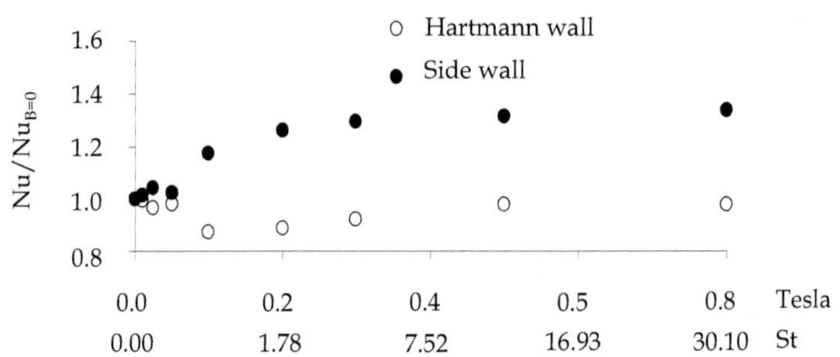

Fig. 18. Low Reynolds number with conducting walls, Rao and Sankar (2010)

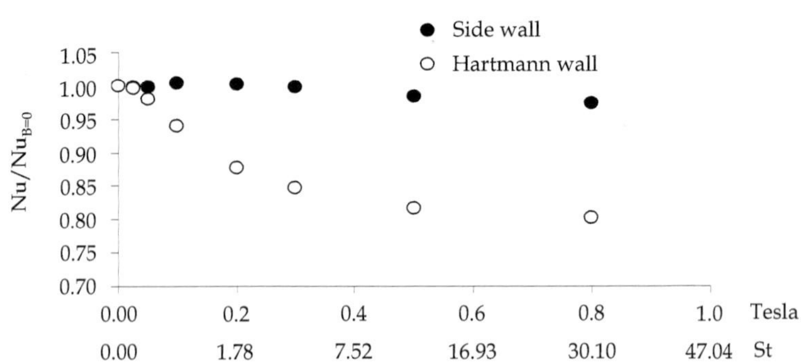

Fig. 19. High Reynolds number with insulating walls, Rao and Sankar (2010)

8. Application of liquid metal MHD studies in nuclear fusion reactors

International Thermo-nuclear Experimental Reactor is an international organization formed in 1985 comprising of researchers from US, EU, China, Japan, India, Korea and Russia working towards development of a test reactor (TOKOMAK) which is expected to be developed by 2020. The test reactor will be installed in France where the head office of ITER is situated. Salient details of the reactor to be developed are shown in figure 20. The reactor height will be close to 100 ft and would weigh around 38000 tons. The cryostat is the external chamber around the TOKOMK which maintains high vacuum inside it to reduce the heat load from atmosphere through conduction and convection. The fusion of Deuterium and Tritium happens inside the plasma chamber. The magnets are used to confine the plasma created inside the plasma chamber using a magnetic field of 4-8 Tesla.

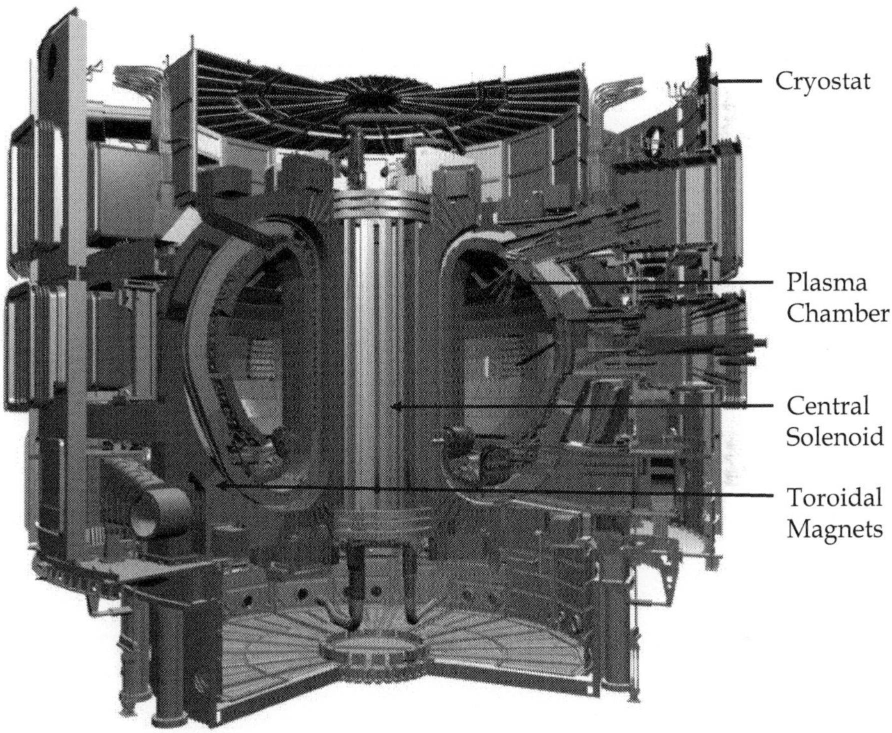

Fig. 20. Details of the TOKOMAK

Tritium breeding modules are used in fusion reactors to produce Tritium by reacting Lithium with neutrons a byproduct of the nuclear fusion reaction. The two basic breeder concepts developed by ITER are liquid breeder and solid breeders. The advantages of liquid breeder over solid breeder are the high Tritium breeding ratio and the Lead-Lithium eutectic can also act as a coolant inside the breeding module which is subjected to high heat from plasma and the heat generated in itself due to bombardment of neutrons. The major disadvantages of liquid breeders over solid breeders is the pressure drop in the form of Lorentz force and the reduction in convective heat transfer characteristics of the liquid metal when it is flowing in the presence of intense magnetic field produced by the cryogenic super-conducting magnets.

Wong *et al.* (2008) has mentioned about the various liquid metal breeders being developed around the world details of which is given in the table 3. All the liquid breeder design uses Lithium as the breeding material though most of them use a eutectic of Lead and Lithium because of the lower electrical conductivity and the neutron multiplication ability of Lead.

Country	Name of TBM	Liquid Metal Used
US	DCLL - Dual Coolant Lead Lithium	PbLi
EU	HCLL - Helium Cooled Lithium Lead	PbLi
Korea	HCML - Helium Cooled Molten Lithium	Li
India	LLCB - Lead-Lithium Cooled Ceramic Breeder	PbLi
China	DFLL - Dual Functional Lithium Lead	PbLi

Table 3. Details of the liquid TBM developed in the various countries, Wong *et al.* (2008)

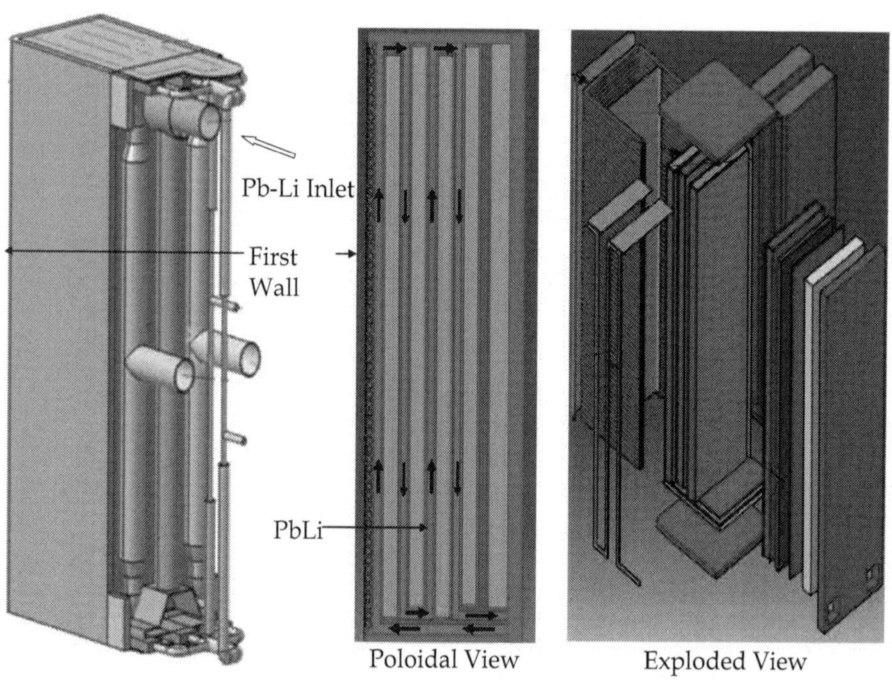

Fig. 21. Details of LLCB- TBM, Wong *et al.* (2008)

Indian Lead lithium Cooled Ceramic Breeder (LLCB) - The design description of LLCB is given in Rao *et al.* (2008). The details of the exploded and cut section views of the LLCB - TBM is shown in figure 21. The two coolants used in LLCB are Helium and a eutectic of Lead-Lithium, Pb-Li. The two coolants are of different molecular properties as Pb-Li has very low Prandtl Number of the order 10^{-2} and Helium gas has Prandtl number of ~0.65. The thermal diffusivity of the two fluids were different as the main temperature difference for Helium in straight ducts were concentrated at the viscous sub layer where as the temperature difference for Pb-Li was also present in the mean core region.
The material of construction of the cooling channels is Ferritic-Martensitic Steel (FMS) having electrical conductivity of the order 10^6 1/'Ω-m, so the pressure drop associated with the flow was very high. Hence a coating of Alumina (Al_2O_3), which has very low electrical conductivity (~10^{-8} 1/ 'Ω-m) is used on the wet surfaces of the cooling channels This makes the configuration similar to the rectangular channel of Shercliff's case with all walls insulating i.e. $d_A = 0$ and $d_B = 0$ and hence as mentioned in 4.2.1, the velocity profiles will not have a high velocity jet near the side walls. So the effect of turbulence modification is more significant on the heat transfer characteristics as mentioned in section 6.2.1. The flow will be electromagnetically laminarized and the heat transfer capacity of the Pb-Li deteriorates at high Hartmann numbers.

9. Nomenclature

$2a$	Distance between Hartmann walls
$2b$	Distance between side walls
B_0	Magnetic field
c	Speed of light
C_p	Specific heat
d_A	Electrical conductivity of wall AA
d_B	Electrical conductivity of wall BB
D	Displacement current
E	Electric field
H	Magnetic field strength
Ha	Hartmann Number
j	Electric charge
k	Thermal conductivity
k_{eff}	Effective thermal conductivity
k_T	Turbulent thermal conductivity
L	Characteristic length
N	Interaction parameter
Nu	Nusselt number
p	Pressure
Pr	Prandtl number
Pr_m	Magnetic Prandtl number
q'''	Volumetric heat generation
S	Source term
Re	Reynolds number

Re_m	Magnetic Reynolds number
t	Time
T	Temperature
U	Axial velocity
U_c	Centre line velocity
U_0	Mean velocity
σ	Electrical conductivity of fluid
η	Non-dimensionalized distance in y direction
ξ	Non-dimensionalized distance in x direction
ν	Kinematic viscosity of fluid
ν_{eff}	Effective viscosity
ν_{τ}	Turbulent viscosity
ρ	Density of fluid
μ	Dynamic viscosity of fluid
μ^*	Magnetic permeability
ρ_c	Electric charge density

10. Acknowledgement

We would like to acknowledge Altair Engineering India Pvt. Ltd., for providing an opportunity to do the associated work

11. References

Alfven, H. (1942). Existence of electromagnetic-hydrodynamic waves. *Nature*, Vol.150, (1942), pp.405-406.

Davidson, H. W. (1968). *Compilation of thermo-physical properties of liquid Lithium.*. NASA Technical Note, Washington. D. C., 1968.

Evtushenko, I. A.; Hua, T. Q.; Kirillov. I. R.; Reed, C. B. & Sidorenkov, S. S. (1995). The effect of a magnetic field on heat transfer in a slotted channel. *Journal of Fusion Engineering and Design*, Vol. 27, (1995), pp. 587-592.

Fink, D. & Beaty, H. W. (October 1999). *Standard handbook for electrical engineers* (14th Edition). McGraw Hill, ISBN 0070220050.

Gardener, R. A. & Lykoudis, P. S. (1971a). Magneto-fluid-mechanic pipe flow in a transverse magnetic field Part 1 Isothermal flow. *Journal of Fluid Mechanics*, Vol.47, (1971), pp 737-764.

Gardener, R. A. & Lykoudis, P. S. (1971b). Magneto-fluid-mechanic pipe flow in a transverse magnetic field Part 1 Heat Transfer. *Journal of Fluid Mechanics*, Vol.48, (1971), pp. 129-141.

Happel, J. & Brenner, H. (1981). *Low Reynolds Number Hydrodynamics*, Springer. ISBN 9001371159.

Hartmann, J. (1937) Theory of the laminar flow of electrically conductive liquid in a homogeneous magnetic field, Hg-Dynamics, *Kgl. Danske Videnskab. Selskab. Mat.-fus. Medd.*, Vol.15, No.6, (1937)

Hartmann, J. & Lazarus. P. (1937). Experimental investigation of flow of Mercury in a homogeneous magnetic field, *Kgl. Damske Videnskabernes Selskab, Math-,Fys. Med*, Vol.14, No. 7, (1937).

Hunt, J. C. R. (1965). Magnetohydrodynamic flow in rectangular ducts. *Journal of fluid mechanics*, Vol. 21, No. 4, (1965), pp. 577-590.

Hunt, J. C. R. & Stewartson, K. (1965). Magnetohydrodynamic flow in rectangular ducts. II. *Journal of fluid mechanics*, Vol. 23, No.3, (1965), pp. 563-581.

Ji, H. C. & Gardener, R. A. (1997). Numerical analysis of turbulent pipe flow in a transverse magnetic field. *International Journal of Heat and Mass Transfer*, Vol.40, No.8, (1997), pp. 1839-1851.

Kirillov, I. R.; Reed, C. B.; Barleon, L. & Miyazaki, K. (1994). Present understanding of MHD and heat transfer phenomenon for liquid metal blankets, *Proceedings of 3rd International Symposium of Fusion Nuclear Technology*, Los Angeles, 1994.

Lielpeteris, J & Moreau, R. (1989). *Liquid metal magnetohydrodynamics*, Kluwer Academic Publishers Group, ISBN 079230344X, Dordrecht, Boston.

Luo, X.; Ying, A. & Abodu, M. (2003). Experimental and computational simulation of free jet characteristics under transverse field gradients. Journal of Fusion Science and Technology, Vol 44, (July 2003), pp. 85-93.

Miyazaki, K.; Inoue, h.; Kimoto, T. ; Yamashita, S.; Inoue, S. & Yamaoka, N. (1986). Heat transfer and temperature fluctuation of lithium flowing under transverse magnetic field. *Journal of Nuclear Science and Technology*, Vol.23, (1986), pp. 582-593.

Miyazaki, K.; Yokomizo, K.; Nakano, M.; Horiba, T. & Inoue, S. *et al.* (1988). Heat Transfer and Pressure Drop of Lithium Flow under Longitudinal Strong Magnetic Field, *Proceedings of LIMET'88, Avignon, 1988.*

Moffatt, H. K. (1967). On the suppression of turbulence by a uniform magnetic field. *Journal of Fluid Mechanics*, Vol. 28, (1967), pp. 571–592.

Muller, U. & Buhler, H. (2001), *Magneto-fluid-dynamics in Channels and Containers* (1st Edition), Springer, ISBN 978-3-540-41253-3.

Rao, J. S. *et al.*(2008). *Design description document for the dual coolant Pb 17Li (DCLL) test blanket module*, Report to the ITER test blanket working group (TBWG), (2008), Institute of Plasma Research, India.

Rao, J. S. & Sankar, H. (2011). Numerical Simulation of MHD Effects on Convective Heat Transfer Characteristics of Flow of Liquid Metal in Annular Tube. *Journal of Fusion Engineering and Design*, Vol.86, No.2-3, (March 2011), pp. 183-191.

Roberts, P. H. (1967). *An Introduction to Magnetohydrodynamics*, Longmans Green and Co Ltd, 1967, ISBN 978-0-582-44728-8.

Shercliff, J. A. (1953). Steady motion of conducting fluids in pipes under transverse magnetic fields, *Proceedings of Cambridge Philosophical Society*, pp. 136-144, 1953.

Uda, N.; Miyazawa, A. ; Inoue, S.; Yamaoka, n.; Horiike, H. & Miyazaki, k. (2001). Forced convection heat transfer and temperature fluctuations of lithium under transverse magnetic field. *Journal of Nuclear Science and Technology*, Vol. 38, (2001), pp. 936-943.

Uda, N.; Miyazawa, A.; Yamaoka, H. N.; Horiike, H. & Miyazaki, K. (2002). Heat transfer enhancement in lithium annular flow under transverse magnetic field. *Energy Conversion and Management*, Vol.43, (2002), pp. 441-447.

Wong, C. P. C. ; Salavy, J. F.; Kim, Y.; Kirillov, I.; Kumar, E. R.; Morley, n. B.; Tanaka, S. & Wu, Y. C. (2008). Overview of liquid metal TBM concepts and programs. *Journal of Fusion Engineering and Design*, Vol.83, (2008), pp. 850-857.

5

Thermal Anomaly and Strength of Atotsugawa Fault, Central Japan, Inferred from Fission-Track Thermochronology

Ryuji Yamada[1] and Kazuo Mizoguchi[2]
[1]*National Research Institute for Earth Science and Disaster Prevention*
[2]*Central Research Institute of Electric Power Industry*
Japan

1. Introduction

Frictional slip induces temperature rise in a fault zone. Abundant frictional heat during an earthquake sometimes produces melt of rocks (i.e. pseudotachylyte; e.g., Sibson, 1975). The amount of heat production along faults provides the essential information to investigate frictional strength of the faults that characterizes the earthquake generation processes. Lachenbruch and Sass (1980) first estimated the coefficient of friction of the San Andreas Fault to be 0.1-0.2 from the measurement of surface heat flow along the fault. Kano et al. (2006) found a temperature rise of ~ 0.06 °C measured in a borehole drilled across the Chelungpu fault six years after the 1999 Chi-Chi, Taiwan earthquake associated with this fault. They found that very low coefficient of friction of 0.04-0.08 can explain the heat anomaly along the Chelungpu fault. The above observations along the natural faults have suggested a very low friction level compared with that of 0.6-0.8 evaluated in laboratory rock friction experiments (Byerlee, 1978).

Fission-track (FT) thermochronology is an effective method to detect heat anomaly caused by past faulting (e.g., Scholz et al., 1979; Camacho et al., 2001; Murakami et al., 2002; Murakami and Tagami, 2004; Yamada et al., 2007a). In order to constrain the frictional properties of faults, d'Alessio et al. (2003) measured apatite FT ages and lengths for samples adjacent to and within the San Gabriel fault zone that is thought to be an abandoned major trace of the San Andreas Fault system active from 13 to 4 Ma. They found no evidence of a localized thermal anomaly in FT data even in samples within just 2 cm of the ultracataclasite, and concluded that either there has never been an earthquake with > 4 m of slip at this locality, or the average apparent coefficient of friction is < 0.4 based on the modelling of heat generation and transport.

In this paper, we estimate the frictional strength of the Atotsugawa fault, central Japan, using the method similar to that used by d'Alessio et al. (2003). In the Atotsugawa fault, Yamada et al. (2009) performed FT thermochronologic analysis at an outcrop without visible pseudotachylyte layers, and revealed a thermal anomaly at a several cm thick gouge whose apatite age is significantly younger than those of other samples in the vicinity. Assuming that the thermal anomaly is cause by frictional heating during a single earthquake, the frictional coefficient and the ancient depth of gouge samples are evaluated by the thermal modelling to satisfy the constraints given by the FT thermochronological data with respect to the geometry and alignment of the gouges in the outcrop.

2. Fission-track thermochronology in Atotsugawa fault zone

The Atotsugawa fault is a right-lateral strike-slip one with a strike of N60°E and almost vertical dip, located in the Hida metamorphic belt, central Japan (Fig. 1). From the trenching surveys, a number of historical large earthquakes were detected along the Atotsugawa fault; the most recent one is the 1858 Hietsu earthquake. The estimate of the magnitude ranges 7.0 (Usami, 1987), 7.3 (Matsu'ura et al., 2006) and 7.9 (Doke and Takeuchi, 2009). Geographical Survey Institute, Japan (GSI; 1997) reported a creeping slip with a rate of 1.5 mm/yr in the central section of the fault.

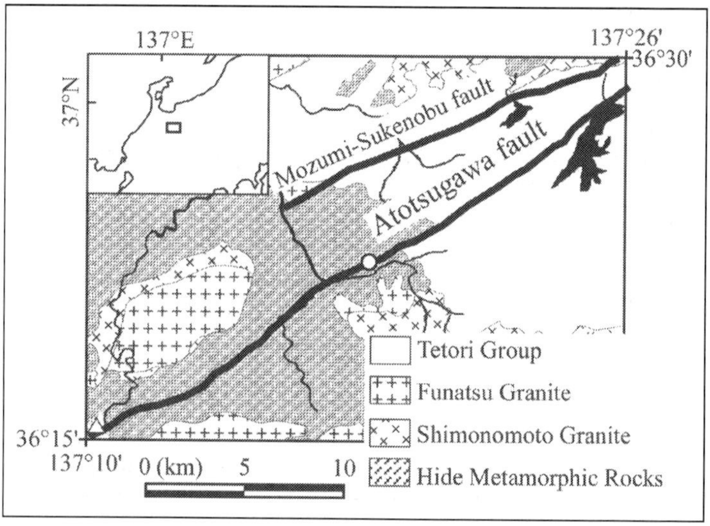

Fig. 1. Distribution of active fault around the Atotsugawa fault. An open circle (centre) and an open triangle (bottom left) symbols indicate the locations of an outcrop and a reference site of R2 in Fig. 2, respectively

In the creeping section, Yamada et al. (2009) performed FT thermochronologic analysis by measuring ages of apatite and zircon grains separated from gouges and fractured rocks at six fracture zones within a 15 m-wide fault zone without visible pseudotachylyte layers (Fig. 2a). This fault zone consists of six fault gouges (name, thickness; gouge-1, 10 cm; gouge-2, 8-20 cm; gouge-3, 8-25 cm; gouge-4, 10 cm; gouge-5, 10-30 cm; gouge-6, 20 cm). FT ages of zircon (c. 120-150 Ma) and apatite (c. 44-60 Ma) for samples except "gouge-1" agree well with emplacement ages for the Funatsu granitic rocks that intruded the Hida Belt (Matsuda et al., 1998). The discordance in zircon and apatite FT ages is interpreted to reflect the rock cooling due to the regional uplift and associated erosion. A thermal anomaly was identified at the gouge sample of "gouge-1" that showed an exceptionally young apatite age (32.1 ± 3.2 Ma, 1σ) with a unimodal FT length distribution, although its zircon age (121 ± 6 Ma, 1σ) was well concordant with other samples (Fig. 2b; after Yamada et al., 2009). The creeping slip observed in the central section of the Atotsugawa fault (GSI, 1997) could be a possible source for this heat anomaly. Such a low slip rate of 1.5 mm/yr, however, causes much smaller increase in temperature (< 20 °C; d'Alessio et al., 2003) in the fault zone. This disagreement can therefore be attributed to the secondary heating induced by frictional slip during an associated earthquake, and the young apatite age possibly gives a younger limit of the initiation of the activity in the Hida Belt.

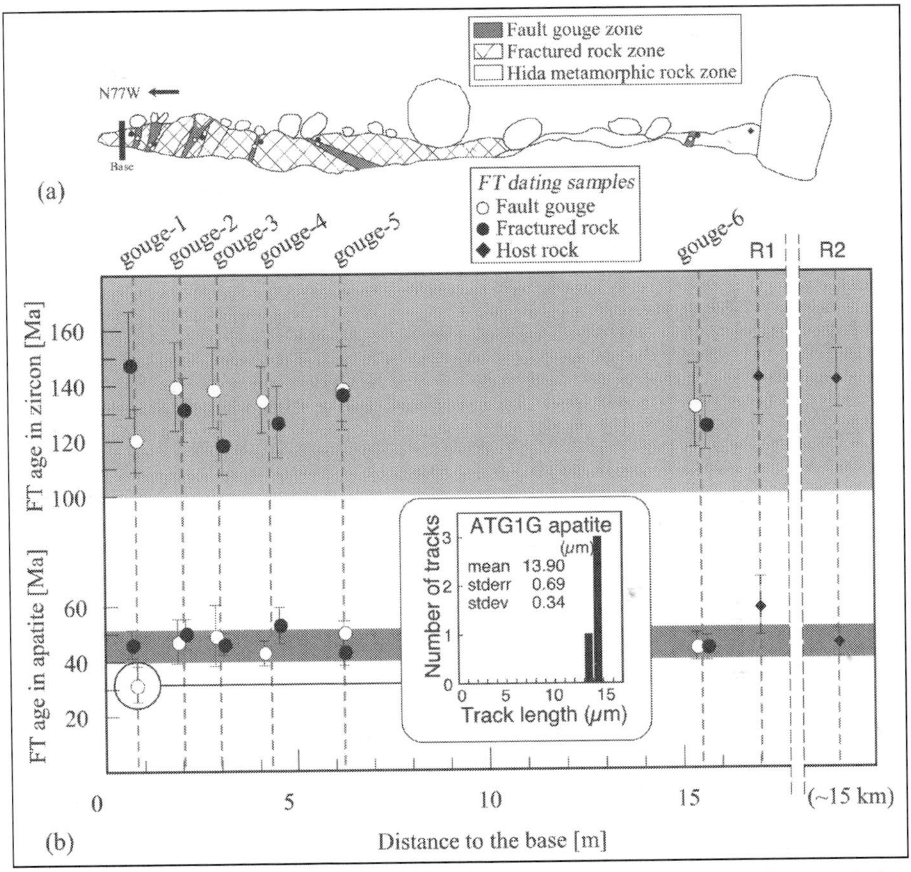

Fig. 2. (a) Sketch of the outcrop of the Atotsugawa fault zone and (b) fission-track age variation in apatite (lower) and zircon (upper) across the outcrop (after Yamada et al., 2009). Open circle, solid circle and square symbols indicate data of fault gouge, fractured rock and host Hida metamorphic rock, respectively. Dashed lines indicate locations of the fault gouge zones. Two reference samples of R1 and R2 (an open triangle in Fig. 1) were also collected where no fractures were observed. Length distribution of apatite FTs for sample ATG1G is also shown. Shaded bands behind the plot indicate the apatite and zircon FT age distributions of the granitic rocks that intrude into the Hida Belt (Matsuda et al. 1998). *Error bars* show 2σ uncertainty in age

3. Thermal modelling associated with frictional heating and estimation of frictional strength

In order to estimate the frictional strength of the Atotsugawa fault based on the FT thermochronological data, we modelled the temporal change in the temperature in and out of the "gouge-1" where an exceptionally young apatite age was found (Yamada et al., 2009). The FT data and the geometry of the occurrence of gouges in the outcrop indicate that the apatite FT age in the 10 cm thick "gouge-1" zone was thermally reset but that in the fractured rock 10 cm apart from "gouge-1" was not. Therefore, the model space for the thermal modelling is composed of a central slip zone of 10 cm thickness and the surrounding rock zone of 10 m thickness with a homogeneous temperature distribution at a certain depth in the initial state (Fig. 3). It is assumed that a single fault slip occurs at a constant rate and all of the frictional work converts into heat. One-dimensional heat transfer model is used to describe the heat diffusion into the surrounding zone at a direction normal to the slip zone. The effect of thermal diffusion by fluid flow is not considered because hydraulic properties of the Atotsugawa fault zone at depth have not yet been investigated.
The equation of thermal diffusion with frictional heat source term for the slip zone is given by

$$\rho_r \cdot C_p \frac{\partial T}{\partial t} = \left(\frac{V \cdot \mu \cdot \sigma_n}{W_c} \right) + \kappa \frac{\partial^2 T}{\partial x^2} \quad (1)$$

where T is temperature, t is the lapse time after a slip occurs, V is the slip rate of the fault, μ is the coefficient of friction, σ_n is the effective normal stress on the fault, C_p is the heat capacity of rock, W_c is the width of a slip zone, k is the thermal conductivity of rock, ρ_r is the density of rock, and x is the distance normal to the fault from the centre of the slip zone. The effective normal stress σ_n is equivalent to the effective overburden pressure given by $(\rho_r - \rho_w) \cdot H \cdot g$, where ρ_w is the density of water, H is depth and g is gravity. For the surrounding zone, Equation (1) without the heat source term (i.e., the first term in the right hand side) is used.

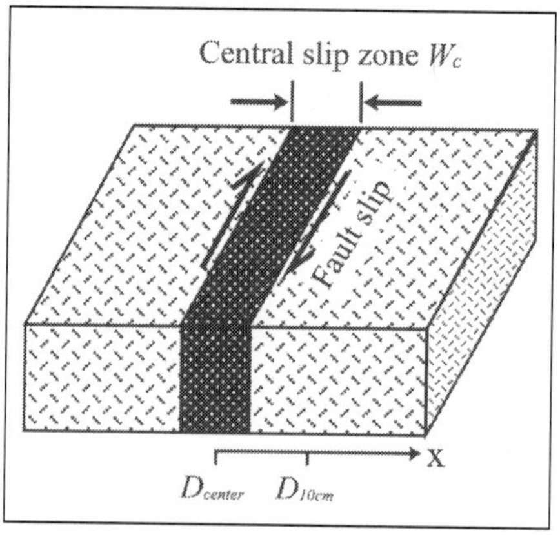

Fig. 3. Fault model for thermal calculation associated with frictional heating. Distance is measured from the centre of the fault

Rock density	2600 kg m^{-3}
Rock heat capacity	1000 J kg^{-1} K^{-1}
Rock thermal conductivity	3.0 W m^{-1} K^{-1}
Water density	1000 kg m^{-3}

Table 1. Parameters used for thermal calculations

Constants of typical physical properties of rocks are used in the thermal modelling as shown in Table 1 (c.f., Schön, 1996). W_c is set as 10 cm that is equivalent to the thickness of the "gouge-1". The initial temperature at a certain depth H is obtained from a geothermal gradient of 30 °C/km (typical value for the upper crust in Japan; e.g., Tanaka et al., 1999) multiplied by H plus a surface temperature of 20 °C, and the temperature distribution over the fault zone is assumed to be uniform. The boundary condition of the temperature at the edge of the model space is fixed at the initial value. Total slip of 5 m long is given for this fault system because the estimate of the magnitude of the associated earthquake ranges from 7.0 to 7.9 (Usami, 1987; Matsu'ura et al., 2006; Doke and Takeuchi, 2009) that corresponds to the total slip of the order of 1-10 m, based on the empirical relationship between the fault displacement and the magnitude (Matsuda, 1975). V is set at 1 m/s (e.g., Heaton, 1990) and therefore the slip duration is 5 sec. Considering the closure temperature of apatite FT (100 ± 20 °C; e.g., Wagner and Van den Haute 1992), H should be shallower than 3 km which corresponds to the environment temperature of 110 °C, and therefore restricted to the three cases of 1, 2 and 3 km for the modelling assuming a geothermal gradient of 30°C/km (e.g., Tanaka et al, 1999).

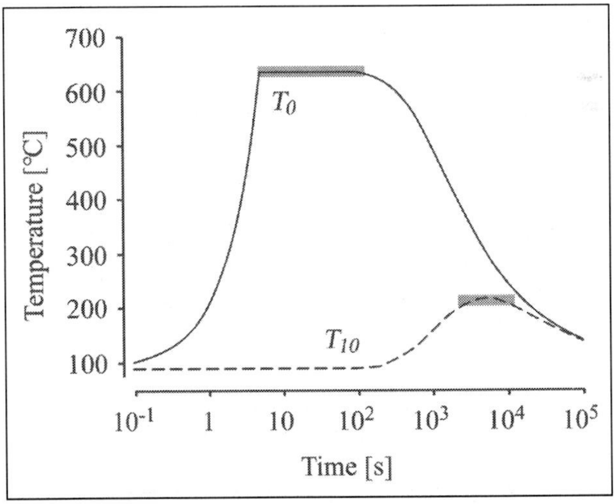

Fig. 4. Temperatures at the centre of the fault (T_0, solid line) and at 10 cm apart from the centre (T_{10}, dashed line) are plotted as a function of time since a slip occurs

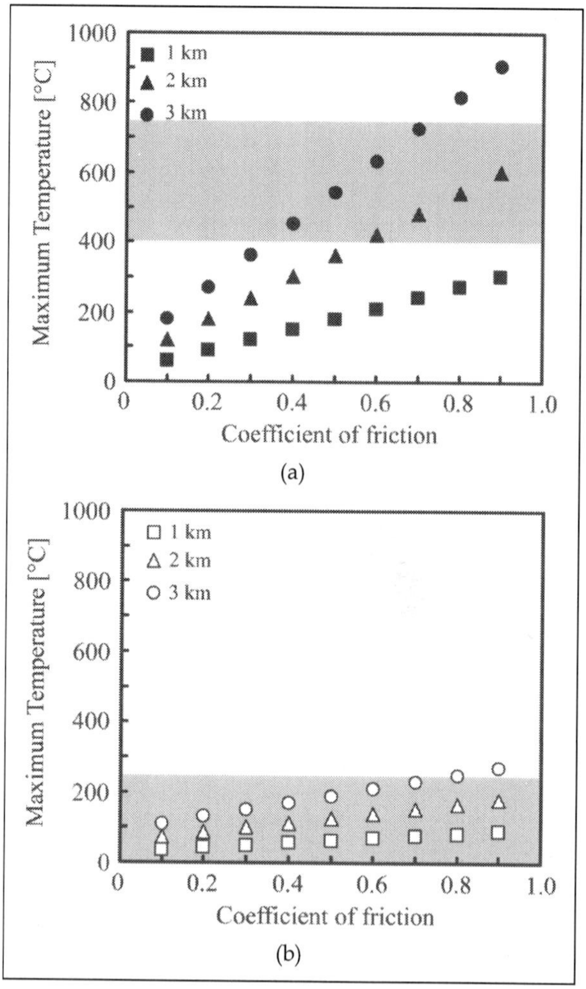

Fig. 5. Maximum temperature at the centre of the fault (T_0; a) and at the location 10 cm apart from the centre (T_{10}; b) during an earthquake are plotted as a function of friction coefficient of 0.1 ~ 0.9 in cases of depth from 1 to 3 km. Square, triangle and circle symbols denote the data at 1, 2 and 3 km, respectively. Shaded areas in (a) and (b) indicate ranges of T_0 (upper) and T_{10} (lower) inferred from apatite and zircon FT thermochronological analyses, respectively

Calculation results of the temporal change in temperature at the two locations of x = 0 cm (D_0) and x = 10 cm (D_{10}) for the combination of μ (0.6) and H (3 km) parameters are shown as representative cases in Fig. 4. These locations are chosen to approximate the positions of the "gouge-1" and a surrounding rock sample, respectively. For any combinations of μ and H parameters, the time during which the temperature in a specific location is maintained at its maximum is almost invariant. The temperatures at D_0 and D_{10} are preserved at the maximums of T_0 and T_{10} (named T_{max0} and T_{max10}) for the order of ~10^2 sec and ~10^4 sec, as indicated by shaded bands in Fig. 4, respectively. Fig. 5 shows the variations of T_{max0} and T_{max10} for the combinations of μ (0.1-0.9) and H (1-3 km) parameters.

Whether FTs in apatite and zircon are annealed or not depends on the temperature and duration of heating. Assuming that the frictional heat caused by an associated single palaeo-earthquake event was responsible for the thermochronologic difference between the "gouge-1" and other samples, calculation results above indicate that the effective heating duration for the samples at D_0 and D_{10} at T_{max0} and T_{max10} are estimated as the order of ~10^2 sec and ~10^4 sec, respectively. Note that the effective heating time is significantly longer than the slip duration, and that FTs in minerals in the distance to the frictional centre are not necessary annealed instantly due to the frictional slip event. The estimates of heating durations and the kinetic relation of time-dependent FT annealing temperature of apatite and zircon (Laslett and Galbraith, 1996; Yamada et al., 2007b) give the following constraints on the T_{max0} and T_{max10} at the secondary heating event (Yamada et al., 2009). For the D_0 sample, the fact that apatite FT age was totally reset although zircon FT age was not indicates that T_{max0} is in the range of 400°C to 750°C (for the heating duration of ~10^2 sec). For the D_{max10} sample, the fact that both apatite and zircon FT ages were not reset indicates that T_{max10} does not exceed 250 °C (for ~10^4 sec). These constraints on T_{max0} and T_{max10} are satisfied with the limited cases of $\mu > 0.6$ for $H = 2$ km, and $0.4 < \mu < 0.7$ for $H = 3$ km, shown as shaded areas in Fig. 5.

4. Discussion

The effect of the pore water is not taken into account in the modelling above. If the pore water exists at the "gouge-1" zone, the temperature will be less than that in the dry condition calculated above, because the pore pressure decreases the stress applied on the fault and the frictional heat is diffused by fluid flow. Therefore, the estimate of the frictional strength in the dry condition can be regarded as a lower bound. The increase in the total amount of slip with the same slip velocity will considerably raise the temperature in the fault zone. If the total amount of slip is doubled compared with the case for the above calculation (= 5 m), the estimated increase in T_0 and T_{10} is almost doubled and the estimate of the coefficient of friction is reduced to almost half. Even in this case, however, the estimated coefficient of friction is still large ($\mu > 0.3$ for H = 2 km; $0.2 < \mu < 0.4$ for H = 3 km) compared with that of 0.1-0.2 (Lachenbruch and Sass, 1980) and 0.04-0.08 (Kano et al., 2006).

Our estimates are obtained by assuming that the thermal anomaly found in "gouge-1" zone is attributed to the frictional heating during a single earthquake associated at ~32 Ma (Yamada et al., 2009). Although a number of earthquakes should have occurred thereafter along the Atotsugawa Fault that remains active to date, an amount of heat generated by each of these quakes might be insufficient to reduce FT length in apatite because the partially annealed tracks are not observed for the "gouge-1" sample. The coefficient of friction for the earthquakes occurred after 32 Ma are, therefore, inferred as < 0.6 for $H = 2$ km, and < 0.4 for $H = 3$ km. In addition, the effect of accumulated residual heat generated by a number of earthquakes should be taken into account if the next heat generation may occur before the temperature in the fault zone is reduced to the ambient temperature due to the thermal diffusion in rocks. This effect should, however, be negligible considering the recurrence interval of general active faults in Japan, ranging from 1000 to 10000 years

(Research Group for Active Faults of Japan, 1991) that would be sufficiently long for the thermal diffusion.

Our modelled estimates of the coefficient of friction are approximately consistent with that obtained in laboratory friction experiments on rocks (Byerlee, 1978). As for the Atotsugawa fault, Mizoguchi et al. (2007) obtained the similar frictional strength of 0.5-0.6 by laboratory friction experiments using fault gouge samples taken from the Atotsugawa borehole core samples at a depth of 326 m, located near to the FT samples of Yamada et al. (2009). This coincidence of frictional strength between the nature and laboratory has rarely reported in the past. In previous studies, frictional strengths of natural faults are estimated much lower than those in the laboratory (Lachenbruch and Sass, 1980; Kano et al., 2006; d'Alessio et al., 2003). At the outcrop of the Atotsugawa fault in Yamada et al. (2009), however, the other five gouges were not heated enough by frictional slip to reset their ages. We suggest that the frictional strengths of the fault during earthquakes when the other gouges were activated were less than that for the "gouge-1". The variety of frictional strength of fault with every event might reflect the complicated earthquake generation processes.

5. Conclusions

The fission-track analysis on fault-related rocks collected from a outcrop of the Atotsugawa fault without visible pseudotachylyte layers revealed that the apatite FT age of the a gouge sample is exceptionally younger than those of the surrounding other rocks (fault gouge, fractured rocks and host rocks) in the vicinity, although the zircon FT age is well concordant with other samples. To explain the thermal anomaly identified at this gouge sample, we modelled the temporal change in the temperature in and out of the gouge after an associated earthquake generated the frictional heat. The calculation results indicate that the effective heating time is significantly longer than the slip duration, and that FTs in minerals in the distance to the frictional centre are not necessary annealed instantly due to the frictional slip event. The estimate of frictional strength for the "gouge-1" is larger than 0.6 for H = 2 km, and between 0.4 and 0.7 for H = 3 km, which is similar to that obtained in laboratory friction experiments using the gouge samples taken from the Atotsugawa Fault.

6. Acknowledgment

We are grateful to Drs. E. Fukuyama, H. Negishi and N. Hasebe for useful discussion and help in preparing the manuscript.

7. References

Byerlee, J. D. (1978). *Friction of rocks*, Pure Appl. Geophys., Vol. 116, pp. 615-626

Camacho, A., McDougall, I., Armstrong, R., Braun, J. (2001). *Evidence for shear heating, Musgrave block, central Australia*, J. Struct. Geol. Vol. 23, pp. 1007-1013

d'Alessio, M. A., Blythe, A. E., Bürgmann, R. (2003). *No frictional heat along the San Gabriel fault, California: Evidence from fission-track thermochronology*, Geology Vol. 31, pp. 541-544

Doke, R., Takeuchi, A. (2009). *The latest event at the eastern part of the Atotsugawa fault, inferred from the outcrops at Sako, Hida City, central Japan (in Japanese with English abstract)*, The Quaternary Res., Vol. 48, pp. 11-17

Geographical Survey Institute, Japan (1997). *Crustal deformation in the Chubu and the Hokuriku regions, Japan,* Rep. Coord. Comm. Earthquake Predict. Vol. 57, pp. 520-524

Heaton, T. H. (1990). *Evidence for and Implications of self-healing pulses of slip in earthquake rupture,* Phys. Earth Planet. Inter. Vol. 64, pp. 1-20

Kano, Y., Mori, J., Fujio, R., Ito, H., Yanagidani, T., Nakao, S., Ma, K. -F. (2006). *Heat signature on the Chelungpu fault associated with the 1999 Chi-Chi, Taiwan earthquake,* Geophys. Res. Lett. Vol. 33, doi:10.1029/2006GL026733

Lachenbruch, A. H., Sass J. H. (1980), *Heat flow and energetics of the San Andreas fault zone,* J. Geophys. Res. Vol. 85, pp. 6185-6222

Laslett, G. M., Galbraith, R. (1996). *Statistical modelling of thermal annealing of fission tracks in apatite,* Geochim. Cosmochim. Acta Vol. 60, pp. 5117-5131

Matsuda, T. (1975). *Magnitude and recurrence interval of earthquakes from a fault (in Japanese with English abstract),* Zishin, J. Seisol. Soc. Japan, Vol. 28, pp. 269-283

Matsuda, T., Goto, A., Kano, T. (1998). *Fission-track thermochronology of Jurassic granitic rocks located in central part of Hida belt (in Japanese),* Abstract of 105th Annual Meeting of Geol. Soc. Jpn., Nagano, Japan

Matsu'ura, R., Nakamura, M., Karakama, I. (2006). *Reexamination of hypocenters and magnitudes for historical earthquakes Part 8 (in Japanese with English abstract),* Programme and Abstracts, The Seismol. Soc. Japan, 2006, Fall Meeting, B024

Mizoguchi, K., Fukuyama, E., Kitamura, K., Takahashi, M., Masuda, K., Omura, K. (2007). *Depth dependent strength of the fault gouge at the Atotsugawa fault, central Japan: A possible mechanism for its creeping motion,* Phys. Earth Planet. Inter. Vol. 161, pp. 115-125

Murakami, M., Yamada, R., Tagami. T. (2002). *Detection of frictional heating of fault motion by zircon fission track thermochronology,* Geochim. Cosmochim. Acta Vol. 66, pp. A537

Murakami, M., Tagami, T. (2004). *Dating pseudotachylyte of the Nojima fault using the zircon fission-track method,* Geophys. Res. Lett. Vol. 31, doi:10.1029/2004GL020211

Ongirad, H., Yasue, K., Takeuchi, A., Nasu, T., Takami, A. (2001). *A newly found fault outcrop at the central part of the Atotsugawa fault, central Japan (in Japanese with English abstract),* Active Fault Res. Vol. 20, pp. 46-51.

Research Group for Active Faults of Japan (1991). *Active faults in Japan: Sheet maps and inventories (in Japanese with English abstract),* (revised ed.) 437 pp., Univ. Tokyo, Tokyo, ISBN 4130607006, Tokyo, Japan

Scholz, C.H., Beavan, J., Hanks, T. C. (1979). *Frictional metamorphism, argon depletion, and tectonic stress on the Alpine fault, New Zealand,* J. Geophys. Res. Vol. 84, pp. 6770-6782.

Schön, J. H. (1996). Physical properties of rocks, 18: fundamentals and principles of petrophysics, In: *Handbook of Geophysical Exploration,* Pergamon, ISBN 0-08-044346-X, New York, USA

Sibson, R. H. (1975). *Generation of pseudotachylyte by ancient seismic faulting,* Geophys. J. R. Astron. Soc. Vol. 43, pp. 775-794

Tanaka, A., Yano, Y., Sasada, M., Okubo, Y., Umeda, K., Nakatsuka, N., Akita, F. (1999). *Compilation of thermal gradient data in Japan on the basis of temperatures in boreholes (in Japanese with English abstract),* Bull. Geol. Surv. Jpn. Vol. 50, pp. 457-487

Usami, T. (1987). *Materials for Comprehensive List of Destructive Earthquakes in Japan (New Edition) (in Japanese),* 434 pp., Univ. Tokyo, ISBN 413060712X, Tokyo, Japan

Wagner, G.A., Van den haute, P. (1992). *Fission Track-Dating*, Kluwer Academic Publishers, 285 pp. ISBN-10: 079231624X, Dordrecht, Netherlands

Yamada, R., Matsuda, T., & Omura, K. (2007a), *Apatite and zircon fission-track dating from the Hirabayashi-NIED borehole, Nojima Fault, Japan: evidence for anomalous heating in fracture zones*, Tectonophysics, Vol. 443, pp. 153-160

Yamada, R., Murakami, M., Tagami, T. (2007b). *Statistical modelling of annealing kinetics of fission tracks in zircon; Reassessment of laboratory experiments*, Chemical Geology, Vol. 236, pp. 75-91

Yamada, R., Ongirad, H., Matsuda, T., Omura, K., Takeuchi, A., Iwano, H. (2009). Fission-track analysis of the Atotsugawa Fault (Hida Metamorphic Belt, central Japan): fault-related thermal anomaly and activation history, In: *Thermochronological Methods: From Paleotemperature Constraints to Landscape Evolution Models*, Lisker, F., Ventura, B., Glasmacher, U. A. (Eds.), 331-337, The Geol. Soc., London, Special Publications, 324, ISBN 978-1-86239-285-4, London, UK

6

Heat Transfer in Freeze-Drying Apparatus

Roberto Pisano, Davide Fissore and Antonello A. Barresi
Dipartimento di Scienza dei Materiali e Ingegneria Chimica, Politecnico di Torino
Italy

1. Introduction

Freeze-drying is a process used to remove water (or another solvent) from a frozen product, thus increasing its shelf-life. It is extensively used in pharmaceuticals manufacturing, to recover the active pharmaceutical ingredient (and the excipients) from an aqueous solution, as well as in some food processes, because of the low operating temperatures that allow preserving product quality. Moreover, the freeze-dried product has a high surface area and can be easily re-hydrated.

In this chapter we focus on pharmaceuticals manufacturing, where the solution containing the product is generally processed in vials, placed over the shelves in a drying chamber. However, it is worth stating that, in industrial practice, other loading configurations can be used to carry out the process, thus this study will be extended also to the case where vials are loaded on trays, or the solution is directly poured in trays. The process consists of three consecutive steps, namely:

1. Freezing: product temperature is lowered below the freezing point and, thus, most of the solvent freezes, forming ice crystals. Part of the solvent can remain bounded to the product, and must be desorbed. Also the product often forms an amorphous glass which can retain a high amount of water.
2. Primary drying: in this step the pressure in the drying chamber is lowered, thus causing ice sublimation. This phase is usually carried out at low temperature (ranging, in most cases, from -40°C to -10°C) and, as sublimation requires energy, heat is transferred to the product through the shelf, by acting on the temperature of the fluid flowing in the coil inserted in the shelf.
3. Secondary drying: when the sublimation of the ice has been completed, shelf temperature is raised (e.g. to 20-40°C) and chamber pressure is further decreased to allow the desorption of the water bounded to the product, thus getting the target moisture in the product.

The freeze-dryer comprises the drying chamber and a condenser where the water vapour is sublimated on some cold surfaces in order to decrease the volumetric flow-rate arriving to the vacuum pump. The pressure in the chamber can be modified either by acting on a valve placed on pump discharge, or using the so called "controlled leakage", i.e. manipulating the flow rate of nitrogen (or another gas) introduced in the chamber. The drying chamber can be isolated from the condenser by means of a valve that is usually placed in the duct connecting the chamber to the condenser (Mellor, 1978; Jennings, 1999; Oetjen & Haseley, 2004; Franks, 2007).

Although it is generally considered a "soft" drying process, because of the low operating temperatures, the heat transfer to the product has to be carefully controlled in order to avoid product overheating. If fact, product temperature has to be maintained below a limit value to avoid the occurrence of undesired events. In case of products that crystallize during freezing, the limit temperature corresponds to the eutectic point: the goal is to avoid the formation of a liquid phase and the successive boiling due to the low pressure. In case the product remains amorphous during freezing, the maximum allowed product temperature is close to the glass transition temperature in order to avoid the collapse of the dried cake: this value can be very low, and is also dependent on the residual moisture. The occurrence of collapse can increase the residual water content in the final product and the reconstitution time, beside decreasing the activity of the pharmaceutical principle; moreover, a collapsed product is often rejected because of the unattractive physical appearance (Pikal & Shah, 1990; Wang, 2000; Rambhatla et al., 2005; Sadikoglu et al., 2006). Beside product temperature, also the residual amount of ice in the product has to be carefully controlled during primary drying, thus identifying the ending point of this phase. In fact, if shelf temperature is increased too early to the value required by the last phase of the cycle, product temperature may exceed the maximum allowed value, thus causing melting or collapse. Also in this case heat transfer to the product plays a key role as, at steady-state, the following equation holds:

$$J_q = \Delta H_s J_w \tag{1}$$

where J_q is the heat flux to the product, J_w is the solvent flux from the product to the chamber, and ΔH_s is the heat of sublimation.
In this chapter, thus, we will focus on heat transfer in vial, as well as bulk, freeze-drying as it is one of the key factor affecting product dynamics and temperature. In particular, the results previously presented by Pikal (2000) for vial freeze-drying are here extended.
The techniques allowing to calculate the heat transfer parameters, or to estimate their values by means of experiments, will be briefly reviewed. We will point out that the heat flux between the heating shelf and the container is the result of several mechanisms that depend on dryer and container geometry, as well as on pressure and temperature of the surrounding gas. Moreover the heat flux to the batch of vials is far from being uniform in a freeze-dryer: the implications on recipe design and scale-up will be finally addressed.

2. Theoretical calculation of heat flux in vial freeze-drying

The heat flux to the product is proportional to the difference between the heating fluid temperature (T_{fluid}) and the product temperature at the vial bottom (T_B):

$$J_q = K_v \left(T_{\text{fluid}} - T_B \right) \tag{2}$$

where K_v is the heat transfer coefficient. It has to be highlighted that in eq. (2) it has been assumed that product temperature is uniform in the radial direction, as it has been demonstrated by means of mathematical simulation and experimental investigations (Pikal, 1985; Sheehan & Liapis, 1998).
Various equations can be found in the literature to calculate the heat exchange coefficient K_v in case the vial is placed directly over the shelf, i.e. if no tray is used. These relationships are still valid in case the solution is directly poured in a tray.

Firstly, the heat transfer coefficient between the heating shelf and the bottom of the vial (K_v') is calculated as the sum of three terms:

$$K_v' = K_c + K_r + K_g \tag{3}$$

corresponding to the various heat transfer mechanisms between the fluid and the vial bottom, namely the direct conduction from the shelf to the glass at the points of contact (K_c), the radiation (K_r), and the conduction through the gas (K_g). In this study, we assume that all these contributions can be referred to the same heat transfer area.
According to Smoluchowski theory, as outlined by Dushman & Lafferty (1962) and reported by Pikal et al. (1984), K_g is a function of the average distance between the bottom of the vial and the shelf (ℓ) and of chamber pressure:

$$K_g = \frac{\alpha \Lambda_0 P_c}{1 + \ell \left(\frac{\alpha \Lambda_0}{\lambda_0} \right) P_c} \tag{4}$$

The parameter α is a function of the energy accommodation coefficient and of the absolute temperature of the gas:

$$\alpha = \frac{a_c}{2 - a_c} \cdot \sqrt{\frac{273.2}{T}} \tag{5}$$

When other gases, beside water vapour, are present in the drying chamber (e.g. nitrogen entering the chamber when controlled leakage is used to regulate the pressure), eqs. (4) and (5) have to be modified to take into account gas composition, as total pressure is due to the sum of the contribution of water and nitrogen partial pressures (Brülls & Rasmuson, 2002). It has to be remarked that the bottom of the vial is not flat, and the curvature depends on the type of vial: thus, the value of K_g is a function of the type of vial considered.
Beside conduction in the gas, there are two radiative heat fluxes towards the product, one from the shelf upon which the vials rest, and the other from the top. Each flux is proportional to the difference in the fourth powers of the absolute temperatures of the two surfaces, and to the effective emissivity for the heat exchange, which depends on the relative areas of the two surfaces, their emissivities, and a geometrical view factor. According to Pikal et al. (1984) the radiative heat transfer coefficient can be written as:

$$K_r = 4(e_s + e_v)\kappa T^3 \tag{6}$$

While the values of Λ_0, λ_0, and e_s can be found in the literature, the values of the parameters K_c, a_c, ℓ, and e_v (in case radiation from upper shelf plays an important role), have to be determined by regression analysis of experimental data, even if some values can be found in the literature for some types of vials (Pikal et al., 1984).
It follows that K_v' depends essentially on the operating conditions at which drying is carried out: chamber pressure, shelf temperature and, in turn, temperature of chamber gas. However, according to literature (Hottot et al., 2005) and to the results shown in the following, the dependence of K_v' on chamber gas temperature is not relevant. On the contrary, the heat transfer between the shelf and the container significantly varies with chamber pressure, and this dependence can be well described by a nonlinear equation with the following structure:

$$K_v^{'} = C_1 + \frac{C_2 \cdot P_c}{1 + C_3 \cdot P_c} \tag{7}$$

where:

$$\begin{cases} C_1 = K_c + 4\kappa \overline{T}^3 \left(e_S + e_v\right) \\ C_2 = \dfrac{a_c}{2 - a_c}\sqrt{\dfrac{273.2}{T}}\Lambda_0 \\ C_3 = \ell\left(\dfrac{\Lambda_0}{\lambda_0}\dfrac{a_c}{2 - a_c}\sqrt{\dfrac{273.2}{T}}\right) \end{cases} \tag{8}$$

At this point, $K_v^{'}$ can be calculated, whichever are the processing conditions, provided that a_c, K_c and ℓ (and thus C_1, C_2 and C_3) are known.

If vials are loaded on a tray, the shelf-vial system can be described as a set of resistors in series, whose total resistance is the sum of their individual resistances. It follows that the resistance for the mentioned system can be expressed as the sum of the resistance due to tray thickness, as well as to the shelf-tray and tray-vial heat transfers that can be both described by equations similar to eq. (7). Provided that the heat transfer between shelf and tray ($K_{v,1}^{'}$), and from tray to vials ($K_{v,2}^{'}$), can be described by the same equations, we can define $K_v^{'}$ as an effective heat transfer coefficient that accounts for both contributions:

$$\frac{1}{K_v^{'}} = \frac{1}{K_{v,1}^{'}} + \frac{1}{K_{v,2}^{'}} \tag{9}$$

It is worth noticing that, as a consequence of this assumption, the value of the global coefficients of eq. (7) (i.e. C_1, C_2 and C_3), as determined by a regression analysis of experimental observations, are effective values that take into account both the heat transfer between shelf and tray and tray and vials.

The coefficient K_v^{*} (from the shelf surface to product bottom) can be calculated by taking into account the conduction in the glass at the bottom of the vial, as well as through the tray (if it is used):

$$K_v^{*} = \frac{J_q}{T_{\text{shelf}} - T_B} = \left(\frac{1}{K_v^{'}} + \frac{s_g}{\lambda_g} + \frac{s_{\text{tray}}}{\lambda_{\text{tray}}}\right)^{-1} \tag{10}$$

In addition, if the considered driving force for the heat transfer is the temperature difference between the heating fluid and the product, we have also to include the resistance to heat transfer through shelf ($1/k_s$). Therefore, eq. (10) has to be substituted by:

$$K_v = \left(\frac{1}{K_v^{'}} + \frac{s_g}{\lambda_g} + \frac{s_{\text{tray}}}{\lambda_{\text{tray}}} + \frac{1}{k_s}\right)^{-1} \tag{11}$$

Finally, it has to be remarked that the product in the vial can receive heat from the walls of the container, as well as from chamber walls, due to radiation. Both contributions do not appear explicitly in eq. (2), but the coefficient K_v can be regarded as an effective heat transfer coefficient, weighing up these additional heat fluxes (Velardi & Barresi, 2008). However, as it will be discussed in the following sections, it is worth remarking that the contribution of radiative heat to the total energy balance of the vial can vary with its position over the shelf. This phenomenon, together with the presence of fluid temperature gradients along the shelf, is responsible for lot unevenness.

3. Experimental methods for K_v measurement

Various methods have been proposed in the past to determine the value of the coefficient K_v (or that of $K_v^{'}$). They are briefly reviewed in the following.

A very simple test that allows determining the coefficient K_v consists of preparing a batch of vials filled with the product, a placebo, or simply water and, then, to carry out the primary drying for a time interval (Δt) and to measure the weight loss (Δm) and the temperature of the ice at the vial bottom (T_B). The coefficient K_v is given by:

$$K_v = \frac{\Delta m \cdot \Delta H_s}{\Delta t \cdot \left(T_{\text{fluid}} - T_B\right) \cdot A_v} \tag{12}$$

As the heat transfer coefficient is a function of chamber pressure, at least three measurements at three different values of P_c are required to calculate the global coefficients of eq. (7).

During the drying, the temperature of the heating fluid can vary along the shelf as the heat is transferred from the technical fluid to the product. However, it must be said that generally the inlet-outlet fluid temperature difference observed under full-load conditions is small (typically lower than 1 K) and of the same order of magnitude of temperature sensor uncertainty. A similar behaviour has been observed in all the experiments whose results are reported in this chapter. It follows that it is not necessary to take into account explicitly the fluid temperature gradient along the shelf in the calculation of K_v, as its effect can be included in the uncertainty of the parameter.

On the other hands, the value of the coefficient K_v is not the same for all the vials of the lot, as a consequence of the different contributions of the various heat transfer mechanisms: as an example, the vials at the edges of the shelf receive also radiant heat from the chamber walls, while the vials located in the central part of the batch are heated only from the heating fluid (Pikal, 2000; Rambhatla & Pikal, 2003; Gan et al., 2005a, 2005b). As a consequence, a different evolution of product temperature can be observed in vials located in various positions of the batch. Therefore, to estimate accurately K_v for each class of vials, product temperature has to be monitored, and thus thermocouples must be inserted both in vials placed in the core and at the edge of the lot.

According to their position over the shelf, the vials of a lot can be classified in various groups depending on the different heat transfer mechanisms involved. In particular, in case vials are confined by a metal or plastic band, we can generally identify five zones (see Figure 1), which are characterized by different heat transfer mechanisms, as reported in Table 1.

A further refinement might be introduced to distinguish edge vials that are exposed to different walls of the drying chamber. This is particularly true for laboratory scale equipment, where the dryer door is not made of steel (like the other chamber walls), but of Plexiglas. Since the two materials have a significantly different value of emissivity (0.36 vs. 0.95), vials close to the door may show a higher value of K_v with respect to the rest of the lot

(Rambhatla & Pikal, 2003). On the contrary, in a manufacturing plant, the dryer door is usually made of steel and, thus, edge vials are less uneven. Nevertheless, it must be said that in a large scale equipment only a limited fraction of vials are radiated, while in a laboratory dryer (where the fraction of edge vials is much greater) the contribution of the radiative heat can be strongly reduced by a proper shielding. These differences, however, must be taken into account during scale-up of a recipe from laboratory to industrial scale dryers.

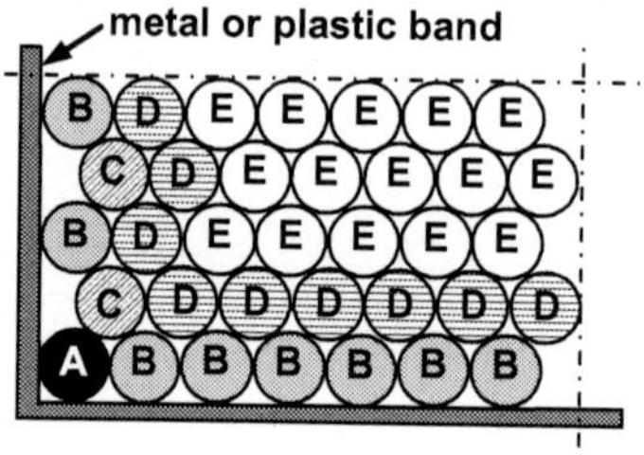

Fig. 1. Scheme of a lot, wherein vials have been classified depending on their position and on the additional heat transfer mechanisms involved

Vial group	**Position over the shelf**	**Additional mechanisms to vial heat transfer**		
		radiation from chamber walls	*contact with the tray band*	*contact with "hot" vials*
A	corner	yes	yes	yes
B	peripheral	yes	yes	yes
C	peripheral	yes	no	yes
D	core	no	no	yes
E	core	no	no	no

Table 1. Characteristics of the various vial families with respect to the position on the shelf and the mechanisms of heat transfer involved

It must be remarked that the gravimetric measurement is the only method that gives a detailed picture of single vial behaviour and of intervial variability, thus allowing to identify the various vial families. A heat flux transducer can also be used for monitoring the dynamics of the single vials composing the lot (Chen et al., 2008), but its use implies that the monitored vials are loaded on the sensor surface altering de facto the heating conditions.
Equation (12) can be still used when the sublimation flux (i.e. $\Delta m/\Delta t \cdot A_v$) is obtained by other means, for example using the Tunable Diode Laser Absorption Spectroscopy (TDLAS) or the pressure rise test technique.
The Doppler-shifted near-infrared absorption spectroscopy is used for measuring the water vapour concentration and gas flow velocity in the duct (that connects drying and condenser chamber). At this point, these variables can be used, known the cross-sectional area of the duct, to estimate the vapour flow rate (Kessler et al., 2006; Gieseler et al., 2007). It must be remarked that TDLAS does not measure directly the mass flow, but only the gas flow velocity in a limited number of points; therefore, the reliability of the vapour flow rate estimation depends on how well the velocity profile is known. This is a critical issue, since

the velocity profile continuously develops in the duct, tightly depends on processing conditions, can be strongly modified by the presence of Clean In Place systems, and even at the duct exit is usually far from the parabolic profile corresponding to fully developed flow. Other drawbacks of this technique are the high investment costs, the difficult to retrofit existing units, and the difficulties in its calibration.

To measure the vapour flow rate (and hence the heat transfer coefficient), an alternative solution, that can be easily implemented also in existing devices, is the pressure rise test technique: the valve in the duct connecting the drying chamber to the condenser is closed for a short time interval, and, thus, pressure in the chamber increases due to vapour accumulation. Once the pressure rise curve has been acquired, the vapour flow rate (J_w) can be calculated by simply evaluating its slope at the beginning of the test (Fissore et al., 2011a):

$$J_w = \frac{V_c M_w}{A_{sub} R T_c} \left. \frac{dp_{w,c}}{dt} \right|_{t=t_0} \tag{13}$$

This measure of J_w is reliable as it is based on the direct measurement of the effect of the sublimation flow, the pressure increase, described by a simple physical law. Unlike TDLAS, the pressure rise technique can be used in both laboratory and industrial scale dryers, provided that drying and ice condenser chamber are separated, without any modifications of existing units. Figure 2 shows an example of pressure rise curves measured (at various time during the primary drying phase) in a laboratory and in an industrial scale dryer. These results were obtained during the drying of a 5% by weight mannitol solution. As stated before, the vapour flow rate is related to the slope of the pressure profile, which, as expected, decreases as the drying proceeds.

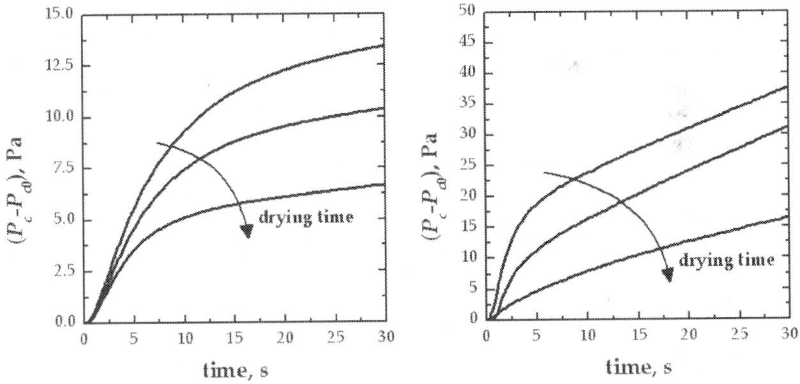

Fig. 2. Examples of pressure rise curves measured at various times during the primary drying stage in (left-side graph) a laboratory and (right-side graph) an industrial dryer

To calculate K_v from the vapour flow rate, the product temperature at the container bottom has to be measured, for example, inserting thermocouples in some vials. For this purpose, the use of wireless technology is promising (Corbellini et al., 2010; Schneid & Gieseler, 2008) as it is compatible with the restrictions of industrial equipment that typically uses automatic loading systems.

The pressure rise curve can be also described by a mathematical model, whose parameters (e.g. product temperature at the beginning of the test, and K_v) are retrieved by looking for the best fit between measured and calculated values of chamber pressure (Milton et al., 1997; Chouvenc et al., 2004; Tang et al., 2006; Velardi et al., 2008). The optimal selection of sampling frequency and duration of the test has been discussed by Fissore et al. (2011a), who also proposed a modification of the Dynamic Parameters Estimation algorithm (Velardi et al., 2008) to cope with the problem of ill-conditioning. It must be said that while the estimation of J_w from the slope of the pressure rise curve can be also carried out with pure ice, the previous algorithms (which take advantage of a model) require that a product, which gives a mass flow resistance, is used. In addition, unlike previous techniques, the pressure rise, coupled with one of the above cited algorithms, does not require an independent measurement of product temperature.
It is worth noticing that the gravimetric measurement is the only method that allows to determine the distribution of the values of K_v in the lot, while both the TDLAS sensor and the pressure rise test based methods allow to determine only a mean value of K_v for the batch. Nevertheless, it must be said that the fraction of edge-vials, and hence its contribution to the mean value of K_v, is becoming less and less important as the size of the dryer becomes larger.
In addition, it must be remarked that the above mentioned methods, differently from the gravimetric test, allow to get an estimation of K_v vs. P_c, even in only one run (Kuu et al., 2009; Pisano et al., 2010) if chamber pressure is properly varied during the test (e.g. as a multi-step constant function), without requiring an excessive effort from the users. Furthermore, these methods allow getting an estimation of K_v whichever is the scale of the equipment. In particular, this eases the problem of K_v measurement for industrial scale apparatus (as well as of process transfer) where the gravimetric procedure cannot be used since the intervention of the user is limited. Therefore, following on from what stated above, the use of global methods is strongly suggested for large scale apparatus.

4. Heat transfer in bulk and vials freeze-drying

In the following the influence of processing conditions, vial and tray characteristics and equipment size will be discussed in detail, presenting a selected set of original experiments planned by the authors to highlight the previous aspects.
In particular, experiments were carried out using both tubing vials (whose characteristics are reported in Table 2) and trays made of stainless steel or polystyrene. In case of freeze-drying in vials, it was also investigated the effect of various loading configurations (e.g. comparing the value of K_v for vials processed directly over the heating shelf or in trays), as well as of different pieces of equipment, not only on the mean value of K_v, but also on intervial variability. With this respect, the gravimetric measurement is the only method that allows to get a complete description of the heat transfer in the lot. However, results obtained by the pressure rise test technique will be also presented, confirming the capacity of this method to get a quick and reliable estimation of the parameter of interest, not only in a laboratory equipment (*LyoBeta 25*™ by Telstar, Terrassa, Spain), but also in an industrial scale dryer where gravimetric experiments cannot be easily carried out. Previous results had already shown this, evidencing at the same time that inaccuracies of the PRT model can affect its reliability (Tang et al., 2006).

Properties	vials V_1	vials V_2	vials V_3	vials V_4
External diameter, mm	14.25	16.20	24.14	14.12
Thickness of side wall, mm	1.00	0.80	1.10	0.90
Thickness of vial bottom wall, mm	0.70	1.41	1.41	0.60
Estimated thickness of bottom vial gap (ℓ), mm	0.20	0.32	0.38	0.34

Table 2. Geometrical characteristics of the various glass vials investigated in this study

4.1 Influence of processing conditions on K_v

Figure 3 (left-side graph) displays the value of K_v vs. P_c measured by gravimetric way for a lot of vials (type V_1), which was directly loaded on the heating shelf and surrounded by a metal band. Vials placed in the central part of the shelf, i.e. those far from the edges where the effect of radiative heat coming from chamber walls can be relevant, are considered. According to Brülls & Rasmuson (2002) and Hottot et al. (2005), K_v increases with P_c, because of a higher heat conduction through the gas trapped between the tray and glass vial bottom.

In Figure 3 (left-side graph), it can be also observed a fairly good agreement between the value of K_v vs. P_c as measured by the pressure rise test technique (coupled with a modified version of Dynamic Parameters Estimation algorithm, DPE$^+$) and by gravimetric way. The values of K_v vs. P_c predicted by the mathematical model described in section §2 is also displayed. The model parameters a_c, K_c and ℓ and, thus, the global coefficients C_1, C_2 and C_3 (that describe the pressure dependence of K_v for vials V_1) were obtained by a non-linear regression of experimental data. The results so obtained are summarized in Table 3.

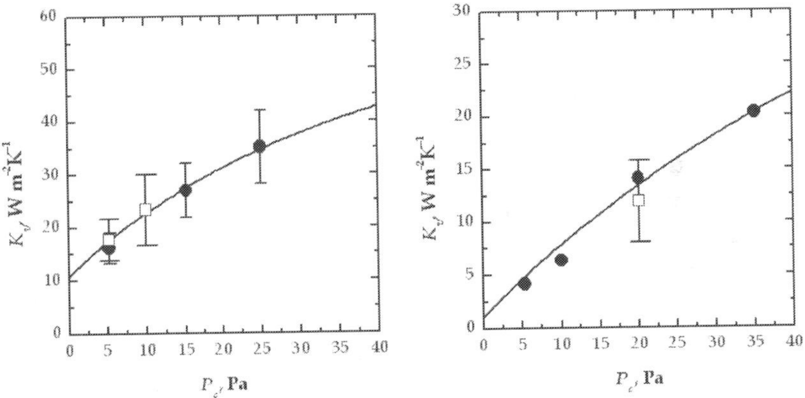

Fig. 3. Effect of chamber pressure on the value of K_v for the vial V_1 (left-side graph) and a stainless steel tray (right-side graph) processed in the laboratory freeze-dryer. The values calculated using the theoretical model, and the parameters of Table 3, are also displayed (solid line) together with the value estimated by pressure rise test technique (□)

In addition, it must be noticed that there is a large variance of K_v, which cannot be explained by the only observational error in the experiment, but is related to the heterogeneity of the lot. This is confirmed by the contour plot of Figure 4, where it can be observed that, because of side radiation and heat conduction though the metal frame, vials at the edge of the batch have a higher value of K_v than those placed in the core. In particular, as already observed by Pisano at al. (2008), these additional contributions increase only the sublimation rate of vials placed just next to the additional heat sources, while their effect is reduced after the first row. However, this issue will be better discussed in the following section.

container	subset	C_1	C_2	C_3	σ_{C_1}
vials V_1	all the lot	10.9	1.4	0.04	4.9
vials V_1	B	21.9	1.4	0.04	6.3
vials V_1	C	13.6	1.4	0.04	1.5
vials V_1	D	9.7	1.4	0.04	0.5
vials V_1	E	7.8	1.4	0.04	0.1
metal tray	-	1.1	0.7	0.01	-

Table 3. Parameters required to calculate the value of $K_v^{'}$ vs. P_c for a metal tray (made of stainless steel) and the various groups of vials V_1

Let's now focus on the temperature influence (of the gas trapped in the gap at the container bottom) on the value of K_v. In this study, as the gas temperature reasonably varies from the product temperature at the interface and T_{fluid}, it can be approximated to their mean. Figure 5 compares the heat transfer coefficient as measured at two different temperatures of the heating fluid, and thus of the chamber gas temperature, while the pressure inside the drying chamber is kept constant. It can be observed that a significant variation in the fluid temperature produces a modification of K_v that is of the same order of magnitude of the parameter uncertainty. This confirms that the dependence on gas temperature of K_v is not significant or, however, is very limited compared with the role of chamber pressure in the heat transfer.

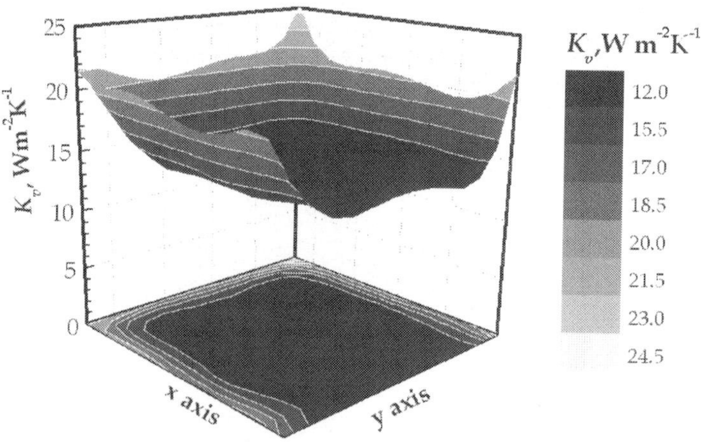

Fig. 4. Contour plot of the heat transfer coefficient for vials V_1 (loaded directly over the heating shelf) as measured over a lot. The primary drying phase was run at T_{fluid}=263 K and P_c=5 Pa. Results refer to a half of the batch

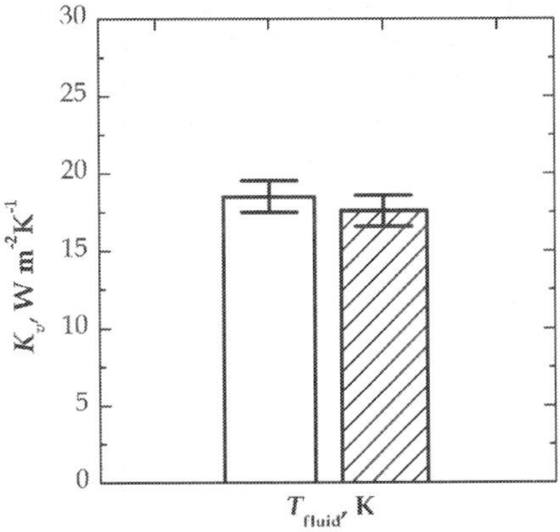

Fig. 5. Effect of T_{fluid}, and in turn of chamber gas temperature, on the value of K_v for the vial V_2 processed in the laboratory freeze-dryer. Both measurements were carried out at P_c=10 Pa, while the fluid temperature was set to: (□) T_{fluid}=275 K; (▨) T_{fluid}=253 K

Figure 3 (right-side graph) displays the value of K_v vs. P_c in case the process is carried out in trays. It can be observed that, similarly to what has been already shown for vials V_1, the heat transfer coefficient is strongly influenced by the total chamber pressure. However, the value of K_v for trays is significantly lower than that of vials. This is probably due a bad contact between the tray bottom and the heating plate, as the resistance to heat transfer of the stainless steel tray is negligible. In addition, as already observed by Bruttini et al. (1991) for bulk freeze-drying of a solution of cloxacillin monosodium salt, it is confirmed that, for freeze-drying in trays, the heat is mostly transferred by the conduction through the thin layer of gas that separates shelf and tray surfaces.

In case of bulk freeze-drying, wherein the solution is poured in a metal tray, the heat transfer between the heating fluid and the product can be still calculated as the sum of the three terms of Equation (3). Therefore, the same mathematical model used for vials freeze-drying is here used to determine (via non-linear regression of experimental data) the optimal combination of a_c, K_c and ℓ that describes the pressure dependence of K_v for trays. The results so obtained are reported in Table 3.

4.2 Non uniformity of the lot

In the previous section it has been observed that the value of K_v significantly varies with the position of the vial in the lot, see Figure 4. Therefore, following on from what stated in section §3 (see Figure 1), the lot of vials has been divided in five zones, which are characterized by different heat transfer mechanisms. The criteria chosen to identify the various zones are summarized in Table 1.

As previously discussed, a further refinement might be introduced to distinguish vials radiated by different walls of the chamber, but it has not been considered here: the correctness of this approach is confirmed by the more detailed statistical analysis that is presented in the following.

Figure 6 compares the value of K_v vs. P_c for the various groups of vials identified in Table 1. Vials A were not considered in the analysis as only four samples can be monitored in each test and, hence, the results are not statistically relevant. As it can be expected, the heat transfer coefficient of edge-vials (i.e. vials B and C) is much higher than that of vials placed in the core of the lot (i.e. vials D and E). In fact, according to Table 1, these vials receive an additional contribution of heat due to radiant energy coming from chamber walls and tray band, as well as, in case of vials B, the contribution due to conduction along the tray band.

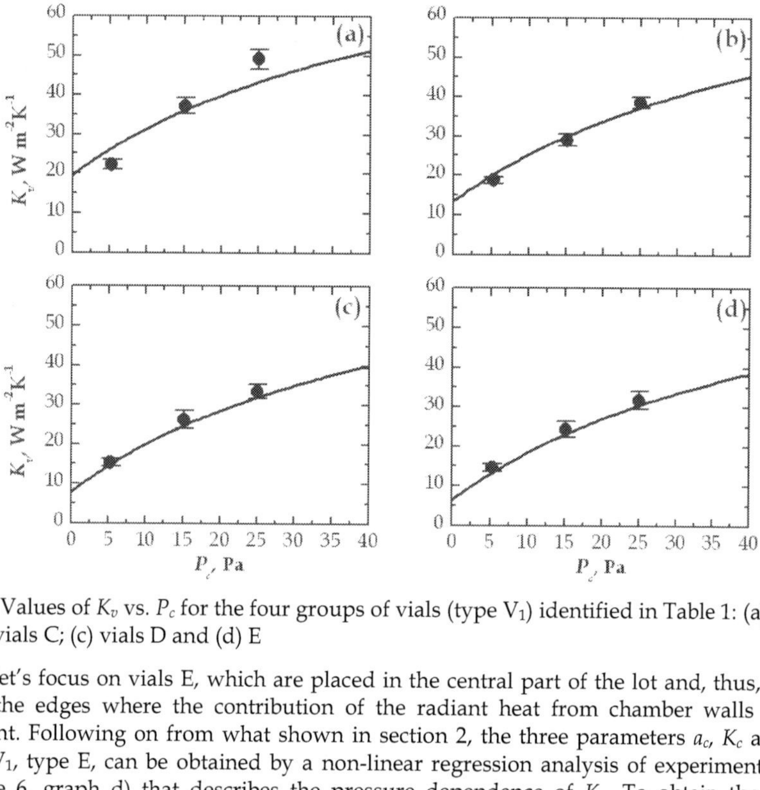

Fig. 6. Values of K_v vs. P_c for the four groups of vials (type V_1) identified in Table 1: (a) vials B; (b) vials C; (c) vials D and (d) E

Now let's focus on vials E, which are placed in the central part of the lot and, thus, are far from the edges where the contribution of the radiant heat from chamber walls can be relevant. Following on from what shown in section 2, the three parameters a_c, K_c and ℓ of vials V_1, type E, can be obtained by a non-linear regression analysis of experimental data (Figure 6, graph d) that describes the pressure dependence of K_v. To obtain the model parameters of the other three groups of vials, we assume that a_c and ℓ, and thus C_2 and C_3, are known, and equal to those determined for vials E. This assumption is reasonable, since these parameters depend on the container and not on the surrounding environment. On the contrary, K_c, and thus C_1, is optimized to get the best fitting between model predictions and experimental observations. Table 3 summarizes the value of global parameters C_1, C_2 and C_3 for the four groups of vials (V_1 type).
Concerning the uncertainty on the parameter K_v, it must be said that a_c and ℓ, thus C_2 and C_3 that define the pressure dependence of K_v, can be estimated, even if with some difficulties. On the contrary, it is almost impossible to estimate C_1 as it is affected by the contact between the shelf surface and the bottom of the container, as well as the view factors for the radiative heat. Furthermore, the effect on the final value of K_v due to the uncertainty on C_2 and C_3 (e.g. deriving from a different geometry of the container bottom) is less important than that due

to the variability associated to transport phenomena involved in C_1. For all these reasons, we assume that the only responsible for the uncertainty on K_v is the parameter C_1. Nevertheless, it is worth noticing that the uncertainty, or variability, on a_c and ℓ (and thus on C_2 and C_3) is implicitly included in the uncertainty on the parameter C_1, even if it is not directly expressed. At this point, once the value of C_2 and C_3 is known, eq. (7) can be used for retrieving the coefficient C_1 of each vial of the lot (under investigation) from the value of K_v that has been previously measured for vials V_1.

Figure 7 displays the distribution of C_1 (of vials V_1) for the entire lot (graph a), as well as for the various families of vials of Table 1 (graphs b–e). It can be observed that the various groups of vials are characterized by a significantly different mean value of C_1 and, as it can be expected, edge-vials have a much higher C_1 than that of vials placed in the core of the batch.

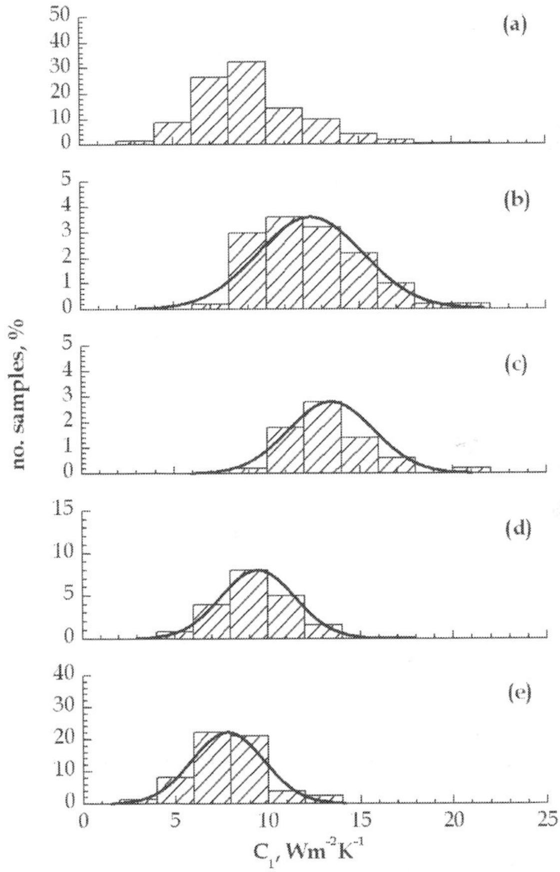

Fig. 7. Distribution of C_1 of vials V_1 for: (a) the entire lot; (b) vials B; (c) vials C; (d) vials D and (e) vials E. The solid line is the Gaussian distribution calculated according to the mean value and standard deviation reported in Table 3

In Figure 7 (graph a), the width of the C_1 distribution for the entire lot is so high that cannot be explained by the only observational error in the experiments but, as already discussed above, is due to the lot unevenness. Thus, the set of data has been divided in the four subsets previously identified and the distribution curve has been recalculated for each group of vials (see graphs b–e). The correctness of this approach is confirmed by the fact that the resulting curves can be well described by Gaussian distributions (that are typically used for describing the observational error in an experiment) and their width is significantly reduced. It can be noticed that the variance of C_1 for vials B is still relevant (even if much smaller than that of the entire lot), but it is now difficult to discern the contribution due to the intrinsic variability of the object of measurement from the uncertainty of the method.

4.3 Role of the vials geometry and material

Following on from what stated in section §2, the shape of the container bottom significantly affects the value of K_v, as well as its dependence on chamber pressure. To better investigate this aspect, the heat transfer coefficient of two different types of vials (i.e. V_1 and V_2), which have a similar geometry (in terms of external diameter and side-wall thickness), but a different shape of the bottom, are compared.

The values of K_v vs. P_c for the two types of containers, and in case they are directly loaded on the heating shelf and processed in the same equipment, are displayed in Figure 8. To discern the contributions due to the characteristics of the vial from those of the environment, the analysis is limited to only those vials that are placed in the centre of the lot. It can be noticed that the two containers have a similar value of C_1 even if vials V_2 have a thicker vial bottom wall.

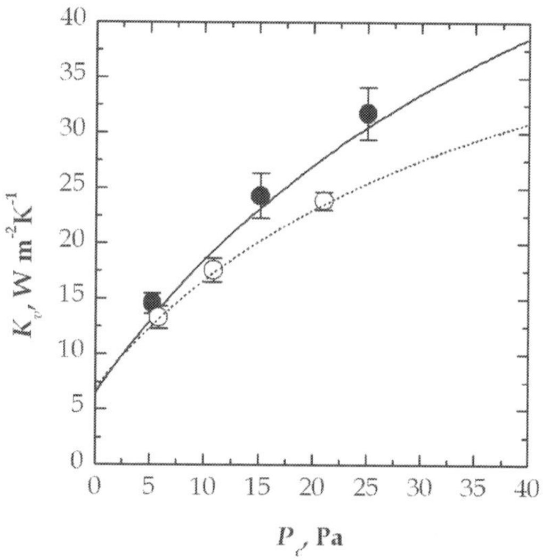

Fig. 8. Comparison between the value of K_v vs. P_c for vials subset E (processed in the laboratory freeze-dryer) in case of two different container geometries: (●) V_1 and (○) V_2 type. The values calculated using the theoretical model are also displayed: (solid line) vials V_1 and (dotted line) V_2

In particular, as the containers have a similar geometry and the contact between the container bottom and the shelf plate tightly depends on the characteristics of the vial surface, this implies that, in this case, the heat conduction (at the points of contact between the container bottom and the shelf surface) and the radiative heat coming from the heating plate give a similar contribution to the energy balance.

On the contrary, it can be observed a significantly different pressure dependence of K_v. This is due to a different contribution of the heat conduction through the gas trapped at the container bottom to the total energy balance of the vial. According to manufacturer specifications, the maximum value of the gap thickness is greater in vials V_1 (ℓ_{max}=4.0x10^{-4} m) and this seems to be in disagreement with experimental results of Figure 8. Nevertheless, it must be said that the heat conduction through the gas is a function of the volume of the gas gap, which depends not only on ℓ_{max}, but also on the shape of the vial bottom. In particular, the visual inspection of the two systems confirms that vials V_1, despite a higher value of ℓ_{max}, are characterized by a flatter gap at the bottom. As a consequence, the heat transfer resistance due to the gas gap is lower in vials V_1, which show a marked pressure dependence of K_v. As further confirmation of the previous evidences, the effective value of ℓ, as obtained by a non-linear regression of experimental data, of vials V_1 (ℓ =2.0x10^{-4} m) is smaller than that of vials V_2 (ℓ =3.1x10^{-4} m).

A final comment concerns the effect of container treatments (e.g. a silane coating of the internal vial walls) on the value of K_v. Figure 9 compares the overall heat transfer coefficient of vials V_2 in case they underwent a silane coating treatment or not. It can be noticed that pre-treated vials show a lower value of K_v and this reduction is of the same order of magnitude for all the four group of vials, thus the coating does not alter the contribution of the radiative heat to the total energy balance of the vial, but probably only the conduction through the vial side-wall.

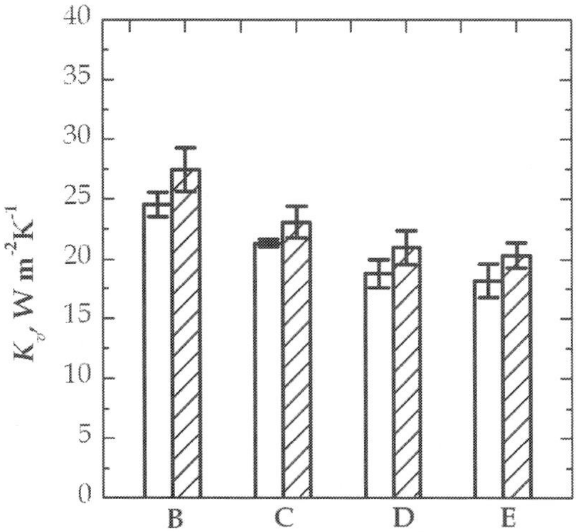

Fig. 9. Heat transfer coefficient of the four groups of vials (V_2 type) in case they underwent a silane coating treatment (void bars) or not (filled bar). The lot of vials is loaded on the heating shelf and enclosed by an external metal band. The measurement was carried out at P_c=10 Pa

4.4 Role of the configuration used for containers loading

In this section, we compare the value of K_v, as well as the heterogeneity of the lot, in case various configurations are used for loading the product into the drying chamber; in particular, the following configurations have been investigated:

- Vials are directly loaded on the heating shelf and enclosed by a metal frame;
- Vials are directly loaded on the heating shelf, but not surrounded by a frame;
- Vials are loaded on a metal tray.

Figure 10 compares the heat transfer coefficient of vials in case they are enclosed by a metal frame or not (the comparison is carried out for vials V_2). In particular, the use of a metal frame can have a different effect on the heat transfer coefficient of the various groups of vials. For example, the mean value of K_v for vials E is not significantly modified by the presence of the surrounding band, while edge-vials (both vials B and C) show a much higher value of K_v in case no metal frame is used. In fact, this frame acts as thermal shield since it reduces the radiative heat coming from chamber walls, which are warmer than the metal frame. The metal band also minimizes the unevenness of edge vials, as they are not anymore exposed to surfaces with different characteristics and temperature. Nevertheless, it must be said that edge-vials continue to receive an additional heat flow (due to the conduction between the vial side-wall and the metal band at the points of contact), but this has a minor effect with respect to chamber walls radiation. All these effects tightly depend upon the type of frame used, as well as the geometry of the drying chamber; hence, it is fundamental to take into account such phenomena during process transfer from one unit to another one, mainly if the loading configuration is modified.

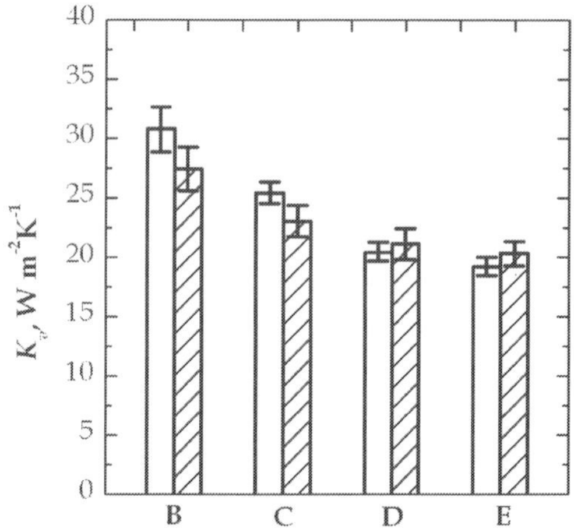

Fig. 10. Heat transfer coefficient (at P_c=10 Pa) of the four groups of vials (V_2 type, silane treated) in case of two different loading configurations: (void bars) the lot of vials is loaded on the heating shelf; (filled bar) the lot of vials is loaded on the heating shelf and enclosed by an external metal band

Figure 11, instead, compares the value of K_v in case vials are loaded on a metal tray or directly loaded on the heating shelf; the comparison was carried out for vials V_3. It can be noticed that if on one side the use of a tray introduces an additional resistance to the heat transfer, on the other side it allows to get a more uniform distribution of the heat over the lot (in Figure 11, the error bar is much smaller in case vials are loaded on a tray), as well as a strong reduction of the value of K_v for edge-vials due to the presence of the tray band. This strongly reduces the difference of K_v between vials D and E, as those vials that are located on the second row of the lot are now in contact with vials that are less warm and, thus, receive a smaller amount of heat from edge-vials.

Figure 12 (a) compares the value of K_v for vials loaded on the heating plate or in trays made of different materials, i.e. stainless steel or polystyrene. Even in this case we can observe a strong reduction of K_v when vials are loaded on a tray. In particular, the value of K_v of central vials (both group D and E) is not modified by the type of tray used; in fact, the observed variations are of the same order of magnitude of the error bar. On the contrary, K_v of edge-vials significantly changes with the type of tray used. This can be explained by a different heat transfer between the tray band and the container side-walls, which, for example, can be due to a different emissivity of the tray material.

4.5 Role of the equipment size

Figure 12 (b) compares the heat transfer coefficient of vials processed in laboratory and manufacturing equipment. Results refer to vials loaded on a plastic tray and in case the drying is carried out at P_c=10 Pa. It can be observed that the value of K_v of vials located in the central part of the lot is not significantly modified, whereas edge-vials processed in the laboratory freeze-dryer have a higher value of K_v than that observed in the large scale freeze-dryer. This is due to a different contribution of the radiative heat (that, for example, can be caused by a different geometry of the drying chamber) and has to be taken into account when we have to transfer a process from one unit to another one, as well as during scale-up operations, as the fraction of edge-vials into the lot varies with the scale of the equipment.

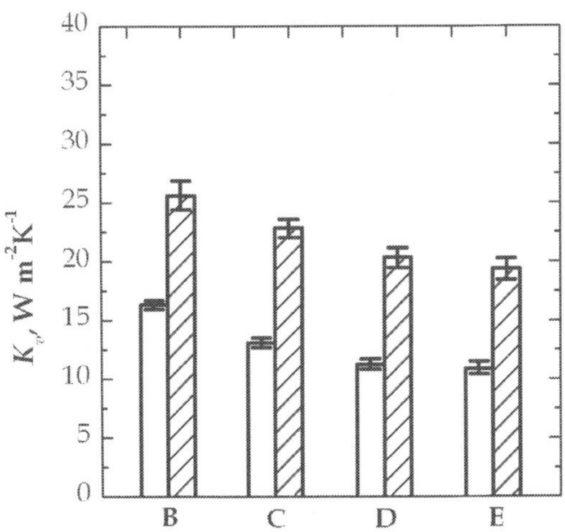

Fig. 11. Heat transfer coefficient (at P_c=10 Pa) of the four groups of vials (V_3 type) in case of two different loading configurations: (void bars) the lot of vials is loaded on a metal tray; (filled bar) the lot of vials is loaded on the heating shelf and enclosed by an external metal band

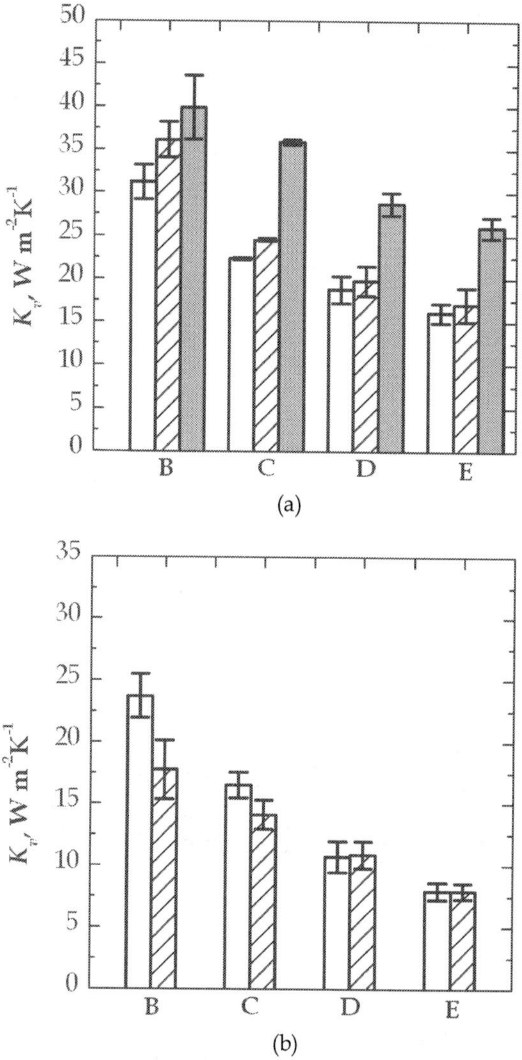

Fig. 12. Comparison between experimental values of K_v for the various groups of vials (V_4 type). (a) Influence of various loading configurations: vials loaded on a stainless steel tray (□), a polystyrene tray (▨) and directly on the heating shelf (■); (b) values measured in two pieces of equipment of different size: (□) laboratory and (▨) industrial scale dryer

For the two pieces of equipment investigated in this study, central vials have almost the same value of K_v, but, in general, it can happen to observe significant differences. In fact, even if the value of ℓ is independent of the equipment used, the parameters a_c and K_c can be affected by the characteristics of the shelf surface. Thus, once the value of ℓ has been determined for the laboratory freeze-dryer (using the procedure described in section §2), and assuming that a_c does not change, we can calculate the effective value of K_c for the four groups of vials in case they are processed in the industrial freeze-dryer. It follows that the global coefficients C_2 and C_3, which describe the pressure dependence of K_v, are the same for both pieces of equipment, while the parameter C_1 of peripheral vials is slightly different (as it depends on the vial-tray and shelf-tray contact, and on radiation effect). The knowledge of the pressure dependence of K_v in the two pieces of equipment is fundamental to make decisions during the phase of scale-up of a recipe obtained in a laboratory scale equipment to an industrial scale freeze-dryer.

5. Conclusion

Following on from what stated in the introduction, the heat transfer to the product has to be carefully controlled in order to avoid product overheating. This result can be obtained in-line, using a control system (Pisano et al., 2010, 2011), or building off-line the design space, i.e. identifying the set of operating conditions (heating fluid temperature and chamber pressure) that allows preserving product quality. Recently, the use of mathematical modelling, coupled with few experiments to determine model parameters (e.g. the value of K_v vs. P_c), has been proposed to get the design space quickly, taking also into account the effect of parameters uncertainty and/or variability (Giordano et al., 2011).

Due to batch non-uniformity, vials have to be divided in various subsets (as shown above), which are characterized by the same value of K_v, and the design space (that is strongly dependent on the value of K_v) has to be built for each group (Fissore et al., 2011b). Of course, since the design space of edge-vials is more contracted than that of vials placed in the core, processing conditions has to be chosen according to the design space of those vials that can be more easily damaged by product overheating (i.e. vials B). The recipe so designed allows preserving product quality of the entire lot, but the resulting drying time might be quite long. Therefore, as edge-vials are a small fraction of the lot, it might be worth referring to central vials to reduce the duration of the process. Of course, this is advantageous only if the energy saving deriving from a shorter drying time compensate the loss due to product waste (that derives from the fact that edge vials did not meet product quality criteria).

It must be remarked that whichever is the tool used to control the process, the resulting recipe is strongly dependent on the value of K_v, which can vary with operating conditions, type of container, equipment and configuration used for loading the product.

The experimental investigation carried out in this study has evidenced a strong dependence of the heat transfer coefficient on processing conditions used for the drying. In particular, it has been proved that while the gas temperature has a minor effect on the value of K_v, the role of chamber pressure cannot be neglected. To be used by the model at the basis of the various tools recently developed for the recipe design, the pressure dependence of K_v has been described by a mathematical model, whose parameters have been obtained, for various types of container (both glass vials and trays), by regression analysis of experimental data.

In addition, the heat transfer of glass vials has been measured in case of various loading configurations. It has been shown that, as expected, the introduction of a tray significantly reduces the value of the heat transfer coefficient of the entire lot and, in particular, a more marked reduction was observed for edge-vials. On the contrary, in case vials are directly loaded on the shelf, but surrounded by a metal frame, the value of K_v of central vials is not modified, while that of edge vials is strongly reduced. This is due to the contribution of the tray band, which acts as thermal shield for the radiative heat coming from chamber walls. Therefore, it must be remarked that, during the phase of process development, the user has to take into account that the pressure dependence of K_v does not depend only on the type of vials, but also on the configuration used for loading the product into the drying chamber.

In general, the gravimetric procedure gives the best accuracy and robustness, even if it is more time demanding with respect to other global methods available. However, the use of the pressure rise test technique is strongly suggested in case of industrial apparatus, where the gravimetric procedure is not practicable as the intervention of the user (to place temperature sensors over the lot of vials) is limited. Therefore, it has been shown that the pressure rise test technique (and in particular the latest developments like the DPE+ algorithm) can be effectively used for measuring the value of K_v, whichever is the scale of the equipment, without requiring an excessive effort from the users. In addition, an estimation of the mean value of K_v is more than enough for an effective description of the heat transfer of the lot, as the effect of batch non-uniformity in a manufacturing process is less marked. A further advantage of the pressure rise test technique is that, with respect to other global methods like TDLAS, it requires no modifications of the equipment and its hardware.

A final comment concerns the problem of scale-up, or process transfer, of a recipe from one unit to another one. It has been proved that the heat transfer coefficient of a specific container can varies significantly (mainly for edge vials) with the type of equipment used, even if the same loading configuration is used. Therefore, if this difference is relevant, the recipe, which is usually developed in laboratory and has to be transferred on manufacturing equipment, should be adapted to take into account the different heat transfer of the containers.

6. Acknowledgment

Development of PRT methods for industrial apparatus has been continuously supported by Telstar S.A. (Terrassa, Spain), whose contribution of data obtained in large scale apparatus and financial support for this chapter is gratefully acknowledged. The authors would like to acknowledge Giovanni Accardo, Salvatore Genco e Daniele Sorce for their valuable support in the experimental investigation.

7. Nomenclature

a_c	energy accommodation coefficient
A_v	cross sectional area of the vial, m^2
A_{sub}	total sublimation area, m^2
C_1	parameter expressing the dependence of $K_v^{'}$ from radiation and the contact between vial bottom and tray surface, $J\ s^{-1}m^{-2}K^{-1}$
C_2	parameter expressing the pressure dependence of $K_v^{'}$, $J\ s^{-1}m^{-2}K^{-1}Pa^{-1}$
C_3	parameter expressing the pressure dependence of $K_v^{'}$, Pa^{-1}
e_s	emissivity for radiation heat exchange from the shelf to the bottom of the vial
e_v	emissivity for radiation heat exchange from the shelf to the top of the vial
ΔH_s	heat of sublimation, $J\ kg^{-1}$
J_q	heat flux to the product, $J\ s^{-1}m^{-2}$
J_w	solvent flux, $kg\ s^{-1}m^{-2}$

k_s	heat transfer coefficient between the technical fluid and the shelf, J $s^{-1}m^{-2}K^{-1}$
K_c	heat transfer coefficient due to direct conduction from the shelf to the glass at the points of contact , J $s^{-1}m^{-2}K^{-1}$
K_g	heat transfer coefficient due to conduction in the gas between the shelf and the vial bottom , J $s^{-1}m^{-2}K^{-1}$
K_r	heat transfer coefficient between the shelf and the vial due to radiation, J $s^{-1}m^{-2}K^{-1}$
K_v	overall heat transfer coefficient between the heating fluid and the product at the bottom of the vial, J $s^{-1}m^{-2}K^{-1}$
$K_v^{'}$	overall heat transfer coefficient between the heating shelf and the vial bottom (or between shelf and tray, and tray and vials), J $s^{-1}m^{-2}K^{-1}$
K_v^{*}	overall heat transfer coefficient between the heating shelf and the product at the bottom of the vial, J $s^{-1}m^{-2}K^{-1}$
ℓ	constant effective distance between the bottom of the vial and the shelf, m
m	mass, kg
M_w	molar mass of water, kg $kmol^{-1}$
$p_{w,c}$	partial pressure of water in the drying chamber, Pa
P_c	chamber pressure, Pa
R	ideal gas constant, J $kmol^{-1}K^{-1}$
s_g	thickness of the glass at the bottom of the vial, m
s_{tray}	thickness of the tray bottom, m
t	time, s
T	temperature, K
T_B	temperature of the product at the vial bottom, K
T_c	temperature of the chamber gas, K
T_{fluid}	temperature of the heating fluid, K
T_{shelf}	temperature of the heating shelf, K
V_c	volume of the drying chamber, m^3

Greeks

α	parameter used to calculate K_g
κ	Stefan-Boltzman constant, J $s^{-1}m^{-2}K^{-4}$
Λ_0	free molecular heat conductivity at 0°C, J $s^{-1}m^{-1}K^{-1}$
λ_0	heat conductivity of the water vapour at ambient pressure, J $s^{-1}m^{-1}K^{-1}$
λ_g	heat conductivity of the glass, J $s^{-1}m^{-1}K^{-1}$
λ_{tray}	heat conductivity of the tray, J $s^{-1}m^{-1}K^{-1}$
σ_{C_1}	standard deviation of the parameter C_1, J $s^{-1}m^{-2}K^{-1}$

8. References

Brülls, M., & Rasmuson, A. (2002). Heat transfer in vial lyophilization. *International Journal of Pharmaceutics*, Vol. 246, pp. 1-16, ISSN 0378-5173.

Bruttini, R., Rovero, G., & Baldi, G. (1991). Experimentation and modelling of pharmaceutical lyophilization using a pilot plant. *The Chemical Engineering Journal*, Vol. 45, pp. B67-77, ISSN 1385-8947.

Chen, R., Slater, N. K. H., Gatlin, L. A., Kramer, T., & Shalaev, E. Y. (2008). Comparative rates of freeze-drying for lactose and sucrose solutions as measured by photographic recording, product temperature and heat flux transducer. *Pharmaceutical Development and Technology*, Vol. 13, pp. 367-374, ISSN 1083-7450.

Chouvenc, P., Vessot, S., Andrieu, J., & Vacus P. (2004). Optimization of the freeze-drying cycle: a new model for pressure rise analysis. *Drying Technology*, Vol. 22, pp. 1577-1601, ISSN 1532-2300.

Corbellini, S., Parvis, M., & Vallan, A (2010). In-process temperature mapping system for industrial freeze dryers. *IEEE Transactions on Instrumentation and Measurement*, Vol. 59, pp. 1134-1140, ISSN 0018-9456.

Dushman, S., & Lafferty, J. M. (1962). *Scientific foundations of vacuum technique*, Wiley, ISBN 978-047-1228-03-5, New York, USA.

Fissore, D., Pisano, R., & Barresi, A. A. (2011a). On the methods based on the Pressure Rise Test for monitoring a freeze-drying process. *Drying Technology*, Vol. 29, pp. 73-90, ISSN 1532-2300.

Fissore, D., Pisano, R., & Barresi, A. A. (2011b). Advanced approach to build the design space for the primary drying of a pharmaceutical freeze-drying process. Submitted to *Journal of Pharmaceutical Sciences*, ISSN 0022-3549.

Franks, F. (2007). *Freeze-drying of pharmaceuticals and biopharmaceuticals*, Royal Society of Chemistry, ISBN 978-085-4042-68-5, Cambridge, UK.

Gan, K. H., Bruttini, R., Crosser, O. K., & Liapis, A. A. (2005a). Freeze-drying of pharmaceuticals in vials on trays: effects of drying chamber wall temperature and tray side on lyophilization performance. *International Journal of Heat and Mass Transfer*, Vol. 48, pp. 1675-1687, ISSN 0017-9310.

Gan, K. H., Crosser, O. K., Liapis, A. I., & Bruttini, R. (2005b). Lyophilisation in vials on trays: effects of tray side. *Drying Technology*, Vol. 23, pp. 341-363, ISSN 1532-2300.

Gieseler, H., Kessler, W. J., Finson, M., Davis, S. J., Mulhall, P. A., Bons, V., Debo, D. J., & Pikal, M. J. (2007). Evaluation of Tunable Diode Laser Absorption Spectroscopy for in-process water vapor mass flux measurement during freeze drying. *Journal of Pharmaceutical Sciences*, Vol. 96, pp. 1776-1793, ISSN 0022-3549.

Giordano, A., Barresi, A. A., & Fissore, D. (2011). On the use of mathematical models to build the design space for the primary drying phase of a pharmaceutical lyophilization process. *Journal of Pharmaceutical Sciences*, Vol. 100, pp. 311-324, ISSN 0022-3549.

Hottot, A., Vessot, S., & Andrieu, J. (2005). Determination of mass and heat transfer parameters during freeze-drying cycles of pharmaceutical products. *PDA Journal of Pharmaceutical Science and Technology*, Vol. 59, pp. 138-53, ISSN 1079-7440.

Jennings, T. A. (1999) *Lyophilization: introduction and basic principles*, CRC Press, ISBN 978-157-4910-81-0, Boca Raton, USA.

Kessler, W. J., Davis, S. J., Mulhall, P. A., & Finson, M. L. (2006). System for monitoring a drying process. United States Patent No. 0208191 A1.

Kuu, W. Y., Nail, S. L., & Sacha, G. (2009). Rapid determination of vial heat transfer parameters using tunable diode laser absorption spectroscopy (TDLAS) in response to step-changes in pressure set-point during freeze-drying. *Journal of Pharmaceutical Sciences*, Vol. 98, pp. 1136-1154, ISSN 0022-3549.

Mellor, J. D. (1978). *Fundamentals of freeze-drying*, Academic Press, ISBN 978-012-4900-50-9, London, UK.

Milton, N., Pikal, M. J., Roy, M. L., & Nail, S. L. (1997). Evaluation of manometric temperature measurement as a method of monitoring product temperature during lyophilisation. *PDA Journal of Pharmaceutical Science and Technology*, Vol. 5, pp. 7-16, ISSN 1079-7440.

Oetjen, G. W., & Haseley, P. (2004). *Freeze-Drying*, Wiely-VHC, ISBN 978-352-7306-20-6, Weinheim, Germany.

Pikal, M. J. (1985). Use of laboratory data in freeze-drying process design: heat and mass transfer coefficients and the computer simulation of freeze-drying. *Journal of Parenteral Science and Technology*, Vol. 39, pp. 115-139, ISSN 0279-7976.

Pikal, M. J. (2000). Heat and mass transfer in low pressure gases: applications to freeze-drying. In: *Transport processes in pharmaceutical systems*, Amidon, G. L., Lee, P. I., & Topp, E. M., pp. 611-686, Marcel Dekker, ISBN 0-8247-66105, New York, USA.

Pikal, M. J., & Shah, S. (1990). The collapse temperature in freeze drying: dependence on measurement methodology and rate of water removal from the glassy phase, *International Journal of Pharmaceutics*, Vol. 62, pp. 165–186, ISSN 0378-5173.

Pikal, M. J., Roy, M. L., & Shah, S. (1984). Mass and heat transfer in vial freeze-drying of pharmaceuticals: role of the vial. *Journal of Pharmaceutical Sciences*, Vol. 73, pp. 1224-1237, ISSN 0022-3549.

Pisano, R., Fissore, D., & Barresi, A. A. (2011). Freeze-drying cycle optimization using Model Predictive Control techniques. *Industrial & Engineering Chemistry Research*, Vol. 50, pp. 7363-7379, ISSN 0888-5885.

Pisano, R., Fissore, D., Velardi, S. A., & Barresi, A. A. (2010). In-line optimization and control of an industrial freeze-drying process for pharmaceuticals. *Journal of Pharmaceutical Sciences*, Vol. 99, pp. 4691-4709, ISSN 0022-3549.

Pisano, R., Rasetto, V., Petitti, M., Barresi, A. A., & Vallan, A. (2008). Modelling and experimental investigation of radiation effects in a freeze-drying process, *Proceedings of EMMC– 5th Chemical Engineering Conference for Collaborative Research in Eastern Mediterranean Countries*, pp. 394-398, Cetraro (CS), Italy, May 24-29, 2008.

Rambhatla, S., Obert, J. P., Luthra, S., Bhugra, C., & Pikal, M. J. (2005). Cake shrinkage during freeze drying: a combined experimental and theoretical study, *Pharmaceutical Development & Technology*, Vol. 1, pp. 33–40, ISSN 0265-2048.

Rambhatla, S., & Pikal, M. J. (2003). Heat and mass transfer scale-up issues during freeze-drying, I: atypical radiation and edge vial effect. *AAPS PharmSciTech*, Vol. 4, Article No. 14, ISSN: 1530-9932.

Sadikoglu, H., Ozdemir, M., & Seker, M. (2006). Freeze-drying of pharmaceutical products: research and development needs. *Drying Technology*, Vol. 24, pp. 849-861, ISSN 0737-3937.

Schneid, S. & Gieseler, H. (2008). Evaluation of a new wireless temperature remote interrogation system (TEMPRIS) to measure product temperature during freeze-drying. *AAPS PharmSciTech*, Vol. 9, pp. 729-739, ISSN 1530-9932.

Sheehan, P., & Liapis, A. I. (1998). Modeling of the primary and secondary drying stages of the freeze-drying of pharmaceutical product in vials: numerical results obtained from the solution of a dynamic and spatially multi-dimensional lyophilisation model for different operational policies. *Biotechnology & Bioengineering*, Vol. 60, pp. 712-728, ISSN 1097-0290.

Tang, X. C., Nail, S. L., & Pikal, M. J. (2006). Evaluation of manometric temperature measurement (MTM), a process analytical technology tool in freeze-drying, part III: heat and mass transfer measurement. *AAPS PharmSciTech*, Vol. 7, Article No. 97, ISSN 1530-9932.

Velardi, S. A., & Barresi, A. A. (2008). Development of simplified models for the freeze-drying process and investigation of the optimal operating conditions. *Chemical Engineering Research and Design*, Vol. 86, pp. 9-22, ISSN 0263-8762.

Velardi, S. A., Rasetto, V., & Barresi A. A. (2008). Dynamic Parameters Estimation Method: advanced Manometric Temperature Measurement approach for freeze-drying monitoring of pharmaceutical solutions. *Industrial Engineering Chemistry Research*, Vol. 47, pp. 8445-8457, ISSN 0888-5885.

Wang, W. (2000). Lyophilization and development of solid protein pharmaceuticals, *International Journal of Pharmaceutics*, Vol. 203, pp. 1–60, ISSN 0378-5173.

7

Radiant Floor Heating System

Byung-Cheon Ahn
Department of Building Equipment System Engineering, Kyungwon University
Korea

1. Introduction

The radiant floor heating system controls indoor air temperature by heat transfer from heated surface to indoor air, after the heat has been applied into a floor structure mass by using hot water heating coil buried under the floor. In this case, hot water is provided by a boiler, then, conveyed to indoor floor heating coil through pipe network.

The radiant floor heating system can operate transmitting power quietly and efficiently with no noise at low costs of the initial investment and with low maintenance. However, since hot water heating coil is buried under the floor, the system has a defect that it has large thermal inertia by heavy heat capacitance of the floor structure mass. In addition, response characteristics with long time delay will be caused due to certain amount of time needed to heat up the structure mass(Ahn, 2010). Thus, saving energy and maintaining comfortable indoor thermal environment would be possible only if a proper control method is applied into the system, considering its thermal inertia.

This chapter introduces system features and mathematical background of radiant floor heating system. Especially, it covers theoretical background of analysis on heat transfer characteristics in pipes and indoor heat flow characteristics to help understand dynamic characteristics of energy in the system. In addition, explanation is given on types and characteristics of automatic thermostatic valves in the system that supplies hot water with on-off or proportional control, and more information is demonstrated on heat flow characteristics and heating performance of the radiant floor heating system in applying various kinds of control systems to comfort indoor heat and save energy.

2. Heat transfer in pipes

In case of radiant floor heating system, hot water from the boiler will be streamed into households through pipes, and these pipes can be distinguished into two types; outdoor exposed pipe covered with heat insulator, and pipe buried under the floor structure mass. Thus, separate mathematical analyzing method is suggested to explain two types of pipes. Firstly, fig. 1 depicts pipe covered with heat insulator. In this case, the pipe has exposed outdoor structure and constant outdoor temperature. Assuming that there is no superheating or subcooling of the fluid that changes phase, and its pressure does not change, the LMTD(Log Mean Temperature Difference) applies and in combination with a heat balance(Stoecker, 1980) gives

$$q = UA\frac{(T_{ao} - T_i) - (T_{ao} - T_o)}{ln[(T_{ao} - T_i)/(T_{ao} - T_o)]} = \rho W C_p(T_o - T_i) \quad (1)$$

where,

q	Heat transfer
U	Heat transfer coefficient
ρ	Water density
C_p	Specific heat
V	Flow velocity
T_i, T_o	Inlet and outlet temperature
T_{ao}	Ambient temperature
W	Flow rate($\pi R_1^2 \cdot V$)
A	Outside surface area of pipe($2\pi R_1 L$)

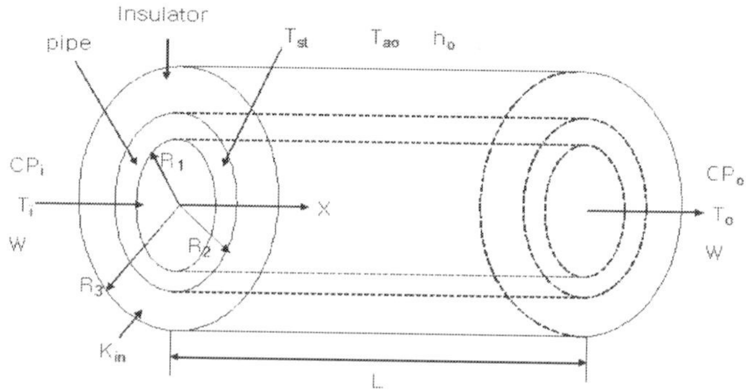

Fig. 1. Insulated pipe

Therefore, T_o, outlet temperature of the pipe, can be indicated from the formula (2)

$$T_o = T_i + (T_{ao} - T_i) \cdot [1 - exp(-\frac{-UA}{\rho W C_p})] \tag{2}$$

T_L, average temperature of hot water considering the length of pipe, can be found as

$$T_L = \frac{1}{L}\int_0^L T_x \, dx = T_i + (T_{ao} - T_i) \cdot [1 + (exp(-E) - 1)/E] \tag{3}$$

Where $E = UA/\rho W C_p$, and equivalent heat transfer coefficient, U, between fluid flow and outdoor is

$$U = \frac{1}{\frac{1}{h_i} + \frac{R_1}{K_{st}} + ln\left(\frac{R_2}{R_1}\right) + \frac{R_1}{K_{in}} ln\left(\frac{R_3}{R_2}\right) + \frac{R_1}{R_2 h_o}} \tag{4}$$

Where,

K_{st},	Thermal conductivity of pipe
K_{in},	Thermal conductivity of insulator
R_1 and R_2	Inside and outside diameter of pipe
R_3	Outside diameter of insulator

The heat transfer coefficient of hot water inside the pipe, (h_i), is

$$h_i = 0.023 \cdot Re^{0.8}Pr^{0.3} \cdot \frac{k_w}{D_i} \quad (Red > 2300) \tag{5}$$

$$h_i = 4.36 \cdot \frac{k_w}{D_i} \quad (Red \leq 2300) \tag{6}$$

Where Re is Reynolds number($V \cdot D_i/\upsilon$) and Pr is Prantl number (υ/α).
Also, heat transfer coefficient of exterior of the pipe, (h_o), is

$$h_o = \frac{k_f \cdot N_{\mu D}}{D_o} \tag{7}$$

Where k_f is thermal conductivity of air.
Nusselt number (Nu_D) can be solved differently regarding horizontal pipe and vertical pipe (Holman, 1981).
In case of horizontal pipe,

$$N_{\mu D} = 0.53(Gr_D \cdot Pr_f)^{\frac{1}{4}} \ \left(10^4 \leq Gr_D \cdot Pr_f \leq 10^9\right) \tag{8}$$

$$N_{\mu D} = 0.13(Gr_D \cdot Pr_f)^{\frac{1}{3}} \ \left(10^9 \leq Gr_D \cdot Pr_f \leq 10^{12}\right) \tag{9}$$

And in case of vertical pipe,

$$N_{\mu D} = 0.59(Gr_D \cdot Pr_f)^{\frac{1}{4}} \ (10^4 \leq Gr_D \cdot Pr_f \leq 10^9) \tag{10}$$

$$N_{\mu D} = 0.1(Gr_D \cdot Pr_f)^{\frac{1}{3}} \ (10^9 \leq Gr_D \cdot Pr_f \leq 10^{13}) \tag{11}$$

Where Pr_f is Prantl number for air and Gr_D is Grashof number.
Considering outdoor temperature (T_{ao}), and temperature difference (ΔT) between outdoor and pipe's external surface, Grashof number can be expressed as below.

$$Gr_D = \frac{g \cdot \beta \cdot \Delta T \cdot {D_o}^3}{{\nu_f}^2} = \frac{g \cdot \Delta T \cdot {D_o}^3}{{\nu_f}^2 \cdot T_{ao}} \tag{12}$$

ΔT value is needed in order to figure out heat transfer coefficient, h_o, while U value must be solved to find ΔT. Thus, accurate value, ΔT, can be measured through repeated calculation, assuming ΔT as a proper number.
Concerning that structure of hot water heating coil pipe is buried under the floor in radiant floor heating system in general, heat transfer phenomena from hot water pipe to floor and ceiling surface must be reviewed. Fig. 2 is a diagram of pipe buried under the household floor. Considering thermal behavior from hot water through pipe gives the following.
Very small volume ($A \cdot dx$) of the amount of heat in the hot water (Δq) can be formulated:

$$\Delta q = \rho \cdot A \cdot dx \cdot C_p \cdot T_x \tag{13}$$

where,

ρ	Water density
A	Cross sectional area of the pipe
C_p	Specific heat of hot water

T_x Temperature of hot water

Hot water in a very small volume has a heat transfer loss after a very short time as follows.

$$d\Delta q = -\rho \cdot A \cdot dx \cdot C_p \cdot dT_x \tag{14}$$

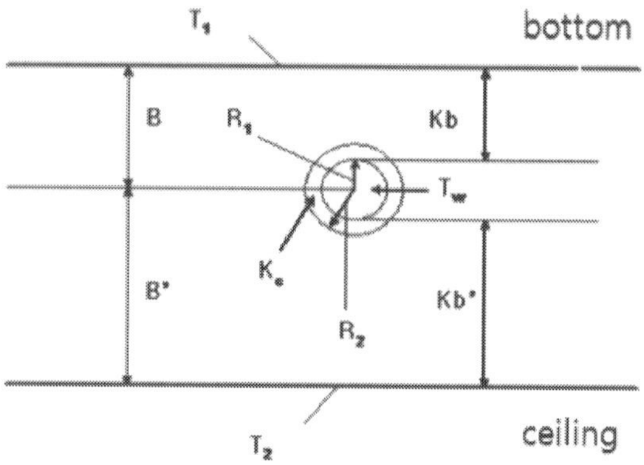

Fig. 2. Pipe buried in semi-infinite medium having isothermal surface

This value is a sum of the heat amounts emitting to the room floor (dΔqb) and to the ceiling surface of the room below (dΔqc).

This occurs because an amount of heat from the heated water is transferred to the floor and ceiling surface below.

$$d\Delta q = d\Delta q_b + d\Delta q_C \tag{15}$$

Where

$d\Delta q_b = U_1 \cdot A_p \cdot (T_k - T_1) \cdot dx \cdot dt$

$d\Delta q_c = U_2 \cdot A_p \cdot (T_k - T_2) \cdot dx \cdot dt$

$$U_1 = \frac{1}{\frac{1}{h_i} + \frac{R_1 \ln(R_2/R_1)}{K_c} + \frac{R_1 \cosh^{-1}(B/R_2)}{K_b}}$$

$$U_1 = \frac{1}{\frac{1}{h_i} + \frac{R_1 \ln(R_2/R_1)}{K_c} + \frac{R_1 \cosh^{-1}(B'/R_2)}{K_{b'}}}$$

where,

U_1 Heat transfer coefficient from pipe surface to the floor surface

U_2 Heat transfer coefficient from pipe surface to the ceiling surface of the bottom layer

h_i Heat transfer coefficient of pipe inner surface

T_1 Floor surface temperature

T_2 Ceiling surface temperature of the bottom layer

K_p Thermal conductivity of pipe

K_b Equivalent thermal conductivity from pipe surface to the floor surface

K_b' Equivalent thermal conductivity from pipe surface to the ceiling surface of the bottom layer
B Distance from the middle of pipe to the floor surface
B' Distance from the middle of pipe to the ceiling surface of the bottom layer
A_p Girth of pipe
R_1 Inside radius of pipe
R_2 Outside radius of pipe

After substituting equation (15) for equation (14), categorized according to hot water temperature, assuming that T_1 and T_2 are steady for a very short time (dt), and integrating for pipe length, the outlet temperature of hot water for length L can be expressed as the following equation (16) for hot water inlet temperature

$$T_O = \frac{U_1T_1 + U_2T_2}{U_1+U_2} + \left(T_i - \frac{U_1T_1+U_2T_2}{U_1+U_2}\right) exp\left(-\frac{(U_1+U_2)\cdot A_p \cdot L}{\rho \cdot A \cdot C_p \cdot v}\right) \tag{16}$$

where,
T_o Hot water temperature of pipe outlet
T_i Hot water temperature of pipe inlet
L The length of pipe
v Mean flow velocity(dx/dt)

If we replace T_o with T_x and L with x, then integrate this for an entire length, the mean temperature of hot water could be achieved as in the following equation (17)

$$T_W = \frac{1}{L}\int_0^L T_X\,dx = \frac{U_1T_1+U_2T_2}{U_1+U_2} + \left(T_i - \frac{U_1T_1+U_2T_2}{U_1+T_2}\right)\left(\frac{1-e^{-DL}}{DL}\right) \tag{17}$$

Where $D = \frac{(U_1+U_2)\cdot A_1}{\rho\cdot A\cdot C_p\cdot v}$

Fig. 3 depicts pipe network buried in a house of apartment building. Hot water is supplied from supply header through 5 distinguished pipes separately. Average temperature of each room can be found using formula (17)

3. Indoor heat transfer

Figure 4 shows heat amounts and temperatures of each part of the room; floor, ceiling, wall and window. There are 3 routes for heat transfer; conduction in the floor, ceiling and wall, convection with indoor air and radiant heat transfer between the heated floor and ceiling surface, and a non-heated wall in the house.

The amount of thermal conduction (q1) from the heated water in the pipe to the floor surface, and the convective amount (q2) from the floor to the indoor air can be shown as equation (18), (19).

$$q_1 = \frac{A_p(T_W T_1)\cdot L}{\frac{1}{h_i} + \frac{R_1 ln\,(R_2/R_1)}{K_c} + \frac{R_1 cosh^{-1}(B/R_2)}{K_b}} \tag{18}$$

$$q_2 = \frac{2.41\cdot A_1(T_1 - T_a)^{1.31}}{De^{0.08}} \tag{19}$$

Where T_a is the temperature of indoor air and D_e is the equivalent diameter of floor slab.

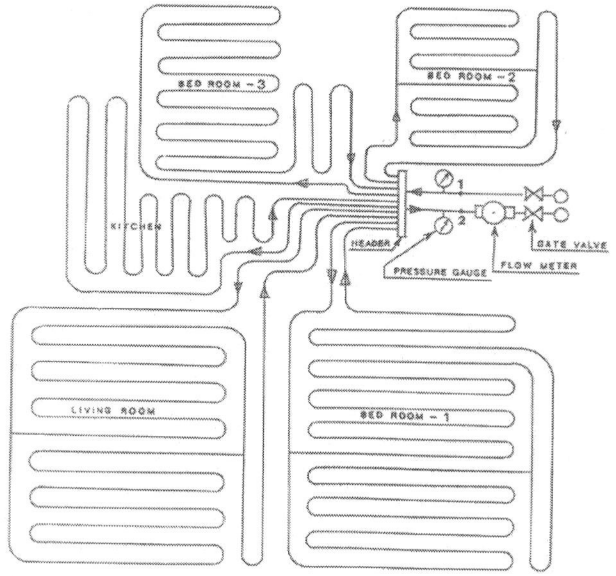

Fig. 3. Pipe network buried in a house of apartment building

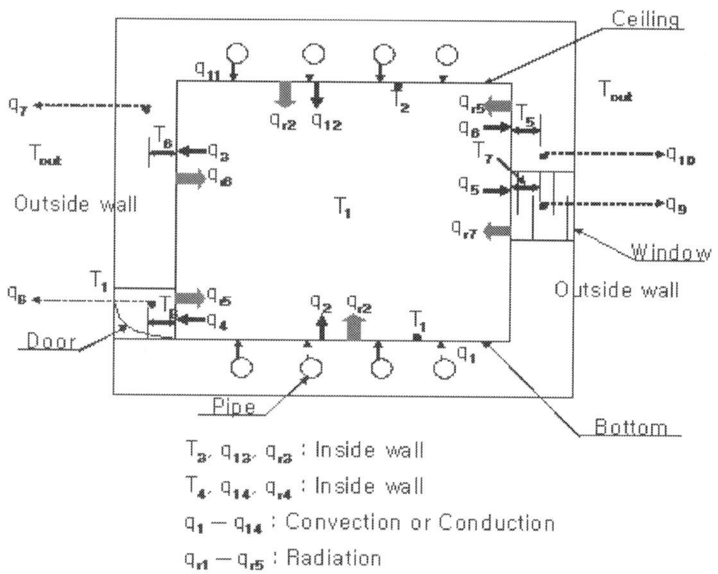

Fig. 4. Schematic diagram for heat flow for the heat transfer analysis in the room

The amount of the thermal convection to wall, window, and door surrounding indoor air(ASHRAE, 2004)

$$q_n = \frac{1.87 \cdot A_k (T_a - T_k)^{1.32}}{H_k^{0.05}} \tag{20}$$

Where H is indoor wall height.

The amount of thermal conduction (q11) from hot water pipe buried under the floor of upper level can be shown as equation (21).

$$q_{11} = \frac{A_p (T_w - T_2) \cdot L}{\frac{1}{h_i} + \frac{R_1 \ln\left(\frac{R_2}{R_1}\right)}{K_c} + \frac{R_1 \cos^{-1}\left(\frac{B'}{R_2}\right)}{K_{b'}}} \tag{21}$$

In addition, the amount of radiant heat transfer (q_{rk}) in each surface is calculated with the Gebhart's enclosure analysis method(Segel, 1981) as in the following equation (22).

$$q_{rk} = A_k \cdot \varepsilon_k \cdot \sigma \cdot T_k^4 - \sum_{j=1}^{n} A_j \cdot \varepsilon_j \cdot \sigma \cdot T_j^4 \cdot G_{jk}$$

$$G_{jk} = F_{j-k} \cdot \varepsilon_k + F_{j-k} \cdot \rho_1 \cdot G_{1k} + \cdots + F_{j-n} \cdot \rho_n \cdot G_{nk} \qquad (j = 1,2,\cdots,n) \tag{22}$$

where,

q_{rk} Radiant heat transfer for k_{th} surface

A_k k_{th} surface area

ε_k k_{th} surface emissivity

σ boltzmann constant

T_k k_{th} surface temperature

ρ_{1-n} Reflectivity of surface

F_{j-k} Coefficient of form between the inside and outside surface

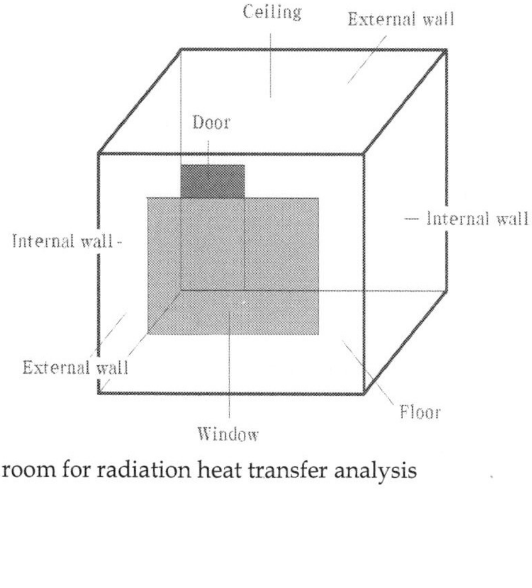

Fig. 5. Schematic of room for radiation heat transfer analysis

Temperatures of indoor air and each part of the room can be determined by an analysis of these 3 heat transfers: conduction, convection and radiation. In this study, each temperature is measured using the electrical resistance-capacitance circuit method(Sepsy, 1972) as shown in Fig. 6. It is based on assumptions that heat capacity for each wall is concentrated to one point in the wall, and that the temperature from the point to the wall surface is steady. Equivalent heat resistances of either side from each central node in the wall are the same by selecting the point. Heat loss to outdoor air is considered by setting the new central node(Chang, 1996), which is the point at which both equivalent heat resistances of the existing node and the surface are the same energy equation of each point is as follows

$$C_a \frac{dT_a}{dt} = q_2 - q_3 - q_4 - q_5 - q_6 - q_{12c} - q_{13} - q_{14}$$

$$C_p \frac{dT_p}{dt} = \sum q_{in} - \sum q_{out} - q_{rk} \qquad (23)$$

where,

C_p Capacitance of each part
T_p Temperature of each part
q_{in}, q_{out} Heat transfer by convection or conduction

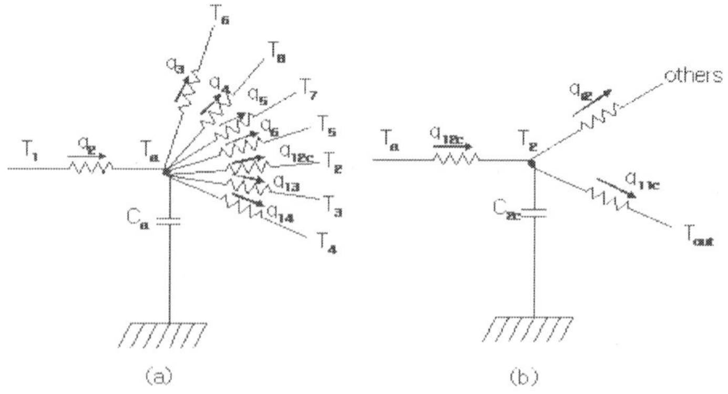

Fig. 6. An equivalent R-C circuit for unsteady energy analysis

4. Automatic thermostatic valves

In case of radiant floor heating system, automatic temperature control valve is used in order to consume energy effectively and maintain pleasant indoor temperature. This valve has similar function as those of gate or glove valve, but it can be separated into electric powered type and non-electric powered type in terms of the source of power that moves valve disk. Electric powered type uses external force such as electricity, while non-electric powered type uses only internal driving element such as shape memory alloy and spring. In general, electric powered type operates by motors is composed of room temperature controller and automatic thermostatic valve, and non-electric powered type only consists of automatic thermostatic valve.

Furthermore, Automatic thermostatic valve itself can be differed into on-off type and proportional control type in terms of control method. Fig. 7 shows control method of automatic thermostatic valve expressed with flux supply method from temperature change.

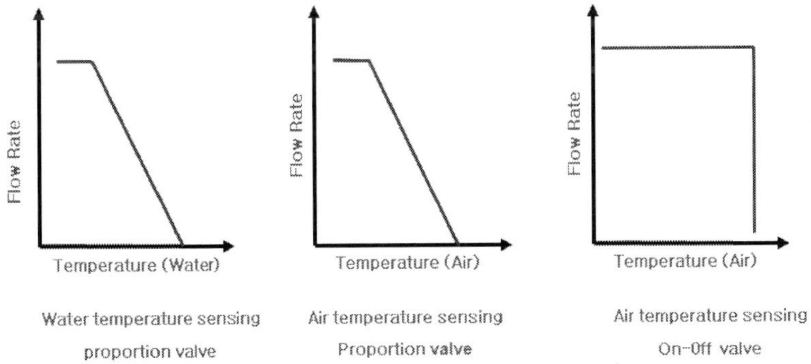

Fig. 7. Characteristics of Flow rates vs temperature in control methods of automatic thermostatic valves

Fig. 8 contains types of heating automatic thermostatic valve, and table 1 is about the characteristics of each type of valves.

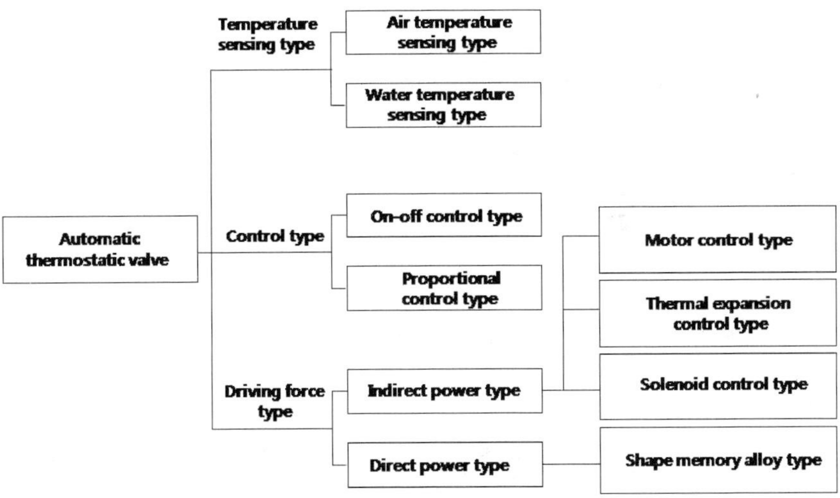

Fig. 8. Various classifications for automatic thermostatic valves

First, in case of electric powered automatic thermostatic valve, control part senses indoor air temperature of each room and transmits signal to driving part (thermostatic valve) to control the amount of heating hot water flowing through the pipe. Temperature sensor called thermister is mostly used to sense temperature from indoor temperature controller, and it is thermally sensitive semiconductor resistance thermometer. Thermister is widely used as temperature sensor for household electric appliances and industrial machineries, since its temperature usage range is -50~500°C, which can be applied to every possible range that ordinary temperature control is needed, and it is not only cheap and tiny but highly sensitive. Driving part of electric powered type automatic thermostatic valve can be separated into 3 different methods using ball valve, cone valve, and solenoid. Solenoid valve has 2 seconds of on-off response time, while ball valve has 10 seconds and cone valve has several ten seconds to minutes

Non-electric powered type automatic thermostatic valve using shape memory alloy actively controls on-off state of valve by sensing shape memory alloy element; closes valve proportional to temperature due to the increase of returning water temperature, and opens valve by returning spring due to decrease of returning water temperature.

As a merit, power supply is not necessary and response time is faster than thermal expansion. Also, structure only consisting of thermal static valve is very simple and endurance is superior, because Ti-Ni shape alloy spring is used as an operational element.

However, it has a demerit that it has to passively decide flux amount that fits to hot water temperature amount considering consumer's thermal surroundings after construction. Also, it is significant to choose appropriate controlling components for types and characteristics of installing heating system.

	Indirect power type			**Direct type**	
	Motor control type	**Solenoid control type**	**Thermal expansional control type**	**Shape memory alloy type**	**Capillary tube Type**
Control method	- Temperature sensing type - On-off type	- Temperature sensing type - On-off type	- Temperature sensing type - On-off type	- water temperature sensing type - Proportional control type	- Water temperature sensing type - Proportional control type
Operating source	Electric motor	Electromagnet	Liquid thermal expansion	Shape memory alloy spring	Liquid thermal expansion bellows
Response time	Within several ten seconds	Within several seconds	Within several minutes	Within several seconds	Within several ten minutes
Merit	- Fast response - Easy installation	- Fastest response - Simple Structure - Low cost		- Electric power isnecessary - Good endurance - Proportional type	- Electric power is unnecessary - Proportional Type
Demerit	-Electric power is necessary - High cost	- High pressure loss	Relatively slow response	Manual valve Setting	-Liquid quality variation -Difficult installation

Table 1. Characteristics of classified automatic thermostatic valves

5. Heating control system performance analysis

If indoor temperature were controlled by applying automatic thermostatic valve for radiant floor heating system, response characteristics of long time delay on control response would occur due to large heat capacitance of floor structure mass. In order to examine response characteristics of radiant floor heating system, simulation was performed and was compared and verified with experiment results by using mathematical analysis model of a radiant floor heating system explained in previous 2nd, 3rd paragraphs.

Fig. 9 shows the results of the experiment and simulation (Ahn, 2010).

We measured temperature changes for 5 hours natural cooling after supplying hot-water for 3 hours. For the floor temperature, two temperatures (one at the nearest part to the pipe and the other between the pipes), were measured and compared with temperatures from the simulation data. The chiller to maintain temperature of artificial chamber in the test house was set up to maintain an outdoor air temperature of 8°C. The reason for cooling is to secure the constant temperature around the room for the indoor heating test.

Data obtained from the entering supply and outdoor temperatures into the simulation for operation, were contrasted to the experimental data.

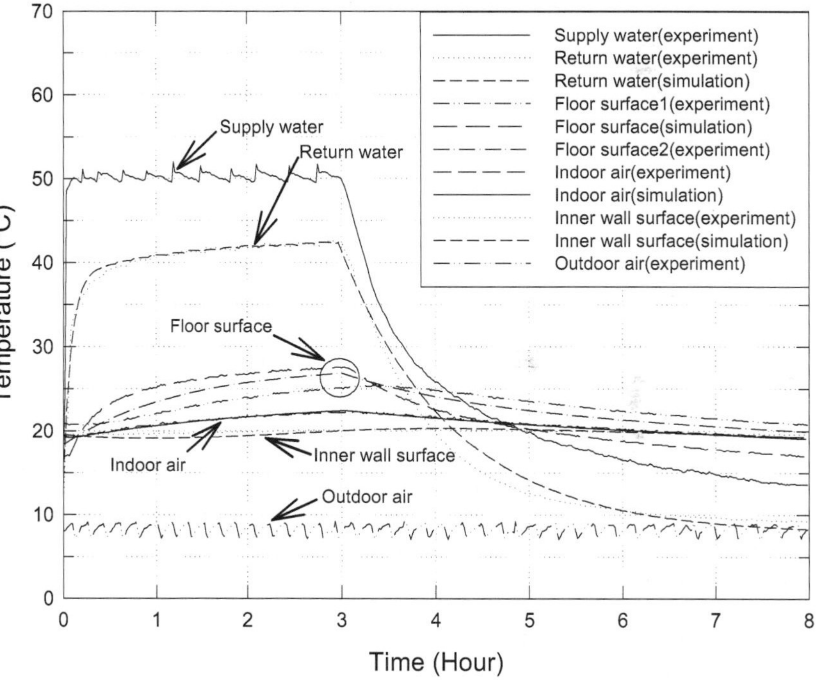

Fig. 9. Comparison of the simulation and experimental data (supply water temperature: 50°C, flow rate: 0.05L/s, outdoor temperature: 8°C)

Results showed that the simulation data agreed well with experimental data. Response characteristics will be examined through using simulation model of radiant floor heating system and applying various kinds of automatic thermostatic valves.

First of all, In terms of proportional control and On-Off control, which are types of automatic thermostatic valves, each is classified according to sensing methods; water temperature sensing and air temperature sensing. They are performed by testing simulation and defining their features.

Each control of each flow rate and temperature has its own peculiarities as outdoor air changes from the daily lowest temperature -5°C to the highest temperature 5°C over 24hours. Table 2 summarizes control characteristics for each of the four studied cases. For the 4 controls: Case 1 does not adjust the flow rate, Case 2 controls the flow rate in proportion to the difference between room temperature and setting point. Case 3 adjusts the flow rate in proportion to the difference between returned-water temperature and setting point, and Case 4 controls the On-off for the supply water by the differential gap according to the difference between the room temperature and setting point.

Classification	Description
Case1	No control method
Case2	Proportional valve control with air temperature feedback
Case3	Proportional valve control with water temperature feedback
Case4	On-off valve control with air temperature feedback

Table 2. Classification of control methods

Fig. 10 summarizes the results of changes in temperatures of return water, the floor and indoor air over 24hours. The outdoor air is vibratory from -5°C to 5°C, the return water increases to 43.3°C from the set point, and the floor and the indoor air rapidly increases to 20°C for 3hours, and then steadily to 32°C and then decreases to 27.5°C. At this time, the mean temperatures of return water and indoor air are 42.3°C and 24.9°C, respectively.

Fig. 11 shows the temperature responses and a flow rate for 24hours as a result of the proportional control for the indoor air temperature (Case 2) designated from 22.3°C to 23.3°C to maintain 22.8°C, the mean indoor temperature. The maximum flow rate helps adjust flow and maintain an indoor air temperature of 23°C before reaching the lower limit of 22.3°C. Controlling flow can offset the change of outdoor air temperature so that indoor air temperature can be maintained.

With the exception of the first stage, the temperature of the return water shows a gentle slope, increasing, and that of floor surface is 27.9°C continuously. In this case, each mean temperature of return water and indoor air is 31.8°C, and 22.8°C. This air-temperature proportional control maintains indoor temperature through light control regardless of changes in outdoor air temperature.

Fig. 12 shows temperature responses and a flow rate for 24hours as a result of the proportional control for return water temperature (Case 3) designated from 30.5°C to 34.5°C to maintain 22.8°C, the mean indoor temperature. The indoor air maintains a temperature of 33.1°C after initially increasing to 24.1°C. The flow rate steadily decreases but its drop dwindles except for the first stage. In this case, the mean temperatures of return water and indoor air are 33°C, and 22.8°C, respectively.

Water proportional control does not cope with changes in indoor air temperature, but maintains return water temperature, and if the indoor air temperature were at the best condition, it would be difficult to find an upper and lower limit that can maintain it.

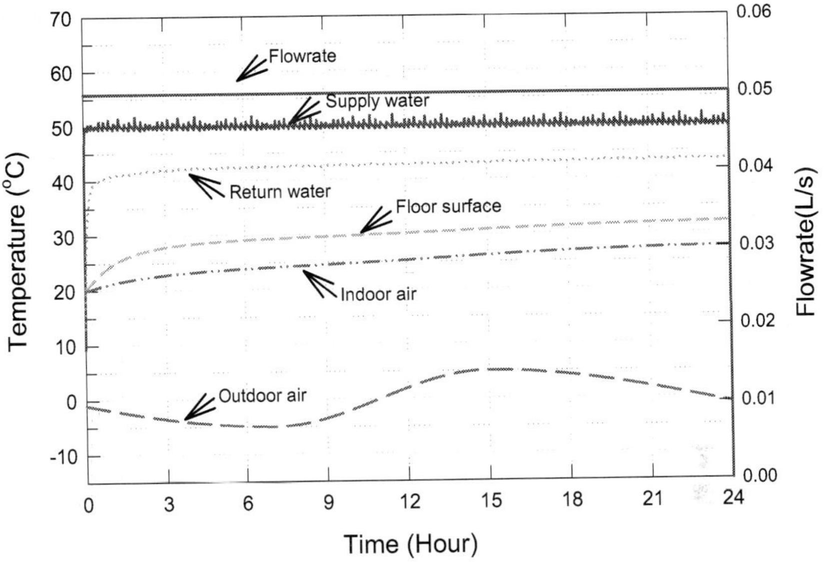

Fig. 10. Various temperature responses for outdoor temperature change with no control (Case 1)

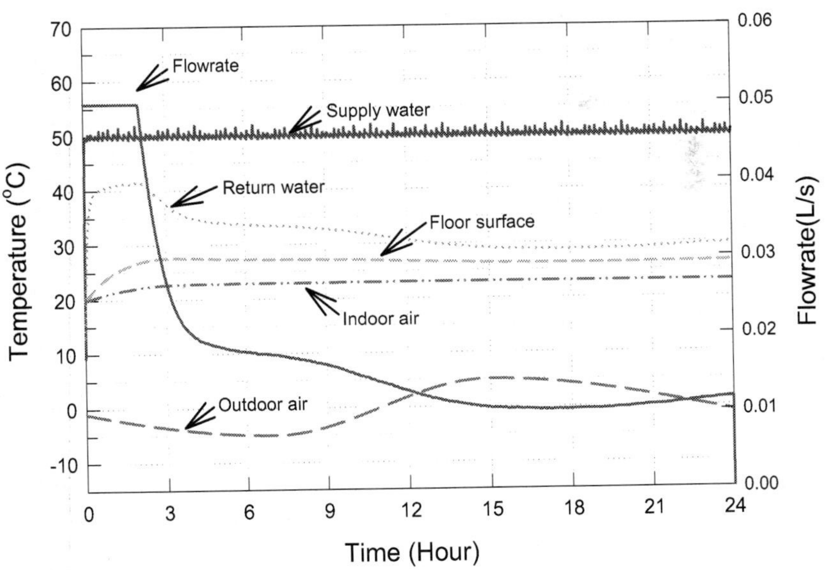

Fig. 11. Various temperature responses with proportional valve control with air temperature feedback (Case 2)

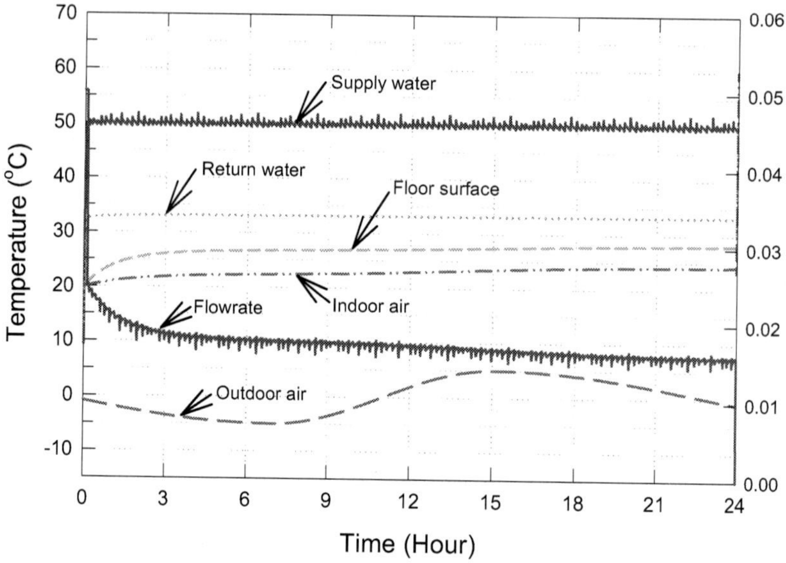

Fig. 12. Various temperature responses with proportional valve control with return water temperature feedback (Case 3)

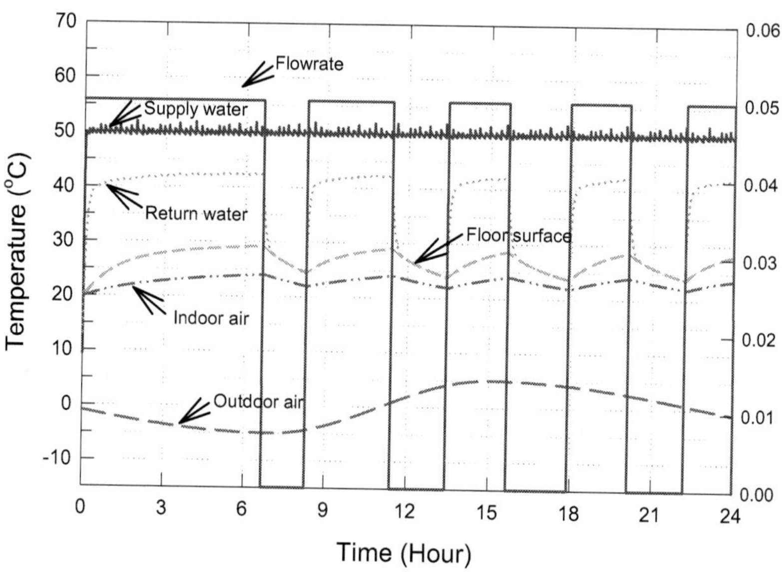

Fig. 13. Various temperature responses with on-off valve control with air temperature feedback (Case 4)

Fig. 13 shows changes to temperature and flow rate for 24hours in flow-rate supply on–off at a range of indoor air temperatures (Case 4), 22.3°C to 23.5°C.
The operation time is long due to the set point of indoor temperature of 20°C, and then shortens due to intervals of 1°C of On-Off. The process is more time consuming when the outdoor temperature is lower and the relatively non-operational time shortens. The indoor air temperature vibrates from 22.5°C to 23.5°C according to the control, and the return water temperature almost reduces to the floor temperature when the flow supply is stopped. At this time, the mean temperatures of return water and indoor air are 34.4°C and 22.8°C, respectively.
For the On-Off control, time differences are created between the hot water supply and non-supply as the outdoor air temperature changes, and the indoor air temperature affects the control. But the control does not maintain the indoor air and creates a fluctuation in temperature. It is therefore important to set the proper point and use a differential gap.
Fig. 14 shows the mean temperature of the room air and total amount of heat supply according to controls. No-Control (Case 1) is 2°C and 14,000Kcal higher than other controls (Case 2~4). All controls, Cases 2~4, have a similar mean temperature but the water temperature-sensing control (Case 3) uses about 8% more heat than the air temperature sensing control (Case 2, Case 4).

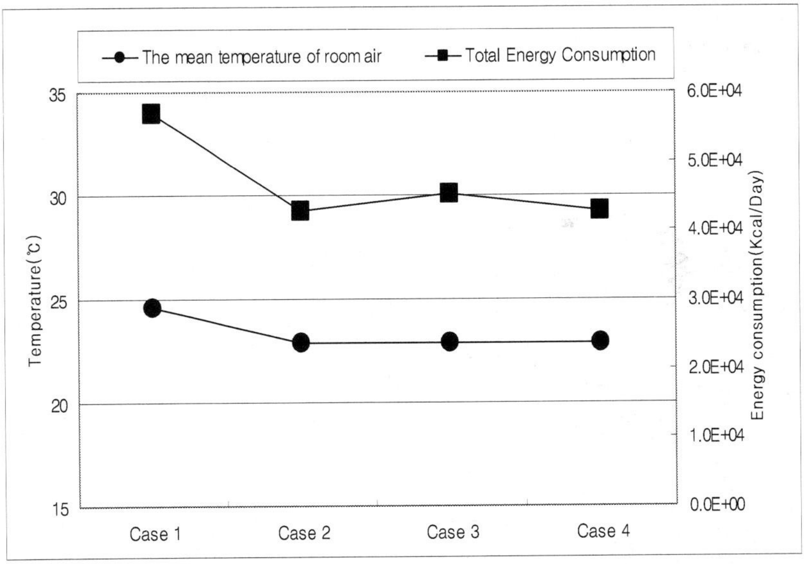

Fig. 14. Mean temperature of indoor air and total energy consumptions with different control methods (supply water temperature of 50°C)

Control variables of radiant floor heating system can usually be considered as supply hot water temperature and set value of indoor temperature. These control variables must be set to a proper value to improve indoor thermal environment and save energy, and variable values reflecting amount of outdoor temperature can be considered.

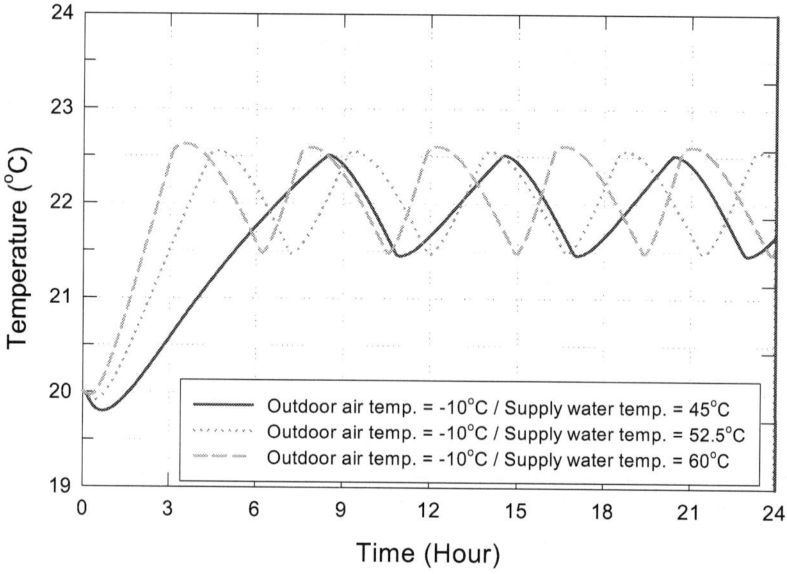

Fig. 15. Indoor air temperature responses for various supply water temperatures

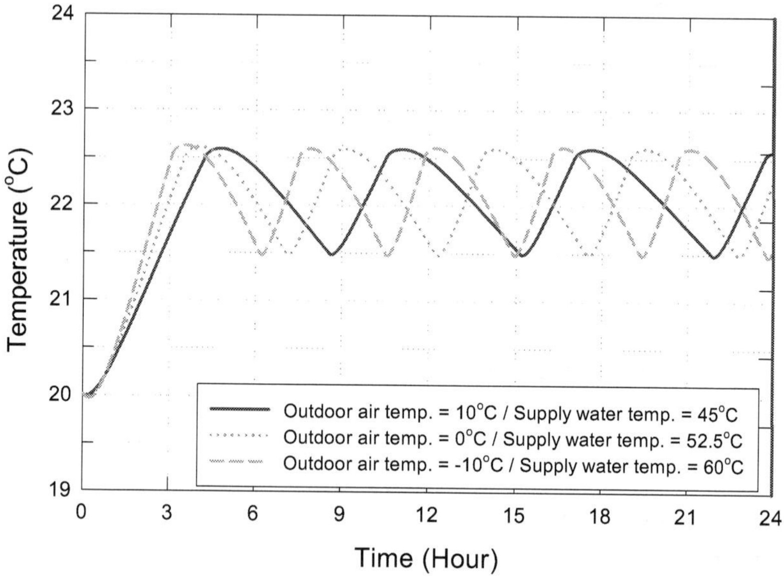

Fig. 16. Indoor air temperature responses for various supply water and outdoor air temperatures

First of all, Fig. 15 shows characteristics of indoor air temperature alternation when supply hot water temperature is 45°C, 52.5°C, 60°C in case of setting outdoor temperature condition as -10°C. The control method here is the result of applying on-off control method (the on-off for the supply water by the differential gap according to the difference between the room temperature and setting point), using indoor air temperature sensor.

The diagram indicates that the time to reach the setting point 22°C is radically increasing when using relatively low temperature supply water in case of decreasing outdoor temperature. Analyzing from the data that more than 6 hours is needed for supply hot water temperature at 45°C to reach 22°C, considerable amount of time is consumed in order to improve indoor thermal environment to a comfortable range in case when outdoor temperature is low and low temperature supply water is provided. Thus, relatively warmer hot water than cooler hot water is appropriate for heating in case of low outdoor temperature (Ahn, 2005).

Fig. 16 shows dynamic characteristics of indoor air temperature change when supply water temperature is 60°C, 52.5°C, and 45°C at outdoor temperature condition of -10°C, 0°C, and 10°C. Overshoot of each 3 separate cases depicted similar amount on indoor air temperature change, and the rise times to reach the set indoor air temperature 22°C were close to 3 hours without difference. By correctly determining supply hot water temperature in regard to amount of outdoor temperature, transient response characteristics for indoor environment can be improved.

Fig. 17 depicts characteristics of energy consumption and indoor air temperature average considering temperature alternation of outdoor and supply hot water.

Energy consumption and indoor air temperature average entirely rises as temperature of supply hot water or outdoor increases on the graph. The energy consumption gap was moderate in terms of supply hot water temperature change, while it was relatively large for outdoor temperature change. Therefore, changing supply hot water temperature in terms of outdoor temperature variation will minimize overheating and benefit energy savings.

Furthermore, if high temperature hot water were provided to make indoor air temperature reach its set temperature, supplying time of hot water would be lessened due to rapid increase of indoor air temperature despite of large consumption of heat amount depending on temperature difference. On the other hand, when low temperature hot water is provided, heat amount consumption will decrease, but supplying time will be delayed due to slow increase rate of indoor air temperature so that amount difference of energy consumption will be small.

Fig. 18 depicts result measured from the experiment about flow rate change characteristics and temperature change of radiant floor heating system. This is to demonstrate general thermal change features of radiant floor heating system from each boiler-installed household that is run by individual heating method.

Fig. 18 indicates temperatures of supply water, returning water, floor surface, indoor air, outdoor air, and flow rate. Temperature of supply water from supply header was at maximum 50°C due to heat loss from pipe, while it is being operated as on-off type control by boiler system.

Therefore, heat loss must be considered when determining set value of the supply water temperature. In addition, while indoor air temperature is being controlled up to 23±0.5°C by

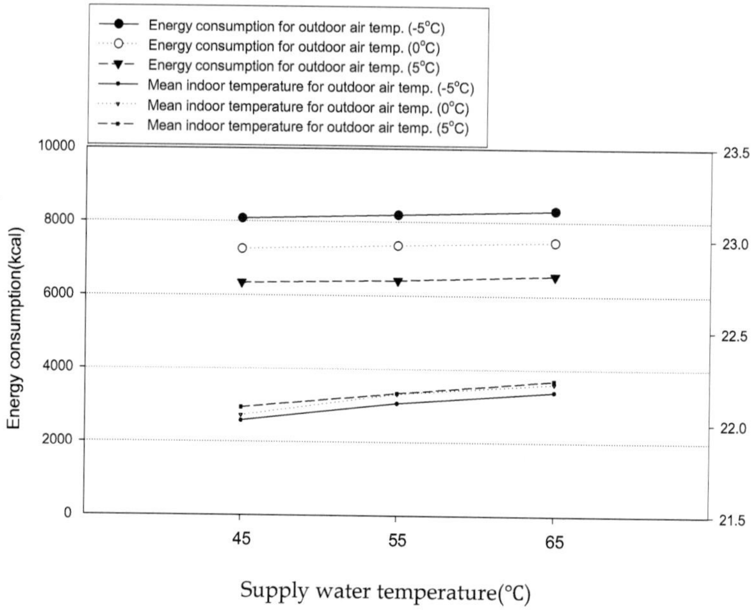

Fig. 17. Energy consumption and mean indoor air temperature responses for various supply water and outdoor air temperatures

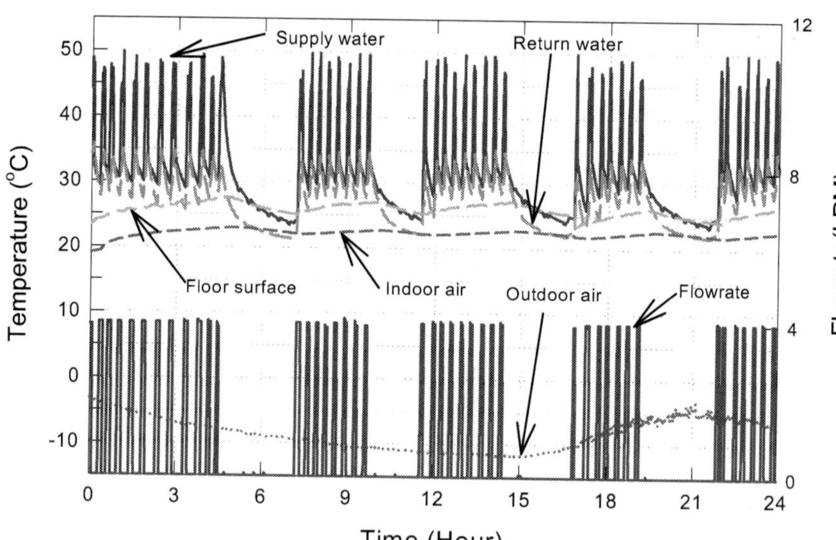

Fig. 18. Experimental results for the response characteristics of temperature and flow rate

on-off type, initial operation time period to reach set indoor temperature is relatively slower than that of after second period, and temperature alternation of indoor air and floor surface seems severely slow even after hot water is being supplied. Thus, thermal environment would be improved if indoor air temperature increasing period were reduced by initially supplying comparably hotter water temperature, and cooling it down afterwards.

6. Conclusions

Radiant floor heating system is mainly applied to residential buildings. Recent trend shows increasing rate of countries applying the system to their residential buildings after considering various profits of the system.
However, the radiant floor heating system may contain long time delay as inappropriate response characteristics on indoor temperature control due to significant influence of thermal inertia when performing indoor air temperature control.
In order to improve thermal environment and save energy by effectively controlling the indoor air temperature, temperature value of supply hot water from boiler and indoor air set temperature can be altered by forecasting the value of outdoor temperature and characteristics variation.
Therefore, application of more effective system design and advanced control technique for radiant floor heating system like HVAC system(Ahn, 2001) is required through consistent technology research and development.

7. References

Ahn, B.C. and Mitchell, J.W. (2001). Optimal Control Development for Chilled Water Plants using a Quadratic Representation, *Energy and Buildings* 33, pp.371-378, Elsevier, ISSN 0378-7788

Ahn. B.C. and Lee, T.W. (2005). Strategy of Energy Saving and Thermal Environmental Improvement for Intermittent Heating System in Apartment Buildings, *Journal of the SAREK*, Vol.17, pp.88-93, ISSN 1229-6422

Ahn, B.C. and Song, J.Y. (2010). Control Characteristics and Heating Performance Analysis of Automatic Thermostatic Valves for Radiant Slab Heating System in Residential Apartment, *Energy* 35, pp.1615-1624, Elsevier, ISSN 0360-5442

ASHRAE Handbook, HVAC Systems and Equipment. (2004). Panael Heating and Cooling, Chapter 6, 6.1-6.22. ISBN 1-931862-48-6, Atlanta, USA

Chang, H.W. and Ahn, B.C. (1996). The Energy Analysis and Control Characteristics of a Ho Water Heating System for Apartment Houses, *Journal of the SAREK*, Vol.8, pp.76-87, ISSN 1229-6422

Holman, H.J. (1981). Heat Transfer, 5th Ed., *McGraw-Hill.*, ISBN 0-07-029618-9, MI, USA

Segel Robert and Howell, J.R. (1981). Thermal Radiation Heat Transfer, *McGraw-Hill.*, ISBN 0-07-057316-6, New York, USA

Sepsy, C.F. (1972). A Thermal Analysis of the Building and the Heating and Cooling Systems Selected for the Field Validation Test, *ASHRAE Sym. Bulletine* No. 72, pp.5-9.

Stoecker, W.F. (1980). Design of Thermal Systems, *McGraw-Hill*, ISBN 0-07-061618-3, Tokyo, Japan

8

Variable Property Effects in Momentum and Heat Transfer

Yan Jin and Heinz Herwig
Institute for Thermo-Fluid Dynamics, Hamburg University of Technology (TUHH)
Germany

1. Introduction

"Variable property effects" is a theoretical construct since a real fluid always is subject to variable properties when changes in temperature or pressure occur. Their influence compared to a corresponding situation but with artificially constant properties may be small and thus neglected in a first approximation. Those artificial "constant property results" may then be corrected with respect to the initially neglected effects due to the variability of the fluid properties. This concept assumes small variable property effects and therefore is not applicable when the flow itself is basically generated by a variable property (like natural convection, generated by density variations) or strongly affected by it (like strongly compressible flow, determined by density variations).

Therefore we define "variable property effects" as those (small) artificial effects that would be present if fluid properties could change from constant to variable. They have to be added to a solution which is gained under the assumption of constant properties in order to account for the fact that real fluids always have properties which are temperature and pressure dependent. The variable property correction of a constant property solution can be accomplished in different ways. Basically there are three methods which are widely used in this context. They are

(1) Property ratio method

Results in terms of the nondimensional friction factor f and Nusselt number Nu are gained by multiplying the constant property results f_{cp} and Nu_{cp} with a property ratio correction factor, i.e.

$$f = f_{cp}\left(\frac{a_1}{a_2}\right)^{m_a}, \quad \mathrm{Nu} = \mathrm{Nu}_{cp}\left(\frac{a_1}{a_2}\right)^{n_a} \tag{1}$$

Here a_1, a_2 are properties (ρ, μ, k, c_p) at two different temperatures and m_a, n_a are empirical exponents. There may be more than one factor, depending on which properties are assumed to contribute to the variable property effects. Studies with this approach are Li et al. (2007) and Mahmood et al. (2003), for example.

(2) Reference temperature method

The constant property results in terms of f and Nu are evaluated at a certain temperature for the properties that appear in f and Nu. This so-called reference temperature,

$$T_R = T_1 + j(T_2 - T_1) \tag{2}$$

between two characteristic temperatures T_1 and T_2 of the problem must be chosen such that f and Nu determined under the assumption of constant properties (and with T_1 or T_2 as reference temperature) give the results for variable properties. For that purpose the factor j in equ (2) must be determined properly. Basically, j is an empirical parameter like the exponents in equ (1) are in the first method. Studies using this method are Jayari et al. (1999) and Debrestian & Anderson (1994), for example.

(3) Asymptotic method

Since the effects of variable properties are assumed to be small the variable property solution can be taken as a perturbation of the constant property solution. In a systematic approach the problem is treated as a regular perturbation problem with a (small) perturbation parameter ε linked to the transfer rate. The constant property solution is that for $\varepsilon = 0$ and variable property effects are described as 1st, 2nd, ... nth order effects of an asymptotic expansion with respect to the perturbation parameter ε. This approach in an early paper has been suggested by Carey and Mollendorf (1980) and afterwards has been adopted by the second author for a wide variety of problems, see for example Herwig and Wickern (1986), Herwig and Bauhaus (1986), Herwig et al. (1989) and Herwig and Schäfer (1992).

The advantage of this method over the first two methods to account for variable properties is its systematic generalization of the correction terms that appear within this approach. These terms hold for all Newtonian fluids and all (small) transfer rates that are the reason for variable property effects. Furthermore, with this approach the empirical parameters in the first two method (m_α, n_α and j) can be embedded in the expansion approach and thus be interpreted on a sound physical basis, as is demonstrated in Herwig and Wickern (1986), for example.

For a long time all three methods have been applied to certain standard situations like pipe, channel or boundary layer flows. Only recently it was shown how the asymptotic method can also be applied to complex problems, such as turbulent mixed convection in a room that is ventilated and heated by several heating elements, see Bünger and Herwig (2009), Jin and Herwig (2010) and Jin and Herwig (2011).

This extension to complex convective heat transfer problems will be presented and illustrated by some examples in the following.

2. The asymptotic approach: State of the art

Variable properties means that all fluid properties involved in a problem depend on temperature and pressure. Whenever temperature and pressure variations occur in a problem these variabilities affect the solution. Assuming the effects to be small, a perturbation approach is straight forward. Since pressure effects almost always are negligibly small, we only account for the temperature impact. The starting point for the temperature dependence is the Taylor series expansion of all properties with a^* representing ρ^* (density), μ^* (viscosity), k^* (thermal conductivity), and c_p^* (specific heat capacity), which is with $a_R^* = a^*(T_R^*)$

$$a := \frac{a^*}{a_R^*} = 1 + \varepsilon K_{a1}\theta + \frac{1}{2}\varepsilon^2 K_{a2}\theta^2 + \cdots + \frac{1}{n!}\varepsilon^n K_{an}\theta^n + O\left(\varepsilon^{n+1}\right) \tag{3}$$

$$\varepsilon := \frac{\Delta T_R^*}{T_R^*} \tag{4}$$

$$\theta := \frac{T^* - T_R^*}{\Delta T_R^*} \tag{5}$$

$$K_{a1} := \left[\frac{\partial a^*}{\partial T^*}\frac{T^*}{a^*}\right]_R \tag{6}$$

$$K_{a2} := \left[\frac{\partial^2 a^*}{\partial T^{*2}}\frac{T^{*2}}{a^*}\right]_R \tag{7}$$

$$K_{an} := \left[\frac{\partial^n a^*}{\partial T^{*n}}\frac{T^{*n}}{a^*}\right]_R \tag{8}$$

Here, ε is a small quantity which basically represents the strength of heat transfer ($\varepsilon = 0$: isothermal situation) and formally serves as perturbation parameter. The dimensionless temperature is θ which then is an $O(1)$ quantity. The fluid behavior is covered by K_{aj} with $j = 1,2,\cdots n$, which are $O(1)$ quantities, representing the partial derivatives of a^* with respect to temperature.

The same expansion can be made with respect to the pressure p^*. Since, however, for most cases the corresponding $\hat{K}_{an}$ values $\left(\left(\partial^n a^*/\partial p^{*n}\right)/\left(p^{*n}/a^*\right)\right)$ and/or pressure variations are very small, we restrict ourselves to the temperature effects.

In order to get solutions in terms of expansions which correspond to the Taylor-series expansions of the properties, all dependent variables V^* are expanded as

$$V := \frac{V^*}{V_R^*} = 1 + \varepsilon\left[K_{\rho 1}V_{\rho 1} + K_{\mu 1}V_{\mu 1} + K_{k1}V_{k1} + K_{c_p 1}V_{c_p 1}\right] + O\left(\varepsilon^2\right) \tag{9}$$

Here, V^* represents the velocity components u^*, v^*, w^* and the temperature T^* which appear in the basic equations of a problem. For many problems these are the Navier-Stokes and the thermal energy equations, here shown for a steady flow $\partial/\partial t = 0$, i.e:

$$\nabla \cdot \left(\boxed{\rho}\vec{u}\right) = 0 \tag{10}$$

$$\left(\boxed{\rho}\vec{u}\cdot\nabla\right)\vec{u} = \frac{1}{\mathrm{Re}_R}\nabla\cdot\left(\boxed{\mu}\left(\nabla\vec{u} + \nabla\cdot\vec{u}^T\right) - \frac{2}{3}\nabla\cdot\vec{u}I\right) - \nabla p + \boxed{\rho}\vec{g} \tag{11}$$

$$\boxed{\rho c_p}\left(\vec{u}\cdot\nabla\right)T = \frac{1}{\mathrm{Re}_R \mathrm{Pr}_R}\nabla\cdot\left(\boxed{k}\nabla T\right) + p\nabla\cdot\vec{u} \tag{12}$$

In the thermal energy equation (12) viscous dissipation is neglected since it contributes very little to the energy balance. Equs (10)-(12) are nondimensional equations in which all fluid properties $a = a^*/a_R^*$ are marked by a box (note that a = 1 for constant properties).

The nondimensional groups are

$$\mathrm{Re}_R = \frac{\rho_R^* u_R^* L^*}{\mu_R^*} \,,\ \mathrm{Pr}_R = \frac{\mu_R^* c_{pR}^*}{k_R^*} \tag{13}$$

It is an important point that here they are formed with the reference fluid properties (index R). When turbulent flows are treated by the RANS-approach, equs (10) - (12) must be time-averaged and subject to turbulence modeling. Then, with a $k-\varepsilon$ model, for example, two more differential equations have to be solved in which fluid properties appear as well.
In the asymptotic method to account for variable property effects, all properties a (i.e. ρ, μ, k, c_p) and all variables υ (i.e. $\vec{u}$, T) are replaced by their expansions (3), (9) in the set of equs (10) - (12). Then two different ways to proceed are possible which will be explained in the next two subsections.

2.1 Higher order equations method (HOE)

In this method the equations are subject to the expansion process, i.e. all terms of $O(1)$, $\varepsilon K_{\rho 1}$, $\varepsilon K_{\mu 1}$, ..., $\varepsilon^2 K_{\rho 2}$, $\varepsilon^2 K_{\rho 1} K_{\rho 1}$, ... are collected separately and form the zero order, first order, second order, ... sets of equations. In the complete approach with all four properties involved there is one zero order set of equations (corresponding to constant properties), four first order sets and fourteen sets of second order. Only the zero order set of equations is a nonlinear one, all subsequent equation sets have linear equations. All equations have to be solved sequentially starting with the zero order equations for constant properties.
The final result in terms of the friction factor f and the Nusselt number Nu gains an asymptotic form as well. For example, the Nusselt number for the thermal boundary condition $T_w = const$, reads, with Nu_{cp} for the constant property case:

$$\frac{\mathrm{Nu}}{\mathrm{Nu}_{cp}} = 1 + \varepsilon \left[K_{\rho 1} \frac{T'_{\rho 1}}{T'_0} + K_{\mu 1} \frac{T'_{\mu 1}}{T'_0} + K_{k1} \frac{T'_{k1}}{T'_0} + K_{c_p 1} \frac{T'_{c_p 1}}{T'_0} \right] + O\left(\varepsilon^2\right) \tag{14}$$

Here, T'_{a1} are the temperature wall gradients since $\dot{q}_w^*$ in the Nusselt number is $\dot{q}_w^* = -\left(k^* T^{*\prime}\right)_w$.
In the second order 14 additional terms in equ (14) appear so that altogether 19 sets of equations have to be solved when second order accuracy is required. Then, however, the result (14) is quite general, since it holds for all Newtonian fluids (different values for $K_{\rho 1}$, $K_{\mu 1}$, ... in (14)) and all small heating rates (different values for ε in equ (14)).
This method is straight forward for all laminar flows and has been applied in Herwig and Wickern (1986) and Herwig and Bauhaus (1986), for example.
When turbulent flows are considered, the higher order equation method needs higher order turbulence model equations. That may be a problem which does not occur in the alternative method described next.

2.2 Higher order coefficients method (HOC)

Instead of solving higher order equations from which the influence of the single properties can be deduced in the final results, like in equ (14) for example, the final results are taken in its asymptotic form as the starting point. Again, shown for the Nusselt number, it reads up to the second order:

$$\frac{\mathrm{Nu}}{\mathrm{Nu}_{cp}} = 1+\varepsilon\left[K_{\rho 1}A_{\rho}+K_{\mu 1}A_{\mu}+K_{k1}A_{k}+K_{c_p 1}A_{c_p}\right]+\varepsilon^2\left[K_{\rho 2}A_{\rho 2}+K_{\mu 2}A_{\mu 2}\right.$$
$$+K_{k2}A_{k2}+K_{c_p 2}A_{c_p 2}+K_{\rho 1}^2 A_{\rho\rho}+K_{\mu 1}^2 A_{\mu\mu}+K_{k1}^2 A_{kk}++K_{c_p 1}^2 A_{c_p c_p} \tag{15}$$
$$+K_{\rho 1}\left(K_{\mu 1}A_{\rho\mu}+K_{k1}A_{\rho k}+K_{c_p 1}A_{\rho c_p}\right)+K_{\mu 1}\left(K_{k1}A_{\mu k}+K_{c_p 1}A_{\mu c_p}\right)$$
$$\left.+K_{k1}K_{c_p 1}A_{kc_p}\right]+O\left(\varepsilon^3\right)$$

with the A-values A_{ρ}, A_{μ} ... as the crucial part of the result. Up to the second order there are 4 + 14 = 18 different A-values.
With the nondimensional property a according to equ (3) is rewritten as

$$a = 1 + h_{a1}\theta + \frac{1}{2}h_{a2}\theta^2 + \cdots + \frac{1}{n!}h_{an}\theta^n + O\left(\varepsilon^{n+1}\right) \tag{16}$$

introducing

$$h_{aj} \;:=\; \varepsilon^j K_{aj} = \left[\frac{\partial^j a^*}{\partial T^{*j}}\frac{\Delta T^{*j}}{a^*}\right]_R, \quad j = 1,2,\cdots n. \tag{17}$$

one immediately recognizes the A-values to be

$$A_a := \frac{1}{\mathrm{Nu}_{cp}}\left[\frac{\partial \mathrm{Nu}}{\partial h_{a1}}\right]_{h=0} \tag{18}$$

$$A_{aa} := \frac{1}{2\mathrm{Nu}_{cp}}\left[\frac{\partial^2 \mathrm{Nu}}{\partial h_{a1}^2}\right]_{h=0} \tag{19}$$

$$A_{a2} := \frac{1}{\mathrm{Nu}_{cp}}\left[\frac{\partial \mathrm{Nu}}{\partial h_{a2}}\right]_{h=0} \tag{20}$$

$$A_{ab} := \frac{1}{\mathrm{Nu}_{cp}}\left[\frac{\partial^2 \mathrm{Nu}}{\partial h_{a1}\partial h_{b1}}\right]_{h=0} \tag{21}$$

where $a,b \in \left\{\rho,\mu,k,c_p\right\}$, $a \neq b$.
According to equs (18) – (21) the 18 A-values in (15) can be determined as first and second derivatives of Nu with respect to the h-values in (16). These h-values quite generally can be interpreted as combinations of ε and fluid properties ($K_{\rho 1},\cdots$). Since the method holds for all Newtonian fluids it also holds for arbitrarily chosen **fictitious fluids** for which only one property is temperature dependent (in some arbitrary way) with all other properties being constant. Such fluids serve as "dummy fluids" from which the general A-values can be determined.
This is done by solving the full equations (10) – (12) or their turbulent version for these dummy fluids with certain values for h_{a1}, h_{b1} and h_{a2}. Then the A values can be determined by the following numerical approximations of (18) – (21):

$$A_a = \frac{1}{\mathrm{Nu}_{cp}} \lim_{s \to 0} \left\{ \frac{\mathrm{Nu}(h_{a1} = s) - \mathrm{Nu}(h_{a1} = -s)}{2s} \right\}_0 \tag{22}$$

$$A_{aa} = \frac{1}{2\mathrm{Nu}_{cp}} \lim_{s \to 0} \left\{ \frac{\mathrm{Nu}(h_{a1} = s) - 2\mathrm{Nu}_{cp} + \mathrm{Nu}(h_{a1} = -s)}{s^2} \right\}_0 \tag{23}$$

$$A_{a2} = \frac{1}{\mathrm{Nu}_{cp}} \lim_{s \to 0} \left\{ \frac{\mathrm{Nu}(h_{a2} = s) - \mathrm{Nu}(h_{a2} = -s)}{2s} \right\}_0 \tag{24}$$

$$\begin{aligned} A_{ab} = \frac{1}{\mathrm{Nu}_{cp}} \lim_{s \to 0} \frac{1}{4s^2} \{ &\mathrm{Nu}(h_{a1} = h_{b1} = s) + \mathrm{Nu}(h_{a1} = h_{b1} = -s) \\ &- \mathrm{Nu}(h_{a1} = -h_{b1} = s) - \mathrm{Nu}(-h_{a1} = h_{b1} = s) \}_0 . \end{aligned} \tag{25}$$

Here the limiting process $\lim\limits_{s \to 0} \ldots$ means that s must be small enough to avoid a higher order influence, but not too small since otherwise truncation errors appear. The index 0 at the right curly brackets indicates that all other variables h_{c1}, h_{c2} are set to zero. Each of the eight terms

- $\mathrm{Nu}(h_{a1} = s)_0, \mathrm{Nu}(h_{a1} = -s)_0$
- $\mathrm{Nu}(h_{a2} = s)_0, \mathrm{Nu}(h_{a2} = -s)_0$
- $\mathrm{Nu}(h_{a1} = h_{b1} = s)_0, \mathrm{Nu}(h_{a1} = h_{b1} = -s)_0,$ $\mathrm{Nu}(h_{a1} = -h_{b1} = s)_0, \mathrm{Nu}(h_{a1} = -h_{b1} = -s)_0$

corresponds to exactly one CFD-simulation. Hence, the calculation of (22), (23), and (24) needs two CFD-simulations for a fixed small value of s, while the determination of A_{ab} according to (25) needs four CFD-simulations. With one further CFD-calculation the constant property solution Nu_{cp} is determined.

In order to determine the complete set of all 18 A-values (second order accuracy of the result) thus 40 variable property solutions and the one for constant properties are needed. Since no higher order equations are determined, all kinds of turbulence modeling can be incorporated.

Details of the approach and an application of the HOC-method to a complex geometry and flow can be found in Bünger and Herwig (2009) and Jin and Herwig (2010), respectively. In Fig. 1, as an example taken from Jin and Herwig (2010), the influence of variable properties on the Nußelt number at two heating elements in a complex geometry is shown, determined with the HOC-method.

There can be, however, a further improvement of the HOC-method which considerably reduces the amount of numerical solutions that are needed for the determination of all A-values. This will be described next.

3. A more efficient (asymptotic) approach

For a more efficient method, we first rewrite the basic equations (10)-(12). Instead of referring the properties ρ^*, μ^*, k^* and c_p^* to their values at a reference temperature T_R^* we incorporate them in the nondimensional groups Re and Pr (which then are no longer Re_R and Pr_R since they now are local, variable quantities).

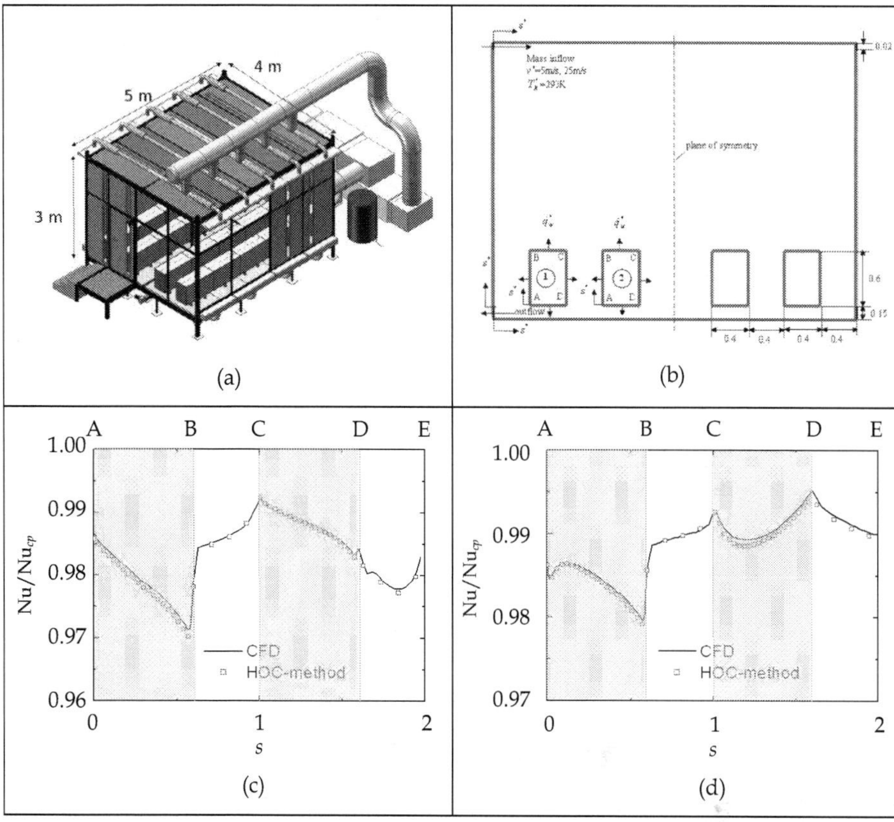

Fig. 1. Complex benchmark geometry, example from (Jin & Herwig, 2010); (a) 3D room for experiments, (b) 2D approximation for simulations, ①, ②: heating elements ($\dot{q}_w^* = 300\,\mathrm{W/m^2}$, $\mathrm{Re} = 1.64 \times 10^6$) s^*: coordinate around the heating elements and along the lower, upper, and side walls *A, B, C, D*: corners of the heating elements (c) Heating element ① (d) Heating element ② CFD: numerical solution, based on equs (10)-(12) HOC: asymptotic result, eq. (15)

Since ρ^*, μ^*, and k^* appear within a spatial derivative in (10)-(12), additional terms appear according to the product rule of differentiation. Instead of (10)-(12) one gets:

$$\nabla \cdot (\vec{u}) + P_\rho \nabla T \cdot \vec{u} = 0 \tag{26}$$

$$(\vec{u} \cdot \nabla)\vec{u} = \left(\frac{1}{\mathrm{Re}}\nabla \cdot + P_\mu \nabla T \cdot\right)\left(\left(\nabla\vec{u} + \nabla\vec{u}^T\right) - \frac{2}{3}\nabla \cdot \vec{u} I\right) - \frac{\nabla p}{1+(\rho-1)} + \vec{g} \tag{27}$$

$$\left(\vec{u} - \frac{P_k}{\mathrm{PrRe}}\nabla T\right) \cdot \nabla T = \frac{1}{\mathrm{PrRe}}\nabla \cdot (\nabla T) \tag{28}$$

with the parameters P_ρ, P_μ, P_k as

$$P_a = \frac{\Delta T^*}{a^*}\frac{\partial a^*}{\partial T^*}; \quad a^* = \rho^*, \mu^*, k^* \tag{29}$$

The only property that is left explicitly is ρ in the term $-\frac{\nabla p}{1+(\rho-1)}$ in equ (27), all other properties are "hidden" in Re, Pr and P_a. Asymptotically the parameters P_a are of order $O(\varepsilon)$, since

$$P_a = \varepsilon K_{a1} + O(\varepsilon) \tag{30}$$

accounting for the fact that a^* in K_{a1} is $a^*\left(T_R^*\right)$ but a^* in P_a is $a^*\left(T^*\right)$.
Based on equs (26)-(28) there are two simplified approaches to account for variable properties which are motivated by the fact that the dimensionless groups account for the variability of the properties leaving additional effects in the asymptotically small terms with P_a and $(\rho-1)$.

3.1 High Reynolds number HOC-method (HR-HOC)

For high Reynolds numbers the energy equation (28) can be approximated by setting $P_k = 0$, since $P_k/\mathrm{RePr} \to 0$ for $\mathrm{Re} \to \infty$. This reduces the number of A-values in the final result for the Nusselt number. Instead of equ (15) one now has for the Nusselt number with

$$K_{\sigma 1} = K_{k1} - K_{c_p 1}, \quad K_{\sigma 2} = K_{k2} - K_{c_p 2} - 2K_{k1}K_{c_{p1}} + 2K_{c_p 1}^2 \tag{31}$$

$$\begin{aligned}\frac{\mathrm{Nu}}{\mathrm{Nu}_{cp}} = {} & 1 + \varepsilon\left[K_{\rho 1}A_\rho + K_{\mu 1}A_\mu + K_{\sigma 1}A_\sigma\right] + \varepsilon^2\left[K_{\rho 2}A_{\rho 2} + K_{\mu 2}A_{\mu 2} + K_{\sigma 2}A_{\sigma 2}\right. \\ & \left. + K_\rho^2 A_{\rho\rho} + K_\mu^2 A_{\mu\mu} + K_\sigma^2 A_{\sigma\sigma} + K_{\rho\mu}A_{\rho\mu} + K_{\rho\sigma}A_{\rho\sigma} + K_{\mu\sigma}A_{\mu\sigma}\right] + O\left(\varepsilon^3\right)\end{aligned} \tag{32}$$

The new h_{aj}-values (c.f. equ (17)) for the determination of the A-values are

$$h_{\sigma 1} = \varepsilon K_{\sigma 1} = \varepsilon\left(K_{k1} - K_{c_p 1}\right) \tag{33}$$

$$h_{\sigma 2} = \varepsilon^2 K_{\sigma 2} = \varepsilon^2\left(K_{k2} - K_{c_p 2} - 2K_{k1}K_{c_{p1}} + 2K_{c_p 1}^2\right) \tag{34}$$

Table 1 shows the reduction in terms and numerical solutions necessary for the determination of the A-values in equ (32), compared to those for equ (15).

Method	A-values		Numerical solutions	
	1. order	2. order	1. order	2. order
HOC	4	14	8	32
HR-HOC	3	9	6	18

Table 1. Number of A-values and numerical solutions

The complex benchmark geometry in Fig. 1 is now treated with the simplified HR-HOC approach. Fig. 2 shows that the quality of the results is as good as in Fig. 1 for the full HOC-method.

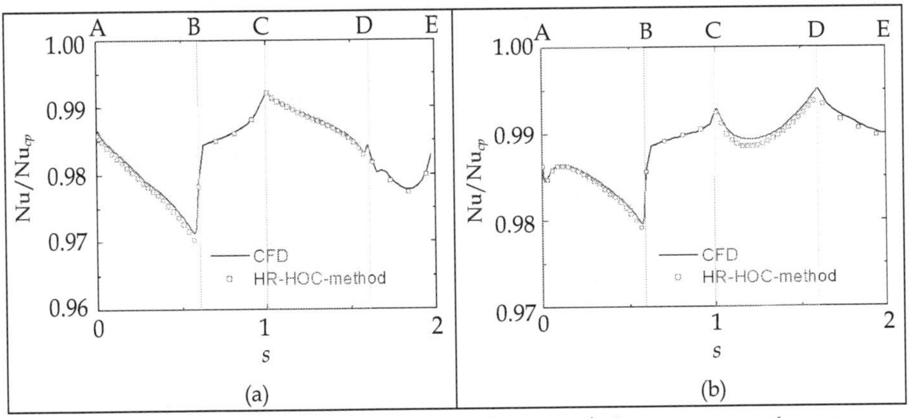

Fig. 2. Complex benchmark geometry, s. Fig. 1, ($\dot{q}_w^* = 300\,\mathrm{W/m^2}$, $\mathrm{Re} = 1.64 \times 10^6$) (a) heating element ① (b) heating element ②

3.2 Locally constant property HOC-method (LCP-HOC)

Without a real asymptotic justification all variable property effects outside the nondimensional groups are now neglected by setting all $P_a = 0$ and $\rho = 1$. Equations (26)-(28) then are the equations for constant properties, but with locally variable properties in *Re* and *Pr*.

Solutions of these equations can be used as an easy to get answer with respect to the question how strong the influence of variable properties is and if it makes sense to apply the more sophisticated methods.

In this LCP-HOC approach Re and RePr in the reduced set of equations (26)-(28) are treated like variable properties, being called α and β, i.e.

$$\alpha = \mathrm{Re} = \frac{\rho^* u_R^* L^*}{\mu^*} \tag{35}$$

$$\beta = \mathrm{RePr} = \frac{\rho^* c_p^* u_R^* L^*}{k^*} \tag{36}$$

Instead of equ (15) one now has for the Nusselt number with

$$\begin{aligned} K_{\alpha 1} &= K_{\rho 1} - K_{\mu 1} \\ K_{\alpha 2} &= K_{\rho 2} - K_{\mu 2} - 2K_{\rho 1}K_{\mu 1} + 2K_{\mu 1}^2 \end{aligned} \tag{37}$$

$$\begin{aligned} K_{\beta 1} &= K_{\rho 1} + K_{c_p 1} - K_{k1} \\ K_{\beta 2} &= K_{c_p 2} + K_{\rho 2} - K_{k2} + 2K_{c_p 1}K_{\rho 1} - 2K_{k1}K_{c_p 1} - 2K_{\rho 1}K_{k1} + 2K_{k1}^2 \end{aligned} \tag{38}$$

the form:

$$\begin{aligned} \frac{\mathrm{Nu}}{\mathrm{Nu}_{cp}} &= 1 + \varepsilon\left[K_{\alpha 1}A_\alpha + K_{\beta 1}A_\beta\right] \\ &+ \varepsilon^2\left[K_{\alpha 2}A_{\alpha 2} + K_{\beta 2}A_{\beta 2} + K_{\alpha 1}^2 A_{\alpha\alpha} + K_{\beta 1}^2 A_{\beta\beta} + K_{\alpha 1}K_{\beta 1}A_{\alpha\beta}\right] + O\left(\varepsilon^3\right) \end{aligned} \tag{39}$$

The new h_{aj}-values (c.f. equ (17)) for the determination of the A-values are

$$h_{\alpha 1} = \varepsilon K_{\alpha 1}; \quad h_{\alpha 2} = \varepsilon^2 K_{\alpha 2} \tag{40}$$

$$h_{\beta 1} = \varepsilon K_{\beta 1}; \quad h_{\beta 2} = \varepsilon^2 K_{\beta 2} \tag{41}$$

Table 2 shows the reduction in terms and numerical solutions necessary for the determination of the A-values in equ (39) compared to those for equ (15).

Method	A-values		Numerical solutions	
	1. order	2. order	1. order	2. order
HOC	4	14	8	32
LCP-HOC	2	5	4	8

Table 2. Number of A-values and numerical solutions

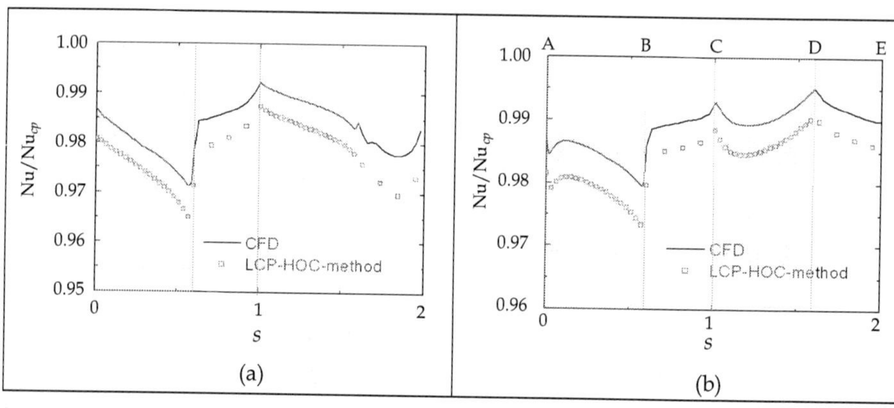

Fig. 3. Complex benchmark geometry, s. Fig. 1, ($\dot{q}_w^* = 300\,\mathrm{W/m^2}$, $\mathrm{Re} = 1.64 \times 10^6$) (a) heating element ① (b) heating element ②

Again, the complex benchmark geometry in Fig. 1 is chosen as a test case, now for the LCP-HOC approach. Fig. 3 shows that the results are of the same order as with the more comprehensive methods and thus can serve as a simple first approximation. So far, there is no explanation for the obviously constant offset in the results.

4. Alternative determination HOC-method (AD-HOC)

So far the RANS-approach has been used to cope with turbulent flows. Computational times are moderate and the relatively large number of numerical solutions necessary for the various HOC-methods is available in acceptable time.

This is no longer true, however, when LES or DNS solutions are aimed at. Then, finding 32 solutions for the second order HOC method (c.f. table 1 or 2) is not a reasonable procedure. In table 3 CPU-times for RANS, LES and DNS solutions of a simple benchmark problem are compared to illustrate the discrepancy in typical CPU-times for the three methods. The benchmark problem is that of a differentially heated cavity with adiabatic top and bottom walls shown in Fig. 4. In x_1-direction there is a periodic boundary condition L1 apart corresponding to an overall 2D geometry. As a result for constant properties (Boussinesq-approximation) the Nusselt number and shear stress distribution is shown for three different Rayleigh numbers at the hot wall in Fig. 5. Whereas DNS results from (Trias et al, 2010a; Trias et al, 2010b; Trias et al, 2007) and our own LES results are close to each other, RANS results show larger deviations. Some numerical details are given in table 3.

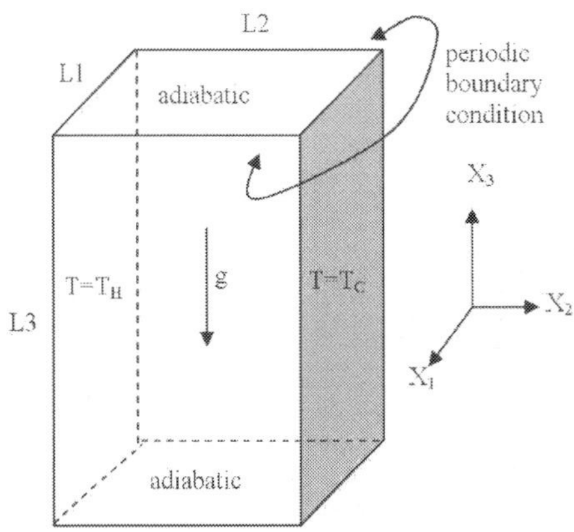

Fig. 4. Differentially heated cavity of aspect ratio $L_3/L_1 = 4$, $L_2/L_1 = 1$

NUMERICAL METHOD	CPU-TIME in hours			NUMERICAL DETAILS
	Ra = 2×10^9	Ra = 4×10^9	Ra = 10^{10}	
RANS (own results)	1	1	1	k - ω SST model; mesh resolution: 100×218; steady simulation. Computer: Linux cluster (4×2.66GHz)
LES (own results)	330	330	330	k -equation eddy-viscosity model; mesh resolution: 64×100×218; time step: 2.4×10^{-2} Computer: Linux cluster (4×2.66GHz)
DNS (Trias et al, 2010a; Trias et al, 2010b; Trias et al, 2007)	6200 (estimated)		26000 (estimated)	Mesh resolution Ra = 2×10^9 : 64×144×318 Ra = 10^{10} : 128×190×462 time step Ra = 2×10^9 : 1.27×10^{-3} Ra = 10^{10} : 6.25×10^{-4} Computer: Nec-Sx5 vectorial parallel machine.

CPU-TIME: sum of computational time of all processors in hours

Table 3. CPU-time and numerical details for different numerical methods applied to the benchmark problem (heated cavity), see Fig. 4

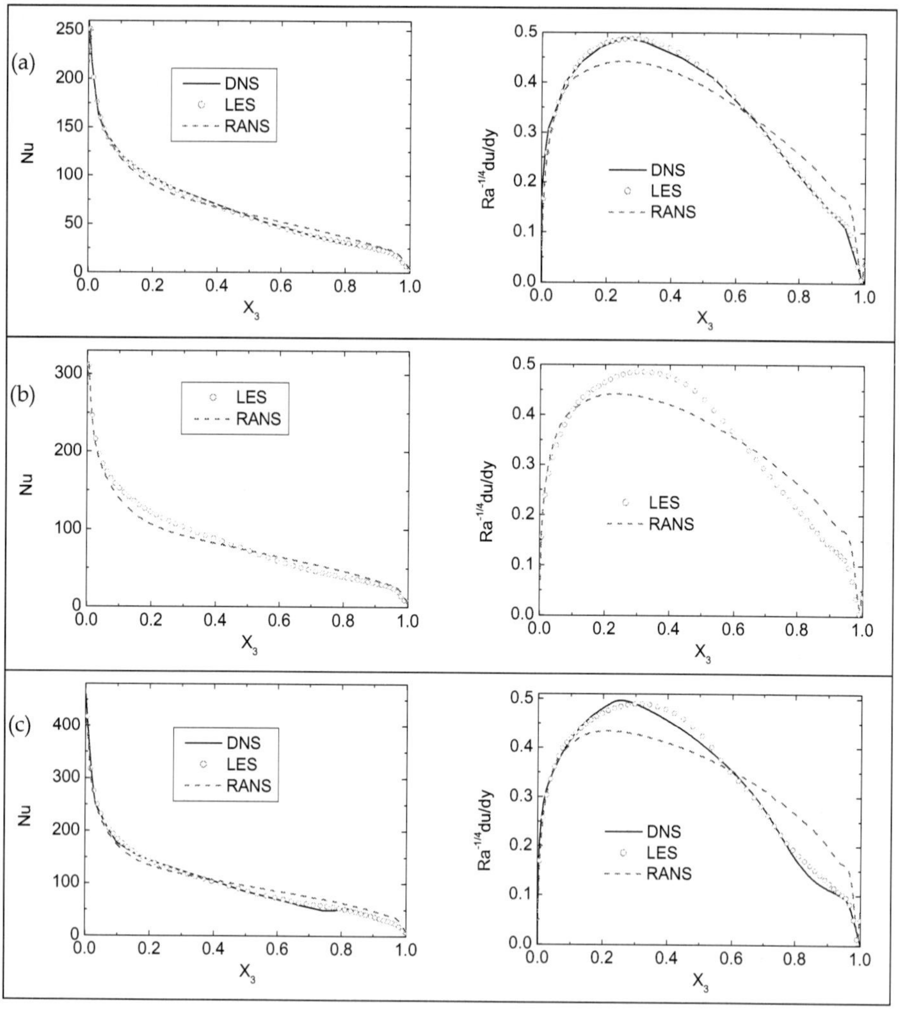

Fig. 5. Constant property results of the flow and heat transfer in a differentially heated cavity, see Fig. 4. (a) $Ra = 2 \times 10^9$ (b) $Ra = 4 \times 10^9$ (c) $Ra = 1 \times 10^{10}$

If now variable property effects are of interest, the A-coefficients should be determined with a RANS-approach. It is assumed that the variable and the constant property RANS results show basically the same errors (compared to the "exact" DNS results) and thus have systematic errors that cancel when the difference is taken in the procedure to determine the A-values.

DNS and LES constant property results can thus be corrected with respect to variable property effects by "alternatively determined" (i.e. AD) A-values. This procedure is called AD-HOC-method, but is not really an ad-hoc-method!

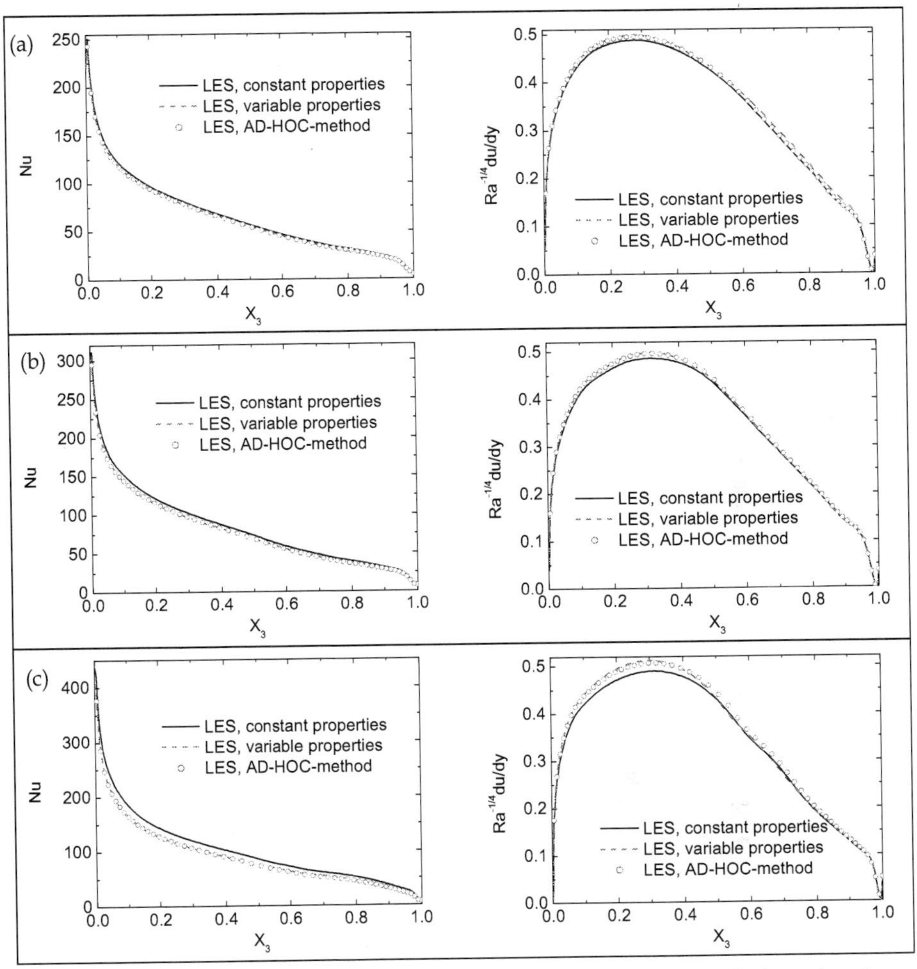

Fig. 6. Variable property results of the flow and heat transfer in a differentially heated cavity, see Fig. 4. (a) $Ra = 2\times10^9$ (b) $Ra = 4\times10^9$ (c) $Ra = 1\times10^{10}$ Fluid: air with properties (42)-(45)

In Fig. 6 these AD-HOC-results that account for variable properties are compared to LES results that account for variable properties from the beginning. They are results for air with variable ρ^*, μ^*, k^* and c_p^* asymptotically accounted for to the first order and for the LES calculations taken as

$$\rho^* = 1.188\cdot\left(1-\gamma^*\left(T^*-T_R^*\right)\right) \tag{42}$$

$$\mu^* = 1.18185\times10^{-5}\cdot\left(1+0.775\frac{T^*-T_R^*}{T_R^*}-0.176\left(\frac{T^*-T_R^*}{T_R^*}\right)^2\right) \tag{43}$$

$$k^* = 0.025721 \cdot \left(1 + 0.891 \frac{T^* - T_R^*}{T_R^*} - 0.1285 \left(\frac{T^* - T_R^*}{T_R^*}\right)^2\right) \tag{44}$$

$$c_p^* = 1014 \cdot \left(1 + 0.068 \frac{T^* - T_R^*}{T_R^*} - 0.038 \left(\frac{T^* - T_R^*}{T_R^*}\right)^2\right) \tag{45}$$

where the reference temperature $T_R^* = 293K$ and the expansion coefficient $\gamma^* = 0.003413$. The density variations are accounted for by the Boussinesq approximation like in (Trias et al, 2010a; Trias et al, 2010b; Trias et al, 2007). Fig. 6 shows that with the influence of variable properties determined by the AD-HOC-method there is a very good coincidence with LES results that account for variable property effects from the beginning. Thus the AD-HOC-method is an attractive approach when LES and DNS results have to be corrected with respect to the influence of variable properties.

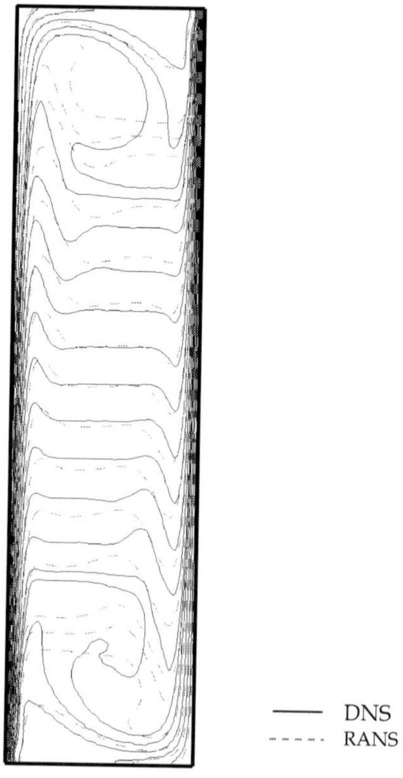

Fig. 7. Constant property results of the temperature field in a differentially heated cavity, see Fig. 4, $\mathrm{Ra} = 2 \times 10^8$

This is also true with respect to the field variables of these cases. For example, Fig. 7 shows that there is an appreciable difference when the temperature field is calculated by DNS compared to the RANS results.
However, as shown in Fig. 8, the iso-temperature lines for variable properties, calculated by DNS are well represented by iso-lines from the AD-HOC method, i.e. those lines from constant property DNS results corrected by A-values from RANS solutions for variable properties.

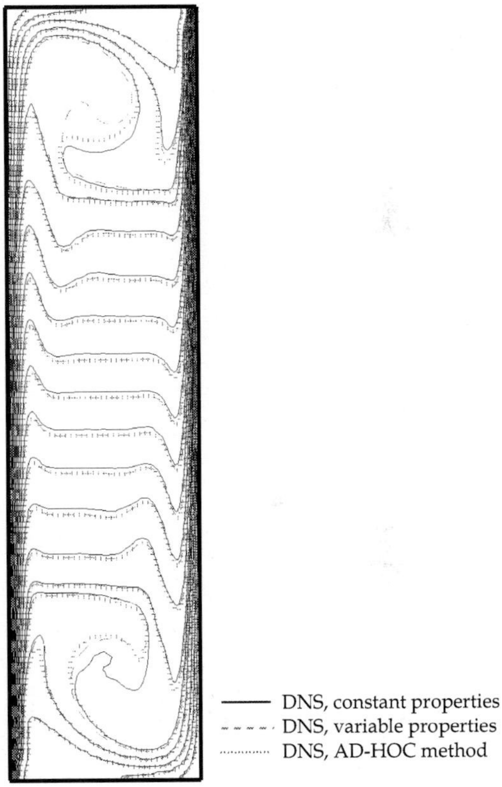

Fig. 8. Variable property results of the temperature field in a differentially heated cavity, see Fig. 4, $\mathrm{Ra} = 2 \times 10^8$

Distribution of the first order A-values A_γ, A_μ, A_k and A_{c_p}, computed by RANS, are shown in Fig. 9. The variable properties behave differently in the core region, where a quasi-laminar flow prevails and in the large vortex region near the bottom and top walls. Also, signs within one region are different. For example A_γ and A_μ are negative in the core region, whereas A_k and A_{c_p} are positive in the same region. Altogether there is a non-negligible effect of variable properties on the temperature distribution.

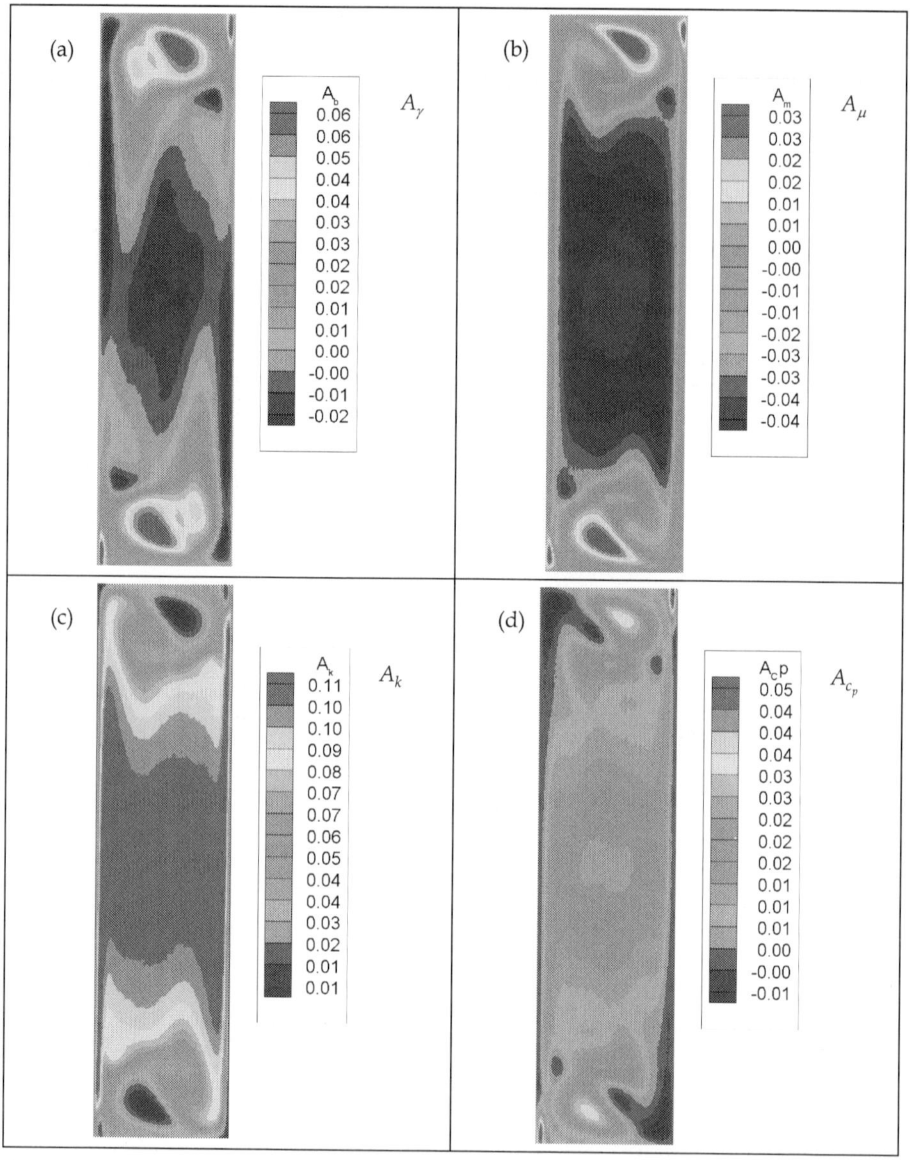

Fig. 9. A-values of temperature computed by RANS in a differentially heated cavity, see Fig. 4, $\mathrm{Ra} = 2 \times 10^8$ (a) A_γ; (b) A_μ; (c) A_k; (d) A_{c_p}

5. Conclusions

Various methods to account for variable property effects in complex geometries and highly sophisticated numerical methods have been proposed. Due to the Taylor series expansions of all properties, which are the starting point for all methods, the influence of variable properties can be accounted for in a general manner, i.e. for all (small) heating rates and for all Newtonian fluids. For a special problem with A-values determined once, the effect of variable properties can be found in the final result by fixing ε (the heat transfer rate) and all K-values (from the fluid of interest).
This way of treating variable property effects is much closer to the physics than empirical methods like the property ratio and the reference temperature methods are.

6. Nomenclature

$A_a, A_{aa}, A_{ab}, A_{a2}$	A-values, $a,b \in \{\rho,\gamma,\mu,k,c_p\}, a \neq b$
c_p	specific heat capacity
f	friction factor
$\vec{g}$	gravity vector
h_{aj}	h-values, $\varepsilon^j K_{aj}$
j	empirical parameter
K_a, K_{a2}, K_{an}	K-values, $a = \rho,\gamma,\mu,k,c_p$
k	heat conductivity
L^*	characteristic length
m_a, n_a	empirical exponents
Nu	Nußelt number
p	pressure
Pr	Prandtl number
P_a	nondimensional properties about variable fluid property
$\dot{q}$	heat flux
Re	Reynolds number
T	temperature
$\vec{u}$	velocity vector
V	variables
Greek symbols	
$\alpha,\beta,\sigma,\delta$	variables composed of fluid properties
γ	expansion coefficient
ΔT^*	temperature difference
ε	nondimensional temperature difference
μ	dynamic viscosity
ρ	density
subscripts	
cp	constant properties
R	reference state
*	dimensional

7. Acknowledgement

This study was supported by the DFG (Deutsche Forschungsgemeinschaft).

8. References

Bünger F. & Herwig H. (2009). An extended similarity theory applied to heated flows in complex geometries. *ZAMP*, Vol. 60, (2009), pp. 1095-1111.

Carey V. P. & Mollendorf J. C. (1980). Variable viscosity effects in several natural convection flows. *Int. J. Heat Mass Transfer*, Vol. 23, (1980), pp. 95-109

Debrestian D. J. & Anderson J. D. (1994). Reference Temperature Method and Reynolds Analogy for Chemically Reacting Non-equilibrium Flowfields. *J. of Thermophysics and Heat Transfer*, Vol. 8, (1994), pp. 190-192

Herwig H. & Wickern G. (1986). The Effect of Variable Properties on Laminar Boundary Layer Flows. *Wärme- und Stoffübertragung*, Vol. 20, (1986), pp. 47-57

Herwig H. & Bauhaus F. J. (1986). A Regular Perturbation Theory for Variable Properties Applied to Compressible Boundary Layers. *Proceedings of 8th International Heat Transfer Conference*, San Fransisco, Vol. 3, 1095-1101, 1986

Herwig H., Voigt M. & Bauhaus F. J. (1989). The Effect of Variable Properties on Momentum and Heat Transfer in a Tube with constant Wall Temperature. *Int. J. Heat Mass Transfer*, Vol. 32, (1989), pp. 1907-1915

Herwig H. & Schäfer P. (1992). Influence of variable properties on the stability of two-dimensional boundary layers. *J. Fluid Mechanics*, Vol. 243, (1992), pp. 1-14.

Jayari S.; Dinesh K. K., & Pillai K. L. (1999). Thermophoresis in natural convection with variable properties. *Heat and Mass Transfer*, Vol. 35, (1999), pp. 469-475

Jin Y. & Herwig H. (2010). Application of the Similarity Theory Including Variable Property Effects to a Complex Benchmark Problem, *ZAMP*, Vol. 61, (2010), pp. 509-528

Jin Y. & Herwig H. (2010). Efficient method to account for variable property effects in numerical momentum and heat transfer solutions, *Int. J. Heat Mass Transfer*, (2011), in press

Li Z.; Huai X.; Tao Y. & Chen H. (2007). Effects of thermal property variations on the liquid flow and heat transfer in micro-channel heat sinks. *Applied Thermal Engineering*, Vol. 27 (2007), pp. 2803–2814

Mahmood G. I.; Ligrani P. M. & Chen K. (2003). Variable property and Temperature Ratio Effects on Nusselt Numbers in a Rectangular Channel with 45 Deg Angle Rib Turbulators. *J. Heat Transfer*, Vol. 125 (2003), pp. 769-778

Trias F. X.; Soria M.; Oliva A. & Pérez-Segarra C. D. (2007). Direct numerical simulations of two- and three-dimensional turbulent natural convection flows in a differentially heated cavity of aspect ratio 4, *J. Fluid Mech.*, Vol. 586, (2007), pp. 259–293

Trias F. X.; Gorobets A.; Soria M. & Oliva A. (2010a). Direct numerical simulation of a differentially heated cavity of aspect ratio 4 with Rayleigh numbers up to 1011 – Part I: Numerical methods and time-averaged flow, *International Journal of Heat and Mass Transfer*, Vol. 53, (2010), pp. 665–673

Trias F. X.; Gorobets A.; Soria M. & Oliva A. (2010b). Direct numerical simulation of a differentially heated cavity of aspect ratio 4 with Rayleigh numbers up to 1011 – Part II: Numerical methods and time-averaged flow, *International Journal of Heat and Mass Transfer*, Vol. 53, (2010), pp. 674–683

9

Bioheat Transfer

Alireza Zolfaghari[1] and Mehdi Maerefat[2]
[1]*Department of Mechanical Engineering, Birjand University, Birjand,*
[2]*Department of Mechanical Engineering, Tarbiat Modares University, Tehran,*
Iran

1. Introduction

Heat transfer in living tissues is a complicated process because it involves a combination of thermal conduction in tissues, convection and perfusion of blood, and metabolic heat production. Over the years, several mathematical models have been developed to describe heat transfer within living biological tissues. These models have been widely used in the analysis of hyperthermia in cancer treatment, laser surgery, cryosurgery, cryopreservation, thermal comfort, and many other applications. The most widely used bioheat model was introduced by Pennes in 1948. Pennes proposed a new simplified bioheat model to describe the effect of blood perfusion and metabolic heat generation on heat transfer within a living tissue. Since the landmark paper by Pennes (1948), his model has been widely used by many researchers for the analysis of bioheat transfer phenomena. And, also a large number of bioheat transfer models have been proposed to overcome the shortcomings of Pennes' equation. These models include the continuum models which consider the thermal impact of all blood vessels as a global parameter and the vascular models which consider the thermal impact of each vessel individually.

Although, several bioheat models have been developed in the recent years, the thermoregulatory control mechanisms of the human body such as shivering, regulatory sweating, vasodilation, and vasoconstriction have not been considered in these models. On the other hand, these mechanisms may significantly influence the thermal conditions of the human body. This causes a serious limitation in using the bioheat models for evaluating the human body thermal response. In order to remove this limitation, Zolfaghari and Maerefat (2010) developed a new Simplified Thermoregulatory Bioheat (STB) model based on the combination of the well-known Pennes' equation and Gagge's thermal comfort model.

The present chapter aims at giving a concise introduction to bioheat transfer and the mathematical models for evaluating the heat transfer within biological tissues. This chapter is divided into six sections. The first section presents an introduction to the concept and history of bioheat transfer. The structure of living tissues with blood perfusion is described in section 2. Next, third section focuses on the mathematical modelling of heat transfer in living tissues. In the mentioned section, a brief description of some of the most important bioheat models (i.e. Pennes (1948) model, Wulff (1974) model, Klinger (1974) model, Chen and Holmes (1980) model and so on) is presented. Afterwards, section 4 explains the complexity of evaluating heat transfer within the tissues that thermally controlled by thermoregulatory mechanisms such as shivering, regulatory sweating, vasodilation, and vasoconstriction. Then, the Simplified Thermoregulatory Bioheat (STB) model is introduced for evaluating heat transfer within the segments of the human body. Finally, section 5 outlines the main conclusions and recommendations of the research. Moreover, the selected references are listed in the last section.

2. Structure of blood perfused tissues

Before we discuss the bioheat models, let us have a brief look at the structure of blood perfused tissues. The biological tissues include the layers of skin, fat, muscle and bone. Moreover, the skin is composed of two stratified layers: epidermis and dermis. Fig. 1 shows a schematic geometry of the tissue structure. Furthermore, the thermophysical properties of the human body tissue are provided in Table 1 (Lv & Liu, 2007; Sharma, 2010).

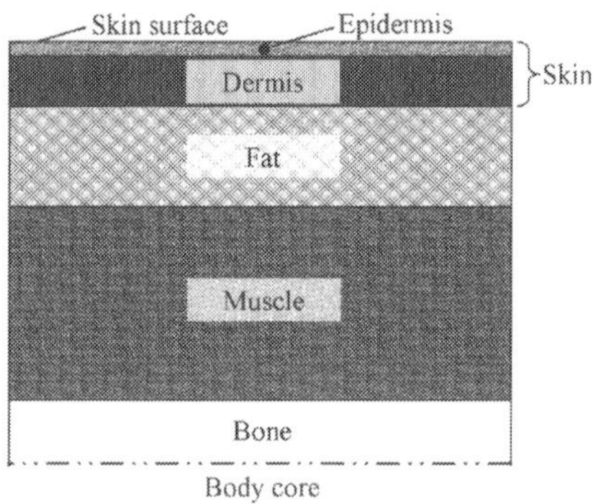

Fig. 1. Schematic geometry of the tissue structure (figure not to scale)

	Thickness l (m)	Density ρ (kg/m³)	Specific heat C (J/kg K)	Blood perfusion rate W_{bl} (m³/s m³)	Thermal conductivity k (W/m K)
Epidermis	80×10^{-6}	1200	3589	0	0.24
Dermis	0.002	1200	3300	0.00125	0.45
Fat	≈ 0.010	937	3258	0.00125	0.21
Muscle	≈ 0.020	1000	4000	0.00125	0.5
Bone	≈ 0.008	1920	1440	-	0.44
Blood	-	937	3889	-	0.64

Table 1. Geometrical information and thermal properties of the human body tissue (Lv & Liu, 2007; Sharma, 2010)

Blood circulation is a key mechanism for regulating the body temperature. The circulatory system of the human body comprises of two sets of blood vessels (arteries and veins) which carry blood from the heart and back. Blood leaves the heart through the aorta, which is the largest artery (diameter ≈ 5000 μm). Vessels supplying blood to muscles are known as main supply arteries and veins (SAV, 300-1000 μm diameter). They branch into primary arteries, (P, 100-300 μm diameter) which feed the secondary arteries (s, 50-100 μm diameter). These vessels deliver blood to the arterioles (20-40 μm diameter) which supply blood to the smallest vessels known as capillaries (c, 5-15μm diameter). Blood is returned to the heart through a system of vessels known as veins. Fig. 2 shows a schematic diagram of a typical vascular structure (Jiji, 2009).

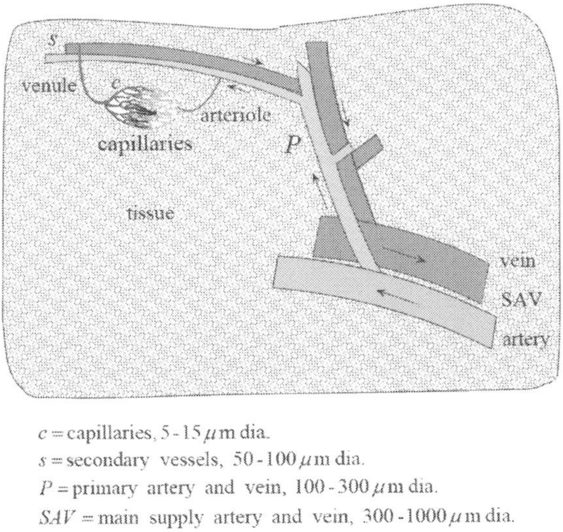

Fig. 2. Schematic diagram of the vascular system (Jiji, 2009)

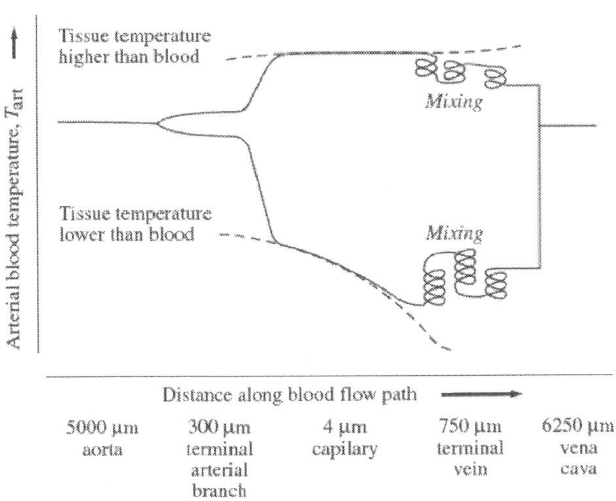

Fig. 3. Schematic of temperature equilibration between the blood and the tissue (Datta, 2002)

Blood leaves the heart at the arterial temperature T_{art}. It remains essentially at this temperature until it reaches the main arteries where equilibration with surrounding tissue begins to take place. Equilibration becomes complete prior to reaching the arterioles and capillaries. Beyond this point, blood temperature follows the solid tissue temperature (T_{ti}) through its spatial and time variations until blood reaches the terminal veins. At this point the blood temperature ceases to equilibrate with the tissue, and remains virtually constant, except as it mixes with other blood of different temperatures at venous confluences. Finally, the cooler blood from peripheral regions and warmer blood from internal organs mix within the vena cava and the right atrium and ventricle. Following thermal exchange in the pulmonary circulation and remixing in the left heart, the blood attains the same temperature it had at the start of the circuit (Datta, 2002). Fig. 3 shows a schematic of temperature equilibration between the blood and the solid tissue.

3. Mathematical models of bioheat transfer

3.1 Pennes model

Over the years, the effects of blood flow on heat transfer in living tissue have been studied by many researchers and a large number of bioheat transfer models have been developed on the basis of two main approaches: the continuum approach and the discrete vessel (vascular) approach. In the continuum approach, the thermal impact of all blood vessels models with a single global parameter; and the vascular approach models the impact of each vessel individually (Raaymakers et al., 2009). The most widely used continuum model of perfused tissue was introduced in 1948 by Harry Pennes. The Pennes (1948) model was initially developed for predicting heat transfer in the human forearm. Due to the simplicity of the Pennes bioheat model, it was implemented in various biological research works such as for therapeutic hyperthermia for the treatment of cancer (Minkowycz et al., 2009).

Pennes bioheat model is based on four simplifying assumptions (Jiji, 2009):

1. All pre-arteriole and post-venule heat transfer between blood and tissue is neglected.
2. The flow of blood in the small capillaries is assumed to be isotropic. This neglects the effect of blood flow directionality.
3. Larger blood vessels in the vicinity of capillary beds play no role in the energy exchange between tissue and capillary blood. Thus the Pennes model does not consider the local vascular geometry.
4. Blood is assumed to reach the arterioles supplying the capillary beds at the body core temperature. It instantaneously exchanges energy and equilibrates with the local tissue temperature.

Based on these assumptions, Pennes (1948) modeled blood effect as an isotropic heat source or sink which is proportional to blood flow rate and the difference between the body core temperature and local tissue temperature. Therefore, Pennes (1948) proposed a model to describe the effects of metabolism and blood perfusion on the energy balance within tissue. These two effects were incorporated into the standard thermal diffusion equation, which is written in its simplified form as:

$$\rho_{ti} C_{ti} \frac{\partial T_{ti}}{\partial t} = \nabla . k_{ti} \nabla T_{ti} + \rho_{bl} C_{bl} W_{bl} (T_{art} - T_{ti}) + q_m \quad (1)$$

where ρ_{ti}, C_{ti}, T_{ti} and k_{ti} are, respectively, the density, specific heat, temperature and thermal conductivity of tissue. Also, T_{art} is the temperature of arterial blood, q_m is the metabolic heat generation and ρ_{bl}, C_{bl} and W_{bl} are, respectively, the density, specific heat and perfusion rate of blood. It should be noted that metabolic heat generation is assumed to be homogeneously distributed throughout the tissue. Also, it is assumed that the blood perfusion effect is homogeneous and isotropic and that thermal equilibration occurs in the microcirculatory capillary bed. In this scenario, blood enters capillaries at the temperature of arterial blood, T_{art}, where heat exchange occurs to bring the temperature to that of the surrounding tissue, T_{ti}. There is assumed to be no energy transfer either before or after the blood passes through the capillaries, so that the temperature at which it enters the venous circulation is that of the local tissue (Kreith, 2000).

Pennes (1948) performed a series of experimental studies to validate his model. Validations have shown that the results of Pennes bioheat model are in a reasonable agreement with the experimental data. Although Pennes bioheat model is often adequate for roughly describing the effect of blood flow on the tissue temperature, there exist some serious shortcomings in his model due to its inherent simplicity. The shortcomings of Pennes bioheat model come from the basic assumptions that are introduced in this model. These shortcomings can be listed as follows (Jiji, 2009):

1. Thermal equilibration does not occur in the capillaries, as Pennes assumed. Instead it takes place in pre-arteriole and post-venule vessels having diameters ranging from 70-500 μm.
2. Directionality of blood perfusion is an important factor in the interchange of energy between vessels and tissue. The Pennes equation does not account for this effect.
3. Pennes equation does not consider the local vascular geometry. Thus significant features of the circulatory system are not accounted for. This includes energy exchange with large vessels, countercurrent heat transfer between artery-vein pairs and vessel branching and diminution.
4. The arterial temperature varies continuously from the deep body temperature of the aorta to the secondary arteries supplying the arterioles, and similarly for the venous return. Thus, contrary to Pennes' assumption, pre-arteriole blood temperature is not equal to body core temperature and vein return temperature is not equal to the local tissue temperature. Both approximations overestimate the effect of blood perfusion on local tissue temperature.

To overcome these shortcomings, a considerable number of modifications have been proposed by various researchers. Wulff (1974) and Klinger (1974) considered the local blood mass flux to account the blood flow direction, while Chen and Holmes (1980) examined the effect of thermal equilibration length on the blood temperature and added the dispersion and microcirculatory perfusion terms to the Klinger equation (Vafai, 2011). In the following sections, a brief review of the modified bioheat models will be given.

3.2 Wulff continuum model

Due to the simplicity of the Pennes model, many authors have looked into the validity of the assumptions used to develop the Pennes bioheat equation. Wulff (1974) was one of the first researchers that directly criticized the fundamental assumptions of the Pennes bioheat equation and provided an alternate analysis (Cho, 1992). Wulff (1974) assumed that the heat transfer between flowing blood and tissue should be modeled to be proportional to the temperature difference between these two media rather than between the two bloodstream temperatures (i.e., the temperature of the blood entering and leaving the tissue). Thus, the energy flux at any point in the tissue should be expressed by (Minkowycz et al., 2009)

$$q = -k_{\text{ti}} \nabla T_{\text{ti}} + \rho_{\text{bl}} h_{\text{bl}} v_{\text{h}} \tag{2}$$

where v_h is the local mean blood velocity. Moreover, h_{bl} is the specific enthalpy of the blood and it is given by

$$h_{\text{bl}} = \int_{T_{\text{o}}}^{T_{\text{bl}}} C_{\text{bl}}(T_{\text{bl}}^*) \text{d}T_{\text{bl}}^* + \frac{P}{\rho_{\text{bl}}} + \Delta H_{\text{f}}(1-\phi) \tag{3}$$

where P is the system pressure, ΔH_f is the enthalpy of formation of the metabolic reaction, and ϕ is the extent of reaction. Also, T_o and T_{bl} are the reference and blood temperatures, respectively. Thus, the energy balance equation can be written as

$$\rho_{\text{ti}} C_{\text{ti}} \frac{\partial T_{\text{ti}}}{\partial t} = -\nabla . q \tag{4}$$

Therefore,

$$\rho_{\text{ti}} C_{\text{ti}} \frac{\partial T_{\text{ti}}}{\partial t} = -\nabla . \left[-k_{\text{ti}} \nabla T_{\text{ti}} + \rho_{\text{bl}} v_{\text{h}} \left(\int_{T_{\text{o}}}^{T_{\text{bl}}} C_{\text{bl}}(T_{\text{bl}}^*) \text{d}T_{\text{bl}}^* + \frac{P}{\rho_{\text{bl}}} + \Delta H_{\text{f}}(1-\phi) \right) \right] \tag{5}$$

Neglecting the mechanical work term (P/ρ_{bl}), setting the divergence of $\rho_{\text{bl}} v_{\text{h}}$ to zero, and assuming constant physical properties, Eq. (5) can be simplified as follows (Minkowycz et al., 2009):

$$\rho_{\text{ti}} C_{\text{ti}} \frac{\partial T_{\text{ti}}}{\partial t} = k_{\text{ti}} \nabla^2 T_{\text{ti}} - \rho_{\text{bl}} C_{\text{bl}} v_{\text{h}} \nabla T_{\text{bl}} + \rho_{\text{bl}} v_{\text{h}} \Delta H_{\text{f}} \nabla \phi \tag{6}$$

Since blood is effectively microcirculating within the tissue, it will likely be in thermal equilibrium with the surrounding tissue. As such, Wulff (1974) assumed that T_{bl} is equivalent to the tissue temperature T_{ti}. In this condition, the metabolic reaction term ($\rho_{\text{bl}} v_{\text{h}} \Delta H_{\text{f}} \nabla \phi$) is equivalent to q_m. Therefore, the final form of the bioheat equation that was derived by Wulff (1974) is

$$\rho_{\text{ti}} C_{\text{ti}} \frac{\partial T}{\partial t} = k_{\text{ti}} \nabla^2 T_{\text{ti}} - \rho_{\text{bl}} C_{\text{bl}} v_{\text{h}} . \nabla T_{\text{ti}} + q_{\text{m}} \tag{7}$$

It should be noted that the main challenge in solving this bioheat equation is in the evaluation of the local blood mass flux $\rho_{\text{bl}} v_{\text{h}}$ (Minkowycz et al., 2009).

3.3 Klinger continuum model

In 1974, Klinger presented an analytical bioheat model that was conceptually similar to Wulff bioheat model. Klinger (1974) argued that in utilizing the Pennes model, the effects of nonunidirectional blood flow were being neglected and thus significant errors were being introduced into the computed results. In order to correct this lack of directionality in the formulation, Klinger (1974) proposed that the convection field inside the tissue should be modeled based upon the *in vivo* vascular anatomy (Cho, 1992). Taking into account the spatial and temporal variations of the velocity and heat source, and assuming constant physical properties of tissue and incompressible blood flow, the Klinger bioheat equation was expressed as:

$$\rho_{ti}C_{ti}\frac{\partial T}{\partial t} = k_{ti}\nabla^2 T_{ti} - \rho_{bl}C_{bl}\mathrm{v}.\nabla T_{ti} + q_m \tag{8}$$

This equation is similar to that derived by Wulff (1974), except it is written for the more general case of a spatially and temporally nonuniform velocity field (v) and heat source (q_m).

3.4 Chen-Holmes (CH) continuum model

Among the continuum bioheat models, the Chen-Holmes model (Chen & Holmes, 1980) is the most developed (Kreith, 2000). Chen and Holmes (1980) showed that the major heat transfer processes occur in the 50 to 500μm diameter vessels. Consequently, they proposed that larger vessels be modeled separately from smaller vessels and tissue. Therefore, in the Chen-Holmes bioheat model, the total tissue control volume is subdivided to the solid-tissue subvolume (V_s) and blood subvolume (V_b) as shown in Fig. 4. By using this concept, Chen and Holmes (1980) proposed a new modified relationship for calculating the blood perfusion term (q_{bl}) in their bioheat model:

$$q_{bl} = \rho_{bl}C_{bl}W_{bl}^*(T_{art}^* - T_{ti}) - \rho_{bl}C_{bl}\mathrm{v}.\nabla T_{ti} + \nabla.k_p\nabla T_{ti} \tag{9}$$

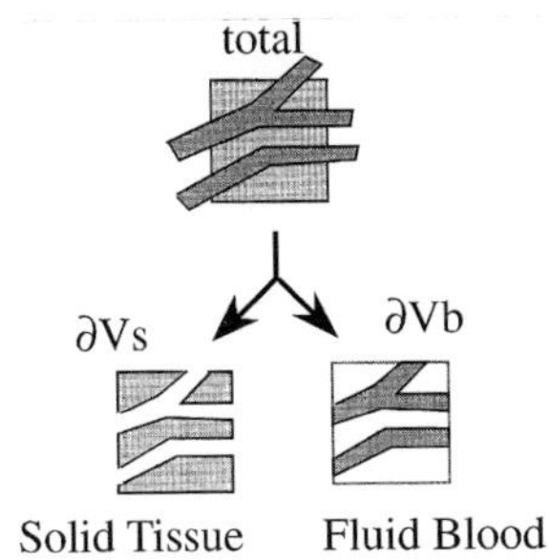

Fig. 4. Schematic representation of tissue control volume as used in Chen-Holmes model (Kreith, 2000)

In Eq. (9), the first term on the right hand side appears similar to Pennes' perfusion term except the perfusion rate (W_{bl}^*) and the arterial temperature (T_{art}^*) are specific to the volume being considered. It should be noted that T_{art}^* is essentially the temperature of blood upstream of the arterioles and it is not equal to the body core temperature. The second term in Eq. (9) accounts for energy convected due to equilibrated blood. Directionality of blood flow is described by the vector **v**, which is the volumetric flow rate per unit area. The third term in Eq. (9) describes conduction mechanisms associated with small temperature fluctuations in equilibrated blood. The symbol k_p denotes "perfusion conductivity". It is a function of blood flow velocity, vessel inclination angle relative to local temperature gradient, vessel radius and number density.

Using a simplified volume-averaging technique, the Chen-Holmes bioheat equation can be written as follows:

$$\rho_{ti,eff}C_{ti,eff}\frac{\partial T_{ti}}{\partial t} = \nabla.k_{ti,eff}\nabla T_{ti} - \left\{\rho_{bl}C_{bl}W_{bl}^*(T_{art}^* - T_{ti}) - \rho_{bl}C_{bl}\mathrm{v}.\nabla T_{ti} + \nabla.k_p\nabla T_{ti}\right\} + q_m \tag{10}$$

where

$$\rho_{\mathrm{ti,eff}} = (1-\varepsilon_{\mathrm{bl}})\rho_{\mathrm{ti}} + \varepsilon_{\mathrm{bl}}\rho_{\mathrm{bl}} \tag{11}$$

$$C_{\mathrm{ti,eff}} = (1-\varepsilon_{\mathrm{bl}})C_{\mathrm{ti}} + \varepsilon_{\mathrm{bl}}C_{\mathrm{bl}} \tag{12}$$

$$k_{\mathrm{ti,eff}} = (1-\varepsilon_{\mathrm{bl}})k_{\mathrm{ti}} + \varepsilon_{\mathrm{bl}}k_{\mathrm{bl}} \tag{13}$$

where $\varepsilon_{\mathrm{bl}}$ is the porosity of the tissue where blood flows and T_{ti}^{*} is the local mean tissue temperature expressed as (Minkowycz et al., 2009)

$$T_{\mathrm{ti}}^{*} = \frac{(1-\varepsilon_{\mathrm{bl}})\rho_{\mathrm{ti}}C_{\mathrm{ti}}T_{\mathrm{ti}} + \varepsilon_{\mathrm{bl}}\rho_{\mathrm{bl}}C_{\mathrm{bl}}k_{\mathrm{bl}}}{\rho_{\mathrm{ti,eff}}C_{\mathrm{ti,eff}}} \tag{14}$$

Since $\varepsilon_{\mathrm{bl}}$ <<1, it follows that $k_{\mathrm{ti,eff}}$ is independent of blood flow and equal to the conductivity of the solid tissue (k_{ti}).

Although the Chen-Holmes model represents a significant improvement over Pennes' equation, it is not easy to implement since it requires detailed knowledge of the vascular network and blood perfusion. Furthermore, the model does not explicitly address the effect of closely spaced countercurrent artery-vein pairs (Kreith, 2000).

3.5 Weinbaum, Jiji and Lemons (WJL) bioheat model

The modeling of countercurrent vascular system, which was not explicitly addressed by the Chen-Holmes model, developed separately from that of the continuum models. In 1984, Weinbaum, Jiji and Lemons presented a new vascular bioheat model by considering the countercurrent blood flow. This model was obtained based on a hypothesis that small arteries and veins are parallel and the flow direction is countercurrent, resulting in counterbalanced heating and cooling effects (Fig. 5). It should be noted that this assumption is mainly applicable within the intermediate tissue of the skin (Minkowycz et al., 2009). In an anatomic study performed on rabbit limbs, Weinbaum et al. (1984) identified three vascular layers (deep, intermediate, and cutaneous) in the outer 1cm tissue layer. For the countercurrent structure of the deep tissue layer, they proposed a system of three coupled equations:

$$\rho_{\mathrm{bl}}C_{\mathrm{bl}}\,\pi r^2\,\overline{\mathrm{v}}.\frac{\mathrm{d}T_{\mathrm{art}}}{\mathrm{d}s} = -q_{\mathrm{art}} \tag{15}$$

$$\rho_{\mathrm{bl}}C_{\mathrm{bl}}\,\pi r^2\,\overline{\mathrm{v}}.\frac{\mathrm{d}T_{\mathrm{v}}}{\mathrm{d}s} = -q_{\mathrm{v}} \tag{16}$$

$$\rho_{\mathrm{ti}}C_{\mathrm{ti}}\frac{\partial T_{\mathrm{ti}}}{\partial t} = \nabla.k_{\mathrm{ti}}\nabla T_{\mathrm{ti}} + \left\{ n\,g\,\rho_{\mathrm{bl}}C_{\mathrm{bl}}(T_{\mathrm{art}}-T_{\mathrm{v}}) - \rho_{\mathrm{bl}}C_{\mathrm{bl}}\,n\pi r^2\,\overline{\mathrm{v}}.\frac{\mathrm{d}(T_{\mathrm{art}}-T_{\mathrm{v}})}{\mathrm{d}s} \right\} + q_{\mathrm{m}} \tag{17}$$

where q_{art} is the heat loss from the artery by conduction through its wall, q_{v} is the heat gain by conduction per unit length through the vein wall into the vein, T_{art} and T_{v} are the bulk mean temperatures inside the blood vessel, r is the vessel radius, $\overline{\mathrm{v}}$ is the mean velocity in either the artery or vein, n is the number of arteries or veins, and g is the perfusion bleed-off velocity per unit vessel surface area (*s*). The first two equations describe the heat transfer of the thermally significant artery and vein, respectively. The third equation refers to the tissue surrounding the artery–vein pair. In Eq. (17), the middle two right-hand-side terms represent the capillary bleed-off energy exchange, and the net heat exchange between the tissue and artery–vein pair, respectively. The capillary bleed-off term is similar to Pennes' perfusion term except the bleed-off mass flow (*g*) is used. Their analysis showed that the major heat transfer is due to the imperfect countercurrent heat exchange between artery–vein pairs. They quantified the effect of perfusion bleed-off associated with this vascular structure, and showed that Pennes' perfusion formulation is negligible due to the temperature differential (Kreith, 2000).

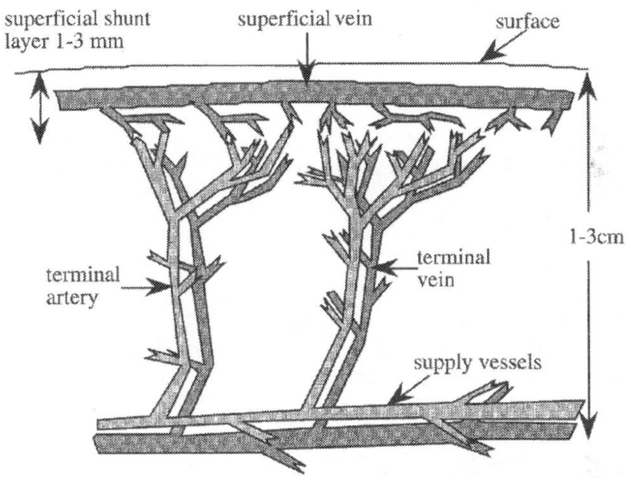

Fig. 5. Schematic of artery and vein pair in peripheral skin layer (Kreith, 2000)

Assumptions of the Weinbaum-Jiji-Lemons model include the following (Kreith, 2000):

1. Neglecting the lymphatic fluid loss, so that the mass flow rate in the artery is equal to that of the vein.
2. Spatially uniform bleed-off perfusion.
3. Heat transfer in the plane normal to the artery–vein pair is greater than that along the vessels (in order to apply the approximation of superposition of a line sink and source in a pure conduction field).
4. A linear relationship for the temperature along the radial direction in the plane normal to the artery and vein.
5. The artery–vein border temperature equals the mean of the artery and vein temperature.
6. The blood exiting the bleed-off capillaries and entering the veins is at the venous blood temperature.

The last assumption has drawn criticism based on studies that indicate the temperature to be closer to tissue. Limitations of this model include the difficulty of implementation, and that the artery and vein diameters must be identical (Kreith, 2000).

3.6 Simplified Weinbaum-Jiji (WJ) model

Since both T_{art} and T_v are unknowns in Equation (17), the tissue temperature T_{ti} cannot be determined. Therefore, Weinbaum and Jiji (1985) derived a simplified single equation to study the influence of blood flow on the tissue temperature distribution. In order to eliminate the artery and vein temperatures from their previous formulation (Weinbaum et al., 1984), two major assumptions were used:

1. Tissue temperature T_{ti} is approximated by the average of the local artery and vein temperatures. That is

$$T_{ti} \approx \frac{T_{art} + T_v}{2} \tag{18}$$

2. Heat from a paired artery is mostly conducted to the corresponding vein:

$$q_{art} \approx q_v \approx \sigma_\Delta k_{ti} \left(T_{art} - T_v\right) \tag{19}$$

where σ_Δ is a geometrical shape factor and it is associated with the resistance to heat transfer between two parallel vessels embedded in an infinite medium (Jiji, 2009). For the case of vessels at uniform surface temperatures with center to center spacing l, the shape factor is given by (Chato, 1980)

$$\sigma_\Delta = \frac{\pi}{\cosh^{-1}(l / 2r)} \tag{20}$$

By using the mentioned assumptions and substituting the Eqs. (18) and (19) in Eqs. (15), (16) and (17), Weinbaum and Jiji (1985) proposed a simplified equation for evaluating the tissue temperature distribution:

$$\rho_{ti} C_{ti} \frac{\partial T_{ti}}{\partial t} = \nabla . k_{eff} \nabla T_{ti} + q_m \tag{21}$$

where k_{eff} is the effective conductivity, defined as

$$k_{eff} = k_{ti}\left[1 + Pe_i V(\xi)\right] \tag{22}$$

where ξ is a dimensionless distance and it defines as x/L and L is the tissue layer thickness. Also, $V(\xi)$ is dimensionless vascular geometry function and it can be calculated if the vascular data are available. Furthermore, Pe_i is the inlet Peclet number; which is defined as (Jiji, 2009)

$$Pe_i = \frac{2\rho_{bl} C_{bl} r_i v_i}{k_{bl}} \tag{23}$$

where r_i and v_i are the vessel radius and the blood velocity at the inlet to the tissue layer at x=0.

The main limitations of the Weinbaum-Jiji bioheat equation are associated with the importance of the countercurrent heat exchange. It was derived to describe heat transfer in peripheral tissue only, where its fundamental assumptions are most applicable. In tissue area containing a big blood vessel (>200 μm in diameter), the assumption that most of the heat leaving the artery is recaptured by its countercurrent vein could be violated; thus, it is not an accurate model to predict the temperature field. Furthermore, unlike the Pennes bioheat equation, which requires only the value of local blood perfusion rate, the Weinbaum-Jiji bioheat model requires many detailed anatomical and vascular data such as the vessel number density, size, and artery-vein spacing for each vessel generation, as well as the blood perfusion rate. These anatomic data are normally not available for most blood vessels in the thermally significant range (Kutz, 2009).

4. Bioheat transfer in physiologically controlled tissues

4.1 Thermoregulatory control mechanisms of the human body

Thermoregulation is the ability of an organism to keep its body temperature within certain boundaries, even when the surrounding temperature is very different. The hypothalamus regulates the body temperature by the thermoregulatory mechanisms such as vasomotion, shivering and regulatory sweating. It receives inputs from central and peripheral temperature receptors situated in the 'core' and in the outer 'shell'. Temperature-sensitive receptors in the core are found in the hypothalamus, spinal cord, abdominal viscera and the great veins. They respond to temperatures between 30°C and 42°C. Peripheral receptors are located in the skin and they contain two types of thermoreceptors: warm receptors and cold receptors. The hypothalamic response to a thermal stimulus depends on the integration of both central and peripheral stimuli (Campbell, 2008). The intensity of these stimuli depends on the difference between the temperature of each compartment of the body (core or skin) and its related neutral temperature. In 1988, Doherty and Arens named the mentioned temperature difference as the thermal signals. The thermoregulatory mechanisms of the body are controlled by these thermal signals. The cold and warm signals of human body for skin and core compartments are defined as follows (Doherty & Arens, 1988)

$$WSIG_{\mathrm{cr}} = \mathrm{Max}\{0, T_{\mathrm{cr}} - T_{\mathrm{cr,n}}\} \tag{24}$$

$$CSIG_{\mathrm{cr}} = \mathrm{Max}\{0, T_{\mathrm{cr,n}} - T_{\mathrm{cr}}\} \tag{25}$$

$$WSIG_{\mathrm{sk}} = \mathrm{Max}\{0, T_{\mathrm{sk}} - T_{\mathrm{sk,n}}\} \tag{26}$$

$$CSIG_{\mathrm{sk}} = \mathrm{Max}\{0, T_{\mathrm{sk,n}} - T_{\mathrm{sk}}\} \tag{27}$$

where *CSIG* and *WSIG*, respectively, represent cold and warm signals of the human body, $T_{\mathrm{sk,n}}$ is neutral skin temperature (≈ 33.7°C), and $T_{\mathrm{cr,n}}$ is neutral core temperature (≈ 36.8°C). The thermal neutrality state of the human body occurs when the body is able to maintain its thermal equilibrium with the environment with minimal regulatory effort (Yigit, 1999).

The thermoregulatory mechanisms of the human body are related to the aforementioned thermal signals of the body. One of these thermoregulatory mechanisms is vasomotion. Vasomotion of blood vessels (vasoconstriction and vasodilation) is caused by cold/warm thermal conditions and it changes the rate of blood flow ($\dot{m}_{\mathrm{bl}}$) and also the fraction of body mass concentrated in skin compartment (α). These parameters can be calculated as follows (Kaynakli & Kilic, 2005)

$$\alpha = 0.0418 + 0.745 / (3600\dot{m}_{\mathrm{bl}} + 0.585) \tag{28}$$

where

$$\dot{m}_{\mathrm{bl}} = \frac{6.3 + 200WSIG_{\mathrm{cr}}}{3600(1 + 0.5CSIG_{\mathrm{sk}})} \tag{29}$$

The other thermoregulatory mechanism of the human body is shivering under cold sensation. Shivering is an increase of heat production during cold exposure due to increased contractile activity of skeletal muscles (Wan & Fan, 2008). Shivering and muscle tension may generate additional metabolic heat. Total metabolic heat production of body includes the metabolic rate due to activity (M_{act}) and the shivering metabolic rate (M_{shiv}). Therefore

$$M = M_{\mathrm{act}} + M_{\mathrm{shiv}} \tag{30}$$

and

$$M_{\mathrm{shiv}} = 19.4CSIG_{\mathrm{sk}}CSIG_{\mathrm{cr}} \tag{31}$$

Another thermoregulatory mechanism of the body is regulatory sweating. Sweating causes the latent heat loss from the skin. The rate of the sweat production per unit of skin area can be estimated by the following equation (Kaynakli & Kilic, 2005)

$$\dot{m}_{\mathrm{rsw}} = 4.7 \times 10^{-5} WSIG_{\mathrm{b}} \exp(WSIG_{\mathrm{sk}} / 10.7) \tag{32}$$

where

$$WSIG_{\mathrm{b}} = \mathrm{Max}\{0, T_{\mathrm{b}} - T_{\mathrm{b,n}}\} \tag{33}$$

and

$$T_{\mathrm{b}} = \alpha T_{\mathrm{sk}} + (1-\alpha)T_{\mathrm{cr}} \tag{34}$$

$$T_{\mathrm{b,n}} = \alpha T_{\mathrm{sk,n}} + (1-\alpha)T_{\mathrm{cr,n}} \tag{35}$$

where $WSIG_{\mathrm{b}}$ is warm signal of body, T_{b} is body temperature (°C), and $T_{\mathrm{b,n}}$ is the neutral temperature of body (°C).

The regulatory sweating leads to an increase in the skin wettedness. The total skin wettedness is composed of wettedness due to diffusion through the skin (w_{dif}) and regulatory sweating (w_{rsw}). Therefore

$$w_{\mathrm{skin}} = w_{\mathrm{dif}} + w_{\mathrm{rsw}} \tag{36}$$

where

$$w_{\mathrm{dif}} = 0.06(1 - w_{\mathrm{rsw}}) \tag{37}$$

$$w_{\mathrm{rsw}} = \frac{\dot{m}_{\mathrm{rsw}} h_{\mathrm{fg}}}{q_{\mathrm{evap,max}}} \tag{38}$$

where h_{fg} is the heat of vaporization of water and $q_{\mathrm{evap,max}}$ is the maximum evaporative potential and can be estimated by the following equation (Kaynakli & Kilic, 2005)

$$q_{\text{evap,max}} = \frac{P_{\text{sk(s)}} - P_{\text{a}}}{R_{\text{e,t}}} \tag{39}$$

where $P_{sk(s)}$ is water vapor pressure in the saturated air at the skin temperature (kPa) and $R_{e,t}$ is the total evaporative resistance between the body and the environment (m^2kPa/W).

4.2 Simplified thermoregulatory bioheat (STB) model

The human body thermal response may be significantly affected by thermoregulatory mechanisms of the human body such as shivering, regulatory sweating and vasomotion. But, these thermoregulatory mechanisms have not been considered in the well-known Pennes model and also in the other modified bioheat models. In addition, although the body core temperature could be changed depending on personal/environmental conditions, it is commonly assumed as a constant value in Pennes model. Therefore, it seems that the well-known Pennes bioheat model must be modified for using in human thermal response applications. In 2010, Zolfaghari and Maerefat (2010) developed a new simplified thermoregulatory bioheat model (STB model) on the basis of two main objectives: the first is to supplement the thermoregulatory mechanisms to Pennes bioheat model, and the second is to consider the body core temperature as a variant parameter depending on personal/environmental conditions. In order to reach the mentioned objectives, Zolfaghari and Maerefat (2010) developed their bioheat model by combining Pennes' equation and Gagge's two-node model. By using this concept, they presented an energy balance equation for core compartment of the human body as follows

$$(1-\alpha)\rho_{\text{b}}C_{\text{b}}\frac{\text{d}T_{\text{cr}}}{\text{d}t} = r_{\text{m}}q_{\text{m}} - \frac{(K_{\text{eff}} + C_{\text{bl}}\dot{m}_{\text{bl}})(T_{\text{cr}} - T_{\text{sk}})}{\ell_{\text{b}}} \tag{40}$$

where ρ_b is specific heat of the body (kg/m^3), C_b is specific heat of body (J/kg°C), r_m is remaining metabolic coefficient, Q_m is the volumetric metabolic heat generation (W/m^3), K_{eff} is the effective conductance between core and skin compartments, C_{bl} is specific heat of blood (J/kg°C), and ℓ_b is characteristic length of the body (m) and it is defined as follows

$$\ell_{\text{b}} = \frac{V_{\text{b}}}{A_{\text{D}}} \tag{41}$$

where V_b is the volume of the human body (m^3) and A_D is the nude body surface area (m^2). A_D is described by the well-known DuBois formula (DuBois and DuBois, 1916)

$$A_{\text{D}} = 0.202m^{0.425}l^{0.725} \tag{42}$$

where m and l are body mass (kg) and height (m). Also, remaining metabolic coefficient (r_m) is defined as follows

$$r_{\text{m}} = 1 - \eta - 0.0014(34 - T_{\text{a}}) - 0.0173(5.87 - P_{\text{a}}) \tag{43}$$

where η is external mechanical efficiency (Fanger, 1970) and

$$\eta = \frac{W}{M} \tag{44}$$

It should be noted that the external mechanical efficiency (η) is insignificant in many human thermal response applications. Also, the respiratory heat losses are negligible compared to the total metabolic rate. Thus, the value of r_m can be approximately estimated as unity.
Discretizing Eq. (40) gives

$$T_{cr}^{new} = T_{cr}^{old} + \frac{\Delta t}{(1-\alpha)\rho_b C_b}\left[r_m q_m - \frac{(K_{eff} + C_{bl}\dot{m}_{bl})(T_{cr}^{old} - T_{sk}^{old})}{\ell_b} \right] \quad (45)$$

Eq. (45) is used as a thermal boundary condition for the human body core in the STB model. By implementing this approach, the core temperature is not treated as a constant value and it varies depending on personal/environmental conditions. Therefore, the main governing equation of the STB model is

$$\rho_{ti} C_{ti} \frac{\partial T_{ti}}{\partial t} = k_{ti} \frac{\partial^2 T_{ti}}{\partial x^2} + \rho_{bl} C_{bl} W_{bl} (T_{art} - T_{ti}) + q_m \quad (46)$$

with time-dependent boundary conditions for skin surface and body core

$$\begin{cases} -k_{ti}\dfrac{\partial T_{ti}}{\partial x} = h(T_{ti} - T_a) + \sigma\varepsilon((T_{ti}+273)^4 - (T_a+273)^4) \\ \quad + (3.054 + 16.7hw_{skin})(0.256T_{ti} - 3.37 - P_a), & \text{at skin surface} \\ T_{cr}^{new} = T_{cr}^{old} + \dfrac{\Delta t}{(1-\alpha)\rho_b C_b}\left[r_m q_m - \dfrac{(K_{eff} + C_{bl}\dot{m}_{bl})(T_{cr}^{old} - T_{sk}^{old})}{\ell_b} \right], & \text{at body core} \end{cases} \quad (47)$$

where T_a is the surrounding air temperature, P_a is the water vapor pressure in the air, h is convective heat transfer coefficient, ε is the skin emissivity, and σ is Stefan-Boltzmann constant ($5.67\times10^{-8}\mathrm{W/m^2K^4}$).
It should be noted that some physiological parameters in Eqs. (46) and (47) such as q_m, w_{skin}, α and $\dot{m}_{bl}$ are influenced by thermoregulatory mechanisms. Also, the metabolic heat production is related to the physical activity of the human body and it can be increased by shivering against cold. Hence,

$$q_m = q_{m,act} + q_{m,shiv} \quad (48)$$

and

$$q_{m,shiv} = \frac{19.4 CSIG_{sk} CSIG_{cr}}{\ell_b} \quad (49)$$

Also, skin wettedness (w_{skin}) can be calculated from Eqs. (36) to (39). In addition, α and $\dot{m}_{bl}$ are, respectively, estimated by Eq. (28) and Eq. (29).
Fig. 6 illustrates calculation steps of the STB model (Zolfaghari & Maerefat, 2010). At the beginning, personal and environmental variables are input. Then, the initial temperature distribution in tissue is computed by solving the steady-state bioheat equation under the initial conditions. It should be noted that the steady-state bioheat equation can be obtained by eliminating time dependent derivations in the well-known Pennes bioheat equation.

Afterwards, at each time step, skin and core control signals are calculated and the thermoregulatory parameters are subsequently computed. Then, the temperature distribution in tissue can be obtained by solving the bioheat equation which expressed by Eq. (46) and its related boundary conditions in Eq. (47). Because of the non-linearity of the mentioned equations, they must be solved numerically. Zolafaghari and Maerefat (2010) used the implicit finite difference scheme to find out the temperature distribution of the human body.

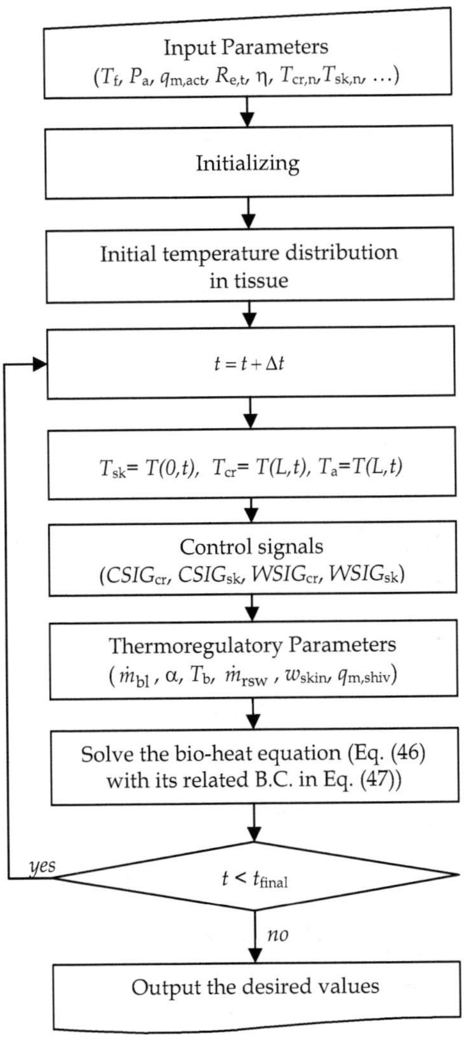

Fig. 6. Flow chart of the STB model calculations

The STB model has been validated against the published experimental and analytical results, where a good agreement has been found. Zolfaghari and Maerefat (2010) showed that the thermal conditions of the human body may be significantly affected by the thermoregulatory mechanisms. Therefore, neglecting the control signals and thermoregulatory mechanisms of the human body can cause a significant error in evaluating the body thermal conditions. Zolfaghari and Maerefat (2010) compared the results of STB model with the results of Pennes bioheat model. This comparison was performed against Stolwijk and Hardy (1966) measured data for a step change in ambient temperature from 30°C/40%RH to 48°C/30%RH for an exposure period of 2 hours followed by 1 hour of environment at 30°C/40%RH. Fig. 7 shows the measured skin temperature data of Stolwijk and Hardy (1966) and the simulation results of Zolfaghari and Maerefat (2010) for STB and Pennes bioheat models. It can be clearly seen that the Pennes bioheat model is not able to accurately estimate the skin temperature under hot environmental conditions. As shown in Fig. 7, the Pennes bioheat model overestimates the value of the skin temperature more than 3.5°C under the mentioned extremely hot conditions. This inaccuracy may be caused by neglecting the thermoregulatory mechanisms such as regulatory sweating and vasomotion in Pennes bioheat model. However, as can be seen in Fig. 7, the results of the STB model are in a good agreement with the experimental results.

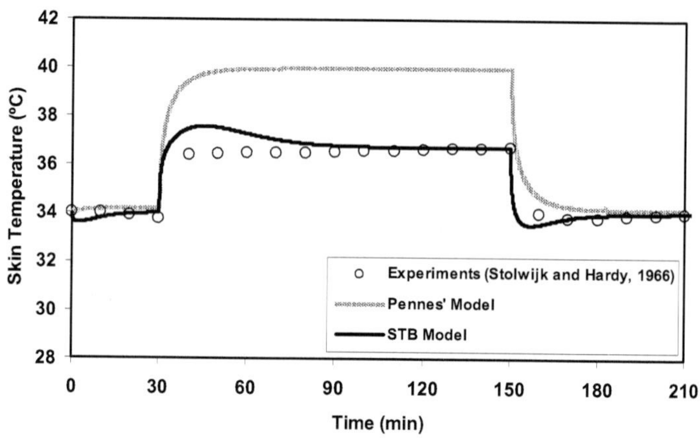

Fig. 7. Comparison of measured (Stolwijk & Hardy, 1966) and simulated skin temperature (Pennes' model and the STB model) during temperature step change

Despite the simplicity of the STB model, it is able to accurately predict the temperature in the cutaneous layer under a wide range of personal/environmental conditions. Also, Zolfaghari and Maerefat (2010) showed that, because of the simplicity and reasonable accuracy of the STB model, it can be widely used in predicting the thermal response of the human body under both transient and steady-state environments. Therefore, the STB model can be utilized for evaluating thermal comfort of the human body in variant personal/environmental conditions. Furthermore, the model validations show that the model results are sufficiently reliable under extremely hot/cold conditions (Zolfaghari & Maerefat, 2010).

5. Conclusion

In this chapter, a brief review of bioheat transfer models (e.g. Pennes' bioheat equation, Wulff Continuum Model, Klinger Continuum Model, Chen-Holmes model, Weinbaum-Jiji-Lemons vascular model and simplified Weinbaum-Jiji vascular Model) has been presented. Also, a new simplified thermoregulatory bioheat (STB) model (Zolfaghari & Maerefat, 2010) has been briefly introduced in the present chapter. The STB model has been developed by combining the well-known Pennes' equation with Gagge's two-node model for evaluating the temperature in the cutaneous layer under a wide range of personal/environmental conditions. This model considers the effects of thermoregulatory mechanisms of the human body by defining the thermal control signals of the body. The STB model has been validated against the published experimental and analytical results, where a good agreement has been found (Zolfaghari & Maerefat, 2010). Therefore, the results of the STB model are sufficiently reliable for estimating the cutaneous themperature under both transient and steady-state thermal conditions.

6. References

Chato, J.C. (1980). Heat Transfer to Blood Vessels, *ASME Journal of Biomechanical Engineering*, Vol. 102, pp. 110-118, ISSN 0148-0731

Chen, M.M. & Holmes, K. R. (1980). Microvascular Contributions in Tissue Heat Transfer, *Annals of the New York Academy of Sciences*, Vol. 335, pp. 137–150, ISSN 0077-8923

Cho, Y.I. (1992). *Bioengineering Heat Transfer*, In: *Advances in Heat Transfer*, J.P. Hartnett & T.F. Irvine, (Ed.), Academic Press, Inc., ISBN 978-0-12-020022-8, San Diego, USA

Campbell, I. (2008). Body Temperature and its Regulation. *Anaesthesia & Intensive Care Medicine*, Vol. 9, No. 6, pp. 259-263, ISSN 1472-0299

Datta, A.K. (2002). *Biological and Bioenvironmental Heat and Mass Transfer*, Marcel Dekker, Inc., ISBN 978-0-8247-0775-3, New York, USA

Doherty, T. & Arens, E.A. (1988). Evaluation of the Physiological Bases of Thermal Comfort Models. *ASHRAE Transaction*, Vol. 94, pp. 1371-1385, ISSN 0001-2505

DuBois D. & DuBois E.F. (1916) A Formula to Estimate Approximate Surface Area, if Height and Weight are Known. *Archives of Internal Medicine*, Vol. 17, pp. 863-871, ISSN 0003-9926

Fanger P.O. (1970). *Thermal Comfort Analysis and Applications in Environmental Engineering*, McGraw-Hill, ISBN 0-07-019915-9, New York, USA

Jiji, L.M. (2009). *Heat Conduction*, Third Edition, Springer, ISBN 978-3-642-01266-2, Berlin, Germany

Kaynakli O. & Kilic M. (2005). Investigation of indoor thermal comfort under transient conditions. *Building and Environment*, Vol. 40, No. 2, pp. 165-174, ISSN 0360-1323

Klinger, H.G. (1974). Heat transfer in perfused biological tissue. I. General theory. *Bulletin of Mathematical Biology*, Vol. 36, pp. 403-415, ISSN 1522-9602

Kreith, F. (2000). *The CRC Handbook of Thermal Engineering*, CRC Press, ISBN 978-0-8493-9581-9, Boca Raton, USA

Kutz, M. (2009). *Biomedical Engineering and Design Handbook*, Second Edition, McGraw-Hill, ISBN 978-0-07-170472-4, New York, USA

Lv Y.G. & Liu J. (2007). Effect of transient temperature on thermoreceptor response and thermal sensation. *Building and Environment*, Vol. 42, pp. 656-64, ISSN 0360-1323

Minkowycz, W.J., Sparrow, E.M. & Abraham, J.P. (2009). *Advances in Numerical Heat Transfer: Volume 3*, CRC Press, ISBN 978-1-4200-9521-0, Boca Raton, USA

Pennes, H.H. (1948). Analysis of Tissue and Arterial Blood Temperatures in the Resting Forearm, *Journal of Applied Physiology*, Vol. 1, pp. 93-122, ISSN 1522-1601

Raaymakers, B.W., Kotte, A.N.T.J. & Lagendijk, J.J.W. (2009). Discrete Vasculature (DIVA) Model Simulating the Thermal Impact of Individual Blood Vessels for In Vivo Heat Transfer, In: *Advances in Numerical Heat Transfer: Volume 3*, Minkowycz, W.J., Sparrow, E.M. & Abraham, J.P., pp. 121-148, CRC Press, ISBN 978-1-4200-9521-0, Boca Raton, USA

Sharma, K.R. (2010). *Transport Phenomena in Biomedical Engineering*, McGraw-Hill, ISBN 978-0-07-166398-4, New York, USA

Stolwijk J.A.J. & Hardy J.D. (1966). Temperature regulation in man — A theoretical study. *Pflügers Archiv European Journal of Physiology*, Vol. 291, No. 2, pp. 129-62, ISSN 0031-6768

Vafai, K. (2011). *Porous Media: Applications in Biological Systems and Biotechnology*, CRC Press, ISBN 978-1-4200-6541-1, Boca Raton, USA

Wan, X. & Fan, J. (2008). A Transient Thermal Model of the Human Body-Clothing-Environment System. *Journal of Thermal Biology*, Vol. 33, pp. 87-97, ISSN 0306-4565

Weinbaum, S., Jiji, L.M. & Lemons, D.E. (1984). Theory and experiment for the effect of vascular microstructure on surface tissue heat transfer. Part I. Anatomical foundation and model conceptualization. *ASME Journal of Biomechanical Engineering*, Vol. 106, pp. 321-330, ISSN 0148-0731

Weinbaum, S. & Jiji, L.M. (1985). A new simplified bioheat equation for the effect of blood flow on local average tissue temperature. *ASME Journal of Biomechanical Engineering*, Vol. 107, pp. 131-139, ISSN 0148-0731

Wulff, W. (1974). The Energy Conservation Equation for Living Tissues. *IEEE Transactions-Biomedical Engineering*, vol. 21, pp. 494-495, ISSN 0018-9294

Yigit, A. (1999). Combining Thermal Comfort Models. *ASHRAE Transactions*, Vol. 105, pp. 149-158, ISSN 0001-2505

Zolfaghari, A. & Maerefat, M. (2010). A New Simplified Thermoregulatory Bioheat Model for Evaluating Thermal Response of the Human Body to Transient Environment. *Building and Environment*, Vol. 45, No. 10, pp. 2068-2076, ISSN 0360-1323

10

The Manufacture of Microencapsulated Thermal Energy Storage Compounds Suitable for Smart Textile

Salaün Fabien[1,2]
[1]*Univ Lille Nord de France,*
[2]*ENSAIT, GEMTEX;*
France

1. Introduction

Smart textiles are able to sense electrical, thermal, chemical, magnetic, or other stimuli from the environment and adapt or respond to them, using functionalities integrated into the textile structure. As an important consideration in active wear, clothing comfort is closely related to microclimate temperature and humidity between clothing and skin.

Since the end of the 80's, functional textiles have been developed to enhance textile performances according to the consumers' demand and to include a large range of properties with a higher added value. One of the possible ways to manufacture functional or intelligent textile products is the incorporation of microcapsules or the use of microencapsulation processes for textile finishing. Thermal storage by latent heat was early recognised as an attractive alternative to sensible heat storage to improve the thermal performance of clothing during the modifications of environmental temperature conditions. Early efforts in the development of latent heat storage used organic phase change materials (PCMs) for this purpose. PCMs are entrapped in a microcapsule of a few micrometers in diameter to protect them and to prevent their leakage during its liquid phase. In the two past decades, microencapsulated Phase Change Materials have drawn an increasing interest to provide enhanced thermal functionalities in a wide variety of applications. When the encapsulated PCMs is heated above its phase change temperature, it absorbs heat as it goes from a solid state to a liquid state or during a solid to solid transition. It can be applied to clothes technology, building insulation, energy storage as well as to coolant liquids. On a more general basis, it can be used to design a broad variety of thermal transient regimes. Recent research has investigated the incorporation of organic PCMs directly into the fabric fibers (Zhang et al. (2006)) or coated on the substrate surface (Choi et al. (2004), Shin et al. (2005)), creating functional and effective textile elements which can significantly affect thermal insulation. Moreover, thermal comfort sensation is closely related to microclimate temperature and humidity. Thus, Fan & Cheng (2005) denoted that a lower moisture absorption rate is beneficial to thermal comfort. Therefore, the thermal functional performance of a thermoregulated fabric is not only influenced by the latent heat of PCMs but also by the design of the textile structure.

2. Classification of heat storage materials to melting temperature range and textile application

Among the various heat storage technologies available, i.e. sensible heat based on increasing the temperature without changing the phase of the material, latent heat based on the transition of a material according to the temperature and thermo-chemical heat (or heat of reaction) based on the thermophysics of the reactions; thermal heat storage in the form of latent heat of phase change seems to be particularly attractive in textile fields. A variety of PCMs are well-known for their thermal characteristics relating to their phase change stage. These compounds possess the ability to absorb and store large amounts of latent heat during the heating process and release this energy during the cooling process. Thus, materials being converted from solid to liquid, from liquid to solid or solid 1 to solid 2 states are suitable to be used in the manufacture of thermoregulated textiles. The selection of a PCMs formulation depends typically on the required phase change temperature depending on end use. Indeed, PCMs should react to changes in temperature of both the body and the outer layer of the garment when they are incorporated in the textile substrate. Thus, for textile applications, PCMs with a phase change within the ambient temperature and comfort range of humans are suitable, i.e. in a temperature range from 15°C to 35°C.

Among the various ways to store energy, the most attractive form is latent heat storage in phase change material, because of the advantages of high storage capacity in a small volume and charging/discharging heat from the system at a nearly constant temperature (Abhat, 1983). Thermal storage by latent heat was recognised early as an attractive alternative to sensible heat storage to improve the thermal performance of clothing during the changes of environmental temperature conditions. The latent heat associated with a first-order phase transition provides a mechanism for the thermal energy storage. The phase changes comprise predominantly solid-liquid transitions for thermal storage applications in textiles.

The use of PCMs is linked to its latent heat of fusion for thermal storage. The latent heat of fusion of a material is substantially greater than its sensible heat capacity. Stated differently, the amount of energy that a material absorbs upon melting or releases upon freezing is much greater than the amount of energy which it absorbs with a weak variation of temperature. Upon melting or freezing, a PCM absorbs and releases substantially more energy than a sensible heat storage material which is heated or cooled to the same temperature range. The storage capacity of PCMs (Q) equals the phase change and the sensible heat stored at the phase change temperature (1). Thus, the latent storage is always increased by a significant extent by their sensible storage capacity. Furthermore, during the complete phase change process, the temperature of the PCMs as well as the surrounding media remains nearly constant (Feldman et al., 1986).

$$Q = m.\int_{T_{initial}}^{T_{Tr}} C_p(T).dT + m.\Delta H + m.\int_{T_{Tr}}^{T_{end}} C_p(T).dT \qquad (1)$$

For a textile application, the following PCMs properties should be required, i.e. a high value of the heat of fusion and specific heat per unit volume and weight; a melting point in the application range (between 15°C to 35°C); a high thermal conductivity; a chemical stability and non-corrosiveness; PCMs should not be hazardous, non-flammable or poisonous; PCMs should have a reproducible crystallisation without decomposition; PCMs should present a small supercooling degree and high rate of crystal growth; they should have a small volume variation during the phase change process; and PCMs should be sufficiently abundant at a low cost.

The most common PCMs, with a phase change temperature suitable for textile application, can be divided into two groups, i.e. organic compounds such as paraffins or linear alkyl hydrocarbon and non paraffinic materials (hydrocarbon alcohol, hydrocarbon acid, polyethylene or polytetramethylene glycol, aliphatic polyester...), and inorganic compounds such as hydrated inorganic salts, eutectics or polyhydric alcohol-water solution (Zhang, 2001).

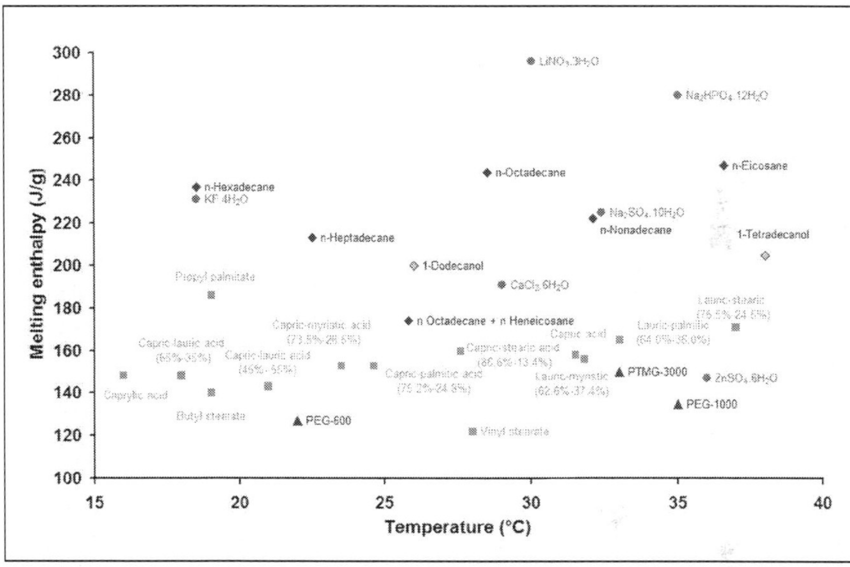

Fig. 1. Temperature ranges and corresponding melting enthalpy of suitable PCMs for textile applications

2.1 Organic phase change materials

Organic solid–liquid PCMs include paraffin, alkyl esters and acids, polyethylene glycol and its derivatives.

2.1.1 Polyhydric alcohols

Polyhydric alcohol, or plastic crystal, are solid-solid phase change materials, and even if they are not suitable for textile application since their phase change temperature is higher than the upper end-use limit, they present small volume change, lower undercooling, no phase separation and no leakage. Amongst them, pentaerythritol ($C(CH_2OH)_4$), trimethylol ethane ($(CH_3)C(CH_2OH)_3$), neopentyl glycol ($(CH_3)_2C(CH_2OH)_2$), their NH_2-substituted compounds and 2-amino-2-methyl-1,3-propanediol ($(CH_2CH_2OH)_2CCH_3NH_2$), 2-amino-2-hydroxymethyl-1,3-propanediol ($(CH_2CH_2OH)_3CNH_2$), and their binary mixtures have an endothermic or exothermic effect under their melting point and are cited as potential candidate for thermal energy storage. Furthermore, the phase change temperatures and heat of binary mixture of polyhydric alcohol are lower than of the raw material. Thus, Wang et al. have found that the phase change of neopentylglycol/pentaerythritol and trihydroxy methylaminomethane were between 30 and 40°C (Wang et al., 2000).

2.1.2 Polyethylene glycols

In the field of organic PCMs, polyethylene glycols (PEGs) (-O-CH_2-CH_2-)n and their derivatives, such as poly(tetra methylene glycol) (PTMG), have been widely used as solid-liquid thermal energy storage materials since they present high transition enthalpy depending on their molecular weight, congruent melting behaviour, wide range of transition temperature to obtain suitable melting point, in contrast to inorganic compounds. Furthermore, their melting points and enthalpies depend on their molecular weight. Thus, PEGs with a molecular weight comprised between 600 and 1500 $g.mol^{-1}$ have melting points in the temperature range between 17°C to 35°C and a melting enthalpy between 120 to 140 $J.g^{-1}$. The use of PTMG as phase change material in textile field is limited due to a supercooling phenomena even if the PTMG 3000 show a melting point about 33°C for a melting enthalpy about 150 $J.g^{-1}$.

Furthermore, great efforts were made to prepare new PEG solid-solid PCMs with another compound to keep the material in solid state during the phase change; and thus prevent its leakage. Nevertheless, these compounds lose their thermal properties after several heating-cooling cycles.

2.1.3 Linear chain hydrocarbons

Linear hydrocarbons, C_nH_{2n+2}, with a melting temperature between 16°C to 38°C, are the most common PCMs used for textiles. Thus, *n*-alkanes were preferred due to their high latent heat, and they are chemically inert, non-toxic, non-corrosive and non-hygroscopic. Thus, *n*-eicosane (Sarier & Onder, 2007a), *n*-octodecane (Pause, 2001; Shim et al., 2001; Choi et al., 2004; Li & Zhu, 2004; Chung & Cho, 2004; Zhang et al., 2005b, 2006; Sarier & Onder; 2007a, 2007b) and *n*-hexadecane (Shim et al., 2001; Sarier & Onder, 2007a, 2007b) are the most PCMs chosen to be applied. In these studies, pure compound was used as PCM material and therefore the phase change occurs in a narrow temperature change at the melting temperature of the *n*-alkane. The use of binary mixture of *n*-alkanes allows to adjust the melting point (or phase change temperature) by modifying the composition, that also leads to a decrease in the overall latent heat of fusion (Kumano et al., 2005). Furthermore, some of the binary mixture, i.e. *n*-hexadecane/ *n*-octadecane (Métivaud, 1999), *n*-octadecane/ *n*-hexadecane (Métivaud, 1999), or *n*-tetradecane/ *n*-hexadecane (Bo et al., 1999) have a solid/liquid or a solid/solid transition which takes place in a relatively narrow temperature range. The use of the binary mixture of *n*-hexadecane/ *n*-eicosane allows to widen this temperature change and can be suitable for textiles application (Salaün et al., 2008a, 2010a). Nevertheless, this widening is accompanied by a decrease of the latent heat which can be caused by the formation of a concentration distribution in the liquid film due to small scale segregation (Kumano et al., 2005). The binary mixture of *n*-hexadecane/ *n*-eicosane was also studied by Sarier et al., they concluded that to enhance the thermal capacity of fabrics, it should be better to use a combination of microcapsules containing different types of PCMs rather than those including a mixture of them (Sarier & Onder, 2007a). They have observed that the mixture was satisfactory to provide a buffering effect against temperature changes and to regulate the temperature at the target of desired value.

2.1.4 Fatty acids

Fatty acids are friendly PCMs and show several advantages such as a high phase change enthalpy, congruent melting and freezing behaviour and good chemical and thermal stability. Even if capric, lauric, myristic, palmitic or stearic acids have good thermal and heat transfer characteristics, their melting points between to 32°C to 61°C are quite high for textile energy storage (Feldman et al., 1989; Sari, 2003). Nevertheless, the eutectics of capric-lauric, capric-palmitic, lauric-myristic, lauric-palmitic, and lauric-stearic binary mixture, with melting points of 18.0, 21.8, 32.7, 34.2 and 34°C, and heats of melting of 120, 171, 150, 167 and 160 $J.g^{-1}$, respectively seem to be suitable PCMs for textile applications (Sari, 2005; Sari & Karaipekli, 2008).

Zeng et al. have analyzed the thermal properties of a binary mixture of palmitic acid and 1-tetradecanol. The eutectic composition mixture (point at 29°C and melting enthalpy of 143 $J.g^{-1}$) could be viewed as a new phase change material with large thermal energy storage capacity (Zeng et al., 2009).

Although, the thermal properties of fatty acid ester were few studied, these compounds show a solid-liquid phase transition within a narrow temperature range, which should be promising for thermal energy storage (Li & Ding, 2007a, 2007b; Alkan et al., 2008; Sari et al., 2009a). Thus, for isopropyl stearate, butyl steareate, and vinyl stearate and, the thermal characteristics determined by Feldman et al. are as follow : melting point : 14, 19, and 27°C , respectively and the melting enthalpy range from 120 to 142 $J.g^{-1}$ (Feldman et al., 1986; Feldman et al., 1995).

2.2 Hydrated inorganic salts

Hydrated inorganic salts, e.g. salt with 'n' water molecules, are attractive materials with their relatively high latent heat per unit weight (from 150 to 300 $J.g^{-1}$), high thermal conductivity (~0.5 $W.M^{-1}.K^{-1}$), non flammability property and moderate costs. Nevertheless, their use is mainly limited by their incongruent melting, supercooling behaviour when there are incorporated in the textile substrate, which can occur after several phase change cycle. If $LiNO_3.3H_2O$ is stable during at least 25 cycles, phase segregation of $CaCl_2.6H_2O$ or $Na_2SO_4.10H_2O$ should be observed after only few cycles. Thus, it was suggested to overcome these problems to use an extra water amount to limit the formation of heavy anhydrous salt, which results in the reduction of the storage density (Salaün et al., 2010b; Biswas, 1977). The use of thickening to overcome phase segregation, such as betonite, or nucleating agent to limit the supercooling phenomena, have also be suggested, but in all cases the thermal properties of the inorganic PCMs are reduced. Furthermore, when inorganic PCMs are encapsulated, the process as well as the morphology of the resulting capsules affects the thermal properties of the materials (Salaün et al., 2010b).

2.3 Conclusion

Many substances have been studied as potential PCMs with regard to their typical range of melting and melting enthalpy, within the human comfort range of 18°C to 35°C. This kind of materials absorbs and emits heat while maintaining a nearly constant temperature, and thus they are very effective since they can store 5 to 14 times more heat per unit volume than sensible storage materials. Each type of PCMs presents some advantages or disadvantages. On the one hand, organic PCMs have excellent stability concerning the thermal cycling, but they are flammable and their melting enthalpy and storage density are relatively low compared to salt hydrates. On the other hand, the main problem with inorganic PCMs are their corrosiveness, their cycling stability and the existence of a subcooling phenomena which be reduced by adding into the materials some nucleating agents.

3. Microencapsulation

For a textile use, PCMs should be surrounded by a protective shell to avoid any liquid migration during the phase change step and therefore to preserve PCMs in the polymer shell as long as possible through the heating/cooling cycles. Besides, the encapsulation step allows to increase the heat transfer area, to decrease the reactivity of the core materials, to limit the interference with other materials parameters, to enhance the low heat conductivity and the manipulation of PCMs becomes easier. Microencapsulated PCMs have attracted more and more attention since the end of the 1980s (Kaska & Chen, 1985; Bryant & Colvin, 1988). It was progressively established as an interesting alternative to the traditional materials where the imprisoned air was the principal insulating element. The choice of the polymer for the membrane synthesis of the microcapsules is linked to the considered application and the required material processes. In the textile field, the polymers used should have good thermomechanical properties to resist to the thermal and mechanical requests during of the implementation processes, and during the daily use by the wearer. The use of microencapsulated phase change material in textile requires a high thermal stability of these particles that is influenced by the shell and the core materials. A review on the various techniques available for the microencapsulation of Phase Change Materials is presented. The methods can be classified as being either physical or chemical based.

3.1 Principle and generalities

Microencapsulation refers to the formation of polymeric particles with a range of diameters from nanometers to millimeters, and which exhibits several kinds of morphologies, in which the mains are so called microcapsules and microspheres (Benita, 1996). Microspheres consist of polymeric network structure in which an active substance is enclosed, whereas in microcapsules or core/shell structures exhibit a reservoir structure, i.e. the core substance is surrounded by a polymeric layer. Even if, microencapsulation principle was firstly introduced by Green and Schleicher in the 1950s to create microcapsules for carbonless copying paper (Green & Schleicher, 1956 a & b, as cited in Arshady, 1990), textile applications were mainly developed in labs since the early 1990's and actually we can see much more applications in the textile industry for developing new properties and added values (Nelson, 2002).

Although, there are more than 200 microencapsulation methods described in scientific literature and patent, most of them include three basic steps, i.e. enclosure of core component, formation of the microparticles, and hardening of it. These methods are generally divided or classified in three main groups, which are based on the mechanisms of microparticle formation, such as mechanical, chemical and physico-chemical processes. The choice of a method rather than another is based on the cost of processing, the use of organic solvents for health and environment considerations. Nevertheless, the microencapsulation processes used in textile industry are generally focused on chemical and physico-chemical methods, because mechanical processes lead to the formation of particles with a mean diameter higher than 100 μm. Even if microparticles with mean size distribution in the range of 20 to 40 μm should be useful for textile application (Nelson, 2001), very small microparticles (mean diameter ranging from 1 to 10 μm) are preferably incorporated within textile fibers (Cox, 1998; Bryant, 1999) but according to Colvin (Colvin, 2000) particles with a mean diameter less than 1 μm are suitable for fiber fabrication; and larger microcapsules (mean diameter up to 100 μm) should be used into foams or coated formulation (Colvin & Bryant, 1996; Pushaw, 1997).

Therefore, the most suitable methods include interfacial, *in-situ* and suspension polymerization methods for chemical processes and simple or complex phase coacervation for physico-chemical processes, and spray drying for the mechanical processes (Borreguero et al., 2011; Hawlader et al., 2000 & 2003; Teixeira et al., 2004). In all cases, the microencapsulation process includes two main steps, e.g. emulsification step which determines the size and the size distribution of the microcapsules; and the formation of the capsules. The emulsification step may be influenced at once by physical parameters such as apparatus configuration, stirring rate and volume ratio of the two phases, and by physicochemical properties such as interfacial tension, viscosities, densities and the chemical compositions of the two phases. The formation of microcapsules is greatly affected by the surfactant, which influences not only the mean diameter but also the stability of the dispersion. The surfactants used in the system have two roles, one to reduce the interfacial tension between oil and aqueous phases allowing formation of smaller microcapsules and one to prevent coalescence by its adsorption on the oil-water interface and therefore by forming a layer around the oil droplets. The second step, or the synthesis of a core/shell particle or other possible morphologies, is mainly governed by the kinetic factors, i.e. the ability of the monomers, pre-polymer or polymer to react or to cross-link, and thermodynamic factors, i.e. the minimum total free energy exchange in the system. The spreading coefficient should be used to predict the engulfment of PCMs within polymer in a continuous immiscible phase (Torza & Mason, 1970; Sundberg & Sundberg, 1993, Loxley & Vincent, 1998; Sánchez et al., 2007; Salaün et al., 2008a).

3.2 State of the art in microencapsulated PCM's

The microencapsulation of phase change materials has been widely investigated since the late 1970s to provide significantly enhanced thermal management for various application fields. Thus, preliminary work on microencapsulation of phase change materials was begun at General Electric under the sponsorship of the U.S Department of Energy. This was the first such attempt and was directed toward the development of a two-component heat transfer fluid for use in solar systems for residential applications (Colvin & Mulligan, 1986).

The microencapsulation of PCMs provides some benefit effects on the heat storage, since the shape of the storage material is arbitrary, temperature gradient is more favourable and PCMs lifetime increases. Furthermore, the product is always in a dried state since the phase change occurs within the coating material. The microcapsules' embedment in heat storage is relatively simple, and the various PCMs formulation with different melting temperatures should be used to obtain the desired temperature in which phase change occurs. The use of microcapsules increase the specific exchange area, which is provided for effective heat transfer; and thus large quantities of thermal energy can be stored and released at a relatively constant temperature without significant volume change since in the particles, the size remains constant and sufficient space or free volume exist. Finally, the encapsulation step allows to reduce or to avoid problems as such supercooling and phase separation (Hawlader et al., 2003). Therefore, to maintain the reversibility of the PCMs to 25.2 wt.%) and therefore the enthalpy heat (28 J.g-1), they proposed to synthetize styrene-divinylbenzene (DVB) co-polymer as shell material rather than melamine-formaldehyde (You et al., 2010).

4.3.2 Coating

For a coating process, the wetted mPCMs are dispersed throughout a polymeric binder, a surfactant, a dispersant, an antifoam agent and a thickener, as preconized by Zuckerman et al. (1997). All the common coating processes, such as knife over roll (Pause, 1995), screen printing (Lottenbach & Sutter, 2002; Choi et al., 2004), pad-dry-cure (Shin et al., 2005), knife over air, and gravure and dip coating (Bryant & Colvin, 1996; Zuckerman et al., 2001) should be used. Besides, even if the method for manufacturing coating composition was widely described in the patent literature, nevertheless few papers published in the literature give account of the formulation of coating, finishing of fabrics and therefore the evaluation of their characteristics more specially thermal and durability properties. Furthermore, the use of polymeric binder has some drawnbacks, since the incorporated amount should be enough to obtain permanent linkage which can alter the fabric properties such as drape, air permeability, breathability, thermal resistance, softness and tensile strength can be affected adversely as the percentage of binder add-on increases. Salaün et al. (2009a) have proposed a thermodynamical method based on the comparison of the surface energy component to select the most suitable polymeric binder for melamine-formaldehyde microcapsules. They also determined that the polyurethane binder plays a main role during the thirty seconds of a cold to warm transition allowing to delay the temperature increase. Furthermore, the thermoregulating response depends on the surface deposited weight and the mass ratio binder to microcapsules. Thus, an interesting cooling effect is found for 20 $g.m^{-2}$ of binder and from 40 $g.m^{-2}$ of mPCMs. And a mass ratio binder to mPCMs taken between 1:2 and 1:4 is suitable to manufacture thermoregulating textile (Salaün et al., 2010c). Choi et al. (2004) have coated a polyester fabric by screen printing and knife over role, and they have denoted that the first process allowed to obtain less stiff and hard fabric, and higher air permeability and lower moisture absorption than the other one. Chung et al. (2004) have succeeded in the manufacture of vapour-permeable, water repellent and thermoregulated fabrics, in coating melamine-formaldehyde microcapsules mixed in to polyurethane binder. The amount of heat was kept constant (~14 $J.g^{-1}$) after 30 launderings, and even if the water repellence property was the same with and without mPCMs, they have also observed a decrease for the water resistance with the presence of mPCMs. On the other hand, Sánchez et al. (2010) have manufactured a cotton coating fabric (with an enthalpy heat about 7.6 $J.g^{-1}$) with 35% of polystyrene microcapsules related to the coating formulation. They observed a thermoregulatory effect of 8.8, 6.3, 5.6 and 2.5°C after 6, 12, 44 and 75 s. Furthermore, their results after rubbing and ironing tests were higher than those of Kim & Cho (2002).

4.3.3 Spinning

The incorporation of mPCMs within fibers was first patented in 1988 (Bryant & Colvin, 1988). This kind of manufacture presents some advantages as a permanent incoporation and no need to modify subsequent processing and after treatment. Thus, the process to obtain artificial or synthetic fibers containing mPCMs can be achieved either by wet spinning for e.g. polyacrylonitrile (Lennox, 1998), polyacrylonitrile-vinylidene chloride (Zhang et al., 2005b), acrylic fiber (Cox, 2001), or by melt spinning (Gao et al., 2009) for polypropylene (Leskovšek et al., 2004) or polyethylene fiber (Zhang et al., 2005b). Although, the touch, drape, softness and color were not modified by the spinning process, the thermal heat capacity of the obtained fiber was limited to a low microPCMs loading content. Up to now, the main envisaged way to incorporate microcapsules directly into fibers was solvent spinning rather than melt-spinning because of their tendency to aggregate themselves under dried powder form, and also sometimes because of the high temperatures involved in melt-spinning which could degrade the microcapsules. Thus, since 1997, polyacrylonitrile fibers containing 7 wt.% of microcapsules were put onto market, evenf if their storage capacity is relatively low (about 8-12 $J.g^{-1}$). Since, Leskovšek et al. (2004) have shown the possibility of introducing microcapsules into polypropylene fibers in the presence of a lubricant. In their study, the added amount of microcapsules was relatively small (1%, 2% and 5%) and the PP was in the form of a single monofilament yarn with a large diameter. They denoted that the microcapsules did not impair the mechanical properties of the fibres. More recently, in 2005, Zhang and co-workers have succeeded in increasing the loading mPCMs content up to 24% in PP fibers, and they denoted that these fibers exhibited acceptable mechanical properties as the content in mPCMs does not exceed 20 wt.%, for an enthalpy about 11 $J.g^{-1}$. Wet-spun thermoregulated polyacrylonitrile-vinylidene chloride fibers containing 30 wt.% of mPCMs were also obtained by Zhang et al.(2006). Gao et al. (2009) have shown the possibility to incorporate the mPCMs during the polymerization step of Acrylonitrile–methyl acrylate copolymer before the melt-spinning process. Thus, they obtained thermoregulated fibers containing 20 and 25 wt.% of mPCMs for a melting enthalpy about 21 to 25 $J.g^{-1}$.

5. Conclusion

One of the advantages of microencapsulated PCMs is their use for the fabrication of any kind of active smart textiles, meaning which sense and react to the conditions or simulus and thus this technology is attractive to improve clothing comfort and this for a wide and diversified market. MPCMs were firstly developed for use in space suits and glove to protect astronauts from extreme temperature fluctuation during activities in space, specialy as a glove liner. Actually, mPCMs are widely used to improve the thermal performance of active wear garment (outwear, sportwear, underwear), shoes linning, accessories, bedding products (comforters, pillows, or mattress covers), and in various activity fields such as building, medical, to enhance the thermo-physical comfort of surgical clothing such as gownd, caps and gloves; or as a product which support the effort to keep the patient warm during an operation by providing insulation tailored to the body's temperature; in automotive sector to stabilise the interior temperature or in the seat to reduce unpleasant temperature changes and to keep the skin temperatures more constant. In fact, more than 150 textile companies or brand use mPCMs licensed by Outlast technology.

On the other hand, a new kind of microcapsules, interacting with the absorbed water to promote a cooling effect, has been recently developed to enhance mass transfer. This technology is based on the thermo-chemical heat storage, which present numerous advantages, i.e. high energy density, high discharge power due to high reaction enthalpy,

Deveci, S.S. & Basal, G. (2009). Preparation of PCM Microcapsules by Complex Coacervation of Silk Fibroin and Chitosan. *Colloid Polymer Science*, Vol. 287, No.12, (December 2009), pp.1455-1467, ISSN 0303-402X

Fan, J. & Cheng, X.Y. (2005). Heat and Moisture Transfer with Sorption and Phase Change Through Clothing Assemblies, Part II: Theoretical Modeling, Simulation, and Comparison with Experimental Results. *Textile Research Journal*, Vol. 75, No.3, (March 2005), pp.187-196, ISSN 0040-5175

Fan, Y.F., Zhang, X.X., Wu, S.Z. & Wang, X.C. (2005). Thermal stability and permeability of microencapsulated n-octadecane and cyclohexane. *Thermochimica Acta*, Vol.429, No.1, (May 2005), pp.25-29, ISSN 0040-6031

Fang, Y., Kuang, S., Gao, X. & Zhang, Z. (2008). Preparation and characterization of novel nanoencapsulated phase change materials. *Energy Conversion and Management*, Vol.49, No.12, (December 2008), pp. 3704-3707, ISSN 0196-8904

Feldman, D., Shapiro, M.M. & Banu, D. (1986). Organic phase change materials for thermal energy storage. *Solar Energy Materials*, Vol.13, No.1, (January 1986), pp. 1-10, ISSN 0165-1633

Feldman, D., Shapiro, M.M., Banu, D. & Fuks, C.J. (1989). Fatty acids and their mixtures as phase-change materials for thermal energy storage. *Solar Energy Materials*, Vol.18, No.3-4, (March 1989), pp. 201-216, ISSN 0165-1633

Feldman, D., Banu, D. & Hawes, D. (1995). Low chain esters of stearic acid as phase change materials for thermal energy storage in buildings. *Solar Energy Materials and Solar Cells*, Vol.36, No.3, (March 1995), pp. 311-322, ISSN 0927-0248

Gao, X.Y., Han, N., Zhang, X.X. & Yu, W.Y. (2009). Melt-processable acrylonitrile–methyl acrylate copolymers and melt-spun fibers containing MicroPCMs. Journal *of Materials Science*, Vol.44, No.21, (November 2009), pp. 5877-5884, ISSN 1573-4803

Hawlader, M.N.A., Uddin, M.S. & Zhu, H.J. (2000). Preparation and evaluation of a novel solar storage material: Microencapsulated paraffin. International Journal of Sustainable Energy, Vol.20, No. 4, (September 2000), pp. 227-238, ISSN 0142-5919

Hawlader, M.N.A., Uddin M.S., & Khin, M.M. (2003). Microencapsulated PCM thermal-energy storage system. Applied Energy, Vol.74, No.1-2, (January-February 2003), pp. 195-202, ISSN 0306-2619

Jahns, E. (n.d.). Microencapsulated phase change material, available from: http://www.fskab.com/annex10/WS_pres/Jahns.pdf

Jin, Y., Lee, W., Musina, Z. & Ding, Y. (2010). A one-step method for producing microencapsulated phase change materials. *Particuology*, Vol.8, No.6, (December 2010), pp. 588-590, ISSN 1674-2001

Kaska, K.E. & Chen, M.M. (1985). Improvement of the Performance of Solar Energy of Waste Heat Utilization Systems using Phase Change Slurry as Enhanced Heat Transfer Storage Fluid. *Journal of Solar Energy Engineering*, Vol.107, No.3, (August 1985), pp. 229-236, ISSN 0199-6231

Kim, J. & Cho, G. (2002).Thermal storage/release, durability, and temperature sensing properties of thermostatic fabrics treated with octadecane-containing microcapsules. *Textile Research Journal*, Vol.72, No.12, (December 2002), pp. 1093-1098, ISSN 0040-5175

Kumano, H., Saito, A., Okawa, S., Takeda, K. & Okuda, A. (2005). Study of direct contact melting with hydrocarbon mixtures as the PCM. *International Journal of Heat and Mass Transfer*, Vol.48, No.15, (July 2005), pp. 3212-3220, ISSN 0017-9310

Lan, X., Tan, Z., Zou, G., Sun L. & Zhang T. (2004). Microencapsulation of n-eicosane as energy storage material. *Chinese Journal of Chemistry*, Vol.22, No.5, (May 2004), pp. 411–414, ISSN 1614-7065

Lane, G.A. (1980). Low temperature heat storage with phase change materials. *International Journal of Ambient Energy*, Vol.1, No.3, (July 1980), pp. 155–168, ISSN 0143-0750

Lennox, K.P. (1998). Outlast Technologies adapts Space-age technology to keep us comfortable, *Technical Textile International*, Vol.7, No.7, (July-August 1998), pp. 25-26, ISSN 0964-5993

Leskovšek, M., Jedrinovic, G. & Stankovic-Elseni, U. (2004). Properties of propylene fibres with incorporated microcapsules, *Acta Chimica Slovenica*, Vol.51, No.4, (December 2004), pp.699-715, ISSN 1318-0207

Li, W.D. & Ding, E.Y. (2007a). Preparation and characterization of a novel solid–liquid PCM: Butanediol di-stearate. *Material Letters*, Vol.61, No.7, (March 2007), pp. 1526–1528, ISSN 0167-577X

Li, W.D. & Ding, E.Y. (2007b). Preparation and characterization of a series of diol di-stearates as phase change heat storage materials. *Material Letters*, Vol. 61, No.21, (August 2007), pp. 4325–4328, ISSN 0167-577X

Li, W., Zhang, X.X., Wang, X.C. & Niu, J.J. (2007). Preparation and characterization of microencapsulated phase change material with low remnant formaldehyde content. *Materials Chemistry and Physics*, Vol.106, No.2-3, (December 2007), pp.437-442, ISSN 0254-0584

Li, Y. & Zhu, Q. (2004). A Model of Heat and Moisture Transfer in Porous Textiles with Phase Change Materials. *Textile Research Journal*, Vol.74, No.5, (May 2004), pp. 447-457, ISSN 0040-5175

Liang, C., Lingling, X., Hongbo, S. & Zhibin, Z. (2009). Microencapsulation of butyl stearate as a phase change material by interfacial polycondensation in a polyurea system. Energy *Conversion and Management*, Vol.50, No.3, (March 2009), pp. 723-729, ISSN 0196-8904

Lottenbach, R. and Sutter, S. (2002). Method for producing temperature-regulating surfaces with phase change material. *WO Patent* 02095314

Loxley, A. & Vincent, B. (1998). Preparation of Poly(methylmethacrylate) Microcapsules with Liquid Cores. *Journal of Colloid and Interface Science*, Vol.208, No.1, (December 1998), pp. 49-62, ISSN 0021-9797

Ma, S., Song, G., Li, W., Fan, P. & Tang, G. (2010). UV irradiation-initiated MMA polymerization to prepare microcapsules containing phase change paraffin. *Solar Energy Materials and Solar Cells*, Vol.94, No.10, (October 2010), pp. 1643-1647, ISSN 0927-0248

Mehling, H. & Cabeza, L.F. (2008). *Heat and cold storage with PCM – An up date to introduction into basics and applications*, Springer, ISBN 978-3-540-68556-2

Métivaud, V. (1999). Systemes multicomposants d'alcanes normaux dans la gamme C14H30–C25H52: aliances structurales et stabilite des echantilons mixtes.

Applications pour la protection thermique d'instalations de telecomunications et de circuits optoelectroniques, *European thesis*, Université Bordeaux I, France.

Mills, A., Farid, M., Selman, J.R. & Al-Hallaj, S. (2006). Thermal conductivity enhancement of phase change materials using a graphite matrix, *Applied Thermal Engineering*, Vol.26, No.14-15, (October 2006), pp. 1652-1661, ISSN 1359-4311

Nelson, G. (2001). Microencapsulation in textile finishing. Review of Progress in Coloration and Related Topics, Vol.31, No.1, (June 2001), pp. 57-64, ISSN 1478-4408

Nelson, G. (2002). Application of microencapsulation in textiles. *International Journal of Pharmaceutics*, Vol.242, No.1-2, (August 2002), pp.55-62, ISSN 0378-5173

Onder, E., Nihal, S. & Cimen E. (2008). Encapsulation of phase change material by complex coacervation to improve thermal performances of woven fabrics. *Thermochimica Acta*, Vol.467, No.1-2, (January 2008), pp. 63-72, ISSN 0040-6031

Pause, B.H. (1994). Investigation of the heat insulation of protective textiles with microencapsulated PCM. *Techtextil-Symposium*, 245, pp. 1-9

Pause, B.H. (1995). Development of heat and cold membrae structures with phase change material. *Journal of Coated Fabrics*, Vol.25, No.7, (July 1995), pp.59-68, ISSN 0093-4658

Pause, B.H. (2001). Interactive thermal insulating system having a layer treated with a coating of energy absorbing phase change material adjacent a layer of fibers containing energy absorbing phase change material. *US Patent* 6,217, 993, available from http://patft.uspto.gov/

Pushaw, R.J. (1997). Coated skived foam and fabric article containing energy absorbing phase change material. *US Patent* 5, 677, 048, available from http://patft.uspto.gov/

Roy, S.K. & Sengupta, S. (1991). An evaluation of phase change microcapsles for use in enhanced heat transfer fluids. *International Communications in Heat and Mass Transfer*, Vol.18, No.4, (July-August 1991), pp. 495-507, ISSN 0735-1933

Salaün, F., Devaux, E., Bourbigot & S., Rumeau, P. (2008a). Preparation of multinuclear microparticles using a polymerization in emulsion process. *Journal of Applied Polymer Science*, Vol.107, No.4, (February 2008), pp. 2444-2452, ISSN 0021-8995

Salaün, F., Devaux, E., Bourbigot, S. & Rumeau, P. (2008b). Development of a precipitation method intended for the entrapment of hydrated salt. *Carbohydrate Polymers*, Vol.73, No.2, (July 2008), pp. 231-240, ISSN 0144-8617

Salaün, F., Devaux, E., Bourbigot, S., Rumeau, P., Chapuis, P.O, Saha, S.K., & Volz, S. (2008c). Polymer nanoparticles to decrease thermal conductivity of phase change materials. *Thermochimica Acta*, Vol.477, No.1-2, (October 2008), pp. 25-31, ISSN 0040-6031

Salaün, F., Devaux, E., Bourbigot, S. & Rumeau, P. (2009a). Application of Contact Angle Measurement to the Manufacture of Textiles containing Microcapsules. *Textile Research Journal*, Vol.79, No.13, (September 2009), pp.1202-1212, ISSN 0040-5175

Salaün, F., Devaux, E., Bourbigot, S. & Rumeau, P. (2009b). Influence of process parameters on microcapsules loaded with n-hexadecane prepared by in situ polymerization, *Chemical Engineering Journal*, Vol.155, No. 1-2, (December 2009), pp. 457-465, ISSN 1385-8947

Salaün, F., Devaux, E., Bourbigot, S. & Rumeau, P. (2010a). Development of Phase Change Materials in Clothing Part I: Formulation of Microencapsulated Phase Change. *Textile Research Journal*, Vol.80, No.3, (February 2010), pp.195-205, ISSN 0040-5175

Salaün, F., Devaux, E., Bourbigot, S. & Rumeau, P. (2010b). Influence of the solvent on the microencapsulation of an hydrated salt. *Carbohydrate Polymers*, Vol.79, No.4, (March 2010), pp. 964-974, ISSN 0144-8617

Salaün, F., Devaux, E., Bourbigot, S. & Rumeau, P. (2010c). Thermoregulating response of cotton fabric containing microencapsulated phase change materials. *Thermochimica Acta*, Vol.506, No.1-2, (July 2010), pp. 82-93, ISSN 0040-6031

Salaün, F., Bedek, G., Devaux, E. & Dupont, D. (2011). Influence of the washings on the thermal properties of polyurea-urethane microcapsules containing xylitol to provide a cooling effect. *Materials Letters*, Vol.65, No.2, (January 2011), pp.381-384, ISSN 0167-577X

Sánchez, L., Sánchez, P., de Lucas, A., Carmona, M. & Rodriguez J.F. (2007). Microencapsulation of PCMs with a polystyrene shell. Colloid polymer science, Vol.285, No. 12, (July 2007), pp.1377-1385, ISSN 0303-402X

Sánchez-Silva, L., Rodriguez, J.F., Romero, A., Borreguero, A.M., Carmona, M. & Sánchez P. (2010). Microencapsulation of PCMs with a styrene-methyl methacrylate copolymer shell by suspension-like polymerisation. *Chemical Engineering Journal*, Vol.157, No.1, (February 2010), pp. 216-222, ISSN 1385-8947

Sánchez, P., Sanchez-Fernandez, M.V., Romero, A., Rodriguez, J.F. & Sanchez-Silva, L. (2010) Development of thermo-regulating textiles using paraffin wax microcapsules. *Thermochimica Acta*, Vol.498, No.1-2, (January 2010), pp. 16-21, ISSN 0040-6031

Sari, A. (2003). Thermal reliability test of some fatty acids as PCMs used for solar thermal latent heat storage applications. *Energy Conversion and Management*, Vol.44, No.14, (August 2003), pp. 2277-2287, ISSN 0196-8904

Sari, A. (2005). Eutectic mixtures of some fatty acids for low temperature solar heating applications: Thermal properties and thermal reliability. *Applied Thermal Engineering*, Vol.25, No. 14-15, (October 2005), pp. 2100-2107, ISSN 1359-4311

Sari, A. & Karaipekli, A. (2008). Preparation and thermal properties of capric acid/palmitic acid eutectic mixture as a phase change energy storage material. *Materials Letters*, Vol.62, No.6-7, (15 March 2008), pp. 903-906, ISSN 0167-577X

Sari, A., Biçer, A. & & Karaipekli, A. (2009a). Synthesis, characterization, thermal properties of a series of stearic acid esters as novel solid–liquid phase change materials, *Materials Letters*, Vol.63, No. 13-14, (May 2009), pp. 1213-1216, ISSN 0167-577X

Sari, A., Alkan, C., Karaipekli, A. & Uzun, O. (2009b). Microencapsulated n-octacosane as phase change material for thermal energy storage. *Solar Energy*, Vol.83, No.10, (October 2009), pp. 1757-1763, ISSN 0038-092X

Sari, A., Alkan, C. & Karaipekli, A. (2010). Preparation, characterization and thermal properties of PMMA/n-heptadecane microcapsules as novel solid-liquid microPCM for thermal energy storage. *Applied Energy*, Vol.87, No.5, (May 2010), pp. 1529-1534, ISSN 0306-2619

Sarier, N. & Onder, E. (2007a). The manufacture of microencapsulated phase change materials suitable for the design of thermally enhanced fabrics. *Thermochimica Acta*, Vol.452, No.2, (January 2007), pp. 149-160, ISSN 0040-6031

Sarier, N. & Onder, E. (2007b). Thermal characteristics of polyurethane foams incorporated with phase change materials. *Thermochimica Acta*, Vol.454, No.2, (March 2007), pp. 90-98, ISSN 0040-6031

Shim, H., McCullough, E.A. & Jones, B.W. (2001). Using Phase Change Materials in Clothing. *Textile Research Journal*, Vol.71, No.6, (June 2001), pp.495-502, ISSN 0040-5175

Shin, Y., Yoo, D. & Son, K. (2005). Development of thermoregulating textile materials with microencapsulated phase change materials (PCM). II. Preparation and application of PCM microcapsules. *Journal of Applied Polymer Science*, Vol.96, No.6, (June 2005), pp. 2005-2010, ISSN0021-8995.

Song, Q., Li, Y., Xing, J., Hu, J.Y. & Marcus, Y. (2007). Thermal stability of composite phase change material microcapsules incorporated with silver nano-particles. *Polymer*, Vol.48, No.11, (May 2007), pp. 3317-3323, ISSN 0032-3861

Su, J.F., Ren, L. & Wang, L.X. (2005). Preparation and mechanical properties of thermal energy storage microcapsules, *Colloid and Polymer Science*, Vol.284, No.2, (November 2005), pp. 224–228, ISSN 0303-402X

Su, J.F., Wang, L.X. & Ren, L. (2007) Synthesis of polyurethane microPCMs containing n-octadecane by interfacial polycondensation: Influence of styrene-maleic anhydride as a surfactant. *Colloids and Surfaces A: Physicochemical and Engineering Aspects*, Vol.299, No.1-3, (May 2007), pp. 268-275, ISSN 0927-7757

Šumiga, B., Knez, E., Vrtačnik, M., Savec, V.F., Starešinič, M. & Boh, B. (2011). Production of Melamine-Formaldehyde PCM Microcapsules with Ammonia Scavenger used for Residual Formaldehyde Reduction. *Acta Chimica Slovenica*, Vol.58, No.1, (March 2011), pp.14-25, ISSN 1318-0207

Sundberg, E.J. & Sundberg, D. C. (1993). Morphology development for three-component emulsion polymers: Theory and experiments. *Journal of Applied Science*, Vol.47, No.7, (February 1993), pp.1277-1294, ISSN 0021-8995

Teixeira, M.I., Andrade, L.R., Farina, M. & Rocha-Leao, M.H.M. (2004). Characterization of short chain fatty acid microcapsules produced by spray drying. *Materials Science and Engineering C: Biomimetic and Supramolecular Systems*, Vol.24, No.5, (November 2004), pp. 653–65, ISSN 0928-4931

Torza, S. & Mason, S.G. (1970). Three-phase interactions in shear and electrical fields. *Journal of colloid and Interface Science*, Vol. 33, No.1 (May 1970), pp. 67-83, ISSN 0021-9797

Uddin, M.S., Zhu, H.J. & Hawlader, M.N.A. (2002). Effects of cyclic operation on the characteristics of a microencapsulated PCM storage material. *International Journal of Solar Energy*, Vol.22, No.3-4, (September-December 2002), pp.105-114, ISSN 0142-5919

Wang, X., Lu, E., Lin, W., Liu, T., Shi, Z., Tang, R. & Wang, G. (2000). Heat storage performance of the binary systems neopentyl glycol/pentaerythritol and neopentyl glycol/trihydroxy methyl-aminomethane as solid-solid phase change materials. *Energy Conversion and Management*, Vol.41, No.2, (January 2000), pp. 129-134, ISSN 0196-8904

Yamagishi, Y., Takeuchi, H., Pyatenko, A.T. & Kayukawa, N. (1999). Characteristics of microencapsulated PCM slurry as a heat-tranfer fluid. *AIChE Journal*, Vol.45, No.4, (April 1999), pp. 696-707, ISSN 0001-1541

You, M., Zhang, X.X., Li, W. & Wang, X.C. (2008) Effects of MicroPCMs on the fabrication of MicroPCMs/polyurethane composite foams. *Thermochimica Acta,* Vol.472,No.1-2, (June 2008), pp. 20-24, ISSN 0040-6031

You, M., Zhang, X.X., Wang, X.C., Li, W. & Wen, W. (2010). Effects of type and contents of microencapsuled n-alkanes on properties of soft polyurethane foams. *Thermochimica Acta,* Vol. 500, No.1-2, (March 2010), pp. 69-75, ISSN 0040-6031

Zeng, J., Cao, Z., Yang, D., Xu, F., Sun, L., Zhang, L. & Zhang, X. (2009). Phase diagram of palmitic acid-tetradecanol mixtures obtained by DSC experiments. *Journal of Thermal Analysis and Calorimetry*, Vol.95 , No.2, (February 2009), pp. 501-505, ISSN 1388-5150

Zhang, X. (2001). Heat-storage and thermoregulated textiles and clothing, IN: *Smart fibres, fabrics and clothing*, Tao, pp. 34-58, Woodhead Publishing Ltd., ISBN 1 85573 546 6, Cambridge.

Zhang, H. & Wang, X. (2009a). Synthesis and properties of microencapsulated n-octadecane with polyurea shells containing different soft segments for heat energy storage and thermal regulation. *Solar Energy Materials and Solar Cells,* Vol.93, No.8, (August 2009), pp. 1366-1376, ISSN 0927-0248

Zhang, H. & Wang, X. (2009b). Fabrication and performances of microencapsulated phase change materials based on n-octadecane core and resorcinol-modified melamine-formaldehyde shell. *Colloids and Surfaces A: Physicochemical and Engineering Aspects*, Vol.332, No.2-3, (January 2009), pp. 129-138, ISSN 0927-7757

Zhang, H., Wang, X. & Wu, D. (2010). Silica encapsulation of n-octadecane via sol-gel process: A novel microencapsulated phase-change material with enhanced thermal conductivity and performance. *Journal of Colloid and Interface Science*, Vol.343, No.1, (March 2010), pp. 246-255, ISSN 0021-9797

Zhang, X.X., Tao, X.M., Yick, K.L. & Wang, X.C. (2004a). Structure and thermal stability of microencapsulated phase-change materials. *Colloid and polymer science,* Vol. 282, No.4, (February 2004), pp. 330-336, ISSN 0303-402X

Zhang, X.X., Fan, Y.F., Tao, X.M. & Yick, K.L. (2004b). Fabrication and properties of microcapsules and nanocapsules containing n-octadecane. *Materials Chemistry and Physics*, Vol.88, No.2-3, (December 2004), pp. 300-307, ISSN 0254-0584

Zhang, X.X., Fan, Y.F., Tao X.M. & Yick, K.L. (2005a). Crystallization and prevention of supercooling of microencapsulated n-alkanes. *Journal of Colloid and Interface Science,* Vol.281, No.2, (January 2005), pp. 299-306, ISSN 0021-9797

Zhang, X.X., Wang, X.C., Tao, X.M. & Yick, K.L. (2005b). Energy storage polymer/MicroPCMs blended chips and thermo-regulated fibers. *Journal of Materials Science*, Vol.40, No.14, (July 2005), pp. 3729-3734, ISSN 0022-2461

Zhang, X.X., Wang, X.C., Tao, X.M. & Yick, K.L. (2006). Structures and Properties of Wet Spun Thermo-regulated Polyacrylonitrile-Vinylidene Chloride Fibers. *Textile Research Journal*, Vol.76, No.5, (May 2006), pp. 351-359, ISSN 0040-5175

Zou, G.L., Tan, Z.C., Lan, X.Z., Sun, L.X. & Zhang, T. (2004). Preparation and characterization of microencapsulated hexadecane used for thermal energy storage. *Chinese Chemical Letters*, Vol.15, No.6, (2004), pp. 729–732, ISSN 1001-8417

Zuckerman, J.L., Pushaw, R.J., Perry, B.T. & Wyner, D.M. (1997). Fabric containing anergy absorbing phase change material and method of manufacturing same. *US Patent* 5, 514,362, available from http://patft.uspto.gov/

Zuckerman, J.L., Pushaw, R.J., Perry, B.T. & Wyner, D.M. (2001). Fabric coating composition containing energy absorbing phase change material. *US Patent* 6,207,738, available from http://patft.uspto.gov/

11

Heat Transfer and Thermal Air Management in the Electronics and Process Industries

Harvey M. Thompson
Institute of Engineering Thermofluids, Surfaces & Interfaces (iETSI)
School of Mechanical Engineering, University of Leeds, Leeds,
United Kingdom

1. Introduction

Rising energy costs and important legislative drivers are making the achievement of efficient heat transfer and thermal management of crucial importance in energy intensive industries. In the electronics industry, inexorable increases in microprocessor performance due to the use of multiple cores on a single chip are creating an enormous challenge for the cooling infrastructure, since almost all of the electrical energy consumed by the chip package is released as heat (Anandan & Ramalingam, 2008). This is particularly relevant to the rapidly increasing number of large scale data centres, see Figure 1, which form the backbone of the digital society on which the world's population is becoming increasingly reliant. The power consumption of data centres is rising sharply, having doubled in the last five years and is likely to double again in the next five years to over 100 billion KWh (Scofield & Weaver, 2008). These enormous energy requirements are presenting governments and industry with a serious energy supply problem (Shehabi et al., 2011) and the importance of data centres' energy efficiency has now been recognised at the international level with the formation of several industry consortia such as the Green Grid, the Uptime Institute and the Data Centre Alliance to promote energy efficiency and best practices in the data centre industry.

Since most enterprise data centres run significant quantities of redundant power and cooling systems to produce higher levels of resiliency, this had led to significant power consumption inefficiencies. The latter are exacerbated by the inefficiencies in the Information Technology (IT) hardware and cooling requirements, each accounting for roughly 40% of the total energy usage. This results in each KWh of energy for data processing requiring a further KWh for cooling (Almoli et al., 2011). In a typical data centre, electrical energy is drawn from the main grid to power an uninterruptible power supply (UPS) which then powers the IT equipment, supply power to offices and to power the cooling infrastructure: computer room air conditioning (CRAC) units, building chilled water pumps and water refrigeration plant. The IT load inefficiencies can be improved by server virtualisation and improved semi-conductor technologies, while the chiller plant is generally the biggest energy cooling component and increasing the set point temperature of the chilled water leaving the chiller evaporator offers significant potential reductions in the overall cooling plant energy consumption.

Fig. 1. A large scale data centre with several rows of server racks

A key strategy for efficient thermal air management in a data centre, as recommended in the EU Code of Conduct on Data Centre Energy Efficiency, is to separate hot and cold air via a layout of alternating hot and cold aisles (Rasmussen, 2006; Niemann, 2008). These are shown schematically in Figure 2. In the *cold aisle containment* strategy, cold air is supplied from the CRAC units through floor tiles or diffusers into cold aisles and the racks are arranged so that all server fronts/intakes face cold aisles. This counteracts the problem that arises if all rows are arranged with intakes facing the same way, when equipment malfunction is inevitable due to server overheating (Cho et al., 2009). In the *hot aisle containment* strategy, it is the hot air that is contained and this approach can have advantages in terms of obviating the need for raised floor tiles and providing hotter air to the CRAC units, increasing their overall efficiency of performance. The importance of good air flow management in data centres has led to increasing use of Computational Fluid Dynamics (CFD) (Versteeg & Malalasekera, 1995) to design data centre operations to ensure the thermal environment within data centres conforms to narrow, acceptable bands. Care must, however, be taken to ensure that CFD predictions are properly validated and the limitations of its key assumptions (for example on the coupling between the small-scale server air flows and the larger scale data centre air flows) are understood (Almoli et al., 2011). Once validated, CFD models can be very useful for data centre air flow management in enabling a large number of design scenarios to be investigated and optimal server rack configurations to be identified much more quickly than would be possible experimentally.

centre cooling is achieved through recycling cold air through server racks and CFD is now an integral means of improving thermal air flow management in data centres. Despite having achieved several successes there are still important areas of weakness of current CFD methods and a need for greater transparency in terms of describing the all-important boundary conditions and greater access to validation case study data.

The energy consumption of process industries is also receiving greater attention in the scientific literature. This chapter has focussed on one important aspect of this enormous subject, namely convective heat transfer from impinging air jets in forced convection ovens used in the coating, converting and bread baking industries. Advances in CFD methods are now being exploited within these industries and have shown, for example, how the required heat flux into products can be achieved more efficiently by optimising the air flow velocity and temperature conditions. Outstanding issues for the CFD modelling of these systems include the variation of turbulence levels throughout the oven, the validity of popular turbulence models used to model them and, once again, the need for a comprehensive database of experimental data for validation purposes. In order for industry to derive the maximum benefit from the improving capabilities of CFD modelling, CFD models will need to be incorporated into formal design optimization frameworks that are capable of minimising physically meaningful objective functions.

5. Acknowledgements

The author would like to thank several colleagues at the University of Leeds for their contribution and support. Thanks are particularly due to Dr Nikil Kapur, Dr Jon Summers, Dr Malcolm Lawes, Professor Phil Gaskell and Professor Vassili Toropov, to industrial collaborators: Airedale International Air Conditioning Ltd, Spooner Industries Ltd and Warburton's Ltd and the UK's Engineering and Physical Sciences Research Council (EPSRC) for financial support in these research areas.

6. References

Almoli, A; Thompson, A., Kapur, N., Summers, J.L, Thompson, H.M. & Hannah, G. (2011). Computational Fluid Dynamic Investigation of Liquid Rack Cooling in Data Centres, to appear in *Applied Energy*, available from: doi:10.1016/j.apenergy.2011.02.003

Anandan, S.S. & Ramalingam, V. (2008). Thermal management of electronics: a review of literature, *Thermal Science*, vol. 12, no. 2, (April 2008), pp 5-26, ISSN 0354-9836.

Ansys Inc. *Ansys Fluent 12*, 18th April 2011, Available from: http://www.ansys.com/Products/Simulation+Technology/Fluid+Dynamics/ANSYS+FLUENT/

Apollonov, V.V. (1999). Intensification of heat transfer in high-power laser diode bars bymeans of a porous metal heat-sink, *Optics Express*, vol. 4, pp 27-32, (June 1999), ISSN 1094-4087.

Apollonov, V.V. (2000). Highly efficient heat exchangers for laser diode arrays, *Proceedings of SPIE*, vol. 3889, pp 71-81, ISBN 1094-4087, Osaka, Japan.

Boulet, M; Marcos, B., Dostie, M & Moresoli, C. (2010). CFD modeling of heat transfer and a flow field in a bakery pilot oven, *J. Food Engineering*, vol. 97, no. 3, (June 2010), pp 393-402, ISSN 0260-8774.

Wolf, D;, Incropera, F.P. & Viskanta, R. (1993). Jet Impingement Boiling, *Advances in Heat Transfer*, vol. 23, (March 1993), pp 1-132, ISSN 0065-2717.

Zareifard, M.R;, Marcotte, M. & Dostie, M. (2006). A method for balancing heat fluxes validated for a newly designed pilot plant oven. *Journal of Food Engineering*, vol. 76, no. 3, (October 2006), pp 303-312, ISSN 0260 8774.

Zheleva, I & Kambourova, V. (2005). Identification of heat and mass transfer processes in bread during baking, *Thermal Science*, vol. 9, no. 2, (September 2005), pp 73-86, ISSN 0354-9836.

Zhou, W. & Therdthai, N. (2007). Three-dimensional CFD modeling of a continuous industrial baking process, in: *Da-Wen Sun (Ed.), Computational Fluid Dynamics in Food Processing*, pp 287-312, ISBN 0849392861, retrieved from http://www.crcnetbase.com/isbn/9781420009217

12

Unsteady Mixed Convection Flow in the Stagnation Region of a Heated Vertical Plate Embedded in a Variable Porosity Medium with Thermal Dispersion Effects

S. M. Alharbi[1] and I. A. Hassanien[2]
[1]*Department of Mathematics, Faculty of Science, Umm AL-Qura University, P.O. 8140, KSA*
[2]*Department of Mathematics, Faculty of science, Assiut University, Assiut, Egypt*

1. Introduction

The mixed convection flow finds applications in several industrial and technical processes such as nuclear reactors cooled during emergency shutdown, solar central receivers exposed to winds, electronic devices cooled by fans and heat exchanges placed in a low-velocity environment. The mixed convection flow becomes important when the buoyancy forces increase due to the temperature difference between the wall and the free stream. The mixed convection flow in the stagnation region of a vertical plate has been investigated by Ramachandra et al. [16].

When there is an impulsive change in the velocity field the inviscid flow is developed instantaneously, but the flow in the viscous layer near the wall is developed slowly which becomes fully developed steady flow after a while. For small period the flow is dominated by the viscous forces and the unsteady acceleration, but for runtime it is dominated by the viscous forces, the pressure gradient and the convective acceleration. The unsteady mixed convection flow in the stagnation region of a heated vertical plate due to impulsive motion has been studied by Schadri et al. [17]. The boundary layer flow development of a viscous fluid on a semi-infinite flat plate due to impulsive motion of the free stream have been investigated by Hall [5], Dennis [3] and Watkins [22]. The corresponding problem over a wedge has been studied by Simth [18], Nanbu [11] and Williams & Rhyne [23].

The problem of unsteady free convection flow in the stagnation-point region of a rotating sphere embedded in a porous medium has been analyzed by Hassanien et al. [7]. The unsteady flow and heat transfer of a viscous fluid in the stagnation region of a three-dimensional body embedded in a porous medium was investigated by Hassanien et al. [8]. The problem of thermal radiation and variable viscosity effects on unsteady mixed convection flow in the stagnation region on a vertical surface embedded in a porous medium with surface heat flux has been studied by Al-Arabi and Hassanien [6].

Motivated by all of the above referenced work and the significant possible applications of porous media in industries, it is of interest in this paper to consider the unsteady mixed convection flow in the region of a heated vertical plate embedded in a porous medium having porosity distribution in the presence of the thermal dispersion with the effect of the buoyancy force. The unsteadiness in the flow field is caused by impulsively creating motion in the free stream and at the same time by suddenly increase in the surface temperature. The partial differential equations governing the flow and the heat transfer have been solved numerically using the finite difference scheme by Pereyra [14]. Particular cases of the present results are compared with previously numerical work by Ramachandra et al. [16] and Scshadri et al. [17]. The problem is formulated in such way that it represented by Rayleigh type of equation at t=0 and for $t \to \infty$ it represented by type of Hemennz equation.

2. Mathematical analysis

Let us consider a semi-infinite vertical plate embedded in a variable porosity porous medium with thermal dispersion effect and uniform temperature T_∞. At $t = 0.0$ the ambient fluid is impulsively moved with a velocity U_e and at the same time the surface temperature is suddenly raised. Figure (1) shows a flow field over a heated vertical surface where the upper half of the field is assisted by the buoyancy force, but the lower part is opposed by the buoyancy force. The surface of the plate is assumed to have an arbitrary temperature. All the physical properties of the fluid are assumed to be constant except the density variation in the buoyancy force term. Both the fluid and the porous medium are assumed to be in local thermal equilibrium. Under above assumptions along with Boussinesq approximation, the unsteady laminar boundary layer equations governing the mixed convection flow are given by Vafai and Tien [20].

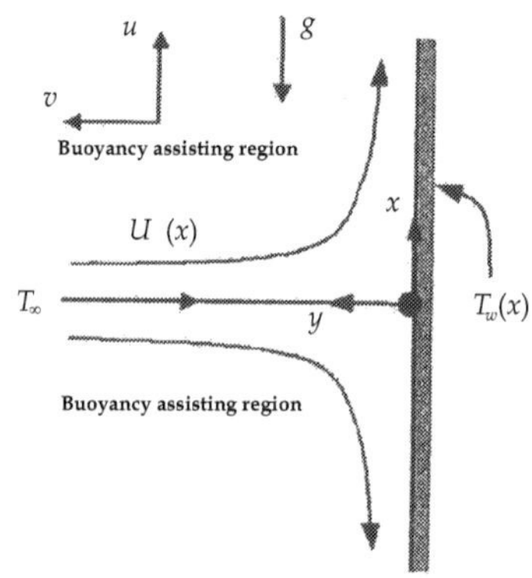

Fig. 1. The flow model and the coordinate system

$$\frac{\partial u}{\partial x} + \frac{\partial v}{\partial y} = 0 \tag{1}$$

$$\frac{1}{\varepsilon}\frac{\partial u}{\partial t} + \frac{1}{\varepsilon^2}\left(u\frac{\partial u}{\partial x} + v\frac{\partial u}{\partial y}\right) = U_e\frac{\partial U_e}{\partial x} + \frac{\nu}{\varepsilon}\frac{\partial^2 u}{\partial y^2} + g\beta(T - T_\infty) + \frac{\nu}{k(y)}(u_e - U_e) \tag{2}$$

$$\frac{\partial T}{\partial t} + u\frac{\partial T}{\partial x} + v\frac{\partial T}{\partial y} = \frac{\partial}{\partial y}\left(\alpha_c\frac{\partial T}{\partial y}\right) \tag{3}$$

The initial conditions are given by

$$u(x,y) = v(x,y) = 0, \quad T(x,y) = T_\infty, \text{ for } t < 0. \tag{4}$$

The boundary conditions for $t \geq 0$ are given by

$$\begin{aligned} &u(x,0) = v(x,0) = 0, \quad u(x,\infty) = U_e = ax, \quad a > 0, \\ &T(x,\infty) = T_\infty, T(x,0) = T_w(x) - bx^n, \quad b > 0, \quad n \geq 0. \end{aligned} \tag{5}$$

The indices $n = 0$ and $n = 1.0$ correspond the constant surface temperature and the linear surface temperature respectively. The variable x is measured along the surface and y is measured normal to it. The fluid velocity u, v is in x and y direction respectively as shown in figure (1). The fluid density, the fluid dynamical viscosity, the gravitational fluid acceleration, the thermal expansion coefficient and the temperature will be denoted by ρ, μ, g, β, T respectively. K(y) is the porous medium permeability, α_c is the effective thermal diffusivity and ε is the porous medium porosity. Equations (1) through (3) are supplemented by constitutive equations for the variations of the porosity permeability and thermal conductivity of the porous medium. It has been shown by Vafai [21] that the results obtained experimentally by Nithiarasu et al. [12] in their study on void fraction distribution in packed beds gives the functional dependence of the porosity on the normal distance from the boundary and so the porosity can be represented by the exponential form

$$\varepsilon = \varepsilon_0 (1 + b \exp\ (-cy / d)) \tag{6}$$

where ε_0 is the free-stream porosity, d is the particle diameter and b, c are empirical constants that depend on the ratio of the bed to particle diameter. The values for ε_0, b and c chosen to be 0.38, 1, and 2 respectively. These values were found to give good approximation to the variable porosity data given by Nithiarasu et al. [12] for a particle diameter d=5 mm. The type of decay of porosity as the normal distance increases given by Equation (4) is well established and has been used extensively in studies on flow in porous media with variable porosity. It is also established that k(y) varies with the porosity as follows

$$K(y) = \frac{d^2 \varepsilon^3}{150(1-\varepsilon)^2} \tag{7}$$

The effective thermal conductivity of the porous medium is given by Al-Arabi and Hassanien [6]

$$\alpha_c = \alpha_m + \gamma\ \mathrm{d}|u| \tag{8}$$

where α_m and γ are the molecular thermal diffusivity and mechanical dispersion coefficients, respectively. Equations (1) through (3) can be transformed into a set of ordinary differential equations by using the following transformations given by Williams and Rhyne [23].

$$\eta = (a/\nu)^{1/2}\, y\, \xi^{-1/2}, \qquad \xi = 1 - \exp(-t^*),$$

$$t^* = at, \quad a > 0, \qquad u(x,y,t) = ax\, f'(\eta,\xi),$$

$$v(x,y,t) = -(a\nu)^{1/2} \xi^{1/2} f(\eta,\xi),$$

$$T(x,y,t) = T_\infty + (T_w(x) - T_\infty)\theta(\eta,\xi),$$

$$\Pr = \nu / \alpha, \qquad \lambda = Gr_x / \mathrm{Re}_x^2,$$

$$Gr_x = g\beta(T - T_\infty)x^3 / \nu^2, \quad \mathrm{Re}_x = ax^2 / \nu. \tag{9}$$

By using the transformations (9), equations (2-5) may be transformed to

$$\begin{aligned} & f''' + 2^{-1}\eta(1-\xi)f'' + \frac{\xi}{\varepsilon}ff'' + \frac{\xi}{\varepsilon}(1-f'^2) - Da^2\frac{150(1-\varepsilon)^2}{\varepsilon}(1-f') \\ & -\frac{\xi}{\varepsilon}(1-f'^2) + \lambda\xi\varepsilon\theta = \varepsilon\xi(1-\xi)\frac{\partial f}{\partial \xi}, \end{aligned} \tag{10}$$

$$\Pr^{-1}\theta'' + +Ds(f'\theta')' + 2^{-1}\eta(1-\xi)\theta' + \xi(f\theta' - nf'\theta) = \xi(1-\xi)\frac{\partial \theta'}{\partial \xi}, \tag{11}$$

where $Da = \dfrac{\nu}{ad^2}$ is the Darcian parameter , $Ds = \dfrac{\gamma d}{a}$ is the dispersion parameter , $\lambda > 0$ for the buoyancy assisting flow and $\lambda < 0$ for the buoyancy opposing flow. The transformed form of the variable porosity function becomes

$$\varepsilon = \varepsilon_0(1 + b\exp(-c\eta\sqrt{\xi}\ Da))$$

The boundary conditions (5) may be reduce to

$$\begin{aligned} & f(0,\xi) = f'(0,\xi) = 0, \qquad f'(\infty,\xi) = 1.0, \\ & \theta(\infty,\xi) = 0, \qquad \theta(0,\xi) = 1.0. \end{aligned} \tag{12}$$

It may be noted that the buoyancy parameter λ is a function of streetwise distance x unless the surface temperature $(T_w = T_\infty)$ for $n = 1.0$, λ is constant. Equations (10) and (11) are coupled nonlinear partial differential equations, but for $\xi = 0$, $(t^* = 0)$ and $\xi = 1.0$, $(t^* \to \infty)$ and they can be reduced to ordinary differential equations. For the case $\xi = 0.0$, these equations take the form

$$f''' + 2^{-1}\eta f'' = 0, \tag{13}$$

$$\theta'' + 2^{-1}\Pr\eta\theta' = 0. \tag{14}$$

For the case $\xi = 1$, equations (7), (8) can be reduced to

$$f''' + ff'' + (1 - f'^2) - \gamma(1 - f') - \Delta(1 - f'^2) + \lambda\theta = 0, \tag{15}$$

$$\Pr^{-1}\theta'' + (f\theta' - nf'\theta) = 0. \tag{16}$$

For the above two special cases the boundary conditions (5) may be reduce to

$$\begin{aligned} & f(0) = f'(0) = 0, \qquad f'(\infty) = 1.0, \\ & \theta(\infty) = 0, \qquad \theta(0) = 1.0 \end{aligned} \tag{17}$$

Equations (13), (14) are uncoupled linear equations while equations (15), (16) are coupled nonlinear equations. Equations (13) and (14) under conditions (17) admit closed form solutions which are given by,

$$f = \eta\ erfc(\eta / 2) - (\pi)^{1/2}[1 - \exp(-\eta^2 / 4)], \tag{18}$$

$$\theta = erfc(\mathrm{Pr}^{1/2}\eta / 2) \tag{19}$$

hence we have

$$f''(0) = (\pi)^{1/2}, \qquad \theta'(0) = -(\mathrm{Pr}/\pi)^{1/2}. \tag{20}$$

Equations (15), (16) do not admit closed form solutions. Equations (10), (11) under conditions (12) for $\xi = 1$ (steady case) and non porous media are identical to those of Ramachandra et al. [16] in nonporous medium ($\varepsilon = 1$. Also equations (10-11) under the condition $\lambda = 0$ (forced convection flow) and $\varepsilon = 1$ are the same as that of Williams and Rhyne [23] if we put m=1 in their equation. The physical quantities of interest in this problem are the skin friction coefficient and the Nusselt number, which are defined by Pop et al. [15].

$$C_f = 2\tau_w / \rho U_\infty^2, \qquad Nu = xq_w / \alpha(T_w - T_\infty), \tag{21}$$

where $\tau_w = \mu(\partial u / \partial y)_{y=0}$ and $q_w = -\alpha(\partial T / \partial y)_{y=0}$. Using equations (9), the quantities in Eqs. (21) can be expressed in the form

$$C_f = 2\xi^{-1/2}\mathrm{Re}_x^{-1/2} f''(\xi,0), \quad \xi > 0. \tag{22}$$

$$Nu = -\mathrm{Re}_x^{1/2}\xi^{-1/2}\theta''(\xi,0), \quad \xi > 0. \tag{23}$$

3. Method of solution

We are going now discuss the local non-similarity method to solve equations (10), (11). Since it was already seen by Pereyra [14], and Sparrow et al. [19] that for the problem of coupled local non-similarity equations, the considerations of equation up to the second level of truncation gives almost accurate results comparable with the solutions from other methods. We will consider here the local non-similar equations (10), (11) only up to the second level of truncation. To do this, we introduce the following new functions

$$g = \partial f / \partial \xi, \qquad \phi = \partial\theta / \partial\xi \tag{24}$$

Introducing these functions into equations (10) and (11) we get

$$f''' + 2^{-1}\eta(1-\xi)f'' + \frac{\xi}{\varepsilon}ff'' + \frac{150(1-\varepsilon)^2}{\varepsilon^2}Da^2(1-f') + \frac{\xi}{\varepsilon}(1-f'^2) + \lambda\xi\varepsilon\theta \;=\; \varepsilon\xi(1-\xi)g, \tag{25}$$

$$\mathrm{Pr}^{-1}\theta'' + +Ds(f'\theta')' + 2^{-1}\eta(1-\xi)\theta' + \xi(f\theta' - nf'\theta) = \xi(1-\xi)\phi' \tag{26}$$

Differentiating the above equations with respect to ξ one may easily neglect the terms involving the derivative functions of g and ϕ with respect to ξ as follows

$$
\begin{aligned}
&g''' + 2^{-1}\gamma[(1-\xi)g'' - f''] + \frac{\varepsilon}{\varepsilon^2}\left[\varepsilon f g'' + \varepsilon g f'' + f f'' - \frac{\varepsilon f f'' \varepsilon_\xi}{\varepsilon}\right] + \\
&\frac{\varepsilon}{\varepsilon^2}\left[(1 - f'^2 + \varepsilon(-2f'g) - \frac{(1-f'^2)\varepsilon_\xi}{\varepsilon}\right] + \\
&150Da^2\frac{1}{\varepsilon^2}\left[(1-\varepsilon^2)^2(-g') + (1-f')\left\{(2)(1-\varepsilon)(-\varepsilon_\xi) - 2\frac{(1-\varepsilon)^2\varepsilon_\xi}{\varepsilon}\right\}\right] \\
&-\frac{1}{\varepsilon}\left[(1-f'^2) + \xi 2f'g' - \xi\frac{(1-f)\varepsilon_\xi}{\varepsilon}\right] + \lambda\varepsilon\xi\phi + \lambda\theta\varepsilon + \lambda\xi\theta\varepsilon_\xi = \varepsilon\phi(1-2\xi) + \xi(1-\xi)\varepsilon_\xi\phi
\end{aligned}
\tag{27}
$$

$$
\begin{aligned}
&pr^{-1}\phi'' + ps(g'\theta'' + \theta' g'') + 2^{-1}\gamma(1-\xi)\phi' \\
&-2^{-1}\gamma\theta' + (f\theta' - nf'\theta) + \xi(f\phi' + g\theta' - ng'\theta - nf'\phi) \\
&= (1-2\xi)\phi'
\end{aligned}
$$

where ε_ξ is the derivative of ε with respect to ξ. The appropriate boundary conditions satisfied by the above equations are given by

$$
\begin{aligned}
&f(0,\xi) = f'(0,\xi) = 0, \qquad f'(\infty,\xi) = 1.0, \\
&\theta(\infty,\xi) = 0, \qquad \theta(0,\xi) = 1.0.
\end{aligned}
$$

$$
\begin{aligned}
&g(0,\xi) = g'(0,\xi) = \phi(0,\xi) = 0, \\
&g'(\infty,\xi) = \phi(\infty,\xi) = 0.
\end{aligned}
\tag{28}
$$

4. Results and discussion

In order to validate our numerical solutions, we have compared the surface shear stress $f''(\xi,0)$ and the surface heat transfer $-\theta'(\xi,0)$ for the prescribed surface temperature with those of Ramachandra et al. [16] and Scshadri et al. [17]. The results are found to be almost compatible to a reasonable degree.. The comparison is shown in Figures (2) and (3), which corresponding Figures (2) and (3) in Scshadri [17].

The variation of the surface shear stress $f''(\xi,0)$, the surface heat transfer rate $-\theta'(\xi,0)$ with time $\xi(0 \le \xi \le 1)$ for the Darcy parameter Da, the dispersion parameter Ds in the presence of the buoyancy assisting flow $(\lambda = 1)$ and buoyancy opposing flow $(\lambda = -1)$ for the non-isothermal surface $(n = 1)$ are shown in Figures (2) through (6).

At the start of motion $(\xi = 0)$, the buoyancy force parameter (λ), the Darcy parameter Da and the dispersion parameter Ds have no effects on both the surface shear stress and the surface heat transfer and these effects become pronounced with increasing time ξ. The steady state is reached at $\xi = 1$ $(t^* \to \infty)$. The surface shear stress $f''(\xi,0)$ and surface heat transfer $-\theta'(\xi,0)$ decrease with the Darcy and dispersion parameters increasing for the two cases (the buoyancy assisting flow $(\lambda = 1)$ and the buoyancy opposing flow $(\lambda = -1)$). It is also clear from these figures that the surface shear stress and heat transfer for buoyancy assisting flow are greater than those of the buoyancy opposing flow. Also, the surface shear stress and the heat transfer rate increase with increasing the Darcy parameter Da and the dispersion parameter Ds.

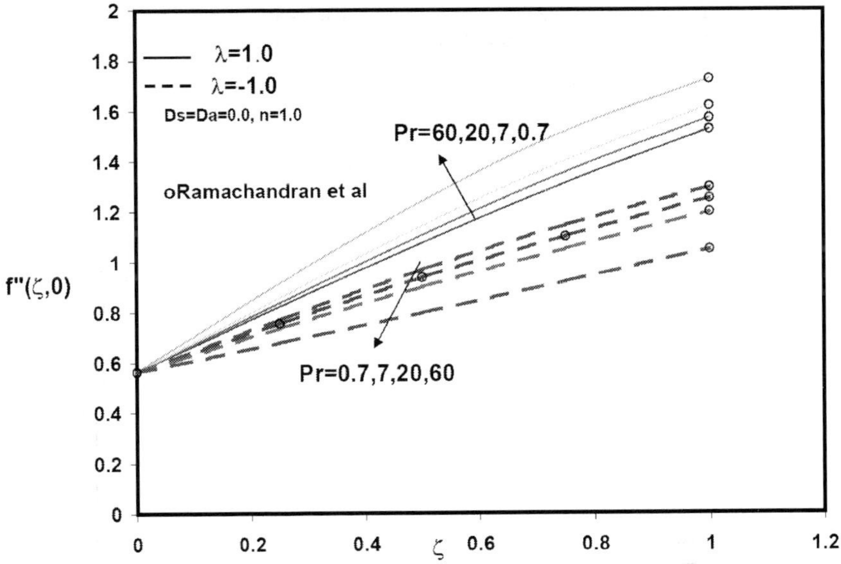

Fig. 2. Variation of shear stress $f''(\xi,0)$ with time ξ

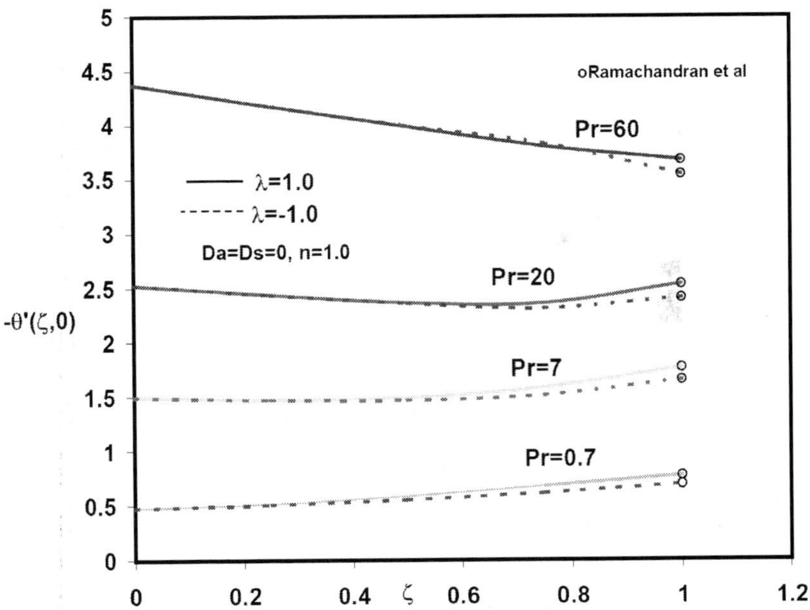

Fig. 3. Variation of surface heat transfer $\theta'(\xi,0)$ with time ξ

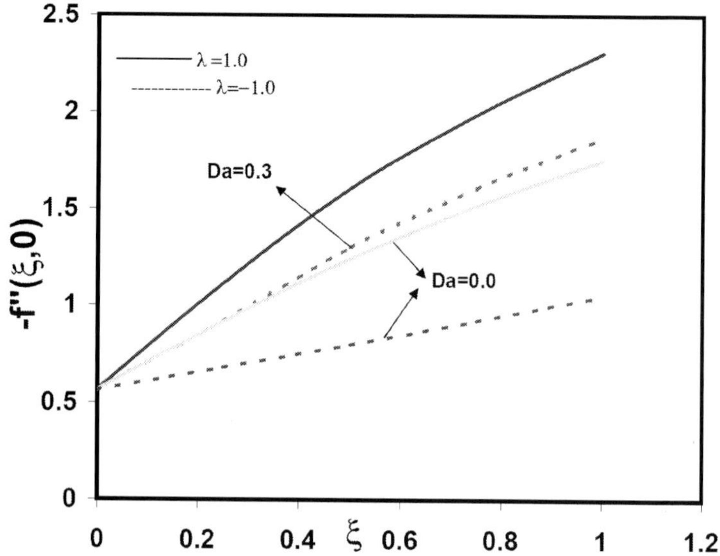

Fig. 4. Variation of surface shear stress with time for $\varepsilon_0 = 0.38$, b=1.0, c= 2.0, Ds=0.0 and Pr=0.7

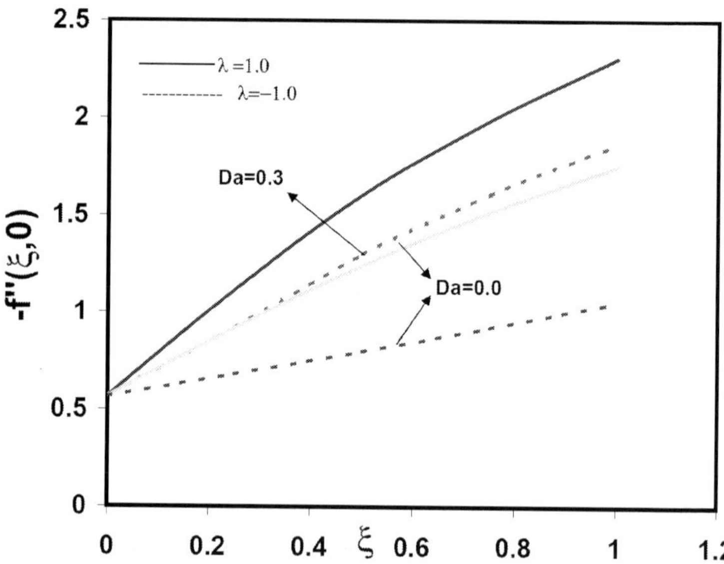

Fig. 5. Variation of the surface heat transfer rate $\theta`(\xi,0)$ with time ξ for ε_0= 0.38, b=1.0, c= 2.0, d=5.0, *Ds*=0.0 and Pr=0.7

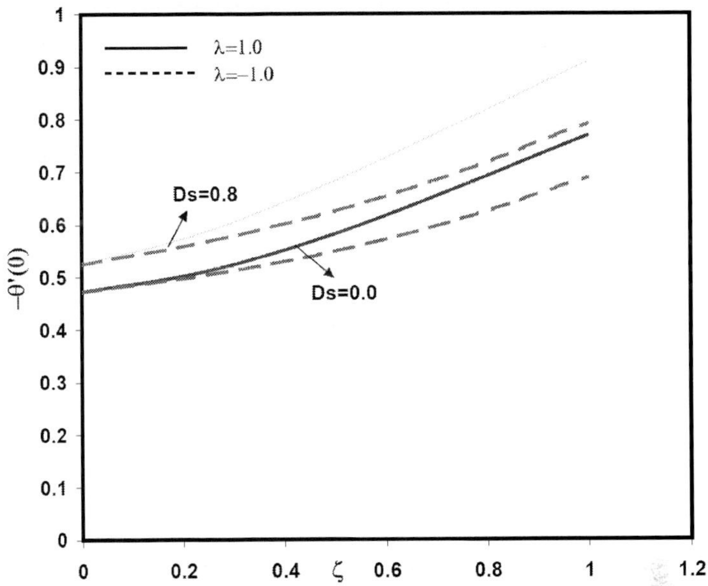

Fig. 6. Variation of the surface heat transfer rate $\theta`(\xi,0)$ with time ξ for $\varepsilon_0 = 0.38$, b = 1.0, c = 2.0, d = 5.0, $Ds = 0.0$ and Pr = 0.7

Figures (5), (6) display the variation of the surface shear stress and the surface heat transfer with time ξ for the prescribed surface temperature when the buoyancy parameter $\lambda = 1.0, -1.0$, $pr = 0.7, n = -0.5$ and for the Darcy parameter Da = 0.0, 0.3, and the dispersion parameter Ds = 0.0, 0.8. The surface shear stress and the heat transfer increase with λ because positive buoyancy force acts like favorable pressure gradient which accelerates the motion and reduces both momentum and thermal boundary layers. Hence both the surface shear stress and the surface heat transfer are increased.

5. Acknowledgement

This research has been supported by a grant from the Institute of Scientific Research at Umm Al-Qura University, Saudi Arabia.

6. References

[1] Amin N. and Riley N., (1995), "Mixed convection at a stagnation point", Quart. J. Mech. App. Maths. Vol. 48, pp. 111-121.
[2] Brown S. N. and Riley N, (1973), "Flow past a suddenly heated vertical plate", J. Fluid Mech. Vol. 59, pp. 225-237.
[3] Dennis S. C. R., (1972), "The motion of a viscous fluid past an impulsively started semi-infinite flat plate", IMA J. Appl. Math. Vol. 10, pp. 105-117.
[4] Ece M. C., (1992), "An initial boundary layer flow past a translating and spinning rotational symmetric body", J. Eng. Math. Vol. 26, pp. 415-428.

[5] Hall M. G., (1969), "The boundary layer over an impulsively started flat plate", Proceedings of the Royal Society of London, Series A, Mathematical and physical science, Vol. 310A, pp. 401-414.

[6] Hassanien, I.A. and Al-Arabi, T.H., (2008), "Thermal Radiation and variable viscosity effects on unsteady mixed convection flow in the stagnation region on a vertical surface embedded in a porous medium with surface heat flux", Far East journal of Mathematical sciences (FJMS), Volume 29, pp. 187 - 207.

[7] Hassanien, I.A., Ibrahim, F.S., Omer, Gh.M., (2004), "Unsteady free convection flow in the stagnation-point region of a rotating sphere embedded in a porous medium", Mechanics and Mechanical Engineering,Vol.7, No.2, pp. 89-98.

[8] Hassanien, I.A., Ibrahim, F.S., Omer, Gh.M., (2006), "Unsteady flow and heat transfer of a viscous fluid in the stagnation region of a three-dimensional body embedded in a porous medium", Journal of Porous Media,Vol.9,No. 4, pp. 357-372.

[9] Ingham D. B., (1985), "Flow past a suddenly heated vertical plate", Proceedings of the Royal Society of London, Series A, Mathematical and physical science, Vol. 402A, pp. 109-134.

[10] Kumari M., (1997), "Development of flow and heat transfer on a wedge with a magnetic field", Archives of Mechanics. Vol. 49, No.5, pp. 977-990.

[11] Nanbu K., (1971), "Unsteady Falkner Skan flow", Zeitschrift Fur Angewandte Mathematik und Physik (ZAMP) , Vol. 22, No. 6 , pp. 1167-1172.

[12] Nithiarasu, P., Setharamu, K. N., and Sundararajin, T., (1997),"Natural convection heat transfer in a fluid saturated variable porosity medium", Int. J. Heat Mass Transfer, (1981), vol. 40, pp. 3955-3967.

[13] Ozturk A. and Ece M. C., (1995), "Unsteady forced convection heat transfer from a translating and spinning body", ASME Journal of Energy Resources Technology, Vol. 117, No. 4, pp. 318-323.

[14] Pereyra V., PASVA3, (1978), "An adaptive finite difference Fortran program for first order non-linear boundary value problems", Lecture Note in Computer Science, vol. 76, Springer, Berlin.

[15] Pop I., Gorla R. S. and Rashidi M., (1992), "The effect of variable viscosity on flow and heat transfer to a continuous moving flat plate", International Journal of Engineering and science, vol. 30, No. 1, pp. 1-6.

[16] Ramachandra N., Chen T., and Armaly B. F., (1988), "Mixed convection in the stagnation flows adjacent to vertical surface", J. Heat Transfer vol. 110, pp. 173-177.

[17] Scshadri R., Srccshylan N. and Nath G., (2002), "Unsteady mixed convection flow in the stagnation region of a heated vertical plate due to impulsive motion", Int. J. Heat and Mass Transfer vol. 45 pp. 1345-1352.

[18] Smith S. H., (1957), "The impulsive motion of a wedge in a viscous fluid", Zeitschrift Fur Angewandte Mathematik und Physik (ZAMP) Vol. 18, No.4pp. 508-522.

[19] Sparrow E.M., Quack H. and Boerner J., (1970), "Local non-similarity boundary solutions, AIAA vol. 8, pp. 1936-1942.

[20] Vafai, K. and Tien, C.L., "Boundary and inertia effects on flow and mass transfer in porous media", Int. J. Heat Mass Transfer, (1981), vol. 24, pp. 492-484.

[21] Vafai, K., "Heat transfer in variable porosity media", J. Fluid Mech., (1984), vol. 147, pp. 233-259.

[22] Watkins C. B., (1975), "Heat transfer in the boundary layer over an impulsively started flat plate", J. Heat Transfer vol. 97, pp. 492-484.

[23] Williams J. C. and Rhyne T. H., (1980), "Boundary layer development on a wedge impulsively set into motion", SIAM J. Appl. Math. Vol. 38, pp. 215-224.

13

Heat Generation and Transfer on Biological Tissues Due to High-Intensity Laser Irradiation

Denise M. Zezell[1], Patricia A. Ana[2],
Thiago M. Pereira[1], Paulo R. Correa[1] and Walter Velloso Jr.[3]
[1]Centro de Lasers e Aplicacoes, Instituto de Pesquisas Energeticas e Nucleares, IPEN – CNEN/SP, Sao Paulo- SP
[2]Centro de Engenharia, Modelagem e Ciencias Sociais Aplicadas, Universidade Federal do ABC, Santo Andre- SP
[3]Faculdade de Zootecnia e Engenharia de Alimentos de Pirassununga, Universidade de Sao Paulo, Pirassununga- SP
Brazil

1. Introduction

The use of high intensity laser irradiation is already used clinically for many dental procedures, since the literature findings confirm its efficacy. However, for the establishment of a laser procedure for any clinical application, it is strongly necessary to study the thermal effects on irradiated tissues, considering that most of biological tissues is highly connective and can easily transmit the received heat (Nyborg & Brännström, 1968; Baldissara et al, 1997; Brown et al, 1970).

The interactions of high intensity laser irradiation on dental hard tissues are commonly induced by thermal action (Seka et al, 1996; Niemz, 1997). For the clinical application of lasers, a precise irradiation parameter must be chosen in order to avoid morphological damage, such as surface carbonization or cracking, which could produce structural, esthetic damages and post-operative complaints. Moreover, the energy densities used must be safe as regards pulp and periodontal tissue vitality (Ana et al, 2006; Nammour et al, 2004; Ana et al, 2007).

Considering a clinical application of high intensity lasers, parameters such as wavelength, energy density, intensity, peak power, average power, repetition rate and pulse length are extremely important to heat generation due to irradiation on any biological tissue. The amount of heat inside the tissue is highly dependent of its optical properties, such as absorption and scattering coefficients. Also, the heat transfer is dependent of some properties of biological tissues, such as thermal diffusivity, thermal conductivity and others (Brown et al, 1970; Seka et al., 1996; Yu et al., 1993).

The knowledge of some basic concepts of temperature and heat are necessary to understand and to evaluate the effects of lasers on biological tissues. These concepts are described below.

1.1 Temperature

The thermal state of a body normally is described by subjective words like "hot", "cold" or "warm". Therefore, the temperature is a physical parameter which is used to characterize the thermal state of the body in a less subjective and more quantitative way; in other words, assigning values.

The temperature is defined as the measurement of the thermal vibration (average kinetic energy) of the body particles. Considering that the thermal vibration of atoms and molecules is difficult to measure, it is necessary to assign numeric values to some macroscopic properties of the body. For example, the Celsius scale was assigned 100 for water boiling temperature and zero to water solidification temperature at the sea level. From this scale it is possible to graduate and assign values to the different thermal states.

1.2 Heat

Heat is the energy associated to the temperature variation. When a hot body gets together with a cold body, the hot body transfers heat (thermal energy) to the cold one. The energy flow stops when both bodies have the same temperature, reaching thermal equilibrium. Microscopically the hot body molecules vibrate more intensely than the cold body, transferring this vibration (energy) to the cold body molecules, until both bodies reach the equilibrium temperature which means they have the same mean kinetic energy. The equation that related the heat with temperature variation is:

$$\Delta Q = m.c.\Delta T \tag{1}$$

where ΔQ is the heat amount of energy (cal) acquired or lost. In this equation, $\Delta T = \left(T_f - T_i\right)$ is the temperature variation (T_f = final temperature, T_i = initial temperature); m is the mass (g) and c is the specific heat (cal/g°C).

1.3 Heat propagation

The heat is energy transferred in association to a temperature variation. The heat propagation occurs spontaneously from the hot body to cold body. There are three different mechanisms for heat propagation into materials:

a. **Conduction:** this is the typical heat propagation mechanism that occurs in solids. The vibrating particles inside the body transmit part of their kinetic energy to their neighbors. As a consequence, he heat flows from the higher temperature regions (where molecules have higher mean kinetic energy) to the low temperature ones (where molecules have lower mean kinetic energy). This process, defined as heat flow, is directly proportional to the cross-sectional area, to the temperature difference between the two points and inversely proportional to the thickness (Figure 1).

   The heat conduction equation is also known as Fourier´s law. For the one-dimensional plane wall shown in Figure 1, the heat flow can be defined as:

$$heat\ flow = \frac{\Delta Q}{\Delta t} = k \cdot \frac{A \cdot (T_i - T_f)}{\Delta z} \tag{2}$$

   where ΔQ is the heat amount that propagates through the area A and thickness Δz, Δt is the time interval, T_i e T_f are temperatures initial and final and k is a proportionality constant, called thermal conductivity coefficient. This coefficient is a characteristic of each material and expresses the facility that the heat flows through the material. The materials with a high k value are called good thermal conductors and with low k is a good thermal insulating.

b. **Convection:** this mechanism describes how the heat propagates in fluids (liquids or gases). When heating a container with a fluid, the part closer to the heat source became warmer, decreasing its density. Whence hot fluid goes up and the colder one goes down, and these movements are called convection currents.

c. **Irradiation:** this heat propagation mechanism corresponds to the emission or absorption of electromagnetic heat waves; for instance, the same way that the Sun heats the Earth.

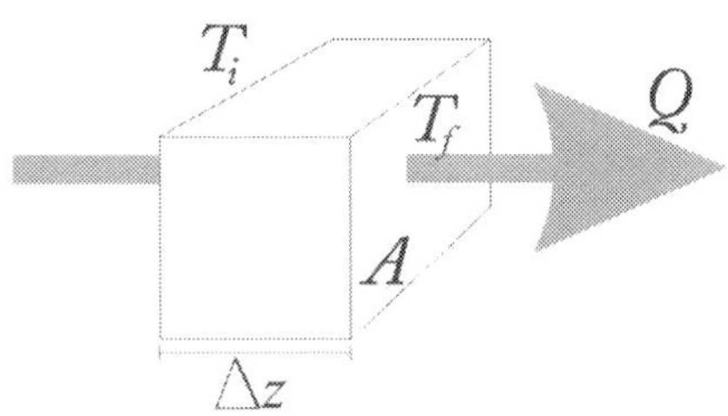

Fig. 1. Schematic diagram of heat propagation in a block having cross section area A and thickness Δz

1.4 Heat generation in a laser-tissue interaction

Considering the interaction of a laser source with the biological tissue, there are some models that should be described: the microscopic and the macroscopic model.

a. Microscopic model

When a photon with energy $h\upsilon$ reach a molecule A, the energy could be absorbed, bringing A to excited state A^*:

$$h\upsilon + A \rightarrow A^* \tag{3}$$

In this way, the molecule A* suffers an inelastic collision (equal energy loss) with the neighbor B (an electron, atom or a molecule), transferring part of its energy and decay to an energy lower state:

$$A^* + B(\varepsilon_{cin}) \rightarrow A + B(\varepsilon_{cin} + \Delta\varepsilon) \tag{4}$$

This process changes part of the laser energy (high frequency electromagnetic waves) into thermal energy (low frequency electromagnetic waves).

The amount $(\varepsilon_{cin} + \Delta\varepsilon)$ represents the increase of the thermal vibration of the particle B and, as a consequence, there is a microscopic temperature increment by the energy absorbed from the photon.

The macroscopic effect of the temperature increase is observed due to the high number of type-A molecules, which absorbed the energy from a high number of photons presented in a laser beam, and transform this energy in thermal vibration. Figure 2 represents the phenomenon described above.

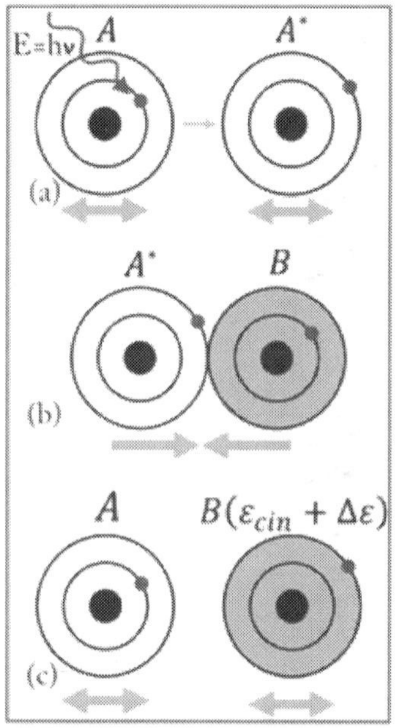

Fig. 2. Schematic diagram of the heat generation in a biological tissue: (a) a photon with the energy hv is applied in the molecule A, then goes to an excited state A^*. (b) The molecule A* collides with B. (c) A* transfers its energy to B; B becomes $B(\varepsilon_{cin} + \Delta\varepsilon)$ and starts to vibrate more intensely

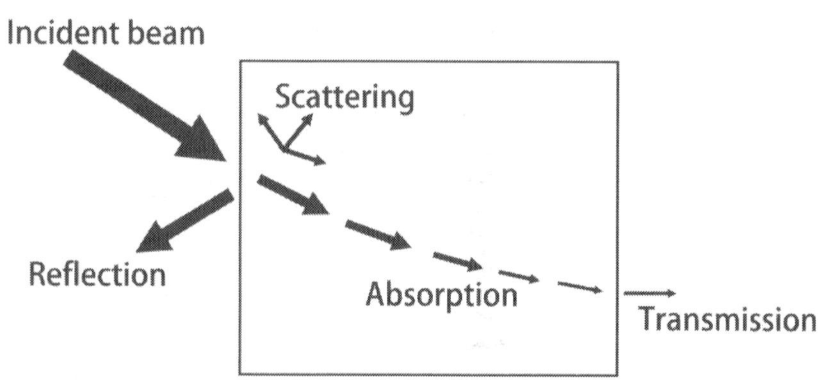

Fig. 3. Schematic diagram showing the laser-tissue interaction: reflection, scattering, absorption and transmission

b. Macroscopic model

In a macroscopic approach it could be observed that the heat generated is directly related with the laser propagation in the tissue. For that, it is convenient to remember how the heat

propagation occurs. When a laser beam irradiates a sample (Figure 3), a part of the beam is reflected and the other part penetrates in the surface. That part which penetrates is attenuated mainly in two different ways: the absorption and scattering - as long as the beam penetrated in the sample.
The absorption and the scattering are characterized for absorption coefficient (μ_a) and scattering coefficient (μ_s), which represents, respectively, the rate of radiation energy loss per penetration length unit, due the absorption and the photons scattering. These two coefficients are specific to each tissue and depend on the laser wavelength.

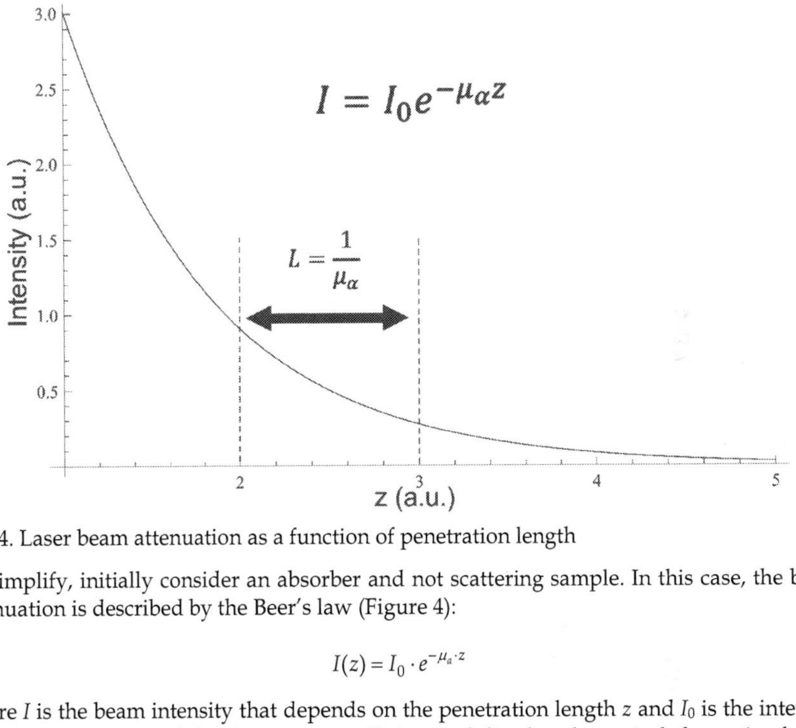

Fig. 4. Laser beam attenuation as a function of penetration length

To simplify, initially consider an absorber and not scattering sample. In this case, the beam attenuation is described by the Beer's law (Figure 4):

$$I(z) = I_0 \cdot e^{-\mu_a \cdot z} \tag{5}$$

where I is the beam intensity that depends on the penetration length z and I_0 is the intensity for $z = 0$. The inverse of the absorption coefficient is defined as the optical absorption length (L) (Figure 4):

$$L = \frac{1}{\mu_a} \tag{6}$$

The generated heat per area unit and per time unit, in a very small thickness Δz, is given by:

$$S(z) = \frac{I(z) - I(z + \Delta z)}{\Delta z} = -\frac{\partial I(z)}{\partial z} = \mu_a \cdot I(z) \tag{7}$$

The equation 7 expresses that the generated heat in the tissue is equal to the absorbed energy and can be described as the absorption coefficient multiplied by the local intensity.

In the most cases, the light is both absorbed and scattered into the sample simultaneously. The beam attenuation continues to be described by a similar law from the Beer's Law, but now the attenuation coefficient is the sum of the absorption and scattering coefficients, which is called total attenuation coefficient $(\mu_T = \mu_a + \mu_s)$.

1.5 Heat propagation in biological tissues

The heat conduction equation in a material medium is given by:

$$\frac{\partial T}{\partial t} = \frac{k}{\rho \cdot c} \nabla^2 T + \frac{S}{\rho \cdot c} \tag{8}$$

where T is the temperature (°C), t is the time (s), k is the thermal conductivity, ρ is the tissue density (g/cm^3), c is the specific heat (cal/g.°C) and S is the generated heat per area and per time (cal/s.cm^2).

This equation can be deduced from the diffusion general equation, but it requires a specific Physics and Mathematical knowledge. Therefore it is important to know that it describes a strong correlation among the temperature temporal variation $\left(\frac{\partial T}{\partial t}\right)$, the temperature spatial variation $(\nabla^2 T)$ and the laser source S. It is also important to say that this same equation also works when the sample is not being irradiated. In order to calculate how the heat propagates after an exposure time, when the laser beam is off, it is only necessary to solve the equation 8 with $S = 0$

There are some other thermal parameters related to the heat propagation. The thermal penetration length is a parameter that describes the propagation extension per time, and it is given by:

$$z_{thermal}(t) = \sqrt{4 \cdot \alpha \cdot t} \tag{9}$$

where $\alpha = \frac{k}{\rho \cdot c}$ is the tissue thermal diffusivity and t is the time. For instance, the thermal diffusivity of water is $\alpha = 1.4 \times 10^{-7}\, m^2/s$.

Other important parameter is the thermal relaxation time, which is obtained mathematically correlating the optical penetration length with thermal penetration length:

$$L = z_{thermal}$$

$$\frac{1}{\mu_a} = \sqrt{4 \cdot \alpha \cdot \tau_{thermal}}$$

$$\tau_{thermal} = \frac{1}{\mu_a^2 \cdot 4 \cdot \alpha} \tag{10}$$

The thermal relaxation time (equation 10) describes the necessary time to the heat propagates from the surface of irradiation until the optical penetration length and is particularly important when the intention is to cause a localized thermal damage, with minimal effect in adjacent structures. This parameter can be interpreted as follows: if the time of the laser pulse is smaller than the relaxation time, the heat would not propagate until a distance given by the optical penetration length L. So the thermal damage will happen only in the first layer where the heat is generated. On the other hand, if the time of the laser pulse is higher than the relaxation time, the heat would propagate for multiple of the optical penetration length, resulting in a thermal damage in a bigger volume to the adjacent structures.

2. Characteristics of dental tissues and their influence on heat propagation

The tooth is composed basically for enamel, dentin, pulp and cementum. Enamel, dentin and cementum are called "dental hard tissues", and the main constituent is represented by the hydroxyapatite (Chadwick, 1997; Gwinnett, 1992) (Figure 5). Dentin and cementum have higher water and organic compound percentage when compared to the enamel and, due to this composition, they are more susceptible to heat storage than the enamel. Dentin has low thermal conductivity values and offers more risk when lasers irradiate in deeper regions, considering that dentinal tubules area and density increase at deepest regions, and subsequently, can easily propagate the generated heat (Srimaneepong et al., 2002). As an example, considering the use of CO_2 lasers in dentistry (wavelength of 9.6 μm or 10.6 μm), the absorption coefficient for dentin tissue is lower than enamel due to its low inorganic content; also, the thermal diffusivity is approximately three times smaller, which can lead a less heat dissipation amount and, as a consequence, can induce higher pulp heating (Fried et al., 1997).

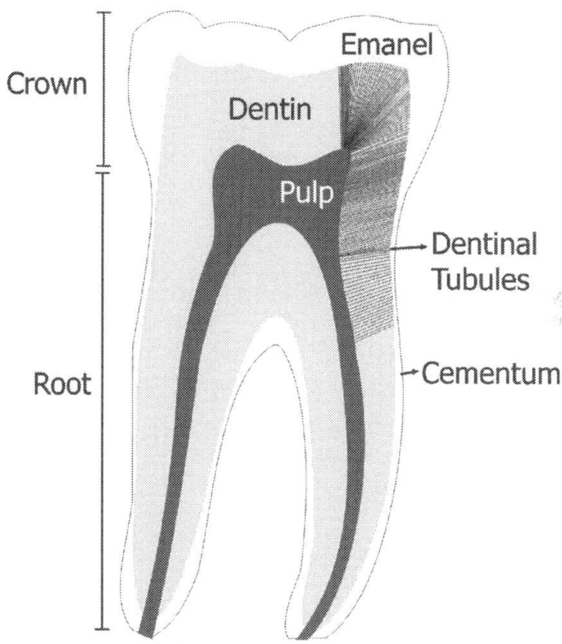

Fig. 5. Representation of a molar tooth, evidencing the macroscopic structures

Dental pulp is a connective and vital tissue, and the higher vascularization makes this tissue strong susceptible to thermal changes. The minimal change in pulp temperature ($\Delta T \leq 5$ °C) is sufficient to alter the microvascularization, the cellular activation and their capacity of hydratation and defense (Nyborg & Brännström, 1968; Zach & Cohen, 1965).
The majority of high intensity lasers used for dental hard tissues cause photothermical and photomechanical effects. Photons emitted at wavelength of visible and near infrared regions of electromagnetic spectrum are poorly absorbed by dental hard tissues (Seka et al., 1996) and, due to this fact, the heat diffusion to the pulp is easy. In this way, in order to choose a parameter of laser irradiation for a clinical application, it is necessary to establish limit energy densities that promote a significant temperature increment on enamel and dentin surface, in order to produce mechanical and/or thermal effects on these structures (Ana et al., 2007). Also, the temperature increment inside the pulp tissue must be bellowing a temperature threshold.
Previous studies have indicated that temperature increments above 5.6 °C can be considered potentially threatening to the vitality of the pulp (Zach & Cohen, 1965) and increments in excess of 16 °C can result in complete pulpal necrosis (Baldissara et al., 1997). Further studies showed levels of 60% and 100% of pulp necrosis when pulp tissue was heated about 11 °C and 17 °C, respectively (Powell et al., 1993). The pulpal temperature rise due to laser-tissue interaction has also been investigated and most of lasers systems promoted an increase in pulpal temperature dependent on the power setting (Ana et al., 2007; Yu et al., 1993; Zezell et al., 1996; Boari et al., 2009).
As well as the knowledge of laser wavelength, energy density and pulse duration, another point to be considered in heat transfer is the tissue characteristics and the influence of the oral environment. Although the calculation of heat transmission and dissipation is performed using hole sound teeth at *in vitro* studies, in clinical situations several characteristics of tissue can change, such as the type of teeth, the remaining thickness, the presence of saliva and the presence of demineralization (Ana et al., 2007; Powell et al., 1993). For instance, due to the great amount of water in carious lesions, the heat transfer to the pulp can be more excessive in decayed teeth. Relating the influence of tissue thickness, White *et al.(1994)* determined that Nd:YAG laser irradiation with a power output of 0.7 W (approximately 87 J/cm^2) induces an increase of 43.2 °C in a remaining dentin thickness of 0.2 mm and induces an increment of 5.8 °C in a dentin thickness of 2.0 mm. Considering that the human teeth present a big variation in volume and weight, and taking into account the low thermal conductivity of dentin, the operator must judge the physical conditions of dental hard tissue in order to adequate the exposition time to avoid dangerous thermal effect on pulp.

3. Changes in tissue thermal characteristics during laser irradiation

Considering the laser irradiation in dental hard tissues, it is necessary to know and to understand the thermal behavior of these tissues when submitted to heating. For that, the evaluation of the heat conduction phenomenon is extremely necessary.
Teeth are mainly composed by hydroxyapatite that, in principle, has high heat capacity value and low heat conduction value (Pereira et al., 2008). The main reason of the changes of thermal parameters of hydroxyapatite can be explained by the complexity of the photon diffusion into the material due to the ionic bond between the chemical elements.
Several studies about thermal parameters measurement in hard dental tissues have been published (Brown et al, 1970; Incropera et al., 2006). Results of these studies are summarized in table 1.

Thermal parameter	Enamel	Dentin	Water
Specific Heat (J/g°C)	0.71 (Brown et al, 1970)	1.59 (Brown et al, 1970)	4.18 (Incropera et al., 2006)
Thermal conductivity (10^{-3}W/cm °C)	9.34 (Brown et al, 1970)	5.69 (Brown et al, 1970)	6.1 (Incropera et al., 2006)
Thermal diffusivity ($10^{-3}cm^2/s$)	4.69 (Brown et al, 1970)	1.86 (Brown et al, 1970)	1.3 (Incropera et al., 2006)

Table 1. Thermal parameters of dental hard tissues (enamel and dentin) and water

Although these thermal values are well-established in literature and can be used for supporting clinical applications, it is important to consider that all parameters were measured at room temperatures. In the moment of laser irradiation of dental hard tissues, the temperature increase can lead several chemical and ultra-structural changes on enamel and dentin (Bachmann et al., 2009; Fowler & Kuroda, 1986); as a consequence, the tissue thermal characteristics of tissue may change during laser irradiation.
Several studies have been developed in order to propose theoretical models of heat propagation in dental hard tissues (Craig R.G & Peyton, 1961; Braden et al., 1964). These models assumed that thermal parameters are constant in function of temperature, which seems to be not true according to the discussed above. Thus, we have to assume that the determination of laser irradiation parameters based only by theoretical calculation that consider thermal properties as constant can be wrong. Figure 6 shows experimental data (Pereira et al., 2008), obtained by infrared thermography, of the thermal diffusivity changes as function of temperature changes.

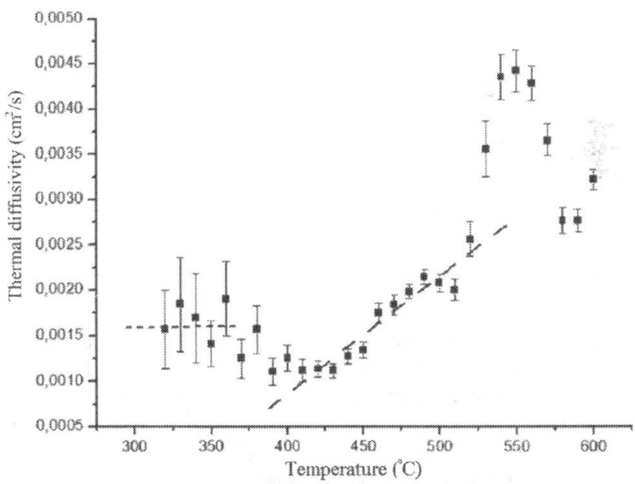

Fig. 6. Thermal diffusivity of dentin as function of temperature (Pereira et al., 2008)

Figure 7 shows the changes on heat penetration on dentin in function of time of exposure. It can be seen that data obtained vary among the related studies due to the fact that some of them consider the thermal diffusivity values always constant, while the present study (Pereira et al., 2008) consider the changes in thermal diffusivity according to the temperature (Figure 7). This fact has significant relevance mainly for clinical procedures using laser irradiation, when it is necessary temperature increases up to 800 °C for cutting dental hard tissues and for caries prevention, for example (Fried et al., 1996; Ana et al., 2007). When a tooth is submitted to this temperature elevation, the heat spreads more quickly than calculated by theoretical models that considered thermal diffusivity values as constant, which can represent a problem mainly for the deeper tissues (pulp tissue).

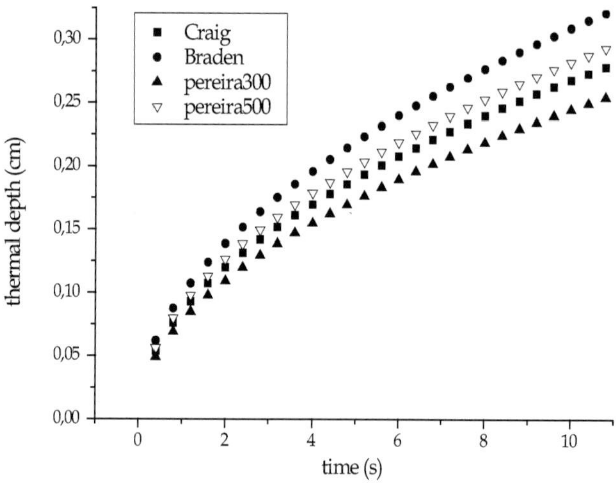

Fig. 7. Calculated thermal depth (cm) in function of time (t) for dentin tissue, obtained by four different literature studies (Pereira et al., 2008)

4. Considering temperature to determine clinical protocols using lasers

As it was stated previously, for the determination of clinical protocols it is demanding to consider the safety and efficacy of lasers, also the characteristics and properties of target tissues. Besides that, literature studies clearly show that laser features, such as wavelength, mode of operation (continuous *versus* pulsed modes), temporal pulse length and repetition rate are characteristics directly related with pulp heating. In this way, among the optical properties, the transmission is the most important property to be considered for preserving pulp vitality.

Among high intensity lasers with high absorption and low transmission through enamel and dentin, erbium lasers seems to be the most appropriated wavelength to be used in dentistry. However, some studies point out that, even with this laser, the repetition rate and pulse duration are decisive on determining clinical parameters; for example, the longer pulse duration is, the higher is the heat generated in pulp (Yu et al., 1993).

Taking into account the clinical application of high intensity lasers on dental hard tissues, some strategies may be useful to control the heat generation and transmission on these tissues. In order to restrict the heat dissipation through the teeth tissues, the application of a photosensitizer is frequently applied over the enamel and dentin surfaces before laser irradiation, and this application can avoid pulpal damages even when laser irradiation occurs with high energy densities (Tagomori & Morioka, 1989; Jennett et al., 1994). The application of a photosensitizer before laser irradiation is commonly used in order to enhance surface tissue absorption in the near-infrared range for ablation and caries prevention actions in dental tissues, considering that some lasers, such as Nd:YAG and Ho:YAG, are poorly absorbed by enamel and dentin. The absorption of the laser beam is increased at the surface of the enamel and the heat produced due to laser absorption in the coating material is transmitted into the adjacent enamel. This technique certifies the deposit of a short laser pulse energy to a small volume of tissue, avoiding the excessive laser beam penetration in deeper dental structures and consequently with less risk of damages in dental pulp (Boari et al., 2009).

The use of Indian Ink is a well-recognized and efficient technique to reduce beam transmission on dental hard tissues. However, because of the difficulty in its removal, which can prejudice the aesthetics of remaining teeth, it has been suggested the application of a coal paste, a mixture of triturated vegetal coal in 50% ethanol, which is biocompatible, easy to remove and presented important results in previous *in vitro* (Boari et al., 2009) and *in vivo* (Zezell et al., 2009) studies. In an *in vitro* study performed by our group, it was demonstrated that the enamel recovering with the coal paste promoted an increase of surface temperatures, which confirmed the absorption of laser beam at the surface (Ana et al., 2007) (Figure 8). Also, the coal paste significantly decreased the heat transfer into the teeth when enamel was irradiated with Nd:YAG and Er,Cr:YSGG lasers, and can assure the pulpal safety when laser irradiation is performed for a long period of time. The

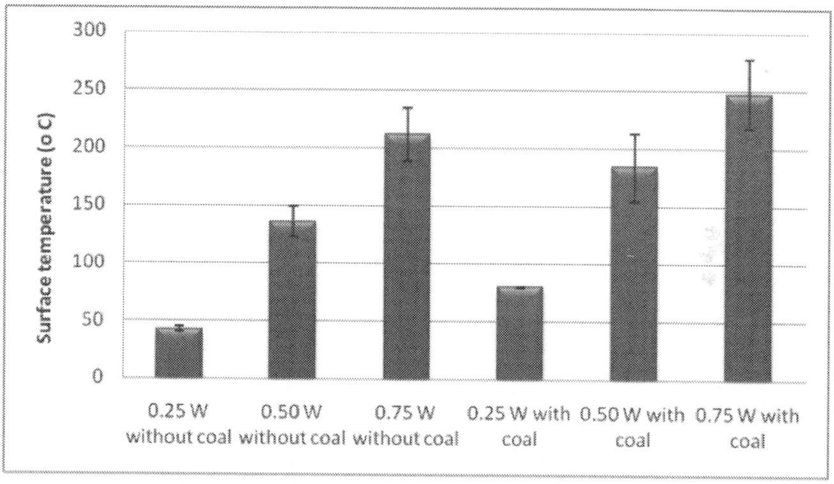

Fig. 8. Surface temperature increase on enamel surface during Er,Cr:YSGG (λ = 2078 nm) laser irradiation with and without the application of coal paste (Ana et al., 2007). It can be noted that, even at three different average powers, the presence of the photosensitizer significantly increased the surface temperature during laser irradiation. Bars mean standard deviation

morphological changes promoted on enamel surface are similar than those promoted by the recovering with Indian Ink, showing evidences of surface heating that promoted melting and recrystallization of enamel (Boari et al., 2009) (Figure 9).

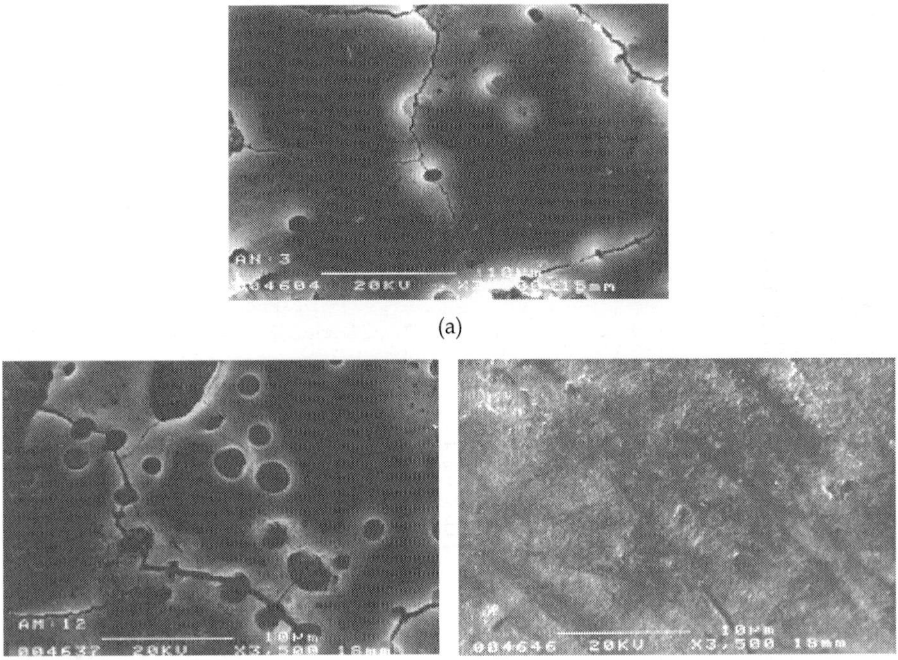

(a)

(b) (c)

Fig. 9. Scanning electron micrography of dental enamel after irradiation with Nd:YAG (λ = 1064 nm) laser irradiation at energy density of 84.9 J/cm^2 after surface recovering with Indian Ink (a) or coal paste (b) or no recovering (c) (Boari et al., 2009). It is possible to note the presence of melting and recrystallization of enamel after recovering with coal paste and Indian Ink. These characteristics are not observed when enamel is irradiated without the presence of a photosensitizer. Original magnification = 3500 X. (a) enamel + Indian Ink; (b) enamel + coal paste; (c) sound enamel

The presence of air-water spray during laser irradiation is another strategy used for clinicians to avoid excessive heat generation on the pulp. The water coolant allows the cleaning of surfaces to be irradiated and increases the efficacy of ablation phenomenon, in a process called "water augmentation" (Fried et al., 2002). When dental hard tissues are irradiated with Er:YAG in addiction to a thin water layer, studies relate that the cutting efficiency increases at the same time that the pulp temperature decreases. However, the thickness of water layer should be well-controlled, considering that erbium lasers interacts primary with water and an thick water layer over the tissue can restrict the laser interaction with the enamel bellow it and, as a consequence, the absorption by the target tissue can decrease.

5. How to determine temperature variations in biological tissues?

Among physical methods to determine the temperatures on materials, the thermocouples (Ana et al., 2007; Boari et al., 2009), elliptical mirrors, HgCdZnTe detectors (Fried et al., 1996) and infrared cameras (Ana et al., 2007) are the most used ones to measure temperature changes in pulp, periodontal tissues and dental hard tissues surfaces. These techniques present good accuracy and efficacy, and can be easily adapted to experimental conditions. However, it should be considered that all experimental methods present some difficulties, such as sample standardization (considering the large variation in volume, size, thickness, and hydratation degree of tissues), the exact duplication of the thermal load, accuracy, availability and cost of equipments (Ana et al., 2008).

The finite element method model (FEM model) is another method that has been popular among researchers, taking into account that this technique is a good analytical tool to model and simulate the thermal or mechanical behavior of dental structures (Toparli et al., 2003). The FEM model can be used to simulate the effects of laser on enamel and dentin, but not on gums or inside the pulp cavity, which is filled with blood vessels and innervated tissues, since these materials are soft and highly inhomogeneous. However, the effects of laser irradiation could be difficult to simulate even on the dental hard tissues (only enamel and dentine), since the thermal characteristics of these materials may not have been well determined.

It must be pointed out that all the *in vitro* methods do not reproduce exactly all the interferences of the oral environment in the photothermal response of enamel and dentin tissues, such as the influence of surrounding saliva, pulpal and periodontal tissues, presence of biofilm, body temperature and other factors (Ana et al., 2008).

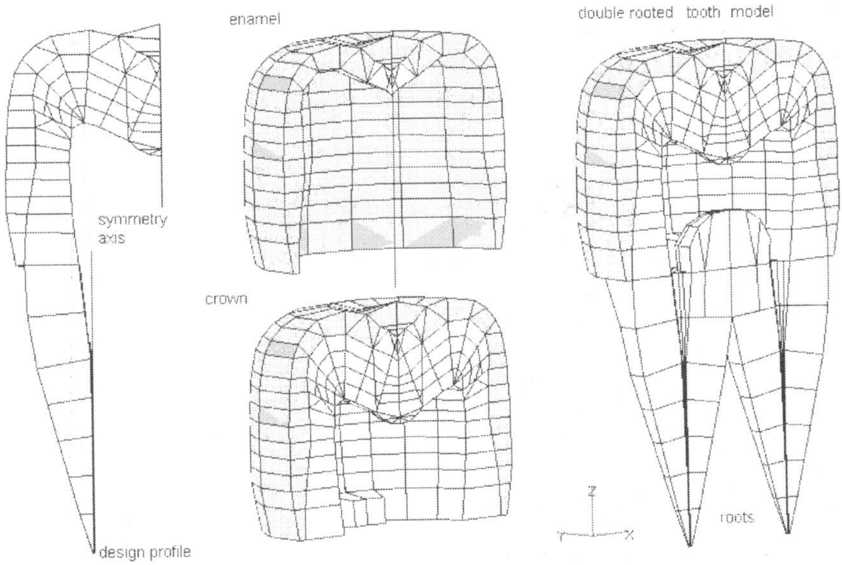

Fig. 10. FEM model used to simulate the heat generation and transmission at dental pulp during Er,Cr:YSGG laser irradiation (Ana et al., 2008)

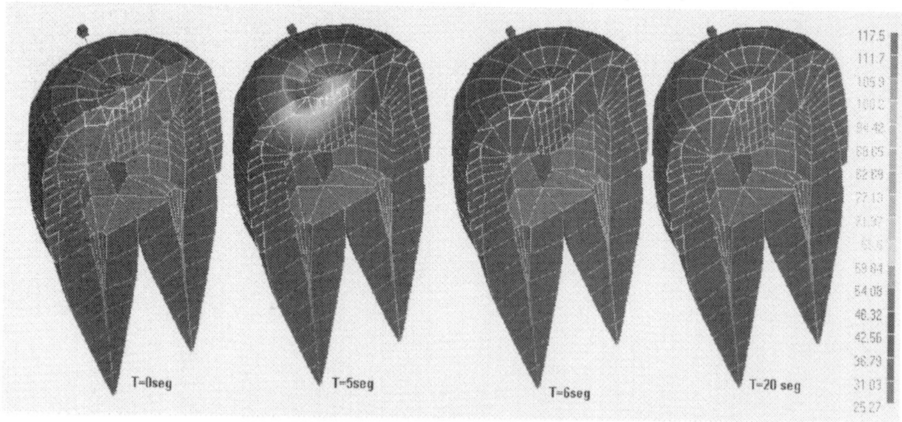

Fig. 11. Calculated temperature distribution: The gray scale represents the temperature in Celsius. The sequence of images illustrates the effects of irradiation with a laser beam and the propagation of heat through the tooth after the beam is turned off

6. Pre-clinical studies

As affirmed previously, the pre-clinical studies are necessary to predict the biological effects of high-intensity lasers and to establish possible parameters and conditions for a future application in dental practice. For preventing dental caries, for example, infrared lasers such as Nd:YAG and Er,Cr:YSGG can be indicated.

The Er,Cr:YSGG laser is emitted in 2.78 μm wavelength, which is better absorbed by water and OH^- contents of hydroxyapatite (Seka et al., 1996), and promotes surface temperatures up to 800 °C at the ablation threshold (Fried et al., 1996). Due to this fact, Er,Cr:YSGG laser is applied for cutting of enamel, dentin and root surfaces, and also for caries prevention.

In a first *in vitro* study, surface temperature measurements were performed in order to verify if this laser had potential to promote chemical and crystalline changes on dental enamel, which can occur in temperatures above 100 °C (Bachmann et al., 2009). For that, the temperature changes in enamel surface during and immediately after laser irradiation were monitored using an infrared high resolution fast thermographic camera (ThermaCam FLIR SC 3000 Systems, USA), which stores infrared images and data at rates up to 900 Hz. This experiment was performed at a controlled room temperature of 24.6 °C, 47 % air relative humidity and considering teeth emissivity as 0.91. The thermographic camera was positioned at 0.1 m distance of samples and the obtained infrared images were recorded at rates of 900 Hz for later analysis (Ana et al, 2007).

For laser irradiation, laser handpiece was positioned at focused beam, at 1 mm distance from the enamel surface. This assembly was kept in optical supports and the area of interest was isolated at a focal length of 0.1 m using an internal macro lens.

The results of surface temperature obtained in this study (Ana et al., 2007) are shown in Figure 12. It is possible to evidence that the surface temperature rises with the increase of energy density, and the presence of photosensitizer (coal paste) propitiated higher temperature values when compared to the surfaces in which were not previously recovered with the coal paste. In this way, even using laser wavelengths highly absorbed by dental enamel, the application of a photosensitizer in the enamel surface can potentiate the absorption phenomenon. This can reflect on temperature rise and, in this way, crystalline changes at this superficial enamel may occur and can favor the caries preventive effect.

Another point to be considered is that the temperature rise of 247.6 °C found when teeth were irradiated with 8.5 J/cm² is lower than that temperature reported by literature studies, who found approximately 400 °C measured by an elliptical mirror and a HgCdZnTe detector with a time resolution of 1 μs (Fried et al., 2006). In this way, the temperature elevation during Er,Cr:YSGG laser irradiation could be higher than those detected by infrared camera. Taking into account that the pulse width of Er,Cr:YSGG laser is 140 μs, even the 900 Hz recording rate of the infrared thermographic method seems to be unable to detect the highest temperature peaks during laser irradiation. In this way, the infrared thermographic camera gives an idea of average temperature changes when teeth are irradiated with high intensity lasers. However more accurate systems are required to precisely determine the maximum temperature peaks, such as the use of integrating sphere.

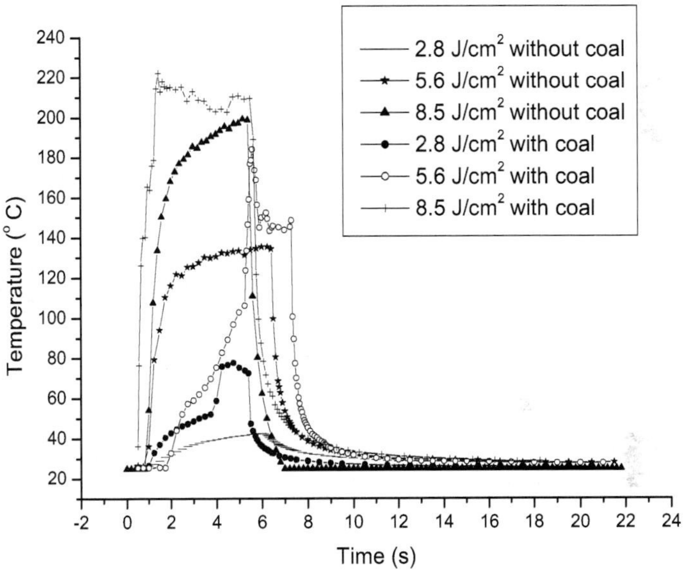

Fig. 12. Dental enamel surface temperature measurement, by infrared thermography, during Er,Cr:YSGG laser irradiation at parameters aimed at caries prevention (Ana et al., 2007)

For measurement of heat transfer to the pulp chamber, the use of fast-response thermocouples seems to be more accurate since it is not possible to see the heat transfer from enamel to pulp by infrared thermography, unless the teeth are half-sectioned.

In this way, calibrated K-type chromel-alumel thermocouples (Omega Engineering, Stanford, USA) were inserted inside the pulp chamber of sound molar human teeth, which were previously filled with a thermally-conductive paste (thermal conductivity of 0.4 cal s^{-1} m^{-1} K^{-1} - Implastec, Votorantim, Brazil) in order to keep thermal contact between the probe end and dentin surface. These thermocouples had 0.05 mm diameter probe and were sensitive to temperature variations between 0.1 °C and 100 °C. The temperature sensitive end of the probe was placed at the closest distance to the area to be irradiated, and its location was controlled radiographically for each sample (Romano et al., 2011). The thermocouple apparatus was connected to an analogue-to-digital converter (SR lock-in amplifier, Stanford Research System, USA) linked to a computer, and time and temperature data were recorded at sampling rate of 20 Hz, with temperature resolution of 0.1° C. During laser irradiation, samples were fixed and immersed in a water-filled heating circulator at standardized temperature of 37 °C, with only the coronal part of the tooth not being submerged in order to simulate body temperature in the oral environment.

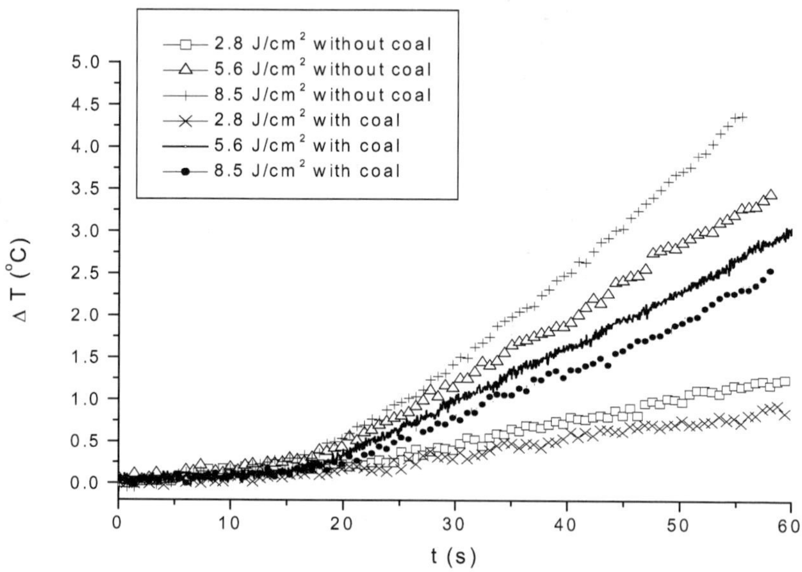

Fig. 13. Pulp chamber temperature variation during enamel surface irradiation with Er,Cr:YSGG laser, measured by fast-response thermocouples (Ana et al., 2007)

Figure 13 shows the results of temperature evaluation inside the pulp chamber during enamel irradiation by Er,Cr:YSGG laser. It is possible to observe that the device can detect minimal temperature variations and, although surface temperatures detected increased up to 230 °C, the pulp temperature variations were up to 4.5 °C. This fact evidences that dental enamel and dentin are good thermal insulating tissues. Also, the presence of the photosensitizer on enamel surface was important to effectively reduce the heat transfer through the pulp chamber, increasing the safety of a future clinical procedure. It must to be emphasized that the time of exposure is important and further histological *in vivo* studies are also necessary to confirm this hypothesis.

All studies described previously were performed to evaluate just one possible clinical application of Er,Cr:YSGG laser. Considering that this laser can be used for multiple applications, the development of a FEM model could be a faster tool to evaluate the heat generation and transfer to dental hard tissues. In this way, a further study was performed to develop this model and to compare the simulation results with those obtained experimentally (Ana et al., 2008).

For that, a geometric FEM model of a half-sectioned double rooted molar tooth was constructed using a typical profile of a tooth root. Since the goal was to calculate the temperature distribution on the surface and inside the tooth, in each element of the model the total heat is given by the internal heat, determined by the material density (ρ) and specific heat (**c**), and the heat flux, determined by the element material thermal conductivity (α). In this study, the external heat source is due to almost instantaneous light absorption converted into heat. Moreover, in spite of not considering wavelength dependence with absorption, reflection, transmission and scattering, the FEM model predicts accurate values for temperatures inside the teeth, mainly because 2.78 μm is strongly absorbed by the dental hard tissue. The optical penetration is very small (to the order of few micrometers) since the optical absorption coefficient of enamel is about 7000 cm^{-1}.

The *in vitro* experiment aimed to compare the results of FEM simulation was performed using half-sectioned teeth, irradiated with Er,Cr:YSGG laser on enamel surface and monitored by infrared imaging. This set-up allowed the visual evaluation of temperature changes and heat diffusion from enamel into dentin and the pulp chamber. Taking into account that the variation in temperature is also dependent on the volume and weight of a tooth, the use of sectioned teeth does not correspond to an *in vivo* condition inside the mouth, but gives a reasonable idea of heat transfer inside the tooth and gives the exact value of surface temperature during laser irradiation. Moreover, in a clinical protocol, laser irradiation is performed by scanning all over the enamel surface, and it was not possible to reproduce this situation in the present study. However, in a clinical application of lasers, the time of irradiation in just one region of tissue is always less than the time considered during experiments, which increases the safety of evaluated parameters.

Energy Density	**Presence of photosensitizer**	**Infrared thermographic camera**		**Finite element method model**	
		Surface	Pulp	Surface	Pulp
	yes	79.6 °C	0.5 °C	94.2 °C	0.95 °C
2.8 J/cm²	no	42.7 °C	1.0 °C	66.0 °C	0.9 °C
	yes	184.1 °C	1.5 °C	178.0 °C	1.9 °C
5.6 J/cm²	no	136.1 °C	1.2 °C	136.0 °C	1.4 °C
	yes	247.6 °C	2.1 °C	231.0 °C	2.1 °C
8.5 J/cm²	no	211.8 °C	1.9 °C	188.0 °C	2.3 °C

Table 2. Comparison of means of temperature rise on surface and in pulp obtained by FEM model and by thermographic camera (Ana et al., 2008)

Table 2 shows the results of temperature monitoring by infrared thermography and the simulation by FEM model. It is possible to observe a good correlation between the *in vitro* and computational method, indicating that FEM model can be used as an alternative to determine heat generation on the enamel surface as well as inside the pulp chamber according to the several laser energies and the presence or not of a photosensitizer. Although it is not possible to simulate the influence of oral environment on FEM model, under the given conditions the simulated model was shown to have a good approximation to the physical reality.

7. Conclusion

The effect of high intensity lasers irradiation on biological tissues and consequently their clinical uses are based on heat generation, which is necessary to assure effective clinical procedures such as faster cutting, good homeostasis and desired chemical changes on target tissues. The comprehension of heating generation and transmission through these tissues is essential to determine safe irradiation parameters of lasers and, for that, the knowledge of optical and thermal properties of tissues and their changes due to heat are strongly necessary. Also, the interaction of laser wavelength with the tissues is necessary to avoid deleterious effects in target tissues, as well in surrounding ones.

The pre-clinical experiments give us important information about laser-tissue interaction, and help to suggest laser parameters and conditions for development of a further clinical protocol. For that, methods such as thermocouples, infrared thermography and finite element simulation are good tools that demonstrated to be useful on predicting the clinical results.

8. References

Ana PA, Bachmann L, Zezell DM. Laser effects on enamel for caries prevention. *Laser Phys* 2006; 16(5): 865 – 875.

Ana PA, Blay A, Miyakawa W, Zezell DM. Thermal analysis of teeth irradiated with Er,C:YSGG laser at low fluences. *Laser Phys Lett* 2007; 4: 827 - 830.

Ana PA, Velloso WF Jr, Zezell DM. Three-dimensional finite element thermal analysis of dental tissues irradiated with Er,Cr:YSGG laser. *Rev Sci Instrum*. 2008; 79(9): 093910.

Bachmann L, Rosa K, Ana PA, Zezell DM, Craievich AF, Kellermann G. Crystalline structure of human enamel irradiated with Er,Cr:YSGG laser. *Laser Phys Lett* 2009; 6: 159-162.

Baldissara P, Catapano S, Scotti R. Clinical and histological evaluation of thermal injuty thresholds in human teeth: a preliminary study. *J Oral Rehabilitation* 1997; 24: 791-801.

Boari HGD, Ana PA, Eduardo CP, Powell GL, Zezell DM. Absorption and thermal study of dental enamel when irradiated with Nd:YAG laser with the aim of caries prevention. *Laser Phys* 2009; 19(7): 1463-1469.

Braden M. heat conduction in teeth + effect of lining material. *J Dent Res* 1964, 43 (3): 315.

Brown WS, Dewey WA, Jacobs HR. Thermal properties of teeth. *J Dent Res* 1970; 49(4): 752-755.

Chadwick DJ, Cardew G. Dental Enamel. London UK: Chichester-Wiley 1997.

Craig R.G, peyton F.A. Thermal conductivity of Tooth Structure, Dental cements, and amalgam. *J Dent Res* 1961, 40 (3): 411.

Fowler BO, Kuroda S. Changes in heated and in laser-irradiated human tooth enamel and their probable effects on solubility. *Calcif Tissue Int* 1986; 38: 197-208.

Fried D, Feathestone JDB, Visuri SR, Seka WW. Walsh JT The caries inhibition potential of Er:YAG and ErCr:YSGG laser irradiation. In: Wigdor HA, Featherstone JDB, White JM, Neev J. *Lasers in Dentistry* II Bellinghan WA Proc SPIE 1996; 2672: 73-77.

Fried D, Zuerlein MJ, Featherstone JDB, Machule D. Thermal and chemical modification of dentin by pulsed CO2 laser irradiation at 9-11 μm. In: Wigdor HA, Featherstone JDB, Rechmann P. *Lasers in Dentistry* III Bellinghan WA: Proc SPIE 1997; 2973: 94-100.

Fried D, Ashouri N, Breunig T, Shori R. Mechanism of water augmentation during IR laser ablation of dental enamel. *Lasers Surg Med* 2002; 31: 186 – 193.

Gwinnett AJ. Structure and composition of enamel. *Oper Dent* 1992; Supl 5: 10-17.

Incropera F P., DeWitt D P, Bergman T L, Lavine A S. Fundamentals of Heat and Mass Transfer. 6th ed., New York: Wiley, 2006.

Jennett E, Motameli M, Rastegar S, Frederickson C, Arcoria C, Powers JM. Dye-enhanced ablation of enamel by pulsed lasers. *J Dent Res* 1994; 73: 1841-1847.

Nammour S, Kowalyk K, Valici C, Zeinoun T, Rocca JP, Powell GL, Van Reck J. Safety parameters for pulp temperature during selective ablation of caries by KTP laser in vitro. *J Clin Laser Med Surg* 2004; 22(2): 99-104.

Niemz MH. Laser-tissue interactions: fundamentals and applications. 3rd ed., Berlin: Springer, 1997.

Nyborg H, Brännström M. Pulp reaction to heat. *J Prosthet Dent* 1968; 19: 605 – 612.

Pereira T. M.; Miyakawa, W; Goulart VP; Zezell, DM. In-plane thermal diffusivity of dentin by infrared thermography. In: *VII Congresso da SBPMAT*, 2008.

Powell GL, Morton TH, Whisenant BK. Argon laser oral safety parameters for teeth. *Lasers Surg Med.* 1993; 13(5):548-52.

Romano AC, Aranha AC, Silveira BL, Baldochi SL, Eduardo CP. Evaluation of carbon dioxide laser irradiation associated with calcium hydroxide in the treatment of dentinal hypersensitivity. A preliminary study. *Lasers Med Sci.* 2011; 26(1): 35-42.

Seka W, Featherstone JDB, Fried D, Visuri SR, Walsh JT. Laser ablation of dental hard tissue: from explosive ablation to plasma-mediated ablation. In: Wigdor HA, Featherstone JDB, White JM, Neev J. *Lasers in Dentistry* II Bellinghan WA. SPIE 1996; 2672: 144-158.

Srimaneepong V, Palamara JEA, Wilson PR. Pulpal space pressure and temperatura changes from ND:YAG laser irradiation of dentin. *J Dent* 2002; 30: 291-296.

Tagomori S, Morioka T. Combined effects of laser and fluoride on acid resistance of human dental enamel. *Caries Res* 1989; 54: 15-17.

Toparli M, Sasaki S. Finite element analysis of the temperature and thermal stress in a postrestored tooth. *J Oral Rehabil.* 2003; 30(9): 921-6.

White JM, Mark CF, Goodis HE. Intrapulpal temperature during pulsed Nd:YAG laser treatment on dentin in vitro. *J Periodontol* 1994; 65: 255-259.

Yu D, Powell GL, Higuchi WI, Fox JL. Comparison of three lasers on dental pulp chamber temperature change. *J Clin Laser Med Surg* 1993; 11(3): 119-122.

Zach L, Cohen G. Pulp response to externally applied heat. *Oral Surg* 1965; 19: 515-530.

Zezell DM, Cecchini SCM, Pinotti M, Eduardo CP. Temperature changes under Ho:YLF irradiation. In: Wigdor H, Featherstone JDB, White JM, Neev J. *Lasers in Dentistry* II Bellinghan WA: Proc SPIE 1996; 2672: 34-39.

Zezell DM, Boari HG, Ana PA, Eduardo CP, Powell GL. Nd:YAG laser in caries prevention: a clinical trial. *Lasers Surg Med.* 2009; 41(1): 31-5.

14

Entransy Dissipation Theory and Its Application in Heat Transfer

Mingtian Xu
Shandong University
P. R. China

1. Introduction

In this chapter, our recent work on the entransy dissipation theory and its application in heat convection and the heat exchanger design is reviewed. In that work, the thermodynamic basis of the entransy dissipation theory is established. It is shown that the entransy is a state variable and the second law of thermodynamics can be described by the entransy and entransy dissipation. Then this entransy dissipation theory is applied to the convective heat transfer and heat exchanger design. The local expression of the entransy dissipation rate is obtained. The extremum principle of entransy dissipation rate is proposed. The expressions of the entransy dissipation caused by heat conduction and fluid friction in heat exchanger are derived. The optimization design of heat exchanger is discussed.

1.1 The second law of thermodynamics in terms of entransy

The second law of thermodynamics is one of the most important fundamental laws in physics, which originates from the study of the efficiency of heat engine and places constraints upon the direction of heat transfer and the attainable efficiencies of heat engines (Kondepudi & Prigogine, 1998). The concept of entropy introduced by Clausius for mathematically describing the second law of thermodynamics has stretched this law across almost every discipline of science. However, in the framework of the classical thermodynamics the definition of entropy is abstract and ambiguous, which was noted even by Clausius (Clausius, 1865). This has induced some controversies for statements related to the entropy. Recently, Bertola and Cafaro found that the principle of minimum entropy production is not compatible with continuum mechanics (Bertola & Cafaro, 2008). Herwig showed that the assessment criterion for heat transfer enhancement based on the heat transfer theory contradicts the ones based on the second law of thermodynamics (Herwig, 2010). The entropy generation number defined by Bejan (Bejan, 1988) is not consistent with the exchanger effectiveness which describes the heat exchanger performance (Guo et al., 2009b). Shah and Skiepko (2004) found that the heat exchanger effectiveness can be maximum, minimum or in between when the entropy generation achieves its minimum value for eighteen kinds of heat exchangers, which does not totally conform to the fact that the reduction of entropy generation leads to the improvement of the heat exchanger performance. These findings signal that the concepts of entropy and entropy generation may not be perfect for describing the second law of thermodynamics.

Although there has been effort to modify the expression of the second law of thermodynamics (Bizarro, 2008; Ben-Amotz & Honig, 2003, 2006) and to improve the classical thermodynamics by considering the Carnot construction cycling in a finite time (den Broek, 2005; Esposito & Lindenberg, 2009; Esposito et al., 2010), the eminent position of entropy in thermodynamics has not been questioned. Recently, Guo et al. (2007) defined two new physical quantities called entransy and entransy dissipation for describing the heat transfer ability and irreversibility of heat conduction, respectively. Guo et al. (2009a) have introduced a dimensionless method for the entransy dissipation and defined an entransy dissipation number which can serve as the heat exchanger performance evaluation criterion. Based on the concept of entransy dissipation, an equivalent thermal resistance of heat exchanger was defined which is consistent with the exchanger effectiveness (Guo et al., 2010). Cheng and Liang (2011) defined the entransy flux and entransy function for the thermal radiation in enclosures with opaque surfaces, and the minimum principle of radiative entransy loss was established. Chen et al. (2011) proposed an entransy dissipation rate minimization approach for the disc cooling system and the influence of various system parameters on the entransy dissipation rate of the cooling system has been investigated.

Although the concepts of entransy and entransy dissipation have been applied to heat transfer and demonstrate some advantages in comparison with the entropy and entropy generation, how to define these concepts from the thermodynamic point of view is still an open question. In this section we place the concepts of entransy and entransy dissipation on the solid thermodynamic basis.

1.2 Carnot's theorem in terms of entransy

We start with the Carnot cycle. In this cycle, the heat engine absorbs heat Q_1 from the hot reservoir with the temperature T_1 (absolute temperature is always assumed in the following discussion), converts part of heat to work W and discards the rest of the heat to the cold reservoir with the temperature T_2. For this cycle, Carnot's theorem states that (Kondepudi & Prigogine, 1998)

$$Q_1 / T_1 \leq Q_2 / T_2 \tag{1}$$

where the equality and inequality correspond to the reversible and irreversible heat engines, respectively. The efficiency of a reversible engine is defined as

$$\eta = 1 - \frac{T_2}{T_1} \tag{2}$$

Carnot's theorem dictates that reversible engines have the maximum efficiency (Kondepudi & Prigogine, 1998).

Equivalently, Inequality (1) can be rewritten as follows

$$Q_1(T_1 - T_2) \geq (Q_1 - Q_2)T_1 \tag{3}$$

Inspired by Inequality (3), we define $E = Q_1(T_1 - T_2)$ as the entransy gained by the heat engine from the hot reservoir in the Carnot cycle. From this definition, one can see that the larger the amount of heat Q_1 and the temperature difference between the hot and cold reservoir, the greater the entransy obtained by the heat engine in the Carnot cycle. Obviously, the larger entransy means higher ability for heat engine to perform work. Guo et al. (2007) shown that the entransy can describe the heat conduction ability. Therefore, we may say that the entransy defined here quantifies the energy transfer ability including the ability to deliver work and transfer heat.

To obtain the efficiency of a reversible heat engine, firstly we know that $W = Q_1 - Q_2$, which is the statement of the law of conservation of energy. We define $E_W = (Q_1 - Q_2)T_1 = WT_1$ and regard it as the entransy consumed by the heat engine for delivering work W to the system's exterior. For the reversible heat engine, Inequality (3) reduces to:

$$Q_1(T_1 - T_2) = (Q_1 - Q_2)T_1 = WT_1 \tag{4}$$

which says that the entransy gained by the reversible heat engine in the Carnot's cycle is completely converted to work W. While for the irreversible heat engine, Inequality (3) becomes

$$Q_1(T_1 - T_2) > (Q_1 - Q_2)T_1 \tag{5}$$

which tells us that only part of the entransy obtained by the heat engine in the Carnot's cycle is utilized to deliver work W, the rest is consumed by the irreversibility in the heat engine. In order to quantify the irreversibility occurring in the heat engine, we define the entransy generation in parallel with the entropy generation in the following way:

$$E_g = (Q_1 - Q_2)T_1 - Q_1(T_1 - T_2) \tag{6}$$

While $E_{diss} = -E_g$ is called the entransy dissipation which represents the entransy consumed by the irreversibility in heat engine. Since the reversible heat engine converts the total entransy gained during the Carnot cycle to perform work, we may define the efficiency of the heat engine in terms of entransy as follows

$$\eta_E = \frac{WT_1}{Q_1(T_1 - T_2)} = \frac{E_W}{E} \tag{7}$$

Notice that the reversible heat engine achieves the maximum value of η_E which is equal to 100%. This is the statement of Carnot's theorem in terms of entransy.

1.3 The second law of thermodynamics

Carnot's theorem has played a pivotal role on the development of the classical thermodynamics. By generalizing Inequality (1) to an arbitrary cycle, Clausius introduced the concept of entropy which is a physical quantity as fundamental and universal as energy. In this section, we aim to generalize Inequality (3) to an arbitrary cycle. Firstly, Inequality (3) can be rewritten as follows:

$$Q_1\Delta T + T_1\Delta Q \geq 0 \tag{8}$$

where $\Delta T = T_1 - T_2$, $\Delta Q = -(Q_1 - Q_2)$. Note that since the system delivers energy to its exterior, therefore $\Delta Q < 0$.

An arbitrary cycle can be decomposed into a group of Carnot's cycles denoted as $C_i\,(i = 1,2,\cdots n)$ (Kondepudi & Prigogine, 1998). Applying Inequality (8) to the i-th Carnot's cycle yields:

$$Q_{a,i}\Delta T_i + \Delta Q_i T_{h,i} \geq 0 \tag{9}$$

where $Q_{a,i}$ is the heat absorbed from the hot reservoir with the temperature $T_{h,i}$, $\Delta Q_i = -(Q_{a,i} - Q_{r,i})$ ($Q_{r,i}$ is the heat discarded into the cold reservoir), $\Delta T_i = T_{h,i} - T_{c,i}$ ($T_{c,i}$ is the temperature of the cold reservoir). If the number of Carnot's cycles under consideration tends to the infinity, the temperature difference between the hot and cold reservoirs approaches to infinitesimal, Inequality (9) becomes

$$\oint QdT + T\delta Q \geq 0 \tag{10}$$

For a reversible cycle, there is no heat conduction, Inequality (10) reduces to

$$\oint QdT + T\delta Q = 0$$

Equation (11) suggests that we can define a quantity E called the entransy as follows

$$dE = QdT + T\delta Q \tag{11}$$

which only depends on the initial and final states of a reversible process. Thus the entransy is a state variable. If E_A and E_B are values of this variable in the initial state A and final state B, respectively, we have

$$E_B - E_A = \int_A^B QdT + T\delta Q \tag{12}$$

If the temperature remains fixed, it follows from Eq. (12) that for a reversible flow of heat δQ, the change in entransy is $T\delta Q$, while the change in entropy is $\delta Q / T$. For this case we get the following relationship between the entransy and entropy,

$$dE = T^2 dS \tag{13}$$

Consider an irreversible process I which starts from the equilibrium state A and ends at the equilibrium state B. In order to form a cycle, we add a reversible process R from the state B to state A. Then from Inequality (10), we have

$$\int_{A_I}^{B_I} QdT + T\delta Q > \int_{A_R}^{B_R} QdT + T\delta Q \tag{14}$$

The subscripts R and I represent the reversible and irreversible processes, respectively. The application of Eq. (12) on Inequality (14) yields

$$\int_{A_I}^{B_I} QdT + T\delta Q > E_B - E_A \tag{15}$$

Thus in parallel with the entropy generation we can define the entransy generation E_g of the irreversible process as follows

$$E_g = (E_B - E_A) - \int_{A_I}^{B_I} QdT + T\delta Q < 0 \tag{16}$$

Note that for the reversible process $E_g = 0$. We define $E_{diss} = -E_g$ as the entransy dissipation. Therefore, the entransy dissipation quantifies the entransy consumed by the irreversibility in the irreversible process. For an irreversible process, the second law of thermodynamics states that the entransy generation is always negative or the irreversible process always decreases the system's ability to do work and transfer heat. From the entransy's definition (12), it is evident that the entransy is an extensive quantity. Subsequently, Inequality (16) is also valid under the local equilibrium assumption.
We may express the system's change in entransy as a sum of two parts

$$dE = d_eE + d_iE \text{, and } d_iE \le 0 \tag{17}$$

In which d_eE is the entransy change due to exchange of matter and energy with the exterior of the system and d_iE is the entransy generation produced by the irreversible processes occurring in the system. For the closed systems that exchange energy, but not matter, we have

$$d_eE = QdT + T\delta Q \text{ and } d_iE \le 0 \tag{18}$$

For open systems that exchange both matter and energy:

$$d_eE = QdT + T\delta Q + (d_eE)_{matter} \text{ and } d_iE \le 0 \tag{19}$$

Finally, for the isolated systems, we have

$$d_eE = 0 \text{, and } d_iE \le 0 \tag{20}$$

Therefore, in the isolated systems the entransy never increases, namely the energy transfer ability can not increase.

1.4 Entransy dissipation due to heat conduction

Now let us consider the heat conduction process discussed by Kondepudi & Prigogine (1998). The system under consideration is an isolated system and consists of two parts, each having a well-defined temperature. Let the temperatures of two parts be T_1 and T_2 ($T_1 > T_2$), respectively. δQ is the amount of heat flow from the hotter part to colder part in a time period *dt*. Since this isolated system does not exchange entransy with its exterior, $d_eE = 0$. Assume the volume of each part is constant, thus $dW = 0$. The energy change in each part is solely due to the flow of heat: $dU_i = \delta Q_i$ $(i = 1, 2)$. In accordance with the first law, the heat gained by one part is equal to the heat lost by the other. Therefore, $-\delta Q_1 = \delta Q_2 = \delta Q$. The total change in entransy d_iE of the system is the sum of the changes of entransy in each part due to the flow of heat

$$d_iE = -T_1\delta Q + T_2\delta Q = -(T_1 - T_2)\delta Q \tag{21}$$

In terms of the rate of flow of heat $\delta Q / \delta t$, the rate of entransy generation can be written as follows

$$\frac{d_iE}{dt} = -(T_1 - T_2)\frac{\delta Q}{\delta t} \tag{22}$$

The rate of heat flow $J_Q = \delta Q / \delta t$ is given by the Fourier law of heat conduction $J_Q = k(T_1 - T_2)$, in which *k* is the coefficient of heat conductivity, therefore

$$\frac{d_i E}{dt} = -k(T_1 - T_2)^2 \le 0 \tag{23}$$

Accordingly, the rate of entransy dissipation is written as

$$\frac{d_i E_{diss}}{dt} = k(T_1 - T_2)^2 \ge 0 \tag{24}$$

This equation is consistent with the expression of the entransy dissipation function obtained by Guo et al. (2007) from the heat conduction equation. Due to the flow of heat from the hot part to cold part, the temperatures of both parts eventually become equal, and the entransy dissipation rate tends to zero. Then the system reaches the equilibrium state. Therefore, the entransy dissipation rate must vanish at the state of equilibrium, which is called as the principle of minimum entransy dissipation rate. This is the counterpart of the principle of the minimum entropy production. Generally, for three-dimensional steady heat conduction without heat source, the entransy dissipation rate is expressed as

$$\frac{d_i E_{diss}}{dt} = \int_V \frac{1}{2} k(\nabla T)^2 dV \tag{25}$$

where V is the volume of the heat conduction medium. The principle of minimum entransy dissipation rate is mathematically formulated as

$$\delta \int_V \frac{1}{2} k(\nabla T)^2 dV = 0 \tag{26}$$

which is consistent with the least entransy dissipation principle established by Guo et al. (2007). By the variational method, Eq. (26) is equivalent to the following Euler-Lagrange equation:

$$\frac{\partial}{\partial x}\left(\frac{\partial F}{\partial(\frac{\partial T}{\partial x})} \right) + \frac{\partial}{\partial y}\left(\frac{\partial F}{\partial(\frac{\partial T}{\partial y})} \right) + \frac{\partial}{\partial z}\left(\frac{\partial F}{\partial(\frac{\partial T}{\partial z})} \right) = 0 \tag{27}$$

with

$$F = \frac{1}{2} k(\nabla T)^2 \tag{28}$$

Substituting Eq. (28) into Eq. (27) gives

$$\frac{\partial^2 T}{\partial x^2} + \frac{\partial^2 T}{\partial y^2} + \frac{\partial^2 T}{\partial z^2} = 0 \tag{29}$$

which is exactly the governing equation of the steady heat conduction without heat sources based on the Fourier law. While Bertola and Cafaro (2008) found that the principle of minimum entropy production can not lead to the governing equation of the steady Fourier heat conduction. In this point, the principle of minimum entransy dissipation rate demonstrates an obvious advantage.

1.5 Concluding remarks

Note that the concept of entropy can be replaced with entransy for describing the second law of thermodynamics. In comparison with entropy, the entransy has a definite physical meaning and the principle of minimum entransy dissipation rate avoids the contradiction between the principle of minimum entropy production and the classical Fourier heat conduction theory. Therefore, we anticipate that the concepts of entransy and entransy dissipation may help us to gain more profound insight on thermodynamics in particular and on science in general.

2. The application of entransy dissipation theory in heat convection

In this section, the entransy dissipation theory developed in Section 1 are applied to analyze the heat convection.

2.1 Introduction

The entropy and entropy generation can help us to deeply understand the momentum and heat transfer (Bejan, 1982; Herwig, 2010). Bejan (1982) realized that in order to improve the performance of the heat transfer enhancement or thermal insulation equipments, one need to reduce the entropy generation rate. Similarly, according to the definition of entransy dissipation given in Section 1.3, it is required to minimize the entransy dissipation rate for achieving the best heat transfer enhancement and thermal insulation. Therefore, it is of great value to derive the expression of the local rate of entransy dissipation rate for heat convection. Xu et al. (2009) have managed to get an expression of the local rate of entransy dissipation rate for heat convection. However, it lacks the theoretical basis in this derivation. In Section 2.2 with the help of the second law of thermodynamics in terms of entransy and entransy dissipation established in Section 1.3 we will make the derivation more rigorously.

2.2 Local thermodynamic entransy dissipation in heat convection

The infinitesimal element as shown in Fig. 1 is an open thermodynamic system, where $[v_x,v_y]^T$ is the velocity, $[q_x,q_y]^T$ is the heat flux. For this system, we assume that the thermodynamic state is irrelevant with the position, but relevant with time. By the second law of thermodynamics for the open system expressed as Eqs. (17) and (19) the rate of local thermodynamic entransy dissipation generation $\dot{E}_g$ per unit volume in the infinitesimal element is expressed as

$$\begin{aligned}\dot{E}_g dxdy &= (T+\frac{\partial T}{\partial x}dx-T)q_x dy+(T+\frac{\partial T}{\partial y}dy-T)q_y dx\\&+T(q_x-q_x-\frac{\partial q_x}{\partial x}dx)dy+T(q_y-q_y-\frac{\partial q_y}{\partial y}dy)dx\\&+(e+\frac{\partial e}{\partial x}dx)(v_x+\frac{\partial v_x}{\partial x}dx)(\rho+\frac{\partial \rho}{\partial x}dx)dy\\&+(e+\frac{\partial e}{\partial y}dy)(v_y+\frac{\partial v_y}{\partial y}dy)(\rho+\frac{\partial \rho}{\partial y}dy)dx\\&-\rho v_x e dy-\rho v_y e dx+\frac{\partial(\rho e)}{\partial t}dxdy\end{aligned} \tag{30}$$

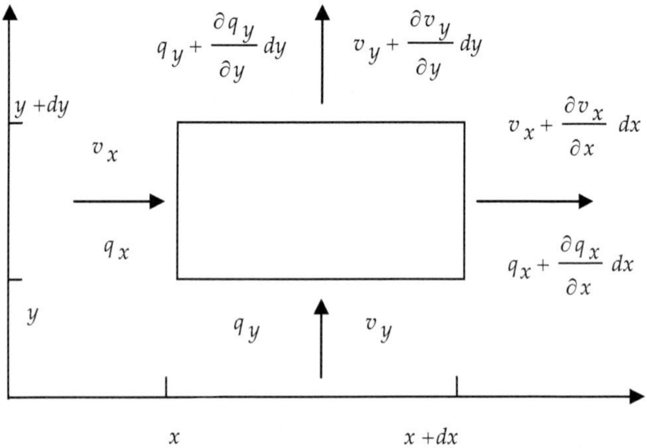

Fig. 1. Two-dimensional infinitesimal element for heat convection

where e is the specific entransy, ρ is the density of fluid. The first four terms in the right side of Eq. (30) account for the entransy exchanged with the environment through the element boundary, the following four terms represent the entransy transported by convection, the last term stands for the variation of the entransy with respect to time. Rearranging Eq. (30) yields,

$$\begin{aligned}\dot{E}_g &= (q_x \frac{\partial T}{\partial x} + q_y \frac{\partial T}{\partial y}) - T(\frac{\partial q_x}{\partial x} + \frac{\partial q_y}{\partial y}) + \rho(\frac{\partial e}{\partial t} + v_x \frac{\partial e}{\partial x} + v_y \frac{\partial e}{\partial y}) \\ &+ e[\frac{\partial \rho}{\partial t} + \rho(\frac{\partial v_x}{\partial x} + \frac{\partial v_y}{\partial y}) + v_x \frac{\partial \rho}{\partial x} + v_y \frac{\partial \rho}{\partial y}]\end{aligned} \tag{31}$$

Making use of the following mass balance equation,

$$\frac{\partial \rho}{\partial t} + v_x \frac{\partial \rho}{\partial x} + v_y \frac{\partial \rho}{\partial y} + \rho(\frac{\partial v_x}{\partial x} + \frac{\partial v_y}{\partial y}) = 0 \tag{32}$$

where ρ is the density of fluid, Eq. (31) is simplified to

$$\dot{E}_g = \mathbf{q} \cdot \nabla T - T \nabla \cdot \mathbf{q} + \rho \frac{De}{Dt} \tag{33}$$

where $D/DT = \partial/\partial t + v_x \partial/\partial x + v_y \partial/\partial y$ is the material derivative. By the following canonical relation of thermodynamics,

$$\rho \frac{De}{Dt} = -T\rho \frac{Du}{Dt} + \frac{PT}{\rho} \frac{D\rho}{Dt} \tag{34}$$

where u is the specific internal energy, P is the pressure,
and the energy balance equation for heat convection

$$\rho \frac{Du}{Dt} = -\nabla \cdot \mathbf{q} - P(\nabla \cdot \mathbf{v}) + \mu\Phi \tag{35}$$

where μ is the viscosity, Φ is the dissipation function and can be expressed as

$$\Phi = 2[(\frac{\partial v_x}{\partial x})^2 + (\frac{\partial v_y}{\partial y})^2] + (\frac{\partial v_x}{\partial y} + \frac{\partial v_y}{\partial x})^2 \tag{36}$$

By Eqs. (34) and (35), Eq. (33) becomes

$$\dot{E}_g = \mathbf{q} \cdot \nabla T - \mu T \Phi \tag{37}$$

The application of the Fourier law, $\mathbf{q} = -k \cdot \nabla T$, on Eq. (37) yields,

$$\dot{E}_{diss} = -\dot{E}_g = k(\nabla T)^2 + \mu T \Phi \tag{38}$$

where k is the thermal conductivity. Thus we obtained the formula for calculating the local entransy dissipation rate in heat convection. It is evident that the expression of the local entransy dissipation rate in the three-dimensional case shares the same form as Eq. (38).
From Eq. (38), one can see that if the temperature and velocity fields in heat conduction problems are known, the local entransy dissipation rate can be obtained exactly. Unfortunately, it is seldom to obtain the exact solution of heat convection problems. In many cases only the heat transfer correlations are available. Therefore it is of importance to develop expressions of the entransy dissipation rate by the knowledge of heat tranfer correlations for heat convection. In this section we attempt to establish such a formula for internal flow in a duct with arbitrary cross section. Consider a flow passage enclosed by two cross sections with distance dx. The temperatures of the duct wall and fluid inside the duct are $T + \Delta T$ and T, respectively. The heat transfer rate from the duct wall to the fluid per unit length is $\dot{q}$. The fluid inside the duct flows with friction in x-direction, thus the pressure gradient $dP / dx \leq 0$. In this heat transfer process, assume there are two irreversibilities. One is the heat conduction from duct wall to the fluid. The other irreversibility is flow friction. As discussed in Section 1.4 the entransy dissipation rate per unit length $(\partial \dot{E}_{diss} / \partial x)_T$ induced by the heat conduction is

$$\left(\frac{\partial \dot{E}_{diss}}{\partial x} \right)_T = \dot{q} \Delta T \tag{39}$$

Assume the fluid is incompressible. The entransy dissipation rate per unit length $(\partial \dot{E}_{diss} / \partial x)_f$ induced by the flow friction is expressed as

$$\left(\frac{\partial \dot{E}_{diss}}{\partial x} \right)_f = -\frac{\dot{m} T}{\rho} \left(\frac{dP}{dx} \right) \tag{40}$$

Where ρ is the density of the fluid, $\dot{m}$ is the mass flow rate. Thus the total entransy dissipation rate per unit length $\partial \dot{E}_{diss} / \partial x$ is

$$\frac{\partial \dot{E}_{diss}}{\partial x} = \dot{q} \Delta T - \frac{\dot{m} T}{\rho} \left(\frac{dP}{dx} \right) \tag{41}$$

The Stanton number correlation reads

$$S_t = \frac{\dot{q}}{p\Delta T c_p R} \tag{42}$$

with

$$R = \dot{m} / A \tag{43}$$

where A is the cross sectional area of the duct, c_p is the specific heat at constant pressure, p is the wetted perimeter. And the friction factor correlation is

$$f = -\frac{\rho D}{2R^2}\frac{dP}{dx} \tag{44}$$

Where D is the hydraulic diameter. By Eqs. (42) and (44), Eq. (41) arrives at

$$\frac{\partial \dot{E}_{diss}}{\partial x} = \frac{q^2}{4\dot{m}c_p}\frac{D}{S_t} + \frac{2\dot{m}T}{\rho^2}\frac{f}{DA^2} \tag{45}$$

Eq. (45) is thus the formula for calculating the derivative of the entransy dissipation rate with respect to the position variable for the forced convection in duct flow. Note that the higher Standon number means the smaller entransy dissipation rate, while a high friction factor leads to the increase of the entransy dissipation rate.

2.3 Entransy dissipation for external flow

Consider an external steady flow that a solid body with arbitrary shape is suspended in a uniform stream with velocity U_∞ and absolute temperature T_∞. For the convenience of analysis, the stream tube with cylindrical shape and large enough volume around the solid body as shown in Fig.2 is taken as the thermodynamic system. The surface temperature of the solid body is T_w. In the regions near the stream tube surface and the external part, the influence of the solid body on the flow is neglected. There is no heat transfer between the solid surface and its external environment. The irreversible dissipation is only caused by the heat conduction and flow friction. For the considered open thermodynamic system, the mass and energy conservations are expressed as,

$$\dot{m}_{in} = \dot{m}_{out} = \dot{m} \tag{46}$$

$$\dot{m}h_{in} + \iint_A \dot{q}ds - \dot{m}h_{out} = 0 \tag{47}$$

where the subscripts "in" and "out" respectively indicate the inlet and outlet, h is the specific enthalpy, $\dot{q}$ is the local heat transfer rate between the solid body and the stream, A is the surface area of the solid body, $\dot{m}$ is the mass flow rate. For the heat conduction across the nonzero temperature difference $T_w - T_\infty$, the entransy dissipation rate $\dot{E}_{diss,T}$ is written as

$$\dot{E}_{diss,T} = \iint_A \dot{q}(T_w - T_\infty)dA$$

where A is the surface area of the solid body. For the flow friction, the entransy dissipation rate $\dot{E}_{diss,P}$ is

$$\dot{E}_{diss,P} = -\frac{\dot{m}T_\infty}{\rho_\infty}(P_{out} - P_{in})$$

where ρ_∞ is the density of the fluid, P is the pressure. Here we assume the fluid is incompressible. Therefore, the total entransy dissipation rate is written as

$$\dot{E}_{diss} = \iint_A \dot{q}(T_w - T_\infty)dA - \frac{\dot{m}T_\infty}{\rho_\infty}(P_{out} - P_{in}) \tag{48}$$

Since

$$\dot{m} = A_{tube}\rho_\infty U_\infty$$

and from the force balance on the control volume, we have

$$F_D = A_{tube}(P_{in} - P_{out})$$

where F_D is the drag force, A_{tube} is the cross sectional area of the cylindrical tube. Therefore,

$$\dot{E}_{diss} = \iint_A \dot{q}(T_w - T_\infty)dA + T_\infty F_D U_\infty \tag{49}$$

If the body temperature is uniform, then the temperature difference $T_w - T_\infty$ is constant, thus Eq. (49) becomes

$$\dot{E}_{diss} = (T_w - T_\infty)\iint_A \dot{q}dA + T_\infty F_D U_\infty \tag{50}$$

Since

$$\iint_A \dot{q}dA = Q = \bar{h}A(T_w - T_\infty)$$

where $\bar{h}$ is the average heat transfer coefficient based on A, Eq. (50) arrives at

$$\dot{E}_{diss} = \bar{h}A(T_w - T_\infty)^2 + T_\infty F_D U_\infty \tag{51}$$

Note that when the body-ambient temperature difference is fixed, the only way to minimize the heat transfer contribution to the entransy dissipation rate is by decreasing the thermal conductance $\bar{h}A$.

The other extreme is that the heat transfer rate $\dot{q}$ is uniform around the solid body. In this case Eq. (50) becomes

$$\dot{E}_{diss} = \dot{q}^2 \iint_A \frac{1}{h}dA + T_\infty F_D U_\infty \tag{52}$$

where h is the local heat transfer coefficient. Note that in order to reduce the heat transfer entransy dissipation rate, we must increase the heat transfer coefficient. In other words, the enhancement of the heat transfer from the solid body to the external fluid can decrease the entransy dissipation rate.

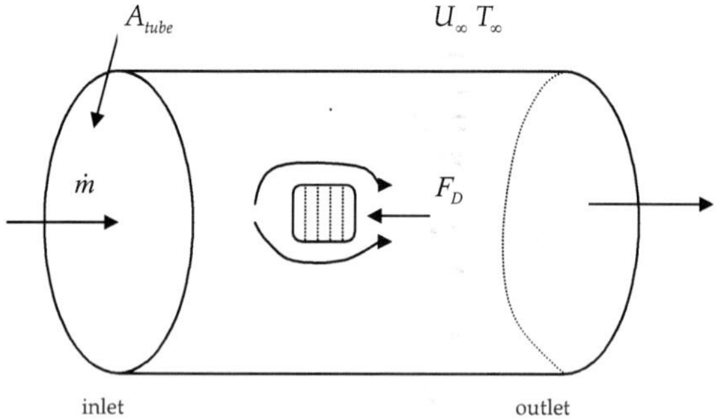

Fig. 2. Schematic diagram for external flow

2.4 Concluding remarks

In this section we have derived the expressions of the entransy dissipation rate for heat convection. These results lay a foundation for the application of entransy dissipation theory in the heat convection problems.

3. From the extremum principle of entransy dissipation to balance equations of fluid mechanics

In this section, our work (Xu & Cheng 2010) is reviewed. In this work, the extremum principle of entransy dissipation is developed by using the entransy dissipation theory developed in Section 1. We further show that this principle is compatible with the Fourier law and Newton's law of viscosity.

3.1 Introduction

The modeling and optimization design of heat exchanger based on the second law of thermodynamics has attracted lots of attention in past decades. In the early 1951, McClintock (1951) recognized that the irreversibility in heat exchanger could not be neglected in heat exchanger design. Bejan's creative work led to the establishment of the minimization entropy generation approach which is widely applied in optimization designs of thermal equipments (Bejan, 1982, 1988, 1995, 1996). However some paradoxes and contradiction appear when the minimization entropy generation approach is applied to the heat transfer problems (Hesselgreaves, 2000; Bertola & Cafaro, 2008). Although a non-dimensionalisation method for the entropy generation in heat exchanger was proposed by Hesselgreaves (2000) that avoids the 'entropy generation paradoxes', it has induced other contradictions.

In fact the thermodynamic basis of the minimization entropy generation approach is the minimum entropy production principle which reads "A steady state has the minimum rate of entropy production with respect to other possible states with the same boundary condition". Actually even for this principle itself, there are lots of controversies (Bertola & Cafaro, 2008; Jaynes, 1980; Mamedov, 2003; Landauer 1975; Ziman, 1956). Landauer (1975) found that the principle of the minimum entropy production is not valid even for some simple thermodynamic systems such as electric resistances. A variational principle developed by Ziman (1956) demonstrates that the entropy production assumes the maximum value in the steady state for transport process. Bertola and Cafaro (2008) found that generally the principle of minimum entropy production does not agree with the conventional Fourier law as well as mass, energy and momentum balance equations in continuum mechanics. In order to remedy the defects of the principle of the minimum entropy production, Guo et al. (2007) developed the extremum principle of entransy dissipation in which only the irreversibility due to heat conduction under finite temperature difference was considered. In Section 3.2, we review the work which extends the current extremum principle of entransy dissipation by taking into account the influence of fluid friction and demonstrate the advantage of this principle over the principle of minimum entropy production in dealing with fluid flow problems (Xu & Cheng, 2010).

3.2 Extremum principle over entransy dissipation and balance equations of fluid mechanics

In order to seek some clues for establishing the extremum principle of entransy dissipation, we first consider natural convection problems, such as the Rayleigh-Benard convection. In this problem, when the temperature difference increases and reaches some critical point, the stationary fluid loses its stability and begins to move. In terms of the entransy, the Rayleigh-Benard convection can be viewed as the entransy related to conduction is converted to the entransy for driving the movement of fluid. The entransy consumed by performing work can be quantified as $-\mu\Phi\delta T$ according to the derivation in Section 2.2. The entransy dissipation associated with the heat conduction in the Rayleigh-Benard convection is quantified by $-\mathbf{q}\cdot\nabla T$. Therefore, we obtain the following extremum principle of entransy dissipation

$$\int_V [-\mathbf{q}\cdot\delta(\nabla T) - \mu\Phi\delta T]dV = 0 \tag{53}$$

The application of the Fourier law on Eq. (53) gives

$$\delta\int_V [\frac{1}{2}k(\nabla T)^2 - \mu\Phi T]dV = 0 \tag{54}$$

This is the principle of minimum entransy dissipation. Note that in this principle, the entransy dissipation comes from two irreversibilities: one is the heat conduction under finite temperature difference, another is the work against the fluid friction.

According to fluid mechanics the governing equation of imcompressible fluid under steady state with constant shear viscosity and thermal conductivity reads

$$\frac{\partial v_i}{\partial x_i} = 0 \tag{55}$$

$$\rho v_j \frac{\partial v_i}{\partial x_j} + \frac{\partial p}{\partial x_i} - \mu\frac{\partial}{\partial x_j}(\frac{\partial v_i}{\partial x_j} + \frac{\partial v_j}{\partial x_i}) = 0 \tag{56}$$

$$\rho c_v v_j \frac{\partial T}{\partial x_j} - k\frac{\partial^2 T}{\partial x_j^2} = \frac{1}{2}\mu(\frac{\partial v_i}{\partial x_j} + \frac{\partial v_j}{\partial x_i})(\frac{\partial v_i}{\partial x_j} + \frac{\partial v_j}{\partial x_i}) \tag{57}$$

where $\mathbf{x}=(x_1,x_2,x_3)^T$ is the position vector, $\mathbf{v}=(v_1,v_2,v_3)^T$ is the velocity field, p is the pressure. The density of entropy production is given by:

$$\sigma=\frac{k}{T^2}\frac{\partial T}{\partial x_i}\frac{\partial T}{\partial x_i}+\frac{1}{2}\frac{\mu}{T}(\frac{\partial v_i}{\partial x_j}+\frac{\partial v_j}{\partial x_i}-\frac{2}{3}\frac{\partial v_k}{\partial x_k}\delta_{ij})(\frac{\partial v_i}{\partial x_j}+\frac{\partial v_j}{\partial x_i}-\frac{2}{3}\frac{\partial v_k}{\partial x_k}\delta_{ij}) \tag{58}$$

Now we consider the one-dimensional steady imcompressible shear flow, $\mathbf{v}=(0,v(x),0)^T$, the continuity equation, momentum balance equation and energy conservative equation reduce to

$$\frac{\partial v}{\partial y}=0 \tag{59}$$

$$\mu\frac{d^2v}{dx^2}=-\frac{\partial p}{\partial x} \tag{60}$$

$$\frac{d^2T}{dx^2}+\frac{\mu}{k}(\frac{dv}{dx})^2=0 \tag{61}$$

From Eq. (59), one can see that the continuity equation is satisfied automatically for the flow under consideration. By Eq.(54), we define the following function

$$F=\frac{1}{2}k\frac{\partial T}{\partial x_i}\frac{\partial T}{\partial x_i}-\frac{\mu T}{2}(\frac{\partial v_i}{\partial x_j}+\frac{\partial v_j}{\partial x_i})(\frac{\partial v_i}{\partial x_j}+\frac{\partial v_j}{\partial x_i})-2\lambda T\frac{\partial v_i}{\partial x_i} \tag{62}$$

In order to satisfy the continuity equation, the Lagrange multiplier λ is introduced in Eq. (62). From the variational method, Eq. (54) is equivalent to the following Euler-Lagrange equations

$$\frac{\partial}{\partial x_i}\left(\frac{\partial F}{\partial(\frac{\partial T}{\partial x_i})}\right)-\frac{\partial F}{\partial T}=0 \tag{63}$$

$$\frac{\partial}{\partial x_i}\left(\frac{\partial F}{\partial(\frac{\partial v_j}{\partial x_i})}\right)-\frac{\partial F}{\partial v_j}=0 \tag{64}$$

Substituting Eq. (62) into Eqs. (63) and (64) yields

$$k\frac{\partial^2 T}{\partial x_i^2}+\frac{\mu}{2}(\frac{\partial v_i}{\partial x_j}+\frac{\partial v_j}{\partial x_i})(\frac{\partial v_i}{\partial x_j}+\frac{\partial v_j}{\partial x_i})+2\lambda\frac{\partial v_i}{\partial x_i}=0 \tag{65}$$

$$\mu\frac{\partial T}{\partial x_i}(\frac{\partial v_i}{\partial x_j}+\frac{\partial v_j}{\partial x_i})+\mu T(\frac{\partial^2 v_i}{\partial x_i\partial x_j}+\frac{\partial^2 v_j}{\partial x_i\partial x_j})=-\lambda\frac{\partial T}{\partial x_i}-\frac{\partial\lambda}{\partial x_i}T \tag{66}$$

The application of the continuity equation (59) on Eq. (65) gives

$$k\frac{\partial^2 T}{\partial x_i^2}+\frac{\mu}{2}(\frac{\partial v_i}{\partial x_j}+\frac{\partial v_j}{\partial x_i})(\frac{\partial v_i}{\partial x_j}+\frac{\partial v_j}{\partial x_i})=0 \tag{67}$$

For the one-dimensional steady shear flow, Eqs. (65) and (66) become

$$\mu\frac{\partial T}{\partial x}\frac{\partial v}{\partial x}+\mu T\frac{\partial^2 v}{\partial x^2}=-\lambda\frac{\partial T}{\partial x}-\frac{\partial \lambda}{\partial x}T \tag{69}$$

$$k\frac{\partial^2 T}{\partial x^2}+\mu\left(\frac{\partial v}{\partial x}\right)^2=0 \tag{70}$$

By setting $\lambda = p = -\mu dv / dx$, Eqs. (69) and (70) reduce to Eqs. (60) and (61), respectively. Therefore, the extremum principle of entransy dissipation yields the same governing equation of the one-dimensional steady shear flow as given in fluid mechanics. Since Eqs. (60) and (61) are determined by the Fourier law and Newton's law of viscosity, we may say that the extremum principle of entransy dissipation is compatible with these two classical laws. While the principle of the minimum entropy production contradicts with the standard balance equations of the one-dimensional steady shear flow (Bertola & Cafaro, 2008). In this point, the extremum principle of entransy dissipation demonstrates the obvious advantage.

3.3 Concluding remarks

In summary, a new type of extremum principle of entransy dissipation is established by taking into account the viscosity resistance of the fluid. The analysis shows that this extremum principle of entransy dissipation gives rise to the standard governing equations of one-dimensional steady shear flow in fluid mechanics. Therefore, the entremum principle of entransy dissipation is consistent with the Fourier law and Newton's law of viscosity.

4. The application of entransy dissipation theory in heat exchanger design

In this section the expressions of the entransy dissipation related to the irreversibility in heat exchanger are developed. The entransy dissipation number which can be employed to evaluate the performance of heat exchangers is defined by a non-dimensionalisation method for the entransy dissipation. The optimization of heat exchanger design is analyzed.

4.1 Introduction

With the sharp decline of fossil fuels such as petroleum and coal, to use energy efficiently is one of effective ways to face the increasing energy demand. Heat exchanger as an important device in thermal system is widely applied in power engineering, petroleum refineries, chemical industries, and so on. Hence, it is of great importance to develop technologies which enable us to reduce the unnecessary energy dissipation and improve the performance of heat exchanger.

The evaluation criteria for heat exchanger performance are generally classified into two groups: the first is based on the first law of thermodynamics; the second is based on the combination of the first and second law of thermodynamics. The heat transfer in heat exchangers usually involves the heat conduction under finite temperature difference, the fluid friction under finite pressure drop and fluid mixing. These processes are characterized as irreversible non-equilibrium thermodynamic processes. Hence, in recent decades the study of the second group has attracted a lot of attention (Yilmaz et al., 2001). Inspired by the principle of the minimum entropy production advanced by Prigogine (1976), Bejan (1982, 1995) developed the entropy generation minimization (EGM) approach to heat exchanger optimization design. In this approach Bejan (1982) took into account two types of the irreversibilities in heat exchanger, namely the heat conduction under the stream-to-stream temperature difference and the frictional pressure drop that accompanies the circulation of fluid through the apparatus. Therefore, the total entropy production rate denoted by $\dot{S}_{gen}$ is the sum of entropy productions associated with heat conduction and fluid friction. However, among all the variational principles in thermodynamics, Prigogine's minimum entropy generation principle is still the most debated one as mentioned in Section 3.1. Accordingly, the entropy generation minimization approach, widely applied to modeling and optimization of thermal systems that owe their thermodynamic imperfection to heat transfer, mass transfer, and fluid flow irreversibilities, demonstrates some inconsistencies and paradoxes in applications of heat exchanger designs (Hesselgreaves, 2000).

Recently based on the concept of entransy Guo et al. (2010), the heat transfer efficiency and equivalent thermal resistance in a heat exchanger were defined. Wang et al. (2009) derived an entransy transfer equation describing the entransy transfer processes of a multi-component viscous fluid subjected to heat transfer by conduction and convection, mass diffusion and chemical reactions. Chen and Ren (2008) defined a ratio of temperature difference to heat flux as the generalized thermal resistance of convective heat transfer processes, and developed the minimum thermal resistance theory for convective heat transfer optimization, it was found that the minimum thermal resistance principle is equivalent to the entransy dissipation extremum principle. Xia et al. (2009) studied the optimum parameter distributions in two-fluid flow heat exchanger by using optimal control theory under the fixed heat load condition and taking the entransy dissipation minimization as the optimization objective. Liu et al. (2009) investigated the applicability of the extremum principles of entropy generation and entransy dissipation for heat exchanger optimization, and found that the former is better for the heat exchanger optimization when it works in the Brayton cycle, while the latter gives better results when heat exchanger is only for the purpose of heating and cooling. Recently the entransy dissipation extremum principle was extended to the radiative heat transfer by Wu and Liang (2008). Guo et al. (2009) introduced a non-dimensionalisation method for the entransy dissipation in heat exchanger and an entransy dissipation number which can be used to evaluate the heat exchanger performance was defined.

However, all the work mentioned above is based on the entransy defined by Guo et al. (2007) which is only related to the heat conduction in heat exchanger. In Section 1 we have presented a new definition of entransy which not only describes the ability to transfer heat, but also the ability to perform work. Subsequently, the entransy dissipation can quantify not only the irreversibility associated with the heat conduction, but also the irreversibility induced by fluid friction. In this section, we first derive the expressions of the entransy dissipation induced by heat conduction, fluid friction and fluid mixing in heat exchanger. Then a non-dimensionalisation of the total entransy dissipation will be introduced.

4.2 Entransy dissipation in heat exchanger

According to the definition of entransy, we have the following relationship between the entransy and enthalpy

$$de = Tdh \tag{71}$$

where e and h are the specific entransy and enthalpy, respectively. Assume that the heat exchanger under consideration is an adiabatic open system. By making use of Eq. (71) and the following relationship

$$h = \dot{m}c_pT \tag{72}$$

the entransy dissipation rate induced by heat conduction in the heat exchanger is written as

$$\dot{E}_1 = \int_o^i (\dot{m}Tdh)_1 = \int_o^i (\dot{m}c_pTdT)_1 = \frac{1}{2}(\dot{m}c_p)_1(T_{1,i}^2 - T_{1,o}^2) \tag{73}$$

where $\dot{E}_1$ is the entransy dissipation rate of the fluid *1*, c_p is the specific heat of fluid at constant pressure and assumed to be a constant, the subscript *1* denotes the fluid *1*, the subscripts *i* and *o* denote the inlet and outlet, respectively. Similarly, for the fluid 2 we have

$$\dot{E}_2 = \int_o^i (\dot{m}Tdh)_2 = \int_o^i (\dot{m}c_pTdT)_2 = \frac{1}{2}(\dot{m}c_p)_2(T_{2,i}^2 - T_{2,o}^2) \tag{74}$$

Therefore, the total entransy dissipation rate induced by heat conduction in the heat exchanger is expressed as

$$\dot{E}_T = \dot{E}_1 + \dot{E}_2 = \frac{1}{2}(\dot{m}c_p)_1(T_{1,i}^2 - T_{1,o}^2) + \frac{1}{2}(\dot{m}c_p)_2(T_{2,i}^2 - T_{2,o}^2) \tag{75}$$

Now we consider the entransy dissipation rate caused by flow friction in heat exchangers. Assume that the fluid flow in the heat exchanger is driven by the finite pressure drop between the inlet and outlet, and the flow is stationary and adiabatic. Then from the definition of entransy, we have

$$de = -Tdp / \rho \tag{76}$$

where p is the pressure. Then the entransy dissipation rate induced by flow friction is written as

$$\dot{E}_p = \int_i^o \dot{m}dg_0 = -\int_i^o \dot{m}T / \rho dp \tag{77}$$

If the fluid is the ideal gas, applying its state equation in Eq. (77) gives

$$\dot{E}_p = -\int_i^o \dot{m}T^2R / pdp \tag{78}$$

where R is the ideal gas constant. If replacing the temperature in Eq. (78) with the logarithmic mean temperature, we obtain

$$\dot{E}_p = -T_{lm}{}^2 \int_i^o \dot{m}R / p dp \tag{79}$$

where $T_{lm} = (T_o - T_i) / (\ln T_o - \ln T_i)$. Similarly, for the incompressible fluid we have

$$\dot{E}_p = \frac{\dot{m} T_{lm} \Delta p}{\rho} \tag{80}$$

where $\Delta p = p_i - p_o$. If both fluids in the heat exchanger are idea gases, the entransy dissipation rate induced by fluid friction is written as

$$\dot{E}_p = -\dot{m}_1 (\frac{T_{1,o} - T_{1,i}}{\ln T_{1,o} - \ln T_{1,i}})^2 R \ln \frac{p_{1,o}}{p_{1,i}} - \dot{m}_2 (\frac{T_{2,o} - T_{2,i}}{\ln T_{2,o} - \ln T_{2,i}})^2 R \ln \frac{p_{2,o}}{p_{2,i}} \tag{81}$$

When both fluids in the heat exchanger are incompressible, we obtain

$$\dot{E}_p = \frac{\dot{m}_1 \Delta p_1}{\rho_1} \frac{T_{1,o} - T_{1,i}}{\ln T_{1,o} - \ln T_{1,i}} + \frac{\dot{m}_2 \Delta p_2}{\rho_2} \frac{T_{2,o} - T_{2,i}}{\ln T_{2,o} - \ln T_{2,i}} \tag{82}$$

When the fluid *1* is idea gas, fluid *2* is impressible, we have

$$\dot{E}_p = -\dot{m}_1 (\frac{T_{1,o} - T_{1,i}}{\ln T_{1,o} - \ln T_{1,i}})^2 R \ln \frac{p_{1,o}}{p_{1,i}} + \frac{\dot{m}_2 \Delta p_2}{\rho_2} \frac{T_{2,o} - T_{2,i}}{\ln T_{2,o} - \ln T_{2,i}} \tag{83}$$

It is well known that when two or more streams of dissimilar fluids are mixing, the irreversibility occurs, which is detrimental to the performance of heat exchanger. In the cross flow exchanger, the mixing of thermal dissimilar fluids (with different temperatures) frequently happens. In the following we focus on the entransy dissipation rate induced by this type of mixing. Assume that n streams of thermal dissimilar fluids enter into the heat exchanger from its inlet. Their temperatures are denoted as $T_{j,i} (j = 1, 2, \cdots, n)$. In the flow process from the inlet to outlet, the streams of thermal dissimilar fluids are adequately mixing. At the outlet, they have the same temperature T_o. Assume that the flow is stationary and the mixing process is adiabatic. Then making use of the law of energy conservation, the entropy balance equation and Eq. (71), we obtain

$$\dot{m}_o = \sum_{j=1}^{n} \dot{m}_{j,i} \tag{84}$$

$$\dot{E}_m = \dot{m}_o e_o - \sum_{j=1}^{n} (\dot{m}e)_{j,i} = \sum_{j=1}^{n} \dot{m}_j \Delta e_j \tag{85}$$

The similar derivation as leading to Eq. (72) yields

$$\dot{E}_m = \sum_{j=1}^{n} \frac{1}{2} (\dot{m}c_p)_j (T_{j,i}^2 - T_o^2) \tag{86}$$

Finally, the total entransy dissipation rate caused by heat conduction, fluid friction and fluid mixing is expressed as

$$\dot{E} = \dot{E}_T + \dot{E}_p + \dot{E}_m \tag{87}$$

4.3 Entransy dissipation number

In this section our work (Guo et al. 2009, Li et al. 2011) is reviewed. In this work, a dimensionless method of the entransy dissipation of the heat exchanger is established. Subsequently a quantity called the entransy dissipation number of heat exchanger is defined. Guo et al. (2009) examined the validity of the entransy dissipation number as the performance evaluation criterion of heat exchanger. Furthermore, for the water-water balanced counter-flow heat exchanger Li et al. (2011) took as the minimum overall entransy dissipation number as an objective function, under certain assumptions it was proved that there is a corresponding optimum in duct aspect ratio or mass velocity since the variation in the duct aspect ratio or mass velocity has opposing effects on the two types of entransy dissipations caused by heat conduction under finite temperature difference and the flow friction under finite pressure drop, respectively. We also develop analytically expressions for the optimal duct aspect ratio and mass velocity of a heat exchanger that are useful for design optimization.

Now we consider a water-water balanced counter-flow heat exchanger. Assume that both the hot and cold fluids are imcompressible. The inlet temperatures and pressures of the hot and cold fluids are denoted as $T_{1,i}$, $P_{1,i}$ and $T_{2,i}$, $P_{2,i}$ respectively. Similarly the outlet temperatures and pressures are $T_{1,o}$, $P_{1,o}$ and $T_{2,o}$, $P_{2,o}$. For the balanced heat exchanger, the heat capacity rate ratio satisfies condition $C^* = (\dot{m}c)_2 / (\dot{m}c)_1 = 1$ (where $\dot{m}$ is the mass flow rate). For the one-dimensional heat exchanger considered in this section, the usual assumptions such as steady flow, no heat exchange with environment, and ignoring changes in kinetic and potential energies as well as the longitudinal conduction are made. In the heat exchanger, there mainly exist two kinds of irreversibility: the first is heat conduction under finite temperature differences and the second is flow friction under finite pressure drops.

Note that according to the definition of entransy given in Section 1 the maximum entransy in a heat exchanger is $Q(T_{1,i} - T_{2,i})$, where Q is the actual heat transfer rate, $(T_{1,i} - T_{2,i})$ is the maximum temperature difference in the heat exchanger. Therefore the entransy dissipation rate caused by heat conduction under a finite temperature difference can be non-dimensionalized in the following way (Guo et al. 2009):

$$\dot{E}_T^* = \frac{\dot{E}_T}{Q(T_{1,i} - T_{2,i})} = \frac{\dot{E}_T}{\varepsilon(\dot{m}c)_1 (T_{1,i} - T_{2,i})^2} \tag{88}$$

where ε is the heat exchanger effectiveness which is defined as the ratio of the actual heat transfer rate to the maximum possible heat transfer rate. As discussed by Guo et al (2009), the entransy dissipation number can be employed as the performance evaluation criterion of heat exchangers which avoids the "entropy generation paradox".

From Eq. (82), the entransy dissipation due to flow friction under a finite pressure drop is expressed as

$$\dot{E}_P = \frac{\dot{m}_1 \Delta P_1}{\rho_1} \frac{T_{1,o} - T_{1,i}}{\ln T_{1,o} - \ln T_{1,i}} + \frac{\dot{m}_2 \Delta P_2}{\rho_2} \frac{T_{2,o} - T_{2,i}}{\ln T_{2,o} - \ln T_{2,i}} \tag{89}$$

where ΔP_1 and ΔP_2 refer to the pressure drops of the hot and cold water, respectively; ρ_1 and ρ_2 are their corresponding densities. Putting in dimensionless form leads to

$$\dot{E}_P^* = \frac{-\Delta P_1}{(\rho c)_1 (T_{1,i} - T_{2,i})} \frac{1}{\ln \frac{T_{2,i}}{T_{1,i}} \left[1 - (1-\varepsilon) \frac{T_{2,i} - T_{1,i}}{T_2} \right]} + \frac{\Delta P_2}{(\rho c)_2 (T_{1,i} - T_{2,i})} \frac{1}{\ln \frac{T_{1,i}}{T_{2,i}} \left[1 + (1-\varepsilon) \frac{T_{2,i} - T_{1,i}}{T_{1,i}} \right]} \tag{90}$$

which is called the entransy dissipation number due to flow friction. Assuming that the heat exchanger behaves as a nearly ideal heat exchanger, then $(1-\varepsilon)$ is considerably smaller than unity (Bejan, 1982). For a water-water heat exchanger under usual operating conditions, the inlet temperature difference between hot and cold water, $\Delta T = T_{1,i} - T_{2,i}$, is less than *100K*, hence $\Delta T / T_i < 100/273 \approx 0.366\ (i = 1,2)$. Therefore, Eq. (90) can be simplified to

$$\dot{E}_P^* = \frac{\Delta P_1}{(\rho c)_1 (T_{1,i} - T_{2,i})} \frac{1}{\ln \frac{T_{1,i}}{T_{2,i}}} + \frac{\Delta P_2}{(\rho c)_2 (T_{1,i} - T_{2,i})} \frac{1}{\ln \frac{T_{1,i}}{T_{2,i}}} \tag{91}$$

Accordingly, the overall entransy dissipation number becomes

$$\dot{E}^* = \dot{E}_T^* + \dot{E}_P^* = (1-\varepsilon) + \frac{\Delta P_1}{(\rho c)_1 (T_{1,i} - T_{2,i})} \frac{1}{\ln \frac{T_{1,i}}{T_{2,i}}} + \frac{\Delta P_2}{(\rho c)_2 (T_{1,i} - T_{2,i})} \frac{1}{\ln \frac{T_{1,i}}{T_{2,i}}} \tag{92}$$

For a typical water-water balanced heat exchanger, the number of heat transfer units *Ntu* can be introduced, which approaches infinity as the effectiveness tends to unity. Since $C^* = 1$, the effectiveness is (Bejan, 1982):

$$\varepsilon = \frac{Ntu}{1 + Ntu} \tag{93}$$

where the number of heat transfer units is defined as

$$Ntu = \frac{UA}{\dot{m} c_p}$$

here *U* is the overall heat transfer coefficient, and *A* is the heat transfer area. Assume that the heat conduction resistance of the solid wall can be neglected, compared with the convective heat transfer, then it is appropriate to replace U with the convective heat transfer coefficient h. Therefore,

$$\frac{1}{UA}=\frac{1}{(hA)_1}+\frac{1}{(hA)_2} \tag{94a}$$

or

$$\frac{1}{Ntu}=\frac{1}{Ntu_1}+\frac{1}{Ntu_2} \tag{94b}$$

where h_1 and h_2 are the convective heat transfer coefficients of the hot and cold fluids, respectively, and $Ntu_i=(hA)_i/(\dot{m}c)_i(i=1,2)$. In the nearly ideal heat exchanger limit, $Ntu>>1$, that is (Bejan, 1982)

$$1-\varepsilon\approx\frac{1}{Ntu} \tag{95}$$

From Eq.(95) the overall entransy dissipation number is

$$\dot{E}^*=\left[\frac{1}{Ntu_1}+\frac{\Delta P_1}{(\rho c)_1(T_{1,i}-T_{2,i})}\frac{1}{\ln\frac{T_{1,i}}{T_{2,i}}}\right]+\left[\frac{1}{Ntu_2}+\frac{\Delta P_2}{(\rho c)_2(T_{1,i}-T_{2,i})}\frac{1}{\ln\frac{T_{1,i}}{T_{2,i}}}\right] \tag{96}$$

The two terms on the right of Eq. (96) correspond to the entransy dissipations of two sides of heat transfer surfaces. For each side, the entransy dissipation number can be written as follows:

$$\dot{E}_i^*=\frac{1}{Ntu_i}+\frac{P_i}{(\rho c)_i(T_{1,i}-T_{2,i})\ln\frac{T_{1,i}}{T_{2,i}}}\frac{\Delta P_i}{P_i}\ (i=1,2) \tag{97}$$

It is evident that the first term accounts for the entransy dissipation from the heat conduction under finite temperature difference and the second for the entransy dissipation from flow friction under finite pressure drop. For simplicity, we now use $\dot{E}^*$ instead of $\dot{E}_i^*$ to denote the entransy dissipation number for each side of the heat exchanger surface.

4.4 Parameter optimization

Theoretically, the exchanger effectiveness increases when the irreversible dissipation in the heat exchanger decreases. From Section 1.3 one can see that the entransy dissipation can be used to describe the irreversible dissipations, therefore we seek optimums in duct aspect ratio and mass velocity by minimizing the entransy dissipation number $\dot{E}^*$ based on Eq. (97). Although the entransy dissipation number on one side of a heat transfer surface can be expressed as the sum of the contributions of the heat conduction under the finite temperature difference and flow friction under the finite pressure drop, the effects of these two factors on heat exchanger irreversibility are strongly coupled through the geometric

parameters of the heat exchanger tube residing on that side. Therefore, based on entransy dissipation minimization, it is possible to obtain optimal geometric parameters of the heat exchanger such as the optimal duct aspect ratio.
Recall the definition of the Stanton number $St\left((\mathrm{Re})_D, \mathrm{Pr}\right)$ and friction factor $f\left((\mathrm{Re})_D\right)$:

$$Ntu = \frac{4L}{D} St \tag{98}$$

$$\frac{\Delta P}{P} = f \frac{4L}{D} \frac{G^2}{2\rho P} \tag{99}$$

where $G = \dot{m} / A$ is the mass velocity, L is the flow path length and D is the duct hydraulic diameter. Introducing the dimensionless mass velocity, $G_* = G / \sqrt{2\rho P}$

$$\tau^2 = \frac{P}{(\rho c)(T_{1,i} - T_{2,i}) \ln \frac{T_{1,i}}{T_{2,i}}}$$

and substituting Eqs. (98) and (99) into Eq. (97), we obtain

$$\dot{E}^* = \frac{1}{\frac{4L}{D} St} + \tau^2 f \frac{4L}{D} G_*^2 \tag{100}$$

Clearly, the duct aspect ratio $4L / D$ has opposing effects on the two terms of the right side of Eq. (100). Therefore, there exists an optimal duct aspect ratio to minimize the entransy number. When the Reynolds number and mass velocity are fixed, minimizing the entransy dissipation number leads to the following expression for this optimum:

$$\left(\frac{4L}{D}\right)_{opt} = \frac{1}{\tau G_* (fSt)^{1/2}} \tag{101}$$

The corresponding minimum entransy dissipation number is

$$\dot{E}^*_{\min} = 2\tau G_* \left(\frac{f}{St}\right)^{1/2} \tag{102}$$

From Eqs.(101) and (102), one can see that the optimal duct aspect ratio decreases as the mass velocity G_* increases, and the minimum entransy dissipation number is directly proportional to the dimensionless mass velocity. Note that the minimum entransy dissipation number is also dependent on the Reynolds number via f and St. However, the impact of the Reynolds number on the minimum entransy dissipation number is very weak since for many heat transfer surfaces the ratio of the friction factor to the Stanton number does not have a significant change as the Reynolds number varies (Bejan, 1982). Therefore, the minimum entransy dissipation number is mainly determined by the selected dimensionless mass velocity. Obviously, the smaller the mass velocity, the longer the working fluid remains on the heat transfer surface and the lower the irreversible dissipations in the heat exchanger.

In designing a heat exchanger, the heat transfer area is an important consideration when it accounts for most of the total cost of a heat exchanger. Thus in the following, we discuss design optimization of the heat exchanger with a fixed heat transfer area.
From the definition of the hydraulic diameter, the heat transfer area for one side is

$$A = \frac{4L}{D} A_c$$

where A_c is the duct cross-section area. This expression can be put in dimensionless form as

$$A_* = \frac{4L}{D} G_*^{-1} \tag{103}$$

where A_* is the dimensionless heat transfer area, $A_* = (2\rho P)^{1/2} A / \dot{m}$. Substituting Eq. (103) into Eq. (100) yields

$$\dot{E}^* = \frac{1}{A_* St} G_*^{-1} + \tau^2 f A_* G_*^3 \tag{104}$$

Obviously, the dimensionless mass velocity has an opposing effect on the two terms of the right side of Eq.(104). Thus there exists an optimal dimensionless mass velocity which allows the entropy dissipation number to reach a minimum value when A_* and Reynolds number $(\mathrm{Re})_D$ are given. Solving this optimization problem yields:

$$G_{*,opt} = \left(\frac{1}{3A_*^2 \tau^2 f St} \right)^{1/4} \tag{105}$$

$$\dot{E}^*_{\min} = 4 \left(\frac{\tau^2 f}{27 A_*^2 St^3} \right)^{1/4} \tag{106}$$

Eqs. (105) and (106) respectively give the optimal dimensionless mass velocity and the minimum entransy dissipation number under fixed A_* and Reynolds number $(\mathrm{Re})_D$. From these two equations, the larger heat transfer area clearly corresponds to the smaller mass velocity and lower entransy dissipation rates. Hence, to reduce the irreversible dissipation occurring in a heat exchanger, the largest-possible heat transfer area should be adopted under the allowable conditions.
If $\dot{E}^*$ and $(\mathrm{Re})_D$ are given, the minimum heat transfer area is

$$A_{*,\min} = \frac{16}{3^{3/2}} \frac{\tau f^{1/2}}{\dot{E}^{*2} St^{3/2}} \tag{107}$$

with

$$\left(\frac{4L}{D} \right)_{opt} = \frac{4}{3} \frac{1}{\dot{E}^* St} \tag{108}$$

From Eqs. (107) and (108), one can see that a lower entransy dissipation rate corresponds to larger heat transfer area or duct aspect ratio. Eqs. (106) and (107) are identical, providing an expression for the minimum attainable value for the product $A_*^{1/2}\dot{E}^*$ under the given Reynolds number.

4.5 Concluding remarks

As exemplified by the water-water counter-flow heat exchanger, the present work shows that there exists an optimal duct aspect ratio for heat exchangers under the fixed Reynolds number and mass velocity when the entransy dissipation number is taken as the performance evaluation criterion. Furthermore, the formula for the optimal duct aspect ratio was obtained analytically. Under constraints of the fixed heat transfer area (or duct volume) and Reynolds number, it was shown that there is an optimal dimensionless mass velocity; for which an analytical expression was also given. The results indicated that to reduce irreversible dissipations in heat exchangers, largest-possible heat transfer areas and lowest-possible mass velocities should be adopted. This conclusion is in agreement with numerical results obtained by design optimization of the shell-and-tube heat exchanger based on the entransy dissipation number as the objective function (Guo et al., 2010).
From the results obtained in this study, it can be seen that the traditional heat exchanger design optimizations based on total cost as an objective function usually sacrifice heat exchanger performance. This issue has been demonstrated by numerical results (Guo et al., 2009). Guo et al. (2009) found that a little improvement in heat exchanger performance can lead to large gains in terms of energy saving and environmental protection. Hence, in heat exchanger design, reduction in total cost and improvement in heat exchanger performance should be treated equally. The present work will be useful to drive new research in this direction.

5. Acknowledgements

The support of our research by National Basic Research Program of China (Project No. 2007CB206900) is greatly appreciated.

6. References

Bejan, A. (1982). *Entropy Generation through Heat and Fluid Flow*. New York, Wiley
Bejan, A. (1988). *Advanced Engineering Thermodynamics*. New York, Wiley
Bejan, A. (1995). *Entropy Generation Minimization*. New York, CRC Press
Bejan, A. (1996). Entropy generation minimization: the new thermodynamics of finite-size devices and finite-time processes. *J Appl Phys*, Vol. 79, pp.1191-1218
Ben-Amotz, D. & Honig, J. M. (2006). Average entropy dissipation in irreversible mesoscopic processes. *Phys Rev Lett*, Vol. 96, pp. 020602.
Ben-Amotz, D. & Honig, J. M. (2003). Rectification of thermodynamic inequalities. *J Chem Phys*, Vol.118, pp. 5932.
Bertola, V. & Cafaro, E. (2008). A critical analysis of the minimum entropy production theorem and its application to heat and fluid flow. *Int J Heat Mass Transfer*, Vol. 51, pp.1907-1912.
Bizarro, J. P. S. (2008). Entropy production in irreversible processes with friction. *Phys Rev E*, Vol. 78, pp. 021137

Chen, L.; Wei, S. & Sun, F. (2011). Constructal entransy dissipation rate minimization of a disc. *Int J Heat Mass Transfer*, Vol. 54, pp. 210.

Chen, Q. & Ren, J. X. (2008). Generalized thermal resistance for convective heat transfer and its relation to entransy dissipation. *Chinese Science Bulletin*, Vol. 53, pp. 3753-3761

Cheng, X. & Liang, X. (2011). Entransy flux of thermal radiation and its application to enclosures with opaque surfaces. *Int J Heat Mass Transfer*, Vol. 54, pp.269.

Cheng, X. G. (2004). Entransy and its Application in Heat Transfer Optimization. PhD thesis, Tsinghua University, Beijing

Clausius, R. (1865). The Mechanical Theory of Heat-With its Applications to the Steam Engine and to Physical Properties of Bodies. London: John van Voorst, 1 Paternoster Row, MDCCCLXVII.

den Broek, C. V. (2005). Thermodynamic efficiency at maximum power. *Phys Rev Lett*, Vol. 95, pp.190602.

Esposito, M. & Lindenberg, K. (2009). Universality of efficiency at maximum power. *Phys Rev Lett*, Vol.102, pp.130602.

Esposito, M. et al. (2010). Efficiency at maximum power of low-dissipation Carnot engines. *Phys Rev Lett*, Vol. 105, pp. 150603.

Guo, J. F.; Cheng, L. & Xu, M. T. (2009a). Entransy dissipation number and its application to heat exchanger performance evaluation. *Chinese Sci Bull*, Vol. 54, pp. 2708.

Guo, J. F.; Xu, M. T. & Cheng, L. (2009b). The application of field synergy number in shell-and-tube heat exchanger optimization design. *Appl Energ*, Vol. 86, pp. 2079-2087

Guo, J. F.; Li, M. X. & Xu M. T. (2010). The application of entransy dissipation theory in optimization design of heat exchanger, *Proceedings of the 14th International Heat Transfer Conference*, IHTC14-23348, Washington, DC, USA, August 8-13, 2010

Guo, Z. Y.; Zhu, H. Y. & Liang, X. G. (2007). Entransy-A physical quantity describing heat transfer ability. *Int J Heat Mass Transfer*, Vol. 50, pp. 2545.

Guo, Z. Y. et al. (2010). Effectiveness-thermal resistance method for heat exchanger design and analysis. *Int J Heat Mass Transfer*, Vol. 53, pp. 2877.

Hesselgreaves, J. E. (2000). Rationalisation of second law analysis of heat exchanger. *Int. J. Heat Mass Transfer*, Vol. 43, pp. 4189-4204

Herwig, H. (2010). The Role of Entropy Generation in Momentum and Heat Transfer, *Proceedings of the 14th International Heat Transfer Conference*, IHTC14-23348, Washington, DC, USA, August 8-13, 2010

Jaynes, E. T. (1980). The minimum entropy production principle. *Annu Rev Phys Chem*, Vol. 31, pp. 579-601

Kondepudi, D & Prigogine, I. (1998). *Modern Thermodynamics-From Heat Engines to Dissipative Structures*, John Wiley & Sons, Chichester

Landauer R. (1975). Inadequacy of entropy and entropy derivatives in characterizing the steady state. *Phys Rev A*, Vol. 12, pp.636-630

Li, X. F.; Guo, J. F., Xu, M. T. & Cheng, L. (2011). Entransy dissipation minimization for optimization of heat exchanger design. *Chinese Sci. Bull.* (Accepted)

Liu, X. B.; Meng, J. A. & Guo, Z. Y. (2009). Entropy generation extremum and entransy dissipation extremum for heat exchanger optimization. *Chinese Science Bulletin*, Vol. 54, pp. 943-947

Mamedov, M. M. (2003). On the incorrectness of the traditional proof of the principle of minimum production. *Technical Physics Letters*, Vol. 29, pp. 69-71

McClintock, F. A. (1951). The design of heat exchangers for minimum irreversibility. *ASME Paper*, No. 51-A-108, 1951, presented at the 1951 ASME Annual Meeting

Prigogine, I. (1967). *Introduction to Thermodynamics of Irreversible Processes*, Third ed., Wiley, New York, pp. 76-77.

Shah, R. K. & Skiepko, T. (2004). Entropy generation extrema and their relationship with heat exchanger effectiveness-number of transfer unit behavior for complex flow arrangement. *ASME J Heat Transfer*, Vol. 126, pp. 994.

Wang, S. P.; Chen, Q. L. & Zhang, B. J. (2009). An equation of entransy transfer and its application, *Chinese science bulletin*, Vol.54, pp. 3572-3578.

Wu, J. & Liang, X. G. (2008). Application of entransy dissipation extremum principle in radiative heat transfer optimization. *Science in China Series E: Technological Sciences*, Vol. 51, pp. 1306-1314

Xia, S. J.; Chen, L. G. & Sun, F. R. (2009). Optimization for entransy dissipation minimizatino in heat exchanger. *Chinese Science Bulletin*, Vol.54, pp. 3587-3595

Xu, M. T. & Cheng L. (2010). From the extremum principle of entransy dissipation to steady balance equations of fluid mechanics, *Proceedings of the 14th International Heat Transfer Conference*, IHTC14-23348, Washington, DC, USA, August 8-13, 2010

Xu, M. T.; Guo, J. F. & Chen, L. (2009). Application of entransy dissipation theory in heat convection. *Front. Energy Power Eng China*, Vol.3, pp. 402-405

Xu, M. T. (2011). The thermodynamic basis of entransy and entransy dissipation. *Energy*, Vol. 36, 4272-4277

Yilmaz, M.; Sara, O. N. & Karsli, S. (2001). Performance Evaluation Criteria for Heat Exchangers Based on Second Law Analysis. *Exergy, an International Journal*, Vol.1, pp. 278-294.

Ziman, J. M. (1956). The general variational principle of transport theory. *Can J Phys*, Vol. 34, pp.1256-1273

15

Inverse Space Marching Method for Determining Temperature and Stress Distributions in Pressure Components

Jan Taler[1], Bohdan Węglowski[1],
Tomasz Sobota[1], Magdalena Jaremkiewicz[1] and Dawid Taler[2]
[1]*Cracow University of Technology*
[2]*University of Science and Technology*
Poland

1. Introduction

Thermal stresses can limit the heating and cooling rates of temperature changes. The largest absolute value of thermal stresses appears at the inner surface. Direct measurements of these stresses are very difficult to take, since the inner surface is in contact with water or steam under high pressure. For that reason, thermal stresses are calculated in an indirect way based on measured temperatures at selected points, located on an outer thermally insulated surface of a pressure element. First, time-space temperature distribution in pressure element is determined using the inverse space marching method.

High thermal stresses often occur in partially filled horizontal vessels. During operation under transient conditions, for example, during power plant start-up and shut-down, there are significant temperature differences over the circumference of the horizontal pressure vessels (Fetköter et al., 2001; Rop, 2010). This phenomenon is caused by the different heat transfer coefficients in the water and steam spaces. This takes place in large steam generator drums, superheater headers and steam pipelines. High thermal stresses caused by nonuniform temperature distribution on vessel circumference also occur in emergency situations such as fire of partially filled fuel tanks. The upper part of the horizontal vessel is heated much faster than the lower part filled with liquid.

Similar phenomenon occurs in inlet nozzles in PWR nuclear reactor, at which high temperature differences on the circumference of the feed water nozzles are observed.

The study presents an analysis of transient temperature and stress distribution in a cylindrical pressure component during start-up of the steam boiler and shut-down operations. Thermal stresses are determined indirectly on the basis of measured temperature values at selected points on the outer surface of a pressure element. Having determined transient temperature distribution in the entire component, thermal stresses are determined using the finite element method. Measured pressure changes are used to calculate pressure caused stresses. The calculated temperature histories were compared with the experimental data at selected interior points.

The presented method of thermal stress control was applied in a few large conventional power plants. It can also be used successfully in nuclear power plants. The developed method for monitoring thermal stresses and pressure-caused stresses is also suitable for nuclear power plants, since it does not require drilling holes for sensors in pressure element walls. Measurements conducted over the last few years in power plants demonstrate that the presented method of stress monitoring can be applied in systems for automatic power boiler start-up operations and in systems for monitoring the fatigue and creep usage factor of pressure components.

2. Present methods used to determine temperature and thermal stress transient in boiler pressure elements

The simplest and most frequently used method for reducing thermal stresses in pressure elements is to limit heating and cooling rates. The allowable rates of fluid temperature changes can be determined using German Boiler Codes TRD 301 (TRD 301, 2001) or European Standard EN 12952-3 (EN 12952-3, 2001). The allowable heating rate of temperature changes is determined from the following condition (EN 12952-3, 2001; TRD 301, 2001)

$$\alpha_m \frac{(p-p_o)d_m}{2s} + \alpha_T \phi_w \phi_f \frac{v_T s^2}{a} = \sigma_{\min} \,. \tag{1}$$

The maximum allowable cooling rate is calculated in a similar way:

$$\alpha_m \frac{(p-p_o)d_m}{2s} + \alpha_T \phi_w \phi_f \frac{v_T s^2}{a} = \sigma_{\max} \,. \tag{2}$$

In Equations (1-2), the following nomenclature is used: $a = k/(c \cdot \rho)$ - thermal diffusivity, m^2/s, r_{in}, r_{out} - inner and outer radius, m, $d_m = r_{out} + r_{in}$ - mean diameter, m, p - absolute pressure, MPa, p_o - ambient pressure, MPa, $s = r_{out} - r_{in}$ - thickness of cylindrical element, m, v_T - rate of temperature changes of fluid or pressure element wall, K/s, α_m - pressure caused stress intensity factor, α_T - thermal stress intensity factor, σ_{min} - allowable stress during start-up (heating), MPa, σ_{max} - allowable stress during shutdown (cooling), MPa.
Coefficients ϕ_w and ϕ_f are defined as follows

$$\phi_w = \frac{E\beta}{1-\nu} \,, \tag{3}$$

$$\phi_f = \frac{\left(3u^2-1\right)\left(u^2-1\right)-4u^4 \ln u}{8\left(u^2-1\right)\left(u-1\right)^2} \,, \tag{4}$$

where: E - Young's modulus, MPa, β - linear coefficient of thermal expansion, 1/K, ν - Poisson's ratio, $u = r_{out}/r_{in}$ - ratio of outer to inner surface radius.
Equations (1-2) can also be used to calculate the maximum total stress at hole edges. The heating or cooling rate of temperature changes in pressure elements v_T can be calculated using the moving average filter (Taler, 1995)

$$\begin{aligned} v_T = \left.\frac{df}{dt}\right|_{t=t_i} &= \frac{1}{693\Delta t}\left(-63 f_{i-4} + 42 f_{i-3} + 117 f_{i-2} + 162 f_{i-1} + \right. \\ &\left. +177 f_0 + 162 f_{i+1} + 117 f_{i+2} + 42 f_{i+3} - 63 f_{i+4}\right), \end{aligned} \tag{5}$$

where f_i are medium or wall temperatures at nine successive time points with Δt time step.

Equations (1-2) and (5) are not only valid for a quasi-steady state (Taler, 1995), but also when the temperature change rate v_T is the function of time: $v_T = v_T(t)$. For pressure components with complex geometry, the stress value in the stress concentration areas can be calculated using the finite element method (FEM). By determining the so called influence function with the use of the FEM, one is subsequently able to carry out an on-line stress calculation with known heat transfer coefficient and fluid temperature transient (Taler et al., 2002).

This chapter formulates the problem of determining the transient temperature in a pressure element as an inverse transient heat conduction problem. The temperature distribution is determined on the basis of temperature histories measured at selected points at the outer insulated surface of a pressure component. After determining transient temperature distribution in the entire pressure component, thermal stresses are computed using the FEM. Inverse problem is solved using the finite volume method (FVM). Thermal and pressure caused stresses are calculated using the FEM.

3. Mathematical formulation of inverse problem

In the following, two dimensional inverse heat conduction problem (IHCP) will be solved (Fig. 1). The analyzed domain is divided into two subdomains: direct and inverse. Boundary and initial conditions are known for the direct region so that the transient temperature distribution is obtained for the solution of the boundary-initial problem. The temperature distribution on the inner closed surface S_m which is located inside the analyzed area (Fig. 1) is known from measurements. Based on the solution of the direct problem the heat flux on the boundary S_m can be evaluated. Thus, the two boundary conditions are known on the surface S_m:

$$T(s,t)\big|_{S_m} = f(s,t), \tag{6}$$

$$-k\frac{\partial T}{\partial n}\bigg|_{S_m} = q(s,t), \tag{7}$$

while on the inner surface S_{in} of the body, the temperature and heat flux are unknown.

In order to evaluate the transient temperature distribution in the inverse region, this region is divided into control volumes (Fig. 2). The method marches in space towards the inner surface of the body S by using the energy balance equations for the finite volumes placed on the boundary S_m to determine the temperatures in adjacent nodes. In this way of proceeding the time derivatives of the measured temperature changes have to be calculated. The accurate calculation of the time derivatives of the measured temperature histories is difficult since the measured temperature values are burdened with random measurement errors. Thus, time-temperature charts have to be smoothed before evaluating the time derivative. In the present section this was achieved by using the local polynomial approximation. The successive nine temperature data points were approximated using the polynomial of the 3rd degree and then the derivatives in the middle of each interval (time coordinate of the point 5) were calculated.

The space-marching method will be illustrated by an example showing the evaluation of the temperature distribution in the cylindrical wall using the temperature measurement points equally distributed on the surface S_m (Fig. 2). It is assumed that the temperature and heat flux distributions are known on the surface S_m from temperature measurements at points 16-22 and from the solution of the direct problem.

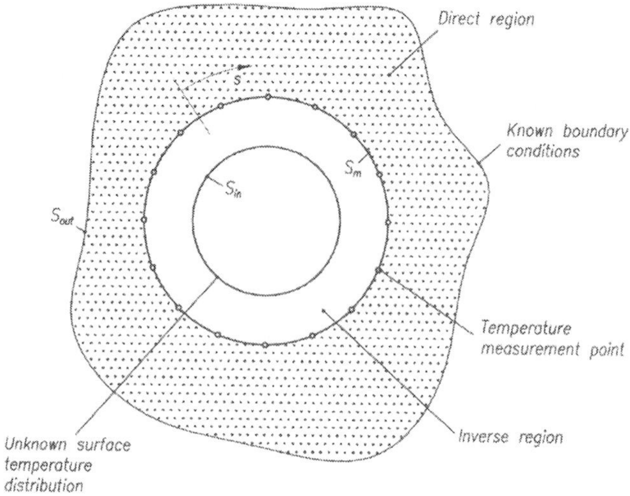

Fig. 1. The analyzed body divided into the inverse and direct regions

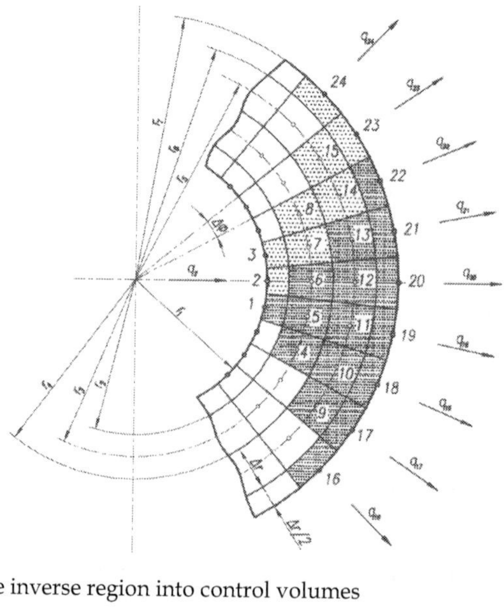

Fig. 2. Division of the inverse region into control volumes

For the ideally insulated outer surface the heat fluxes q_{16}-q_{22} and their time derivatives are equal to zero. The heat balance equation for the finite volume with the node (i, j) is as follows (Fig. 3):

$$\frac{1}{2}\left[\left(r_{i,j}+\frac{\Delta r}{2}\right)^2-\left(r_{i,j}-\frac{\Delta r}{2}\right)^2\right]\Delta\varphi c\left(T_{i,j}\right)\rho\left(T_{i,j}\right)\frac{dT_{i,j}}{dt}=$$
$$=\left(r_{i,j}+\frac{\Delta r}{2}\right)\Delta\varphi\frac{k\left(T_{i,j}\right)+k\left(T_{i+1,j}\right)}{2}\frac{T_{i+1,j}-T_{i,j}}{\Delta r}+$$
$$+\left(r_{i,j}-\frac{\Delta r}{2}\right)\Delta\varphi\frac{k\left(T_{i,j}\right)+k\left(T_{i-1,j}\right)}{2}\frac{T_{i-1,j}-T_{i,j}}{\Delta r}+\Delta r\frac{k\left(T_{i,j+1}\right)+k\left(T_{i,j}\right)}{2}\frac{T_{i,j+1}-T_{i,j}}{\Delta\varphi r_{i,j}}+ \tag{8}$$
$$+\Delta r\frac{k\left(T_{i,j-1}\right)+k\left(T_{i,j}\right)}{2}\frac{T_{i,j-1}-T_{i,j}}{\Delta\varphi r_{i,j}},$$

where: c – specific heat, J/(kg·K), ρ – density, kg/m³, k – thermal conductivity, W/(m·K), t – time, s, T – temperature, °C, Δr – space step in radial direction, m, $\Delta\varphi$ – angular step, rad. Transforming the heat balance Equation (8) for $T_{i-1,j}$, we obtain:

$$T_{i-1,j}=T_{i,j}+\Delta r\frac{\left(r_{i,j}+\frac{\Delta r}{2}\right)^2-\left(r_{i,j}-\frac{\Delta r}{2}\right)^2}{r_{i,j}-\frac{\Delta r}{2}}\cdot\frac{c\left(T_{i,j}\right)\rho\left(T_{i,j}\right)}{k\left(T_{i,j}\right)+k\left(T_{i-1,j}\right)}\cdot\frac{dT_{i,j}}{dt}-$$
$$-\frac{r_{i,j}+\frac{\Delta r}{2}}{r_{i,j}-\frac{\Delta r}{2}}\cdot\frac{k\left(T_{i,j}\right)+k\left(T_{i+1,j}\right)}{k\left(T_{i,j}\right)+k\left(T_{i-1,j}\right)}\cdot\left(T_{i+1,j}-T_{i,j}\right)-\frac{\left(\Delta r\right)^2}{r_{i,j}-\frac{\Delta r}{2}}\cdot\frac{k\left(T_{i,j+1}\right)+k\left(T_{i,j}\right)}{k\left(T_{i,j}\right)+k\left(T_{i-1,j}\right)}\cdot\frac{T_{i,j+1}-T_{i,j}}{\Delta\varphi^2 r_{i,j}}- \tag{9}$$
$$-\frac{\left(\Delta r\right)^2}{r_{i,j}-\frac{\Delta r}{2}}\cdot\frac{k\left(T_{i,j-1}\right)+k\left(T_{i,j}\right)}{k\left(T_{i,j}\right)+k\left(T_{i-1,j}\right)}\cdot\frac{T_{i,j-1}-T_{i,j}}{\Delta\varphi^2 r_{i,j}}.$$

Since Eq. (9) is nonlinear, the fixed-point iterative technique is used to determine the temperature $T_{i,j-1}$:

$$T_{i-1,j}^{(n+1)}=T_{i,j}+\Delta r\frac{\left(r_{i,j}+\frac{\Delta r}{2}\right)^2-\left(r_{i,j}-\frac{\Delta r}{2}\right)^2}{r_{i,j}-\frac{\Delta r}{2}}\frac{c\left(T_{i,j}\right)\rho\left(T_{i,j}\right)}{k\left(T_{i,j}\right)+k\left(T_{i-1,j}^{(n)}\right)}\frac{dT_{i,j}}{dt}-$$
$$-\frac{r_{i,j}+\frac{\Delta r}{2}}{r_{i,j}-\frac{\Delta r}{2}}\frac{k\left(T_{i,j}\right)+k\left(T_{i+1,j}\right)}{k\left(T_{i,j}\right)+k\left(T_{i-1,j}^{(n)}\right)}\left(T_{i+1,j}-T_{i,j}\right)-\frac{\left(\Delta r\right)^2}{r_{i,j}-\frac{\Delta r}{2}}\frac{k\left(T_{i,j+1}\right)+k\left(T_{i,j}\right)}{k\left(T_{i,j}\right)+k\left(T_{i-1,j}^{(n)}\right)}\frac{T_{i,j+1}-T_{i,j}}{\Delta\varphi^2 r_{i,j}}- \tag{10}$$
$$-\frac{\left(\Delta r\right)^2}{r_{i,j}-\frac{\Delta r}{2}}\frac{k\left(T_{i,j-1}\right)+k\left(T_{i,j}\right)}{k\left(T_{i,j}\right)+k\left(T_{i-1,j}^{(n)}\right)}\frac{T_{i,j-1}-T_{i,j}}{\Delta\varphi^2 r_{i,j}}.$$

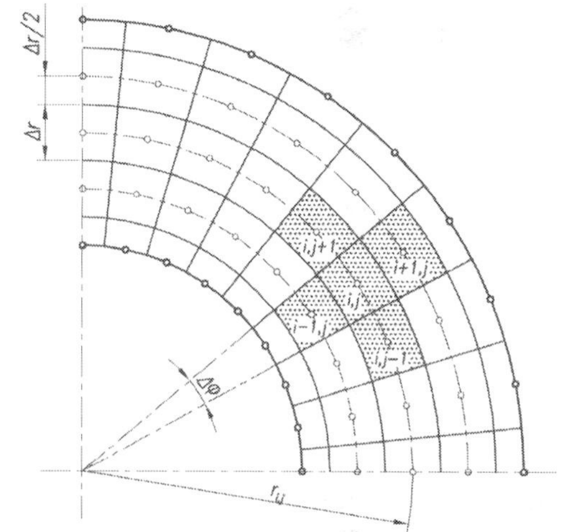

Fig. 3. Space marching in the inverse region

The iteration process continuous until the condition $\left|T_{i-1,j}^{(n+1)} - T_{i-1,j}^{(n)}\right| \le \varepsilon$ is satisfied. Symbol n denotes the iteration number and ε is some small number (tolerance), for example, ε = 0.0001 K. If the thermal properties are constant, the iterations are not required.

To determine the wall temperature at the node 1 (Fig. 2) we proceed as follows. First the heat balance equations are written for the nodes 16 to 22 from which the temperature at the nodes 9 to 13 is determined. Then the heat balance equations are set for the nodes 9 to 13 from which the temperature at the nodes 4 to 6 is evaluated. From the heat equations for the nodes 4, 5 and 6, we can determine the temperature at node 1 located at the inner surface of the pressure component. Based on the measured temperatures at the nodes 17 to 23, the temperature at the node 2 is calculated in similar way. Using the measured temperature at nodes 18 to 24 the temperature at the node 3 is estimated.

Repeating the procedure described above for all the nodes, located at the outer surface, at which the wall temperature is measured, the temperature at the nodes situated at the inner component surface are determined. After determining the temperature distribution at the whole cross-section at the time point t, the temperature distribution at time $t + \Delta t$ is computed. After calculating the temperature distribution the thermal stresses were calculating using the FEM. The finite element mesh is constructed so that the FEM nodes are coincident with the nodes used in the finite volume method.

Random measurement errors of the temperature $f_j(t)$ have great influence on the estimated temperature and thermal stress distributions. If the temperature data are burdened with random errors, least squares smoothing is used to reduce the effect of the measurement errors on the calculated time derivatives df_j/dt or $dT_{i,j}/dt$. The Gram orthogonal polynomials were used for smoothing the measured time-temperature history $f_j(t)$ and estimated temperatures $T_{i-1,j}(t)$ (Taler, 1995). For linear IHCP, when thermal properties are temperature independent, node temperatures can be expressed in explicit form (Taler & Zima, 1999; Taler et al., 1999). In the following subsection the linear IHCP will be presented in detail.

4. Linear inverse problem

The typical way of evaluation of inverse method accuracy is to use the direct method to generate the "measured data" on the outer surface assuming the values of fluid temperature and heat transfer coefficients on the inner surface. The generated data are then used as input data for the inverse method. The results from the inverse solution are accurate when they agree with assumptions made for the direct problem.
First, a direct problem of heat conduction, where the boundary conditions on the drum inner and outer surfaces as well as initial conditions are known, will be discussed. Then the temperature distribution will be determined on the basis of measured temperature changes on the outer, thermally insulated surface of the drum with unknown boundary condition on the inner surface (the inverse problem). The problem formulated in this way, as ill-posed, is more difficult to solve than the first one.
Both direct and inverse problems of heat conduction allow the determination of temperature distribution in the drum cross-section. However, the procedure for thermal stresses determination is slightly different in the two cases. In the direct method both temperature distribution and thermal stresses are calculated using FEM. In the inverse method, unlike the direct problem solved by FEM, the initial and boundary conditions on the inner surface are not set because the transient temperature distribution is obtained from temperature measurements on the outer surface of the drum. Temperature values used for stress calculations are obtained from the inverse solution presented in this section.

4.1 Direct problem of heat conduction

The direct problem has been solved with the following assumptions:
Temperature distribution is described by equation:

$$\frac{1}{r}\frac{\partial}{\partial r}\left(r\frac{\partial T}{\partial r}\right)+\frac{1}{r^2}\frac{\partial^2 T}{\partial \varphi^2}=\frac{1}{a}\frac{\partial T}{\partial t} \tag{11}$$

with boundary conditions:

$$k\left.\frac{\partial T}{\partial r}\right|_{r=r_o}=0\,, \tag{12}$$

$$k\left.\frac{\partial T}{\partial r}\right|_{r=r_{in}}=h_w\left(T_w-\left.T\right|_{r=r_{in}}\right) \tag{13}$$

$0\le\varphi\le\varphi_w$,

$$k\left.\frac{\partial T}{\partial r}\right|_{r=r_{in}}=h_s\left(T_s-\left.T\right|_{r=r_{in}}\right) \tag{14}$$

$\varphi_w\le\varphi\le\pi$,
and initial condition:

$$\left.T\right|_{t=0}=T_0\left(r,\varphi\right). \tag{15}$$

The symbols h_w and h_s stand for heat transfer coefficients at water and steam region, respectively. The initial temperature $T_0(r, \varphi)$ is known from measurements or calculations at steady state.

Since the temperature field is symmetrical with respect to the vertical plane passing through the drum longitudinal axis, the temperature field is analyzed only in half of the drum. The division of drum cross-section into control volumes and finite elements has been shown in Figs. 4 and 5, respectively.

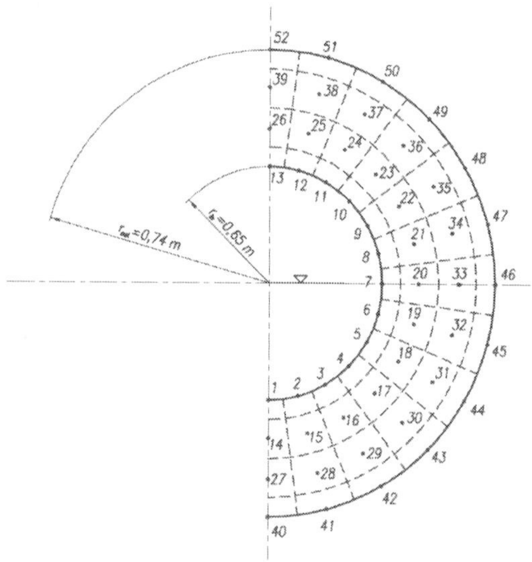

Fig. 4. Division of half of drum cross section into control volumes (52 control volumes)

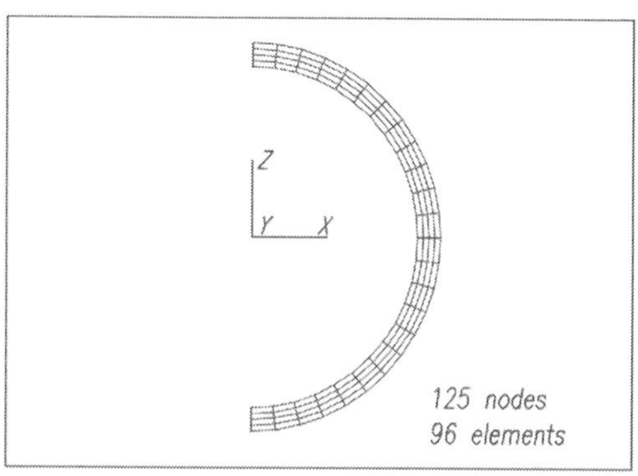

Fig. 5. Division of half of drum cross-section into finite elements (125 nodes, 96 elements)

The direct problem will be solved separately by means of two methods: the method of control volume (CVM) and finite elements. The following data have been used for calculations: r_{in} = 0.65 m, r_{out} = 0.74 m, k = 47 W/(m·K), ρ = 7850 kg/m^3, c = 500 J/(kg·K), $\varphi_w = \pi/2$, h_w = 1000 W/(m^2·K) and h_s = 2000 W/(m^2·K). The changes of the fluid temperature in time ($T_w = T_s$) have been shown in Fig. 6.

In the control volumes method, following planar discretization, 52 ordinary differential equations are obtained, which have been integrated by Runge-Kutta method.

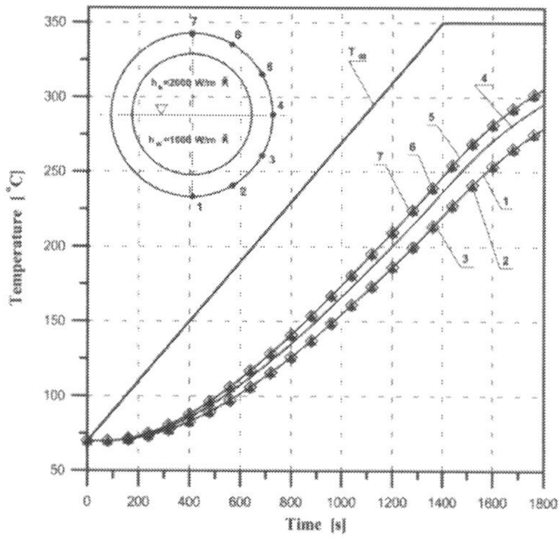

Fig. 6. Fluid temperature changes in drum as function of time and temperature at nodes on the outer surface obtained from the solution of the direct problem using FEM; 1 – 6 numbers of curves refer to the numbers of temperature measuring points on the drum outer surface

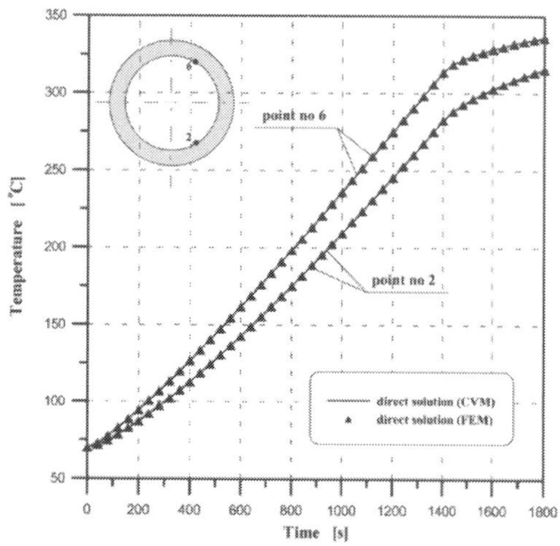

Fig. 7. Comparison of temperature changes at points 2 and 6 calculated by control volume method and FEM

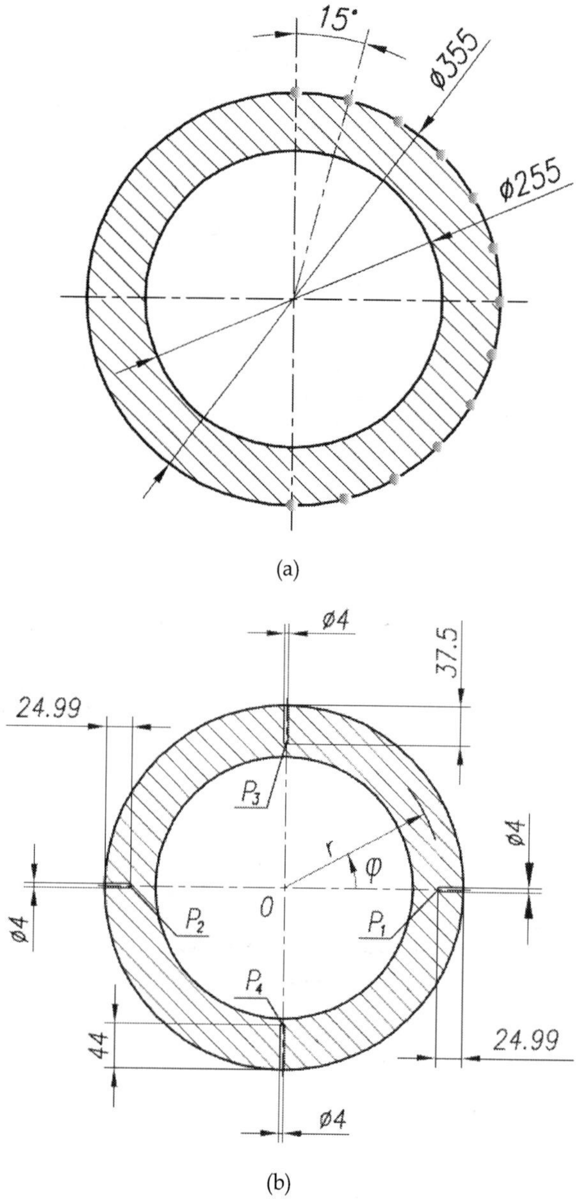

Fig. 12. Location of temperature sensors on the steam header (a) thermocouples at the outer surface on the steam header; (b) location of interior thermocouples used for experimental validation of the inverse procedure

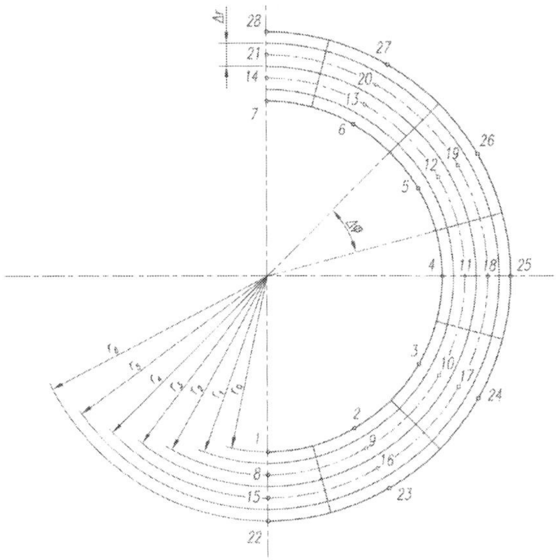

Fig. 13. Division of the steam header cross-section into finite volumes

The largest stresses (Figs. 16-18) occur at the beginning of the heating process, when the ratio of wall temperature change is significant. The stresses at the outer surface are lower in comparison to the inner surface. The greatest absolute value has the longitudinal stress at the inner surface (Figs. 16-17). The influence of the inner pressure on the calculated stresses is negligible since the inner pressure is very low (Fig. 14a).

6. Conclusions

The chapter presents a method of transient temperature field identification in the drum on the basis of measured temperature changes at locations on the outer insulated surface of the pressure component. The method's accuracy was demonstrated by comparison of measured and calculated temperature at a few internal points. In order to reduce the sensitivity of the inverse method to random errors, the measured temperature histories were smoothed by the moving average filter based on the Gram's polynomials approximating nine successive measurement points. The smoothed value of measured temperature and its time derivatives are determined only in the middle of the interval, i.e., in time $t_i + 4(\Delta t)$, where t_i is time coordinate of the first point in the analyzed time interval: $[t_i, t_i + 8(\Delta t)]$. From the measured temperature histories on the outer surface of a pressure component, the temperature distribution can be determined at different time points in the whole cross section. Based on nodal temperatures determined in such a way, thermal stresses are calculated using FEM.

Values of fluid temperature and heat transfer coefficient on the inner surface of the component are not needed because the temperature distribution in the component cross-section is determined by the developed method for solving the inverse heat conduction problem. Thus, the developed method is more accurate than the direct temperature distribution methods, because the fluid temperatures and heat transfer coefficients are unknown in operating components.

2. Thermal design of the heat exchangers

The suitable use of heat transfer knowledge in the design of practical heat transfer equipment is an art. Designers must be constantly aware of the differences between the idealized conditions under which the fundamental knowledge was obtained and the real conditions of their design and its environment. The result must satisfy process and operational requirements and do so cost-effectively. An important element of any design process is to consider and compensate the consequences of error in the basic knowledge, in its subsequent incorporation into a design method. Heat exchanger design is not a extremely accurate procedure under the best of conditions (Shilling et al., 1999).

The design of a heat exchanger usually consists the subsequent steps:

1. Specification of the process conditions, e.g. flow compositions, flow rates, temperatures, pressures.
2. Obtaining of the required physical properties over the temperature and pressure ranges of interest obtained.
3. Choosing the type of heat exchanger that is going to be used.
4. An initial estimation of the size of the heat exchanger that is made, using a heat transfer coefficient appropriate to the fluids, the process, and the equipment.
5. A first design is chosen, complete in all details necessary to carry out the design calculations.
6. Evaluation of ability to perform the process specifications with respect to both heat transfer and pressure drop as the design of heat exchanger is chosen.
7. Described in point above procedure can be repeated to new heat exchanger design if it is necessary. The final design should meet process requirements within reasonable error expectations.

The calculation of convective heat transfer coefficients constitutes a crucial issue in designing and sizing any type of heat exchange device. Thus its correct determining permits for the proper selection of heat transfer area during designing of heat exchangers and calculation of the fluids outlet temperature. A lot efforts have been made during experimental investigations of pressure drop and heat transfer in different types of heat exchangers to obtain proper heat transfer correlation formulas.

2.1 The Wilson plot technique to determine heat transfer correlations in heat exchangers

One of the widely used methods for calculations of heat transfer coefficient is the Wilson plot technique. This approach was developed by E.E. Wilson in 1915 in order to evaluate the heat transfer coefficients in shell and tube condensers for the case of a vapour condensing outside by means of a cooling liquid flow inside (Viegas et al., 1998; Kumar et al., 2001; Rose, 2004; Fernández-Seara et al., 2007). It is based on the separation of the overall thermal resistance into the inside convective thermal resistance and the remaining thermal resistances participating in the heat transfer process. The overall thermal resistance R_{overal} of the condensation process in a shell-and-tubes heat exchanger can be expressed as the sum of three constituent thermal resistances: R_{in} – the internal convection, R_{wall} – the tube wall and R_o – the external convection, presented in Eq. (1).

$$R_{total} = R_{in} + R_{wall} + R_o \, . \tag{1}$$

The thermal resistances of the fouling in Eq. (1) was neglected. Employing the expressions for the thermal resistances in Eq. (1), the overall thermal resistance can be rewritten as follows:

$$R_{total} = \frac{1}{h_{in}A_{in}} + \frac{\ln\left(\frac{d_o}{d_{in}}\right)}{2\pi\lambda_{wall}L_{wall}} + \frac{1}{h_o A_o}. \quad (2)$$

where h_{in} and h_o is the internal and outer heat transfer coefficients, d_{in} and d_o – the inner and outer tube diameters, λ_{wall} is the tube material thermal conductivity, L_{wall} is the tube length and A_i and A_o are the inner and outer tube surface areas, respectively.
On the other hand, the overall thermal resistance can be written as a function of the overall heat transfer coefficient referred to the inner or outer tube surfaces and the corresponding areas. Assuming this the overall thermal resistance is expressed as a function of the overall heat transfer coefficient referred to the inner or outer surface $U_{in/o}$ and the inner or outer surface area $A_{in/o}$ (Eq. 3)

$$R_{total} = \frac{1}{U_{in/o}A_{in/o}}. \quad (3)$$

Taking into account the specific conditions of a shell and tube condenser Wilson assumed that if the mass flow of the cooling liquid was modified, then the change in the overall thermal resistance would be mainly due to the variation of the in-tube heat transfer coefficient, while the remaining thermal resistances remained nearly constant. Therefore, as specified in Eq. (4) the thermal resistances outside of the tubes and the tube wall could be regarded as constant:

$$R_{wall} + R_o = C_1. \quad (4)$$

Wilson determined that for the case of fully developed turbulent flow inside a tube of circular cross-section, the heat transfer coefficient was proportional to a power of the reduced velocity w_r which describes the variations of the fluid property and the tube diameter. Thus, the heat transfer coefficient could be written in form:

$$h_{in} = C_2 w_r^n, \quad (5)$$

where C_2 is a constant, w_r – the reduced fluid velocity and n – velocity exponent. In this case the convective thermal resistance related to the inner tube flow is proportional to $1/w_r^n$. Inserting Eqs. (4) and (5) into Eq. (1), the overall thermal resistance becomes the linear function of $1/w_r^n$, where C_1 is the intercept and $1/(C_2 A_{in})$ is the slope of the straight line.
The overall thermal resistance can be calculated using experimental data using the following formula:

$$Q = U_o A_o \Delta T_{lm}. \quad (6)$$

Substituting Eq. (3) into Eq. (6), and assuming $Q = \dot{m}_1 c_{p1}(T_{1outlet} - T_{1inlet})$, where $\dot{m}_1$ is the mass flow rate of cooling liquid, c_{p1} – average specific heat of cooling liquid, and T_{1inlet}, $T_{1outlet}$, are inlet and outlet temperatures of cooling liquid, respectively, yields to

$$R_{total} = \frac{\Delta T_{lm}}{\dot{m}_1 c_{p1} \left(T_{1outlet} - T_{1inlet} \right)} . \tag{7}$$

As the constants C_1 and C_2 are determined from straight–line approximation of measured data, to evaluation, for a given mass flow rate, the internal heat transfer coefficient can be used Eq. (5) and internal heat transfer coefficient Eq. (8):

$$h_o = \frac{1}{A_o \left(C_1 - R_{wall} \right)} . \tag{8}$$

The original Wilson plot technique depends on the knowledge of the overall thermal resistance, that involves to remain of one fluid flow rate constant and varying flow rate of the another fluid.

Approach of Wilson plot technique to determine constant in heat transfer correlation formula for helically coiled tube-in-tube heat exchanger is presented by Sobota (Sobota, 2011).

3. Experimental prediction of heat transfer correlations in heat exchangers

In this chapter, the experimental and numerical investigations of helically coiled tube-in-tube heat exchanger are presented. Calculations of unknown constants and exponents in correlations formula for Nusselt number have been performed with least squares method using Levenberg-Marquardt algorithm. Presented method allows for determining unknown values of constants and exponents in correlation formulas for Nusselt number. This method enables to obtain values of heat transfer coefficient on both sides of the barrier simultaneously without earlier indirect calculations of the overall heat transfer.

3.1 Mathematical formulation of the inverse problem

The issue consisting of simultaneous determining of the heat transfer coefficient on the cooling and heating liquid is ranked among inverse heat transfer problems (IHCP) (Beck et al., 1985). In discussed methodology the knowledge of correlation formula form for heat transfer coefficient on the both sides of the heat transfer surface, for counter-flow and parallel-flow heat exchanger, was assumed to be known. An unknown value of the parameters in correlation formulas was hidden in equations for outlet temperature of the liquids (Nashchokin, 1980):

a) parallel-flow arrangement of heat exchanger

- heating liquid

$$T''_{1,calc} = T'_{1,meas} - \left(T'_{1,meas} - T'_{2,meas} \right) \frac{1 - e^{-\left(1+\frac{W_1}{W_2}\right)\cdot\frac{k \cdot F_c}{W_1}}}{1 + \frac{W_1}{W_2}} \tag{9}$$

- cooling liquid

$$T''_{2,calc} = T'_{2,meas} + \left(T'_{1,meas} - T'_{2,meas} \right) \frac{W_1}{W_2} \frac{1 - e^{-\left(1+\frac{W_1}{W_2}\right)\cdot\frac{k \cdot F_c}{W_1}}}{1 + \frac{W_1}{W_2}} \tag{10}$$

b) counterflow arrangement of heat exchanger

- heating liquid

$$T''_{1,calc} = T'_{1,meas} - \left(T'_{1,meas} - T'_{2,meas}\right)\frac{1-e^{-\left(1-\frac{W_1}{W_2}\right)\frac{k\cdot F_c}{W_1}}}{1-\frac{W_1}{W_2}e^{-\left(1-\frac{W_1}{W_2}\right)\frac{k\cdot F_c}{W_1}}} \tag{11}$$

- cooling liquid

$$T''_{2,calc} = T'_{2,meas} + \left(T'_{1,meas} - T'_{2,meas}\right)\frac{W_1}{W_2}\frac{1-e^{-\left(1-\frac{W_1}{W_2}\right)\frac{k\cdot F_c}{W_1}}}{1-\frac{W_1}{W_2}e^{-\left(1-\frac{W_1}{W_2}\right)\frac{k\cdot F_c}{W_1}}} \tag{12}$$

where $T'_{1,meas}$ and $T'_{2,meas}$ - measured temperature of the heating and cooling liquid at the inlet of the helically coiled heat exchanger respectively, °C; $T''_{1,calc}$ and $T''_{2,calc}$ - calculated temperature of the heating and cooling liquid at the outlet of the helically coiled heat exchanger respectively, °C; and expression $W = V\cdot\rho\cdot c_v$ is called as water equivalent.
The minimum of the square of the differences between measured and calculated from analytical formula temperatures of the hot fluid and differences between measured and calculated from analytical formula temperatures of the cold fluid at the outlet of heat exchanger was searching for:

$$S(\boldsymbol{\alpha}) = \sum_{i=1}^{n}\left[\left(T''_{1,meas} - T''_{1,calc}\right)^2 + \left(T''_{2,meas} - T''_{2,calc}\right)^2\right] \to \min . \tag{13}$$

In analysed example the solution of the nonlinear least square problem was searching for. Determining the values of the constants, and indirectly the values of heat transfer coefficient, was carried out using least squares method with modified Levenberg-Marquardt algorithm (Visual Numerics, 2007; Press et al., 1996).
In Levenberg-Marquardt algorithm unknown are formed an column vector $\mathbf{x} = (x_1, x_2, ..., x_m)^T$, for which the sum becomes minimum

$$S(\mathbf{x}) = \sum_{i=1}^{n}\left[r_i(\mathbf{x})\right]^2 \to \min , \tag{14}$$

where $r_i(\mathbf{x}) = T''_{meas} - T''_{calc}$.
The method performs the k-th iteration as

$$\mathbf{x}^{(k+1)} = \mathbf{x}^{(k)} + \boldsymbol{\delta}^{(k)} , \tag{15}$$

where

$$\boldsymbol{\delta}^{(k)} = \left[\left(\mathbf{J}_m^{(k)}\right)^T \mathbf{J}_m^{(k)} + \mu^{(k)}\mathbf{I}_n\right]^{-1}\left(\mathbf{J}^{(k)}\right)^T\left[\mathbf{f} - \mathbf{T}_m(\mathbf{x})\right], \qquad k = 0, 1, ... \tag{16}$$

The symbols **f** and $\mathbf{T}_m(\mathbf{x})$ stand for vector of measured and vector of computed temperature, respectively.
Jacobian determinant is described by formula

$$\mathbf{J}_m = \frac{\partial \mathbf{r}_m(\mathbf{x})}{\partial \mathbf{x}^T} = \left[\left(\frac{\partial r_i(\mathbf{x})}{\partial x_j} \right) \right]_{mxn}, \tag{17}$$

where $i = 1, \ldots, n$, $j = 1, \ldots, m$, $\mathbf{D}^{(k)}$ denotes diagonal matrix with positive elements. Quite often $\mathbf{D}^{(k)} = \mathbf{I}_m$, where $\mathbf{I}_m$ is identity matrix.
The value of the parameter $\mu^{(k)} \to 0$ when $\mathbf{x}^{(k)} \to \mathbf{x}^*$. In the proximity of minimum $\mathbf{x}^*$ the iteration step in the Levenberg-Marquardt method is almost the same as in the Gauss-Newton method. The computation programs for solving the non-linear least square problem by the Levenberg-Marquardt method are described in (Lawson & Hanson, 1974) and in the IMSL Library (Visual Numerics, 2007).

3.2 Correlations for Nusselt number

Although curved pipes are used in a wide range of applications, flow in curved pipes is relatively less well known than that in straight ducts. A helical coil can be geometrically described by the coil radius R, the pipe radius r, and the coil pitch $2\pi b$ (Fig. 1).

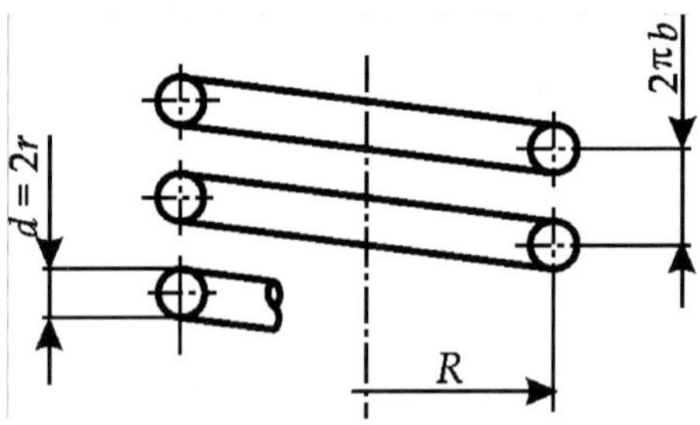

Fig. 1. Schematic representation of a helical pipe with its main geometrical parameters: r - tube radius; R - coil radius; $2\pi b$ - coil pitch

The observations on the complexity of a flow in such a channels allowed to notice the effect of curvature on the fluid flow regime which occurs delaying the transition from laminar to transitional flow to a higher Reynolds number with respect to straight pipes (Ito, 1959; Schmidt, 1967). Using data from his own experiments as well as that from previous investigations, Ito developed the following empirical relation to determine the critical Reynolds number for the range of curvature ratios of 1/15 to 1/860:

$$\mathrm{Re}_{crit} = 20000\left(r/R \right)^{0.32}, \tag{18}$$

whilst Schmidt suggested the form of critical Reynolds number listed below:

$$\mathrm{Re}_{crit} = 2300 \cdot \left[1 + 8.6\left(\frac{r}{R} \right)^{0.45} \right]. \tag{19}$$

For curvature ratios $\delta = r/R$ less than 1/860, the critical Reynolds number was found to correspond with that of a straight pipe.
Equation for Nusselt number that are most commonly found in literature concerning heat transfer in curved or helical tubes can be assumed formula developed by Schmidt (Schmidt, 1967):

- for laminar regime

$$Nu = 3.65 + 0.08 \cdot \left[1 + 0.8\left(\frac{r}{R}\right)^{0.9}\right] \cdot \mathrm{Re}^{\left[0.5+0.2903\left(\frac{r}{R}\right)^{0.194}\right]} \cdot \mathrm{Pr}^{\frac{1}{3}}, \tag{20}$$

where Reynolds number varies $100 < \mathrm{Re} < \mathrm{Re}_{crit}$, where Re_{crit} is described by Eq. (19).

- for turbulent flow

$$Nu = 0.023 \cdot \left[1 + 14.8 \cdot \left(1 + \frac{r}{R}\right) \cdot \left(\frac{r}{R}\right)^{\frac{1}{3}}\right] \cdot \mathrm{Re}^{\left[0.8+0.22 \cdot \left(\frac{r}{R}\right)^{0.1}\right]} \cdot \mathrm{Pr}^{\frac{1}{3}}, \quad \mathrm{Re}_{crit} < \mathrm{Re} \le 22000, \tag{21}$$

$$Nu = 0.023 \cdot \left[1 + 3.6 \cdot \left(1 - \frac{r}{R}\right)\left(\frac{r}{R}\right)^{0.8}\right] \cdot \mathrm{Re}^{0.8} \cdot \mathrm{Pr}^{1/3}, \quad 22000 < \mathrm{Re} \le 150000. \tag{22}$$

Another formulas for Nusselt number that is valid for turbulent flow was invented by Seban and McLaughlin (Seban & McLaughlin, 1963):

$$Nu = 0.023 \cdot \left(\frac{r}{R}\right)^{0.1} \cdot \mathrm{Re}^{0.85} \cdot \mathrm{Pr}^{0.4} \tag{23}$$

and Rogers and Mayhew (Rogers & Mayhew, 1964)

$$Nu = 0.021 \cdot \left(\frac{r}{R}\right)^{0.1} \cdot \mathrm{Re}^{0.85} \cdot \mathrm{Pr}^{0.4}. \tag{24}$$

Eqs. (23) and (24) have simple structure that makes them easy to use.
Other widely used method is that of Seider and Tate, who recommended the following expression for applications with large property variations from the bulk flow to the wall temperature:

$$Nu = 0.027 \cdot \mathrm{Re}^{0.8} \cdot \mathrm{Pr}^{1/3} \left(\frac{\mu_{bulk}}{\mu_{wall}}\right)^{0.14} \tag{25}$$

for $0.7 < \mathrm{Pr} < 16000$, $\mathrm{Re} > 10000$ and $L/D > 10$.
For more accurate calculations in fully developed turbulent flow it is recommended to use Petukhov heat transfer correlation that is valid for $0.5 < \mathrm{Pr} < 2000$ and $10000 < \mathrm{Re} < 5000000$:

$$Nu = \frac{(f/2)\mathrm{Re}\,\mathrm{Pr}}{1.07 + 12.7(f/2)^{0.5}\left(\mathrm{Pr}^{2/3} - 1\right)} \left(\frac{\mu_{bulk}}{\mu_{wall}}\right)^{0.14}, \tag{26}$$

where friction factor f can be obtained from the Moody diagram or from Petukhov's friction factor correlation that is valid for 3000 < Re < 5000000:

$$f = \frac{1}{1.58\ln(\mathrm{Re}) - 3.28}. \tag{27}$$

Another heat transfer correlation commonly used is that of Gnielinski (Smith, 1997), which extends the Petukhov correlation down into the transition regime:

$$Nu = \frac{\left(f/2\right)(\mathrm{Re}-100)\mathrm{Pr}\left(\mathrm{Pr}/\mathrm{Pr}_w\right)^{0.14}}{1+12.7\left(f/2\right)^{0.5}\left(\mathrm{Pr}^{2/3}-1\right)}\left(\frac{\mu_{bulk}}{\mu_{wall}}\right)^{0.14}, \tag{28}$$

where

$$f_c = \frac{\frac{0.3164}{\mathrm{Re}^{0.25}} + 0.03\sqrt{\frac{r}{R}}\cdot\left(\frac{\eta_{wall}}{\eta_{bulk}}\right)^{0.27}}{4}. \tag{29}$$

Worth to be mentioned are the following heat transfer correlations:

- Mori and Nakayama (Manglik, 2003)

$$Nu = \frac{\left(\frac{r}{R}\right)^{\frac{1}{12}}\left[1+0.061\left(\left(\frac{r}{R}\right)^{2.5}\mathrm{Re}\right)^{0.167}\right]\mathrm{Pr}^{0.4}\,\mathrm{Re}^{\frac{5}{6}}}{41.0}. \tag{30}$$

- Jeschke, which is the oldest one (Rogers & Mayhew, 1964)

$$Nu = 0.045\cdot\left[1+3.54\frac{r}{R}\right]\cdot\mathrm{Re}^{0.76}\cdot\mathrm{Pr}^{0.4}. \tag{31}$$

This correlation was developed as a result of transposition formula that was valid for air flow through helical two loop heat exchanger into water.

- Kirpikov (Nashchokin, 1980)

$$Nu = 0.0456\cdot\left(\frac{r}{R}\right)^{0.21}\mathrm{Re}^{0.76}\cdot\mathrm{Pr}^{0.4},\quad 10000 < \mathrm{Re} \le 45000 \tag{32}$$

and Mikheev (Nashchokin, 1980)

$$Nu = 0.021\cdot\left[1+1.77\cdot\left(\frac{r}{R}\right)\right]\cdot\mathrm{Re}^{0.85}\cdot\mathrm{Pr}^{0.43}. \tag{33}$$

The discussed correlations can be helpful in selecting the form of the heat transfer correlation in which certain coefficients and exponents are to be determined.

3.3 Experimental setup

To determine the Nusselt number correlation for forced convection in helically coiled tube-in-tube heat exchanger an experimental setup was build. It consisted of copper made heat exchanger, electric heater and circulating pumps.

3.3.1 Heat exchanger

Helical coil heat exchangers are one of the most common equipment found in many industrial applications ranging from chemical and food industries, power production, electronics, environmental engineering, air-conditioning, waste heat recovery and cryogenic processes. Helical coils are extensively used as heat exchangers and reactors due to higher heat and mass transfer coefficients, narrow residence time distributions and compact structure. The modification of the flow in the helically coiled tubes is due to the centrifugal forces (Dean, 1927, Dean, 1928). The curvature of the tube produces a secondary flow field with a circulatory motion, which causes the fluid particles to move toward the core region of the tube. The secondary flow increases heat transfer rates as it reduces the temperature gradient across the cross-section of the tube. Thus there is an additional convective heat transfer mechanism, perpendicular to the main flow, which does not exist in conventional heat exchangers. An extensive review of fluid flow and heat transfer in helical pipes has been presented in the literature (Kumar et al. 2008; Shah & Joshi, 1987).

The examined heat exchanger was constructed from copper tubing and typical connections were made of copper also and consisted of 6.5 loop. The outer tube of the heat exchanger had an outer diameter of 35 mm and a wall thickness of 1.5 mm. The inner tube had an outer diameter of 22 mm with wall thickness of 1 mm.

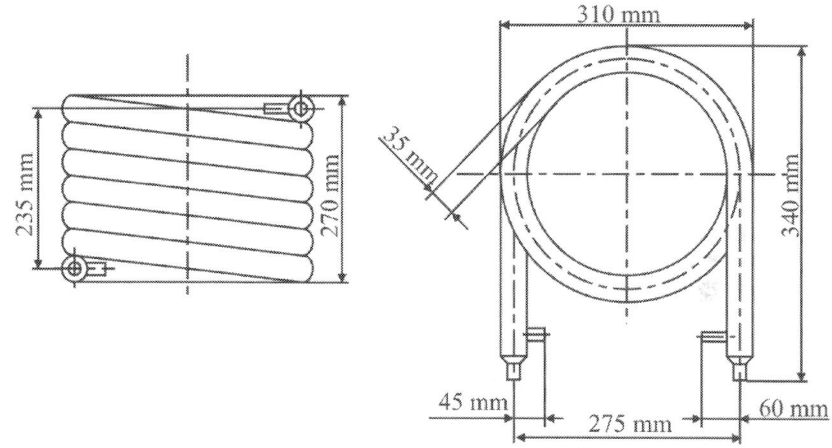

Fig. 2. Schematic of the examined heat exchanger with basic geometry

Coil had a radius of curvature, measured from the centre of the inner tube, of 137.5 mm. Its thermal power Q_n was equal to 14 kW, pressure drop Δp = 0.32 bar and volumetric flow V = 2.3 m^3/h. Calculated heat transfer area F_c of the heat exchanger was 0.3952 m^2. The heat exchanger was very carefully insulated with polyurethane foam to avoid heat losses to the surroundings.

it is shown in Fig. 4. In the following, Eq. (34) will be used for the calculation of the Nusselt number in the circular duct.

Flow type in helically coiled tube-in-tube heat exchanger	
Parallel flow	Counter flow
Equation (36)	
$A_1 = 0.0202$	$A_1 = 0.0188$
Equation (37)	
$A_2 = 30.2301$	$A_2 = 61.8249$

Table 1. Values of the constants in Eq. (36) and Eq. (37) calculated with Levenberg-Marquardt method

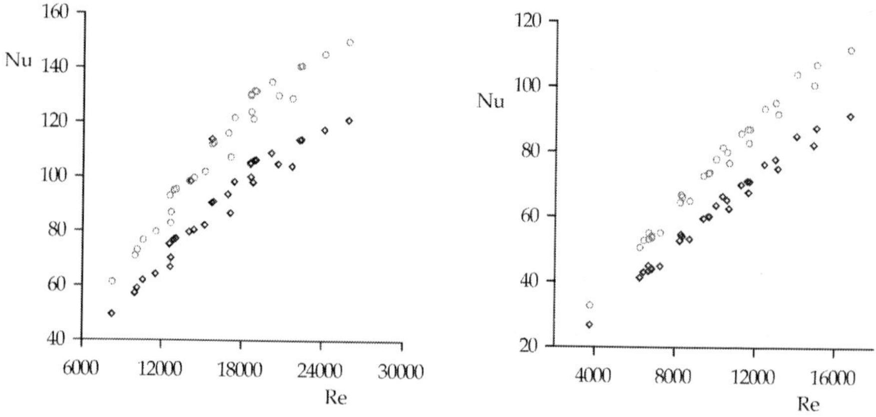

a) b)

Fig. 4. Distribution of the Nusselt number as a function of Reynolds number in circular channel of the helically coiled tube-in-tube heat exchanger; a) parallel-flow, b) counterflow; ○ Eq. (34), ◊ Eq. (36)

For next stage of calculations Eq. (34) and (35) were modified. Constant A_i and exponent B_i, where $i = 1, 2$ of Reynolds number were investigated in this case.

$$Nu_1 = \left(1 + 3.5 \cdot \frac{d_{1in}}{D}\right) \cdot A_1 \cdot \mathrm{Re}_1^{B_1}\, \mathrm{Pr}_1^{0.33} \tag{38}$$

and

$$Nu_2 = \left(1 + 3.5\frac{d_h}{D}\right) \cdot \left[3.66 + 1.2 \cdot \left(\frac{d_1}{d_{2in}}\right)^{-0.8} + A_2 \cdot \left(\mathrm{Re}_2 \cdot \mathrm{Pr}_2 \cdot \frac{d_h}{L}\right)^{B_2}\right]. \tag{39}$$

As a first constant A_2 in formula (39) for Nusselt number in annular channel for parallel flow in examined heat exchanger was determined and then constant A_2 and exponent B_2

simultaneously. During the calculations the flow in circular channel was described by formula (34).

Identical calculations were carried out for Eq. (38), which was modified in order to adapt it to describe heat transfer in annular channel:

$$Nu_1 = \left(1 + 3.5 \cdot \frac{d_h}{D}\right) \cdot A_1 \cdot \mathrm{Re}_1^{B_1}\, \mathrm{Pr}_1^{0.33}. \tag{40}$$

Also in this case as a first was calculated constant A_1, and next constant A_1 and exponent B_1 in Eq. (40).

This procedure was to test, which of the equations will be better to map the set of experimental data of the working fluid in annular channel - less complicated and correct in case of turbulent flow - Eq. (34) or Eq. (35) describing the heat transfer for fluid flow in laminar range.

$Nu_2 = (1+3.5\cdot d_h/D)\cdot\left[3.66+1.2\cdot(d_1/d_{2in})^{-0.8} + A_2\cdot(\mathrm{Re}_2\cdot \mathrm{Pr}_2\cdot d_h/L)^{B_2}\right]$	
$A_2 = 19.7433$	
$A_2 = 3.5754$	$B_2 = 0.8229$
$Nu_1 = (1+3.5\cdot d_h/D)\cdot A_1\cdot \mathrm{Re}_1^{B_1}\, \mathrm{Pr}_1^{0.33}$	
$A_1 = 0.0566$	
$A_1 = 0.0269$	$B_1 = 0.8926$

Table 2. Values of constant and exponents calculated with Levenberg-Marquardt for parallel flow in helically coiled tube-in-tube heat exchanger

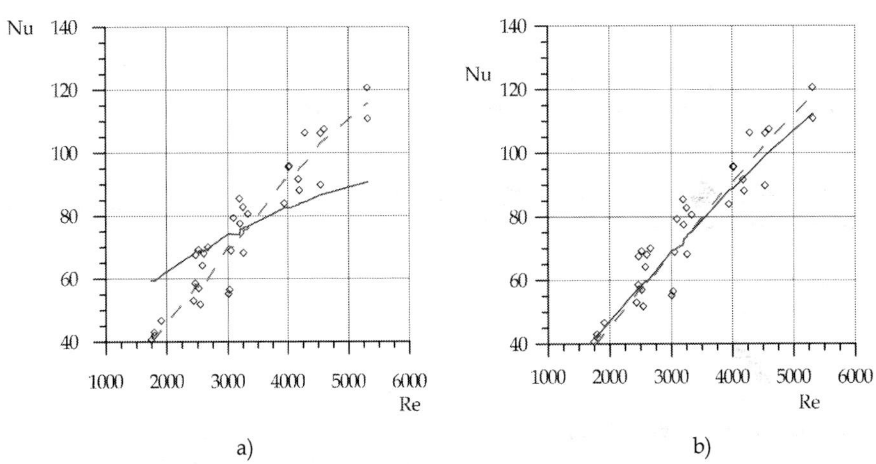

Fig. 5. Comparison of Nusselt number changes in annular channel (parallel flow) for Eqs. (39) and (40). Using Levenberg-Marquardt method was determined: a) one coefficient ——— and b) two coefficients — — —, experimental points ◊

In case when examined helically coiled tube-in-tube heat exchanger was working as a counter flow analogous calculations, using least squares method with modified Levenberg-Marquardt, were carried out to determine constants and exponents in correlation formulas for Nusselt number. Results are shown in Table 3 – forms of correlation formulas and determined values of the constants and exponents. Fig. 6. shows changes of two different formulas for Nusselt number as a function of Reynolds number in annular channel. Also in this case as a first was determined value of a constant, and as a second value of constant and exponent.

$Nu_2 = (1+3.5\cdot d_h/D)\cdot\left[3.66+1.2\cdot(d_1/d_{2in})^{-0.8} + A_2\cdot(\mathrm{Re}_2\cdot \mathrm{Pr}_2\cdot d_h/L)^{B_2}\right]$	
$A_2 = 25.397$	
$A_2 = 14.1959$	$B_2 = 0.4997$
$Nu_1 = (1+3.5\cdot d_h/D)\cdot A_1\cdot \mathrm{Re}_1^{B_1}\,\mathrm{Pr}_1^{0.33}$	
$A_1 = 0.0827$	
$A_1 = 0.3787$	$B_1 = 0.6052$

Table 3. Values of constant and exponents calculated with Levenberg-Marquardt for counter flow in helically coiled tube-in-tube heat exchanger

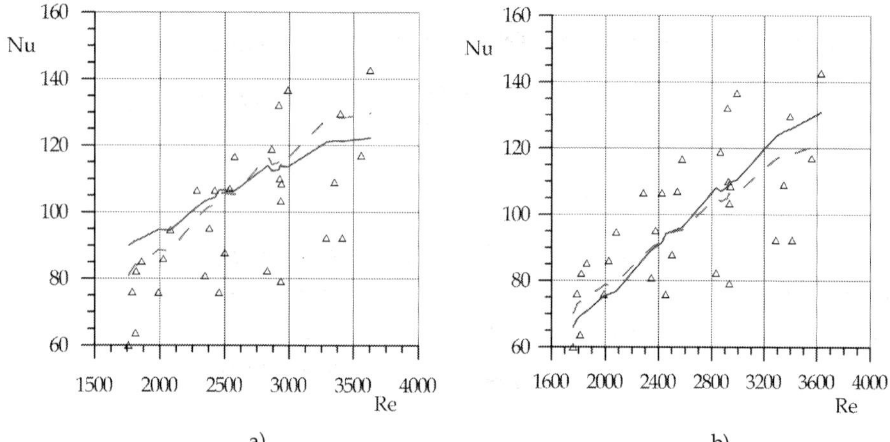

Fig. 6. Comparison of Nusselt number changes in annular channel (counter flow) for Eqs. (38) and (39). Using Levenberg-Marquardt method was determined: a) one coefficient ——— and b) two coefficients — — —, experimental points △

Analyzing changes of the Nusselt number in annular channel (hot fluid) shown on the Fig. 5a and Fig. 6a, drawn as dotted line, the best fit curves to experimental points can be noticed when the Levenberg-Marquardt method is used to determine values of two unknown parameters – constant and exponent in Eq. (39) than in case where only value of the one parameter was investigated.

The differences between calculated values of constants and exponents in heat transfer correlation formulas for counterflow and parallel-flow configuration of the examined heat exchanger are result of the larger average temperature difference between the two fluids.
Comparison of the Nusselt number changes with calculated one and two unknown parameters for turbulent (40) and laminar (39) form of the Nusselt formula leads to the conclusion that Eq. (40) better describes the kind of the fluid flows in annular channel for heat exchanger operating in parallel and counter flow.

4. Conclusion

In this paper methodology which allows for numerical determination unknown parameters in correlation formulas for Nusselt number and heat transfer coefficients on the hot and cold fluid side simultaneously was presented. Calculations were carried out on the basis of gathered experimental data for parallel and counter flow of working fluid in helically coiled tube-in-tube heat exchanger. The changes of Nusselt number in circular channel (turbulent flow) and annular channel (laminar flow) as a function of Reynolds number were presented also.
Described methodology for determining constants and exponents in correlation formulas for Nusselt number can be used in designing of the different heat exchangers types and shapes of heat transfer surface.

5. References

Beck, J. V.; Blackwell, B.; Clair, Ch. R. St. (1985). *Inverse Heat Conduction. Ill-posed Problems.* John Wiley & Sons, ISBN 0-47-108319-4, New York, USA

Dean W.R. (1928). The Streamline Motion of Fluid in a Curved Pipe (second paper), *Philos. Mag.*,Vol. 7, pp. 673–695, ISSN 1941-5982

Dean, W.R. (1927). Note on the Motion of Fluid in a Curved Pipe, *Philos. Mag.*, Vol. 4, pp. 208–223, ISSN 1941-5982

Fernández-Seara, J.; Uhía F.J.; Sieres, J.; Campo, A. (2007). A General Review of the Wilson Plot Method and its Modifications to Determine Convection Coefficients in Heat Exchange Devices. *Applied Thermal Engineering*, Vol.27, pp. 2745–2757, ISSN 1359-4311

Hewitt, G.F.; Shires, G.L.; Bott T.R. (1994). *Process Heat Transfer.* Begell House Publishers, ISBN 0-8493-9918-1, New York, USA

Holman, J.P. (2009). *Heat Transfer.* McGraw-Hill, ISBN 0-07-114320-3, New York, USA

Ito, H. (1959). Friction factors for turbulent flow in curved pipes, *Transactions of ASME D 81*, pp. 123-132, ISSN 0021-9223

Kumar, R.; Varma, H.K.; Agrawal, K.N.; Mohanty, B. (2001). A Comprehensive Study of Modified Wilson Plot Technique to Determine the Heat Transfer Coefficient During Condensation of Steam and R-134a Over Single Horizontal Plain and Finned Tubes. *Heat Transfer Engineering*, Vol.22, pp.3–12, ISSN 0145-7632

Kumar, V.; Faizee, B.; Mridha, M.; Nigam, K.D.P. (2008). Numerical Studies of a Tube-in-Tube Helically Coiled Heat Exchanger. *Chemical Engineering and Processing*, Vol.47, pp. 2287–2295, ISSN 0255-2701

Kuppan, T. (2000). *Heat Exchanger Design Handbook*, Marcel Dekker, Inc, ISBN 0-8247-9787-6, New York, USA

Lawson, C.; Hanson R. (1995). *Solving Least Squares Problems*, ISBN 0-89871-356-0, Prentice Hall, Englewood Cliffs, USA

Manglik, R.M. (2003). HeatTransfer Enhancement, In: *Heat Transfer Textbook,* A. Bejan and Kraus A.D., (Eds.), John Wiley & Sons, ISBN 0-471-39015-1, New York, USA

Nashchokin, V.V. (1980). *Engineering Thermodynamics and Heat Transfer,* Central Books Ltd, ISBN 0-71-471523-9, London, Great Britain

Pitts, D.R.; Sissom, L.E. (1998). *Shaum's Outline of Theory and Problems of Heat Transfer.* McGraw-Hill, ISBN 0-07-050207-2, New York, USA

Press, W.H.; Teukolsky, S.A.; Vetterling W.T.; Flannery B.P. (1996). *Numerical Recipes in Fortran 77: The Art of Scientific Computing.* Cambridge University Press, ISBN 0-521-43064-X, New York, USA

Rennie, T.J.; Raghavan, V.G.S. (2006). Effect of Fuid Thermal Properties on the Heat Transfer Characteristics in a Double-Pipe Helical Heat Exchanger. *International Journal of Thermal Sciences,* Vol.45, pp. 1158–1165, ISSN 1290-0729

Rogers, G.F.C.; Mayhew Y. R. (1964). Heat Transfer And Pressure Loss In Helically Coiled Tubes With Turbulent Flow. *International Journal of Heat Mass Transfer,* Vol. 7, pp. 1207 – 1216, ISSN 0017-9310

Rose, J.W. (2004). Heat-Transfer Coefficients, Wilson Plots and Accuracy of Thermal Measurements. *Experimental Thermal and Fluid Science,* Vol.28, pp.77–86, ISSN 0894-1777

Schmidt, E.F. (1967). Wärmeübergang und Druckverlust in Rohrschlangen. *Chemie Ingenieur - Technik,* 39 Jahrgang, Heft 13, pp. 781 – 789, ISSN 0009-286X

Seban,. R.A.; McLaughlin, E.F. (1963). Heat Transfer in Tube Coils with Laminar and Turbulent Flow. *International Journal of Heat Mass Transfer,* Vol. 6, pp. 387 – 395, ISSN 0017-9310

Serth, R.W. (2007). *Process Heat Transfer. Principles and Applications.* Academic Press, ISBN 978-0-12-373588-1, Burlington, USA

Shah, R.K.; Joshi S.D. (1987). Convective Heat Transfer in Curved Ducts, In: *Handbook of Single-Phase Convective Heat Transfer,* S. Kakac, R.K. Shah and W. Aung (Eds.), ISBN 0-47-181702-3, John Wiley & Sons, New York, USA

Shilling, R.L; Bell, K.J.; Bernhagen, P.M.; Flynn, T.M.; Goldschmidt, V.M.; Hrnjak, P.S.; Standiford F.C.; Timmerhaus K.D. (1999). Heat-Transfer Equipment, In: *Perry's Chemical Engineers' Handbook,* R.H. Perry and Green D.W., (Eds.), McGraw-Hill, ISBN 0-07-049841-5, New York, USA

Shokouhmand, H.; Salimpour, M.R., Akhavan-Behabadi, M.A. (2008). Experimental Investigation of Shell and Coiled Tube Heat Exchangers Using Wilson Plots. *International Communications in Heat and Mass Transfer,* Vol.35, pp. 84–92, ISSN 0735-1933

Smith, E. M. (1997). *Thermal Design of Heat Exchangers: A Numerical Approach: Direct Sizing and Stepwise Rating, and Transients,* John Wiley & Sons, ISBN 0-47-001616-7, New York, USA

Sobota, T. (2011). Determinig of Convective Heat Transfer Coefficients in Heat Exchangers Using the Wilson Plot Technique, In: *Thermal and Flow Proesses in Large Steam Boilers. Modeling and Monitoring,* J. Taler, (Ed.), ISBN 978-83-01-16479-9, Warszawa, Poland (in Polish)

Viegas, R.M.C.M.; Rodríguez, M.; Luque, S.; Alvarez, J.R.; Coelhoso, I.M.; Crespo, J.P.S.G. (1998). Mass Transfer Correlations in Membrane Extraction: Analysis of Wilson-Plot Methodology. *Journal of Membrane Science,* Vol.145, pp.129-142, ISSN 0376-7388

Visual Numerics Inc. (2007). *IMSL Fortran and C Application Development Tools,* Vol. 1 and 2. Houston, Texas, USA

17

High Temperature Thermal Devices for Nuclear Process Heat Transfer Applications

Piyush Sabharwall and Eung Soo Kim
Idaho National Laboratory
United States of America

1. Introduction

Thermal device is a broad term used to describe a component whose main objective is to transport thermal energy across a system. The two thermal devices discussed in this chapter are heat exchangers and thermosyphons.

1.1 Heat exchangers

A heat exchanger is a component used to transfer heat from one medium to another. The media may be separated by a solid wall, so they never mix, or they may be in direct contact (Kakac and Liu 2002). Heat exchangers generally have no external heat nor work interactions and are typically used in the following applications:

- Space heating
- Refrigeration
- Air conditioning
- Power plants
- Chemical plants
- Petrochemical plants
- Petroleum refineries
- Natural gas processing.

1.1.1 High temperature heat exchangers

The high temperature heat exchanger technology is emerging in many industrial applications such as gas turbines, chemical plants, and nuclear power plants. For this reason, many researchers have been studying the development of heat exchangers that can operate at high temperatures. Unlike typical heat exchangers, high temperature heat exchangers require and exhibit some special characteristics to be operated in severe environments (Sunden 2005):

- The transfer of radiative heat may have a significant role in the heat transfer process.
- Tube diameters and pitch should be larger for lower pressure drop. Adding a fan or blower to work at high temperatures might be prohibitive.

- Fins are generally not used in high temperature units because of suspended dirt particles that will foul and low available pressure drops. The advantage of fins is negligible.
- Ceramic and high-temperature alloys (such as: Alloy 800H, 617, 230) are used for construction materials.
- The thickness and mechanical design of selected materials are mainly governed by thermal stress, but the extent of the materials oxidation, thermal shock bearing capability, and erosion from suspended dirt particles, fouling, and corrosion because of metallic salts, sulfates, etc., also need to be considered.
- Differential expansion is an important factor in high temperature units and should be accounted for by using either expansion bellows or by using bayonet-type units. Floating tube sheets cannot generally be used, because sealing gaskets or packing materials do not work effectively at such high temperatures.
- Heat losses from the outside surface to the environment have to be considered in the mechanical design of the unit and design of the foundation.
- Gases, air, liquid metals, or molten salts are preferred over steam for high temperature heat transfer, because the latter require a very thick shell and tubes to contain its high pressure.

Therefore, the thermal stress during startup, shutdown, and load fluctuations can be significant for high temperature heat exchangers. The heat exchanger must be designed accordingly for reliability and long life. The thermal capacitance should therefore be reduced for high temperature heat exchangers for shorter startup time. High temperature heat exchangers also require costly materials contributing to the high cost of balance of power plant. Heat exchanger costs increase significantly for temperatures above 675°C.

1.1.2 Heat exchanger types and classifications

A variety of heat exchanger types with various features are used in industry. This subsection generally explains how to classify and categorize them. According to Kakac and Liu (2002), heat exchangers can be generally classified as follows:

1. Recuperator/Regenerator
   a. Recuperations
   b. Regenerations
2. Transfer Process
   a. Direct contact
   b. Indirect contact
3. Geometry of Construction
   a. Tubular heat exchanger
      i. Double pipe heat exchanger
         - High pressure (in both sides)
      ii. Shell and Tube heat exchanger
      iii. Spiral tube type heat exchanger
   b. Plate heat exchanger
      i. Gasketed plate heat exchanger
      ii. Spiral plate heat exchanger
   c. Extended surface heat exchanger
      i. Plate-fin heat exchanger

ii. Tubular-fin heat exchanger (Gas to Liquid)

4. Heat Transfer Mechanism
   a. Single phase convection on both sides
   b. Single phase convection on one side, two phase convection on other side
   c. Two phase convection on both sides
5. Flow Arrangement
   a. Parallel flow
   b. Counter flow
   c. Cross flow

Table 1 shows the principle features for several types of heat exchangers (Shah and Sekulic 2003). According to this table, shell-and-tube, Bavex (plate heat exchanger), printed-circuit, and Marbond are available for high temperature applications above 700°C.

The **shell and tube heat exchanger** is the most common type found in industry. This exchanger is generally built of a bundle of round tubes mounted in a cylindrical shell with the tube axis parallel to that of the shell. One fluid flows inside the tubes and the other fluid flows across and along the tubes. The major components of this exchanger are tubes (or tube bundles), shell, front-end head, rear-end head, baffles, and tube sheets (Shah and Sekulic 2003). The diameter of the outer shell in a shell and tube heat exchanger is greatly increased, and a bank of tubes rather than a single central tube is used, as shown in Figure 1 (Sherman and Chen 2008). Fluid is distributed to the tubes through a manifold and tube sheet. To increase heat transfer efficiency, further modifications to the flow paths of the outer and inner fluids can be accomplished by adding baffles to the shell to increase fluid contact with the tubes, and by creating multiple flow paths or passes for the fluid flowing through the tubes (Sherman and Chen 2008). These heat exchangers are used for gas-liquid heat transfer applications, primarily when the operating temperature and/or pressure is very high (Shah and Sekulic 2003).

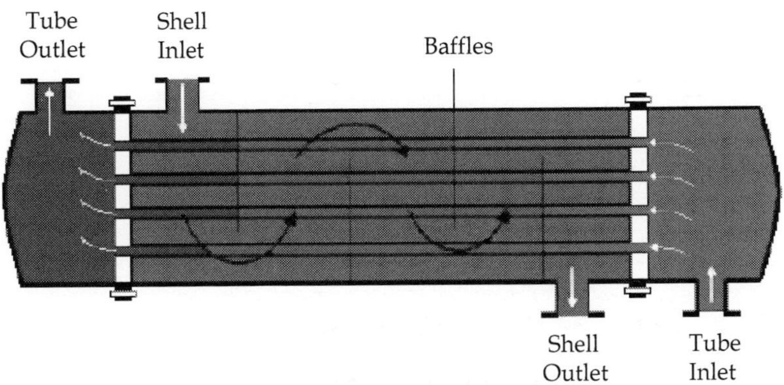

Fig. 1. Shell and tube heat exchanger with baffles (Sherman and Chen 2008)

The **Bevax hybrid welded-plate heat exchanger** is a plate type heat exchanger that deploys metal plates arranged in a stack-wise fashion and sealed with welds as shown in Figure 2. This heat exchanger is reported to be operational at 900°C with pressures to 6 MPa on the

HX Type	Compactness (m^2/m^3)	System Types	Material	Temperature Range (°C)[a]	Maximum Pressure (bar)[b]	Cleaning Method	Corrosion Resistance	Multistream Capability[c]	Multipass Capability[d]
Shell and Tube	~100	Liquid/Liquid, Gas/Liquid, 2Phase	Stainless steel (s/s), Ti, Inoloy, Hastellroy, graphite, polymer	~ +900	~ 300	Mechanical, Chemical	Good	No	Yes
Plate-and-frame (gaskets)	~200	Liquid/Liquid, Gas/Liquid, 2Phase	s/s, Ti, Inoloy, Hastellroy, graphite, polymer	-35 ~ +200	25	Mechanical	Good	Yes	Yes
Partially welded plate	~200	Liquid/Liquid, Gas/Liquid, 2Phase	s/s, Ti, Inoloy, Hastellroy	-35 ~ +200	25	Mechanical, Chemical	Good	No	Yes
Fully welded plate (Alfa Rex)	~200	Liquid/Liquid, Gas/Liquid, 2Phase	s/s, Ti, Ni alloys	-50 ~ + 350	40	Chemical	Excellent	No	Yes
Brazed plate	~200	Liquid/Liquid, 2Phase	s/s	-195 ~ +220	30	Chemical	Good	No	No
Bavex plate	200 ~ 300	Gas/Gas, Liquid/Liquid, 2Phase	s/s, Ni, Cu, Ti, special steels	-200 ~ +900	60	Mechanical, Chemical	Good	Yes	Yes
Platular plate	200	Gas/Gas, Liquid/Liquid, 2Phase	s/s, Hastelloy, Ni alloys	~700	40	Mechanical	Good	Yes	Yes
Packinox plate	~300	Gas/Gas, Liquid/Liquid, 2Phase	s/s, Ti, Hastelloy, Inconel	-200 ~ +700	300	Mechanical	Good	Yes	Yes
Spiral	~200	Liquid/Liquid, 2Phase	s/s, Ti, Incoloy, Hastelloy	~400	25	Mechanical	Good	No	No
Brazed plate fin	800 ~ 1500	Gas/Gas, Liquid/Liquid, 2Phase	Al, s/s, Ni alloy	~650	90	Chemical	Good	Yes	Yes
Diffusion bonded plate fin	700 ~ 800	Gas/Gas, Liquid/Liquid, 2Phase	Ti, s/s	~500	>200	Chemical	Excellent	Yes	Yes
Printed circuit	200 ~ 5000	Gas/Gas, Liquid/Liquid, 2Phase	Ti, s/s	-200 ~ +900	> 400	Chemical	Excellent	Yes	Yes
Polymer (e.g. channel plate)	450	Gas/Liquid	PVDF, PP	~150	6	Water Wash	Excellent	No	No
Plate and shell	–	Liquid/Liquid	s/s, Ti	~350	70	Mechanical, Chemical	Excellent	Yes	Yes
Marbond	~10,000	Gas/Gas, Liquid/Liquid, 2Phase	s/s, Ni, Ni alloys, Ti	-200 ~ +900	>400	Chemical	Excellent	Yes	Yes

a. Heat exchanger operational temperature ranges.
b. Heat exchanger maximum applicable pressure.
c. Capability to connect several independent flow loops in a single heat exchanger.
d. Capability to split flow into several paths in the heat exchanger.

Table 1. Principal features of several types of heat exchangers (Shah and Sekulic 2003)

plate side. It is called a hybrid because one fluid is contained inside the plates while the other flows between the plates from baffled plenums inside a pressure boundary (Fisher and

Sindelar 2008). It is reminiscent of a shell and tube arrangement with substantially greater surface area. Plates can be produced up to 0.35 m wide and 16 m long (Fisher and Sindelar 2008). Other variants of the welded plate-type heat exchanger are produced, some of which do not require external shells.

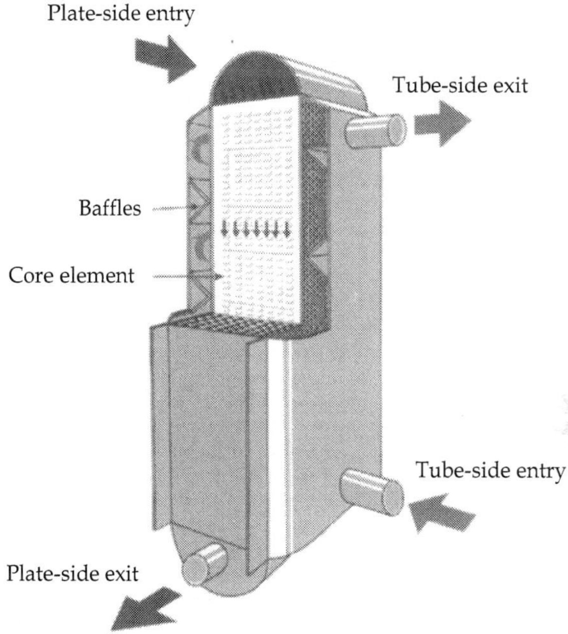

Fig. 2. Bevax welded-plate heat exchanger (Reay 1999)

The **printed circuit heat exchanger (PCHE)** is a relatively new concept that has only been commercially manufactured by Heatric™ since 1985. PCHEs are robust heat exchangers that combine compactness, low pressure drop, high effectiveness, and the ability to operate with a very large pressure differential between hot and cold sides (Heatric™ Homepage 2011). These heat exchangers are especially well suited where compactness is important. The Heatric™ heat exchanger falls within the category of compact heat exchangers because of its high surface area density (2,500 m^2/m^3) (Hesselgreaves 2001). As the name implies, PCHEs are manufactured by the same technique used for producing standard printed circuit boards for electronic equipment. In the first step of the manufacturing process, the fluid passages are photochemically etched into the metal plate (See Figure 3). Normally, only one side of each plate is etched-out. The etched-out plates are thereafter joined by diffusion bonding, which is the second step and results in extremely strong all-metal heat exchanger cores. Plates for primary and secondary fluids are stacked alternately and formed into a module. Modules may be used individually or joined with others to achieve the needed energy transfer capacity between fluids. The diffusion bonding process allows grain growth, thereby essentially eliminating the interface at the joints, which in turn gives the parental metal strength. Because of the use of diffusion bonding, the expected lifetime of the heat exchanger exceeds that of heat exchangers that are based on a brazed structure (Dewson and Thonon 2003).

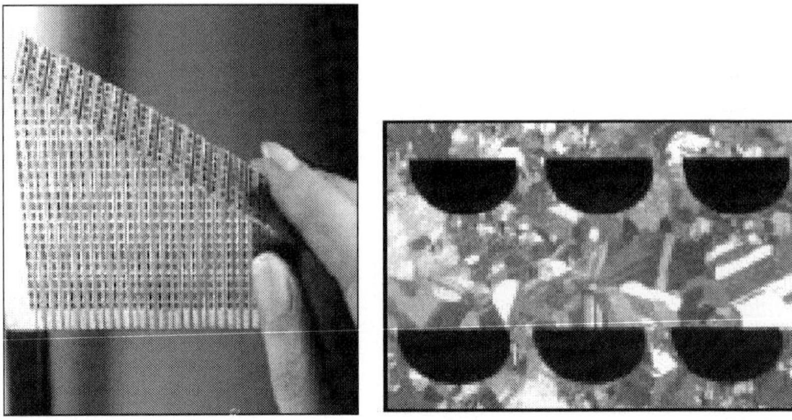

(a) Photo of PCHE (b) Cross-sectional view of PHCE

Fig. 3. Printed circuit heat exchanger (Heatric™ Homepage 2011)

The **Marbond heat exchanger** is a type of compact heat exchanger based on a novel combination of photochemical etching and diffusion bonding (Phillips 1996). The internal construction of this heat exchanger comprises a stack of plates that are etched photochemically to form a series of slots as shown in Figure 4. The plates are stacked with high positional tolerance such that series of slots form discrete flow paths. Adjacent flow paths are separated by means of intervening solid plates. Thus, two or more separate flow paths may be formed across a group of plates, enabling different fluid streams. Injecting a secondary reactant into the flow of the primary reactant may be achieved by means of perforations in the solid separator plate that are aligned exactly with the flow paths of the primary reactant. The use of a positive pressure differential between the secondary and primary reactant streams ensures that the secondary reactants flow in the desired direction.

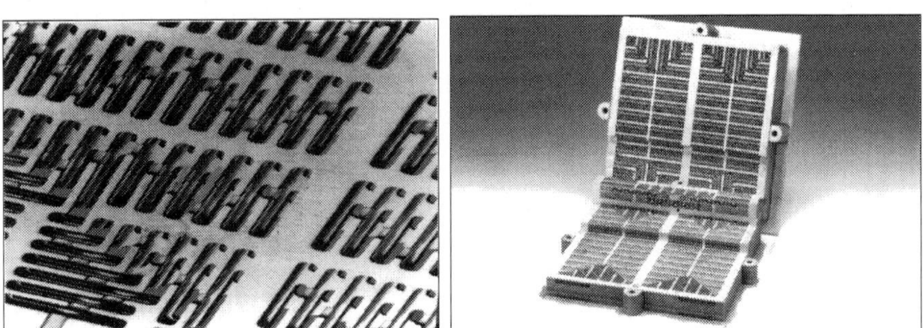

Fig. 4. Marbond heat exchanger (Phillips 1996)

1.1.3 Heat exchanger fluid types and comparisons

A variety of heat transfer fluids are available for high temperature heat exchangers including gases, liquid metals, molten salts, etc. The following lists some general characteristics required for the heat transfer fluid:

- High heat transfer performance to achieve high efficiency and economics
- Low pumping power to improve economics through less stringent pump requirements
- Low coolant volume for better economics
- Low structural materials volume for better economics
- Low heat loss for higher efficiency
- Low temperature drop for higher efficiency.

Characteristics of heat transfer fluids have been extensively investigated by Kim, Sabharwall, and Anderson (2011) for high temperature applications based on the following Figures-of-Merit (FOMs):

- *FOMht* represents the heat transfer performance of the coolant. It measures the heat transfer rate per unit pumping power for a given geometry.
- *FOMp* represents the pumping power of the coolant. It measures the pumping power required to transport the same energy for a given geometry.
- *FOMcv* represents the volume of the coolant. It measures the coolant volume required for transferring heat with the same heat and pumping power.
- *FOMccv* represents the volume of the structural materials. It measures the volume of the coolant structural materials required for transferring heat with the same heat duty and pumping power under given operating conditions (T and P).
- *FOMhl* represents the heat loss of the coolant. It measures the heat loss of the coolant when it is transported the same distance with the same heat duty and pumping power.
- *FOMdt* represents the temperature drop in the coolant while transferring thermal energy with a given heat duty and pumping power.

Table 2 shows the comparisons of the thermal-hydraulic characteristics of the various coolants based on the estimated FOMs (Kim, Sabharwall, and Anderson 2011). In this estimation, the water at 25°C and 0.1 MPa was selected to be the reference coolant. The following summarizes the results:

- Higher FOMht is preferred for better heat transfer performance. According to the comparisons, sodium shows the highest value (=19.05) and argon has the lowest value (0.05). Overall, FOMht is the highest in liquid metal followed by liquid water, molten salt, and gases, respectively.
- Lower FOMp is preferred for better efficiency and economics. According to the comparisons, liquid water has the lowest value (=1.0) and argon has the highest value (=72592). Overall, FOMp is the lowest in molten salt followed by liquid metals and gases, respectively.
- Lower FOMcv is preferred because it requires less coolant volume for providing the same amount of heat transfer performance under the same pumping power. According to the comparisons, the liquid water has the lowest value (=1.0) and argon has the highest (=101.44). Overall, FOMcv is the lowest in molten salt followed by liquid metals and gases, respectively.
- Lower FOMccv is preferred because it requires less structural material volume for both heat transfer pipes and components. Overall, the same result was obtained as the FOMcv. The FOMccv is the lowest in molten salt followed by liquid metals and gases, respectively.

	Coolant	FOM_{ht}	FOM_p	FOM_{cv}	FOM_{ccv}	FOM_{hl}	FOM_{dl}
Ref.	Water (25°C, 1 atm)*	1.00	1.00	1.00	1.00	1.00	1.00
Gas (700°C, 7 MPa)	He	0.12	25407.41	67.74	4741.80	0.40	0.40
	Air	0.07	40096.15	80.10	5607.14	0.26	0.26
	CO_2	0.11	11390.17	47.19	3303.46	0.32	0.32
	H_2O (Steam)	0.11	10012.63	45.10	3157.12	0.32	0.32
	Ar	0.05	72592.09	101.44	7100.53	0.20	0.20
Molten Salt (700°C)	LiF-NaF-KF	0.80	2.87	1.57	1.57	0.92	0.92
	NaF-ZrF_4	0.45	5.02	1.98	1.98	0.56	0.56
	KF-ZrF_4	0.38	8.69	2.49	2.49	0.51	0.51
	LiF-NaF-ZrF_4	0.40	5.36	2.05	2.05	0.50	0.50
	LiCl-KCl	0.55	14.99	3.07	3.07	0.76	0.76
	LiCl-RbCl	0.47	23.03	3.66	3.66	0.70	0.70
	NaCl-$MgCl_2$	0.58	16.26	3.18	3.18	0.81	0.81
	KCl-$MgCl_2$	0.50	14.30	3.02	3.02	0.70	0.70
	NaF-$NaBF_4$	0.71	5.66	2.04	2.04	0.88	0.88
	KF-KBF_4	0.64	8.98	2.47	2.47	0.84	0.84
	RbF-RbF_4	0.54	14.61	3.01	3.01	0.75	0.75
Liquid Metal (700°C)	Sodium	19.05	33.62	4.19	4.19	28.91	28.91
	Lead	6.05	111.64	6.90	6.90	10.82	10.82
	Bismuth	6.61	100.69	6.60	6.60	11.66	11.66
	Lead-Bismuth	4.86	142.94	7.65	7.65	8.95	8.95

Table 2. Principal features of several types of heat exchangers (Shah and Sekulic 2003)

Fluid	h [W/m²K]
Gases (natural convection)	3–25
Engine Oil (natural convection)	30–60
Flowing liquids (nonmetal)	100–10,000
Flowing liquid metal	5000–25,000
Boiling heat transfer:	
Water, pressure < 5 bars, dT<25K	5000–10,000
Water, pressure 5-100, dT = 20K	4000–15,000
Film boiling	300–400
Condensing heat transfer:	
Film condensation on horizontal tubes	9000–25,000
Film condensation on vertical surface	4000–11,000
Dropwise condensation	60,000–120,000

Table 3. Order of magnitude of heat transfer coefficient (Kakac and Liu 2002)

- Lower FOMhl is preferred because it requires less insulation for preventing heat loss. According to the comparisons, argon has the lowest value (0.2), and sodium has the highest (28.9). Overall, the FOMhl is the lowest in gases followed by molten salt and liquid metal, respectively.

- Lower FOMdt is preferred because more thermal energy can be transferred long distances without much of a temperature drop. Same values were obtained for the FOMdt as obtained from FOMhl.

In the heat exchanger design, the heat transfer coefficient is a very important parameter because it determines overall heat exchanger sizes and performance. Table 3 lists some coolant types and the ranges of their heat transfer coefficients (Kakac and Liu 2002). As can be seen, water exhibits the highest heat transfer coefficient in the drop-wise condensation, and gases exhibit the lowest in the natural circulation.

1.1.4 Heat exchanger materials and comparisons

Material selection is one of the most important things in the high temperature application. There are four main categories of high temperature materials: high temperature nickel-based alloy, high temperature ferritic steels and advanced carbon silicon carbide composite, and ceramics (Sunden 2005).

Ohadi and Buckley (2001) extensively reviewed materials for the high temperature applications. High temperature nickel-based material has good potential for helium and molten salts up to 750°C. High temperature ferrite steels shows good performance under fusion and fission neutron irradiation to around 750°C. Advanced carbon and silicon carbide composite has excellent mechanical strength at temperatures exceeding 1000°C. It is currently used for high temperature rocket nozzles to eliminate the need for nozzle cooling and for thermal protection of the space shuttle nose and wing leading edges. Many options are available that trade fabrication flexibility and cost, neutron irradiation performance, and coolant compatibility. Table 4 compares the properties of most commonly used high temperature materials (Ohadi and Buckley 2001). It includes nickel-based alloy, ceramic materials. and carbon and SiC composites. Figure 5 shows the specific strength versus temperature for various composite materials.

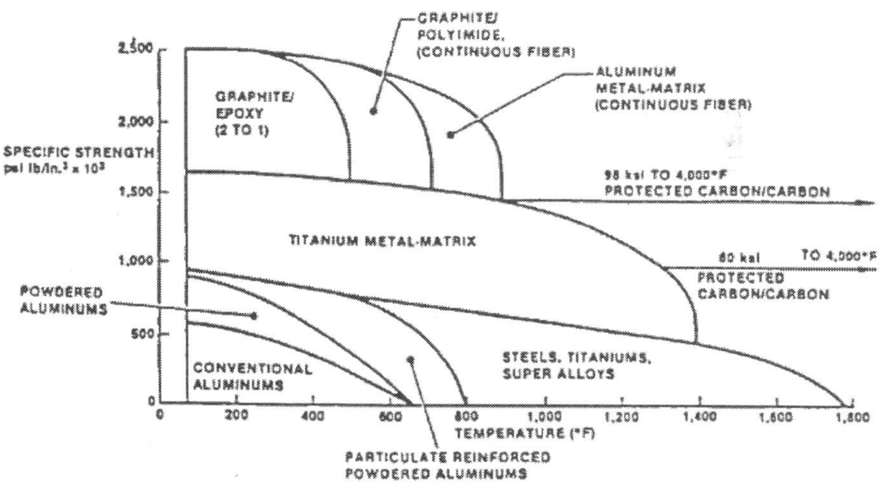

Fig. 5. Specific strength vs. temperature (Brent 1989)

High temp. material/fabrication technology	Metallic Ni alloys (Inconel 718)	Ceramics oxides of Al, Si, Sr, Ti, Y, Be, Zr, B and SiN, AiN, B4C, BN, WC94/C06	Carbon-carbon composite	Carbon fiber-SiC composite
Temperature range	1200 – 1250 °C	1500 – 2500 °C	3300 °C (inert environment) 1400 – 1650 °C (with SiC layer)	1400 – 1650 °C
Density	8.19 g/cm³	1.8 – 14.95 g/cm³	2.25 g/cm³	1.7 – 2.2 g/cm³
Hardness	250 – 410 (Brinell)	400 – 3000 kgf/mm² (V)	0.5 – 1.0 (Mohs)	2400 – 3500 (V)
Elongation	< 15%	N/A	N/A	-
Tensile strength	800 – 1300 MPa	48 – 2000 MPa	33 (Bulk Mod.)	1400 – 4500 MPa
Tensile modules	50 GPa	140 – 600 GPa	4.8 GPa	140 – 720 GPa
Strength of HE	Strength – adequate, but limited due to creep and thermal exp	Strength – not adequate, low mechanical parameters for stress. Good thermal and electrical parameters	Strength – poor, oxidation starts at 300 °C	Highest due to carbon fiber and SiC
Electrical conductivity	125 μΩ cm	2E-06 – 1E+18 Ω cm	1275 μΩ cm	1275 μΩ cm
Thermal conductivity	11.2 W/m K	0.05 – 300 W/m K	80 – 240 W/m K	1200 W/m K
Thermal expansion	13E-06 K⁻¹	0.54 – 10E-06 K⁻¹	0.6 – 4.3E-06 K⁻¹	-
Comments	Metallic expansion joints are the weak link	Often very expensive fabrication cost for conventional applications. Technology proprietary for the most part. Technologically hard to produce	Life-time is low even protected by SiC (adhesion is poor)	Comparatively less expensive, successful proprietary fabrication technologies available.

Table 4. Selected properties of most commonly used high-temperature materials and fabrication technologies (Ohadi and Buckley 2001)

Dewson and Li (2005) carried out a material selection study of very high temperature reactor (VHTR) intermediate heat exchangers (IHXs). They selected and compared the following eight candidate materials based on ASME VIII (Boiler and Pressure Vessel Code): Alloy 617, Alloy 556, Alloy 800H, Alloy 880HT, Alloy 330, Alloy 230, Alloy HX, and 253MA.

Alloys	**UNS No**	**Tmax (°C)**	**S898°C (MPa)**	**UTS (MPa)**	**0.2%PS (MPa)**	**El (%)**	**Nominal compositions (wt%)**
617	N06617	982	12.4	655	240	30	52Ni-22Cr-13Co-9Mo-1.2Al
556	R30556	898	11.0	690	310	40	21Ni-30Fe-22Cr-18Co-3Mo-3W-0.3Al
800HT	N08811	898	6.3	450	170	30	33Ni-42Fe-21Cr
800H	N08810	898	5.9	450	170	30	33Ni-42Fe-21Cr
330	N08330	898	3.3	483	207	30	Fe-35Ni-19Cr-1.25Si
230	N06230	898	10.3	760	310	40	57Ni-22Cr-14W-2Mo-0.3Al-0.05La
HX	N06002	898	8.3	655	240	35	47Ni-22Cr-9Mo-18Fe
253MA	S30815	898	4.9	600	310	40	Fe-21Cr-11Ni-0.2N

Table 5. Candidate materials for VHTR IHXs (Dewson and Li 2005)

Table 5 lists the allowable design stress (S) at 898°C, minimum required mechanical properties (ultimate tensile stress [UTS]), 0.2% proof stress (PS), and elongation (El) at room temperature, together with the nominal compositions of the alloys. They extensively compared the mechanical properties, physical properties, and corrosion resistance for the candidate materials, and finally concluded that Alloy 617 and 230 are the most suitable materials for an IHX.

1.1.5 General heat exchanger design methodology

Once the process requirements are given, the high temperature heat exchanger is designed using the same methods used for the typical heat exchangers. This subsection summarizes the basic logical structure of the process heat exchanger design procedure described by Bell (2004).

Figure 6 shows the logical process for heat exchanger design. The fundamental goal of any such process is to find the optimal design among the infinite set of designs that could satisfy the thermal-hydraulic and mechanical requirements. Because of large number of qualitative factors in the design process, optimal needs are to be considered broadly. Generally, the design process aims at the least costly (which usually means the smallest) heat exchangers that meet the required thermal duty within the allowed pressure drops and satisfy the mechanical requirements.

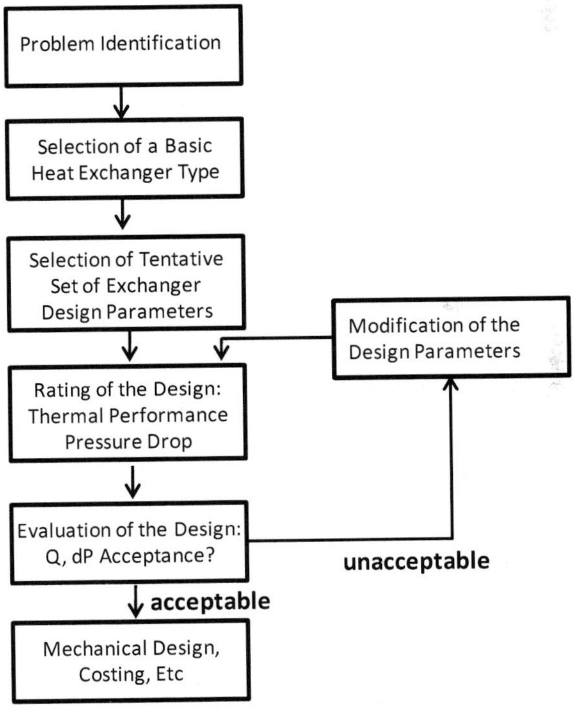

Fig. 6. Basic logical structure of the process heat exchanger design procedure (Bell 2004)

First, a basic heat exchanger type is selected based on the given operational requirements and problems identified. The basic criteria for heat exchanger selection are as follows:

- It must satisfy the process specifications (performance) for temperature and pressure
- It must withstand the service conditions of the plant environment (reliability) for temperature and pressure
- It must be maintainable for cleaning or replacement of a special component
- It should be cost effective (installed operating and maintenance costs)
- It must meet the site requirements or limitations for diameter, length, weight, and tube configurations, and lifting and servicing capability or inventory considerations.

Second, main heat exchanger design parameters are selected based on the followings main design factors:

- Heat duty
- Materials
- Coolant
- Pressure drop
- Pressure level
- Fouling
- Manufacturing techniques
- Cost
- Corrosion control
- Cleaning (with ease).

Third, rating and evaluation of the heat exchanger designs are iteratively conducted to thermally optimize the design. The main purpose of this is to find all possible configurations to meet the process requirements. Generally, the optimum heat exchanger is determined based on the cost.

Once the optimal thermal design is provided, a mechanical design is developed based on the following factors:

- Plate, tube, shell, and header thickness and arrangements
- Corrosion resistance
- Manifold
- Location of pressure and temperature measuring device
- Thermal stress analysis under steady and transient
- Flow vibrations
- Level of velocity to eliminate fouling and erosion
- Maintenance.

Detailed cost analyses are also conducted at this stage including both capital and operational costs. The capital cost includes materials, manufacturing, testing, shipment, and installation. The operational cost includes pumping power, repair, and cleaning.

1.1.6 Heat exchanger cost analyses

The cost of a heat exchanger is an important factor for heat exchanger design and selection. Generally, manufacturers have their own methods for cost estimation. This section introduces a simple heat exchanger costing methodology based on empirical cost data of the ESDU (1994) for various feasible heat exchanger types (Shah and Sekulic 2003). Detailed cost will depend on the operating conditions and materials used. The decision variable in this method is cost of a heat exchanger per unit of its thermal size defined by

$$C_{UA} = \frac{\cos t}{\frac{q}{\Delta T_m}} \tag{1}$$

Where:
cost = cost of the heat exchanger
q = thermal duty
ΔT_m = mean temperature.
Table 6 shows a selection of the cost data represented by the values of C_{UA} (Shah and Sekulic 2003). This table is prepared for an application between gas as a hot fluid at a medium pressure of about 2 MPa and cold fluid as treated water. An extensive set of C_{UA} data for various heat exchangers can be found in ESDU (1994).

$q/\Delta T_m$	C_{UA} ($/(W/K))				Welded Plate	
	Shell-and-Tube U = 484 (W/m² K)	Double Tube U = 484 (W/m² K)	Printed Circuit U = 1621 (W/m² K)	Plate-Fin U = 491 (W/m² K)	U (W/m² K)	C_{UA} ($/(W K))
10^3	3.98	2.5	12	-	349	4.9
5×10^3	1.00	0.75	2.4	3.1	1187	1.22
3×10^4	0.29	0.31	0.6	0.513	1068	0.42
10^5	0.17	0.31	0.42	0.210	1112	0.28
10^6	0.106	0.31	0.28	0.115	1173	0.22
* Original cost data in ESDU are approximated to the US dollar value in 2000.						

Table 6. Cost data CUA vs. UA for various heat exchanger types (Shah and Sekulic (2003), ESDU (1994))

2. Thermal energy transfer for process heat application

Recent technological developments in next generation nuclear reactors have created renewed interest in nuclear process heat for industrial applications. The Next Generation Nuclear Plant (NGNP) will most likely produce electricity and process heat for hydrogen production. Process heat is not restricted to hydrogen production, but is also envisioned for various other technologies such as the extraction of iron ore, coal gasification, and enhanced oil recovery. To utilize process heat, a thermal device is needed to transfer the thermal energy from NGNP to the hydrogen plant in the most efficient way possible. There are several options to transferring multi-megawatt thermal power over such a distance. One option is simply to produce only electricity, transfer it by wire to the hydrogen plant, and then reconvert the electric energy to heat via Joule or induction heating. Electrical transport, however, suffers energy losses of 60 to70% because of the thermal-to-electric conversion inherent in the Brayton cycle. A second option is to transport thermal energy via a single-phase forced convection loop where a fluid is mechanically pumped between heat exchangers at the nuclear and hydrogen plants. High temperatures, however, present unique challenges for materials and pumping. Single phase, low pressure helium is an attractive option for NGNP, but is not suitable for a single-purpose facility dictated to hydrogen production because low pressure helium requires higher pumping power and makes the process very inefficient. A third option is two-phase heat transfer utilizing a high-temperature thermosyphon. Heat transport occurs via evaporation and condensation, and the heat transport fluid is recirculated by gravitational force. Thermosyphons have the ability to transport heat at high rates over appreciable distances, virtually isothermally, and without any requirement for external pumping devices.

Heat pipes and thermosyphons have the ability to transport very large quantities of heat over relatively long distances with small temperature loses. The applications of heat pipes and thermosyphons require heat sources for heating and heat sinks for cooling. The development of the heat pipe and thermosyphon was originally directed towards space applications. However, the recent emphasis on energy conservation has promoted the use of heat pipes and thermosyphons as components in terrestrial heat recovery units and solar energy systems. Thermosyphons have less thermal resistance, wider operating limits (the integrity of the wick material might not hold in heat pipes at very high temperatures), and lower fabrication costs than capillary heat pipes, which makes a thermosyphon a better heat recovery thermal device. Perhaps the most important aspect of thermosyphon technology is that it can easily be turned off when required, whereas a heat pipe cannot be turned off. This safety feature makes the licensing of NGNP process heat transfer systems comparatively easier. This section describes the thermosyphon system and the potential benefits of using it in order to transfer process heat from the nuclear plant to the hydrogen production plant.

2.1 Thermosyphon design

Considerable effort has been invested in thermosyphon and heat pipe development, resulting in broad applications. One significant advantage of heat transfer by thermosyphon is the characteristic of nearly isothermal phase change heat transport, which makes the thermosyphon an ideal candidate for applications where the temperature gradient is limited and high delivery temperatures are required, as in the case of thermochemical hydrogen production (Sabharwall and Gunnerson 2009; Sabharwall 2009). The nature of isothermal heat transport results in an extremely high thermal conductance (defined as the heat transfer rate per unit temperature difference). A schematic diagram of a thermosyphon system is shown in Figure 7.

The controllable thermosyphon, conceptually illustrated in Figure 7, is a wickless heat pipe with a separate liquid return line, which is an intriguing option to traditional pumped fluid heat transfer. Thermosyphons rely on convection to transport thermal energy inside pipes and high-temperature heat exchangers for the evaporation and condensation end processes. Ideally, no pumping power is required in contrast to single-phase gas or liquid loops that require compressors or pumps, both of which are problematic at very high temperatures.

Heat is transported by saturated or superheated vapor expanded from an evaporative heat exchanger, through a long pipe, to a condensation heat exchanger. Liquid condensate returns to the evaporator assisted by gravity through a separate liquid return line with a liquid return control valve. When the thermosyphon is started by applying power (process heat from NGNP) to the evaporator, the working fluid is evaporated and the latent heat of vaporization is transported (~isothermally) along the thermosyphon to the condenser region. Expansion joints are added near the condenser section and at the inlet to the condensate return line in order to accommodate the thermal expansion of the thermosyphon piping at higher temperature. The condensate returns to the evaporator region through a liquid bypass line containing a liquid storage reservoir and a control check valve as shown in Figure 7. The storage reservoir and part of the liquid lines may incorporate electric resistance heating if necessary in order to melt the working fluid and restart the thermosyphon after a long shutdown period. Liquid from the storage reservoir passes into the thermosyphon system evaporator through a control valve which, as needed, plays a role in controlling the rate of heat transfer and shutting off or isolating the thermosyphon. In order for the thermosyphon system to be shut and completely disabled from heat transfer, the control valve is closed wherein all the working fluid is collected in the liquid storage reservoir and the condensing-evaporating cycle is terminated. When it is desired to resume the thermosyphon action, the control valve is opened to again allow the liquid to flow into the evaporator region of the system. The heat input governs the rate of evaporation and the subsequent rate of heat transfer. The rate of thermal energy exchange can be regulated over a spectrum of conditions from "off" to "fully on," hence the term *controllable thermosyphon*.

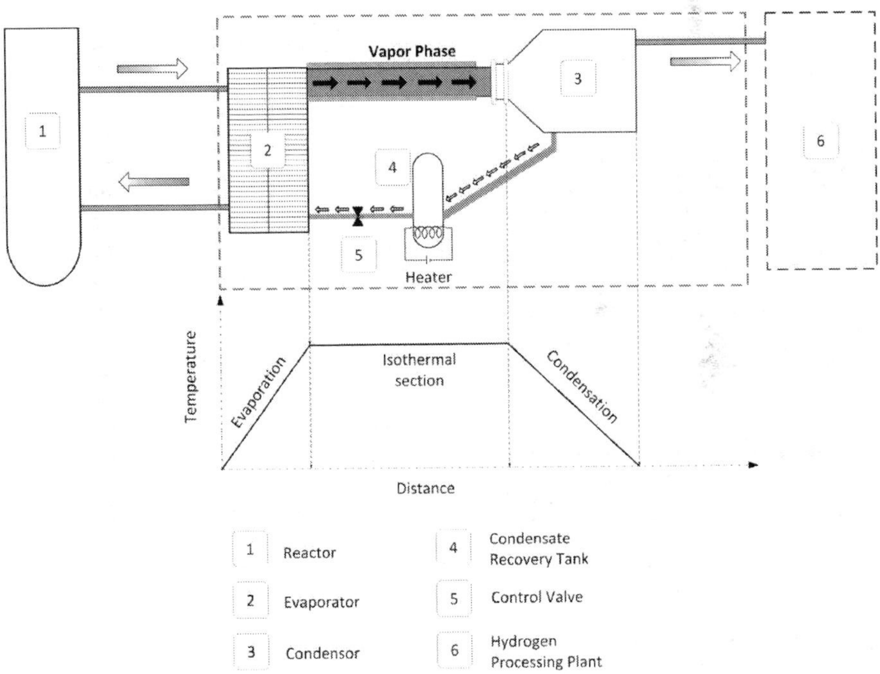

Fig. 7. Schematic of a simple controllable thermosyphon

Depending on the temperature and pressure of operation, favorable working fluids can be identified. Alkaline metals, for example, may be suited for process heat transfer because they have the characteristics of:

- High boiling temperature
- Availability and cost effectiveness
- Good heat transfer properties (latent and specific heat are both high)
- Typically good chemical compatibility (except Li).

Working fluids more suitable than alkali metals may exist, such as molten salts (Sohal et al. 2010). Corrosive behavior at high temperatures or lack of high temperature thermodynamic properties, especially for superheated vapors, rule out fundamental analysis of many possible thermosyphon working fluids.

2.2 Thermosyphon startup

The charging of a thermosyphon requires a transfer station wherein molten working fluid under an inert environment or vacuum is transferred to the evacuated thermosyphon pipe(s). Pure fluids without condensable gases are required for proper thermosyphon operation. Otherwise, the impurities, which are more volatile than the fluid itself, will be driven to the condenser section of the thermosyphon and less volatile impurities will be collected in the evaporator causing hot spots and reducing heat transfer. Noncondensable gases will accumulate within the condenser. Although conceptually simple, the startup of a large, high temperature thermosyphon is difficult to accurately predict.

Figure 8 describes the procedure commonly practiced for filling up the thermosyphon. The liquid sodium valve is opened till 20%-by-volume limit is reached for filling up the thermosyphon. The 20%-by-volume is a good approximation for the coolant as described by Gunnerson and Sanderlin (1994).

For a thermosyphon to startup efficiently and effectively, the working fluid has to initially be in molten state. The liquid reservoir, as shown in Figure 8, has provisions for external heating. During normal startup for both heat pipes and thermosyphons, the temperature of the evaporator section increases by a few degrees until the thermal front reaches the end of the condenser as described by Reay and Kew (2006). At this point, the condenser temperature will increase until the pipe structure becomes almost isothermal.

2.3 Comparison of thermosyphon with convective loop

Alkaline metal thermosyphons and alkaline metal forced convective loops can both deliver comparable rates of heat transfer through a given size pipe. This can be demonstrated by considering the ideal rate of convective heat transport through a pipe without losses, modeled in terms of enthalpy as:

$$Q'' = \Delta h \, \rho \, V = \frac{\dot{m}}{A} \Delta h \tag{2}$$

where:

A -Cross-sectional flow area

Δh -Specific enthalpy change of the transport fluid

$\dot{m}$ -Mass flow rate

Q'' -Rate of thermal energy transport per unit flow area
V -Average flow velocity
ρ -Density of the fluid
Δ -Difference.

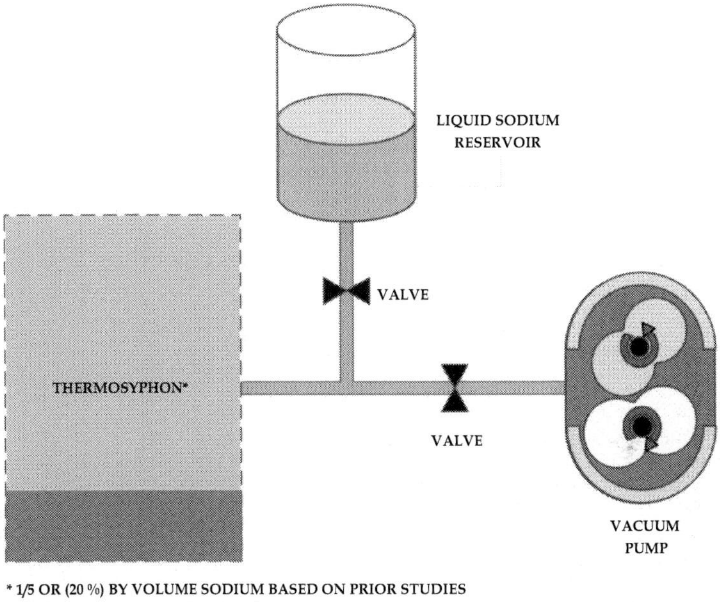

Fig. 8. Na filling procedure for a thermosyphon

Two-phase heat transfer by a thermosyphon has the advantage of high enthalpy transport when compared with single-phase forced convection. Vapor-phase velocities within a thermosyphon can also be much greater than single-phase liquid velocities within a forced convective loop.

Figure 9 exemplifies the enthalpy enhancement in heat transfer afforded by a two-phase thermosyphon versus a single-phase convective loop with sodium as the working fluid as shown by Sabharwall et al. 2009. The specific enthalpy (Δh) of saturated liquid and vapor, relative to the solid at 298.15 K, is illustrated as a function of temperature. Assuming heat transfer from a high temperature gas-cooled reactor to an industrial facility at 1223 K, the maximum single-phase heat transfer is given by the enthalpy gain from points A to B in Figure 9, or approximately 1,190 kJ for each kilogram of sodium. Compared with two-phase heat transfer from points A to B to C, where the enthalpy gain is approximately 3,864 kJ per kilogram with no vapor superheat, over three times more heat per kilogram of sodium is needed than with the single-phase. The saturation pressure of sodium at 1223 K is only 0.188 MPa, thus minimizing pressure and stress forces. Vapor flow through a pipe is limited by compressible choke flow when the vapor reaches its sonic velocity. The sonic velocity for sodium vapor is approximately 737 m/s at 1223 K as given by Bystrov et al. (1990).

The limiting heat transfer rate for an ideal sodium thermosyphon operating at 1223 K can therefore be estimated as

$$Q'' = h_{fg}\,\rho_V\,V_S = (3{,}864)\ \text{kJ/kg}\ (0.47)\text{kg/m}^3(737)\text{m/s} = 1338\ \text{MW/m}^2$$

where:

h_{fg} -Latent heat & vaporization

Similarly, single-phase liquid sodium could transport the same rate of thermal energy with an average flow velocity of about 2.2 m/s, well within the capabilities of advanced liquid metal pumps. This simple analysis (Sabharwall 2009) for sodium as the working fluid theoretically illustrates that both a thermosyphon and a forced convective loop can deliver comparable rates of heat transfer through comparable diameter pipes. The thermosyphon, however, has the luxury of controllable heat transfer without the need for high temperature pumping and can deliver the heat at the same approximate temperature as the source. The enthalpy gain for sodium that can be achieved by two-phase heat transfer versus a single-phase is about 3.7 times greater (Sabharwall 2009).

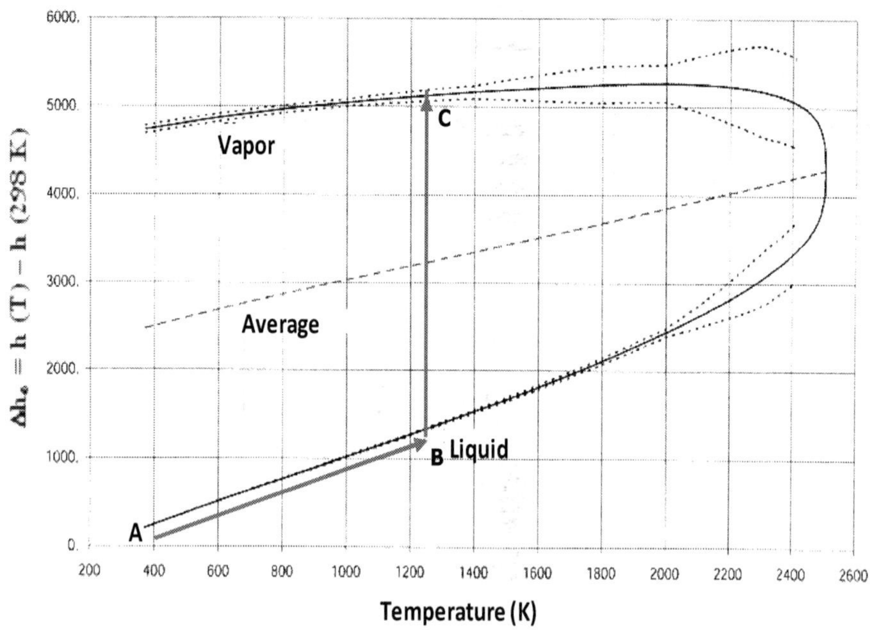

Fig. 9. Enthalpy for saturated sodium: liquid and vapor (thermodynamic data from Fink and Leibowitz [1995]; Gunnerson, Sabharwall, and Sherman [2007])

2.4 Heat transport limitations

Depending on the operational conditions, the heat transport may be limited by one of the following (Sabharwall et al. 2009), as described below:

2.4.1 Sonic limit (choking) of vapor flow

After continuum flow is established, the evaporator-condenser pressure difference accelerates the vapor until it reaches a maximum velocity at the evaporator exit. The maximum vapor velocity that can exist at the evaporator exit corresponds to sonic velocity, or Mach 1. This choked flow condition is a fundamental limit on the axial vapor flow in a thermosyphon.

2.4.2 Viscous limit

At startup for liquid metals, the vapor pressure difference between the evaporator and the condenser is zero or very small. In such cases, the viscous forces may be larger than the vapor pressure gradients, thus preventing vapor flow.

2.5 Flow instabilities

Instabilities are common to both forced and natural circulation systems; the latter is more unstable than forced circulation systems. Instabilities can cause operational problems in process heat transfer. Thus, it is important to classify mechanisms that can lead to unstable operational behavior of the thermosyphon. The following processes can lead to an unstable behavior for the thermosyphon.

2.5.1 Surging (chugging) and geysering instability

Surging and geysering occur mainly because of liquid superheat. Surging occurs when boiling is initiated in the evaporator, but because of nonuniformity in the temperature at the wall and bulk fluid temperature, the vapor being generated becomes trapped, eventually resulting in vapor expulsion as described by Bergles et al. (1981). Geysering is a similar phenomenon that occurs when the heat flux is sufficiently high and boiling is initiated at the bottom. In low pressure systems, this results in a sudden increase in vapor generation because of the reduction in hydrostatic head, usually causing an expulsion of vapor.

2.5.2 Thermosyphon evaporator instability

If the evaporator section of the thermosyphon system is not sufficiently long for vapor superheat, instability can occur such that the fluid at the outlet of the evaporator experiences a static pressure decrease, leading to the onset of fluid condensation within the thermosyphon. Slight vapor superheat from the evaporator should reduce this concern.

2.5.3 Fluid superheating (alkaline metals)

Alkaline metals have relatively high boiling temperatures at atmospheric pressure. If the heater surface does not have enough active nucleation sites, boiling may not occur near saturation temperature but rather require significant superheat. At high superheat temperatures a vapor burst expulsion can be expected upon phase change, which could lead to structural damage and flow excursions.

3. Summary

Thermal device is a component whose main objective is to transport thermal energy across a system. In this chapter, two key thermal devices for high temperature heat transfer applications were introduced and discussed in detail; (1) heat exchanger and

(2) thermosyphon. A heat exchanger is a key component in the thermal systems used for transferring heat from one medium to another. Especially, the high temperature heat exchanger technology is emerging in many industrial applications such as gas turbines, chemical plants, and nuclear power plants with increasing demands. Unlike typical heat exchangers, high temperature heat exchangers require and exhibit some special characteristics to be operated in severe environments.

In this chapter, shell-and-tube, Bevax, PCHE, and Marbond heat exchangers are recommended to be the possible option for high temperature applications.

Working fluids and structural materials are the important design parameters that should be carefully considered for high temperature heat exchangers. A variety of heat transfer fluids are available for high temperature heat exchangers including gases, liquid metals, molten salts, etc. General characteristics required for heat transfer coolant are (1) high heat transfer performance, (2) low pumping power, (3) low coolant volume, (4) low heat loss, and (5) low temperature drop. In this chapter, six figures-of-merit (FOMs) for heat transfer coolants were introduced for evaluating important coolant characteristics. Based on these FOMs, various candidate coolants were compared extensively. Material selection is one of the most important things in the high temperature application. There are four main categories of high temperature materials, high temperature nickel-based alloy, high temperature ferritic steels and advanced carbon silicon carbide composite, and ceramics. Basic design methods and cost analysis for the heat exchangers are also briefly mentioned. Another thermal device are heat pipes and thermosyphons, which have the ability to transport very large quantities of heat over relatively long distances with small temperature loses. In this chapter, the thermosyphon system and the potential benefits of using it were described in the heat transfer perspective from the nuclear plant to the hydrogen production plant (also applicable to any other process application industry). One significant advantage of heat transfer by thermosyphon is the characteristic of nearly isothermal heat transport, which makes the thermosyphon an ideal candidate for applications where the temperature gradient is limited and high delivery temperatures are required, as in the case of thermochemical hydrogen production.

Depending on the temperature and pressure of operation, favorable working fluids can be identified. Alkaline metals, for example, may be suited for process heat transfer because of high boiling temperature, availability and cost effectiveness, good heat transfer properties (latent and specific heat are both high), and typically good chemical compatibility (except for Li).

4. References

Bell, K.J., (2004). "Heat Exchanger Design for the Process Industries," *Journal of Heat Transfer*, Vol. 126, pp. 877-885, 2004

Bergles, A.E., Collier, J.G., Delhaye, J.M., Hewitt, G.F. and Mayinger, F., (1981). *Two-Phase Flow and Heat Transfer in the Power and Process Industries*, Hemisphere Publishing Corporation.

Brent, A., (1989). Fundamentals of Composites Manufacturing Materials Methods and Applications, Society of Manufacturing Engineers, Dearborn, MI.

Bystrov, P.V., Kagan, D.N., Krechetova, G.A. and Shpilrain, E.E., (1990). *Liquid- Metal Coolants for Heat Pipes and Power Plants*, Hemisphere Publishing Corporation.

Dewson, S., Li, X., 2005, "Selection Criteria for the High Temperature Reactor Intermediate Heat Exchanger," *Proceedings of ICAPP'05*, Seoul, Korea, May 15-19, 2005

Dewson, S.J., and Thonon, B., (2003). "The Development of High Efficiency Heat Exchangers for Helium Gas Cooled Reactors," Report No. 3213, *International Conference on Advanced Nuclear Power Plants (ICAP)*, Cordoba, Spain.

ESDU, (1994). "Selection and costing of heat exchangers," *Engineering Science Data*, Item 92013, ESDU, Int. London, UK.

Fink, J.K. and Leibowitz, L., (1995). Thermodynamic and Transport Properties of Sodium Liquid and Vapor, ANL/RE-95/2

Fisher, D.L., and Sindelar, R.L., (2008). *Compact Heat Exchanger Manufacturing Technology Evaluation*, Savannah River Nuclear Solutions, SRNS-STI-2008-00014, Savannah River Site, Aiken, South Carolina.

Gunnerson, F.S., Sabharwall, P., and Sherman, S., (2007). "Comparison of Sodium Thermosyphon with Convective Loop," *Proceedings of the 2007 AIChE Conference*, Salt Lake City, November 2007.

Gunnerson, F.S., and Sanderlin, F.D., (1994). "A Controllable, Wickless Heat Pipe Design for Heating and Cooling," Fundamentals of Heat Pipes, *ASME HTD*, Vol. 278.

Heatric™ Homepage, (Jan 20, 2011). URL: www.heatric.com.

Hesselgreaves, J.E., (2001). Compact Heat Exchangers, Selection, Design and Operation, First edition, Pergamon.

Kakaç, S. and Liu, H. (2002). *Heat Exchangers: Selection, Rating and Thermal Design* (2nd ed.). CRC Press. ISBN 0849309026.

Kim, E.S., Sabharwall, P., Anderson, N., (2011). "Development of Figure of Merits (FOMs) for Intermediate Coolant Characterization and Selection," *ANS Annual Meeting*, Hollywood Florida, USA, June 25–30, 2011.

Ohadi, M.M. and Buckey, S.G., (2001). "High temperature heat exchangers and microscale combustion systems: applications to thermal system miniaturization," *Experimental Thermal and Fluid Science*, Vol. 25, pp. 207–217.

Phillips, C., (1996). *Intensification of Chemical Batch Processes using Integrated Chemical Reactor - Heat Exchangers*, Final Report for JOULE II Contract No. JOU2-CT94-0425 (DG 12 WSME), June 1996.

Reay, D. and Kew, P., (2006). *Heat Pipes: Theory Design and Applications*, Fifth edition, Butterworth-Heinemann, Elsevier.

Reay, D.A., (1999). "Learning from Experiences with Compact Heat Exchangers," *CADDET Analyses Series No. 25*, Centre for the Analysis and Dissemination of Demonstrated Energy Technologies, Sittard, The Nederland.

Sabharwall, P., (2009). "Nuclear Process Heat Transfer for Hydrogen Production," VDM Verlag Publications, August 2009.

Sabharwall, P., and Gunnerson, F., (2009). "Engineering Design Elements of a Two-Phase Thermosyphon for the Purpose of Transferring NGNP Thermal Energy to a Hydrogen Plant," *Journal of Nuclear Engineering and Design*, Vol. 239, June 2009

Shah, R. K., and Sekulic S. P., (2003). "Fundamentals of Heat Exchanger Design," John Wiley and Sons.

Sherman, S. R., and Chen Y., (2008). Heat Exchanger Testing Requirements and Facility Needs for the NHI/NGNP Project, WSRC-STI-2008-00152, April 2008.

Sohal, M., Ebner, M., Sabharwall, P., and Sharpe, P., (2010). *Engineering Database of Liquid Salt Thermo-Physical and Thermo-Chemical Properties,* INL/EXT-10-18297, Idaho National Laboratory, Idaho, March 2010.

Sunden, B., (2005). "High Temperature Heat Exchangers (HTHE)," Proceedings of Fifth International conference on Enhanced, Compact and Ultra-Compact Heat Exchangers: Science, Engineering and Technology, Hoboken, NJ, USA, September 2005.

18

Flow Properties and Heat Transfer of Drag-Reducing Surfactant Solutions

Takashi Saeki
Yamaguchi University
Japan

1. Introduction

The frictional resistance of fluids can be reduced by adding small amounts of certain polymers, a phenomenon first reported (Toms, 1948) known as drag reduction or Toms phenomenon. The added polymers might form thread-like structures in fluids, which interact with turbulent eddies due to their viscoelasticity. Since the polymer synthesis technology and cost effectiveness have been highly improved, polymer drag reduction has been adopted widely in large pipeline systems for crude oils and refined petroleum products. Presently, more than 40% of all the gasoline consumed in the United States has polymer drag reducer in it (Motier, 2002). However, mechanical degradation of polymer chains in high shear rate regions, such as pumps, is frequently observed, which lowers the molecular weight and causes a loss of drag reduction. For that reason, polymer drag reduction cannot be adopted for circulating flow systems.

Drag reduction caused by surfactant solutions was first reported by Gadd (Gadd, 1966). Combinations of certain cationic surfactants with a suitable counter ion are often chosen as the drag-reducing agents. Some nonionic surfactants also show the drag-reducing effects, rendering the use of counter ions dispensable. A number of authors have pointed out that the surfactant molecules come together to form rod-like micelles, which are necessary for drag reduction. Figure 1 shows surfactant molecule and micelle structures. Drag-reducing surfactants form rod-like micelles, and their aggregates might be present in a solution. Figure 2 shows a transmission electron microscope (TEM) image of surfactant micelles (Shikata et al., 1988). Again, aggregates of rod-like micelles might interact with turbulent eddies and cause drag reduction. These aggregates suffer mechanical degradation in high shear rate regions, which is then repaired in lower shear stress regions, such as in flow through pipes.

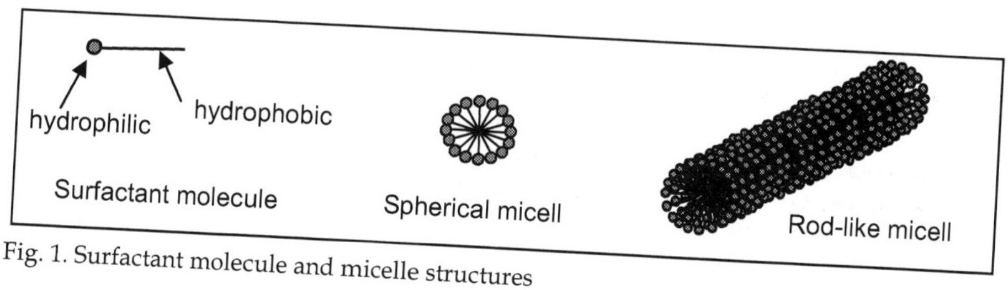

Fig. 1. Surfactant molecule and micelle structures

practical air conditioning system. In this study, drag reduction and heat transfer data measured at our university library building is presented as an example of application in a practical air conditioning system.

4.2 Experimental procedure

The library building of Yamaguchi University (Figure 13) is two stories with the total floor space of 2,400 m^2. Each floor has its own pipeline system with a heat pump and a chiller unit (40 RT, 33 kW, Figure 14). Figure 15 shows a schematic diagram of the air conditioning system. Produced cold or hot water is transported to an air conditioner at the first floor in a machine room to exchange heat with the air. The temperatures of each position mentioned in the figure were measured continuously by using resistance bulb thermometers. The water flow rate was measured by an ultrasonic flow meter (Model UFP-20, Tohki Sangyo, Japan). All the data were stored in a PC and analyzed.

Fig. 13. Yamaguchi University library building

Fig. 14. Heat pump - chiller units

Experiments were conducted for almost one month during both the summer and winter seasons in 2007. In the summer season, operation conditions of the air conditioning system without drag-reducing additives were as follows:

1. The system was started at 7:30 every morning. At this time, the temperature of the circulating water was around 22 °C.
2. The chiller unit produced cold water. When the circulating water temperature became 11 °C, the unit stopped.
3. When the circulating water temperature became 18 °C, the chiller unit started automatically.

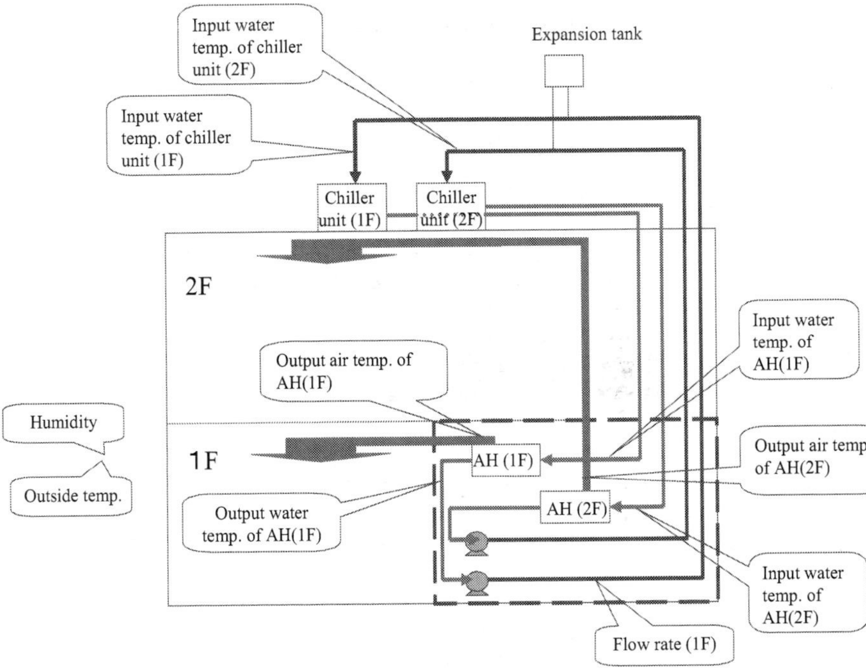

Fig. 15. Schematic diagram of the air conditioning system. (AH: Air Handling Unit)

After accumulating data for two weeks, we added LSP-01M, which included Ethoquad O/12, NaSal, and sodium molybdate as a corrosion inhibitor, to the system (Figure 16). The concentration of LSP-01M was 5000 mg/L, and thus the surfactant concentration was 500 mg/L.

4.3 Experimental results

The water flow rate increased from 5.7 L/s to 6.01 L/s with the addition of 5000 mg/L LSP-01. To adjust the flow rate to the former value (5.7 L/s) by controlling an inverter system installed on the main pump, we changed the electric current of the pump from 5.4 A to 4.5A, and as a consequence, 16.7% reduction of energy consumption was obtained. In the analyses of all the stored temperature data for a month, we found no significant difference between before and after the addition of the drag-reducing additive. The operation of this air conditioning system can be summarized as follows:

1. Measure the output water temperature of the chiller unit.
2. If the measured temperature is higher than an upper setting temperature, run the chiller unit.
3. If the measured temperature is lower than a lower setting temperature, stop the chiller unit.

Fig. 16. Addition of LSP-01 from the discharged side of the main pump

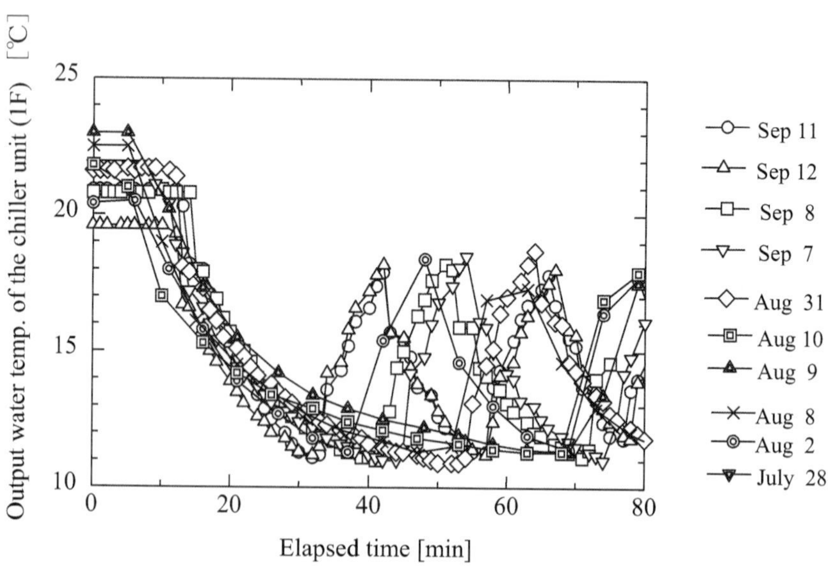

Fig. 17. Output water temperature versus elapsed time for of the chiller unit (1F)

If the heat transfer reduction caused by the drag-reducing additive occurs, the operating time of the chiller unit should be longer than the normal operation without the additive.
Figure 17 shows the output temperatures of the chiller unit (1F). The figure indicates that the temperatures were around 20 °C just before the operation of the air conditioning system. Then, the temperature decreased gradually after the chiller unit was run. After several tens of minutes, the chiller stopped and the temperature suddenly increased. If the temperature reached the upper setting temperature, the chiller started. Therefore, the temperature charts with respect to the elapsed times became zigzag shapes.
To determine the operating intervals of the chiller unit, the following function was used to fit the down side of the zigzag chart:

$$y = y_0 \exp\left(\frac{-t}{t_R}\right) \qquad (6)$$

where y, y_0, t, and t_R are the output water temperature, constant, elapsed time, and relaxation time, respectively. The relaxation time can be used to evaluate the operating interval. Table 2 indicates the relaxation times for the down side of the first zigzag chart of each day. The drag-reducing additive was introduced on August 21st; however, the relaxation times did not show a significant change between before and after the addition.

Date	y_0	$1/t_R$	t_R [min]
July 28	20.0	0.0117	85.5
Aug 8	20.0	0.0155	64.5
Aug 9	19.6	0.0103	97.1
Aug 10	18.9	0.0093	107.5
Aug 31	18.0	0.0142	70.4
Sep 7	19.6	0.0189	52.9
Sep 8	18.1	0.0209	47.8
Sep 11	19.3	0.0317	31.5
Sep 12	17.4	0.0249	40.2

Table 2. Relaxation time for the down side of the first zigzag chart of each day

4.4 Discussion

In section 3, we described the measurement of the heat transfer characteristics for drag-reducing flow by the laboratory experimental apparatus, and a significant amount of heat transfer reduction was observed when the additive conditions were unsuitable. In section 4-3, we evaluated the possibility of a heat transfer reduction occurring and found no evidence of a heat transfer reduction. One of the following may have caused the difference in these results:

1) In the case of the laboratory experiment, the heat transfer characteristics were evaluated based on the reduction of the heat transfer coefficient between the wall of the copper pipe (heat exchanger) and water. Since it was impossible to measure the wall temperature of the

chilling unit, we used the temperature differences of the inlet and outlet of the chilling unit to evaluate the heat transfer characteristics. In other words, the heat transfer characteristics were evaluated based on the overall coefficient of heat transfer, U, which is determined the following equation:

$$\frac{1}{U}=\frac{1}{h_w}+\frac{x}{k}+\frac{1}{h_a} \tag{7}$$

where h_w, x, k, and h_a are the heat transfer coefficient of water, the thickness of the heat exchanger's wall, the thermal conductivity of the wall, and the heat transfer coefficient of air, respectively. Since h_w is two orders of magnitude larger than h_a, the value of U is not significantly changed even if the value of h_w is reduced due to the heat transfer reduction. To took at a quantitative example, let us assume the following values:

h_w=5800 W/m^2·K for water flow

h_w=1200 W/m^2·K for drag-reducing flow ($HTR\%$=80 %)

x=1 mm, k=350 W/mK, h_a=60 W/m^2·K

The calculated overall heat transfer coefficient for water flow is 59.4 W/m^2·K, while that for drag-reducing flow is 57.1 W/m^2·K. As a result, the heat transfer reduction with respect to the overall heat transfer coefficient is only a 3.9% reduction, even though the heat transfer coefficient of fluid was reduced significantly.

2) In the case of the laboratory experiment, the constant heat flux conditions were maintained during the experiment, which means the difference of the inlet and outlet temperatures of the copper pipe did not affect the flow conditions. In this experiment, when the heat transfer occurred, the wall temperature increased. On the other hand, for the practical air conditioning system, the heat transfer phenomenon occurred unsteadily due to the intermittent operation of the chilling unit and inadequate mixing of the fluid. Therefore, even though the heat transfer reduction occurred in response to the drag-reducing additives, it was not easy to monitor the influence of the reduction of heat transfer.

3) Finally, we want to stress that micelle aggregates of drag-reducing surfactants normally suffer mechanical degradation in the chiller unit pipe of practical air conditioning systems. The average velocity inside heat exchangers is usually designed to be in the range of 1.5 m/s to 2.0 m/s. Also, practical pipelines have many elbows, branches, reduction or expansion pipes, valves, and so on, in which the shear rates applied to the fluids are larger than the rated in straight pipes. However, it is quite difficult to evaluate the drag reduction and the heat transfer reduction in the practical heat exchangers while factoring in the associated changes in the micelle structure of drag-reducing surfactants.

5. Conclusion

The drag-reducing technology will be applied to an ever increasing range of applications. There is also enormous potential for application of this approach to other fluids besides water, so we can expect a lot of research in this area. Finally, we observed certain characteristic near-wall momentum and heat transfer behavior in flows that calls for further investigation. This behavior is attributed to the effects of uneven transfer phenomena of micelle motion in pipes for drag-reducing flows, and investigation of this phenomenon should lead to new insights regarding the mechanism involved in drag-reducing flows.

6. Acknowledgment

The author would like to express his thanks to Mr. T. Matsumura, LSP Cooperative Union, and Mr. K. Tokuhara, Shunan Region Local Industry Promotion Center Foundation, for their valuable discussion on the drag reduction by surfactant solutions. The author also would like to acknowledge the experimental assistance of a student colleague (Yusuke Uchiyama).

7. References

Aguilar, G.; Gasljevic, K. & Matthys, E. J. (2001). Asymptotes of Maximum Friction and Heat Transfer Reductions for Drag-reducing surfactant Solutions, *Int. J. of Heat and Mass Transfer*, Vol.44, pp.2835-2843

Gadd, G. E. (1966). Reduction of Turbulent Friction in Liquids by Dissolved Additives, *Nature*, Vol. 212, pp. 874-877

Ohlendorf, D.; Interthal, W. & Hoffmann, H. (1986). Surfactant systems for drag reduction:physico-chemical properties and rheological behaviour, *Rheology Acta*, Vol. 25, pp.468-486

Qi, Y.; Kawaguchi, Y.; Lin, Z.; Ewing, M.; Christensen, R. & Zakin, J. (2001). Enhanced Heat Transfer of Drag Reducing Surfactant Solutions with Fluted Tube-in-tube Heat Exchanger, *Int. J. of Heat and Mass Transfer*, Vol.44, pp.1495-1505

Saeki, T.; De Guzman, M.; Morishima, H.; Usui, H. & Nishimura, T. (2000). A Flow Visualization Study on the Mechanism of Turbulent Drag Reduction by Surfactants, *Nihon Reoroji Gakkaishi*, Vol. 28, No.1, pp.35-40

Saeki, T.; Tokuhara, K.; Matsumura, T. & Yamamoto, S. (2002). Application of Surfactant Drag Reduction for Practical Air Conditioning Systems, *Nihon Kikai Gakkai Ronbunshu*, B, Vol. 68, No. 669, pp.1482-1488 (in Japanese)

Shikata T. et al. (1988). Micelle formation of detergent molecules in aqueous media. II: Role of free salicylate ions on viscoelastic properties of aqueous cetyltrimethylammonium bromide-sodium salicylate solutions, *Langmuir*, Vol. 4, pp. 354-359, 1988

Usui, H. & Saeki T. (1993). Drag Reduction and Heat Transfer Reduction by Cationic Surfactants, *J. of Chem. Eng. Japan*, Vol.26, No.1, pp.103-106

Usui, H. et al. (1998). Effect of surfactant molecular structure on turbulent drag reduction, *Kagaku Kogaku Ronbunshu*, Vol. 24, No.1, pp.134-137 (in Japanese)

Chou, L.-C.; Christensen, R. N. & Zakin, J. L. (1989). The influence of chemical composition of quaternary ammonium salt cationic surfactants on their drag reducing effectiveness, *Proceedings of 4th Int. Conf. on Drag Reduction*, Davos, IAHR/AIRH, Ellis Horwood Pub., pp.141-148

Motier, J. F. (2002). Polymer advances and the dramatic growth in commercial pipeline drag reduction, *Proceedings of 12th European Drag Reduction Meeting*, Herning, Denmark, April 18-20, 2002

Steiff, A.; Klopper, K. & Weinspach, P. (1998). Influence of Drag Reducing Additives in Heat Exchangers, *Proceedings of 11th IHTC*, 6, pp.317-322, Kyongju, Korea

Toms, B. A. (1948). Some observations on the flow of linear polymers solutions through straight pipe at large Reynolds numbers. *Proceedings of International Congress on Rheology*, Vol. 2, pp. 135-141, Scheveningen, Italy, 1948

Li, B.; Xu, Y. & Choi, J. (1996). Applying Machine Learning Techniques, *Proceedings of ASME 2010 4th International Conference on Energy Sustainability*, pp. 14-17, ISBN 842-6508-23-3, Phoenix, Arizona, USA, May 17-22, 2010

19

Entransy - a Novel Theory in Heat Transfer Analysis and Optimization

Qun Chen, Xin-Gang Liang and Zeng-Yuan Guo
Department of Engineering Mechanics, Tsinghua University
China

1. Introduction

It has been estimated that, of all the worldwide energy utilization, more than 80% involves the heat transfer process, and the thermal engineering has for a long time recognized the huge potential for conserving energy and decreasing CO_2 release so as to reduce the global warming effect through heat transfer efficiency techniques (Bergles, 1988, 1997; Webb, 1994; Zimparov, 2002). In addition, since the birth of electronic technology, electricity-generated heat in electronic devices has frequently posed as a serious problem (Arden, 2002; Chein & Huang, 2004), and effective cooling techniques are hence needed for reliable electronic device operation and an increased device lifespan. In general, approaches for heat transfer enhancement have been explored and employed over the full scope of energy generation, conversion, consumption and conservation. Design considerations to optimize heat transfer have often been taken as the key for better energy utilization and have been evolving into a well-developed knowledge branch in both physics and engineering.

During the last several decades and promoted by the worldwide energy shortage, a large number of heat transfer enhancement technologies have been developed, and they have successfully cut down not only the energy consumption, but also the cost of equipment itself. However, comparing with other scientific issues, engineering heat transfer is still considered to be an experimental problem and most approaches developed are empirical or semi-empirical with no adequate theoretical base (Gu et al., 1990). For instance, for a given set of constraints, it is nearly impossible to design a heat-exchanger rig with the optimal heat transfer performance so as to minimize the energy consumption.

Therefore, scientists developed several different theories and methods to optimize heat transfer, such as the constructal theory (Bejan, 1997) and the minimum entropy generation (Bejan, 1982). Then heat transfer processes were optimized with the objective of minimum entropy generation. Based on this method, several researchers (Nag & Mukherjee, 1987; Sahin, 1996; Sekulic et al., 1997; Demirel, 2000; Sara et al., 2001; Ko, 2006) analyzed the influences of geometrical, thermal and flow boundary conditions on the entropy generation in various convective heat transfer processes, and then optimized them based on the premise that the minimum entropy generation will lead to the most efficient heat transfer performance. However, there are some scholars (Hesselgreaves, 2000; Shah & Skiepko, 2004; Bertola & Cafaro, 2008) who questioned whether the entropy generation is the universal irreversibility measurement for heat transfer, or the minimum entropy generation is the general optimization criterion for all heat transfer processes, regardless of the nature of the applications. For instance, by analyzing the relationship between the efficiency and the entropy generation in 18 heat exchangers with different structures, Shah & Skiepko (2004) demonstrated that even when the system entropy generation reaches the extremum, the efficiency of the heat exchangers can be at either the maximum or the minimum, or anything in between. In addition, the so-called "entropy generation paradox" (Bejan, 1996; Hesselgreaves, 2000) exists when the entropy generation minimization is used as the optimization criterion for counter-flow heat exchanger. That is, enlarging the heat exchange area from zero simultaneously increases the heat transfer rate and improves heat exchanger efficiency, but does not reduce the entropy generation rate monotonously - the entropy generation rate increases at first then decreases. Therefore, it was speculated that the optimization criterion of minimum entropy generation is not always consistent with the heat transfer improvement.

Recently, Guo et al. (2007) introduced the concepts of entransy and entransy dissipation to measure, respectively, the heat transfer capacity of an object or a system, and the loss of such capacity during a heat transfer process. Moreover, Guo et al. (2007) proposed the entransy dissipation extremum and the corresponding minimum entransy dissipation-based thermal resistance as alternative optimization criterions for heat transfer processes not involved in thermodynamic cycles, and consequently, developed the minimum entransy dissipation-based thermal resistance principle to optimize the processes of heat conduction (Guo et al., 2007; Chen et al., 2009a, 2011), convective heat transfer (Meng et al., 2005; Chen et al., 2007, 2008, 2009b), thermal radiation (Cheng & Liang, 2011), and in heat exchangers (Liu et al., 2009; Guo et al., 2010).

This chapter summarizes the entransy theory in heat transfer, such as the definitions of entransy, entransy dissipation and its corresponding thermal resistance with multi-temperatures, the minimum entransy dissipation-based thermal resistance principle for heat transfer, etc., and introduces its applications in heat transfer optimization. Finally, we will make comparisons between entransy optimization and entropy optimization to further examine their applicability to heat transfer optimization in applications of different natures.

2. The origin of entransy

After an intensive study, we found that all transport processes contain two different types of physical quantities due to the existing irreversibility, i.e., the conserved ones and the non-conserved ones, and the loss or dissipation in the non-conserved quantities can then be used as the measurements of the irreversibility in the transport process. Taking an electric system as an example, although both the electric charge and the total energy are conserved during an electric conduction, the electric energy however is not conserved and it is partly dissipated into the thermal energy due to the existence of the electrical resistance. Consequently, the electrical energy dissipation rate is often regarded as the irreversibility measurement in the electric conduction process. Similarly, for a viscous fluid flow, both the mass and the momentum of the fluid, transported during the fluid flow, are conserved, whereas the mechanical energy, including both the potential and kinetic energies, of the fluid is turned into the thermal energy due to the viscous dissipation. As a result, the mechanical energy dissipation is a common measure of irreversibility in a fluid flow process. The above two examples show that the mass, or the electric quantity, is conserved during the transport processes, while some form of the energy associated with them is not. This loss or dissipation of the energy can be used as the measurement of irreversibility in these transport processes. However, an irreversible heat transfer process seems to have its own particularity, for the heat energy always remains constant during transfer and it doesn't appear to be readily clear what the non-conserved quantity is in a heat transfer process (Chen et al., 2011). Based on the analogy between electrical and heat conductions, Guo et al. (2007) made a comparison between electrical conduction and heat conduction as shown in table 1. It could be found in the table that there is no corresponding parameter in heat conduction for the electrical potential energy in a capacitor, and hence they defined an equivalent quantity, G, that corresponding to the electrical potential energy in a capacitor

$$G = \frac{1}{2} Q_{vh} T \,, \tag{1}$$

which is called entransy. They further derived Eq. (1) according to the similar procedure of the derivation of the electrical potential energy in a capacitor. Entransy was originally referred to as the heat transport potential capacity in an earlier paper by the authors (Guo et al., 2003).

Electrical charge stored in a capacitor Q_{ve} [C]	Electrical current I [C/s]=[A]	Electrical resistance R_e [Ω]	Capacitance $C_e = Q_{ve}/U_e$ [F]
Thermal energy stored in a body $Q_{vh} = Mc_vT$ [J]	Heat flow rate $\dot{Q}_h$ [W]	Thermal resistance R_h [K/W]	Heat capacity $C_h = Q_{vh}/T$ [J/K]
Electrical potential U_e [V]	Electrical current density $\dot{q}_e$ [A/m²]	Ohm's law $\dot{q}_e = -K_e \nabla U_e$	Electrical potential energy in a capacitor $E_e = Q_{ve}U_e/2$
Thermal potential (temperature) $U_h = T$ [K]	Heat flux density $\dot{q}_h$ [W/m²]	Fourier's law $\dot{q}_h = -K_h \nabla T$	Thermal potential in a body ?

Table 1. Analogy between electrical and thermal conductions (Guo et al., 2007)

Entransy represents the heat transfer ability of an object (Guo et al., 2007). It possesses both the nature of "energy" and the transfer ability. If an object is put in contact with an infinite number of heat sinks that have infinitesimally lower temperatures, the total quantity of "potential energy" of heat which can be output is $\frac{1}{2}Q_{vh}T$. Biot (1955) suggested a similar concept in the derivation of the differential conduction equation using the variation method. Eckert & Drake (1972) pointed out that "Biot in a series of papers beginning in 1955 formulated from the ideas of irreversible thermal dynamics a variational equivalent of the heat-conduction equation that constituted a thermodynamical analogy to Hamilton's principle in mechanics and led to a Lagrangian formulation of the heat conduction problem in terms of generalized coordinates..., Biot defines a thermal potential $E = \frac{1}{2}\iiint_{\Omega} \rho c T^2 dV$ ....

The thermal potential E plays a role analogous to a potential energy...". However, Biot did not further explain the physical meaning of thermal potential and its application was not found later except in the approximate solutions of anisotropic conduction problems.

Accompanying the electric charge, the electric energy is transported during electric conduction. Similarly, along with the heat, the entransy is transported during heat transfer too. Furthermore, when a quantity of heat is transferred from a high temperature to a low temperature, the entransy is reduced and some of entransy is dissipated during the heat transport. The lost entransy is called entransy dissipation. Entransy dissipation is an evaluation of the irreversibility of heat transport ability. For instance, let's consider two bodies A and B with the initial temperatures T_A and T_B, respectively. Their initial entransy is

$$G_1 = \frac{1}{2}\left(M_A c_{vA} T_A^2 + M_B c_{vB} T_B^2\right), \tag{2}$$

where M is the body mass, c_v the specific heat at constant volume, and the subscripts A and B represent the bodies A and B, respectively. When these two bodies contact each other, thermal energy will flow from the higher temperature body to the lower temperature one. After infinite long time, their temperature will be the same and the entransy becomes

$$G_2 = \frac{1}{2}(M_A c_{vA} + M_B c_{vB}) T_2^2 , \tag{3}$$

where T_2 is the equilibrium temperature

$$T_2 = \frac{M_A c_{vA} T_A + M_B c_{vB} T_B}{M_A c_{vA} + M_B c_{vB}} . \tag{4}$$

The entransy difference before and after equilibrium is

$$G_1 - G_2 = \frac{1}{2} \frac{M_A M_B c_{vA} c_{vB} \left(T_A - T_B\right)^2}{M_A c_{vA} + M_B c_{vB}} > 0 . \tag{5}$$

Equation (5) proves that the total entransy is reduced after the two bodies getting equilibrium, i.e. there is an entransy dissipation associating with the heat transport.

3. Entransy balance equation and optimization for heat conduction (Guo et al., 2007; Chen et al., 2011)

3.1 Entransy balance equation for heat conduction

In a heat conduction process, the thermal energy conservation equation is expressed as

$$\rho c_v \frac{\partial T}{\partial t} = -\nabla \cdot \dot{q} + \dot{Q} , \tag{6}$$

where ρ is the density, t the time, $\dot{q}$ the heat flow density, and $\dot{Q}$ the internal heat source. Multiplying both sides of Eq. (6) by temperature T gives an equation which can be viewed as the balance equation of the entransy in the heat conduction:

$$\rho c_v T \frac{\partial T}{\partial t} = -\nabla \cdot \left(\dot{q} T\right) + \dot{q} \cdot \nabla T + \dot{Q} T , \tag{7}$$

that is,

$$\frac{\partial g}{\partial t} = -\nabla \cdot (\dot{q}T) - \phi_h + \dot{g}\ , \tag{8}$$

where g=G/V=$uT/2$ is the specific entransy, V the volume, u the specific internal energy, $\dot{q}T$ the entransy flow density, $\dot{g}$ the entransy change due to heat source, and ϕ_h can be taken as the entransy dissipation rate per unit volume, expressed as

$$\phi_h = -\dot{q} \cdot \nabla T \tag{9}$$

The left term in either Eq. (7) or (8) is the time variation of the entransy stored per unit volume, consisting of three items shown on the right: the first represents the entransy transferred from one (or part of the) system to another (part), the second term can be considered as the local entransy dissipation during the heat conduction, and the third is the entransy input from the internal heat source. It is clear from Eq. (9) that the entransy is dissipated when heat is transferred from high temperature to low temperature. Thus, heat transfer is irreversible from the viewpoint of entransy, and the dissipation of entransy can hence be used as a measurement of the irreversibility in heat transfer.

3.2 The entransy dissipation extremum principle for heat conduction

Heat transfer optimization aims for minimizing the temperature difference at a given heat transfer rate,

$$\delta(\Delta T) = \delta f\left(x, y, z, \tau, T, k, q, \rho, c_{v,} \cdots\right) = 0\ , \tag{10}$$

Or maximizing the heat transfer rate at a given temperature difference,

$$\delta(\dot{Q}) = \delta g\left(x, y, z, \tau, T, k, q, \rho, c_{v,} \cdots\right) = 0\ . \tag{11}$$

In conventional heat transfer analysis, it is difficult to establish the relationship between the local temperature difference, or local heat transfer rate, and the other related physical variables over the entire heat transfer area, so the variational methods in Eqs. (10) and (11) are not practically useable. However, the entransy dissipation in Eq. (9) is a function of the local heat flux and local temperature gradient in the heat transfer area, and thus the variational method will become utilizable if written in terms of the entransy dissipation (Cheng, 2004).

Integrating the balance equation of the entransy Eq. (7) over the entire heat transfer area gives:

$$\iiint_{\Omega} \rho c_v T \frac{\partial T}{\partial t} dV = \iiint_{\Omega} -\nabla \cdot (\dot{q}T) dV + \iiint_{\Omega} \dot{q} \cdot \nabla T dV + \iiint_{\Omega} \dot{Q} T dV\ . \tag{12}$$

For a steady state heat conduction problem, the left term in Eq. (12) vanishes, i.e.

$$0 = \iiint_{\Omega} -\nabla \cdot (\dot{q}T) dV + \iiint_{\Omega} \dot{q} \cdot \nabla T dV + \iiint_{\Omega} \dot{Q} T dV\ . \tag{13}$$

If there is no internal heat source in the heat conduction domain, Eq. (13) is further reduced into:

$$\iiint_{\Omega} -\nabla \cdot (\dot{q}T) dV = \iiint_{\Omega} -\dot{q} \cdot \nabla T dV \cdot \tag{14}$$

By transforming the volume integral to the surface integral on the domain boundary according to the Gauss's Law, the total entransy dissipation rate in the entire heat conduction domain is deduced as

$$\Phi_h = \iiint_{\Omega} -\dot{q} \cdot \nabla T dV = \iint_{\Gamma} -\dot{q} T dS = \iint_{\Gamma^+} \dot{q}_{in} T_{in} dS - \iint_{\Gamma^-} \dot{q}_{out} T_{out} dS \cdot \tag{15}$$

where Γ^+ and Γ^- represent the boundaries of the heat flow input and output, respectively. The continuity of the total heat flowing requires a constant total heat flow $\dot{Q}_t$,

$$\dot{Q}_t = \iint_{\Gamma^+} \dot{q}_{in} dS = \iint_{\Gamma^-} \dot{q}_{out} dS \,. \tag{16}$$

We further define the ratio of the total entransy dissipation and total heat flow as the heat flux-weighted average temperature difference ΔT

$$\Delta T = \frac{\Phi_h}{\dot{Q}_t} = \iint_{\Gamma^+} \frac{\dot{q}_{in}}{\dot{Q}_t} T_{in} dS - \iint_{\Gamma^-} \frac{\dot{q}_{out}}{\dot{Q}_t} T_{out} dS \,. \tag{17}$$

For one-dimensional heat conduction, Eq. (17) is reduced into $\Delta T = (T_{in} - T_{out})$, exactly the conventional temperature difference between the hot and cold ends. Using the heat flux-weighted average temperature difference defined in Eq. (17) and applying the divergence theorem, a new expression for optimization of a steady-state heat conduction at a given heat flow rate can be constructed as:

$$\dot{Q}_t \delta(\Delta T) = \delta \iiint_{\Omega} -\dot{q} \cdot \nabla T dV = \delta \iiint_{\Omega} k |\nabla T|^2 dV = 0 \,. \tag{18}$$

It shows that when the boundary heat flow rate is given, minimizing the entransy dissipation leads to the minimum in temperature difference, that is, the optimized heat transfer. Conversely, to maximize the heat flow at a given temperature difference, Eq. (18) can be rewritten as:

$$\Delta T \delta \dot{Q}_t = \delta \iiint_{\Omega} -\dot{q} \cdot \nabla T dV = \delta \iiint_{\Omega} \frac{1}{k} |\dot{q}|^2 dV = 0 \,, \tag{19}$$

showing that maximizing the entransy dissipation leads to the maximum in boundary heat flow rate.

Likewise, for a steady state heat dissipating process with internal heat source in Eq. (13), the total entransy dissipation rate in the entire heat conduction domain is derived as

$$\Phi_h = \iiint_{\Omega} -\dot{q} \cdot \nabla T dV = \iiint_{\Omega} \dot{Q} T dV - \iint_{\Gamma^-} \dot{q}_{out} T_{out} dS \cdot \tag{20}$$

Since the heat generated in the entire domain will be dissipated through the boundaries, i.e.,

$$\dot{Q}_t = \iiint_{\Omega} \dot{Q} dV = \iint_{\Gamma^-} \dot{q}_{out} dS \cdot \tag{21}$$

Again the heat flux-weighted average temperature is defined as the entransy dissipation over the heat flow rate

$$\Delta T = \frac{\Phi_h}{\dot{Q}_t} = \iiint_{\Omega} \frac{\dot{Q}}{\dot{Q}_t} T dV - \iint_{\Gamma^-} \frac{\dot{q}_{out}}{\dot{Q}_t} T_{out} dS , \tag{22}$$

and thus the optimization of the process is achieved when

$$\dot{Q}_t \delta(\Delta T) = \delta \iiint_{\Omega} -\dot{q} \cdot \nabla T dV = \delta \iiint_{\Omega} k |\nabla T|^2 dV = 0 , \tag{23}$$

which means that in a heat dissipating process, minimizing the entransy dissipation leads to the minimum averaged temperature over the entire domain.
Based on the results from Eqs. (18), (19) and (23), it can be concluded that the entransy dissipation extremum (EDE) lead to the optimal heat transfer performance at different boundary conditions. This extreme principle can be concluded into the minimum thermal resistance principle defined by entransy dissipation. (Guo et al., 2007)

3.3 Application to a two-dimensional volume-point heat conduction

We will apply our proposed approach to practical cases where heat transfer is used for heating or cooling such as in the so-called volume-point problems (Bejan, 1997) of heat dissipating for electronic devices as shown in Fig. 1. A uniform internal heat source distributes in a two-dimensional device with length and width of L and H, respectively. Due to the tiny scale of the electronic device, the joule heat can only be dissipated through the surroundings from the "point" boundary area such as the cooling surface in Fig. 1, with the opening W and the temperature T_0 on one boundary. In order to lower the unit temperature, a certain amount of new material with high thermal conductivity is introduced inside the device. As the amount of the high thermal conductivity material (HTCM) is given, we need to find an optimal arrangement so as to minimize the average temperature in the device.

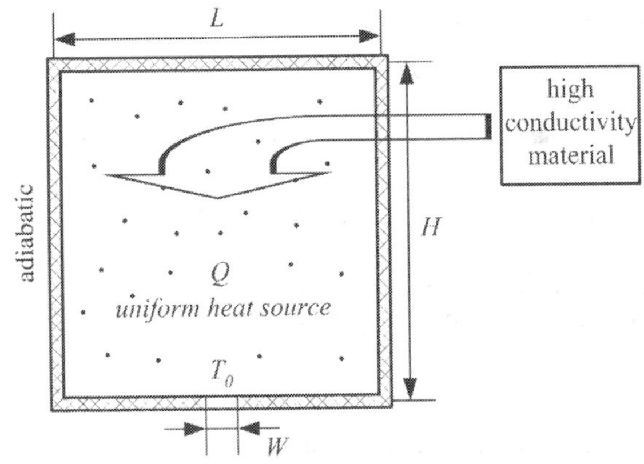

Fig. 1. Two-dimensional heat conduction with a uniformly distributed internal heat source (Chen et al., 2011)

According to the new extremum principle based on entransy dissipation, for this volume-point heat conduction problem, the optimization objective is to minimize the volume-average temperature, the optimization criterion is the minimum entransy dissipation, the optimization variable is the distribution of the HTCM, and the constraints is the fixed amount of the HTCM, i.e.,

$$\iiint_{\Omega} k(x,y)\,dV = const. \tag{24}$$

By variational method, a Lagrange function, Π, is constructed

$$\Pi = \iiint_{\Omega}\left(k|\nabla T|^2 + Bk\right)dV. \tag{25}$$

where the Lagrange multiplier B remains constant due to a given amount of thermal conductivity.

The variation of Π with respect to temperature T gives

$$\iint_{\Gamma} k\nabla T\delta T\cdot\vec{n}\,dS - \iiint_{\Omega}\nabla\cdot(k\nabla T)\delta T\,dV = 0. \tag{26}$$

Because the boundaries are either adiabatic or isothermal, the surface integral on the left side of Eq. (26) vanishes, that is,

$$\iint_{\Gamma} k\nabla T\delta T\cdot\vec{n}\,dS = 0. \tag{27}$$

Moreover, due to a constant entransy output and a minimum entransy dissipation rate, the entransy input reaches the minimum when

$$\delta\iiint_{\Omega}\dot{Q}T\,dV = \iiint_{\Omega}\dot{Q}\delta T\,dV = 0. \tag{28}$$

Substituting Eqs. (27) and (28) into Eq. (26) in fact gives the thermal energy conservation equation based on Fourier's Law:

$$\nabla\cdot(k\nabla T) + \dot{Q} = 0. \tag{29}$$

This result validates that the irreversibility of heat transfer can be measured by the entransy dissipation rate. The variation of Π with respect to thermal conductivity k gives

$$|\nabla T|^2 = -B = const. \tag{30}$$

This means that in order to optimize the heat dissipating process, i.e. to minimize the volume-average temperature, the temperature gradient should be uniform. This in turn requires that the thermal conductivity be proportional to the heat flow in the entire heat conduction domain, i.e. the HTCM be placed at the area with the largest heat flux.

As an example, the cooling process in low-temperature environment is analyzed here. For the unit shown in Fig. 1, $L = H = 5$ cm, $Q = 100$ W/cm^2, $W = 0.5$ cm and $T_0 = 10$ K. The thermal conductivity of the unit is 3 W/(m·K), and that for the HTCM is 300 W/(m·K) occupying 10% of the whole heat transfer area.

Fig. 2(b) shows the distribution of the HTCM according to the extremum principle of entransy dissipation, where the black area represents the HTCM - the same hereinafter. (The implementary steps are as shown in the reference of Chen et al., 2011.) The HTCM with a tree structure absorbs the heat generated by the internal source and transports it to the isothermal outlet boundary - similar in both the shape and function of actual tree roots.

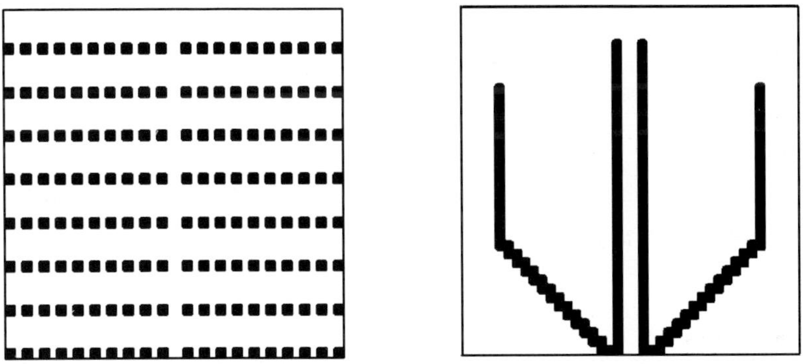

(a) Simple uniform HTCM arrangement (b) HTCM arrangement using the extremum principle of entransy dissipation

Fig. 2. Different arrangements of HTCM (Chen et al., 2011)

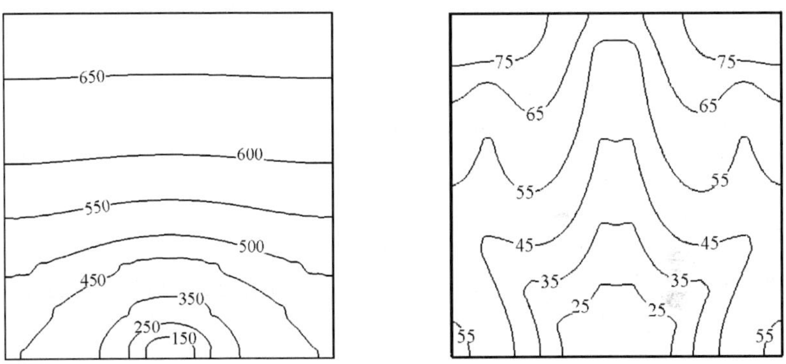

(a) From the uniform HTCM arrangement in Fig. 2(a) (b) From the optimized arrangement of HTCM in Fig. 2(b)

Fig. 3. The temperature fields obtained from different arrangements of HTCM (Chen et al., 2011)

For a fixed amount of HTCM, Figs. 3(a) and 3(b) compare the temperature distributions between a uniform distribution of HTCM shown in Fig. 2(a), and the optimized distribution in Fig. 2(b) based on the extremum principle of entransy dissipation. The average temperature in the first case is 544.7 K while the temperature in the second optimized case is 51.6 K, a 90.5 % reduction! It clearly demonstrates that the optimization criterion of entransy

dissipation extremum is highly effective for such applications. Furthermore, as shown in Fig. 3(b), the temperature gradient field is also less fluctuating in the optimized case.
In addition, based on the concept of the entransy dissipation rate, we (Chen et al. 2009a) introduced the non-dimensional entransy dissipation rate and employed it as an objective function to analyze the thermal transfer process in a porous material.

4. Entransy balance equation and dissipation for convective heat transfer

4.1 Entransy balance equation for convective heat transfer (Cheng, 2004)

For a convection problem, the energy equation is

$$\rho\left[c_p\frac{DT}{Dt}+P\frac{D}{Dt}\left(\frac{1}{\rho}\right)\right]=\phi_\mu-\nabla\cdot\dot{q}+\dot{Q}\,, \tag{31}$$

where P, t and ϕ_μ are the pressure, time, and viscous dissipation, respectively. Multiply Eq. (31) by T and by some subsequent derivation (Cheng, 2004), there is

$$\rho\left(\frac{\partial g}{\partial t}+U\cdot\nabla g\right)=-\nabla\cdot(\dot{q}T)+\dot{Q}T+\phi_\mu T+\dot{q}\cdot\nabla T\,, \tag{32}$$

where $g=c_pT^2/2$, is the entransy per unit volume of the fluid. On the left, the 1st term is time variation of entransy, and the 2nd term is entransy variation accompanying fluid flow; on the right, the 1st term is entransy variation due to boundary heat exchange, the 2nd term is the entransy change due to heat source, the 3rd term is the entransy variation due to dissipation and the 4th term is the entransy dissipation. It is clear from Eq. (32) that the heat transport ability is reduced when heat is transferred from high temperature to low temperature.

4.2 The entransy dissipation extremum principle for convective heat transfer

Similar to the derivation of EDE principle in heat conduction, for a steady-state convective heat transfer process of a fluid with constant boundary heat flux and ignoring the heat generated by viscous dissipation, integrating Eq. (32), and transforming the volume integral to the surface integral on the domain boundary yields:

$$Q_0\delta\left(T_w-\frac{T_{in}+T_{out}}{2}\right)=\delta\iiint\limits_{\Omega}k|\nabla T|^2\,dV\,. \tag{33}$$

Since the inlet and outlet temperatures, T_{in} and T_{out}, of the fluid are fixed for a given boundary heat flux, Equation (33) means that, the minimum entransy dissipation in the domain is corresponding to the minimum boundary temperature, i.e. the minimum boundary temperature difference.
Similarly, from (32) the maximum entransy dissipation is obtain,

$$(T_w-T_{in})\delta Q_0-\frac{Q_0}{\rho c_p}\delta Q_0=\delta\iiint\limits_{\Omega}k|\nabla T|^2\,dV\,. \tag{34}$$

Because Eq. (33) illustrates that for the range of the boundary heat flux from 0 to its maximum value, $\rho Vc_p(T_w-T_{in})$, the entransy dissipation increases monotonically with the increasing boundary heat flux, that means maximum entransy dissipation results in the maximum boundary heat flux. Equation (33) together with Eq. (34) is called entransy dissipation extreme (EDE) principle in convective heat transfer. Thus, the entransy dissipation can be used to optimize the flow field with given viscosity dissipation so that the heat transfer rate can be increased most with given cost of pressure lose (Guo et al., 2007; Meng et al. 2005, Chen et al. 2007).

4.3 Application to laminar heat transfer (Meng et al., 2005)

For a steady laminar flow, the viscosity dissipation is

$$\phi_m = \mu \begin{bmatrix} 2\left(\frac{\partial u}{\partial x}\right)^2 + 2\left(\frac{\partial v}{\partial y}\right)^2 + 2\left(\frac{\partial w}{\partial z}\right)^2 \\ + \left(\frac{\partial u}{\partial y} + \frac{\partial v}{\partial x}\right)^2 + \left(\frac{\partial u}{\partial z} + \frac{\partial w}{\partial x}\right)^2 + \left(\frac{\partial v}{\partial z} + \frac{\partial w}{\partial y}\right)^2 \end{bmatrix}, \tag{35}$$

where u, v and w are the velocity components along x, y and z directions, respectively. The mechanical work maintaining the fluid flow equals to the integral of the viscous dissipation function over the whole domain,

$$\Pi_{\phi m} = \iiint_{\Omega} \phi_m dV . \tag{36}$$

The optimal flow field was established by configuring a Lagrange functional that includes the objective and constraint functions. The established Lagrange function is,

$$\Pi = \iiint_{\Omega} \left[k\nabla T \cdot \nabla T + C_0\phi_m + A\left(\nabla \cdot k\nabla T - \rho c_p U \cdot \nabla T\right) + B\nabla \cdot \rho U \right] dV , \tag{37}$$

where A, B and C_0 are the Lagrange multipliers. Because of the different types of the constraints, A and B vary with position, while C_0 remains constant for a given viscous dissipation. The first term on the right is the entransy dissipation, the second is the constraint on prescribed viscous dissipation, the third and forth terms are the constraints on energy equation and continuum equation. The variational of Eq. (37) with respect to velocity U offers

$$\mu\nabla^2 U + \frac{\rho c_p}{2C_0} A\nabla T + \frac{1}{2C_0}\nabla B = 0 . \tag{38}$$

The variational of Eq. (37) with respect to temperature T is

$$-\rho c_p U \cdot \nabla A = \nabla \cdot k\nabla A - 2\nabla \cdot k\nabla T , \tag{39}$$

and the boundary conditions of the variable A are $A_b = 0$ for given boundary temperatures, and $\left(\partial A/\partial n\right)_b = 2\left(\partial T/\partial n\right)_b$ for given boundary heat flow rates.

Comparing the momentum equations and Eq. (38), gives the following relations

$$B = -2C_0 P , \tag{40}$$

$$F = C_{\Phi} A \nabla T + \rho U \cdot \nabla U \,, \tag{41}$$

where C_{Φ} is related to the viscous dissipation as

$$C_{\Phi} = \frac{\rho c_p}{2C_0} \,. \tag{42}$$

Substituting Eqs. (41) and (42) into the momentum equation gives

$$\rho U \cdot \nabla U = -\nabla P + \mu \nabla^2 U + \left(C_{\Phi} A \nabla T + \rho U \cdot \nabla U \right) . \tag{43}$$

This is the Euler's equation, i.e. field synergy equation, governing the fluid velocity and temperature fields with the EDE principle during a convective heat transfer process. Equation (43) essentially the momentum equation with a special additional volume force defined in Eq. (41), by which the fluid velocity pattern is adjusted to lead to an optimal temperature field with the extremum of entransy dissipation during a convective heat transfer process.

For a given C_{Φ}, solving Eqs. (39) and (43), the continuity equation and the energy equation in combination result in the optimal flow field with the extremum of entransy dissipation with a specific viscous dissipation and fixed boundary conditions. Fig. 4 shows a typical numerical result of the cross-sectional flow field (Re=400, C_{Φ}=-0.01). Compared with the fully-developed laminar convection heat transfer in a circular tube ((fRe)$_s$=64, Nu$_s$=3.66), the flow viscous dissipation is increased by 17%, the Nusselt number is increased by 313% in the case of Fig. 4. Meanwhile, the numerical analysis shows that multiple longitudinal vortex flow is the optimal flow pattern for laminar flow in tube. With the guide of this optimal flow field, the discrete double-inclined ribs tube (DDIR-tube) is designed to improve heat transfer in laminar flow. Fig. 5 is a typical cross-sectional vortex flow of a DDIR-tube and experiments have demonstrated the DDIR-tube have better heat transfer performance but lower resistance increase (Meng et al., 2005).

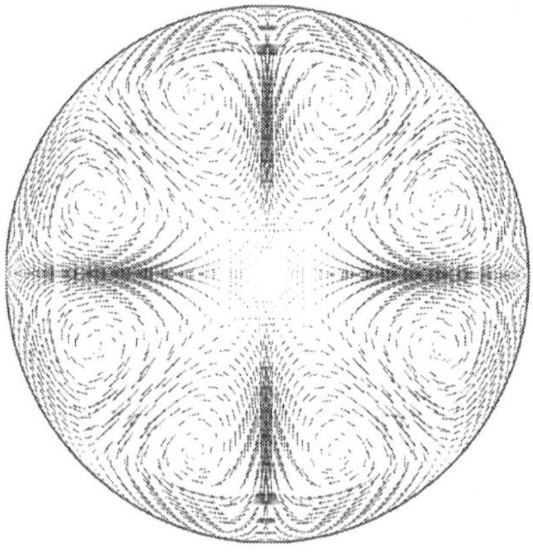

Fig. 4. Optimum flow field of laminar heat transfer in circular tube (Re=400) (Meng et al., 2005)

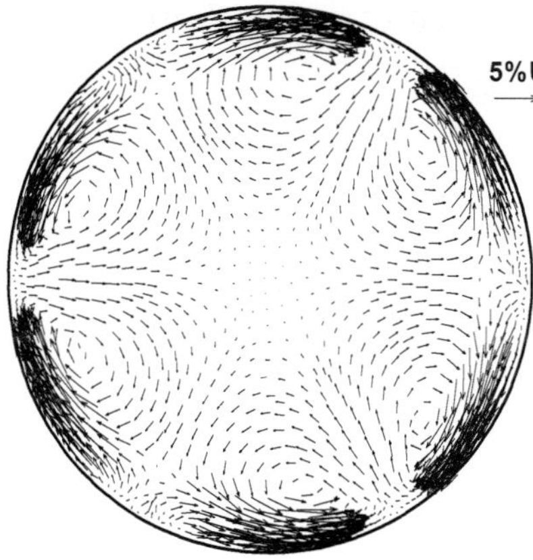

Fig. 5. Numerical solution of cross-sectional flow fields in the DDIR-tube (Meng et al., 2005)

4.4 Application to turbulent heat transfer (Chen et al., 2007)

For turbulent heat transfer, the entransy dissipation function is modified as:

$$\phi_{ht} = k_{eff} \left|\nabla T\right|^2, \tag{44}$$

where k_{eff} is the effective thermal conductivity during turbulent heat transfer.

The optimization objective of turbulent heat transfer is to find an optimal velocity field, which has the extremum of entransy dissipation for a given decrement of the time-averaged kinetic energy, i.e. a fixed consumption of pumping power. Meanwhile, the flow is also constrained by the continuity equation and the energy equation. All these constraints can also be removed by using the Lagrange multipliers method to construct a function

$$\Pi_t = \iiint_{\Omega} \left[k_{eff} \nabla T \cdot \nabla T + C_0 \phi_{mt} + A\left(\nabla \cdot k_{eff} \nabla T - \rho c_p U \cdot \nabla T\right) + B \nabla \cdot \rho U \right] dV \,. \tag{45}$$

where ϕ_{mt} is the decrement function of the mean kinetic energy

$$\phi_{mt} = \left(\mu + \mu_t\right) \begin{bmatrix} 2\left(\dfrac{\partial u}{\partial x}\right)^2 + 2\left(\dfrac{\partial v}{\partial y}\right)^2 + 2\left(\dfrac{\partial w}{\partial z}\right)^2 \\ + \left(\dfrac{\partial u}{\partial y} + \dfrac{\partial v}{\partial x}\right)^2 + \left(\dfrac{\partial u}{\partial z} + \dfrac{\partial w}{\partial x}\right)^2 + \left(\dfrac{\partial v}{\partial z} + \dfrac{\partial w}{\partial y}\right)^2 \end{bmatrix}. \tag{46}$$

The decrement of the mean kinetic energy equals to the viscous dissipation by the viscous forces, plus the work of deformation of the mean motion by the turbulence stresses, which transform the mean kinetic energy to the turbulence-energy. The turbulent viscosity, μ_t, which is a function of the velocity, can be calculated using the Prandtl's mixing-length model, the Prandtl-Kolmogorov assumption or the k-ε model. However, the Prandtl-Kolmogorov assumption and the k-ε model require additional differential equations for the turbulent kinetic energy (k) and the turbulent dissipation rate (ε) to calculate the turbulent viscosity, which severely complicate obtaining the variation of Eq. (45). Thus, a single algebraic function, which has been validated and used for indoor air flow simulation (Chen & Xu, 1998), is used to calculate the turbulent viscosity:

$$\mu_t = 0.03874\rho|U|l\,, \tag{47}$$

where, l is the distance to the nearest wall. As with other zero-equation turbulent models, Eq. (47) is not very theoretically sound, but yields some reasonable results for turbulent flows.

The variational of Eq. (45) with respect to temperature T is:

$$-\rho c_p U \cdot \nabla A = \nabla \cdot k_{eff} \nabla A - 2\nabla \cdot k_{eff} \nabla T\,. \tag{48}$$

The variational of Eq. (45) with respect to velocity component u is:

$$\begin{aligned}
&\frac{1}{2C_0}\frac{\partial B}{\partial x} + \nabla \cdot \left(\mu_{eff} \nabla u\right) + \frac{\rho c_p A}{2C_0}\frac{\partial T}{\partial x} - \frac{\Gamma u}{2C_0 \mathrm{Pr}_t}\left[|\nabla T|^2 + A\nabla \cdot (\nabla T)\right] - \frac{\phi_{mt}\Gamma u}{2\mu_{eff}} \\
&- \frac{A\Gamma}{2C_0 \mathrm{Pr}_t}\frac{\partial T}{\partial x}\left[\frac{u}{l}\frac{\partial l}{\partial x} - \left(u\frac{\partial u}{\partial x} + v\frac{\partial v}{\partial x} + w\frac{\partial w}{\partial x}\right)|U|^{-2} u + \frac{\partial u}{\partial x}\right] \\
&- \frac{A\Gamma}{2C_0 \mathrm{Pr}_t}\frac{\partial T}{\partial y}\left[\frac{u}{l}\frac{\partial l}{\partial y} - \left(u\frac{\partial u}{\partial y} + v\frac{\partial v}{\partial y} + w\frac{\partial w}{\partial y}\right)|U|^{-2} u + \frac{\partial u}{\partial y}\right] \\
&- \frac{A\Gamma}{2C_0 \mathrm{Pr}_t}\frac{\partial T}{\partial z}\left[\frac{u}{l}\frac{\partial l}{\partial z} - \left(u\frac{\partial u}{\partial z} + v\frac{\partial v}{\partial z} + w\frac{\partial w}{\partial z}\right)|U|^{-2} u + \frac{\partial u}{\partial z}\right] \\
&+ \left(\frac{\partial \mu_{eff}}{\partial x}\frac{\partial u}{\partial x} + \frac{\partial \mu_{eff}}{\partial y}\frac{\partial v}{\partial x} + \frac{\partial \mu_{eff}}{\partial z}\frac{\partial w}{\partial x}\right) + \frac{1}{2C_0}\nabla \cdot \left(\frac{A\Gamma u}{\mathrm{Pr}_t}\nabla T\right) = 0
\end{aligned} \tag{49}$$

where $\Gamma = 0.03874\rho l|U|^{-1}$, Pr_t is the turbulent Prantdl number with the value of 0.9. The variationals of Eq. (45) with respect to the velocity components v and w are similar to Eq. (49). There are four unknown variables and four governing equations including Eqs. (48), (49), the continuity equation and the energy equation, so the unknown variables can be solved for a given set of boundary conditions. Meanwhile, the flow must also satisfy the momentum equation,

$$\rho U \cdot \nabla U = -\nabla P + \mu_{eff} \nabla^2 U + F\,. \tag{50}$$

Comparison with Eq. (50), Eq. (49) is actually a momentum equation with a special additional volume force, which is referred to as the field synergy equation for turbulent heat transfer. For a given set of boundary conditions, the optimal velocity field, which has a larger heat transfer rate than any other field flow, can be obtained by solving this field synergy equation.

As an example, flow in a parallel plate channel is studied to illustrate the applicability of the field synergy equation. For simplicity, a repeated segment with the height of 20 mm and the length of 2.5 mm was chosen. Water flowing between the parallel plates is assumed periodically fully developed with a Reynolds number of 20,000. The inlet water temperature is 300 K and the wall temperature is 350 K.

It is well known that for original results before optimization, the velocity vectors and the temperature gradients are nearly perpendicular to each other, leading to a small scalar product between the velocity vector and the temperature gradient, i.e. the field synergy degree (Meng et al. 2003) is relatively poor. In the computational domain, the heat transfer rate is 1782 W and the decrement of the mean kinetic energy is 4.61×10^{-3} W. The thicknesses of the laminar sublayer and the transition sublayer are 0.116 mm and 0.928 mm, respectively.

The optimized velocity and temperature fields near the upper wall for $C_0 = -1.5\times10^{7}$ are shown in Figs. 6 (a) and (b). There are several small counter-clockwise eddies near the upper wall. The distances between eddies centres is about 0.4 mm and the eddy heights perpendicular to the primary flow direction are about 0.2 mm. There are also several small clockwise eddies near the lower wall for symmetry. For this case, the heat transfer rate is 1887 W and the decrement of the mean kinetic energy is 5.65×10^{-3} W. Compared with the original results before optimization, the heat transfer rate is increased by 6 %, while the decrement of the mean kinetic energy is increased by 23 %.

For heat transfer in turbulent flow between parallel plates, the temperature gradients in the laminar sublayer are two to three orders of magnitude larger than the gradients far from the wall, which means that the thermal resistance in the laminar sublayer is the dominate resistance in turbulent flow. The conventional heat transfer enhancement viewpoint is to first reduce the dominate resistance to most effectively increase the heat transfer rate. Eddies and disturbances near the wall will increase the velocities, reduce the thermal resistances, and enhance the heat transfer. The results support that the tubes with micro fins effectively enhance turbulent heat transfer, which is different from multiple longitudinal vertex generation in laminar heat transfer, and further point out the optimal heights of the fins for different Reynolds numbers should be half of the transition sublayer of turbulent flow, which is also validated experimentally (Li et al., 2009).

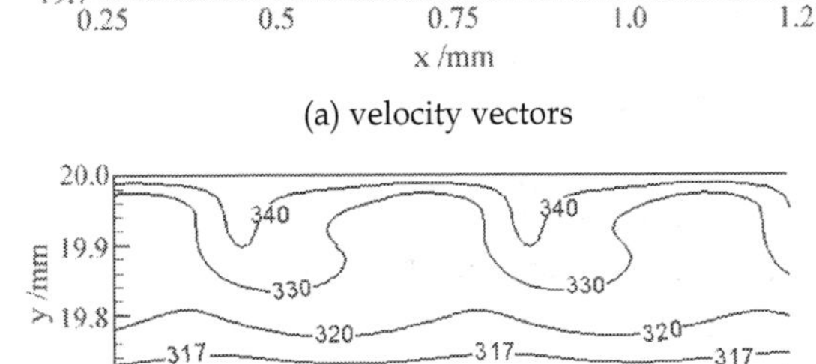

(a) velocity vectors

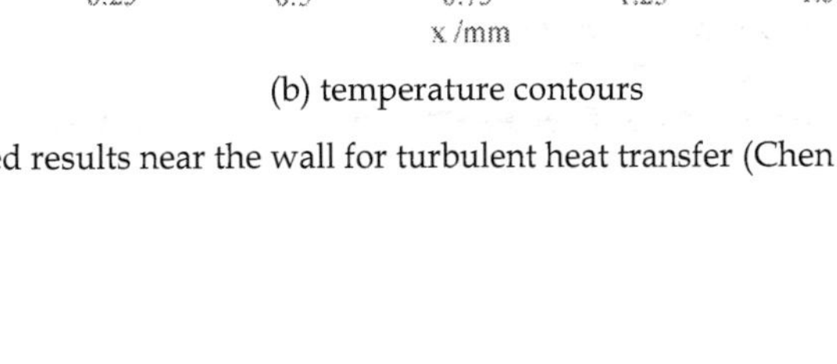

(b) temperature contours

Fig. 6. Optimized results near the wall for turbulent heat transfer (Chen et al., 2007)

5. Entransy dissipation-based thermal resistance for heat exchanger design

In engineering, both designing and checking heat exchanger performance are generally used such four approaches as the logarithmic mean temperature difference method (LMTD), the heat exchanger effectiveness – number of transfer units method (ε-NTU), the P-NTU method, and the ψ-P method. However, these methods are not very convenient when used in some particular situations. For example, in the LMTD method, it is inevitable to introduce a correction factor to adjust the effective temperature difference for cross-flow and multi-pass exchangers. Meanwhile, the use of LMTD method for checking heat exchanger performance has to involve tedious iterations to determine the proper outlet fluid temperatures and thereafter the value of LMTD which satisfy the requirement that heat transferred in the heat exchanger be equal to the heat carried out by the fluid. On the other hand, in the ε-NTU method, the fluid with the minimum heat capacity rate has to be first taken as the benchmark to calculate both heat exchanger effectiveness ε and NTU, so iterations are also unavoidable for the design of fluid flow rates. Besides, different types, e.g. parallel, counter-flow, cross-flow and shell-and-tube, of heat exchangers have their individual diverse and complex relations between heat exchanger effectiveness ε and NTU, which are more or less inconvenient for engineering applications. Therefore, it is necessary to develop a general criterion for the evaluation of heat exchanger performance and, more importantly, develop a common method for heat exchanger performance design and optimization.

For a heat transfer process in a parallel heat exchanger, the heat lost by the hot fluid over a differential element should be the same as that gained by the cold fluid, which both equal to the heat transferred through the elements, that is

$$dq = -m_h dh_h = m_c dh_c \,. \tag{51}$$

where, m is the mass flow rate, h is the specific enthalpy, and the subscripts h and c represent the hot and cold fluids, respectively.

Integrating Eq. (51), we will obtain the total heat transfer rate in the heat exchange from the viewpoint of energy conservation

$$Q = m_h\left(h_{h,a} - h_{h,b}\right) = m_c\left(h_{c,d} - h_{c,c}\right), \tag{52}$$

For a heat exchanger without any phase-change fluid, if the fluid specific heats are constant, Eq. (51) will be rewritten as

$$dT_h = -\frac{1}{m_h c_{p,h}} dq \,, \tag{53}$$

$$dT_c = \frac{1}{m_c c_{p,c}} dq \,, \tag{54}$$

and Eq. (52) is rewritten as

$$Q = m_h c_{p,h}\left(T_{h,a} - T_{h,b}\right) = m_c c_{p,c}\left(T_{c,d} - T_{c,c}\right). \tag{55}$$

Based on Eqs. (53) and (54), Fig. 7 gives the fluid temperature variations versus the heat transfer rate q. As shown, the shaded area is:

$$dS = T_h dq - T_c dq \,. \tag{56}$$

where, the first term on the right-hand side represents the entransy output accompanying the thermal energy dq flowing out of the hot fluid, while the second term represents the entransy input accompanying the thermal energy dq flowing into the cold fluid. Therefore, the shaded area exactly indicates the entransy dissipation rate during the heat transferred from the hot fluid to the cold one:

$$d\phi_h = (T_h - T_c)dq \,. \tag{57}$$

The total entransy dissipation in the heat exchanger is deduced by integrating Eq. (57),

$$\Phi_h = \int_0^{\Phi_h} d\phi_h = \int_0^Q (T_h - T_c)dq = \frac{(T_{h,a} - T_{c,c}) + (T_{h,b} - T_{c,d})}{2} Q = \Delta T_{AM} Q \,. \tag{58}$$

where, ΔT_{AM} is the arithmetical temperature difference between the hot and cold fluids in the heat exchanger.

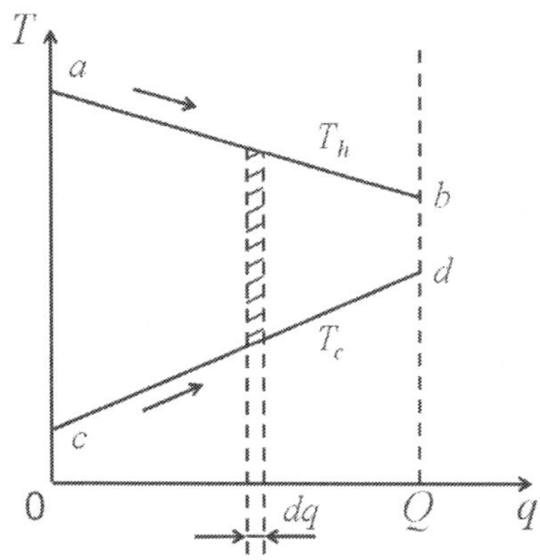

Fig. 7. Sketch of the fluid temperature variations versus the heat transfer rate in a parallel heat exchanger

Substituting Eqs. (55), (58) and the heat transfer equation

$$Q = KA \frac{(T_{h,a} - T_{c,c}) - (T_{h,b} - T_{c,d})}{\ln(T_{h,a} - T_{c,c}) - \ln(T_{h,b} - T_{c,d})} \tag{59}$$

into the definition of entransy dissipation-based thermal resistance, EDTR, for heat exchangers (Liu et al., 2009), $R_h = \Phi_h/Q^2 = \Delta T_{AM}/Q$, we get the formula of such thermal resistance for parallel heat exchangers:

$$R_{h,p} = \frac{\xi_p}{2} \frac{\exp(KA\xi_p) + 1}{\exp(KA\xi_p) - 1} \,, \tag{60}$$

where, $\xi_p = \left(\frac{1}{m_h c_{p,h}} + \frac{1}{m_c c_{p,c}}\right)$, termed the arrangement factor for parallel heat exchangers.

Similarly, for counter-flow or TEMA E-type heat exchangers, we can also deduce the same formula of EDTR as shown in Eq. (60), except that the expression of the arrangement factor ξ is diverse, i.e. $\xi_c = \left(\frac{1}{m_h c_{p,h}} - \frac{1}{m_c c_{p,c}}\right)$ for counter-flow heat exchangers and $\xi_s = \sqrt{\frac{1}{\left(m_c c_{p,c}\right)^2} + \frac{1}{\left(m_h c_{p,h}\right)^2}}$ for TEMA E-type heat exchangers.

This general expression is convenient for us to design the heat transfer performance of heat exchangers. For example, a one shell and two tube passes TEMA E-type shell-and-tube exchanger with the heat transfer coefficient K and area A of 300 W/(m^2·K) and 50 m^2, respectively, is used to cool the lubricating oil from the initial temperature $T_{h,in}$ = 57 °C to the desired temperature $T_{h,out}$ = 45 °C. The mass flow rate and specific heat of the oil are m_h = 10 kg/s and $c_{p,h}$ = 1.95 kJ/(kg·K), respectively. If the cooling water enters the heat exchanger at the temperature of 33 °C, what are its heat capacity rate and outlet temperature, and what is the rate of heat transfer by the exchanger.

According to the relation between EDTR and arithmetical mean temperature difference, the total heat transfer rate in the exchanger is

$$Q = \frac{\Delta T_{AM}}{R_{h,s}} = \frac{\left(T_{h,in} + T_{h,out} - T_{c,in} - T_{c,out}\right)\left(\exp\left(KA\xi_s\right) - 1\right)}{\xi_s\left(\exp\left(KA\xi_s\right) + 1\right)}. \tag{61}$$

Numerically solving Eq. (61) and the energy conservation equation (55) simultaneously, we can easily obtain the heat capacity rate and the outlet temperature of the cooling water are 73.8 kJ/(s·K) and 36.17 °C, respectively, and the total heat transfer rate is 234 kJ.

In this problem, neither the heat capacity rate nor the exit temperature of the cooling water are known, therefore an iterative solution is required if either the LMTD or the ε-NTU method is to be used. For instance, when using the LMTD method, the detail steps are: (1) obtain the required heat transfer rate of the exchanger Q_1 from the energy conservation equation of the oil; (2) assume a heat capacity rate of the cooling water, and then calculate its exit temperature; (3) according to the inlet and outlet temperatures of both the oil and the cooling water, obtain the logarithm mean temperature difference and the correction factor of the shell-and-tube exchanger; (4) based on the heat transfer equation, derive another heat transfer rate of the exchanger Q_2. Because the heat capacity rate of the cooling water is assumed, iteration is unavoidable to make the derived heat transfer rate Q_2 in step 4 close to the required one Q_1. Thus, it is clear that the entransy dissipation-based thermal resistance method can design heat exchanger performance conveniently.

6. Differences between entransy and entropy

Besides entransy, the concepts of entropy and entropy generation are considered to be two important functions in thermodynamic and used for estimating the irreversibility of and optimizing heat transfer based on the premise that the minimum entropy generation (MEG) will lead to the most efficient heat transfer performance. Thus, there exist two optimization

principles, and it is highly desired to investigate the physical essentials and the applicability of, as well as the differences between, them in heat transfer optimization.
The local entropy generation function induced by the heat transfer over finite temperature difference is

$$s_g = \frac{k|\nabla T|^2}{T^2} . \tag{62}$$

Let's reconsider the same volume-point problems as shown in Fig. 1. What we seek is also the optimal HTCM arrangement, except in this case that: 1. the optimization criterion is the minimum entropy generation; 2. the corresponding energy conservation equation should be added as a constraint, because it is not implied in the principle of minimum entropy generation when the thermal conductivity is constant (Bertola & Cafaro, 2008).
Introducing the corresponding Lagrange function

$$\Pi' = \iiint_{\Omega} \left[k\frac{|\nabla T|^2}{T^2} + B'k + C'\left(\nabla \cdot k\nabla T + \dot{Q}\right) \right] dV . \tag{63}$$

where B' and C' are also the Lagrange multipliers. The constraint of thermal conductivity is the isoperimetric condition, and consequently B' is a constant. C' is a variable related to space coordinates. The variation of Π' with respect to temperature T gives

$$-\nabla \cdot \left(k\nabla C'\right) = \frac{2k|\nabla T|^2}{T^3} + \frac{2\dot{Q}}{T^2} . \tag{64}$$

while the variation of Π' with respect to thermal conductivity k yields

$$\nabla C' \cdot \nabla T - \frac{\nabla T^2}{T^2} = B' = const . \tag{65}$$

Likewise, Eq. (65) gives the guideline for optimization based on the criterion of minimum entropy generation. That is, the HTCM be placed at the area with the extreme in absolute value of $\nabla C' \cdot \nabla T - \nabla T^2/T^2$.
Fig. 8(a) shows the distribution of HTCM based on the principle of minimum entropy generation. Comparison of Figs. 2(b) and 8(a) shows that although the distributions of HTCM are similar between the two results in most areas, the root-shape structure from the minimum entropy generation principle is not directly connected to the heat flow outlet, leaving some parts with the original material in between them so that the heat cannot be transported smoothly to the isothermal outlet boundary. Fig. 8(b) gives the optimized temperature distribution obtained by the minimum entropy generation. Because the low thermal conductivity material is adjacent to the heat outlet, the temperature gradient grows larger and thus lowers the entire heat transfer performance. The averaged temperature of the entire area is 150.8 K - 99.2 K higher than that obtained by the extremum principle of entransy dissipation. From the definition, it is easy to find that in order to decrease the entropy generation, we have to both reduce the temperature gradient, and raise the temperature, thus leading to the arrangement of HTCM showed in Fig. 8(a).

In addition, according to the principle of minimum entropy generation, the optimization objective of a steady-state heat dissipating process can be expressed as:

$$\dot{Q}\delta\left(\Delta\frac{1}{T}\right)_m = \delta\iiint_{\Omega}\frac{k|\nabla T|^2}{T^2}dV = 0\,. \tag{66}$$

where $\left(\Delta\frac{1}{T}\right)_m = \left(\frac{1}{T}-\frac{1}{T_0}\right)_m$, is the equivalent thermodynamics potential difference which represents the "generalized force" in the entropy picture for heat transfer. Thus, minimizing the entropy generation equals to minimizing the equivalent thermodynamics potential difference $\left(\Delta\frac{1}{T}\right)_m$, leading to the highest exergy transfer efficiency. That is, the minimum entropy generation principle is equivalent to the minimum exergy dissipation during a heat transfer process.

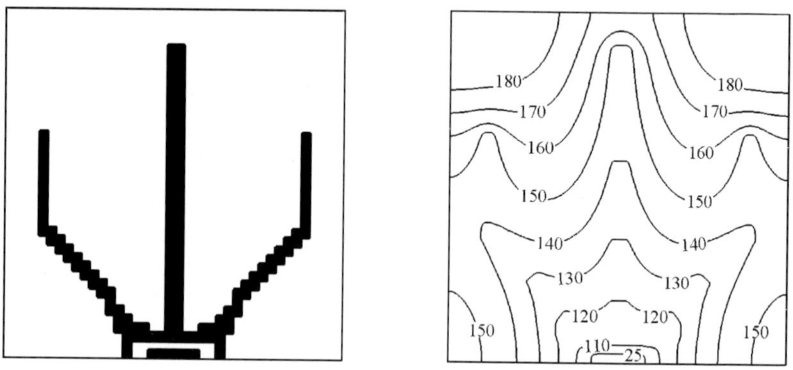

(a) HTCM arrangement (b) Temperature field

Fig. 8. Optimized results using the minimum entropy generation principle (Chen et al., 2011)

To facilitate the comparison between the two results from Figs. 3(b) and 8(b), Table 2 lists the key findings side by side, obtained respectively by the optimization criteria of the minimum entropy generation and the entransy dissipation extremum. It indubitably shows in the table that the proposed entransy based approach is more effective than the entropy based one in heat transfer optimization, for the former leads to a result with significantly reduced mean temperature than that by the latter (51.6 K vs. 150.8 K), and much lower maximum temperature (83.0 K vs. 194.9 K). Whereas the entropy based approach is preferred in exergy transfer optimization, as it results in a significantly lower equivalent thermodynamic potential (7.1×10^{-3} /K vs. 2.2×10^{-2} /K).

Besides heat conduction, we (Chen et al. 2009b) also compared the two criteria in convective heat transfer optimization. Our results indicate that both principles are applicable to convective heat transfer optimization, subject however to different objectives. The minimum entropy generation principle works better in searching for the minimum exergy dissipation during a heat-work conversion, whereas the entransy dissipation extremum principle is more effective for processes not involving heat-work conversion, in minimizing the heat-transfer ability dissipation.

Optimization results / Optimization Criterions	$\Phi_h/(W\bullet K)$	$S_{gen}/(W/K)$	T_m/K	T_{max}/K	$\left(\Delta\frac{1}{T}\right)_m/(1/K)$
Extremum Entransy Dissipation	5.5×10^{4}	100.7	51.6	83.0	2.2×10^{-2}
Minimum Entropy Generation	1.58×10^{5}	81.7	150.8	194.9	7.1×10^{-3}

Table 2. Optimized results obtained by the optimization criterions of minimum entropy generation and entransy dissipation extremum (Chen et al., 2011)

7. Conclusion

The entransy is a parameter that is developed in recent year. It is effective in optimization of heat transfer. Entransy is an evaluation of the transport ability of heat. Both the amount of heat and the potential contribute to the entransy. Entransy will be lost during the heat transportation from a high temperature to lower one and entransy dissipation will be produces. Based on the energy conservation equation, the entransy balance equations for heat conduction and convective heat transfer are developed. The entransy dissipation extreme principles are developed, that is, the maximum entransy dissipation corresponds to the maximum heat flux for prescribed temperature difference and the minimum entransy dissipation corresponds to minimum temperature difference for prescribed heat flux. This extreme principle can be concluded into the minimum thermal resistance principle defined by entransy dissipation.

The entransy dissipation-based thermal resistance of heat exchangers is introduced as an irreversibility measurement, for parallel, counter-flow, and shell-and-tube exchangers, which has a general expression, is the function of heat capacity rates and thermal conductance, and may analyze, compare and optimize heat exchanger performance from the physical nature of heat transfer. Besides, from the relation among heat transfer rate, arithmetical mean temperature difference and EDTR, the total heat transfer rate may easily be calculated through the thermal conductance of heat exchangers and the heat capacity rates of fluids, which is convenient for heat exchanger design.

Finally, we compared the criteria of entransy dissipation extremum to that of entropy generation minimization in heat transfer optimization, the results indicates that the minimum entransy dissipation-based thermal resistance yields the maximum heat transfer efficiency when the heat transfer process is unrelated with heat-work conversion, while the minimum entropy generation leads to the highest heat-work conversion when such is involved in a thermodynamic cycle.

8. Acknowledgments

This work was financially supported by the National Basic Research Program of China (2007CB206901) and the National Natural Science Foundation of China (Grant No. 51006060).

9. References

Arden W. M. (2002). The International Technology Roadmap for Semiconductors - Perspectives and Challenges for the Next 15 Years. *Current Opinion in Solid Sate and Materials Science*, Vol.6, No.5, (October 2002), pp. 371-377, ISSN 1359-0286

Bejan A. (1982). *Entropy Generation through Heat and Fluid Flow*, John Wiley & Sons, ISBN 0-471-09438-2, New York, USA.

Bejan A. (1996). *Entropy Generation Minimization: the Method of Thermodynamic Optimization of Finite-size Systems and Finite-time Processes*, CRC Press, ISBN 0-849-39651-4, Boca Raton, USA

Bejan A. (1997). Constructal-theory Network of Conducting Paths for Cooling a Heat Generating Volume. *International Journal of Heat and Mass Transfer*, Vol.40, No.4, (March 1997), pp. 799-811, ISSN 0017-9310

Bergles A. E. (1988). Some Perspectives on Enhanced Heat Transfer --- 2nd-generation Heat Transfer Technology. *Journal of Heat Transfer-Transactions of the Asme*, Vol.110, No.4B, (November 1988), pp. 1082-1096, ISSN 0022-1481

Bergles A. E. (1997). Heat Transfer Enhancement --- The Encouragement and Accommodation of High Heat Fluxes. *Journal of Heat Transfer-Transactions of the Asme*, Vol.119, No.1, (Feburary 1997), pp. 8-19, ISSN 0022-1481

Bertola V. & Cafaro E. (2008). A Critical Analysis of the Minimum Entropy Production Theorem and Its Application to Heat and Fluid Flow. *International Journal of Heat and Mass Transfer*, Vol.51, No.7-8, (April 2008), pp. 1907-1912, ISSN 0017-9310

Biot M. (1955). Variational Principles in Irreversible Thermodynamics with Application to Viscoelasticity. *Physical Review*, Vol.97, No.6, (March 1955), pp. 1463-1469, ISSN 0031-899X

Chein R., & Huang G. (2004). Thermoelectric Cooler Application in Electronic Cooling. *Applied Thermal Engineering*, Vol.24, No.14-15, (October 2004), pp. 2207-2217, ISSN 1359-4311

Chen Q., Ren J. X. & Meng J. A. (2007). Field Synergy Equation for Turbulent Heat Transfer and Its Application. *International Journal of Heat and Mass Transfer*, Vol.50, No.25-26, (December 2007), pp. 5334-5339, ISSN 0017-9310

Chen Q. & Ren J. X. (2008). Generalized Thermal Resistance for Convective Heat Transfer and Its Relation to Entransy Dissipation. *Chinese Science Bulletin*, Vol.53, No.23, (December 2008), pp. 3753-3761, ISSN 1001-6538

Chen Q., Wang M. R., Pan N. & Guo Z. Y. (2009a). Irreversibility of Heat Conduction in Complex Multiphase Systems and Its Application to the Effective Thermal Conductivity of Porous Media. *International Journal of Nonlinear Sciences and Numerical Simulation*, Vol.10, No.1, (January 2009), pp. 57-66, ISSN 1565-1339

Chen Q., Wang M. R., Pan N. & Guo Z. Y. (2009b). Optimization Principles for Convective Heat Transfer. *Energy*, Vol.34, No.9, (September 2009), pp. 1199-1206, ISSN 0360-5442

Chen Q. & Xu W. (1998). A Zero-equation Turbulence Model for Indoor Airflow Simulation. *Energy and buildings*, Vol.28, No.2, (October 1998), pp. 137-144, ISSN 0378-7788

Chen Q., Zhu H. Y., Pan N. & Guo Z. Y. (2011). An Alternative Criterion in Heat Transfer Optimization. *Proceedings of the Royal Society a-Mathematical Physical and Engineering Sciences*, Vol.467, No.2128, (April 2011), pp. 1012-1028, ISSN 1364-5021

Cheng X. G. (2004). *Entransy and Its Applications in Heat Transfer Optimization.* (Doctoral Dissertation), Tsinghua University, Beijing, China

Cheng X. T. & Liang X. G. (2011). Entransy Flux of Thermal Radiation and Its Application to Enclosures with Opaque Surfaces. *International Journal of Heat and Mass Transfer,* Vol.54, No.1-3, (January 2011), pp. 269-278, ISSN 0017-9310

Demirel Y. (2000). Thermodynamic Analysis of Thermomechanical Coupling in Couette Flow. *International Journal of Heat and Mass Transfer,* Vol.43, No.22, (November 2000), pp. 4205-4212, ISSN 0017-9310

Gu W. Z., Shen S. R. & Ma C. F. (1990). *Heat Transfer Enhancement,* Science Press, ISBN 7-03-001601-7, Beijing, China

Guo Z. Y., Liu X. B., Tao W. Q. & Shah R. (2010). Effectiveness-thermal Resistance Method for Heat Exchanger Design and Analysis. *International Journal of Heat and Mass Transfer,* Vol.53, No.13-14, (June 2010), pp. 2877-2884, ISSN 0017-9310

Guo Z. Y., Zhu H. Y. & Liang X. G. (2007). Entransy --- a Physical Quantity Describing Heat Transfer Ability. *International Journal of Heat and Mass Transfer,* Vol.50, No.13-14, (July 2007), pp. 2545-2556, ISSN 0017-9310

Guo Z. Y., Cheng X. G. & Xia Z. Z. (2003). Least Dissipation Principle of Heat Transport Potential Capacity and Its Application in Heat Conduction Optimization. *Chinese Science Bulletin,* Vol.48, No.4, (Feburary 2003), pp. 406-410, ISSN 1001-6538

Hesselgreaves J. E. (2000). Rationalisation of Second Law Analysis of Heat Exchangers. *International Journal of Heat and Mass Transfer,* Vol.43, No.22, (November 2000), pp. 4189-4204, ISSN 0017-9310

Ko T. H. (2006). Numerical Analysis of Entropy Generation and Optimal Reynolds Number for Developing Laminar Forced Convection in Double-sine Ducts with Various Aspect Ratios. *International Journal of Heat and Mass Transfer,* Vol.49, No.3-4, (Febuary 2006), pp. 718-726, ISSN 0017-9310

Li X. W., Meng J. A. & Guo Z. Y. (2009). Turbulent Flow and Heat Transfer in Discrete Double Inclined Ribs Tube. *International Journal of Heat and Mass Transfer,* Vol.52, No.3-4, (January 2009), pp. 962-970, ISSN 0017-9310

Liu X. B. & Guo Z. Y. (2009). A Novel Method for Heat Exchanger Analysis. *Acta Physica Sinica,* Vol.58, No.7, (July 2009), pp. 4766-4771, ISSN 1000-3290

Meng J. A. (2003). *Enhanced Heat Transfer Technology of Longitudinal Vortices Based on Field-Coordination Principle and Its Application.* (Doctoral Dissertation), Tsinghua University, Beijing, China

Meng J. A., Liang X. G. & Li Z. X. (2005). Field Synergy Optimization and Enhanced Heat Transfer by Multi-longitudinal Vortexes Flow in Tube. *International Journal of Heat and Mass Transfer,* Vol.48, No.16, (July 2005), pp. 3331-3337, ISSN 0017-9310

Nag P. K. & Mukherjee P. (1987). Thermodynamic Optimization of Convective Heat Transfer through a Duct with Constant Wall Temperature. *International Journal of Heat and Mass Transfer,* Vol.30, No.2, (Feburary 1987), pp. 401-405, ISSN 0017-9310

Sahin A. Z. (1996). Thermodynamics of Laminar Viscous Flow through a Duct Subjected to Constant Heat Flux. *Energy,* Vol.21, No.12, (December 1996), pp. 1179-1187, ISSN 0360-5442

Sara O. N., Yapici S., Yilmaz M. & Pekdemir T. (2001). Second Law Analysis of Rectangular Channels with Square Pin-fins. *International Communications in Heat and Mass Transfer,* Vol.28, No.5, (July 2001). pp. 617-630, ISSN 0735-1933

Sekulic D. P., Campo A. & Morales J. C. (1997). Irreversibility Phenomena Associated with Heat Transfer and Fluid Friction in Laminar Flows through Singly Connected Ducts. *International Journal of Heat and Mass Transfer*, Vol.40, No.4, (March 1997), pp.905-914, ISSN 0017-9310

Shah R. K. & Skiepko T. (2004). Entropy Generation Extrema and Their Relationship with Heat Exchanger Effectiveness - Number of Transfer Unit Behavior for Complex Flow Arrangements. *Journal of Heat Transfer - Transactions of the ASME*, Vol.126, No.6, (December 2004), pp. 994-1002, ISSN 0022-1481

Webb R. L. (1994). *Principles of Enhanced Heat Transfer*, Taylor & Francis Group, ISBN 0-471-57778-2, Wiley, New York, USA

Zimparov V. (2002). Energy Conservation through Heat Transfer Enhancement Techniques. *International Journal of Energy Research*, Vol.26, No.7, (June 2002), pp. 675-696, ISSN 0363-907X

20

Transient Heat Transfer and Energy Transport in Packed Bed Thermal Storage Systems

Pei Wen Li[1], Jon Van Lew[1], Wafaa Karaki[1],
Cho Lik Chan[1], Jake Stephens[2] and James. E. O'Brien[3]
[1]*Department of Aerospace and Mechanical Engineering*
The University of Arizona, Tucson, AZ 85721,
[2]*US Solar Holdings LLC., 1000 E. Water Street, Tucson, AZ 85719,*
[3]*Idaho National Laboratory, Idaho Falls, Idaho 83415,*
USA

1. Introduction

Compared to fossil fuel energy resources, the major types of renewable energy—such as wind power, solar energy, ocean currents, and tidal energy—generally possess the innate characteristics of intermittence of availability, fluctuation of magnitude, as well as low energy density (Li, 2008). However, the utilization of energy and power in industry, living, and working often requires high energy densities, and demand may be out of phase with the period of availability of renewable energy. In other words, renewable energy is not always load following. This variability creates a demand for energy storage when people develop renewable energy technologies (Kolb, 1998).

Among the several types of renewable energy, solar energy has the largest proportion of the total available and may be directly used as thermal energy in conventional thermal power plants, or converted into electrical power directly using photovoltaic panels. Although direct electrical energy storage in batteries or capacitors may have a high efficiency, it is still very challenging and expensive—particularly when storing a large quantity of electrical energy (Spiers, 1995). Electricity may be indirectly stored by pumping water to reservoirs, or by compressing air, or by electrolyzing water and making hydrogen fuel, etc. However, these methods often have low round-trip efficiency (from electricity to electricity), or are restricted by the availability of geographical conditions or suitable locations. In comparison, it has been recognized that direct solar thermal energy storage is relatively easy to approach at a reasonably low cost and high efficiency, and the energy storage capacity can be much larger than that of direct electricity storage (Price et al., 2002; Montes et al, 2009). Thermal energy storage systems use materials that can be kept at high temperatures in insulated containers. The heat retrieved can then be used in conventional thermal power plants for power generation at times when sunlight is not available or when weather conditions are not favorable (Singer et al, 2010; Laing et al, 2010).

Researchers worldwide have done a great amount of research and development on concentrated solar thermal power generation technologies in the last ten years (Renewable Energy Policy Network for the 21st Century, 2007). Particularly, with these efforts solar trough and solar tower concentrated thermal power generation technologies have become more and more reliable and matured, and the cost of concentrated solar power systems have been significantly reduced due to increased productivity and demand (Pitz-Paal et al., 2007; Herrmann & Kearney, 2002; Gil et al., 2010.).

It has been widely recognized that further cost reduction of electricity generation using concentrated solar thermal power may be accomplished by adding solar thermal storage systems. Storage provides the heat necessary for operation of thermal power plants when sunlight availability is out of phase, and thus increases the operational capacity (in terms of the daily operational time) of the power plants. In addition, the extended operation of solar thermal power plants using stored thermal energy can significantly improve the power dispatch ability (Herrmann & Kearney, 2002) of the power plant.

Other than electrical power generation, solar thermal energy can also be stored for sundry applications such as house heating, hot water supply, industrial drying processes, as well as heating for greenhouse agriculture and animal husbandry. From an energy efficiency perspective, direct use of solar thermal energy for heating is much more efficient than using electricity for heating, as electricity generation requires much more input of other types of energy than the generated electricity. Therefore, although thermal energy storage is not a new technology, it will receive more and more attention, particularly with the development of renewable energy technologies.

2. Brief review of thermal storage techniques

The ideal scenario for thermal energy storage is such that the energy-carrying fluid is stored in a thermal storage system and can be withdrawn at a temperature of no degradation from that of when the fluid was stored. On the basis of this fundamental distinction, one may classify thermal energy storage systems into two categories: (1) a system with direct storage of heat transport fluid, which may approach the performance of an ideal thermal storage system; (2) two-medium heat storage system, which has a fluid serving as heat- carrying medium, and another medium, either solid materials or a liquid, serving the purpose as of a primary thermal storage material. A two-medium heat storage system uses a reduced amount of expensive heat transfer fluid while sacrificing the energy storage efficiency.

A first generation direct heat transfer fluid storage system usually has two storage tanks, as shown in Fig. 1(a), one for hot fluid and the other for cold fluid (Herrmann at al., 2004). During the energy storage process, fluid from the cold tank is pumped to the solar field to be heated and then stored in the hot fluid tank; while during the energy discharging process, fluid from the hot fluid tank is pumped out to release heat to the power plant and afterwards, flows back to the cold fluid tank. Although there are two tanks in such a thermal storage system, the heat transfer fluid only occupies a volume equivalent to that of one tank at any instant in time. This means that the elimination of one tank in the system is possible. As a consequence, a second generation direct heat transfer fluid storage system has only one tank, as shown in Fig. 1(b). A stratification of fluid, which maintains hot fluid on top of cold fluid, is important to such a single tank thermal storage system (Michel et al., 2009; Abdoly & Rapp, 1982; Krane R.J. & Krane M.J.M., 1992; Brosseau et al., 2005). During a heat charging process, hot fluid is injected from the top of the tank, whereas cold fluid is pumped out from bottom of the tank. The opposite is true for the heat discharge process, during which hot fluid is pumped out from top of the tank and cold fluid is injected into the tank from the bottom (Canada et al., 2006).

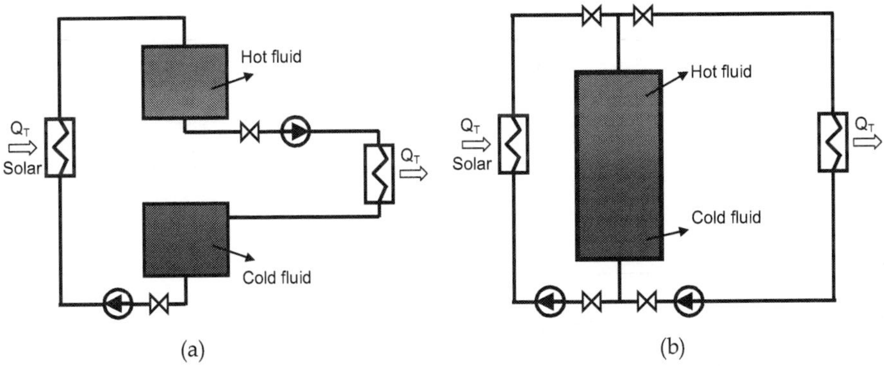

Fig. 1. Thermal storage using heat transport fluid only

Two-medium heat storage systems must have a heat transport fluid and a primary thermal storage material (Laing et al, 2006), either solid or liquid. Also, a two-medium heat storage system typically uses only a single tank. Depending on the contact and heat transfer interaction between the heat-carrying fluid and the primary energy storage material, the storage system may have two types. The first type, as shown in Fig. 2(a), is one that includes loosely packed solid materials (such as rocks, pebbles of metals, and capsules of phase change materials (PCM), etc.) as a porous bed held in a container and through which the heat transport fluid (HTF) flows and transports energy to or from the solid material. In this type of thermal storage system, the heat transfer between thermal storage material and the heat transport fluid is relatively efficient due to the ubiquitous contact between the fluid and storage materials. The second type of two-medium heat storage system, as shown in Fig. 2(b), is such that the heat transfer fluid flows in tubes or pipes that run through thermal storage material, either solid (such as, concrete, wax, sands, soil (Nassar et al., 2006), salts, etc), or liquid (such as oil, or liquid salts, etc). Due to the smaller contact area between the fluid and thermal storage material, the heat transfer between the fluid and the thermal storage material in this case is worse compared to that of the first type.

Cost-effectiveness is always the dictating criterion for selecting a thermal storage system for a specific application. However, there are many factors that can influence the cost of a thermal storage system; for example, the cost of the heat transfer fluid is a key factor which can determine whether a direct HTF storage system or a two-medium thermal storage should be used. A high-pressure tank may be needed if the vapor pressure of the heat transfer fluid is high, or a stainless steel tank may be needed to mitigate corrosion problems. Possible chemical interaction between tank material, the heat transfer fluid, and primary thermal storage material, if applicable, must also be prevented.

The following section provides a survey of suitable heat transport fluids and solid materials for thermal energy storage applications.

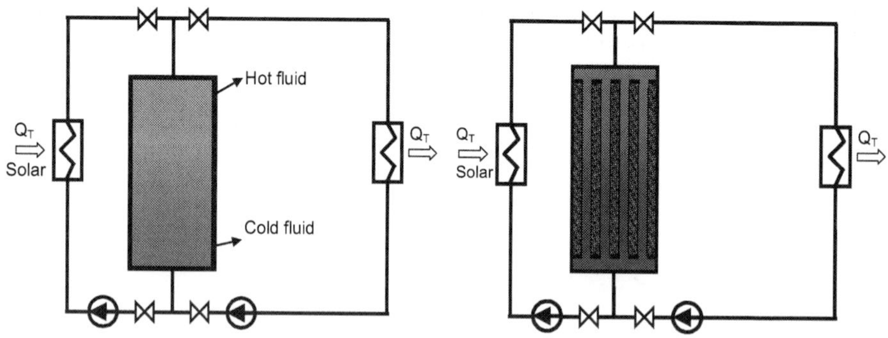

(a) direct contact heat transfer (b) heat transfer through fluid pipes

Fig. 2. Thermal storage using a primary thermal storage material with heat transport fluid

3. Thermal storage materials and heat transfer fluids

The quantity of sensible thermal energy stored in a mass is given by the equation:

$$Q_T = V\rho C(T_H - T_L) \tag{1}$$

where V, ρ, and C are the volume of the mass, the average density of the mass, and the heat capacity of the material, respectively.
Obviously, a high specific heat and a large density are both important for a substance to be a good thermal energy storage material, for that will keep the volume of the storage container small. Other required properties such as high thermal conductivity, low cost, low thermal expansion coefficient, etc. are also important. For a liquid to serve as a heat transport fluid or a thermal storage material, a low solidification point, high boiling point, and low vapor pressure, are all important. Particularly, if the vapor pressure is high, it will require pressurized storage tanks, which can significantly increase the cost of the thermal storage system.

3.1 Properties of solid thermal storage materials

Solid materials, suck as concrete (Zhang et al., 2004), sand, rock, brick, soil, graphite, silicon carbide, taconite, cast iron, and even waste metal chips, have been considered or applied for thermal energy storage purposes. Depending on the formation and granular size of these solid materials, some of them can be used to form a packed bed storage system as illustrated in Fig. 2(a) and others may only be suitable for use in a storage system as shown in Fig. 2(b). The main properties of some of these materials (Tritt, 2005; Hasnain, 1998) are given in Table 1.
Compared to sensible heat storage, the high latent heat associated with phase change of a material offers the potential for higher energy storage densities. There are two approaches using phase change material (PCM) for thermal storage. One is encapsulation of small amounts of PCM in spherical or cylindrical capsules (Wu et al., 2011), which can be arranged to form a packed bed. Heat transport fluid can flow through the packed bed for energy delivery and extraction. This approach needs encapsulation technologies, either coating or canning. The other approach is to embed the PCM in a matrix made of another solid material with high heat conduction, and HTF pipes run through the PCM matrix (Sari

& Kaygusuz, 2001; Regin et al., 2008). The use of a matrix material (e.g. graphite or metal mesh) helps enhance heat conduction in the PCM.

Medium	melting(°C) (or crumbles)	ρ (kg/m³)	C(kJ/kg·°C)	ρ·C(kJ/m³·°C)	k(W/m·°C)
Aluminum	660	2700	0.92	2484.0	250
Brick (common)	1800	1920	1.0	1920.0	1.04
Fireclay	1800	2100-2600	1.0	2100-2600	1-1.5
Soil (dry)	1650	1200-1600	1.26	1512-2016	1.5
Granite	1215	2400	0.79	1896	1.7-4.0
Sand (dry)	1500	1555	0.8	1244	0.15-0.25
Sandstone	1300	2000-2600	0.92	1840-2392	2.4
Rocks	1800	2480	0.84	2086.6	2-7
Concrete	1000 (Crumbles)	2240-2400	0.75	1680-1800	1.7
Graphite	3500	2300-2700	0.71	1633-1917	85
Silicon carbide	2730	3210	0.75	2407.5	3.6
Taconite	1538	3200	0.8	2560	1.0-2.0
Cast iron	1150	7200	0.54	3888	42-55

Table 1. Properties of solid material suitable for thermal energy storage application

There are many suitable PCM for relatively high temperature thermal storage purposes, with melting temperatures ranging from below 100 Celsius up to several hundred Celsius. Paraffin is a typical PCM used for low-grade heat storage. For concentrated high temperature solar thermal energy storage, alloy and molten salts are often considered. Table 2 includes the properties of some alloys and salts (Zalba et al., 2003).

Medium	ρ kg/m³	Heat of fusion (kJ/kg)	Melting temperature (°C)
Mg/Cu/Ca (52/25/23)	2000	184	453
Mg/Cu/Zn (60/25/15)	2800	254	452
Al/Cu/Mg/Zn (54/22/18/6)	3140	305	520
Al/Mg/Zn (59/35/6)	2380	310	443
Al/Cu/Si (65/30/5)	2730	422	571
$NaNO_3$	2260	172	307
KNO_3	2110	266	333
K_2CO_3/Na_2CO_3(51/49)	2400	163	710
CaCl/NaCl(67/33)	2160	281	500

Table 2. Properties of alloy and salt PCM for thermal energy storage application

3.2 Heat transfer fluids for thermal storage application

Heat transport fluid (HTF), either used for direct fluid storage or as a heat-carrying medium, must have favorable properties for heat transfer and at the same time, must be stable and have a low vapor pressure. Molten salts and oils are the two major types of heat transport fluids developed so far.

Most molten salts have low vapor pressures, which is an advantage to the large quantity storage of thermal energy as no pressurized tanks are needed. However, many molten salts

degradation is inevitable due to the heat transfer between the solid thermal storage material and the heat transfer fluid, even if initially the solid thermal storage material is fully charged, or its temperature is exactly equal to T_H.

Considering the need of heat transfer fluid in a power plant, it is always important that during the required operational period of time, $t_{ref,discharge}$, the temperature of the heat transfer fluid have minimum or no degradation from the temperature the fluid is stored at. To meet this requirement in an actual thermocline storage system, one needs to first store a sufficient amount of energy (more than the ideal amount) in the tank. This requires a storage tank to have a sufficiently large thermal energy storage capacity as well as a sufficiently long charge time that allows heat to be charged to the tank. Giving this requirement as a mathematical expression,it is such that:

$$\left\{\left[\rho_s C_s(1-\varepsilon)+\rho_f C_f \varepsilon\right] V_{real}\right\} > \left[\left(\rho_f C_f\right) V_{ideal}\right] \tag{3}$$

In engineering reality, one needs to know, specifically, how large the real thermal storage volume, V_{real}, is and how long a charging time is necessary, if the assumed operation time period of a power plant is $t_{ref,discharge}$. This must be addressed through mathematical analysis.

In the following section, the modeling of the heat transfer and energy transport between the solar thermal storage material and the heat transfer fluid will be described. The goal of the modeling analysis is to predict the size of the storage tank and the period of time required to charge the tank for a given subsequent period of heat discharge from the system, within which minimum or no temperature degradation must be maintained.

5. Analysis of transient heat transfer and energy transport in a packed bed

5.1 Thermal storage/delivery process explained

The following qualitative analysis helps readers better understand the behavior of the fluid temperature variation from a thermocline tank during a discharging operation. When a tank of volume V_{ideal} is filled with thermal storage material at a void fraction of ε, the volume for heat transfer fluid in the tank will be εV_{ideal}, and the volume of the primary thermal storage material must be the remaining portion, $(1-\varepsilon)V_{ideal}$. Due to the existence of solid filler material in the tank, the heat transfer fluid velocity in the charge/discharge processes is higher than that in an ideal thermocline tank as the same mass flow rate of heat transfer fluid is considered. The mass flow rate is a condition determined by the required power output of the power plant, regardless of what type of thermal storage system is used.

Assume that a thermocline tank is initially fully charged. During a thermal discharge process the temperature of the fluid flowing out from top will decrease after a time when the pre-existing hot fluid in the tank is completely discharged; from then on, the hot fluid discharged out from the top is originated from cold fluid that picks up energy from the solid material during the discharging process. The longer the discharge process progresses, the more the temperature of the discharged fluid will decrease. This scenario is illustrated in Fig. 4.

In order to avoid the temperature degradation as show in Fig. 4, one either needs to use an ideal thermal storage system or to have a thermocline system that stores much more energy than is needed so that during the required time period, $t_{ref,discharge}$, the temperature degradation of the discharged fluid is minimal or ignorable.

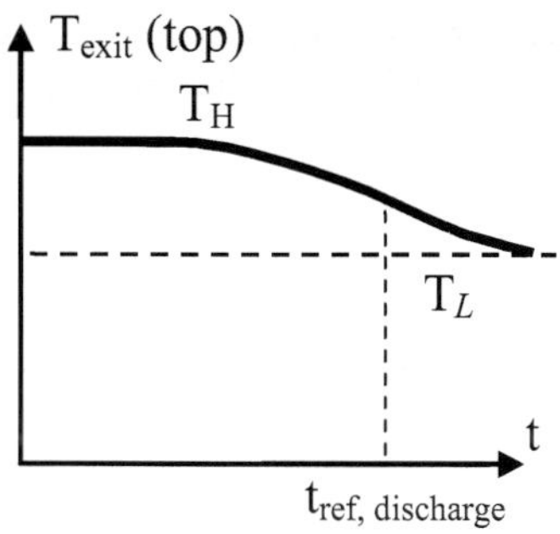

Fig. 4. The temperature variation of heat transfer fluid during discharge from a tank having filler material

5.2 A generalized mathematical modeling

In general, the thermocline thermal storage systems shown in Fig. 2(a) and 2(b) can be regarded as systems that have porous medium, through which a fluid flows and heat transfer between the solid and fluid occurs. Therefore, generalized governing equations for the energy balance in the solid material and fluid can be constructed. As a representative case of a tank filled with porous medium, the system shown in Fig. 2(a) is subjected to analysis for the objective of formulating an analytical model.

Shown in Fig. 5 is a one-dimensional control volume of an element dz in the packed bed. For convenience in analysis, the positive direction of coordinate z is set to be always identical to the fluid flow direction. In the energy charge process hot fluid flows into the tank from the top, and thus $z = 0$ is at the top of the tank. During heat discharge process, cold fluid flows into the tank from bottom to extract heat from the solid material, and this makes $z = 0$ at the bottom of the tank.

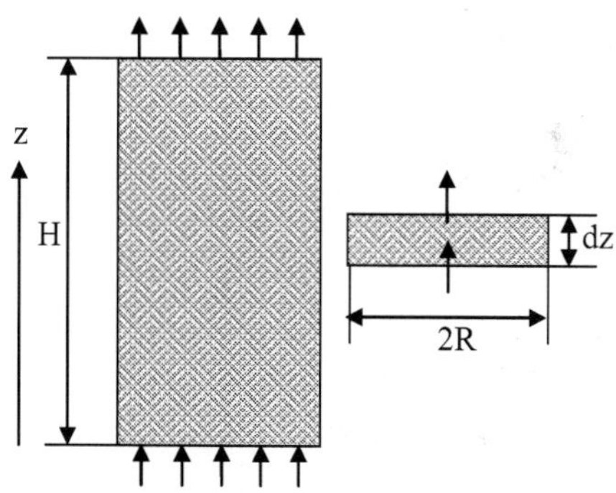

Fig. 5. Schematic of a packed-bed thermal storage system and a control volume for analysis

Several modeling assumptions are made to reasonably simplify the analysis of heat transfer between the heat transfer fluid and the solid packing material:

1. There is a uniform radial distribution of fluid flow and filler material throughout the storage tank. This allows the model to be one-dimensional, only in the z direction.

2. Assume that the particles of filler material have only point contact and therefore axial heat conduction between layers of filler material is negligible.
3. Heat conduction in the axial direction in the fluid is negligible compared to the convective heat transfer.
4. The lumped heat capacitance method is applied to the transient heat conduction in the filler material (particles of size of 0.25-5.0 cm in nominal diameter). When this method is inadequate, due to the large size of solid filler material, a modified lumped capacitance method will be used, which introduces a modified heat transfer coefficient for the convection heat transfer between the fluid and the solid filler material.
5. Assume that there is no heat loss from the storage tank to the surroundings. This assumption applies to both the processes of energy charge and discharge, as well as the resting time between a charge and a discharge.

Assumption (3) is valid when the Peclet number (=RePr) in the HTF is sufficiently large, which is satisfied for most thermal energy storage applications (Kays, 2005). Assumption (4) is valid when the Biot number $(= hL_p / k_s)$ for the thermal storage material is sufficiently small (Incropera, 2002). If the Biot number is large, a correction to the heat transfer accounting for the effects of an internal temperature gradient in the filler material will be considered. Heat loss from a thermal storage tank is inevitable and should be considered. However, from the design point of view, one needs to first decide on the dimensions of a storage tank in order to find the heat loss. To compensate for the heat loss from the tank, a larger volume heat storage tank and a longer heat charge period may be adopted. A simple way of refining this design is to increase both the heat charge time and tank size with a factor that is equal to the ratio of heat loss versus the projected heat delivery. To focus on the main issues, the current work determines the dimensions of a storage tank without considering heat loss. The assumption of no heat loss to the surroundings also provides a basis for using the results from a heat charge process as the initial condition of the following discharge process, and visa versa. By using the end results of one process as the initial conditions of the following process, multiple cyclic energy charges and discharges in the actual operation can be simulated relatively easily.

5.2.1 Energy balance in the heat transfer fluid

Based upon the above modeling assumption (1), the cross-sectional area of the tank seen by the fluid flow is assumed constant at all locations along the axis of the tank, which gives:

$$a_f = \varepsilon \pi R^2 \tag{4}$$

The thermal energy balance of the fluid in the control volume dz is:

$$\rho_f \varepsilon \pi R^2 U(\hbar_z - \hbar_{z+dz}) + hS_s(T_s - T_f)dz = \rho_f C_f \varepsilon \pi R^2 dz \frac{\partial T_f}{\partial t} \tag{5}$$

where the parameter S_s denotes the heat transfer surface area between the filler material and the heat transfer fluid per unit length of the tank; U is the actual fluid velocity in the packed bed:

$$U = \frac{\dot{m}}{\rho_f a_f} \tag{6}$$

The heat transfer coefficient h in Eq. (5) is for the convection between the heat transfer fluid and the packing material. It can be different depending on the flow, packing condition of thermal storage material, fluid properties, and the interaction between packed material and heat transfer fluid, such as is shown in Fig. 2(a) and 2(b). Detailed discussions of S_s and h will be presented following the modeling work.

Using the definition of enthalpy change and a Taylor's series expansion, $\hbar_{z+dz} - \hbar_z = C_f(\partial T_f / \partial z)dz$, the energy balance equation for the heat transfer fluid becomes:

$$\frac{hS_s}{\rho_f C_f \varepsilon \pi R^2}(T_s - T_f) = \frac{\partial T_f}{\partial t} + U\frac{\partial T_f}{\partial z} \tag{7}$$

Introducing the following dimensionless variables,

$$\theta_f = (T_f - T_L)/(T_H - T_L) \tag{8.a}$$

$$\theta_s = (T_s - T_L)/(T_H - T_L) \tag{8.b}$$

$$z^* = z / H \tag{8.c}$$

$$t^* = t / (H / U) \tag{8.d}$$

The dimensionless governing equation for the heat transfer fluid is finally reduced to:

$$\frac{\partial \theta_f}{\partial t^*} + \frac{\partial \theta_f}{\partial z^*} = \frac{1}{\tau_r}(\theta_s - \theta_f) \tag{9}$$

where

$$\tau_r = \frac{U}{H}\frac{\rho_f C_f \varepsilon \pi R^2}{hS_s} = \frac{C_f \dot{m}}{HhS_s} \tag{10}$$

The boundary condition for Eq. (9) is from the fluid inlet temperature, while the initial condition is the temperature distribution in a tank before a charge or a discharge starts.

5.2.2 Energy balance in solid thermal storage material

For the energy balance of the filler material in a control volume dz as shown in Fig. 5, it is understood that the filler material delivers or takes heat to or from the passing fluid at the cost of a change in the internal energy of the filler. The energy balance equation is:

$$hS_s(T_s - T_f)dz = -\rho_s C_s(1-\varepsilon)\pi R^2 dz\frac{\partial T_s}{\partial t} \tag{11}$$

By substituting in the dimensionless variables given in Eq. (8), the above governing equation for filler material is reduced to:

$$\frac{\partial \theta_s}{\partial t^*} = -\frac{H_{CR}}{\tau_r}(\theta_s - \theta_f) \tag{12}$$

where

$$H_{CR} = \frac{\rho_f C_f \varepsilon}{\rho_s C_s (1-\varepsilon)} \tag{13}$$

In the energy charge and discharge processes, the filler material and heat transfer fluid will have a temperature difference at any local location. Once the fluid comes to rest upon the completion of a charge or discharge process, the fluid will equilibrate with the local filler material to reach the same temperature, T_{final}. The energy balance of this situation at a local location is:

$$\varepsilon \rho_f C_f T_{f-initial} + (1-\varepsilon)\rho_s C_s T_{s-initial} = \varepsilon \rho_f C_f T_{final} + (1-\varepsilon)\rho_s C_s T_{final} \tag{14}$$

Here, the initial temperatures of primary thermal storage material and HTF are from the results of their respective charge or discharge processes. The final temperatures of the storage material and the fluid are the same after their thermal equilibrium is reached.
According to the assumption of no heat loss from the storage tank, it can be seen that the equilibrium temperature at the end of one process (charge or discharge) will necessarily be the initial condition of the next process in the cycle. This connects the discharge and charge processes so that overall periodic results can be obtained.
The initial temperatures of filler material and fluid in the storage tank should be known. Also, the inlet fluid temperature is known as a basic boundary condition, with which the filler temperature at inlet location z=0 can be easily solved mathematically from Eq. (12).

5.2.3 Energy delivery efficiency

With the solution of the governing equations for filler material and HTF, the discharged fluid temperature from a storage tank can be obtained. With the required heat discharge period being given as $t_{ref,discharge}$, an energy delivery effectiveness can be obtained from Eq.(2) as discussed before. For convenience of expression, the dimensionless form of the required time period of energy discharge is defined as:

$$\Pi_d = \frac{t_{\text{ref,discharge}}}{H/U} \tag{15}$$

Similarly, a dimensionless form of the time period of energy charge is defined as:

$$\Pi_c = \frac{t_{\text{charge}}}{H/U} \tag{16}$$

Substituting the dimensionless energy discharge period Π_d into Eq. (2), we obtain:

$$\eta = \frac{1}{\Pi_d} \int_0^{\Pi_d} \theta_{f(z^*=1,t^*)} dt^* \tag{17}$$

The energy discharge efficiency will obviously be affected by how much energy is charged into the storage tank. Therefore, it should be noted that η is essentially the function of the following four parameters— Π_c / Π_d, Π_d, τ_r, and H_{CR}. By specifying the dimensionless time period of the discharge process and the mass flow rate, the dimensionless time period of the energy charge can be determined to achieve the objective value of η, which is always desired to approach as close as possible to 1.0.

As has been discussed, a longer energy charging time than energy discharging time is needed in order to achieve an energy delivery effectiveness of approximately 1.0 in a packed-bed system. In addition, the energy storage capacity of a packed-bed tank must be larger than that of an ideal thermal storage tank, as expressed in Eq. (3).

5.2.4 Heat transfer area S_s and heat transfer coefficient h in different types of storage systems

As has been discussed above, the governing equations for the temperatures and energy exchange between the primary thermal storage material and the HTF are generally the same for all the thermal storage systems as schematically shown in Fig. 2. However, the heat transfer coefficients and the heat transfer area between the primary thermal storage material and HTF for different types of storage systems (for example, in Fig. 2(a) and Fig. 2(b)) can be significantly different.

If we consider the use of rocks as a filler material, the heat transfer area between rocks and fluid per unit length of tank was denoted as S_s. Therefore, the unit of S_s is in meters. For spherical filler materials, S_s is obtained through the following steps:

1. The volume of filler material in a unit length Δz of tank is given as $\pi R^2 \Delta z(1-\varepsilon)$.

One sphere of rock has a volume of $V_{sphere} = 4\pi r^3 / 3$, and therefore, in the length of Δz in the tank, the number of rocks is $\pi R^2 \Delta z(1-\varepsilon) / V_{sphere}$. The total surface area of rocks is then determined to be $\pi R^2 \Delta z(1-\varepsilon) \times 4\pi r^2 / V_{sphere}$, which becomes $3\pi R^2 (1-\varepsilon)\Delta z / r$ after V_{sphere} being substituted in.

2. Finally, the heat transfer area of rocks per unit length of tank is:

$$S_s = 3\pi R^2 (1-\varepsilon) / r \tag{18}$$

The above discussion considers the actual volume (assuming ε is known) for solid 'spherical particles' in a packed volume. Depending on the packing scheme, the void fraction ε in a packed bed with spheres of a fixed diameter may range from 0.26 to 0.476 (Conway & Sloane, 1998). The loosest packaging of spherical rocks in a volume is given by the case where each sphere (of diameter $2r$) is packed into a cube with side lengths of $2r$. The densest packing of spheres causes a void fraction of 0.26, which was due to Kepler's conjecture (Hales, 2006). Nevertheless, if the packed bed void fraction ε is known, Eq. (18) should be used for finding S_s.

The heat transfer coefficient h(W/m^2 °C) between the primary thermal storage material (porous media) and HTF can be found from reference (Nellis & Klein, 2009):

$$h = 0.191 \frac{\dot{m} C_f}{\varepsilon \pi R^2} \mathrm{Re}^{-0.278} \mathrm{Pr}^{-2/3} \tag{19}$$

where *Re* is Reynolds number (equal to $4G\, r_{char} / \mu_f$) for porous media, as defined by Nellis (Nellis & Kline, 2009). The mass flux of fluid through the porous bed is G (equal to $\dot{m} / (\varepsilon \pi R^2)$), and r_{char} is defined as the characteristic radius of the filler material (Nellis & Kline, 2009), which is equal to $0.25\varepsilon d_r / (1-\varepsilon)$ for spherical solid filler. Here, d_r is the nominal diameter of a rock, if it is not perfectly spherical.

5.2.5 Modification of lumped capacitance method

The model and equations in section 5.2.1 and 5.2.2 use the lumped capacitance method to determine the heat transfer inside the filler material. This method actually ignores the resistance to heat conduction inside the filler material. This will result in the calculated energy going into, or coming out from, a filler material being higher than that in the actual physical process. It is known (Incropera, 2002) that when the Biot number of the heat transfer of a particle is larger than 0.1, the lumped capacitance assumption will result in increased inaccuracy. In order to correct the lumped capacitance approximation for a spherical 'particle' in fluid, Bradshaw et al. (Bradshaw et al., 1970) and Jeffreson (Jeffreson, 1972) proposed to correct the convective heat transfer coefficient between the solid spherical 'particle' and the fluid. The modified heat transfer coefficient is then used in the equations for the transient temperature in the 'particle' from the standard lumped capacitance method. The modified heat transfer coefficient for a spherical 'particle' is:

$$h_p = h\frac{1}{1+Bi/5} \tag{20}$$

The smaller the Bi number $[=h(d_r/2)/k_s]$ the smaller the correction that is needed. The modified heat transfer coefficient h_p will be used in Eqs. (10), where h should be replaced by h_p when calculating τ_r.

Figure 6 shows a comparison of the results of dimensionless energy (normalized by the ideal maximum energy change of the 'particle') going in or out from a single spherical 'particle' to a fluid during a transient heat transfer. The energy is quantified based on the local change of internal energy in the particle, and the ideal energy is the internal energy change assuming its temperature completely changed from an initially uniform temperature to the fluid temperature around the particle. The Bi of the case shown in Figure 6 is 2.54, which is close

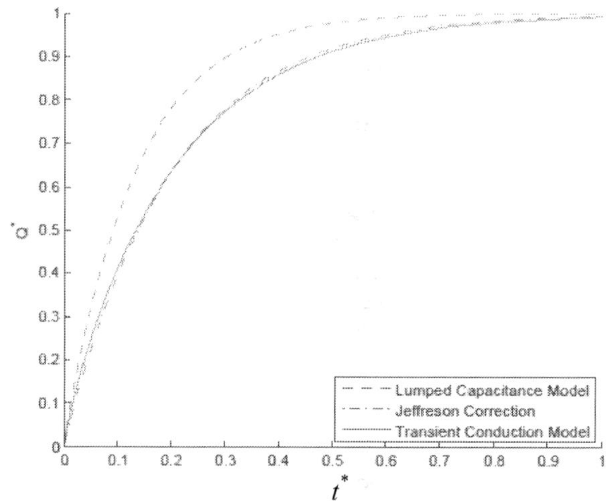

Fig. 6. Jeffreson correction to lumped capacitance heat conduction in a sphere (here only in this figure the dimensionless time is specifically defined as $t^* = t/[(d_r/2)^2/\alpha_s]$)

to the value seen for rocks in typical thermal storage systems. Results from an exact transient heat conduction solution from reference (Incropera, 2002) are compared to the results using the lumped capacitance method, as well as the results obtained using the corrected lumped capacitance method by introducing the modified heat transfer coefficient given in Eq. (20). The heat transfer coefficient used for the comparison was listed in Table 5. Results presented in Fig. 6 indicate that the lumped capacitance method has appreciable discrepancy compared to the exact analytical solution. Results from the Jeffreson correction model agree with the exact transient heat conduction solution very well. Therefore, the Jeffreson correction introducing the modified heat transfer coefficient in the heat transfer model for thermal storage medium and heat transfer fluid is recommended. The Jeffreson correction allows for the thermocline model to remain in a one-dimensional system yet increases the accuracy of the results by accounting for the internal thermal gradient in the packed bed filler material.

Parameters	Value	Unit
d_r	0.04	m
h	355.1	$W/(m^2 \cdot K)$
k_s	2.8	$W/(m \cdot K)$
ρ_s	2630	kg/m^3
C_s	775	$J/(kg \cdot K)$
$\alpha_s = k_s/(\rho_s C_s)$	$1.374 \cdot 10^{-6}$	m^2/s

Table 5. Properties of the solid particle and heat transfer coefficient for the results in Fig. 6

5.2.6 Application of the model to the storage system as shown in Fig. 2(b)

The same governing equations in sections 5.2.1 and 5.2.2 are also applicable to the case shown in Fig. 2(b). However, the porosity in the tank, the heat transfer coefficient h and the heat transfer surface area per unit length, S_s, for the thermal storage tank shown in Fig. 2(b) will be quite different from that in Fig. 2(a). A cross-section of the storage system in Fig. 2(b), including the thermal storage filler material (either solid or liquid) and tubes for the heat transfer fluid, are illustrated in Fig. 7.

An equivalent porosity in the tank is obtained as:

$$\varepsilon = N\frac{d_i^2}{D^2} \tag{21}$$

where N is the number of heat transfer fluid tubes in the storage tank; d_i and D are indicated in Fig. 7.

The heat transfer surface area per unit length in the tank is obtained as:

$$S_s = N\pi d_i \tag{22}$$

The convective heat transfer coefficients inside tubes for either turbulent or laminar fluid flow can be easily obtained from heat transfer textbooks elsewhere. The Reynolds number and Nusselt number for the heat transfer coefficient in tubes are obtained assuming a uniform distribution of total fluid to all the heat transfer fluid tubes.

It is assumed that each heat transfer fluid tube exchanges heat with the storage material in an equivalent control area of diameter D_{eq} as shown in Fig. 7.

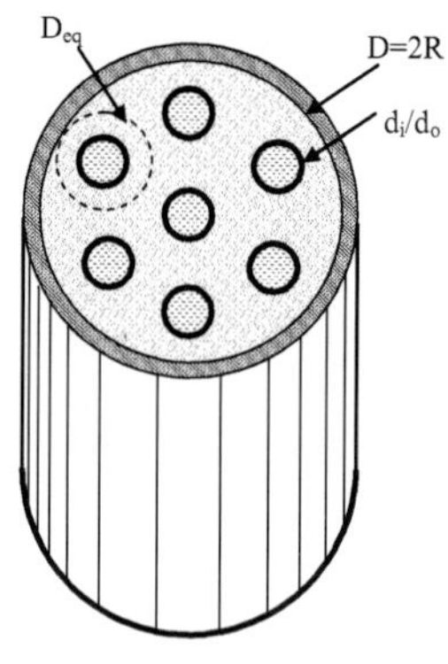

Fig. 7. Schematic of HTF tube and its surrounding thermal storage material (D_{eq} is an equivalent diameter based on the cross sectional area of the container divided by the number of HTF tubes)

The concept of the Jeffreson correction for the lumped capacitance method using a modified heat transfer coefficient is still applicable to the heat conduction in the thermal storage material. However, since the material is not in the form of spherical particles but rather is an integrated solid or liquid, the modified heat transfer coefficient will be different and must be found through analysis.

The following analysis takes one HTF tube and its surrounding thermal storage material for analysis, as shown in Fig. 8. In order to accurately determine the heat going into or out of the filler material in a transient heat conduction process, a transient heat conduction problem is analyzed. This will allow for a comparison of results respectively based on an exact analytical solution, the lumped capacitance method, and the lumped capacitance method with the adoption of a modified heat transfer coefficient.

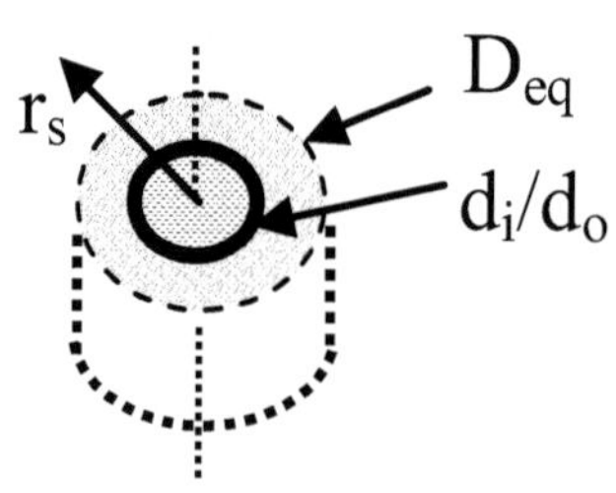

Fig. 8. Physical model of transient heat conduction in a thermally protected "washer"

The analytical solution is for a transient heat conduction problem in a "washer" as shown in Fig. 8. The 'washer' has its outer surface and the two flat surfaces thermally insulated, leaving only the inner surface at $r_s = d_i / 2$ having a convective heat transfer flux.

The mathematical description of this one dimensional problem is

$$\frac{\partial T}{\partial t} = \alpha_s \frac{1}{r_s} \frac{\partial}{\partial r_s} (r_s \frac{\partial T}{\partial r_s}) \tag{23}$$

which is subjected to boundary conditions of:

$$t = 0 \quad T = T_L \tag{24. a}$$

$$t > 0 \text{ at } r_s = d_i / 2;\ k_s \frac{\partial T}{\partial r_s} = h(T - T_f) \tag{24. b}$$

$$t > 0 \text{ at } r_s = D_{eq} / 2;\ \frac{\partial T}{\partial r_s} = 0 \tag{24. c}$$

Solutions for the energy going into the tube based on the standard lumped capacitance method and solutions of the exact analytical model can be easily obtained. Using results from the exact analytical solution, a modified heat transfer coefficient h_p is introduced to correct the lumped capacitance method for the 'washer'.

$$h_p = h\left(\frac{1}{1 + Bi_{washer} / w_c}\right) \tag{25}$$

where Bi_{washer} is a specially defined Biot number $[= h(d_i / 2) / k_s]$ for the "washer"; w_c is a coefficient which depends on the ratio of the washer diameters, D_{eq} / d_i. For a value of $D_{eq} / d_i = 6$ we found w_c = 0.83442 through analysis. Using data for typical thermal storage materials as shown in Table 6, a comparison of results for the dimensionless heat stored in the washer based respectively on the lumped capacitance method, the exact analytical solution, and the modified lumped capacitance method, is shown in Fig. 9. The agreement between the modified lumped capacitance method and the exact analytical solution is very good.

Parameter	Value	Unit
d_i	0.05	m
D_{eq}	0.3	m
h (laminar in tube)	6.88	$W / (m^2 \cdot K)$
k_s	0.57	$W / (m \cdot K)$
ρ_s	1730	kg / m^3
C_s	1.47	$kJ / (kg \cdot K)$
$\alpha_s = k_s / (\rho_s C_s)$	$2.24 \cdot 10^{-7}$	m^2 / s

Table 6. Heat transfer coefficient of HTF-Therminol® VP-1 and properties of primary thermal storage material (molten salt: $Na_2S_2O_3 \cdot 5H_2O$) for the results in Fig. 9

In the energy storage model discussed in sections 5.2.1 and 5.2.2, axial heat conduction in the filler material (solid spherical particles in magnitude of 1.0-10 cm in diameter) is neglected. For integrated solid or liquid thermal storage materials, it is necessary to investigate whether axial heat conduction will be significant. When axial heat conduction in the thermal storage material is considered, the energy balance equation for the thermal storage material is:

$$hS_s(T_s - T_f)dz - (1-\varepsilon)\pi R^2 k_s \frac{\partial^2 T_s}{\partial z^2} dz = -\rho_s C_s (1-\varepsilon)\pi R^2 dz \frac{\partial T_s}{\partial t} \tag{26}$$

Its dimensionless form is:

$$\frac{\partial \theta_s}{\partial t^*} = -\frac{H_{CR}}{\tau_r}(\theta_s - \theta_f) + \frac{k_s}{\rho_s C_s H U}\frac{\partial^2 \theta_s}{\partial z^{*2}} \quad (27)$$

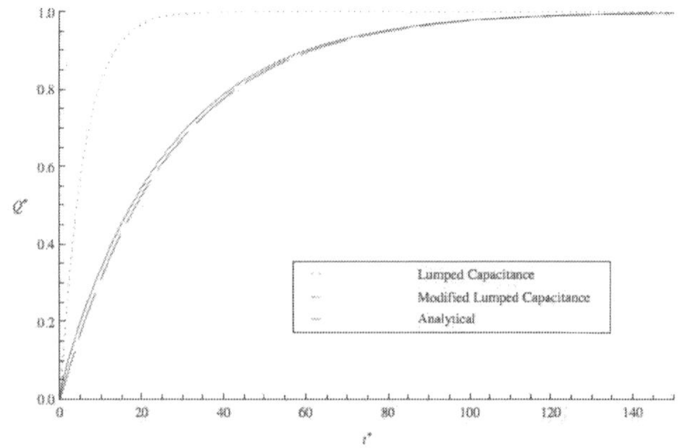

Fig. 9. Comparison of dimensionless energy storage in the 'washer', normalized by the ideal maximum energy change in the 'washer', due to different methods of solution ($Bi_{washer} = h(d_i / 2) / k_s$ =3.0; $D_{eq} / d_i = 6.0$; $t^* = t / [(d_i / 2)^2 / \alpha_s]$; w_c = 0.83442)

The parameter cluster of $k_s / (\rho_s C_s HU)$ is a dimensionless term. If it is sufficiently large, the axial conduction term in Eq. (27) may not be dropped off. A basic effect of significant axial heat conduction is that it will destroy the thermocline effect—a temperature gradient with hot material being on top of cold. This can lower the thermal storage performance in general. Therefore, to take into account the axial heat conduction effect, a similar correction via the introduction of another factor to the modified heat transfer coefficient is proposed. This results in a new modified heat transfer coefficient of:

$$h_p = h\left(\frac{1}{1 + Bi_{washer} / w_c}\right)\left(\frac{1}{1 + \frac{k_s}{\rho_s C_s HU}}\right) \quad (28)$$

For most thermal storage materials, such as rocks, molten salts, concrete, soil, and sands, the value of $k_s / (\rho_s C_s HU)$ is very small (in the order of 1×10^{-6}); while other terms in Eq.(27) are in the order of 1.0. Therefore, the axial heat conduction effect in the thermal storage material in Eq. (27) is negligible.

5.2.7 Application of the model to the storage system with PCM

For thermal storage with phase-change involved, the PCM can be enclosed in capsules to form a packed bed as shown in Fig. 2(a), or simply put in a storage tank that has heat transfer tubes inside as shown in Fig. 2(b). The governing equations discussed above are still

applicable to the heat transfer at locations where either phase change has not yet occurred or has already been completed. However, at locations undergoing phase change, the energy equations must account for the melting or solidification process (Halawa & Saman, 2011; Wu et al., 2011). The key feature in a melting or solidification process is that the temperature of the material stays constant.
Considering the energy balance for the thermal storage material:

$$hS_s(T_m - T_f) = -\Gamma \rho_s (1-\varepsilon)\pi R^2 \frac{d\Phi}{dt} \tag{29}$$

where Γ is the fusion energy of the material, and Φ is the ratio of the liquid mass to the total mass in the control volume of dz. For melting, Φ increases from 0 to 1.0, while for solidification it decreases from 1.0 to 0.
Considering the invariant of the temperature of the material during a phase change process, the energy balance equation for HTF is:

$$\frac{hS_s}{\rho_f C_f \varepsilon \pi R^2}(T_m - T_f) = \frac{\partial T_f}{\partial t} + U\frac{\partial T_f}{\partial z} \tag{30}$$

Equation (29) and (30) can be reduced to dimensionless equations by introducing the same group of dimensionless parameters:

$$\frac{1}{\tau_r}(\theta_m - \theta_f) = \frac{\partial \theta_f}{\partial t^*} + \frac{\partial \theta_f}{\partial z^*} \tag{31}$$

$$-\frac{H_{CR}}{\tau_r}\psi(\theta_m - \theta_f) = \frac{d\Phi}{dt^*} \tag{32}$$

where a new dimensionless parameter $\psi = (T_H - T_L)C_s / \Gamma$ is introduced. Since the phase change temperature is known, Eqs. (31) and (32) can be solved separately.

5.3 Numerical methods and solution to governing equations
5.3.1 Solution for the case of no phase change

A number of analyses and solutions to the heat transfer governing equations of a working fluid flowing through a packed-bed have been presented in the past (Schumann, 1929; Shitzer & Levy, 1983; McMahan, 2006; Beasley, 1984; Zarty & Juddaimi, 1987). As the pioneering work, Schumann (Schumann, 1929) presented a set of equations governing the energy conservation of fluid flow through porous media. Schumann's equations have been widely adopted in the analysis of thermocline heat storage utilizing solid filler material inside a tank. His analysis and solutions were for the special case where there is a fixed fluid temperature at the inlet to the storage system. In most solar thermal storage applications this may not be the actual situation. To overcome this limitation, Shitzer and Levy (Shitzer & Levy, 1983) employed Duhamel's theorem on the basis of Schumann's solution to consider a transient inlet fluid temperature to the storage system. The analysis of Schumann, and Shitzer and Levy, however, still carry with them some limitations. Their method does not consider a non-uniform initial temperature distribution. For a heat storage system, particularly in a solar thermal power plant, heat charge and discharge are cycled daily. The initial temperature field of a heat charge process is dictated by the most recently completed heat discharge process, and vice versa. Therefore, non-uniform and nonlinear temperature distribution is typical for both charge and discharge processes. To consider a non-uniform initial temperature distribution and varying fluid temperature at the inlet in a heat storage system, numerical methods have been deployed by researchers in the past.

To avoid the long mathematical analysis necessary in analytical solutions, numerical methods used to solve the Schumann equations were discussed in the literature by McMahan (McMahan, 2006, 2007), and Pacheco et al. (Pacheco et al., 2002), and demonstrated in the TRNSYS software developed by Kolb and Hassani (Kolb & Hassani, 2006). Based on the regular finite-difference method, McMahan provided both explicit and implicit discretized equations for the Schumann equations. Whereas the explicit solution method had serious stability issues, the implicit solution method encountered an additional computational overhead, thus requiring a dramatic amount of computation time. The solution for the complete power plant with thermocline storage provided by the TRNSYS model in Kolb's work (Kolb & Hassani, 2006) cites the short time step requirement for the differential equations of the thermocline as one major source of computer time consumption. To overcome the problems encountered in the explicit and implicit methods, McMahan et al. also proposed an infinite-NTU method (McMahan, 2006, 2007). This model however is limited to the case in which the heat transfer of the fluid compared to the heat storage in fluid is extremely large.

The present study has approached the governing equations using a different numerical method (Van Lew et al., 2011). The governing equations have been reduced to dimensionless forms which allow for a universal application of the solution. The dimensionless hyperbolic type equations are solved numerically by the method of characteristics. This numerical method overcomes the numerical difficulties encountered in McMahan's work—explicit, implicit, and the restriction on infinite-NTU method (McMahan, 2006, 2007). The current model yields a direct solution to the discretized equations (with no iterative computation needed) and completely eliminates any computational overhead. A grid-independent solution is obtained at a small number of nodes. The method of characteristics and the present numerical solution has proven to be a fast, efficient, and accurate algorithm for the Schumann equations.

The non-dimensional energy balance equations for the heat transfer fluid and filler material can be solved numerically along the characteristics (Courant & Hilbert, 1962; Polyanin, 2002; Ferziger, 1998). Equation (9) can be reduced along the characteristic $t^* = z^*$ so that:

$$\frac{D\theta_f}{Dt^*} = \frac{1}{\tau_r}(\theta_r - \theta_f) \tag{33}$$

Separating and integrating along the characteristic, the equation becomes:

$$\int d\theta_f = \int \frac{1}{\tau_r}(\theta_r - \theta_f)dt^* \tag{34}$$

Similarly, Eq.(12) for the energy balance for the filler material is reposed along the characteristic z^* =constant so that:

$$\frac{d\theta_r}{dt^*} = -\frac{H_{CR}}{\tau_r}(\theta_r - \theta_f) \tag{35}$$

The solution for Eq. (35) is very similar to that for Eq. (33) but with the additional factor of H_{CR}. The term H_{CR} is simply a fractional ratio of fluid heat capacitance to filler heat capacitance. Therefore, the equation for the solution of θ_r will react with a dampened speed when compared to θ_f, as the filler material must have the capacity to store the energy being delivered to it, or vice versa. Finally, separating and integrating along the characteristic for Eq.(35) results in:

$$\int d\theta_r = \int -\frac{H_{CR}}{\tau_r}(\theta_r - \theta_f)dt^* \tag{36}$$

There are now two characteristic equations bound to intersections of time and space. A discretized grid of points, laid over the time-space dimensions will have nodes at these intersecting points. A diagram of these points in a matrix is shown in Fig. 10. In space, there are i = 1, 2, . . . ,M nodes broken up into step sizes of Δz^* to span all of z^*. Similarly, in time, there are j = 1, 2, . . . ,N nodes broken up into time-steps of Δt^* to span all of t^*. Looking at a grid of the ϑ nodes, a clear picture of the solution can arise. To demonstrate a calculation of the solution we can look at a specific point in time, along z^* where there are two points, $\vartheta_{1,1}$ and $\vartheta_{2,1}$. These two points are the starting points of their respective characteristic waves described by Eq. (33) and (36). After the time Δt^* there is a third point $\vartheta_{2,2}$ which has been reached by both wave equations. Therefore, Eq. (34) can be integrated numerically as:

$$\int_{\vartheta_{1,1}}^{\vartheta_{2,2}} d\theta_f = \int_{\vartheta_{1,1}}^{\vartheta_{2,2}} \frac{1}{\tau_r}(\theta_r - \theta_f)dt^* \tag{37}$$

The numerical integration of the right hand side is performed via the trapezoidal rule and the solution is:

$$\theta_{f_{2,2}} - \theta_{f_{1,1}} = \frac{1}{\tau_r}\left(\frac{\theta_{r_{2,2}} + \theta_{r_{1,1}}}{2} - \frac{\theta_{f_{2,2}} + \theta_{f_{1,1}}}{2}\right)\Delta t^* \tag{38}$$

where $\theta_{f_{1,1}}$ is the value of θ_f at $\vartheta_{1,1}$, and $\theta_{f_{2,2}}$ is the value of θ_f at $\vartheta_{2,2}$, and similarly so for θ_r.

The integration for Eq. (36) along z^* =constant is:

$$\int_{\vartheta_{2,1}}^{\vartheta_{2,2}} d\theta_r = \int_{\vartheta_{2,1}}^{\vartheta_{2,2}} [-\frac{H_{CR}}{\tau_r}(\theta_r - \theta_f)]dt^* \tag{39}$$

The numerical integration of the right hand side is also performed via the trapezoidal rule and the solution is:

$$\theta_{r_{2,2}} - \theta_{r_{2,1}} = -\frac{H_{CR}}{\tau_r}\left(\frac{\theta_{r_{2,2}} + \theta_{r_{2,1}}}{2} - \frac{\theta_{f_{2,2}} + \theta_{f_{2,1}}}{2}\right)\Delta t^* \tag{40}$$

Equations (38) and (40) can be reposed as a system of algebraic equations for two unknowns, $\theta_{f_{2,2}}$ and $\theta_{r_{2,2}}$, while θ_f and θ_r at grid points $\vartheta_{1,1}$ and $\vartheta_{2,1}$ are known.

$$\begin{bmatrix} 1+\dfrac{\Delta t^*}{2\tau_r} & -\dfrac{\Delta t^*}{2\tau_r} \\ -\dfrac{H_{CR}\Delta t^*}{2\tau_r} & 1+\dfrac{H_{CR}\Delta t^*}{2\tau_r} \end{bmatrix} \begin{bmatrix} \theta_{f_{2,2}} \\ \theta_{r_{2,2}} \end{bmatrix} = \begin{bmatrix} \theta_{f_{1,1}}\left(1-\dfrac{\Delta t^*}{2\tau_r}\right)+\theta_{r_{1,1}}\dfrac{\Delta t^*}{2\tau_r} \\ \theta_{f_{2,1}}\left(\dfrac{H_{CR}\Delta t^*}{2\tau_r}\right)+\theta_{r_{2,1}}\left(1-\dfrac{H_{CR}\Delta t^*}{2\tau_r}\right) \end{bmatrix} \quad (41)$$

Cramer's rule (Ferziger, 1998) can be applied to obtain the solution efficiently. It is important to note that all coefficients/terms in Eq.(41) are independent of z^*, t^*, θ_f, and θ_r, thus they can be evaluated once for all. Therefore, the numerical computation takes a minimum of computing time, and is much more efficient than the method applied in references (McMahan, 2006, 2007).

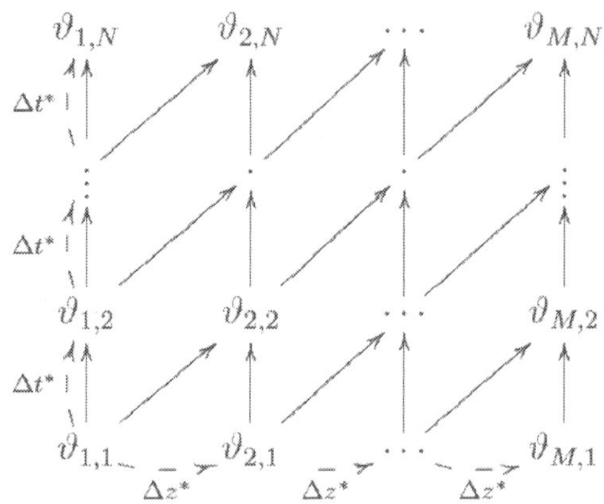

Fig. 10. Diagram of the solution matrix arising from the method of characteristics

From the grid matrix in Fig.10 it is seen that the temperatures of the filler and fluid at grids $\vartheta_{i,1}$ are the initial conditions. The temperatures of the fluid and filler at grid $\vartheta_{1,1}$ are the inlet conditions which vary with time. The inlet temperature for the fluid versus time is given. The filler temperature (as a function of time) at the inlet can be easily obtained using Eq.(12), for which the inlet fluid temperature is known. Now, as the conditions at $\vartheta_{1,1}$, $\vartheta_{1,2}$, and $\vartheta_{2,1}$ are known, the temperatures of the rocks and fluid at $\vartheta_{2,2}$ will be easily calculated from Eq.(41).

Extending the above sample calculation to all points in the ϑ grid of time and space will give the entire matrix of solutions in time and space for both the rocks and fluid. While the march of Δz^* steps is limited to $z^* = 1$ the march of time Δt^* has no limitation.

The above numerical integrations used the trapezoidal rule; the error of such an implementation is not straightforwardly analyzed but the formal accuracy is on the order of $O(\Delta t^{*2})$ for functions (Ferziger, 1998) such as those solved in this study.

5.3.2 Solutions for the case with phase change

For the governing equations of the phase change case, the adopted convention of having the z-direction coordinate always follow the flow direction is preserved, such that for heat charging, z=0 is for the top of a tank, and for heat discharging, z=0 is for the bottom of a tank. The two governing equations (Eq. (31) and Eq.(32)) for the phase change process can be discretized using finite control volume methodology:

$$\frac{1}{\tau_r}(\theta_m - \theta_{f(i)}^{t^*+\Delta t^*}) = \frac{\theta_{f(i)}^{t^*+\Delta t^*} - \theta_{f(i)}^{t^*}}{\Delta t^*} + \frac{\theta_{f(i)}^{t^*+\Delta t^*} - \theta_{f(i-1)}^{t^*+\Delta t^*}}{\Delta z^*} \tag{42}$$

$$-\frac{H_{CR}}{\tau_r}\psi(\theta_m - \theta_{f(i)}^{t^*+\Delta t^*}) = \frac{\Phi_i^{t^*+\Delta t^*} - \Phi_i^{t^*}}{\Delta t^*} \tag{43}$$

From Eq.(42) the fluid temperature $\theta_{f(i)}^{t^*+\Delta t^*}$ can be solved, which is then used in Eq. (43) to solve for the fusion ratio $\Phi_i^{t^*+\Delta t^*}$.

The procedures for finding the solution of phase change problem are as follows:

1. Solve the non-phase-change governing equation analytically using Eq.(12) for the phase change material for the inlet point.
2. Monitor the temperature at each time step as given by Eq.(12), and see if the temperature at a time step is greater than the fusion temperature, if yes, the solution for that and subsequent time steps are to be solved using the phase change equation (Eq. (43)
3. For each time step solved using Eq (43), monitor the fusion ratio, Φ; when it becomes larger than 1.0 then the solution for that and subsequent time steps are to be solved using the non-phase-change governing equation (Eq.(12)) for the remainder of the required time.
4. March a spatial step forward and repeat all of the above steps. However, now in part (1) of this procedure, Eq.(41) must be used to solve the temperatures of both the fluid and PCM for time steps before phase change starts; and also in part (3) of this procedure Eq.(41) should be used to solve the temperatures of the fluid and PCM for steps after the phase change is over. The repetition of parts (1) to (3) of this procedure is to be continued until all the spatial steps are covered.

6. Results from simulations and experimental tests

6.1 Numerical results for the temperature variation in a packed bed

The first analysis of the storage system was done on a single tank configuration of a chosen geometry, using a filler and fluid with given thermodynamic properties. The advantage of having the governing equations reduced to their dimensionless form is that by finding the values of two dimensionless parameters (τ_r and H_{CR}) all the necessary information about the problem is known. The properties of the fluid and filler rocks, as well as the tank dimensions, which determined τ_r and H_{CR} for the example problem, are summarized in Table 7.

The numerical computation started from a discharge process assuming initial conditions of an ideally charged tank with the fluid and rocks both having the same high temperature throughout the entire tank, i.e. $\theta_f = \theta_s = 1$. After the heat discharge, the temperature distribution in the tank is taken as the initial condition of the following charge process. The discharge and charge time were each set to 4 hours. The fluid mass flow rate was determined such that an empty (no filler) tank was sure to be filled by the fluid in 4 hours.

With the current configuration, after five discharge and charge cycles the results of all subsequent discharge processes were identical—likewise for the charge processes. It is therefore assumed that the solution is then independent of the first-initial condition. The data presented in the following portions of this section are the results from the cyclic discharge and charge processes after 5 cycles.

ε	τ_r	H_{CR}	H	R	t
0.25	0.0152	0.3051	14.6	m	7.3 m
Fluid (Therminol® VP-1) properties:					
T_H=395 °C;		T_L=310 °C;	ρ_f =753.75 kg/m³;		C_f=2474.5 J/(kg K);
k_f=0.086W/(m K);		$\dot{m}$ =128.74 kg/s;		μ_f =1.8×10⁻⁴ $Pa \cdot s$;	
Filler material (granite rocks) properties:					
ρ_s = 2630 kg/m3;	C_s=775 J/(kg K);		k_s=2.8 W/(m K);		d_r= 0.04 m;

Table 7. Dimensions and parameters of a thermocline tank (Van Lew et al., 2011)

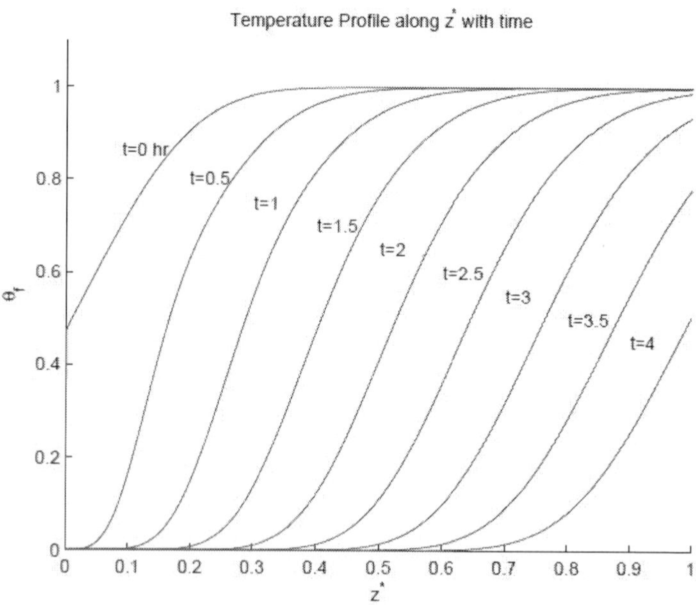

Fig. 11. Dimensionless fluid temperature profile in the tank for every 0.5 hours

Shown in Fig. 11 are the temperature profiles in the tank during a discharge process, in which cold fluid enters into the tank from bottom of the tank. The location of $z^* = 0$ is at the bottom of a tank for a discharge process. The temperature profile evolves as discharging proceeds, showing the heat wave propagation and the high temperature fluid moving out of the storage tank. The fluid temperature at the exit ($z^* = 1$) of the tank gradually decreases after 3 hours of discharge. At the end of the discharge process, the temperature distribution along the tank is shown in Fig. 12. At this time the fluid and rock temperatures, θ_f and θ_s respectively (θ_s is denoted by θ_r when the filler material is rock), in the region with z^* below 0.7 are almost zero, which means that the heat in the rocks in this region has been completely extracted by the passing fluid. In the region from $z^* = 0.7$ to $z^* = 1.0$ the temperature of the fluid and rock gradually becomes higher, which indicates that some heat has remained in the tank.

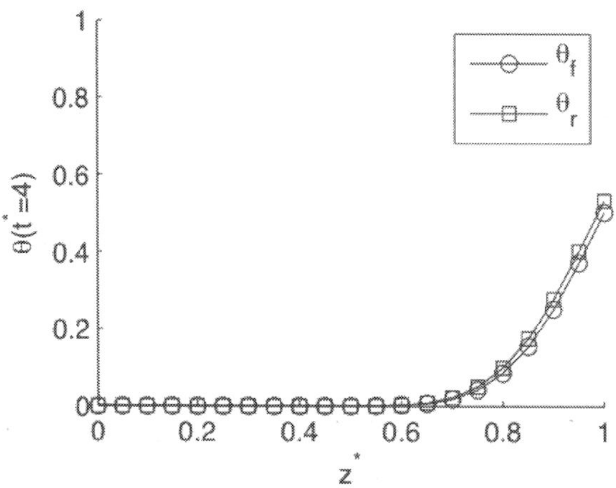

Fig. 12. Dimensionless temperature distribution in the tank after time $t^* = 4$ of discharge (Here θ_r is used to denote θ_s, as rocks are used as the storage material in the example).

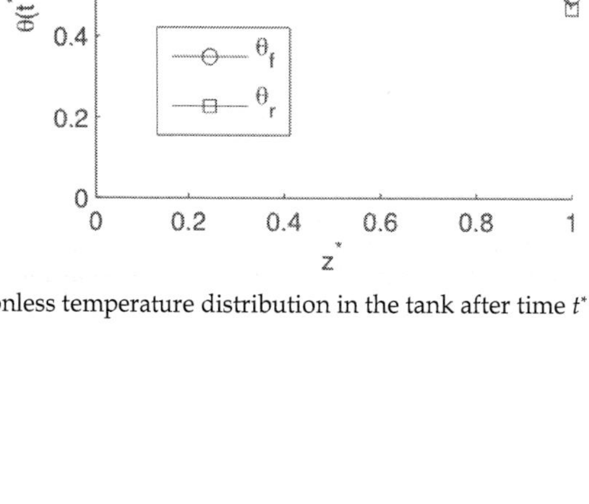

Fig. 13. Dimensionless temperature distribution in the tank after time $t^* = 4$ of charge

(Here θ_r is used to denote θ_s, as rocks are used as the storage material in the example)
A heat charge process exhibits a similar heat wave propagation scenario. The temperature for the filler and fluid along the flow direction is shown in Fig. 13 after a 4 hour charging process. During a charge process, fluid flows into the tank from the top, where z^* is set as zero. It is seen that for the bottom region (z^* from 0.7 to 1.0) the temperatures of the fluid and rocks decrease significantly. A slight temperature difference between heat transfer fluid and rocks also exists in this region.

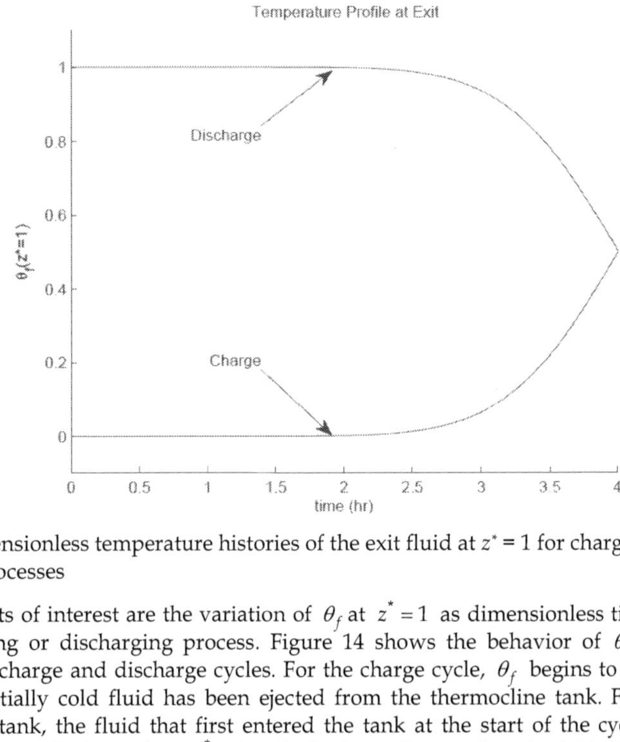

Fig. 14. Dimensionless temperature histories of the exit fluid at $z^* = 1$ for charge and discharge processes

The next plots of interest are the variation of θ_f at $z^* = 1$ as dimensionless time progresses for a charging or discharging process. Figure 14 shows the behavior of θ_f at the outlet during both charge and discharge cycles. For the charge cycle, θ_f begins to increase when all of the initially cold fluid has been ejected from the thermocline tank. For the present thermocline tank, the fluid that first entered the tank at the start of the cycle has moved completely through the tank at $t^* = 1$, which also indicates that the initially-existing cold fluid of the tank has been ejected from the tank. Similarly, during the discharge cycle, after the initially-existing hot fluid in the tank has been ejected, the cold fluid that first entered the tank from the bottom at the start of the cycle has moved completely through the tank at $t^* = 1$. At $t^* = 2.5$, or t=2.5 hours, the fluid temperature θ_f starts to drop. This is because the energy from the rock bed has been significantly depleted and incoming cold fluid no longer can be heated to $\theta_f = 1$ by the time it exits the storage tank.
The above numerical results agree with the expected scenario as described in section 4. To validate the above numerical method, analytical solutions were obtained using a Laplace Transform method by the current authors (Karaki, et al, 2010), which were only possible for cases with a constant inlet fluid temperature and a simple initial temperature profile. Results compared in Fig. 15 are obtained under the same operational conditions—starting from a

fully charged initial state and run for 5 iterations of cyclic discharge and charge processes. The fluid temperature distribution along the tank (z^*=0 for bottom of the tank) from numerical results agrees with the analytical results very well. This comparison essentially proves the effectiveness and reliability of the numerical method developed in the present study.

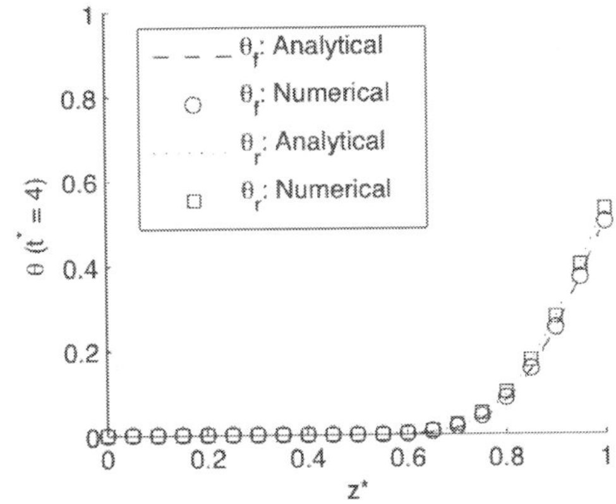

Fig. 15. Comparison of numerical and analytical results of the temperature distribution in the tank after time t^* = 4 of a discharge (Here θ_r is used to denote θ_s, as rocks are used as the storage material in the example)

Based on the results shown in Fig. 15, the temperature distribution along z^* at the end of a charge is nonlinear. This distribution will be the initial condition for the next discharge cycle. Similarly a discharge process will result in a nonlinear temperature distribution, which will be the initial condition for the next charge. It is evident that the analytical solutions developed by Schumann (Schumann, 1929) could not handle this type of situation.

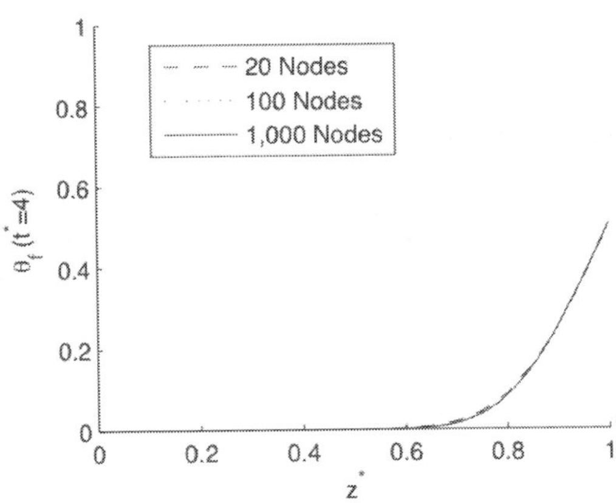

Fig. 16. Comparison of dimensionless temperature distributions in the tank after time t^* = 4 of discharge for different numbers of discretized nodes

Another special comparison was made to demonstrate the efficiency of the method of characteristics at solving the dimensionless form of the governing equations. Shown in Fig. 16 are the temperature profiles at $t^* = 4$ obtained by using different numbers of nodes (20, 100, and 1000) for z^*. The high level of accuracy of the current numerical method, even with only 20 nodes, demonstrates the accuracy and stability of the method with minimal computing time.

6.2 Comparison of modeling results with experimental data from literature

6.2.1 Temperature variations in charge processes

The authors have conducted experimental tests (Karaki at al., 2011). The test conditions for a given heat charge process are listed in Table 8, which also shows the dimensionless parameters. As shown in Fig. 17, at the initial time the thermocline tank has a uniform temperature equal to room temperature. The temperature readings from the thermocouples at the top of the tank provide the inlet fluid temperatures in a charge process.

Tank Length	0.65(m)	Initial temperature	21.9 (°C)			
Tank inner diameter	0.241 (m)	High temperature	79.82 (°C)			
Rock nominal diameter	0.01 (m)	Low temperature	21.9 (°C)			
Oil flow rate	1.0 (Liter/min.)	Porosity	0.324			
Density of rocks	2632.8 kg/m³	$\Pi_c = t_c / (H / U)$	7.2511			
Charging period t_c	70.08 (min.)	τ_r	0.1262			
Rock heat capacity	790 (J/(kg K))	H_{CR}	0.4068			
Time (s)	0	856	1713	2570	3427	4284
Dimensionless time t^*	0	1.45	2.9	4.35	5.80	7.25

Table 8. Conditions of a heat charging test (HTF is Xceltherm 600 by Radco Industries)

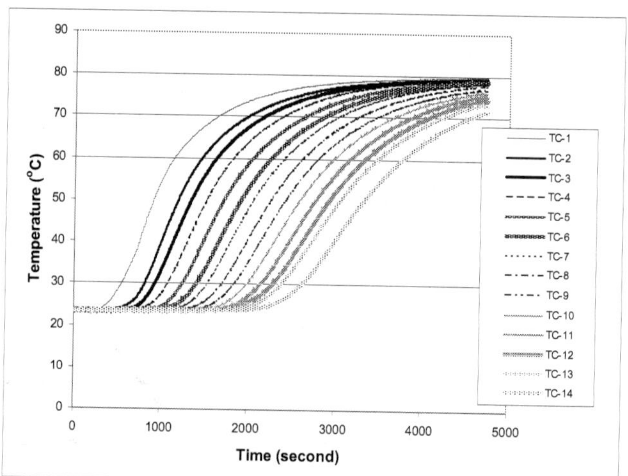

Fig. 17. Temperatures in the center of the tank along the height of 65 cm for a charging process (Thermocouples were set every 5cm)

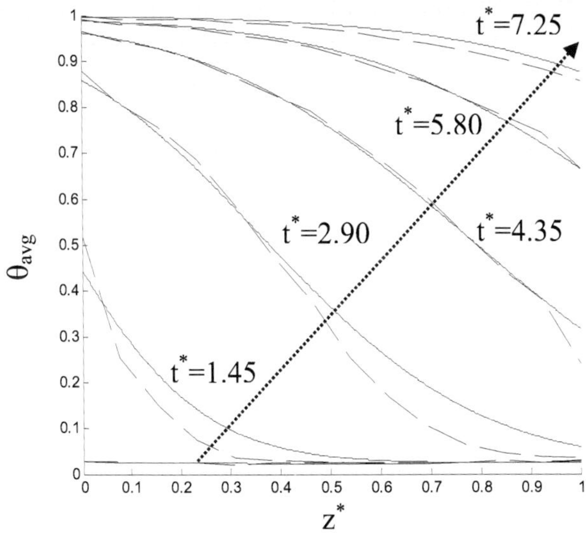

Fig. 18. The temperature distribution along the height in the tank at different time points (Solid lines are from simulation results and dashed lines are from experimental tests)

Based on the temperature measurements, the temperature in the tank increases gradually and at the end of the charging process, the temperatures at all the locations are sufficiently high. The temperature distribution along the height in the tank at different times is shown in Fig. 18. Obviously at the end of the charge, the temperature at the top of the tank (at $z^*=0$) is high. Using the initial temperature distribution and the inlet fluid temperature, together with the properties listed in Table 8, numerical simulation results were obtained and are shown in Fig. 18 for the average temperature of the fluid and rocks. The real time and the dimensionless time are listed in Table 8. The agreement between the experimental data and the modeling simulation is very satisfactory.

The thermal storage performance test results of a thermocline tank reported in the literature (Pacheco et al., 2002) was also referenced to validate the current modeling work. The experimental tests used eutectic molten salt ($NaNO_3$-KNO_3, 50% by 50%) as the heat transfer fluid and quartzite rocks and silica sands as the filler material. Thermocouples in the test apparatus were imbedded in the packed-bed. Temperatures in the tank at different height locations were recorded during a two-hour heat discharge process after the tank was charged for the same length of time. The storage tank dimensions, packed-bed porosity, and properties of the fluid and filler material are listed in Table 9.

Using the modeling of section 5, the heat charge followed by heat discharge was simulated. In Fig. 19, the predicted temperatures at several height locations of the tank at different time instances during the discharge process were compared to the experimental data reported (Pacheco et al., 2002). The trend of temperature curves from the modeling prediction and the experiment is quite consistent. Considering the uncertainties in the experimental test and the properties of materials considered, the agreement between the experimental data and the modeling prediction is quite satisfactory. This comparasion firmly validates the current modeling and its numerical solution method.

ε	τ_r	H_{CR}	H	R	t
0.22	0.0041	0.2733	6.1m	1.5m	2 hr
Molten salt $NaNO_3$-KNO_3 properties:					
T_H=396 °C;	T_L=290 °C;	ρ_f =1733 kg/m³;	C_f=1550 J/(kg K);		
k_f=0.57 W/(m K);	$\dot{m}$ =7.0 kg/s;	μ_f = 0.0021 $Pa \cdot s$;			
Quartzite rocks/sands mixture properties:					
ρ_s = 2640 kg/m3;	C_s=1050 J/(kg K);	k_s=2.5 W/(m K);	d_r= 0.015 m;		

Table 9. Dimensions and parameters of a thermocline tank for the test in literature (Pacheco et al., 2002)

Tank Length measured	0.65(m)		Initial temperature		129.0 (°C)	
Tank inner diameter	0.241 (m)		High temperature		129.0 (°C)	
Rock nominal diameter	0.01 (m)		Low temperature		56.0 (°C)	
Oil flow rate	1.96 (Liter/min.)		Porosity		0.324	
Density of rocks	2632.8 kg/m³		$\Pi_d = t_d / (H / U)$		8.0596	
Charging period t_c	39.74 (min)		τ_r		0.1044	
Rock heat capacity	790 (J/(kg K))		H_{CR}		0.4210	
Time (s)	0	476.2	952.5	1428.7	1905.0	2381.2
Dimensionless time t^*	0	1.61192	3.22384	4.83576	6.44768	8.0596

Table 10. Conditions of a discharge test

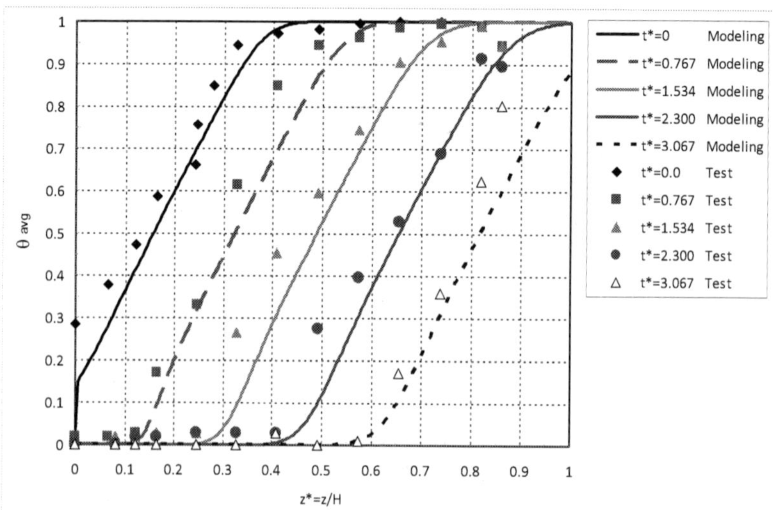

Fig. 19. Comparison of modeling predicted results with experimental data from reference (Pacheco et al., 2002)

6.2.2 Temperature variations in a discharge process

A heat discharge experiment was conducted under the conditions listed in Table 10, which also includes the dimensionless parameters. Shown in Fig. 20 is the temperature variation versus time at different locations in the tank along the tank height. The thermocouples at the top (z^*=1.0) of the tank measure the temperature of the discharged fluid. The degradation of discharged fluid temperature is clearly shown in the figure.

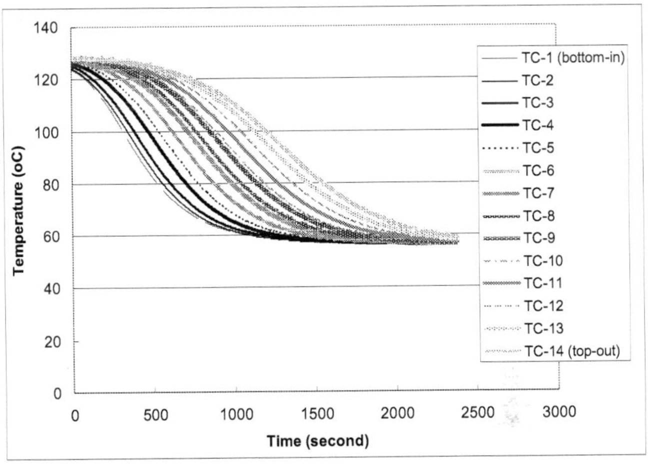

Fig. 20. Temperatures in the center of the tank along the height of 65 cm for a discharging process (Thermocouples were set every 5cm)

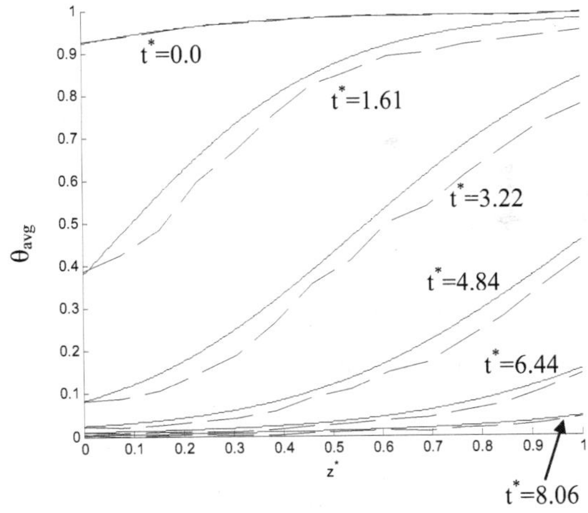

Fig. 21. The temperature distribution along the height in the tank at different time points (Solid lines are from simulation results and dashed lines are from experimental tests)

The temperature distribution along the height in the tank at different times is shown in Fig. 21. At the end of the discharge, the temperature on top of the tank (at z*=1) becomes low, which means that stored energy has been discharged. Using the measured initial temperature distribution and the inlet fluid temperature (at z*=1) together with the properties listed in Table 10, numerical simulation results were obtained and are compared with the test results in Fig. 21. The real time and the dimensionless time are listed in Table 10. Again, the agreement between the experimental data and the modeling simulation is satisfactory.

6.3 Correlation of energy delivery effectiveness to dimensionless parameters

Based on the above discussion and the dimensionless governing equations obtained in section 5.2.3, the energy delivery effectiveness, η, is a function of four dimensionless parameters, Π_c/Π_d, Π_d, τ_r, and H_{CR}. Solutions of the dimensionless governing equations for energy charge and discharge allow us to develop a database so that a series of charts and diagrams for $\eta = f(\Pi_c/\Pi_d, \Pi_d, \tau_r, H_{CR})$ may be prepared for reference by engineers in the design of thermocline storage tanks. Illustrated in Fig. 22 is a configuration of a group of database charts for a given Π_d, in which multiple graphs, each with a specific τ_r, may be provided. In each graph, multiple curves, each with a given H_{CR}, for the energy storage effectiveness η versus Π_c/Π_d are provided. To build a large database, more graphs in the same configuration can be provided covering a large range of Π_d values.

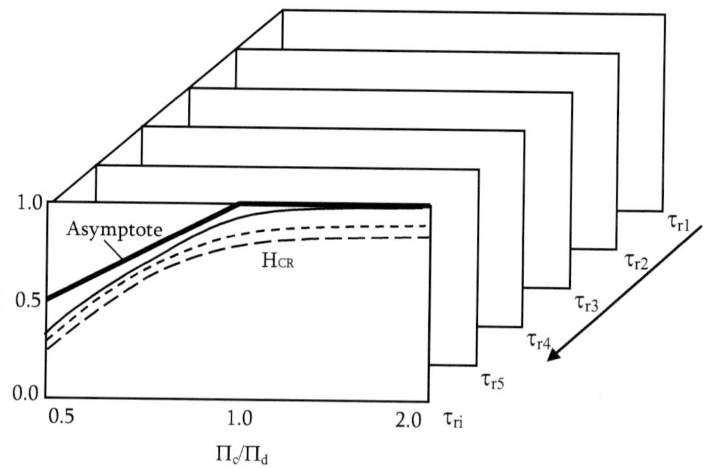

Fig. 22. Configuration of a group of database charts including multiple graphs of η versus Π_c/Π_d at a fixed Π_d

It is easy to understand that under the same mass flow rate and a desired Π_d, the energy delivery effectiveness will increase with the increase of Π_c/Π_d. The asymptote of the energy delivery effectiveness is due to an ideal thermal storage system, in which η can be exactly 1.0 if the energy charge period is equal to or larger than that of the discharge ($\Pi_c/\Pi_d \geq 1.0$). For any non-ideal thermocline system η can only approach 1.0.

Figure 23 shows four charts of η versus Π_c / Π_d at Π_d =4.0. Each chart is for a specific τ_r with multiple curves for different H_{CR}. All the data of energy delivery effectiveness were obtained based on several cyclic operations of the energy charge and discharge, and the results are consequently independent of the number of cycles. More charts with wide range of Π_d, τ_r, and H_{CR} may be developed. However, in actual design practice, the ranges of Π_d, τ_r, and H_{CR} are not very wide and specific charts can be easily prepared.

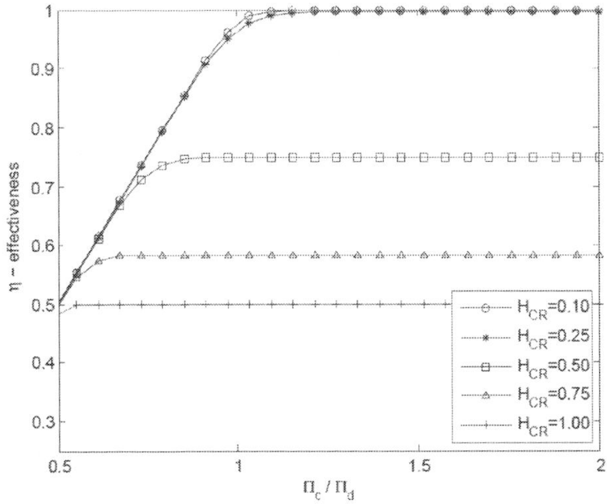

(a) $\tau_r = 0.01$

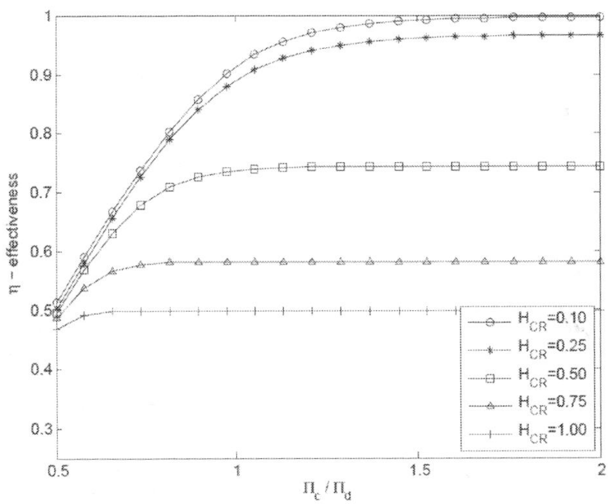

(b) $\tau_r = 0.04$

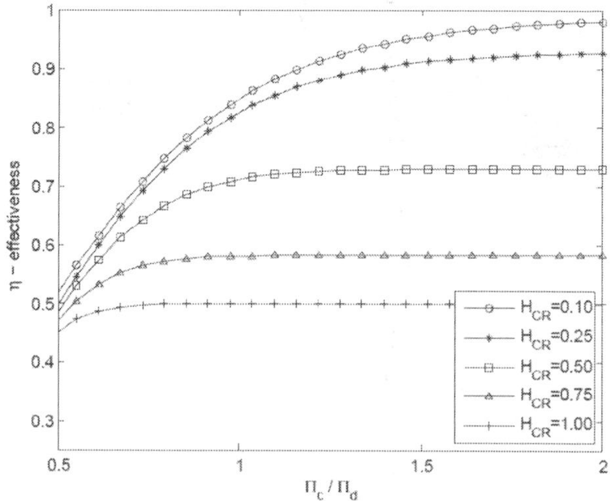

(c) $\tau_r = 0.1$

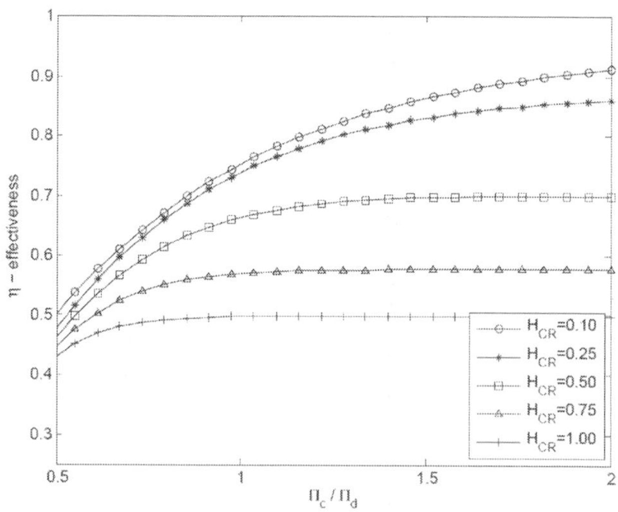

(d) $\tau_r = 0.2$

Fig. 23. Multiple graphs from modeling results for energy storage effectiveness versus Π_c / Π_d at Π_d =4.0

Observing the above four graphs one can easily draw the following conclusions:

1. The energy delivery effectiveness never reaches 1.0 if $\Pi_c / \Pi_d < 1.0$. This proves that only for an ideal thermocline storage tank can $\eta = 1.0$ at $\Pi_c / \Pi_d = 1.0$.

2. With the decrease of τ_r, η will increase. For example, at a ratio of $\Pi_c / \Pi_d = 1.5$ and $H_{CR} = 0.25$, the energy delivery effectiveness approaches 1.0 when τ_r changes from 0.2 to 0.01. This is because a decrease of τ_r is due to an increase in the volume of the storage tank.
3. It is understood that a small H_{CR} corresponds to the case where $(\rho C)_s$ is relatively large when compared to $(\rho C)_f$, and therefore the energy storage capability is improved, and η can approach 1.0 easier. On the other hand, when the void fraction in a packed bed approaches 1.0, it will make $H_{CR} \to \infty$, and the thermal storage effectiveness can approach that of an ideal case. However, in most practical applications, a low void fraction in a thermocline tank is required for the purpose of using less heat transfer fluid, and therefore a smaller H_{CR} value is practical and preferable.
4. For cases where η could never approach 1.0, even at large Π_c / Π_d values, it is obviously attributable to the fact that the storage tank is too small, and reselection of a larger storage tank is needed.

Designers for a thermal storage system often need to calibrate or confirm that a given storage tank can satisfy an energy delivery requirement. Under such a circumstance, the dimensions of the storage tank and the power plant operational conditions are known, which means the values of τ_r, Π_d, and H_{CR} are essentially given. One can easily check whether over a range of values of Π_c / Π_d the energy delivery effectiveness η can approach 1.0.

7. Procedures of sizing and design of a thermal storage system

The required operational conditions of the power plant dictate the size of the thermocline storage tank. The relevant operational conditions include: electrical power, thermal efficiency of the power plant, the extended period of operation based on stored thermal energy, the required high temperature of the heat transfer fluid from the storage tank, and the low temperature of the fluid returned from the power plant, the specific choice of heat transfer fluid and thermal storage material, as well as the packing porosity in a thermocline tank. The design analysis using the general charts provided in the present study will include the following steps:

1. Select a minimum required volume for a thermocline tank using Eqs. (1) and (3).
2. Choose a radius, R, and the corresponding height, H, from the minimum volume decided in step (1). Using these dimensions, the parameters– Π_d, τ_r, H_{CR} for a thermocline tank with filler material, can be evaluated, where Π_d is determined based on the required operational time.
3. Look up the design charts (such as those in Fig. 23) and see if an energy delivery effectiveness of 1.0 can be achieved. Often the energy delivery effectiveness will not approach 1.0 for the first trial design. This is because the first trial uses a minimum volume. However, with the results from the first trial one can predict the required height or volume of the tank necessary to decrease τ_r and Π_d in the same proportion. A couple of trail iterations may be needed to eventually satisfy the criterion of η closing to 1.0.

If the energy delivery effectiveness from step (1) cannot approach 1.0 even if a large Π_c / Π_d is chosen, one actually has two ways to improve the effectiveness during the second trial. These are to decrease H_{CR}, or decrease both τ_r and Π_d in the same proportion. However, H_{CR} is determined by properties of the fluid and filler material, which has very limited options, and therefore a decrease of τ_r and Π_d is more practical. The decrease of τ_r can be done by an increase of the height of a storage tank. This means that to achieve an effectiveness of 1.0 one has to increase the size of the storage tank. When the height of tank is increased, Π_d decreases accordingly since it depends on the height of the tank.

Occasionally, a calibration analysis requires a designer to find a proper time period of energy charge that can satisfy the needed operation time of a power plant. The known parameters will be the tank volume, τ_r, as well as H_{CR} at a required operation period of Π_d. The first step of the calibration should be the examination of the criterion given in Eq. (3), from which a minimum tank volume can be chosen. If the minimum tank volume is satisfied, the second step of calibration will be to find a proper Π_c / Π_d that can make the energy delivery effectiveness approach 1.0. Graphs including curves at the required H_{CR} and the given τ_r and Π_d must be looked up. Conclusions can be easily made depending on whether the energy delivery effectiveness can approach 1.0 for a particular value of Π_c / Π_d.

Two practical examples of thermocline thermal storage tank design are provided to help industrial designers practice the design procedures proposed in this work. Readers can easily repeat the design procedures while using their specific material parameters and operational conditions.

7.1 Design example 1—a system as shown in Fig. 2(a)

A solar thermal power plant has 1.0 MW electrical power output at a thermal efficiency of 20%. The heat transfer fluid used in the solar field is Therminol® VP-1. The power plant requires high and low fluid temperatures of 390 °C and 310 °C, respectively. River rocks are used as the filler material and the void fraction of packed rocks in the tank is 0.33. The required time period of energy discharge is 4 hours, the storage tank diameter is chosen to be 8 m. The rock diameter is 4 cm.

The solution is discussed as follows:

1. Making use of Eqs. (1) to (3) and the above given details on the power plant, as well as the properties of Therminol® VP-1, we can find a necessary mass flow rate of 25.34 kg/m³ and an ideal tank height of 9.59 m. The minimum volume of the storage tank can be determined from Eq. (3). Here, the ideal volume is used in the first trial of the design. Using Eqs. (19) and (20) we find the modified heat transfer coefficient to be 32.05 W/(m² K). With this information, the values of H_{CR}, Π_d, and τ_r are found to be 0.451, 3.03, and 0.0227, respectively. Given in Fig. 24 is a chart for Π_d =3 and τ_r =0.0227 at various values of H_{CR} and Π_c / Π_d. It is seen that on the curve of H_{CR} =0.45, there is no time ratio Π_c / Π_d that allows the energy delivery effectiveness to be close to 1.0. Therefore, the ideal volume chosen will not satisfy the energy storage need.
2. One option to provide the ability to store and deliver more energy and approach an effectiveness of 1.0 is to increase the height of the storage tank. When the height is increased to 12 m, the values of Π_d and τ_r changed to 2.42 and 0.0181, respectively. Figure 25 gives the chart for Π_d =2.42 and τ_r =0.0181. It is seen on the curve of H_{CR} = 0.45 that at the time period ratio, Π_c / Π_d =1.2, the energy delivery effectiveness approaches 0.99. This should be an acceptable design.

In this example, compared to an ideal thermal storage tank, the rock-packed-bed thermocline tank uses about 40.0% of the heat transfer fluid. To avoid the temperature degradation of the heat transfer fluid in a required discharge time period, a 20% longer charging time than discharging time is always applied in every charge and discharge cycle.

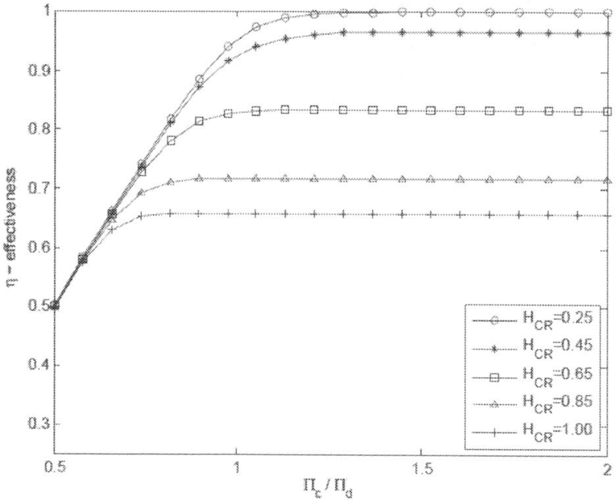

Fig. 24. Energy delivery effectiveness versus Π_c / Π_d at Π_d =3 and τ_r =0.0227

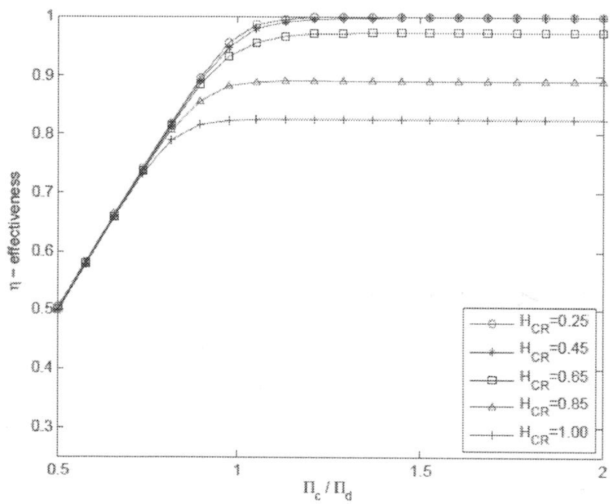

Fig. 25. Energy delivery effectiveness versus Π_c / Π_d at Π_d =2.42 and τ_r =0.0181

7.2 Design example 2—a system as shown in Fig. 2(b)

For the same solar thermal power plant and operational conditions as in Example 1, the thermal storage primary material is molten salt with properties of $\rho = 1680\,kg/m^3$, $C_s = 1560\,J/(kg \cdot K)$, and $k_s = 0.61\,W/(m \cdot K)$. The heat transfer fluid, Hitec (Bradshaw & Siegel, 2009), flows in multiple heat transfer tubes, in the same arrangement as shown in Fig.

2(b). The required time period of energy discharge is 4 hours, the storage tank diameter is 8 m. The study will find the storage tank height.
The solution and design procedures are as follows:

3. Following the same procedure as seen in Example 1, we determine the mass flow rate to be 40.35 kg/m³, and the ideal tank diameter and height to be 8 m and 6.44 m respectively. It is assumed that we have 8448 steel pipes (with inner diameters of 0.025m) in the storage tank, and that the heat transfer fluid flows in all of the pipes with an equal flow rate. The void fraction ε is found to be 0.33, and the heat transfer surface area per unit of length of the tank is found to be S_s =1327 m. The modified heat transfer coefficient inside a pipe for laminar flow heat transfer is used based on the correction using Eqs. (23) to (25), where the correction coefficient for the washer is w_c = 3.69894 for the ratio of $D_{eq} / d_i = 1.74$. The minimum volume of the tank is used in the first trial of the design. The dimensionless values of H_{CR}, Π_d, and τ_r are 0.522, 3.032, and 0.221, respectively. Given in Fig. 26 is a chart for Π_d =3.032 and τ_r =0.221 at various values of H_{CR} and Π_c / Π_d. The figure shows that at an H_{CR} value of 0.50 (close to 0.522) it is impossible to get an energy efficiency of 1.0 for attempted time ratios of Π_c / Π_d up to 2.0.

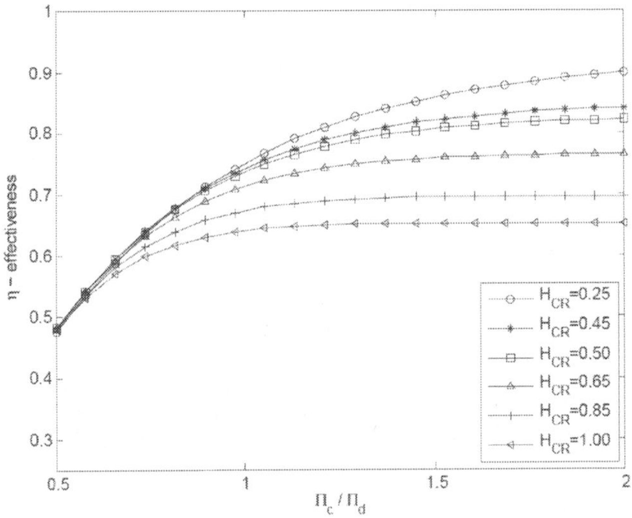

Fig. 26. Energy delivery effectiveness versus Π_c / Π_d at Π_d =3.032 and τ_r =0.221

To increase the energy delivery efficiency, a new tank height of 2.1 times that of the ideal volume tank is used. This makes the value of Π_d and τ_r be 1.444 and 0.105, respectively. Figure 27 shows the curves for Π_d =1.444 and τ_r =0.105. It is seen that at an H_{CR} value of 0.50 (close to 0.522) and a time period ratio, Π_c / Π_d, of 1.2, the energy delivery effectiveness can reach 0.96.
Note that in the finalized storage tank (with a height of 13.5 m), the volume of heat transfer fluid takes 69% of the volume of an ideal thermal storage tank (with a height of 6.44 m). The energy delivery effectiveness reached 0.96 if the charging time period was kept 20% longer than the required discharge time. Obviously, the weak heat transfer between the HTF and

the primary thermal storage material in Example 2 is responsible for the much larger ratio of the actual volume compared to the volume of the ideal storage tank. In Example 1 the ratio of the actual volume of the finalized storage tank is only 1.25 times that of the volume of its corresponding ideal storage tank. Therefore, it is important that in order to improve the energy delivery efficiency, the heat transfer between the fluid and thermal storage material must be improved, for example, by using pipes with fins. Nevertheless, this type of thermal storage system, as in Example 2, still saves 31% of the heat transfer fluid.

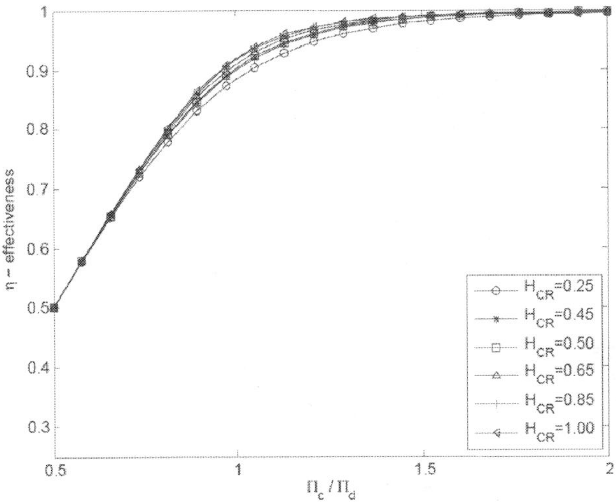

Fig. 27. Energy delivery effectiveness versus Π_c / Π_d at Π_d =1.444 and τ_r =0.105

Thermal energy storage in soil (Nassar et al., 2006), concrete (Zhang et al., 2004; Laing at al., 2006; Reuss et al., 1997), and in sands (Wyman, 1980) has similar features to that of the Example 2. When the heat transfer performance (the multiplication of the heat transfer coefficient and the heat transfer area) between the heat transfer fluid and thermal storage material is poor, the energy delivery effectiveness can be rather low. If the temperature degradation of the discharged fluid is a big concern, for example in a power plant, the thermal storage system in Example 2 may need enhancement of the heat transfer performance between the heat transfer fluid and thermal storage material.

8. Concluding remarks

Thermal energy storage is very important to the development of concentrated solar thermal power technologies. Thermal energy storage is also particularly attractive for large capacity energy storage. It is becoming more and more common for industrial engineers to be required to size and design thermal storage systems. Therefore, it is necessary to provide a general analytical tool for the sizing and design of thermal energy storage systems.

The details of the analytical model provided in this book chapter will help researchers and industrial engineers to understand the behavior of the heat transfer and energy transport between heat transfer fluid and thermal energy storage material. Two examples for the practical design of the size of thermal storage systems were provided for readers to practice the design process concretely. The authors believe that the currently developed charts, relating the energy storage efficiency to the properties and parameters of a thermal storage system, are of great potential to help industrial engineers performing sizing calculations for thermal storage systems. In this regard, the publication of this chapter is a significant and timely event.

Due to space limitations, this chapter did not provide a large number of the developed charts for thermal storage system sizing and design. However, the authors are currently working to provide a handbook including a large number of design charts covering a wide range of design parameters that industry will need. It is hoped that the handbook will be available to readers in the near future.

9. Acknowledgement

The authors are grateful for the support provided by the US Department of Energy and the National Renewable Energy Laboratory, under DOE Award Number DE-FC36-08GO18155, and the US Solar Thermal Storage LLC. The Support from the Idaho National Laboratory under award number 00095573 for phase change thermal storage is also gratefully acknowledged.

10. Nomenclature

a_f	The cross section area of a storage tank (m^2)
Bi	Biot number ($= L_p h / k_s$)
C	Heat capacity ($J / kg \cdot {}^oC$)
d_r	Nominal diameter of a single filler 'particle' (rocks) (m)
D	Diameter of a storage tank (m)
D_{eq}	Equivalent diameter (m)
$\hbar$	Enthalpy of fluid at a location along the axis (J / kg)
h	Heat transfer coefficient ($W / m^2\ {}^oC$)
h_p	Modified heat transfer coefficient ($W / m^2\ {}^oC$)
H	Length or height of a storage tank (m)
H_{CR}	A dimensionless parameter
k	Thermal conductivity ($W / m\ {}^oC$)
Lp	Characteristic length of 'particles' for Biot number.
$\dot{m}$	Mass flow rate (kg/s)
N	Number of tubes for heat transfer fluid in a storage tank
Pr	Prandtl number
$\dot{Q}$	Thermal energy involved per unit of time (W)
Q	Thermal energy involved (J)
r	Average radius of the filler material (rocks) (m)
R	Radius of the storage tank (m)
Re	Modified Reynolds number for porous media
Ss	Surface area of filler material per unit length of the storage tank (m)
t	Time (*sec*)
T_H	High temperature of fluid from solar field (°C)
T_L	Low temperature of fluid from power plant (°C)

U Fluid velocity in the axial direction in the storage tank (*m/s*)
V Volume (m^3)
z Location of a fluid element along the axis of the tank (*m*)

Greek symbols

α_s Thermal diffusivity $(= k_s / (\rho_s C_s))$ [m^2/s]
ε Porosity of packed bed in a storage tank.
η_s Thermal storage efficiency.
η_T Thermal efficiency of a solar power plant.
Γ Fusion energy of thermal storage material (J / kg)
μ Dynamic viscosity ($Pa \cdot s$)
Π Dimensionless charge or discharge time
τ_r A dimensionless parameter
ρ Density (kg / m^3)
θ Dimensionless temperature.

Subscript

c Energy charge process
d Energy discharge process
f Thermal fluid
ref A required reference value
r Rocks
s Filler material (rocks), the primary thermal storage material.
z Location along the axis of the tank

Superscript

* Dimensionless values

11. References

Abdoly, M.A., Rapp, D., 1982, Theoretical and experimental studies of stratified thermocline storage of hot water, Energy Conversion and Management, Vol. 22, No. 3, 1982, pp. 275-285.

Beasley, D.E., and Clark, J.A., 1984, Transient response of a packed bed for thermal energy storage, International Journal of Heat and Mass Transfer, 27(9), pp. 1659 -1669.

Becker, M., 1980, Comparison of heat transfer fluid for use in solar thermal power stations. Electric Power Systems Research, 3 (1980) 139-150.

Bradshaw R.W., Siegel N.P., 2009, Development of molten nitrate salt mixtures for concentrating solar power systems, paper no. 11538, proc. SolarPACES, Berlin, Germany, 2009.

Bradshaw, A.V., Johnson, A., McLachlan, N.H., Chiu, Y-T., 1970, Heat transfer between air and nitrogen and packed beds of non-reacting solids, Trans. Instn Chem. Engrs, Vol. 48, 1970, pp. T77-T84.

Brosseau, D., Kelton, J.W., Ray, D., Edgar, M., Chisman, K., and Emms, B., 2005, Testing of thermocline filler materials and molten-salt heat transfer fluids for thermal energy storage systems in parabolic trough power plants, Journal of Solar Energy Engineering, 127(1), pp. 109–116.

Canada, S., Brosseau, D. A., and Price, H., 2006, Design and Construction of the APS 1MWe parabolic trough power plant, ASME Conference Proceedings, 2006(47454), pp. 91–98.

Conway, J.H. & Sloane, N.J.H., 1998, Sphere Packings, Lattices and Groups, (Third Edition). ISBN 0-387-98585-9.

Courant, R., Hilbert, D., 1962, *Methods of Mathematical Physics, Volume II,* Wiley-Interscience.

Ferziger, J.H., 1998, *Numerical Methods for Engineering Applications.* Wiley-Interscience.

Gil, A., Medrano, M., Martorell, I., Lazaro, A., Dolado, P., Zalba, B., and Cabeza, L.F., 2010, State of the art on high temperature thermal energy storage for power generation, part 1—concepts, materials and modellization, Renewable and Sustainable Energy Reviews, 14(1): 31-55.

Halawa, E., Saman, W., 2011, Thermal performance analysis of a phase change thermal storage unit for space heating, Renewable Energy 36 (2011) 259-264.

Hales, T.C., 2006, Historical overview of the Kepler conjecture, Discrete & Computational Geometry. an International Journal of Mathematics and Computer Science 36 (1): 5–20.

Hasnain S.M., 1998, Review on sustainable thermal energy storage technologies, part I: heat storage materials and techniques. Energy Convers. Mgmt. 1998, Vol. 39: 1127–1138.

Herrmann, U., and Kearney, D.W., 2002, Survey of thermal energy storage for parabolic trough power plants, Journal of Solar Energy Engineering, 124(2), pp. 145–152.

Herrmann, U., Kelly, B., and Price, H., 2004, Two-tank molten salt storage for parabolic trough solar power plants, Energy, 29(5-6), pp. 883 - 893.

Incropera, F.P., and DeWitt, D.P., 2002, Introduction to Heat Transfer, fourth ed. John Wiley and Sons, Inc.

Jeffreson, C.P., 1972, Prediction of breakthrough curves in packed beds: 1. applicability of single parameter models, American Institute of Chemical Engineers, 18(2), pp. 409-416.

Karaki, W., Van Lew, J.T., Li, P.W., Chan, C.L., Stephens, J., 2010, Heat transfer in thermocline storage system with filler materials: analytical model, Proceedings of the ASME 2010 4th International Conference on Energy Sustainability, ES2010-90209, May 17-22, 2010, Phoenix, Arizona, USA.

Karaki, W., Li, P.W., Van Lew, J.T., Valmiki, M.M., Chan, C.L., Stephens, J., 2011, Experimental investigation of thermal storage processes in a thermocline storage tank, Submitted to ASME 5th International Conference on Energy Sustainability, ESFuelCell2011, August 7-10, Washington DC, USA

Kays, W. M., Crawford, M. E., and Weigand, B., 2005, Convective Heat and Mass Transfer, fourth ed. McGraw Hill.

Kolb, G.J., 1998, Economic evaluation of solar-only and hybrid power towers using molten-salt technology, Solar Energy, Vol. 62 (1), January 1998, pp. 51-61.

Kolb, G.J., and Hassani, V., 2006, Performance analysis of thermocline energy storage proposed for the 1 mw saguaro solar trough plant, *ASME Conference Proceedings,* 2006 (47454), pp. 1–5.

Krane, R.J., and Krane, M.J.M., 1992, The optimum design of stratified thermal energy storage systems—part II: completion of the analytical model, presentation and interpretation of the results, Journal of Energy Resources Technology, 114(3), pp. 204–208.

Laing, D., Steinmann, W. D., Viebahn, P., Gräter, F., and Bahl, C., 2010, Economic analysis and life cycle assessment of concrete thermal energy storage for parabolic trough power plants, Journal of Solar Energy Engineering, 132, 041013.

Laing, D, Steinmann, W.D., Tamme, R., Richter, C., 2006, Solid media thermal storage for parabolic trough power plants, Solar Energy, Vol. 80, pp. 1283-1289.

Li, P.W., 2008, Energy storage is the core of renewable energy technologies, IEEE Nanotechnology Magzine, December 2008, pp. 3-18.

Li, P.W., Van Lew, J.T., Chan, C.L., Karaki, W., Stephens, J., 2010, Similarity and generalized analysis of efficiencies of thermal energy storage systems, Submitted to Journal of Renewable Energy, 2010.

McMahan, A. C., 2006, Design and optimization of organic Rankine cycle solar-thermal power plants, Master's thesis, University of Wisconsin, Madison, Wisconsin.

McMahan, A.C., Klein, S. A., and Reindl, D. T., 2007, A finite-time thermodynamic framework for optimizing solar-thermal power plants, Journal of Solar Energy Engineering, 129(4), pp. 355–362.

Michel, Y., Haller, C.A., Cruickshank, W.S., Harrison, S.J., Andersen, E., Furbo, S., 2009, Methods to determine stratification efficiency of thermal energy storage processes – Review and theoretical comparison, Solar Energy, Vol. 83, Issue 10, October 2009, pp. 1847-1860.

Montes, M.J., Abánades, A., Martínez-Val, J.M., and Valdés, M., 2009, Solar multiple optimization for a solar-only thermal power plant, using oil as heat transfer fluid in the parabolic trough collectors, Solar Energy, Vol. 83 (12), Dec. 2009, pp. 2165-2176.

Nassar, Y., ElNoaman, A., Abutaima, A., Yousif, A., Salem, A., 2006, Evaluation of the underground soil thermal storage properties in Libya, Renewable Energy, Volume 31, Issue 5, April 2006, Pages 593-598.

Nellis, G., and Klein, S., 2009, *Heat Transfer,* Cambridge University Press.

Pacheco, J.E., Showalter, S.K., and Kolb, W.J., 2002, Development of a molten salt thermocline thermal storage system for parabolic trough plants, Journal of Solar Energy Engineering, 124(2), pp. 153–159.

Pitz-Paal, R., Dersch, J., Milow, B., Tellez, F., Ferriere, A., Langnickel, U., Steinfeld, A., Karni, J., Zarza, E., Popel, O., 2007, Development steps for parabolic trough solar power technologies with maximum impact on cost reduction, Journal of Solar Energy Engineering, 131, pp. 371-377.

Polyanin, A. D., 2002, *Handbook of Linear Partial Differential Equations for Engineers and Scientists,* Boca Raton: Chapman & Hall/CRC Press, ISBN 1-58488-299-9.

Price, H., Lupfert, E., Kearney, D., Zarza, E., Cohen, G., Gee, R., Mahoney, R., 2002, Advances in parabolic trough solar power technology, Journal of Solar Energy Engineering, Vol. 124, May 2002, pp. 109-125.

Product Resources, http://www.radcoind.com/index.html

Regin, A. F., Solanki, S.C., Saini, J.S., 2008, Heat transfer characteristics of thermal energy storage system using PCM capsules: A review, Renewable and Sustainable Energy Reviews 12 (2008) 2438–2458.

Renewables 2007, 2008, Global Status Report, Issued by Renewable Energy Policy Network for the 21st Century (www.REN21.net).

Reuss, M., Beck, M., Muller, J.P., 1997, Design of a seasonal thermal energy storage in the ground, Solar Energy, Vol. 59, No. 4-6, April-June 1997, pp. 247-257.

Sari, A., Kaygusuz, K., 2001, Thermal performance of myristic acid as a phase change material for energy storage, application. Renewable Energy, 24 (2001) 303–317.

Schumann, T. E.W., 1929, Heat transfer: a liquid flowing through a porous prism. Journal of the Franklin Institute, 208(3), pp. 405 - 416.

Shitzer, A., and Levy, M., 1983, Transient behavior of a rock-bed thermal storage system subjected to variable inlet air temperatures: Analysis and experimentation. *Journal of Solar Energy Engineering,* 105(2), May, pp. 200–206.

Singer, C., Buck, R., Pitz-Paal, R., and Müller-Steinhagen, H., 2010, Assessment of solar power tower driven ultrasupercritical steam cycles applying tubular central receivers with varied heat transfer media. Journal of Solar Energy Engineering, 132, 041010.

Spiers, D.J., and Rasinkoski, A.D., 1995, Predicting the service lifetime of lead/acid batteries in photovoltaic systems, J. Power Sources, vol. 53, no. 2, pp. 245–253.

Therminol VP-1 heat transfer fluid by Solutia, 1999, Technical Bulletin 7239115B, Solutia, Inc., 1999.

Tritt T.M., 2005, Thermal Conductivity: Theory, Properties, and Applications (Physics of Solids and Liquids), Springer, 2005-05-13, ISBN: 0306483270.

Van Lew, J.T., Li, P.W., Chan, C.L., Karaki, W., Stephens, J., 2009, Transient heat delivery and storage process in a thermocline heat storage system, IMECE2009-11701, Proceedings of the ASME 2009 International Mechanical Congress and Exposition, November 13-19, 2009, Lake Buena Vista, Florida, USA.

Van Lew, J.T., Li, P.W., Chan, C.L., Karaki, W., Stephens, J., 2011, Analysis of heat storage and delivery of a thermocline tank having solid filler material, Journal of Solar Energy Engineering, MAY 2011, Vol. 133/021003.

Wu, S.M., Fang, G.Y., Liu, X. 2011, Dynamic discharging characteristics simulation on solar heat storage system with spherical capsules using paraffin as heat storage material, Renewable Energy, 36 (2011) 1190-1195.

Wyman, C., Castle, J., Kreith, F., 1980, A review of collector and energy storage technology for intermediate temperature applications, Solar Energy, Vol. 24, No. 6, 1980, pp. 517-540.

Yang, Z., Garimella, S.v., 2010, Molten-salt thermal energy storage in thermoclines under different environmental boundary conditions, Applied Energy, Vol. 87, No. 11, November 2010, pp. 3322-3329.

Zalba, B., Marin, J.M., Cabeza, L.F., Mehling, H., 2003, Review on thermal energy storage with phase change: materials, heat transfer analysis and applications, Applied Thermal Engineering, 23 (2003) 251–283.

Zarty, O., Juddaimi, A. E., 1987, Computational models of a rock-bed thermal storage unit, Solar and wind Technology, 2(4), pp. 215- 218.

Zhang, D., Li, Z.J., Zhou, J.M., Wu, K.R., 2004, Development of thermal energy storage concrete , Cement and Concrete Research, Vol. 34, No. 6, June 2004, pp. 927-934.

2.1 Plasma channel radius

The extreme high temperature gradient that exists between plasma channel and electrode-workpiece interface enhances heat conduction through the contact surface. Plasma channel with high-energy plasma exhibits spatial and sequential variation at plasma channel-electrode-workpiece interface. This leads to variation in energy flux available for conduction into the electrode and workpiece (Dibitonto et al., 1989). Plasma channel radius is proportional to current, pulse duration, and the proportionality constants and indices depend on inter-electrode gap, electrode and workpiece materials and as well as dielectric fluid (Marafona & Chousal, 2006).

Measurement of plasma channel radius is extremely difficult due to very high pulse frequencies (Vinod Yadav et al., 2002). There are number of different equations used to calculate plasma channel radius according to different authors. It was proposed to use Eq 2 which depends on discharge current and time to calculate plasma channel radius.

$$R(t) = KQ^{m}t^{n} \tag{2}$$

where

R = Plasma channel radius (μm)

Q = Discharge current (Amp)

exponents m, n and K are empirical constant

Eq. 3 obtained by Patel et al, 1989 was used to calculate plasma channel radius in *EDM* process. According to the authors, the equation is time dependent with the empirical constants K and exponent n given as 0.788 and 0.75 respectively

$$R(t) = Kt^{n} \tag{3}$$

In the work of Shuvra et al, 2003 it was pointed out that plasma radius varies with time as per the relationship in Eq. 4

$$a = Rt^{\frac{3}{4}} \tag{4}$$

where t is time in μs, a is radius in μm and R is the proportionality constant.

According to Shuvra et al, because they could not have a good estimated value of proportionality constant R, they took plasma radius (a) to be 0.75 μm which is equal to the radius of the crater. Pandey & Jilani, 1986 in their paper titled, "Plasma channel growth and the resolidified layer in *EDM*", used Eq. 5 to calculate the plasma radius R.

$$T_b = \frac{E_0 R}{K\pi^{0.5}} tan^{-1}\left[\frac{4at}{R^2}\right]^{0.5} \tag{5}$$

where T_b = boiling temperature (^{0}C), E_o is the energy density (Watt/m^2) and a is the thermal diffusivity (m^2/s). Salonitis et al, 2007 in their works used Eq. 6 to obtained equivalent heat input radius which is dependent on the current intensity and pulse on-time duration.

$$r_{sp} = 2040 \times (I)^{0.43} \times (t_{on})^{0.44} \tag{6}$$

where r_{sp} is heat input radius in μm, I is current intensity in A and t_{on} is the spark time in μs. In the work of Bulent et al., 2006 and Ozgedik & Cogun, 2006 they used Eq. 7 to estimate the energy released due to a single spark.

$$E_{sp} = I \times V \times t_{on} \tag{7}$$

where
E_{sp} = Energy released due to one spark (Watt)
I = Pulse current (Amp)
V = Gap voltage (Volt)
t_{on} = on-time (μs)
According to Bulent Ekmekci et al, 2006 the total energy received by any material due to a single spark depends on it thermal conductivity, and is given by Eq. 8.

$$E_M = W_M \times I \times V \times t_{on} \tag{8}$$

where
E_M = Total energy received by the material (Watt)
W_M = Fraction of energy received by the material (Watt)

3. Experimental details

3.1 Electrode and work materials

In the present study copper and brass were taken as the electrode materials. These materials have high electrical and thermal conductivity. High electrical conductivity of these materials facilitates conducting electrical sparks. Their high thermal conductivity allows the absorbed heat to escape easily and thus reduces EW. Moreover, it is easy to machine these materials by conventional machining techniques into complex shapes. The electrodes used in the present study were 70 mm long with a cross-sectional dimension of 15 mm x 15 mm. The major properties of the electrode materials are shown in Table 1.

Electrode materials	Thermal conductivity (W/m-°K)	Melting point (°C)	Electrical resistivity (ohm-cm)	Specific heat capacity (J/g-°C)
Copper	391	1083	1.69	0.385
Brass	159	990	4.7	0.38

Table 1. Major properties of electrode materials

The workpiece materials used in the present study were mild steel and aluminum. Their chemical composition and major properties are shown in the Table 2 and Table 3 respectively.

Work materials	Chemical composition
Aluminum	Al: 99.9%, Cu:0.05%, Fe:0.4%, Mg:0.05%, Mn:0.05%, Si:0.25%, Zn:0.05%
Mild steel	C: 0.14%-0.2%, Fe: 98.81-99.26%, Mn: 0.6%-0.9%, P: 0.04%, S: 0.05%

Table 2. Chemical composition of the work materials

Work materials	Thermal conductivity (W/m-°K)	Melting point (°C)	Electrical resistivity (ohm-cm)	Specific heat capacity (J/g-°C)	Hardness (HB)	Tensile strength (MPa)	Yield strength (MPa)	Percentage elongation at break (%)
Aluminum	227	660	2.9	0.9	35	131	124	8
Mild steel	51.9	1523	1.74	0.472	143	475	275	38

Table 3. Major properties of work materials

3.2 EDM machine

The experimental work were conducted on a die sinking EDM machine of type Mitsubishi EX 22 model C11E FP60E. The machine has a rectangular table for mounting the workpiece. The electrode is fixed above the workpiece on the tool holder and can be given a vertical feed to machine a particular cavity. Usually the electrode is connected to the negative terminal and the workpiece is connected to the positive terminal of the power supply which is called direct polarity. However, depending on the combination of the electrode and the work materials, reverse polarity is also sometimes used. The machine used in the present study is shown in Fig. 3.

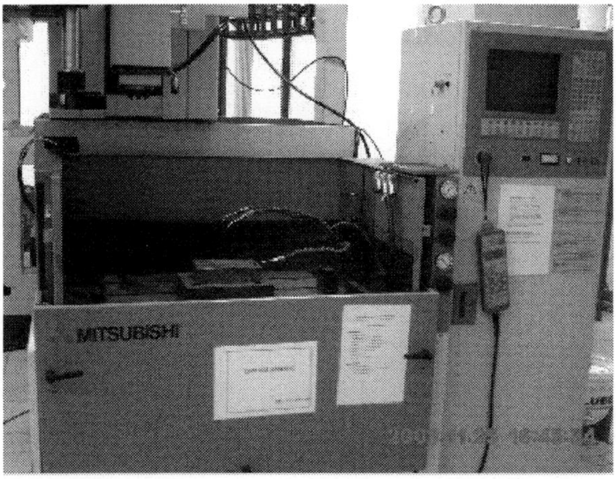

Fig. 3. EDM die sinking machine

3.3 Dielectric material

The common properties of dielectric fluids are: The dielectric fluid should have sufficient and stable dielectric strength; it should de-ionize rapidly after the spark discharge; it should have low viscosity and good wetting capacity; its flash point should be sufficient high to avoid any fire hazard; it should not emit any toxic vapours or have unpleasant odours; it should maintain its properties under all working conditions; it should be chemically neutral to the workpiece, electrode or the work table and it should be easily available at a reasonable price. Considering the above factors, kerosene was selected as the dielectric fluid in the present study.

The product of machining, the tiny debrtis should be removed from the machining zone in order to avoid sparks between the electrode and the debrtis. This is done by effective flushing system of the dielectric fluid to remove the debrits away from the machining zone. Suction flushing of dielectric fluid is preferred when straight vertical walls of the machined cavity are needed. The vertical surfaces are produced with a small taper angle of 0.005mm to 0.05mm per 10mm depth when pressure flushing is employed. However, for machining die cavities a small taper angle is desired. In the present study the machine was incorporated with circulating emission flushing system coupled with jets of three nozzles in order to assure the adequate flushing of the debris from the gap between the tool and the electrode. The experimental set-up is shown in Fig. 4. The current tried in the experiments were 2.5 amps, 3.5 amps and 6.5 amps. The level of voltage used was 10 volts and 5 volts.

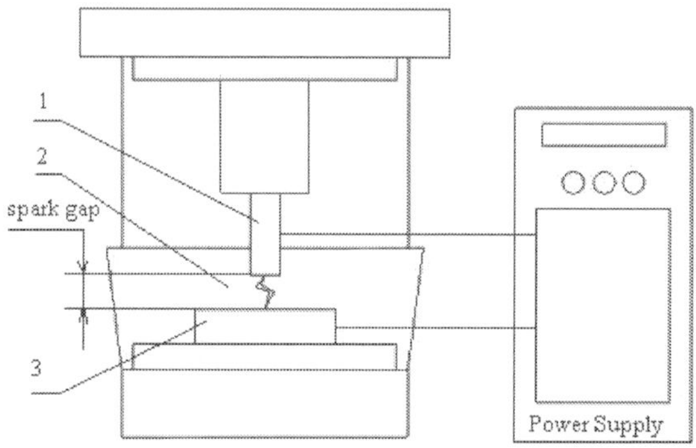

1 - tool; 2 - dielectric fluid; 3 - workpiece
Fig. 4. Set-up for the experiments

Square holes of dimensions of 15 mm x 15 mm were machined with a depth of 3 mm. After machining, the wear (Fig. 5) along the direction of the electrode movement (y-direction) and across the electrode (x-direction) was measured using an optical microscope Mitutoyo Hisomet II.

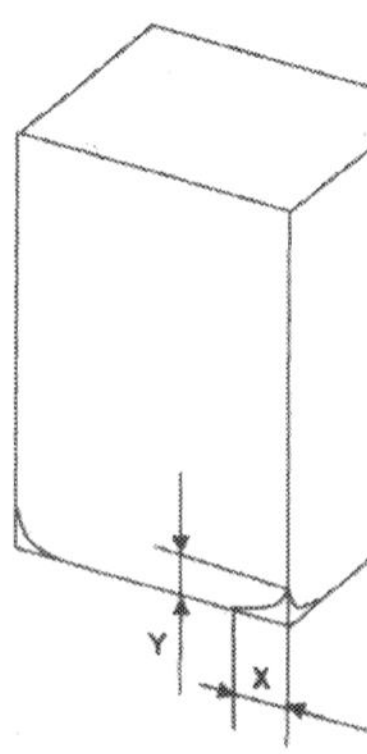

Fig. 5. Electrode wear in x and y directions

4. Results and discussions

4.1 Electrode wear rate

The present *EDM* technology allows for a very small level of *EW* because *EW* cannot be eliminated completely. *EW* is a critical problem in *EDM* since the tool shape degeneration directly affects the final shape of the product. In addition, the cost of a part manufactured by *EDM* is directly determined by the tool cost, which consists of the raw material cost, the production cost and the number of electrodes required for the machining. In most *EDM* operations, the cost of the electrode contributes more than 70% to the total operation cost. Due to these reasons, *EW* should be carefully taken into consideration in planning and designing *EDM* operations. Electrode wear is affected by the precipitation of carbon from the hydrocarbon dielectric onto the electrode surface during sparking. The thickness of the carbon inhibitor layer made a significant improvement on the *EWR* with little effect on the *MRR*.

As it was mentioned before, during *EDM* the electrode is connected to the negative terminal and the workpiece is connected to the positive terminal of the power supply. Generally, the rate of material removal from the cathode is comparatively less than that from the anode due to the following reasons (Ghosh & Malik, 1991):

i. The momentum with which the stream of electrons strikes the anode is much more than that due to the stream of the positive ions impinging on the cathode though the mass of an individual electron is less than that of the positive ions.
ii. The pyrolysis of the dielectric fluid (normally a hydrocarbon) creates a thin film of carbon on the cathode.
iii. A compressive force is developed on the cathode surface.

Therefore, normally, the tool is connected to the negative terminal of the DC source.

The weight of the electrodes before and after machining gives *EW*. It can be expresses as follows:

$$EWR = \frac{volume\ of\ material\ removed\ from\ the\ electrode}{machining\ time}$$

$$EWR = \frac{EV\mathbf{B} - EW\mathbf{A}}{M\mathbf{T}}$$

where $EV\mathbf{B}$ and $EW\mathbf{A}$ are the volumes of the workpiece before and after machining and $M\mathbf{T}$ is the machining time.

In Fig. 6 to Fig. 9 electrode wear along the cross-sectional direction and along the length of the tool has been presented after a machining a cavity of 3 mm depth. Fig. 6 to Fig. 9 shows that electrode wear rate increases both in x-direction and y-direction with increase in current. It is obvious that a higher current will produce a stronger spark which would cause more material to be eroded from the electrodes. It can also be observed that at a higher gap voltage an electrode undergoes more wear compared to that at a low gap voltage.

For example, during machining of aluminum with brass electrode with a current of 3.5 A the *EW* in x-direction is 0.64 mm at a gap voltage of 5 volts (Fig. 6), whereas, under the same condition the wear in x-direction is 0.86 mm at a gap voltage of 10 volts (Fig. 8).

It can be observed that in all cases (Fig. 6 to Fig. 9) wear of the electrodes in y-direction (along the length) is less than that in x-direction (along the cross-section) of the electrodes. As it was mentioned before, the length of the electrode is 70 mm (in y-direction) and the length of each side of the cross-sectional area is 15 mm (in x-direction). As a result the heat generated during the spark cannot be transferred into the body of the electrode easily in x-direction

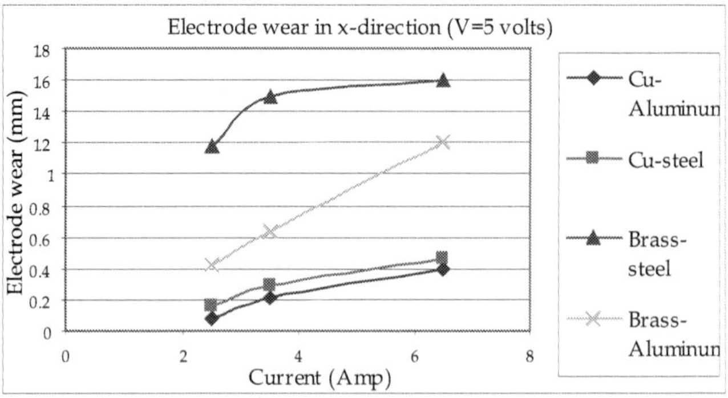

Fig. 6. Relationship of current with electrode wear along the cross-section (v=5 volts)

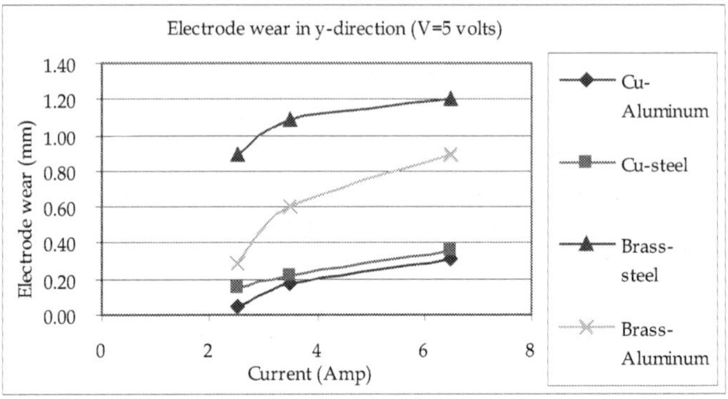

Fig. 7. Relationship of current with electrode wear along the length (v=5 volts)

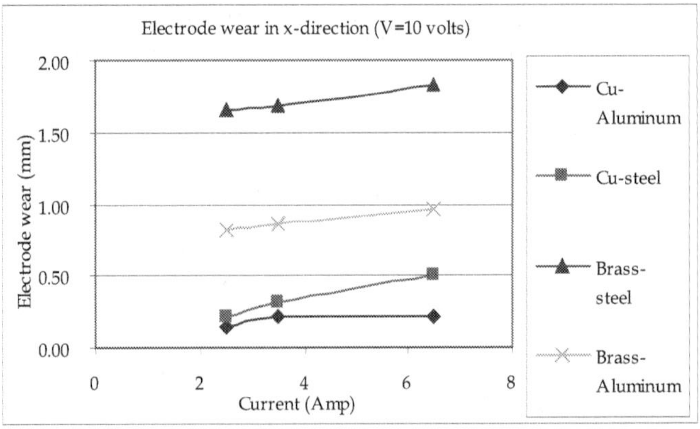

Fig. 8. Relationship of current with electrode wear along the cross- section (v=10 volts)

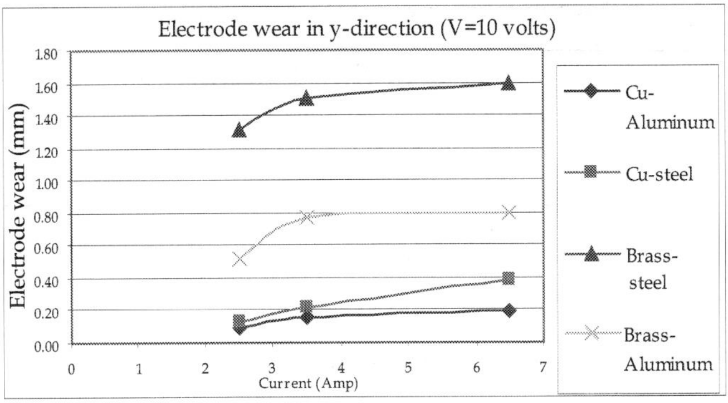

Fig. 9. Relationship of current with electrode wear along the length (v=10 volts)

compared to that in y-direction. This results more wear of electrodes in x-direction compared to that in y-direction. For the same reason it can be expected that that an electrode of a smaller cross-section permits poor heat transfer in x-direction and will undergo more wear compared to that of a larger cross-section. Fig. 10 clearly shows that electrode wear along the cross-section of the electrode is higher compared to the same along its length during machining of aluminum using brass electrodes.

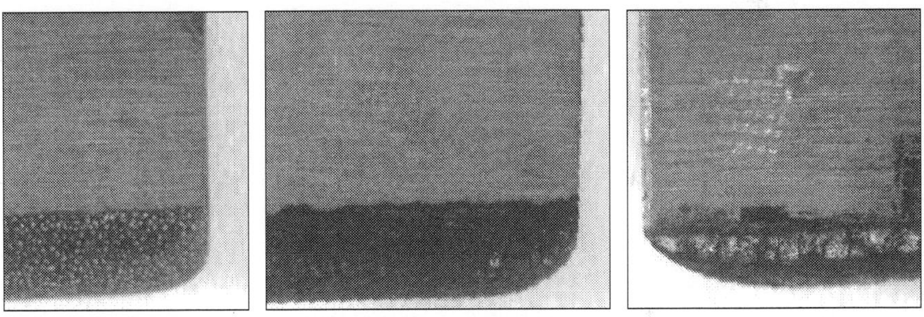

(a) current: 2.5 amps (b) current: 3.5 amps (c) current: 6.5 amps

Fig. 10. Wear of brass electrodes; electrode - brass, workpiece - aluminum

Results of the experiments show that copper electrodes undergo less wear compared to brass electrodes. For example, during machining of steel at a current of 3.5 amps and a voltage of 5 volts, the wear of copper electrode is 0.29 mm in x-direction, whereas the same for brass electrode is 1.5 mm (Fig. 6). This is due to the fact that the thermal conductivity of copper (391 W/m-K) is almost 2.5 times higher than that of brass (159 W/m-K). This facilitates rapid heat transfer through the body of copper electrodes compared to brass electrodes. It can also be noted that melting point of copper (1083°) is higher to that of brass (990°) that causes less melting and wear of copper electrodes.

From Fig. 6 to Fig. 9 it can also be observed that during machining of steel both the copper and the brass electrodes undergo more wear compared to the same during machining of aluminum. It can be mentioned that thermal conductivity of aluminum (227 W/m-K) is almost four times to that of steel (51.9 W/m-K). As a result the heat generated during each spark is easily absorbed by aluminum compared to that absorbed by steel. This causes less wear of electrodes during machining of aluminum compared to the same during machining of steel. For example, during machining of steel with brass electrodes at a current of 6.5 amps and a voltage of 10 volts electrode wear in y-direction is 1.6 mm, while the same is 0.8 mm during machining of aluminum (Fig. 9).

4.2 Wear ratio

Wear ratio (*WR*) is calculated as the ratio of the material removed from the work to the material removed from the electrode. It can be expresses as follows:

$$WR = \frac{volume\ of\ material\ removed\ from\ the\ electrode}{volume\ of\ material\ removed\ from\ the\ workpiece} \times 100\%$$

$$WR = \frac{EV\mathbf{B} - EW\mathbf{A}}{WPV\mathbf{B} - WPW\mathbf{A}} \times 100\%$$

where $EV\mathbf{B}$ and $EW\mathbf{A}$ are the volumes of the workpiece before and after machining.

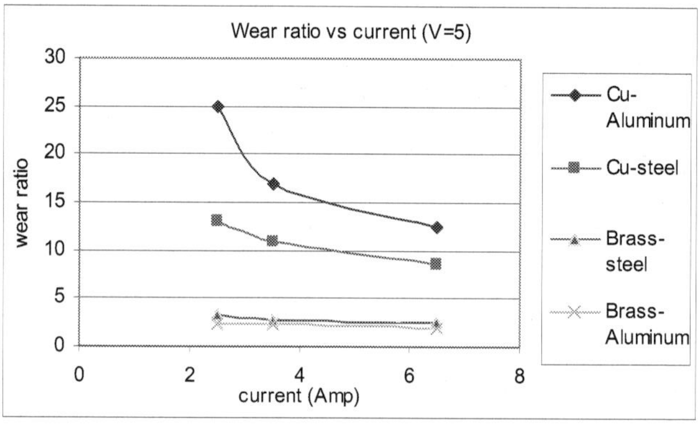

Fig. 11. Relationship of current with wear ratio (v=5 volts)

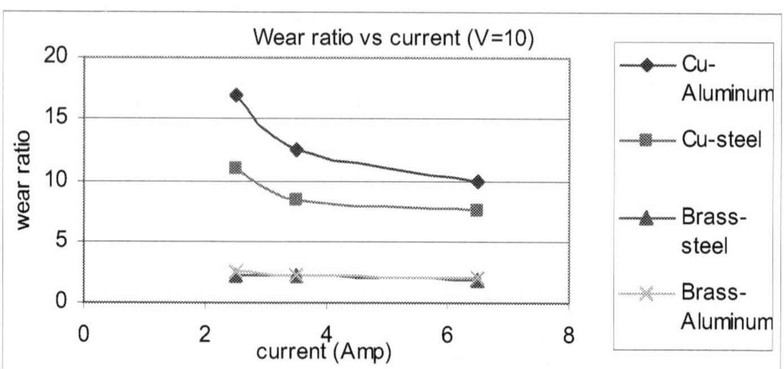

Fig. 12. Relationship of current with wear ratio (v=10 volts)

Fig. 11 and Fig. 12 illustrate the relationship of *WR* with current and voltage. It is obvious from the figures that *WR* decreases with increase in current. That means, though a higher current causes more removal of work material and the electrode, but comparatively more material is removed from the electrode. At a higher current a stronger spark is generated producing more heat. The size of the workpiece is massive and heat is easily dissipated through it. But since the electrode is a smaller one, heat is accumulated in it resulting high temperature and consequently high *EW*.

It can also be observed that *WR* is higher at a lower voltage (Fig. 11) compared to that at a higher voltage (Fig. 12). Highest *WR* of 25.00 was found during machining of aluminum using a copper electrode at a current of 2.5 amps and a voltage of 5 volts. The total heat generated during a spark is absorbed by the workpiece, electrode, dielectric fluid and the machine parts. It is desirable that most of the heat should be absorbed by the work material, while the least quantity of heat should be absorbed by the electrode. High thermal conductivity of copper electrodes facilitates easy heat transfer and its high melting point facilitates low melting of the electrode material. At the same time, high thermal conductivity of aluminum facilitates easy heat absorption and its low melting temperature facilitates fast removal of work material. These factors result the highest *WR* during machining of aluminum using a copper electrode. The lowest WR of 2.00 was found during machining of aluminum using brass electrodes at a current of 6.5 A and a voltage of 5 volts. Thermal conductivity of brass is only 1.4 times higher than that of the aluminum workpiece and its melting point is approximately 1.5 times higher than that of aluminum. As a result, material removal from the aluminum workpiece is comparatively low and the material removed from the electrode is comparatively high. Therefore, the *WR* is only 2.00. From Fig. 11 and Fig. 12 it can be concluded that in order of high to low *WR* of electrode-work pair are: copper-aluminum, copper-steel, brass-steel and brass-aluminum.

4.3 Material removal rate

The mechanism of material removal in *EDM* is a sudden violent splash of the molten material coinciding with the collapse of the plasma channel. As soon as the spark collapses and the hydrostatic pressure of the arc is released and dielectric rushes back to fill the void, the pressurized molten metal splatters from the workpiece surface leaving a crater on the surface and some splattered fragments around the crater (Shuvra et al, 2003). Material removal mechanism can also be explained in terms of the migration of material elements between the workpiece and electrode. This could take place in the form of elements diffusing from the electrode to the workpiece and vice-versa.

The use of CNC in *EDM* has helped to explore the possibility of using alternative methods of tooling to improve the *MRR*. *EDM* is commonly used for producing complex and deep 3-D shaped cavity in hard materials which necessitate the use of 3-D profile electrodes, which are costly and time-consuming to manufacture for the sparking process. However, in order to produce a complex shape a lot of experimental works have been performed with different types of electrode configurations to generating different path movement on workpiece surfaces by means of controlling the electrode motion.

Flushing pressure has a considerable effect on *MRR* and *EW*. When the dielectric fluid is forced at low velocity into the spark gap, short-circuiting becomes less pronounced as a result of the accumulated particles. This helps in improving the efficiency and thereby

increases *MRR*. Higher flushing pressure hinders the formation of ionized bridges across the gap and results in higher ignition delay and decrease discharge energy and reduces *MRR*.
It was found by many researchers that the influential machining factors on *MRR* are the current intensity and voltage.
Usually *EDM* is carried out by electrical sparks between the electrode and the workpiece using a single discharge for each electrical pulse. Some researchers have carried out experiments using a multi-electrode discharging system, delivering additional discharge simultaneously from a corresponding electrode connected serially. The design of electrode was based on the concept of dividing an electrode into multiple electrodes, which are electrically insulated. The energy efficiency were claimed to be better than the conventional *EDM* without any significant difference in work suface finish.
Material removal rate is expressed as the ratio of the difference in volume of the workpiece before and after machining to the machining time, i.e.:

$$MRR = \frac{volume\ of\ material\ removed\ from\ the\ workpiece}{machining\ time}$$

$$MRR = \frac{WPV\mathbf{B} - WPW\mathbf{A}}{M\mathbf{T}}$$

where $WPV\mathbf{B}$ and $WPW\mathbf{A}$ are the volumes of the workpiece before and after machining and $M\mathbf{T}$ is the machining time.

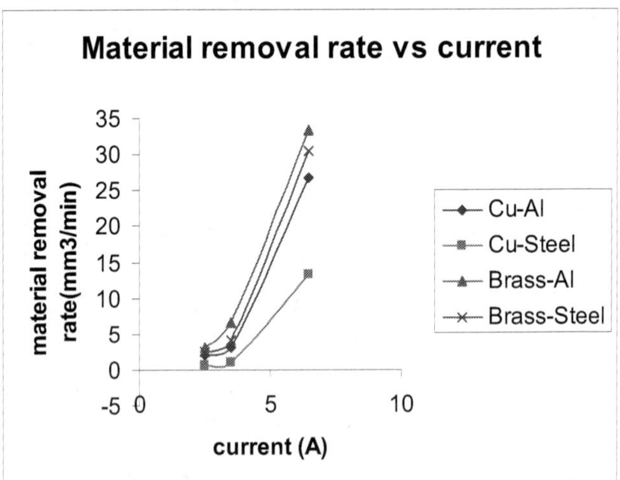

Fig. 13. Relationship between current and MRR

Relationship of *MRR* with current during machining of aluminum and steel using brass and copper electrodes are illustrated in Fig. 13. It is to be noted that at a low current *MRR* is very low, but with increase in current *MRR* increases sharply. At a low current, a small quantity of heat is generated and a substantial portion of it is absorbed by the surroundings and the machine components and the left of it is utilized in melting and vaporizing the work material. But as the current is increased, a stronger spark with higher energy is produced, more heat is generated and a substantial quantity of heat is utilized in material removal. However, the highest material removal rate was observed during machining of aluminum using copper electrodes. Comparatively low thermal conductivity of brass as an electrode material doesn't allow absorbing much of the heat energy and most of the heat is utilized in removal of material from aluminum workpiece of low melting point. But during machining of steel using copper electrodes, comparatively smaller quantity of heat is absorbed by the work material due to its low thermal conductivity. As a result *MRR* becomes very low.

4.4 Micro cracks

During the spark discharge in *EDM* the temperature is usually in the range of 8,000°C to 20,000°C. After the spark the work surface is immediately cooled rapidly by the dielectric fluid. Repeated heating to a very high temperature followed by rapid cooling develops micro-cracks on the work surface. Micro-cracks on the work surface are a major problem in *EDM*. They strongly influence on the fatigue strength of the part machined by *EDM*. Micro-cracks in the surface and loose grains in the subsurface resulted from thermal shock causes surface damage and leads to degradation of both strength and reliability. Comparing the SEM images in Fig. 14 it can be observed that more micro-cracks were formed during *EDM* with a higher current of 6.5 Amp as shown in Fig. 14 (a) compared to that with a low current of 2.5 Amp as illustrated in Fig. 14 (b). More heat is developed during EDM at a higher current heating the work surface to a higher temperature followed by rapid cooling. As a result more micro-cracks are found at a higher current. A larger t_{on} results more cracks as it can be observed comparing the Figs. 14 (c) and 14 (d). However, it was suggested by Lee & Tai, 2003 that when the pulse voltage is maintained at a constant value of 120 V, it is possible to avoid the formation of cracks if machining is carried out with a current in the range of 12-16 A together with pulse duration of 6-9 μs.

4.5 Recast layer

There are three layers created on the top of the base metal which are spattered *EDM* surface layer, recast layer and Heat Affected Zone (*HAZ*). Recast layer consists of dielectric fluid, molten electrode and molten workpiece that are melted during *EDM* machining and solidified. Usually recast layer has a higher hardness when compared to the base metal. The recast layer is also known as white layer because it often appears as a bright white layer in a sectional view under magnification. It occurs as the second layer under the spattered *EDM* surface layer. This layer is formed by the un-expelled molten metal solidifying in the crater. The recast layer is usually very thin and it can be removed by finishing operations. Recast layer can cause problems in some applications due to stress cracking or premature failure. Recast structure greatly affects die fatigue strength and shortens its service life. This is because the recast layers have micro-cracks and discharge craters that cause bad surface quality. HAZ consists of two layers: a hardened layer and the annealed layer. The depth of the hardened layer depends on the machining conditions. Usually the depth is 0.002mm for finish cut and 0.012 mm for rough cut. Below the hardened layer there is a layer which was cooled slowly and as a result, the layer is annealed. Its hardness is 2 to 5 points below the same of the base metal. Its thickness may be 0.05mm for finish cut and 0.2mm for rough cut.

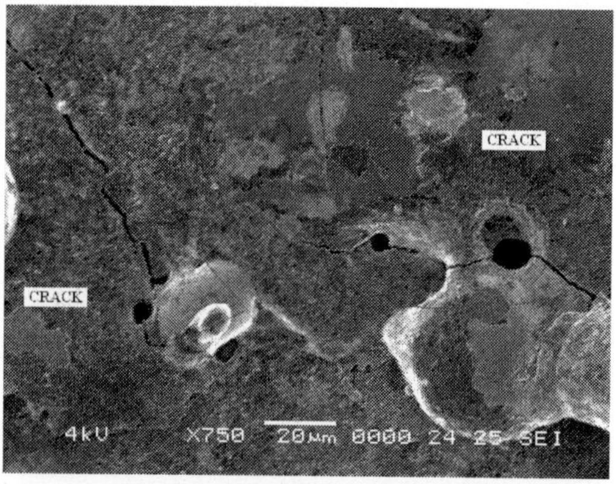

(a) I=6.5 Amp; t_{on}=10 μs

(b) I=2.5 Amp; t_{on}=10 μs

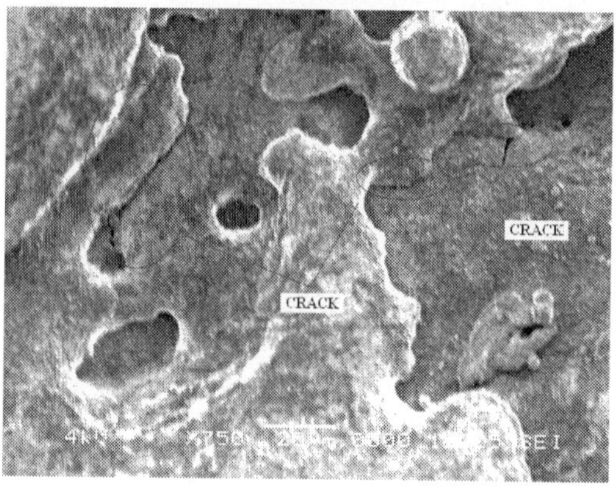

(c) I=2.5 Amp; t_{on}=10 μs

(d) I=2.5 Amp; t_{on}=3 µs

Fig. 14. Influence of current and pulse-on time on micro cracks

As stated above, a higher current and a higher pulse-on time produce a spark with more energy, melt more materials from the workpiece and the electrode. Consequently higher thickness of recast layer is found at a current of 6.5 Amp, Fig. 15 (a) compared to that at a current of 2.5 Amp, Fig. 15 (b). Similarly, thickness of the recast layer was found to be at a higher pulse-on time, Fig. 15 (c) compared to that at a shorter pulse-on time, Fig. 15 (d). Hwa-Teng Lee et al., 2004 also stated that R_a and average white layer thickness tend to increase at higher values of pulse current and t_{on}. However, they found that for extended pulse-on duration *MRR*, R_a and crack density all decrease.

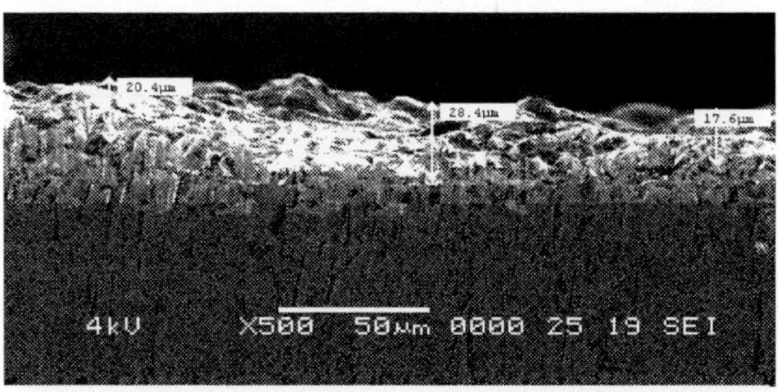

(a) t_{av} 22.1 µm: I= 6.5 Amp; t_{on} =10µs

(b) t_{av} 18.1 µm: I= 2.5 Amp; t_{on}=10 µs

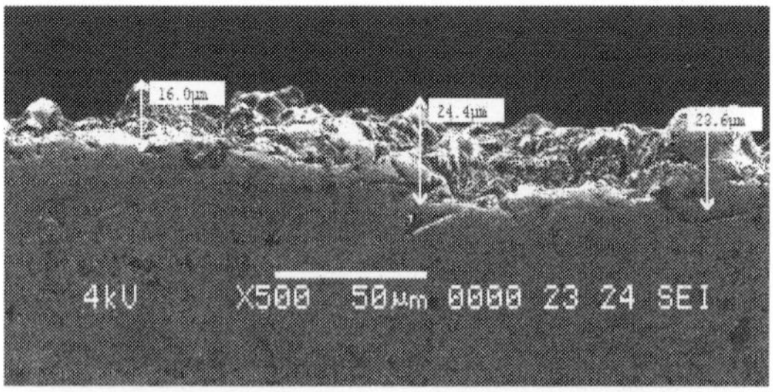

(c) t_{av} 21.3 μm: I= 2.5 Amp; t_{on}=10 μs

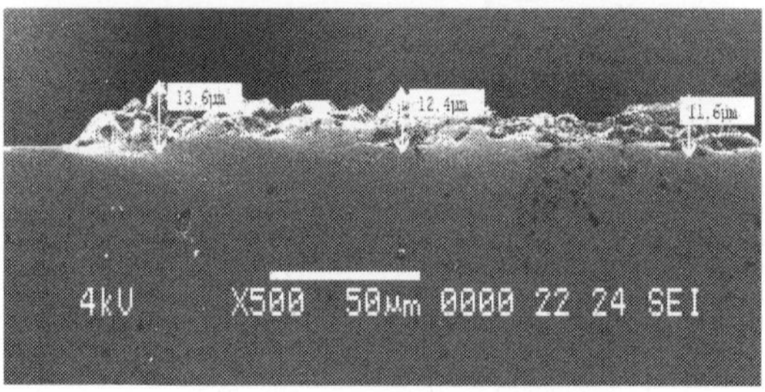

(d) t_{av} 12.5 μm: I= 2.5 Amp; t_{on}=1.5 μs

Fig. 15. Thickness of recast layer at different machining conditions

5. Conclusion

From the above discussions the following conclusions can be drawn:

1. Electrodes undergo more wear along its cross-section compared to that along its length.
2. Electrode wear increases with increase in current and voltage. Wear of copper electrodes is less than that of brass electrodes. This is due to the higher thermal conductivity and melting point of copper compared to those of brass.
3. During machining of mild steel, electrodes undergo more wear than during machining of aluminum. This is due to the fact that thermal conductivity of aluminum is higher to that of mild steel which causes comparatively more heat energy to dissipate into the electrode during machining of mild steel.
4. Wear ratio decreases with increase in current, but decreases with increase in gap voltage. The highest wear ratio was found during machining of aluminum using a copper electrode.
5. MRR increases sharply with increase in current. In the present study, highest MRR was obtained during machining of aluminum using a brass electrode.
6. Micro cracks are found on the machined surface. The tendency of formation of micro cracks increases during EDM with a higher current and larger pulse-on time.
7. A recast layer was found on the machined surface which consists of the molten materials from the workpiece and the electrode that could not be flushed away completely by the dielectric fluid. A thicker layer of recast layer was formed on the work surface machined with a higher current and pulse-on time.

6. Acknowledgement

The author of this work is indebted to the Research Management Center, International Islamic University Malaysia (IIUM) for its continuous help during the research work. Also, the author likes to appreciate the help of the staff and the technicians of the Department of Manufacturing and Materials Engineering, International Islamic University Malaysia.

7. References

Bleys, P.; Kruth, P. & Lauwers, B. (2004). Sensing and compensation of tool wear in milling EDM. *Journal of Materials Processing Technology*, vol.149, No.1-3, pp. 139-146, ISSN 0924-0136

Bulent, E.; Erman, A. & Abdulkadir E. (2006). A semi-empirical approach for residual stresses in electric discharge machining (EDM). *International Journal of Machine Tool & Manufacture*. Vol.46, pp. 858-865, ISSN 0890-6955

Dibitonto, D.; Eubank, T.; Patel, R. & Barrufet, A. (1989).Theoretical models of the electro discharge machining process–a simple cathode erosion model. *Journal of Applied Physics*, vol.69, pp. 4095-4103, ISSN 0021-8979

Ghosh, A. & Mallik, K. (1991). Electrical Discharge Machining, In: *Manufacturing Science*, 383-403, Affiliated East- West Press Private Limited, ISBN 81-85095-85-X, New Delhi, India.

Ho, H. & Newman, T. (2003). State of the art electrical discharge machining (EDM). *International Journal of Machine Tools and Manufacture*, vol.43, No.13, pp. 1287-1300, ISSN 0890-6955

Hu, F. Zhou, C. & Bao, W. (2008). Material removal and surface damage in EDM of Ti3SiC2 ceramic. *Ceramics International*, vol.34, issue 3, pp. 537-541, ISSN 0272-8842

Hwa-Teng, L.; Fu-ChuanHsu & Tzu-Yao, T. (2004). Study of surface integrity using the small area EDM process with a copper–tungsten electrode, *Materials Science and Engineering A*, vol. 364, issues 1-2, pp. 346-356, ISSN 0025-5416

Kalpakjian, S. Schmid, R. (2001). Electrical-discharge machining, in: *Manufacturing Engineering and Technology*, 6th edition, Prentice Hall, 769-774, Singapore

Khan, A. & Mridha, S. (2006). Performance of copper and aluminum electrode during EDM of stainless steel and carbide. *International Journal of Manufacturing and Production*, vol. 7, No.1, pp. 1-7, ISSN 0793-6648

Khanra, K.; Sarker, R.; Bhattacharya, B.; Pathak, C. & Godkhindi, M. (2007). Performance of ZrB_2–Cu composite as an EDM electrode. *Journal of Materials Processing Technology*, vol.183, No.1, pp. 122-126, ISSN 0924-0136

Kunieda, M. & Kobayashi, T. (2004). Clarifying mechanism of determining tool electrode wear ratio in EDM using spectroscopic measurement of vapor density. *Journal of Materials Processing Technology*, vol.149, No. 1-3, pp. 284-288, ISSN 0924-0136

Lee, T. & Tai, Y. (2003). Relationship between EDM parameters and surface crack formation. *Journal of Materials Processing Technology*, vol.142, issue 3, pp. 676-683, ISSN 0890-6955

Marafona, J. & Wykes. C. (2000). A new method of optimizing material removal rate using EDM with copper-tungsten electrodes. *International Journal of Machine Tools and Manufacture*, vol.40, pp. 153-164, ISSN 0890-6955

Marafona, J., & Chousal, G. (2006). A finite element model of EDM based on the Joule effect. *International Journal of Machine Tools & Manufacture*, vol.46, pp. 595-602, ISSN 0890-6955

Pandey, C. & Jilani, T. (1986). Plasma channel growth and the resolidified layer in EDM. *Precision Engineering*, Vol.8, issue 2, pp. 104-110, ISSN 0141-6359

Ozgedik, A. & Cogun, C. (2006). An experimental investigation of tool wear in electric discharge machining. *International Journal of Advance Manufacturing Technology*. Vol.27, pp. 488-500, ISSN 0268-3768

Patel, R.; Barrufet, A.; Eubank, T. & DiBitonto, D. (1989). Theoretical models of the electrical discharge machining process-II: the anode model. *Journal of Applied Physics*, vol.66, pp. 4104-4111, ISSN 0021-8979

Peter, Fonda.; Zhigang, Wang.; Kazuo, Yamazaki. & Yuji, A. (2007). *A* fundamental study on Ti-6Al-4V's thermal and electrical properties and their relation to EDM productivity. *Journal of Materials Processing Technology*, doi:10.1016/j.jmatprotec.2007.09.060, ISSN 0924-0136

Puertas, I.; Luis, J. & Alvarez, L. (2004). Analysis of the influence of EDM parameters on surface quality, MRR and EW of WC-Co. *Journal of Materials Processing Technology*, vol.153-154, No.10, pp. 1026-1032, ISSN 0924-0136

Ramasawmy, H. & Blunt, L. (2001). 3D surface characterization of electropolished EDMed surface and quantitative assessment of process variables using Taguchi Methodology. *International Journal of Machine Tools and Manufacture*, vol.42, pp. 1129-1133, ISSN 0890-6955

Salonitis, K.; Stournaras, A.; Stavropoulos, P. & Chryssolouris, G. (2007). Thermal modeling of the material removal rate and surface roughness for die-sinking EDM. *International Journal of Advance Manufacturing Technology*. DOI 10.1007/s00170-007-1327-y, ISSN 0268-3768

Shuvra, D.; Mathias, K. & Klocke, F. (2003). *EDM* simulation: finite element-based calculation of deformation, microstructure and residual stresses. *Journal of Materials Processing Technology*, vol.142, pp. 434-451, ISSN 0924-0136

Tariq, S. & Pandey, C. (1984). Experimental investigation into the performance of water as dielectric in EDM, *International Journal of Machine Tool Design and Research*, vol.24, pp. 31-43, ISSN 0020-7357

Thomas, N.; Shreyes, M.; Thomas, W.; Rosa, T. & Laura, R. (2009). Investigation of the effect of process parameters on the formation and characteristics of recast layer in wire-EDM of Inconel 718. *Materials Science and Engineering: A*, vol.513-514, pp. 208-215, ISSN 0025-5416

Vinod, Y.; Vijay, K. & Prakash, M. (2002). Thermal stresses due to electrical discharge machining, *International Journal of Machine Tools & Manufacture*. Vol.42, pp. 877-888, ISSN 0890-6955

Wang, J. & Tsai, M. (2001a) Semi-empirical model on work removal and tool wear in electrical discharge machining. *Journal of materials processing technology*, vol.114, No.4, pp. 1-17, ISSN 0924-0136

Wang, J. & Tsai, M. (2001b). Semi-emperical model of surface finish on electrical discharge machining. *International Journal of Machine Tools and Manufacture*, vol.41, pp. 1455-1477, ISSN 0890-6955

Yan, H.; Tsai, C. & Huang, Y. (2005). The effect of EDM of a dielectric of a urea solution in water on modifying the surface of titanium. *International Journal of Machine Tools and Manufacture*, vol. 45, No.2, pp. 194-200, ISSN 0890-6955

Zarepour, H.; Tehrani, A.; Karim, D & Amini, S. (2007). Statistical analysis on electrode wear in EDM of tool steel DIN 1.2714 used in forging dies. *Journal of Material Processing Technology*, vol.187-188, pp. 711-714, ISSN 0924-0136

Zaw, M.; Fuh, H.; Nee, C. & Lu, L. (1999). Fabrication of a new EDM electrode material using sintering techniques. *Journal of Materials Processing Technology*, vol.89-90, pp. 182-186, ISSN 0924-0136

22

Thermal Treatment of Granulated Particles by Induction Thermal Plasma

M. Mofazzal Hossain[1] and Takayuki Watanabe[2]
[1]Department of Electronics and Communications Engineering, East West University
[2]Department of Environmental Chemistry and Engineering
Tokyo Institute of Technology,
[1]Bangladesh
[2]Japan

1. Introduction

After the invention of induction plasma torch by [Reed, 1961], tremendous achievements have been earned by the researchers in the field of thermal treatment of micro particles by induction plasma torch. Induction thermal plasma (ITP) has become very popular in material processing due to several of its inherent characteristics: such as contamination free (no electrode), high thermal gradient (between torch and reaction chamber), wide pressure range and high enthalpy. ITP have extensively been used for the synthesis and surface treatment of fine powders since couple of decades as a clean reactive heat source [Fan, 1997], [Watanabe, 2004]. ITP technology may ensure essentially the in-flight one-step melting, short melting time, and less pollution compared with the traditional technologies that have been using in the glass industries for the vitrification of granulated powders. Moreover ITP technology may be very effective in the thermal treatment of porous micro particles and downsizing the particle size. During in-flight treatment of particles, it is rear to have experimental records of thermal history of particles; only some diagnosis of the quenched particles is possible for the characterization. Thus, the numerical analysis is the only tool to have comprehensive characterization of the particle thermal history and energy exchange during in-flight treatment. Thus, for numerical investigation it is the challenge to predict the trajectory and temperature history of the particles injected into the ITP torch. Among others Yoshida et al [Yoshida, 1977] pioneered the modeling of particle heating in induction plasmas; though their work assumed the particle trajectory along the centerline of the torch only. Boulos [Boulos, 1978] developed a model and comprehensively discussed the thermal treatment of alumina powders in the fire ball of argon induction plasma. Later (Proulx et al) [Proulx, 1985] predicted the trajectory and temperature history of alumina and copper particles injected into ITP torch and discussed the particle loading effects in argon induction plasma.

In this chapter we shall discuss the in-flight thermal treatment mechanism of soda-lime-silica glass powders by ITP and to optimize the plasma discharge parameters, particle size and feed-rate of input powders that affect the quenched powders size, morphology, and compositions. The thermal treatment of injected particles depends mainly on the plasma-particle heat transfer efficiency, which in turn depends to a large extent on the trajectory and temperature history of the injected particles. To achieve that goal, a plasma-particle interaction model has been developed for argon-oxygen plasma, including a nozzle inserted into the torch for the injection of carrier gas and soda-lime-silica glass powders. This model can be used to demonstrate the particle loading effects and to optimize the parameters that govern the particles trajectory, temperature history, quenched particles size and plasma-particle energy exchange efficiency. This model may be used to optimize the plasma and particle parameters for any combination of plasma gases for example argon-oxygen or argon nitrogen etc.

2. Modeling

2.1 Plasma model

The schematic geometry of the ITP torch is presented in Fig.1. The torch dimensions and discharge conditions are tabulated in Table 1. The overall efficiency of the reactor is assumed to be 50%, thus, plasma power is set to 10 kW. The torch dimension, power and induction frequency may vary and can be optimized through the simulation. The model solves the conservation equations and vector potential form of Maxwell's equations simultaneously under LTE (local thermodynamic equilibrium) conditions, including a metal nozzle inserted into the torch. It is assumed that plasma flow is 2-dimensioanl, axi-symmetric, laminar, steady, optically thin, and electromagnetic fields are 2-dimensional. Adding the source terms to the conservation equations, the plasma-particle interaction and particle loading effects have been taken into account. In this model, the conservation equations are as follows:
Mass conservation:

$$\nabla \cdot \rho \mathbf{u} = S_p^C \tag{1}$$

Momentum conservation:

$$\rho \mathbf{u} \cdot \nabla \mathbf{u} = -\nabla p + \nabla \cdot \mu \nabla \mathbf{u} + \mathbf{J} \times \mathbf{B} + S_p^M \tag{2}$$

Distance to initial coil position (L_1)	19 mm
Length of injection tube (L_t)	52 mm
Distance to end of coil position (L_2)	65 mm
Torch length (L_3)	190 mm
Coil diameter (d_c)	5 mm
Wall thickness of quartz tube (T_{wall})	1.5 mm
Inner radius of injection tube (r_1)	1 mm
Outer radius of injection tube (r_t)	4.5 mm
Outer radius of inner slot (r_2)	6.5 mm
Inner radius of outer slot (r_3)	21.5 mm
Torch radius (r_0)	22.5 mm
Coil radius (r_c)	32 mm
Plasma power	10 kW
Working frequency	4 MHz
Working pressure	0.1 MPa
Flow rate of carrier gas (Q_1)	$4 \sim 9$ L/min of Argon
Flow rate of plasma gas (Q_2)	2 L/min of Argon
Flow rate of sheath gas (Q_3)	22 L/min Argon & 2 L/min Oxygen

Table 1. Torch dimensions & discharge conditions

Energy conservation:

$$\rho \mathbf{u} \cdot \nabla h = \nabla \cdot \left(\frac{\kappa}{C_p} \nabla h \right) + \mathbf{J} \cdot \mathbf{E} - Q_r - S_p^E \tag{3}$$

Species conservation:

$$\rho \mathbf{u} \cdot \nabla y = \nabla \cdot \left(\rho D^m \nabla y \right) + S_p^C \tag{4}$$

Vector potential form of Maxwell electromagnetic field equation [Mostaghimi, 1998]:

$$\nabla^2 \mathbf{A_c} = i \mu_0 \sigma \omega \mathbf{A_c} \tag{5}$$

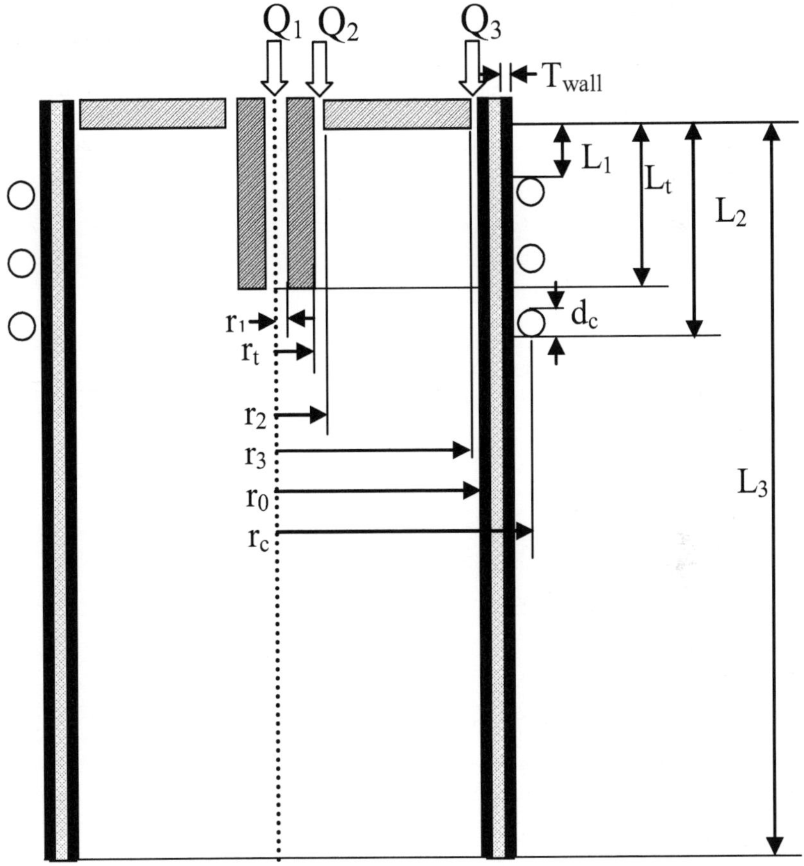

Fig. 1. Schematic geometry of induction thermal plasma torch

2.1.1 Boundary conditions

The boundary conditions for the mass, momentum, energy and species conservation equations are: at the inlet, gas temperature was set to 300 K and uniform velocity profiles are assumed based on the given flow rates; on the axis of symmetry, the symmetry conditions are imposed; on the walls, no-slip condition is assumed; the outer wall temperature is set to 350 K; and, at the exit, axial gradients of all fields are set equal to zero. The inserted nozzle is assumed to be water cooled at 300 K. On the nozzle wall, the velocity is set to zero. The boundary conditions for the vector potential form of Maxwell's equation are the same as those described in reference [Mostaghimi, 1998].

2.1.2 Computational procedure and thermophysical properties

The conservation equations, which are listed in previous section, are solved numerically using the SIMPLER algorithm of Patankar [Patankar, 1980]. The algorithm is based on a control-volume finite-difference scheme for solving the transport equations of incompressible fluids. Calculations are performed for a 44 (in radial direction) by 93 (in axial direction) non-uniform grid system.

Thermodynamic and transport properties of argon and oxygen gases required for the simulation are mass density, specific heat at constant pressure, viscosity, electrical and thermal conductivity and radiative loss coefficient. The transport properties, which are function of temperature, are calculated under LTE conditions using Chapman-Enskog first approximation to Boltzmann equation [Tanaka, 2000]. The effective diffusion coefficient of species is calculated based on the following equations:

$$D_i^m = \frac{(1-y_i)}{\sum\limits_{j\neq i, j=1}^{\upsilon} \frac{x_i}{D_{ij}}} \tag{6}$$

$$D_{ij} = 2.628 \times 10^{-2} \sqrt{\frac{(M_i + M_j)}{2M_i M_j}} \frac{T^{1.5}}{\bar{\Omega}_{ij}^{(1.1)}} \tag{7}$$

The ambipolar diffusion coefficient for ions can be approximated as $D_a = D_{ion}(1+T_e/T_{ion})$. As the thermal equilibrium condition i.e $T_h = T_e = T_{ion}$ was applied thus, $D_a \cong 2D_{ion}$.

2.2 Particle model

The following assumptions are made in the analysis of plasma-particle interactions in the ITP torch; the particle motion is two-dimensional, only the viscous drag force and gravity affect the motion of an injected particle, the temperature gradient inside the particle is neglected, and the particle charging effect caused by the impacts of electrons or positive ions is negligible. The particle charging effects have not been intensively studied yet. However, the electromagnetic drag forces caused by the particle charging of the injected particles are negligible compared with those by neutrals and charged particles due to negligible electrical conductivity of soda-lime-silica powders. Thus, the momentum equations for a single spherical particle injected vertically downward into the plasma torch can be expressed as follows:

$$\frac{du_p}{dt} = -\frac{3}{4} C_D \left(u_p - u\right) U_R \left(\frac{\rho}{\rho_p d_p}\right) + g \tag{8}$$

$$\frac{dv_p}{dt} = -\frac{3}{4} C_D \left(v_p - v\right) U_R \left(\frac{\rho}{\rho_p d_p}\right) \tag{9}$$

$$U_R = \sqrt{\left(u_p - u\right)^2 + \left(v_p - v\right)^2} \tag{10}$$

The particle temperature, liquid fraction and diameter are predicted according to the following energy balances:

$$Q = \pi d_p^2 h_c \left(T - T_p\right) - \pi d_p^2 \sigma_s \varepsilon \left(T_p^4 - T_a^4\right) \tag{11}$$

$$\frac{dT_p}{dt} = \frac{6Q}{\pi \rho_p d_p^3 C_{pp}} \quad \text{for } T_p < T_b \tag{12}$$

$$\frac{d\chi}{dt} = \frac{6Q}{\pi \rho_p d_p^3 H_m} \quad \text{for } 1000 \le T_p \le 1600 \tag{13}$$

$$\frac{dd_p}{dt} = \frac{2Q}{\pi \rho_p d_p^2 H_v} \quad \text{for } 1000 \le T_p \le 1600,\ T_p \ge T_b \tag{14}$$

Drag coefficient C_{Df} is calculated using Eq. (15) and the property variation at the particle surface layer and the non-continuum effects are taken into account by Eq. (16) and (17) [Chen, 1983].

$$C_{Df} = \begin{cases} \dfrac{24}{R_e} & R_e \le 0.2 \\ \dfrac{24}{R_e}\left(1 + \dfrac{3}{16} R_e\right) & 0.2 < R_e \le 0.2 \\ \dfrac{24}{R_e}\left(1 + 0.11 R_e^{0.81}\right) & 2.0 < R_e \le 21.0 \\ \dfrac{24}{R_e}\left(1 + 0.189 R_e^{0.62}\right) & 21.0 < R_e \le 200 \end{cases} \tag{15}$$

$$f_1 = \left(\frac{\rho_\infty \mu_\infty}{\rho_s \mu_s}\right)^{-0.45} \tag{16}$$

$$f_2 = \left\{1 + \left(\frac{2-\alpha}{\alpha}\right)\left(\frac{\gamma}{1+\gamma}\right)\frac{4}{Pr_s} Kn\right\}^{-0.45}, \quad 10-2 < Kn < 0.1 \tag{17}$$

source terms for the mass, momentum and energy conservation equations for each control volume throughout the torch are then predicted. The plasma temperature and flow fields are predicted again incorporating these particle source terms. The new plasma temperature and flow fields are used to recalculate the particles trajectories, temperature and size histories. Calculating the new source terms and incorporating them into conservation equations constitute the effects of plasma-particle interaction, thereby completing the cycle of mutual interaction. The above computation schemes are repeated until convergence. The physical properties of soda-lime-silica glass powders used in the present investigation are listed in Table 3.

Mass density	2300 kg/m^3
Specific heat at constant pressure	800 J/kg-K
Porosity	80%
Fusion temperature	1000~1600 K
Boiling temperature	2500 K
Latent heat of fusion	3.69×10^5 J/kg
Latent heat of vaporization	1.248×10^7 J/kg

Table 3. Physical properties of soda-lime-silica glass powders

3. Simulated results

The calculation has been carrier out for a plasma power of 10 kW, reactor pressure 0.1 MPa and induction frequency 4 MHz. The discharge conditions are tabulated in Table 1. In this study, attention is given to the plasma-particle interaction effects on individual particle trajectory, velocity, and temperature history along the trajectories for different carrier gas flow-rate and powder feed-rates. Attention also paid to investigate how the plasma-particle energy exchange process is affected by the particle loading effects. Two aspects of the thermal treatment are investigated: the behavior of the individual particles, and the global effects of the particles on the plasma fields. The carrier gas flow-rate is very vital in determining the individual particle trajectories, and the allowable powder feed-rate. Figure 2 shows the isotherms in the torch for a carrier gas flow-rate of 6 L/min argon and various powder feed-rates. The other discharge conditions are the same as presented in Table 1. A comparison among the isotherms clearly reveals the intense cooling around the torch centerline that increases with powder feed-rate. However, the plasma temperature away from the centerline of the torch remains almost unaffected by higher powder feed-rates. This is because the

individual particle trajectories are not widely outbound in the radial direction; rather the trajectories are very close to the torch axis. Thus, the plasma-particle interaction around the centerline is very crucial at higher powder feed-rate. The same kind of arguments is proposed by Ye et al [Ye, 2000] to explain the particle trajectories for alumina and tungsten particles. The effects of carrier gas flow-rates on the individual particle trajectories are presented in Fig. 3, for the particle diameter of 50 μm and a feed-rate of 5 g/min. It is comprehended that the higher flow-rate of carrier gas enhances the axial velocity of the particles, because the initial axial velocity of the particles depends on carrier gas flow rate; as a result the trajectories become closer to the torch axis at higher flow-rate. The individual particle temperature history along the trajectory is also influenced by the carrier gas flow-rate and powder feed-rate. Figure 4 shows the effects of carrier gas flow-rate on the particle temperature for a feed-rate of 5 g/min. It is found that the particle temperature along the trajectory decreases at higher carrier gas flow-rate. The main reason is the cooling of plasma at Fig. 2 Effects of powder loading on the isotherms for a carrier gas flow-rate of 6 L/min higher carrier gas flow-rate that leads less heat transfer to particles. Figure 5 describes the effects of powder feed-rate on the particle temperature along the trajectory. Like the flow-rate of carrier gas, the higher feed-rate of powder also causes intense cooling of plasma; thus, the heat transfer to particles decreases what results lower particle temperature. At this stage of investigation, it is indeed necessary to discuss the energy transfer mechanism to particles. The energy transfer is affected by the particles physical properties, plasma temperature, and velocity. The last two parameters are affected to a large extent by the carrier gas flow-rate and powder loading. The net energy transfer to particles is calculated by integrating the energy transfer rate to the particles injected per unit time over the residence time for all the particle trajectories. Mathematically the net energy transfer to particles (Q_{net}) can be expressed as follows:

$$Q_t = \int_{t=0}^{t=t_s} \left\{\pi d_p^2 h_c \left(T - T_p\right) - \pi d_p^2 \sigma_s \varepsilon \left(T_p^4 - T_a^4\right)\right\} dt \tag{32}$$

$$Q_{net} = \sum_l \sum_k N^{(l,k)} Q_t \tag{33}$$

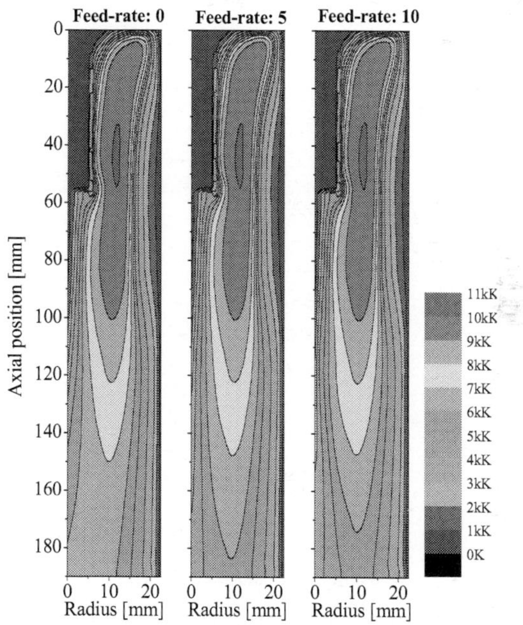

Fig. 2. Effects of powder loading on the isotherms for a carrier gas flow-rate of 6 L/min

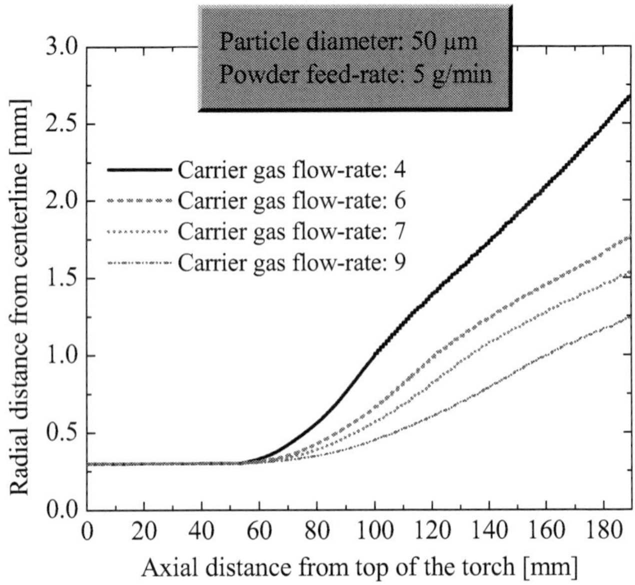

Fig. 3. Effects of carrier gas flow-rate on the particle trajectories for a powder feed-rate of 5 g/min

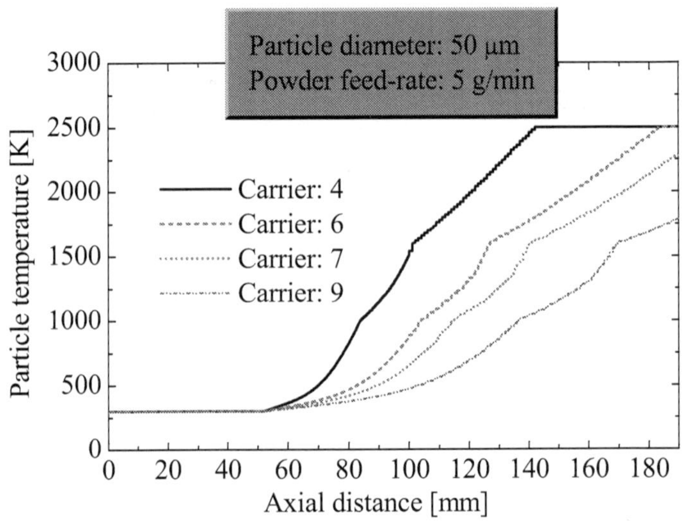

Fig. 4. Dependence of particle temperature history along the trajectory on carrier gas flow-rate

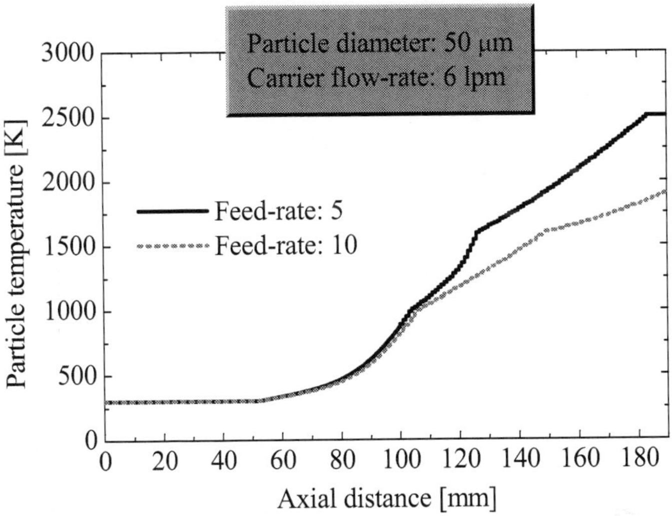

Fig. 5. Dependence of particle temperature history along the trajectory on the powder feed-rate

Figure 6 clearly presents how the net energy transfer to particles is affected by the carrier gas flow-rate under powder loading conditions. Only 5 g/min of powder feeding decreases

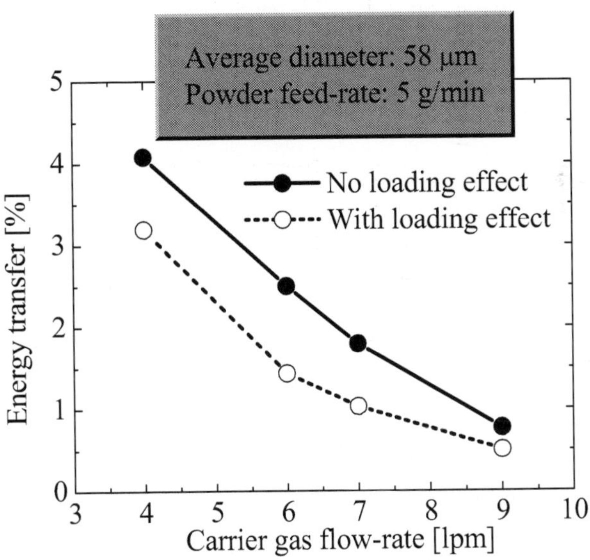

Fig. 6. Effects of powder loading and carrier gas flow-rate on the plasma-particle energy transfer

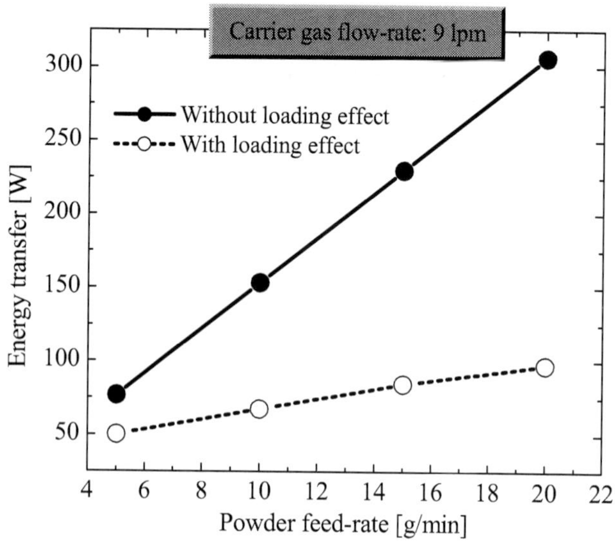

Fig. 7. Particle loading effects on plasma-particle energy transfer at various powder feed-rate

the energy transfer to particle by about 44%. The powder loading effect and the dependence of energy transfer to particles on the powder feed-rate is presented in Fig. 7, for a carrier gas flow-rate of 9 L/min. It can be noticed that energy transfer to particles increases linearly with feed-rate in the absence of particle loading effect; however, when particle loading effect is taken into account, energy transfer to particles yet increases with feed-rate but with a declined slop. The main reason is the intense local cooling of plasma around the torch centerline under dense particle loading. It is also evident that the particle loading effect is pronounced at higher powder feed-rate.

4. Experimental

4.1 Setup

The experimental setup consists of a plasma torch (Fig. 1), a reaction chamber, powder feeder, and a power supply unit (4 MHz, 20 kW). The plasma torch consists of a water-cooled co-axial quartz tube surrounded by a three-turn induction coil. The granulated soda-lime-silica glass powders are prepared by spray-drying method from the reagents of Na_2CO_3, $CaCO_3$ and SiO_2 with the composition of Na_2O:16, CaO:10 and SiO_2:74 in wt%. The mean diameter and porosity of soda-lime-silica glass powders are 58 μm and 80%, respectively. The plasma discharge conditions are the same as those described in Table 1 in the modeling section. The soda-lime-silica glass powders are injected into ITP torch along with the carrier gas at a rate of 5-20 g/min and the quenched powders are collected on a water-cooled ceramic block at 340 mm from the nozzle exit.

4.2 Characterization of plasma-treated particles

The treatment quality of the powders is characterized by the vitrification degree, the surface morphology, cross-sectional structure and composition of the quenched powders. The

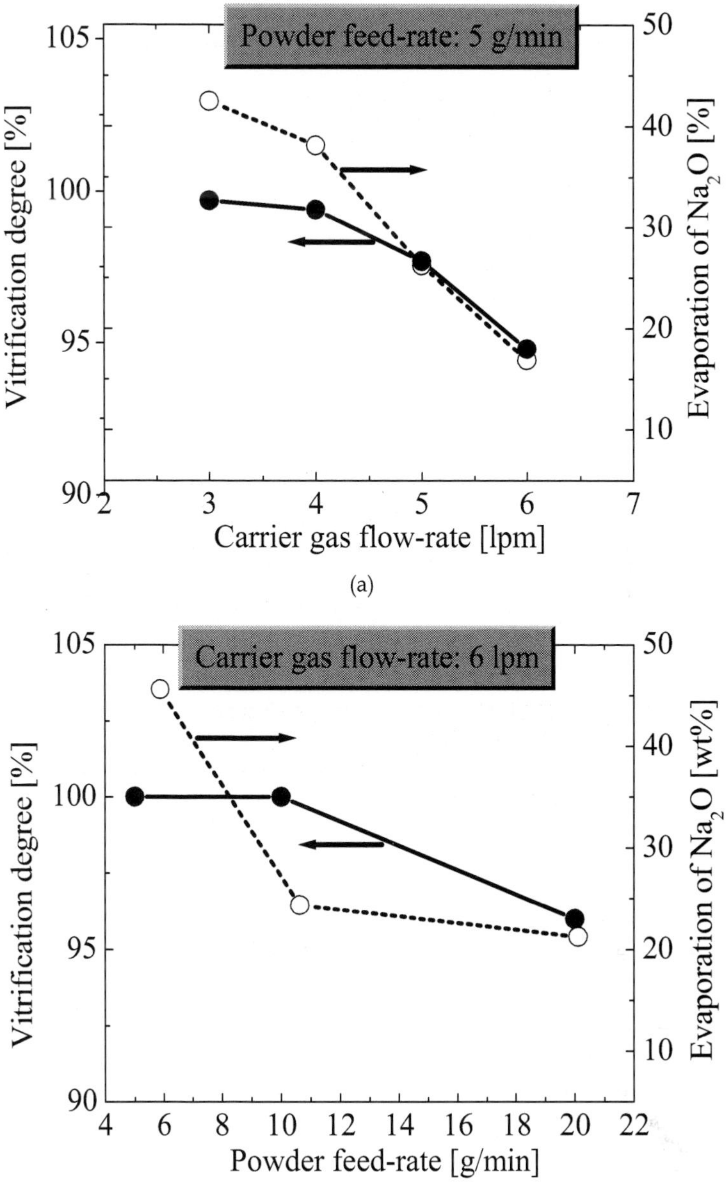

(a)

(b)

Fig. 8. Effects of carrier gas flow-rate (a), and powder feed-rate (b) on the vitrification degree and Na_2O evaporation rate

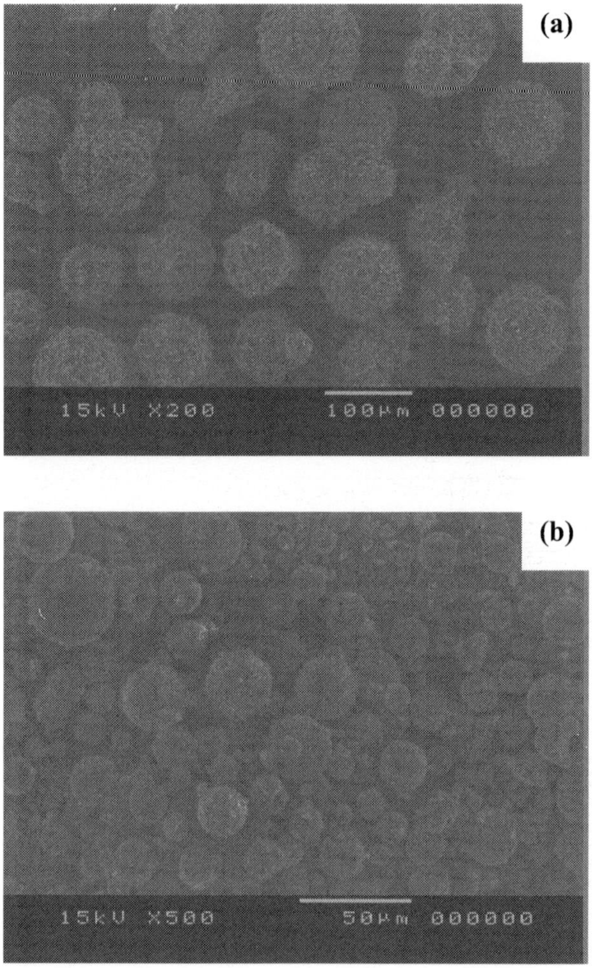

Fig. 9. The SEM photographs of soda-lime-silica glass powders before (a), and after (b) thermal treatment

vitrification degree is defined as the ratio of the converted crystalline phases of SiO_2 in the quenched powder to the crystalline SiO_2 in the raw powders. The vitrification degree of quenched powders is quantitatively determined by X-ray diffractometry (XRD) on Miniflex (Rigaku) with Cu Kα radiation at 30 kV and 15 mA. The data are collected in the 2θ range 3-90° with a step size of 0.02° and a scan speed of 4°/min. The quenched powders collected at the reaction chamber are examined by scanning electron microscopy (SEM) on JSM5310 (JEOL) to observe their surface morphologies and cross-sectional microstructures. The composition of quenched powders is analyzed by inductively coupled plasma (ICP) on ICP-8100 (SHIMADZU).

4.3 Experimental results

In the experiment soda-lime-silica glass powders are injected along with the carrier gas through the nozzle inserted into the plasma torch. Thus, the initial particle velocity is the same as that of carrier gas. When the particles come in contact to high temperature plasma flame, they are heated and their temperature starts to rise. As the particle temperature reaches to its melting temperature, particle porosity decreases drastically; as a result, particle diameter shrinks. When particle temperature reaches to its boiling temperature, vaporization takes place and particle diameter shrinks further. In order to investigate the effects of carrier gas flow-rate and powder feed-rate on energy transfer to particles, the vitrification degree and Na_2O evaporation rate are estimated through XRD and ICP spectrum analysis. Higher vitrification degree and Na_2O evaporation rate indirectly indicate the large energy transfer to particles. Figure 8 shows the XRD and ICP spectrum analysis results. It can be noticed that both the evaporation rate of Na_2O and the vitrification degree decrease with the increase of both carrier gas flow-rate [Fig. 8(a)] and powder feed-rate [Fig. 8(b)]. The vitrification degree and the evaporation of Na_2O depend to a large extent on the particle temperature. Higher carrier gas flow-rate and powder feed-rate cause lower plasma temperature which causes less heat transfer to particles; as a result particle temperature is lower. It is important to underline that after the thermal treatment, the size, composition, and morphology of the particles have been changed significantly. The effects of thermal treatment are visualized in the SEM photograph as shown in Fig. 9. From Fig. 9 (a) and (b), it can be noticed that after treatment, the particle size becomes smaller, quite spherical, smoother and compact surface.

5. Discussions

To validate the modeling and simulated results, a comparative discussion between simulated and experimental results are indeed necessary. From the experimental results it is found that at higher carrier gas flow-rate and powder feed-rate, both the evaporation rate of Na_2O and the vitrification degree decrease. These results indicate that less heat transfer to particles takes place at higher carrier gas flow-rate and powder feed-rate. From the simulated results it is evident that at increased carrier gas flow-rate and powder feed-rate, the energy transfer to particles decreases; as a result, particles temperature becomes lower. It is convinced that the main reason of less heat transfer to particles is the severe local cooling of plasma around the torch centerline at higher carrier gas flow-rate and powder feed-rate. Thus, it may be argued that the simulated results well agree with the experimental findings.

6. Conclusions

In this chapter we basically, discussed the way of thermal treatment of any type of granulated porous particles by induction thermal plasma. A general plasma-particle interactive flow model has been discussed using what it is possible to simulate the particle trajectories, temperature histories, plasma temperature contours etc. The described model can be used to optimize the carrier gas flow-rate, particle size, and powder feed-rate to achieve the maximum treatment efficiency during thermal treatment of granulated powders by argon-oxygen induction thermal plasmas. Numerically, it is found that the heat transfer to particles decreases at increased carrier gas flow-rate and powder feed-rate, and these results well agree with those of experiment. Thus, it can be concluded that, efficient thermal treatment of particles depends not only on the physical properties of the particles, but also on the plasma discharge conditions and particle parameters. Therefore, for a particular type of powder (certain physical properties) both carrier gas flow-rate and powder feed-rate mainly govern the treatment quality.

7. Nomenclature

A_c Complex amplitude of vector potential
B Magnetic field vector
C_p Specific heat at constant pressure
C_{pp} Particle specific heat at constant pressure
d_p Particle diameter
D_{ij} Binary diffusion coefficient between species i and j
D_m Multicomponent diffusion coefficient
E Electric field vector
f Frequency
g Acceleration of gravity
h Enthalpy
h_c Heat transfer coefficient
H_m Latent heat of melting
H_v Latent heat of vaporization
i Complex vector ($\sqrt{-1}$)
J Current density vector
K_n Knudsen number
M_i Molecular weight of species i
M_j Molecular weight of species of j
N_t^0 Total number of particles injected per unit time
n_d Particle size distribution
n_r Fraction of Nt0 injected at each point
Nu Nusselt number
p Pressure
Pr Prandtl number
Q Net heat exchange between the particle and its surroundings
Q_r Volumetric radiation loss
Re Reynold number
S_p^C Particle source term in continuity equation
S_p^M Particle source term in momentum equation
S_p^E Particle source term in energy equation
t Time
T Plasma temperature
T_a Ambient temperature
T_b Boiling point temperature of particles
T_e Electron temperature
T_h Heavy particle temperature
T_{ion} Ion temperature

T_p Particle temperature
t_s Residence time of particle in the plasma
u Velocity vector
u_p Axial velocity component of particle
U_R Relative speed of particles with respect to plasma
v_p Radial velocity component of particle
x_i Mole fraction of species i
y Mass fraction

Greek symbols

Vector operator
κ Thermal conductivity
Mass density
Viscosity
σ Electrical conductivity
κ Thermal conductivity
μ_0 Permeability of free space
ω Angular frequency ($2\pi f$)
$\bar{\Omega}_{ij}^{(1.1)}$ Collision integral between species i and j
ε Particle porosity
σ_s Stefan-Boltzmann constant
Liquid mass fraction of a particle
Thermal accommodation coefficient
γ Specific heat ratio

Subscripts

f Properties corresponding to film temperature
p Particle
s Properties corresponding to particle temperature
∞ Properties corresponding to plasma temperature
(i, j) Location of a control volume or cell

Superscripts

(l, k) Particles having an initial diameter d_l, and injection point r_k.

8. References

Boulos, M. I, 1978, "Heating of powders in the fire ball of an induction plasma," IEEE Trans. on Plasma Sci. PS-6, pp. 93-106.
Crowe, C. T., Sharma, M. P., and Stock, D. E., 1977, "The particle-source-in cell (PSI-CELL) model for gas-droplet flows," J. Fluid Eng., 99, pp. 325-332.
Chen, X., and Pfender, E., 1983, "Effects of Knudsen number on heat transfer to a particle immersed into a thermal plasma," Plasma Chem. Plasma Process., 3, pp. 97-113.
Fan, X., Ishigaki, T., and Sato, Y., 1997, "Phase formation in molybdenum disilicide powders during in-flight induction plasma treatment," J. Mater. Res., 12, pp. 1315-1326.
Mostaghimi, J., Paul, K. C., and Sakuta, T., 1998, "Transient response of the radio frequency inductively coupled plasma to a sudden change in power," J. Appl. Phys., 83, pp. 1898-1908.

Lee, Y. C., Chyou, Y. P., and Pfender, E., 1985, "Particle dynamics and particle heat and mass transfer in thermal plasmas. Part II. Particle heat and mass transfer in thermal plasmas," Plasma Chem. Plasma Process., 5, pp. 391-414.

Patankar, S. V., 1980, Numerical fluid flow and heat transfer, Hemisphere, New York.

Proulx, P., Mostaghimi, J., and Boulos, M. I., 1985, "Plasma-particle interaction effects in induction plasma modeling under dense loading conditions," Int. J. Heat Mass Transfer, 28, pp. 1327-1335.

Reed T. B. "Induction-Coupled Plasma Torch," *Journal of Applied Physics*, 1961, 32(5), p.821

Tanaka, Y., Paul, K. C., and Sakuta, T., 2000, "Thermodynamic and transport properties of N_2/O_2 mixtures at different admixture ratio," Trans. IEE Japan, 120-B, pp. 24-30.

Watanabe, T., and Fujiwara, K., 2004, "Nucleation and growth of oxide nanoparticles prepared by induction thermal plasmas," Chem. Eng. Comm., 191, pp. 1343-1361.

Yoshida, T., and Akashi, K., 1977, "Particle heating in a radio-frequency plasma torch," J. Appl. Phys., 48, pp. 2252-2260.

Ye, R., Proulx, P., and Boulos, M. I., 2000, "Particle turbulent dispersion and loading effects in an inductively coupled radio frequency plasma," J. Phys. D: Appl. Phys., 33, pp. 2154-2162.

23

Method for Measurement of Single-Injector Heat Transfer Characteristics and Its Application in Studying Gas-Gas Injector Combustion Chamber

Guo-biao Cai, Xiao-wei Wang and Tao Chen
School of Astronautics, Beijing University of Aeronautics and Astronautics
P. R. China

1. Introduction

In the development of a Liquid Propellant Rocket Engine (LPRE), injector is always the element which requires the longest development period. An injector always especially in the initial design phase needs hundreds and thousands of tests to get a choice. These numerous tests, in turn, require reliable and accurate measurement method to give the basic support. The strength, the life cycle, and the cooling system effectiveness are highly dependent on heat transfer into and out of the system (Tramecourt et al., 2005). Heat transfer characteristics, combined with combustion efficiency, and combustion instability, are the three key parameters needed to be investigated in developing a new injector in a combustion chamber. The heat transfer characteristics generally contain the temperature and heat flux on the hot-gas-wall in the combustion chamber. The combustion efficiency and combustion instability can be observed easily by measuring the chamber pressure. However, it is always very hard to measure the temperature and heat flux on hot-gas-wall of the chamber in the hot-tests due to the extreme environment in LPRE chambers (temperature > 3000K, pressure > 10atm).

Additionally, considerable efforts have been dedicated to model combustion in combustion chambers to understand and predict the heat transfer to the chamber walls (Zurbach, 2006), whereas the validation of computational fluid dynamics (CFD) design tools requires reliable experimental data assessment. This acquires a comprehensive set of data in the same facility, including wall heat fluxes along with inflow measurements (Tucker et al., 2005). To obtain the exact wall heat flux data, a reliable measurement method, in turn, is also required.

Though a number of efforts had contributed to design new injectors and study on heat transfer of injectors, studies focusing on the measurement method are less well documented. In recent years, Pennsylvania State University, University of Florida and NASA MSFC etc. made many attempts on studying the heat transfer characteristics of some injectors in heat sink chambers (Conley et al., 2007; Jones et al., 2006; Marshall et al., 2005; Santoro & Pal, 2005; Vaidyanathan et al., 2007, 2010b). They applied coaxial thermocouple to measure two point temperatures at the same axial location and difference radial locations in a heat-sink copper combustion chamber, and then used the temperatures to calculate the local heat flux through a simple radial heat conduction analytical calculation. Coy summarized and analyzed the inverse heat transfer methods, and described a method for resolving inverse heat conduction problems using approximating polynomials (Coy, 2010), in which two sensors are also needed in any axial measurement location. All of previous method tried to get the exact time-dependent data of heat flux and surface temperature based on analytical solutions. Though these measurement methods can get the exact real-time heat flux data if right calculation methods are used, they always need two measurement points at the any axial measurement location, which increases the failure probability twice at any axial measurement location, and the current most popular method always put one measurement point right on the inner wall, and the thermocouple plug must be chamber-side geometry contoured to match the internal radius of curvature of the chamber, otherwise it will influence the inner flowfield and could not get the right temperature data. A little influence on the inner flowfield may induce a large deviation of the temperature on the measurement point. The potential influence of the thermocouple plug on the inner flowfield always exists, because the separated plug was put on the inner wall. In summary, all of the previous methods tried to get the exact time-dependent data with sacrificing the success probability and the measurement systems are complicated. In addition, they solved the axisymmetric problems with 1-D assumptions or solved 3-D problem with 2-D assumption.

This study introduces a simple method to obtain the inner wall temperature and heat flux in a heat sink combustion chamber. This method uses numerical calculation method to get the ultimate time-dependent inner wall temperature and heat flux with numerical calculation method. It considers the inner wall heat transfer coefficient as unchanged, which suffers a little change in hot-test, but the change is always less than 5%. Only single temperature measurement point at one axial location is just needed. In current method the measurement points are not on the inner wall surface, thus the potential influence on the inner flowfield doesn't exist anymore and there is not any sealing problem which also may cause severe measurement error. A reliable 2-D axisymmetric numerical calculation was applied to get the ultimate data. The measurement method, the related data processing and error analysis are demonstrated in detail with a specific hot-testing example of a gaseous hydrogen/gaseous oxygen single-element heat sink combustion chamber using this method. This method can be a feasible alternate one to get the inner wall temperature and heat flux, and it was originally developed for single-element axisymmetric chamber, and also can be a reference for non-axisymmetric chambers and multi-element injector chambers.

Furthermore, this method was used to investigate the heat transfer characteristics of a shear-coaxial single-element gas-gas injector combustion chamber. The full flow stage combustion (FFSC) cycle engine is preferable because of its high performance and high reliability, for which gas-gas injector technology is a key technology for this kind of engine and has received extensive studies (Archambault et al., 2002a, 2002b; Davis & Compbell, 1997; Farhangi et al., 1999; Foust et al., 1996; Meyer et al., 1996; Schley et al., 1997; Tucker et al., 1997; Santoro et al 2005; Tucker, 2007a ,2008b; Lin et al., 2005; Cai et al., 2008; Wang et al, 2009a, 2010b, 2010c etc.). Also, gas-gas injectors have been widely used on other engines and combustion devices (Calhoon, 1973; Groot, 1997). In this sense, it is of great value to investigate the heat transfer characteristics of a gas-gas injector combustion chamber. A single-injector heat-sink chamber was designed and hot-fire tested for 17 times at chamber pressure from 0.92MPa to 6.1MPa. Inner hot-gas-wall temperature and heat flux along with the axial direction of the chamber were obtained. The heat flux results were compared with each other qualitatively and quantitatively. The inner combustion flows were also numerically simulated with multi-species turbulence N-S equations at higher chamber pressure from 5MPa to 20MPa to extend the experimental results. Both the combustion flow structures and heat flux profiles on inner wall were obtained and discussed.

2. Measurement method

The heat-sink chamber is preferable (Calhoon, 1973; Marshall et al., 2005; Santoro et al 2005; Conley et al., 2007; Jones et al., 2006) in the initial design phase of a new injector or in the study on the heat transfer characteristics of injector owing to its simple structure, low cost and easy manufacture. Thus, heat-sink combustion chamber was selected in this study. And the temperature measurement scheme is shown in Fig. 1.

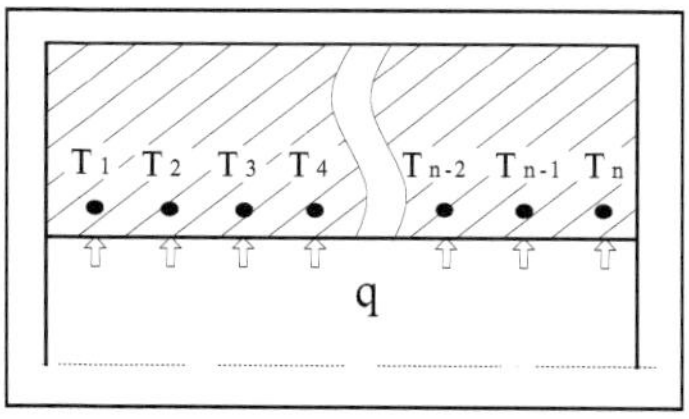

Fig. 1. Temperature measurement scheme in the heat sink chamber

This method applies thermocouples to measure the temperatures of a series of points near the inner wall along the axial direction in the heat-sink chamber. The distances between these points and inner wall surface are the same. The basic theory of this method is stated this way: with the development of combustion flow in the chamber, different heat fluxes are produced on the inner wall axially, further resulting in different temperatures on these measurement points. Therefore, the temperatures of these points can be utilized to obtain the heat flux and temperature on the inner wall.

In view that a heat-sink chamber can not undertake a long hot test, the tests are always completed within several seconds in which, fortunately, enough effective temperature output curves can be produced. In light of the temperatures are measured at those points near the inner surface of the chamber, the temperature profile along measurement point can clearly reflect the inner surface heat flux profile. The longitudinal heat transfer effect will be considered in the heat flux calculation.

This measurement method was employed in a single-element gaseous hydrogen/gaseous oxygen shear-coaxial injector test. The chamber shown in Fig. 2 was a modular heat sink design assembled with two OFHC (oxygen-free high conductivity) copper cylindrical barrel spool sections and an ablation resistance material nozzle. The inner diameter was 26 mm and the cylinder part was 255 mm in length.

Many small holes were designed at the axial locations of the measurement points, with temperatures of those at the bottom measured by the thermocouples. Axial locations for instrumentation are indicated in Fig. 2. The distances between all the measurement points and the chamber inner wall were the same except the distance between the first point and the inner wall. Multiple instrumentations (4 each) were presented at some axial locations, and four thermocouples were separated at 90° to examine the non-axisymmetry which may probably happen due to presumable manufacture and installation errors. Four thermocouples were also fixed in the injector surface to observe the plate surface's heat load.

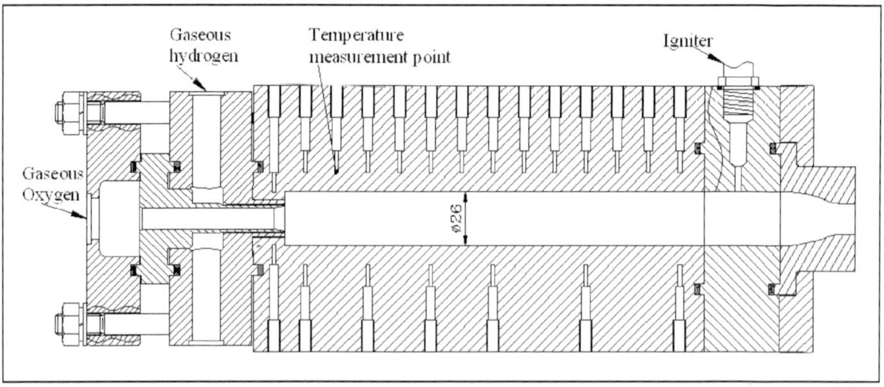

Fig. 2. Schematic of the chamber and thermocouple locations

The timing of the tests was designed as follows: the igniter preceded the injector, and with the chamber pressure got steady, the igniter was shut down. In the light of the fact that the igniter mass flowrate was less than 8.0g/s, the total ignition gas temperature was less than 1000K, and the ignition time lasted less than 0.8s, the influence of the ignition on wall temperature of the long cylinder part could be neglected here. A photo of the calorimeter chamber is shown in Fig. 3.

Fig. 3. The photo of the calorimeter chamber

3. Key design parameters and thermocouples

The test time was set to be 3s for this heat sink chamber. Some key parameters of the testing article, including the chamber wall thickness *H*, measurement hole diameter *D* and the distance *L* from the measurement point to the inner wall surface are listed in Table 1.

The Type-K thermocouples used in this hot-testing example were designed and provided by the Beijing West Zhonghang Technology Ltd. Their measurement ranges were all set to be 0-600°C. The temperature versus voltage response for the Type-K thermocouple is shown in Fig. 4, which shows a near-linear response. Their response time constants were designed less than 100ms and calibrated individually.

H/mm	48
D/mm	1.5
L/mm	8

Table 1. The values of key parameters

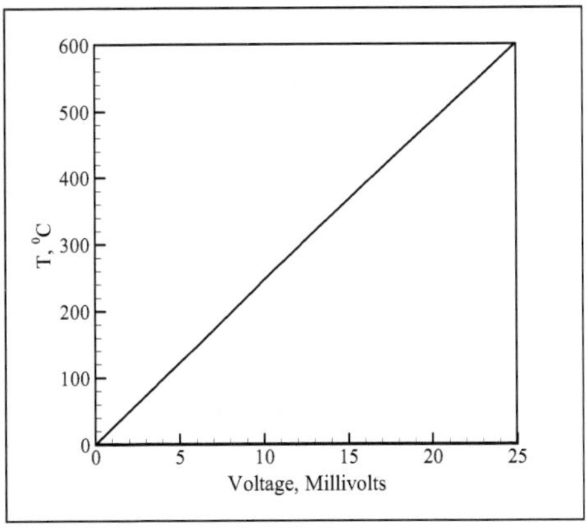

Fig. 4. Type-K thermocouple response

The chamber was fixed on a testing platform at the FFSC Laboratory in BUAA, resulting that the chamber was electronically grounded, which can produce strong interference signal in the output data. Therefore, an isolation module transmitter was designed for each thermocouple to eliminate the interference signal. The temperature responses of a thermocouple without and with isolation module are shown in Fig. 5.

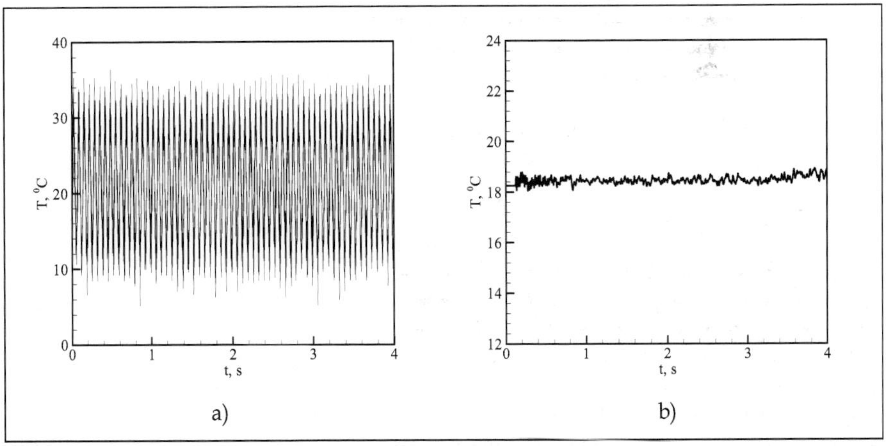

Fig. 5. Temperature outputs of transmitters a) without and b) with insulate module

4. Calculation of inner wall temperature and heat flux

Several hot-fire tests were carried out for this gaseous hydrogen/gaseous oxygen single-element injector. A typical chamber pressure and one typical temperature curve are shown in Fig. 6. The temperature curve is depicted at the measurement point which is located 100mm away from the injector plane axially.

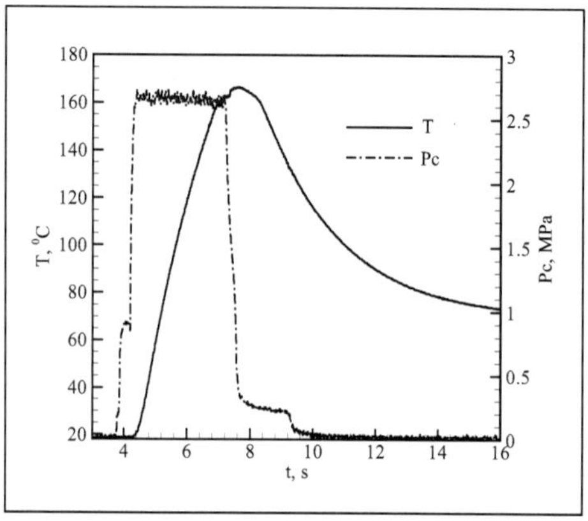

Fig. 6. Typical chamber pressure and wall temperature curve

Two steps are involved in converting these temperature output data into temperature and heat flux on the inner wall: 1. depicting the real measurement point temperature curves based on the output temperature curves, 2. obtaining the temperature and heat flux on the inner wall with 2-D axisymmetric heat transfer simulation. An error analysis will also be conducted in the end for this method.

4.1 Temperature conversion

Any thermocouple has somewhat response delay, that is, when thermocouple of initial temperature T_0 is used to measure an object of constant temperature (T_1), the response curve function can write into equation (1):

$$\frac{T_1 - T}{T_1 - T_0} = \exp(-\frac{1}{\tau_c}\tau) \tag{1}$$

where τ_c is thermocouple delay constant and τ is time. The qualitative relationship between the object temperature and thermocouple output is shown in Fig. 7.

The heat transfer was unsteady in the test. The wall temperature kept rising after the combustion started. Obviously, there was some discrepancy between the real temperature curve and the experimental output curve for the existence of response delay, and it increased with the testing time, as shown in Fig. 8.

In this method, curve fitting and analytical calculation are employed to solve the real temperature curve from the thermocouple output. An example is shown in Fig. 9, T_1 is a thermocouple output curve, and T_2 is the biquadratic polynomial fitting curve of T_1 (almost overlapped with T_1). T_3 is the real temperature result considering the response delay effect, which is higher than T1. The thermocouple delay constant was set 100ms in this case.

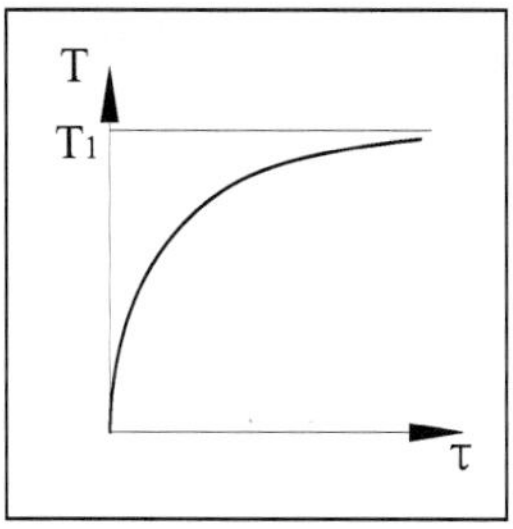

Fig. 7. The thermocouple characteristic curve schematic

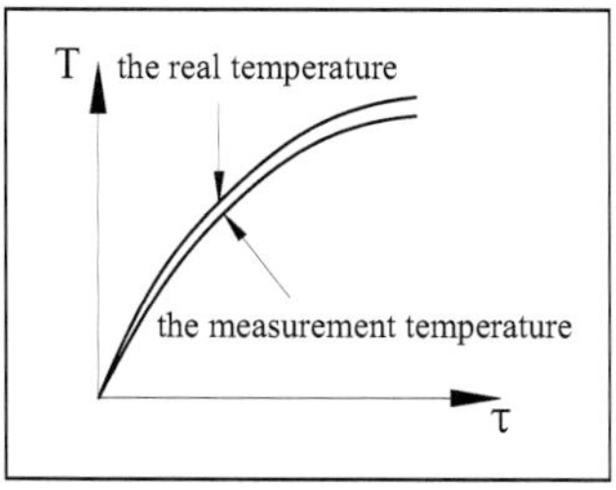

Fig. 8. Comparison of the TC measurement temperature curve and real wall temperament curve

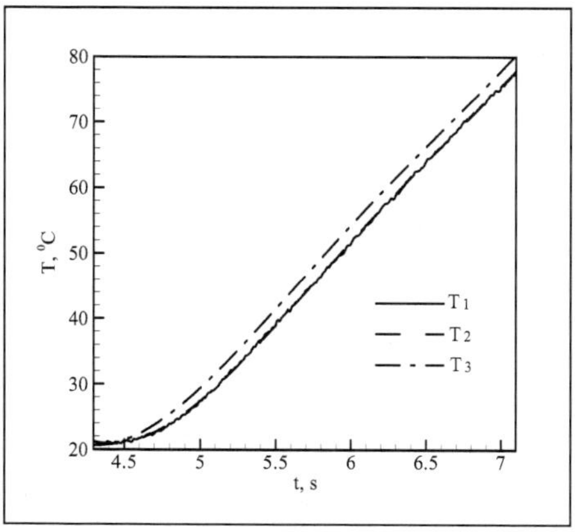

Fig. 9. Example of temperature curve conversion

4.2 Inner wall heat flux and temperature calculation

The influence of the hot-testing start-up on the wall temperature was very limited, which can be observed from Fig. 10, thus, this influence was neglected here. When chamber pressure reached 90% of the stable value was considered as the beginning of the heat transfer and the calculation initial time, as t_0 shown in Fig. 10. The end time of calculation

was chosen at the end of stable combustion, as t_1 shown in Fig. 10. The heat transfer from t_0 to t_1 was calculated and the combustion was assumed to be stable in this period.
The inner wall heat flux can be defined with equation (2), in which the coefficient h can be defined by the Bartz equation as show in equation (3)(4).

$$q = h(T^* - T_w) \tag{2}$$

$$h = const.\frac{1}{d_t^{0.2}}\left(\frac{\eta^{0.2} c_p}{\Pr^{0.6}}\right)\left(\frac{p_c^*}{c^*}\right)^{0.8}\left(\frac{A_t}{A}\right)^{0.9}\sigma \tag{3}$$

$$\sigma = \left[\frac{1}{2}\frac{T_{wg}}{T^*}\left(1+\frac{\kappa-1}{2}Ma^2\right)+\frac{1}{2}\right]^{-0.68}\left(1+\frac{\kappa-1}{2}Ma^2\right)^{-0.12} \tag{4}$$

Here, T^* and T_w are inner flow total temperature and the inner wall temperature. From Equations (3)-(4), it can be seen that the coefficient h is almost unchanged when the chamber pressure p_C^* is constant. Figure 10 shows that the chamber pressure is steady and constant in the test, thus, the coefficient h is considered as unchanged in the hot-test, i.e. time-averaged heat transfer coefficient through the hot-test, which will be used as the boundary condition in the following heat conduction numerical modeling. In fact, in a real engine, the total combustion flow temperature, specific heat ratio and the *Ma* number are about 3600K, 1.2, and 0.1, the inner wall temperature changes from 300K to 600K, the heat transfer coefficient just changes less than 5% according to Equation (3)-(4).

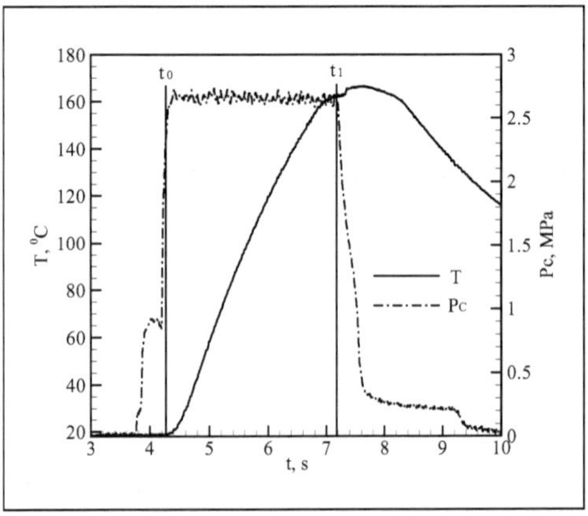

Fig. 10. The calculation time and temperature chosen from the output curve

The original thermocouple output data were obtained from the hot-fire test directly, and then the temperature T_m at the end time of calculation (t_1) were gotten. Then the real temperatures T_r of each measurement point can be obtained through a conversion program described above. Tm and T_r are shown in Table 2. The injector plate was set at zero point axially.

Axial location/mm	T_m/°C	T_r/°C
-3	74.5	78.5
10	51.1	54.2
25	92.4	97.2
40	108.3	113.6
55	115.4	121.4
70	98.6	103.5
85	108.3	113.6
100	144.1	150.8
115	157.9	164.9
130	159.6	166.6
160	161.5	168.7
175	156.4	163.3
190	158.6	165.6
210	165.7	173
240	171.1	178.6

Table 2. The measurement and real temperatures

The calculation method for the inner wall temperature and heat flux from the real temperatures at measurement points are introduced below. The whole calculation procedure can be divided into two parts: 1-D and 2-D modelling. The real temperatures at the axial measurement points were labeled as T_{rk} (k=1, 2, …, n) for the sake of convenience.

4.2.1 1D axisymmetric modeling

Firstly, surface heat transfer coefficients at all measurement points were calculated with 1-D axisymmetric model regardless of the longitudinal heat transfer effect. These 1-D results will be used as the first iteration data for the 2-D calculation to accelerate the convergence. The unsteady axisymmetric 1-D heat conduction equation is as follows:

$$\rho c \frac{\partial T}{\partial t} = \frac{1}{r}\frac{\partial}{\partial r}(\lambda r \frac{\partial T}{\partial r}) + \dot{\Phi} \tag{5}$$

The total temperature of the inner combustion flow was chosen as the adiabatic combustion flame temperature T^*. Two empirical surface heat transfer coefficients h_{k1} and h_{k2} ($h_{k1}<h_{k2}$) (k=1,2,…,n) were chosen firstly for each measurement point. Two temperatures T_{k1} and T_{k2} at time t_1 shown in Fig. 10 were obtained from h_{k1} and h_{k2} with unsteady 1-D axisymmetric heat transfer model respectively. The 1-D mesh contains 60 grid points for the 48mm thickness chamber wall, and the stretching of the grid were used near the inner wall boundary. Each unsteady calculation period is from t_0 to t_1 shown in Fig. 10 (~2.95s in this case). Then the first iteration data h_k for 2-D modeling can be obtained using the linear interpolation with Equation (6). All the h_k (k=1,2,…,n) are shown in Table 3.

$$h_k = h_{k1} + \frac{T_k - T_{k1}}{T_{k2} - T_{k1}}(h_{k2} - h_{k1}) \tag{6}$$

4.2.2 2D axisymmetric modeling

2-D axisymmetric unsteady modeling was applied to calculate the real inner wall temperature and the heat flux from t_0 to t_1, as shown in Fig. 11.

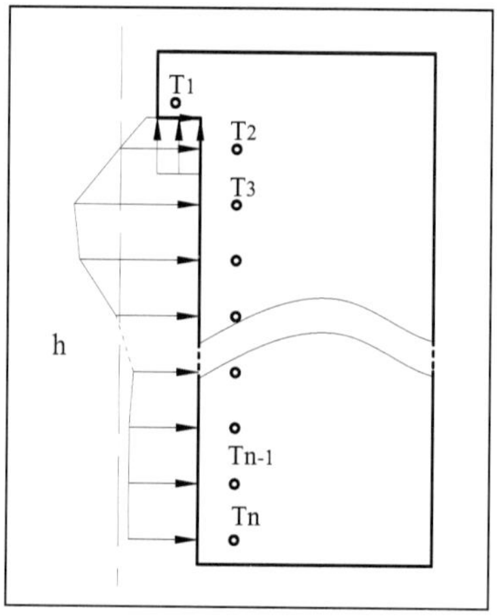

Fig. 11. Schematic of continuous calculation boundary condition of 2-D modeling

In this step, several 2-D unsteady simulations are needed to get a converged inner wall heat transfer coefficients. In each 2-D unsteady simulation, the inner wall of the chamber was treated as the third boundary condition, the calculation time is from t_0 to t_1, and the initial temperature at time t_0 of the whole chamber was set at ambient temperature. The 2-D mesh contains 500×60 grid points, and the stretching of the grid was also used near the inner wall boundary. The unsteady axisymmetric heat conduction equation is as follows (Chapman 1987):

$$\rho c \frac{\partial T}{\partial t} = \frac{1}{r}\frac{\partial}{\partial r}(\lambda r \frac{\partial T}{\partial r}) + \frac{\partial}{\partial z}(\lambda \frac{\partial T}{\partial z}) + \dot{\Phi} \tag{7}$$

Source term $\dot{\Phi}$ is set to $h(T^* - T_w)$ on the inner surface, where h is heat transfer coefficient, a series of heat transfer coefficients h_k (k = 1,2,..,n) obtained from the 1-D modeling were used for the first 2-D simulation at the corresponding inner wall grid points. T_w is the inner wall temperature which keeps changing in each 2-D simulation process, ambient temperature T_{w0} for the initial condition of all the 2-D simulations. After the first unsteady 2-D simulation, a series of measurement point temperatures T_k' (k = 1,2,…,n) at the end of calculation time (t_1) were obtained. Generally, T_k' (k = 1,2,...,n) differs from the real temperatures T_{rk} (k = 1,2,…,n). A correction of each heat transfer coefficients h_k' for the next unsteady 2-D simulation was made with Equation(8). And the unsteady 2-D simulation was repeated until the differences between the simulation results and the real temperatures T_{rk} (k = 1,2,...,n) reach the tolerance.

$$h_k' = h_k \cdot (T_{rk} - T_0) / (T_k' - T_0) \tag{8}$$

Axial location/mm	h_{1D}/MW/(m²K)	h_{2D}/MW/(m²K)
0	368	255
10	933	655
25	1701	1928
40	2038.5	2379
55	2175	2360
70	1838	1450
85	2019	1564
100	2831	3136
115	3028	3280
130	3062	2964
160	3100	3020
175	3000	2800
190	2840	2991
210	3180	2800
240	3296	3060

Table 3. Heat transfer coefficients on the hot-gas-wall

In addition, the temperature-dependent metal (OFHC copper) properties were also considered in the modeling. Table 3 lists the surface heat transfer coefficients at the corresponding axial locations from 1-D and 2-D calculations when the adiabatic flame temperature was considered as the total temperature. Because in the 2-D equation calculation, the inner wall heat flux is the source term and unique, if difference total temperature of the inner flow was set, the coefficients will be changed, but it did not influence the ultimate results of inner wall temperature and heat flux.

The inner wall temperature and the heat flux distributions at the end of calculation time are shown in Fig. 12, where X-coordinate denotes the distance from injector face plate. The results show that heat load on the injector face plate is very low. Temperature and heat flux distributions have similar profiles: They both begin to rise from the injector plate and reach the first peak at around 0.05m, and decrease until 0.075m, and then continue to climb successively to get the maximum values at 0.120m, followed by some small fluctuations and a little decrease in the end section. Though some small fluctuations exits for both the temperature and heat flux at the end, they mainly keep stable. It should be mentioned here that the fluctuations here were somewhat produced by using the same delay constant in the conversion calculation process, which were actually not involved in all thermocouples.

This heat load profile in the chamber was directly produced by the inner combustion flowfield structure. The first peak of the heat flux in the beginning is due to the existence of the strong recirculation zone there. Then the heat flux gets up continuously is because the inner mixing and combustion become more and more intense and sufficient, and the flow becomes faster and faster downstream. When the combustion is mainly completed, the flowfield temperature and velocity reach their maxima. As the flow moves further downstream, the combustion heat release is generally finished, but the wall heat loss still exists, which induces the heat flux goes down a little in the end.

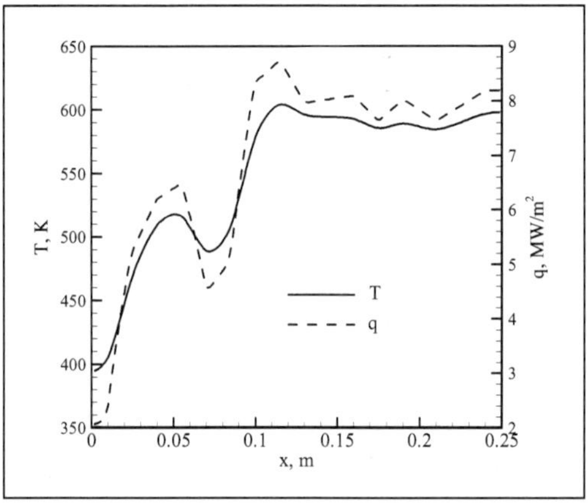

Fig. 12. The hot-gas wall temperature and heat flux distributions

4.3 Error analysis

Errors of this method mainly come from the temperature conversion process and the assumption of the 2-D axisymmetric calculation. Detailed analysis will be conducted in these two parts.

4.3.1 Temperature conversion errors

Errors from the temperature conversion process occurred in the polynomial curve-fitting and application of the same response time constants for all thermocouples. For example as is shown in Fig. 9, the curve-fitting variance was 90.0635K^2 with 2800 temperature data points, indicating that this kind of error was very small and acceptable.

The deviation of response time also caused some errors to the results. Though the response time constants were provided by the manufacturer, analysis should be carried out to estimate its influence on the results.

Fig. 13 shows the different converted temperature curve results from one measurement output data with different response time constants of 100ms and 150ms respectively. Curve T_1 is the original output of thermocouple. T_{100} is the converted curve for response time of 100ms and T_{150} for 150ms. It can be seen that though the difference of the time constants is even 50%, difference of the results is about 1.2 K, just 2% compared with the total temperature increment 60K.

4.3.2 Calculation assumption error

2-D axisymmetric chamber physical model was employed in the heat conduction simulation, whereas real chamber body is three dimensional containing the measurement holes. The 2-D axisymmetric assumption produced somewhat new error, for which, a 3-D calculation and a 2-D axisymmetric calculation for one chamber structure were carried out.

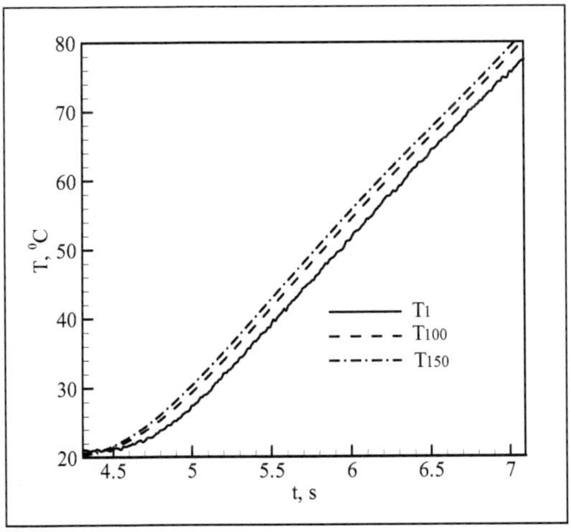

Fig. 13. Comparison of the converted temperature curves for different response times

Real chamber body 3-D physical model for calculations is shown in Fig. 14, where 1/4 of a chamber section with a measurement hole was selected. Uniform convection heat load was imposed on the inner wall surface. Bartz equation was used to get the convection heat coefficient h, 7500W/m^{2}K in this case. Initial temperature was set at 300K. Another 2-D axisymmetric simulation was conducted with a whole chamber body without the measurement hole under the same boundary conditions. The results of these two cases after 3s unsteady calculations are shown in Table 4.

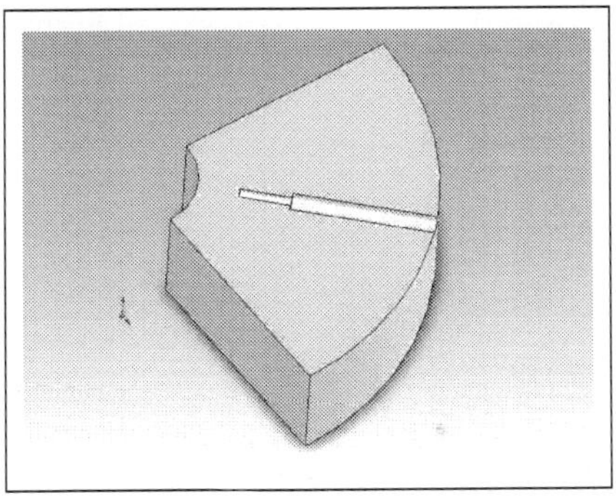

Fig. 14. The physical model of 3-D calculation

Table 4 indicates that the differences of temperature increment on the measurement hole bottom (ΔT_h), the heat flux (q_i) and temperature increment (ΔT_i) at the corresponding inner wall surface were very low. The maximum error was just 3.6%. The reason is that the 3-D character of this chamber body with this measurement method is not evident because measurement hole was rather small compared with the chamber body. The result indicates that for this measurement method, though the chamber structure is three dimensional, the 2-D axisymmetric calculation is feasible and the calculation cost can be sharply saved.

Parameters	Chamber with holes	Chamber without holes	Errors
ΔT_h/K	361	348	3.6%
q_i/MK/m2	17.1	17.5	2.3%
ΔT_i/K	675	665	1.5%

Table 4. Comparison of some key results with 2-D and 3-D calculations

5. Application in studying gas-gas injector combustion chamber

This measurement method was used to investigate the heat transfer characteristics of a single-element gas-gas injector combustion chamber.

5.1 The testing article

The single-element shear-coaxial injector selected from the real engine shown in Fig. 15 was hot-tested here. The configuration of the single-element chamber can be seen in Fig. 2. All of the design parameters of this chamber were the nominal ones from the real engine. The combustor contract ratio was 3.1 and the character length was 800mm. The inner diameter was 26 mm and the cylinder part was 255 mm in length.

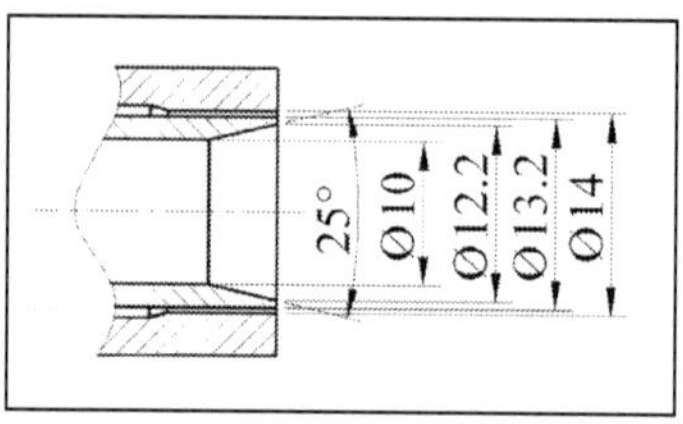

Fig. 15. Schematic of the injector

In the tests of different chamber pressures, with the heat-sink design of the chamber and the heat protection in consideration, the durations of the steady hot-fires were set from 1.5s to 5s for different chamber pressure cases: 0.92MPa—5s, 1.83MPa—4s, 2.69MPa—4s, 3.63MPa—3s, 4.52MPa—2s, 5.42MPa—2s, 6.1MPa—1.5s.

5.2 Test conditions

A total of 17 hot-fire tests were conducted steadily. The operation conditions and typical chamber pressure profiles are summarized in Table 5 and Fig. 16. The nominal Mixture Ration (*MR*) of all conditions were 6.0, the uncertainties of the *MR* of tests were -2.8%-2.3%. The injection velocities and the propellant temperatures for all the cases were kept unchanged.

Chamber Pressure (MPa)	OX Flowrate (g/s)	Ox. injection velocity/(m/s)	Fuel Flowrate (g/s)	Fuel injection velocity/(m/s)	*MR*	Repeat Times
0.92	66.7	~70	11.2	~760	5.96	3
1.83	135.1	~70	22.0	~760	6.14	2
2.69	195.4	~70	32.8	~760	5.95	2
3.63	258.3	~70	44.0	~760	5.87	3
4.52	327.2	~70	54.4	~760	6.01	2
5.42	397.8	~70	65.6	~760	6.06	3
6.1	446.9	~70	76.6	~760	5.83	2

Table 5. Test conditions summary

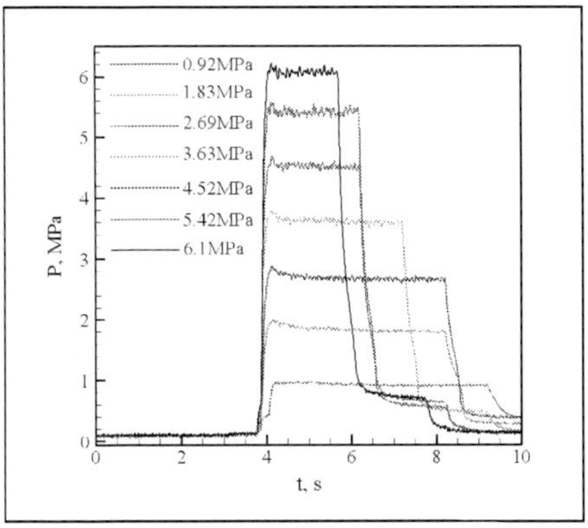

Fig. 16. The typical chamber pressure profiles of 7 cases

5.3 Results and discussion

The time traces of some thermocouples for a representative 2.69MPa chamber pressure test are shown in Fig. 17. A total of eight sets of thermocouple temperature measurements are shown. In terms of nomenclature in the figure, for example, the first trace labeled TC-10-00 denotes that the thermocouple was at the 10mm axial location, at 00 degrees (angle was defined with respect to major array of thermocouple). Except the curve TC-25-00, it can be seen that all temperature traces had the same response characteristic, were all well behaved and not noisy. The TC-25-00 had an obvious longer response time than others, so it could not be utilized.

During the steady state portion of the firing, the temperatures rose steadily owing to the heat sink nature of the chamber design. The curves of two thermocouples located respectively at 40mm and 100mm are nearly identical suggesting that the chamber flow was concentric. According to theory of heat transfer, higher heat flux on the inner wall at axial location of measurement point consequentially induces higher temperature raise at this point. Picture of the raises of temperatures at these measurement points versus the axial distance for 2.69MPa chamber pressure case is shown in Fig. 18, manifesting that the results of 2 repetitive tests were nearly identical.

All temperature curves were obtained for all the pressure cases, and then an axisymmetric heat conduction numerical calculation was conducted to obtain the hot-gas-wall heat flux for each pressure case. Inspection of empirical heat transfer correlations available in the literature such as the Bartz (Bartz, 1957), all the heat flux data were scaled by $1/p^{0.8}$, and the results are shown in Fig.19. It can be seen that all the heat flux distribution curves collapse to a single profile, and all the cases show the same qualitative distribution trends and the almost same quantitative local values, which means that the heat flux q of a gas-gas injector combustor correlates well with the pressure p as $q \sim p^{0.8}$. A valuable suggestion can thus be drawn that the heat flux data at high pressure condition can be predicted from that at a low pressure condition.

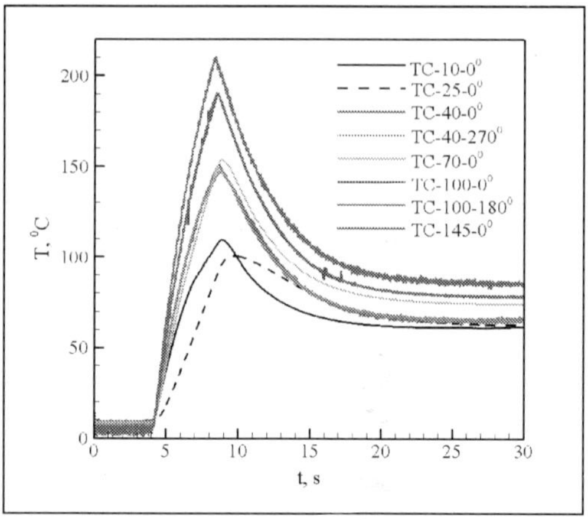

Fig. 17. Thermocouple temperature traces (representative) for a 2.69MPa test

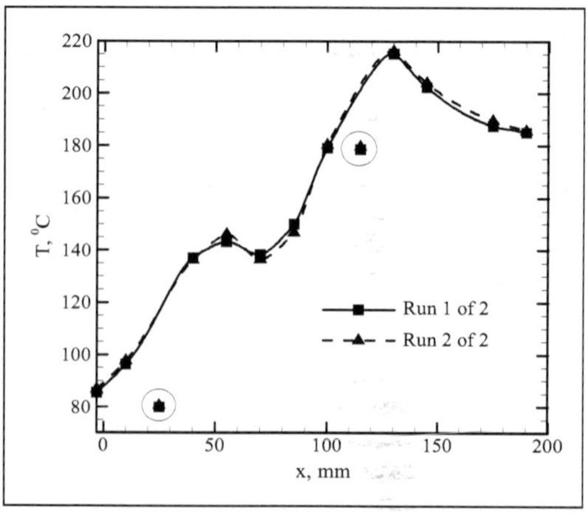

Fig. 18. Wall temperature versus axial distance for 2.69MPa chamber pressure

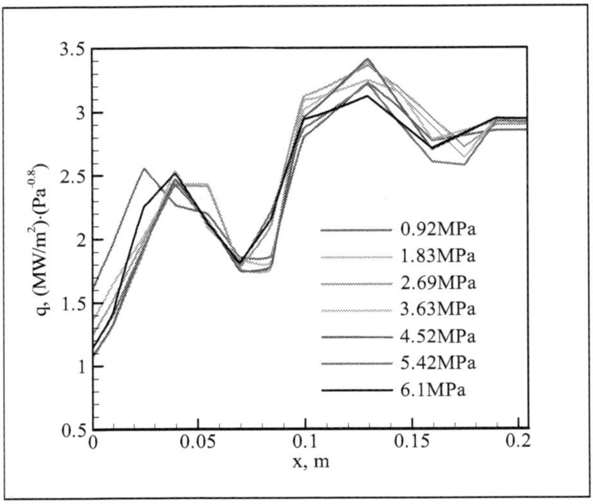

Fig. 19. Heat flux (scaled with respect to $(1/Pc^{0.8})$) versus axial distance for each chamber pressure case

5.4 Numerical study

In order to investigate the heat transfer characteristics at the high pressure condition unavailable in the experimental hot-test, and further examine the inner combustion flowfields at different chamber pressures, numerical simulations were conducted on this combustion chamber.

5.4.1 Numerical models

A great effort has been made to perform the CFD simulation of gas-gas combustion flow at Pennsylvania State University, NASA Marshall Space Flight Center, University of Michigan and Beihang University et al. (Foust et al., 1996; Schley et al., 1997; Lin et al., 2005; Tucker et al., 2007a, 2008b; Cai et al., 2008; Sozer et al., 2009; Wang, 2009a, 2010b, 2010c) And the results indicated that the steady Reynolds Average Navier-Stokes (RANS) method combined with a $k-\varepsilon$ turbulence model could effectively simulate the whole combustion flow and obtain the statistical average solutions that can match the experimental results. In reference (Wang, 2010), difference RANS models were used to simulate a hot-testing chamber, and a feasible $k-\varepsilon$ turbulence model was obtained. Here, the RANS method combined with this $k-\varepsilon$ turbulence model was used.

Constant pressure specific heat of each species was calculated as a function of temperature

$$C / R = a_0 + a_1 T + a_2 T^2 + a_3 T^3 + a_4 T^4 + a_5 T^5 \tag{9}$$

Coefficients of laminar viscosity and heat conduction of single component were calculated by molecular dynamics. The compressibility of the gas propellants at high pressure was considered. The R-K equation was substituted for the ideal state equation to take the real gas effect into account.

5.4.2 Numerical method and boundary condition

The entire system was solved by a strongly coupled implicit time-marching method with ADI factorization for the inversion of the implicit operator. Convective terms were 2-order flux split upwinding differenced, whereas diffusion terms were centrally differenced.
The calculation domain only occupied half the chamber. The radial and axial stretchings of the grid were used near the wall boundary and in the shear layer domain. The grid consisted of 29,028 cells, and the grid of half the cylinder was 43×350.
The inlets were fixed mass flowrate, and the inlet turbulence intensities both set to be 5%. The centerline was an axisymmetric boundary, and the nozzle exit was specified as a supersonic outlet. Non-slip wall boundaries were used on the chamber walls. The temperature of the combustor wall was set at environment temperature of 300K to achieve a steady heat flux.

5.5 Results and discussion

The dimensions of the chambers were kept unchanged, and a total of 4 numerical cases under different pressures from 5MPa to 20 MPa were chosen and shown in Table 6. The combustion flowfields and heat flux along with the combustor wall were obtained. The temperature contours are shown in Fig. 20, which shows that all the temperature contours of 4 pressure conditions are similar. And the similarity of the inner combustion flowfield structures leads to the same inner wall heat flux distribution shown in Fig. 21. From the time-mean inner flowfield results, the wall heat flux distribution can be clearly explained. The little peak of the heat flux in the beginning originates from the existence of the strong recirculation zone there. Then the heat flux gets up continuously with the increasing intensity and sufficiency of the inner mixing and combustion and the increasing velocity of the downstream flow. With the combustion mainly completed at the end of the combustor, the flowfield temperature and velocity both reach their maximum values. As the flow moves further downstream, the combustion heat release is generally finished, but the wall heat loss still exists, inducing a little downward movement of the heat flux in the end. In Fig. 21 all the heat flux data were scaled by $p_c^{0.8}$. It can be seen that all the curves almost collapse to a

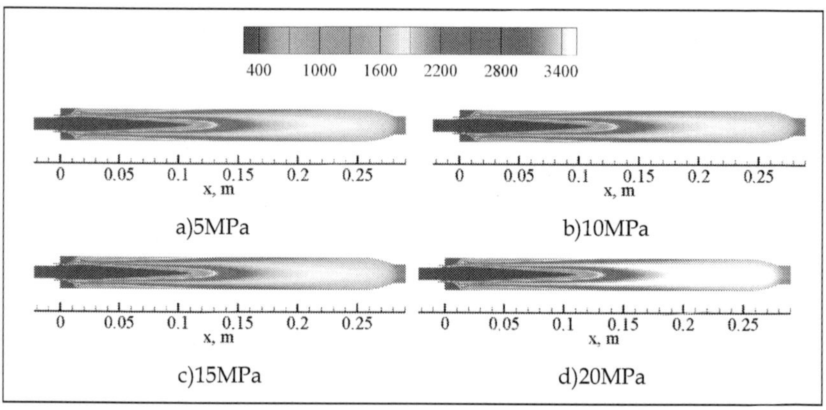

Fig. 20. Temperature contours of the five different pressure cases

single profile, which indicates that in the high pressure conditions, the heat flux in gas-gas injector combustors of different pressures also have the same qualitative distribution, and in a good agreement with $q \sim p_c^{0.8}$ quantitatively.

Chamber pressure /MPa	H2 flowrate /(kg/s)	H2 temperature /K	H2 injection velocity /(m/s)	O2 flowrate /(kg/s)	O2 temperature /K	O2 Injection velocity /(m/s)
5	0.054	300	~760	0.324	300	~70
10	0.108	300	~760	0.648	300	~70
15	0.162	300	~760	0.972	300	~70
20	0.216	300	~760	1.296	300	~70

Table 6. Parameters of pressure scaling conditions

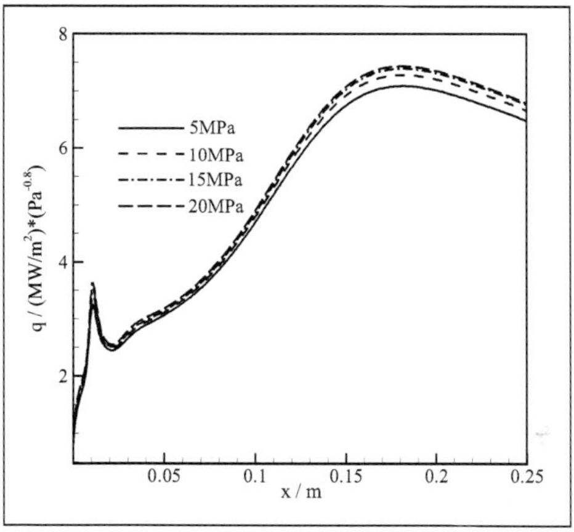

Fig. 21. Heat flux (scaled with respect to $1/Pc^{0.8}$) versus axial distance for four chamber pressure cases

6. Conclusion

A method for measurement of single-injector heat transfer characteristics in a heat sink chamber was expound in this chapter. A series of measurement points are designed in the chamber with the same axial intervals and the same distance from the inner wall surface. This method measures the temperatures at these measurement points and then converts these temperatures into inner wall temperatures and heat flux with 2-D axisymmetric calculation. A hot-testing of a single-element gas-gas shear-coaxial injector chamber applying this method was introduced to explain this method. And the inner wall temperature and heat flux for this case were obtained and demonstrated. The basic principle and design, data processing and the corresponding error analysis were described in detail. And the error analysis showed that the accuracy of this method is sufficient for engineering application, and the 2-D axisymmetric calculation can substitute for the expensive 3-D calculation with its cost-saving advantage. The method was originally developed for single-element axisymmetric chamber, and can also serve as a reference for non-axisymmetric chambers and multi-element injector chambers.

Furthermore, this method was used to investigate the heat transfer characteristics of a single-element shear-coaxial gas-gas injector combustion chamber. A single-injector heat-sink chamber was designed and hot-fire tested for 17 times at chamber pressure from 0.92MPa to 6.1MPa. Inner hot-gas-wall temperature and heat flux along with the axial direction of the chamber were obtained. The results show that heat flux in gas-gas injector combustors of different pressures not only have the same distribution qualitatively, also show a good agreement with $q \sim p_c^{0.8}$ quantitatively. The inner combustion flows were also numerically simulated with multi-species turbulence N-S equations at higher chamber pressure from 5MPa to 20MPa to extend the experimental results. Both the flows structures and heat flux profiles on inner wall were obtained and discussed, and the results of numerical simulations indicated that the combustion flowfield of different pressures are similar and the heat flux is also proportional to pressure to the power 0.8.

7. Acknowledgments

The authors acknowledge the support of the state high-tech research and development fund. The authors also thank W. Zhang and Sh. Li from Beijing West Zhonghang Technology Ltd. for helps in designing the thermocouples. Finally, the authors thank all the people who made contribution and gave much help to this paper.

8. References

Archambault, M. R., Peroomian, O., "Characterization of a Gas/Gas Hydrogen/Oxygen Engine," *AIAA Paper* 2002-3594, 2002a.

Archambault, M. R., Talley,R. D., Peroomian, O., "Computational Analysis of a Single-Element Shear-Coaxial GH2/GO2 Engine," *AIAA Paper* 2002-1088, 2002b.

Bartz, D.R., "A Simple Equation for Rapid Estimation of Rocket Nozzle Convective Heat Transfer Coefficients," *Jet Propulsion*, Vol.27, No.1, Jan. 1957. pp: 49-51.

Cai G B, Wang X W, Jin P, Gao Y S. Experimental and Numerical Investigation of Large Mass Flow Rate Gas-Gas Injectors .*AIAA Paper* 2008-4562.

Calhoon, D., Ito, J., and Kors, D., "Investigation of Gaseous Propellant Combustion and Associated Injector-Chamber design Guide- lines," *NAS 3-13379*, Aerojet Liquid Rocket Company, Sacramento, California, 1973.

Chapman A. J., *Fundamentals of Heat Transfer*, Macmillan, New York, 1987.

Conley, A., Vaidyanathan, A., and Segal, C., "Heat Flux Measurements for a GO2/GH2 Single-Element, Shear Injector," *Journal of Spacecraft and Rockets*, Vol. 44, No. 3, May-June 2007. pp. 633-639.

Coy E., "Measurement of Transient Heat Flux and Surface Temperature Using Embedded Temperature Sensors", *Journal of Thermophysics and Heat Transfer*, Vol.24, No.1. January–February 2010. pp. 77-84.

Davis, J. A., Campbell, R. L., "Advantages of A Full-flow Staged Combustion Cycle Engine System", *AIAA Paper* 1997-3318, 1997.

Farhangi, S., Yu, T., Rojas, L., and Sprouse, K., "Gas-Gas Injector Technology for Full Flow Stage Combustion Cycle Application," *AIAA Paper* 1999-2757, 1999.

Foust, M. J., Deshpande, M., Pal, S., Ni, T., Merkle, C. L., Santoro, R. J., "Experimental and Analytical Characterization of a Shear Coaxial Combusting GO2/GH2 Flow field," *AIAA Paper* 1996-0646, 1996.

Groot, W., A., McGuire, T., J., and Schneider, S., J., "Qualitative Flow Visualization of an 110N Hydrogen/Oxygen Laboratory Model Thruster", *AIAA Paper* 1997-2847, 1997.

Jones G., Protz C., Bullard B., and Hulka J., "Local Heat Flux Measurements with Single Element Coaxial Injectors," *AIAA Paper* No. 2006-5194, July 2006.

Lin, J., West, J. S., Williamst, R. W., and Tucker, P. K., "CFD Code Validation of Wall Heat Fluxes for a GO2/GH2 Single Element Combustor," *AIAA Paper* 2005-4524, 2005.

Marshall W. M., Pal S., and Santoro R. J., "Benchmark Wall Heat Flux Data for a GO2/GH2 Single Element Combustor," *AIAA Paper* No. 2005-3572, July 2005.

Meyer, L., Nichols, J., Jones, J. M., "Integrated Powerhead Demonstrator (booster hydrogen oxygen rocket engines)," *AIAA Paper* 1996-4264, 1996.

NASA Space Vehicle Design Criteria. "Liquid rocket engine injectors," *NASA SP-8089*, 1976.

Santoro R. J. and Pal S., "Validation Data for Full Flow Staged Combustion Injectors," *Final Report for NASA Contract Grant NAG8-1792*, Pennsylvania State University, 2005.

Schley, C-A., Hagemann, G., Tucker, P. K., "Comparison of Calculation Codes for Modeling Hydrogen-Oxygen Injectors," *AIAA Paper* 1997-3302, 1997.

Sozer E, Vaidyanathan A, Segal C, and Shyy W, Computational Assessment of Gaseous Reacting Flows in Single Element Injector, *AIAA Paper* 2009-449.

Tramecourt, N., Masquelet, M., and Menon, S., "Large-Eddy Simulation of Unsteady Wall Heat Transfer in a High Pressure Combustion Chamber," *AIAA Paper* No. 2005-4124, July 2005.

Tucker, K., West, J., Williams, R., Lin, J., Rocker, M., Canabal, F., Robles, B., and Garcia, R., "Using CFD as a Rocket Injector Design Tool: Recent Progress at Marshall Space Flight Center," *NASA NTRS 20050217148*, Jan. 2005.

Tucker, P. K., Klemt, M. D., and Smith, T. D., "Design of Efficient GO2/GH2 Injectors: a NASA, Industry and University Cooperative Effort," *AIAA Paper* 1997-3350, 1997.

Tucker, P. K., Menon, S., Merkle, C. L., Oefelein, J. C., and Yang, V., "An Approach to Improved Credibility of CFD Simulations for Rocket Injector Design," *AIAA Paper* 2007-5572, 2007.

Tucker, P. K., Menon, S., Merkle, C. L., Oefelein, J. C., and Yang, V., "Validation of High-Fidelity CFD Simulations for Rocket Injector Design," *AIAA Paper* 2008-5226, 2008.

Vaidyanathan A., Gustavsson J., and Segal C., "Heat Fluxes/OH-PLIF Measurements in a GO2-GH2 Single-Element Shear Injector," *AIAA Paper* No. 2007-5591, July 2007.

Vaidyanathan A., Gustavsson J. and Segal C., "One- and Three-Dimensional Wall Heat Flux Calculations in a O_2-H_2 System," *Journal of Propulsion and Power*, Vol. 26, No. 1, January-February 2010.

Wang X W, Cai G B, Gao Y S. Large Flow Rate Shear-Coaxial Gas-Gas Injector. *AIAA Paper* 2009-5042.

Wang X W, Cai G B, Jin P. Scaling of the flowfield in a combustion chamber with a gas-gas injector. *Chinese Physics B*, Vol. 19, No.1 (2010). SCI DOI: 10.1088/1674-1056/19/1/019401.

Wang X W, Jin P, Cai G B . Method for investigatio n of combustion flowfield characteristics in single-element gas/gas injector chamber. *Journal of Beijing University of Aeronautics and Astronautics,* 35(9), (2009). pp.1095-1099

Zurbach, S. (ed.), Rocket Combustion Modeling, *3rd International Symposium,* Centre National D'Etudes Spatiales, Paris, March 2006.

24

Heat Transfer Related to Gas Hydrate Formation/Dissociation

Bei Liu, Weixin Pang, Baozi Peng, Changyu Sun and Guangjin Chen
*State Key Laboratory of Heavy Oil Processing,
China University of Petroleum, Beijing 102249,
P. R. China*

1. Introduction

Gas hydrates are ice-like crystalline compounds comprised of small guest molecules, such as methane or other light hydrocarbons, which are trapped in cages of a hydrogen-bonded water framework. It has drawn attention in the gas and oil industry since 1930s because it was found that the formation of gas hydrates may block oil/gas pipelines (Sloan and Koh, 2007). However, with the gradual discovery of huge reserve of natural gas hydrates in the earth as well as the understanding of the peculiar properties of gas hydrates, more and more studies have focused on how to benefit from gas hydrates in recent decades. The most important aspect of gas hydrates research is attributed to the exploration and exploitation of natural gas hydrates. Additionally, people also try a lot in the development of novel technologies based on hydrates, such as separation of gas mixture via forming hydrates, storage of natural gas or hydrogen in the form of solid hydrates, and sequestration of CO_2, etc. As the formation of gas hydrates is an exothermic process, heat transfer always accompanies hydrate formation or dissociation. The understanding of heat transfer mechanism is critical to the modeling of formation/dissociation kinetic process of gas hydrates, which favors the best exploitation of natural gas hydrates and the best design of reactor for hydrate production or decomposer for hydrate dissociation with respect to different kinds of hydrate application objects.
In recent years, a variety of experimental and theoretical works focused on heat transfer involved in formation/dissociation of gas hydrates have been reported. They are summarized in this chapter accompanying presentation of our new work relevant to this topic. This chapter is organized as follows. In section 2, we present progresses in experimental measurement of the thermal conductivities of different kinds of gas hydrates, including pure gas hydrates and hydrate-bearing sediments. The achievements on mechanism and modeling of heat transfer occurring in the growth of hydrate film at the guest/water interface, as well as its influence upon the hydrate film growth rate are summarized in section 3. Our new experimental study on heat transfer in stirring or flowing hydrate system is given in section 4. Section 5 presents our recent work on the experimental and modeling studies on heat transfer in quiescent reactors for producing or decomposing big blocks of hydrates, and the formulation of the influence of heat transfer upon the hydrate formation/dissociation rate. In section 6, the mechanism of heat transfer in hydrate

bearing-sediment are analyzed and discussed. Finally, some concluding remarks are given in section 7.

2. Thermal conductivity of gas hydrate

Thermal conductivity is a kind of basic data for studying the heat transfer of hydrates involved systems. In recent decades, a number of researchers have made their efforts to measure the thermal conductivities of different types of gas hydrates at different conditions. Regarding to measurement technique, the most widely adopted ones are standard needle probe technique and transient plane source (TPS) technique (Gustafsson et al., 1979, 1986). For example, thermal conductivity of methane hydrate has been determined by deMartin (2001), Krivchikov et al. (2005), and Waite et al. (2007) using the needle probe technique. With same technique, thermal conductivities of several other gas hydrates, such as tetrohydrofuran (THF) hydrate (Cortes et al., 2009), xenon hydrate (Krivchikov et al., 2006), HCFC-141b hydrate (Huang et al., 2004), and CFC-11 hydrate (Huang et al., 2004) have been measured. Transient plane source (TPS) technique in double- and single-sided configurations has been used more recently to measure thermal conductivity of gas hydrates (Huang and Fan, 2004; Li et al., 2010; Rosenbaum et al., 2007). This technique is based on the transient method and the needle probe, but it has a very small probe (Gustafsson et al., 1979, 1986). It allows measurements without any disturbance from the interfaces between the sensor and the bulk samples. In addition, it is possible to measure thermal conductivity, thermal diffusivity, and heat capacity per unit volume simultaneously (Gustafsson et al., 1979). It is hard to draw a definite conclusion that which technique is better for pure gas hydrate samples synthesized in laboratory; however, for in-situ determination of the thermal properties of hydrate-containing sediments, the single-sided TPS technique may be more suitable as the needle probe and double-sided TPS techniques need the probe to be surrounded by the hydrates (English and Tse, 2010).

There are several factors, such as the porosity of the samples, temperature, pressure, and measurement time, that influence thermal conductivity of gas hydrates. As pointed out by English and Tse (2010), for relatively pure hydrates, reducing the porosity of the samples by compacting them is critical for obtaining the reliable thermal conductivity in the intermediate temperature range. For hydrate-bearing sediments, Tzirita (1992) concluded that porosity is also a critical factor in controlling the thermal conductivity. More recently, Cortes et al. (2009) carried out a systematic measurement of the thermal conductivity of THF-hydrate saturated sand and clay samples. They found the influence is a complex interplay among particle size, effective stress, porosity, and fluid-versus-hydrate filled pore spaces, not only porosity. With respect to temperature effect, many studies found that hydrates exhibit a glass-like temperature dependence of thermal conductivity (Andersson and Ross, 1983; Handa and Cook, 1987; Krivchikov et al., 2005, 2006; Ross et al., 1981; Ross and Andersson, 1982; Tse and White, 1988). Among these studies, the works of Krivchikov et al. (2005, 2006) are interesting as they found that both methane and xenon hydrates show crystal-like temperature dependence below 90 K, while exhibiting glass-like behavior above 90 K. The effect of pressure has also been investigated by many groups (Andersson and Ross, 1983; Rosenbaum et al., 2007; Waite et al., 2007). Only weak pressure dependency was observed by them. Finally, the relationship between thermal conductivity and measurement time for methane hydrate

has been studied by Li et al. (2010) very recently. They found that in 24h, thermal conductivity increases 5.45% at 268.15 K; however, at 263.15 K, the increment is 196.29%. From their results we may say that measurement time needs to be considered for thermal conductivity studies at relatively low temperatures.
To give readers a clear picture of measured thermal conductivities of different kinds of gas hydrates, the results of pure gas hydrates and hydrate-bearing sediments are listed in Table 1.

compound	T/K	P/MPa	λ/W m-1 K-1	Ref
	pure gas hydrates			
methane	253 - 290	31.5	~ 0.62	Waite et al., 2007
methane (compacted samples)	263.05 – 277.97	6.6	~ 0.57	Huang and Fan, 2004
methane (compacted samples)	261.5 – 277.4	3.8 – 14.2	~ 0.68	Rosenbaum et al., 2007
tetrahydrofuran	15 -100		0.04 - 0.12	Tse and White, 1988
tetrahydrofuran ·17 H_2O	261		0.5	Waite et al., 2005
tetrahydrofuran ·17 H_2O	261	0.05 - 1	0.58	Cortes et al., 2009
xenon	245	~ 0.05	0.36	Handa and Cook, 1987
ethylene oxide	263		0.49	Cook and Laubitz, 1983
cyclobutanone	260	100	~ 0.47	Andersson and Ross, 1983
1,3 – dioxolane	260	100	~ 0.51	Andersson and Ross, 1983
propane	275	1	~ 0.4	Stoll and Bryan, 1979
sodium sulphide ·9 H_2O	295	0.001	0.12	Lunden et al., 1986
HCFC-141b	~ 250	0.1	~ 0.5	Huang et al., 2004
CFC-11	~ 250	0.1	~ 0.5	Huang et al., 2004
	hydrate-bearing sediments			
natural methane hydrate-layer sand	263 -283	0.1	3.8 – 5.8	Yamamoto et al., 2008
sand with 100% THF ·17 H_2O hydrate	261	0.05 - 1	4.1 – 4.5	Cortes et al., 2009
clay with 100% THF ·17 H_2O hydrate	261	0.05 - 1	2.8 – 3.0	Cortes et al., 2009

Table 1. Thermal conductivities, λ, of pure gas hydrates and hydrate-bearing sediments

3. Heat transfer in growth of hydrate film

Generally, because most of hydrate formers (guests) are water insoluble, the initial formation of hydrate occurs at the guest/water interface, taking the form of thin porous crystalline film. The further growth of hydrate is controlled by mass transfer of water or hydrate former through the film. Many experimental and/or theoretical studies on the growth of hydrate film have been carried out by several groups (Freer et al., 2001; Ma et al., 2002; Mochizuki and Mori, 2006; Mori, 2001; Ohmura et al., 2000, 2005; Peng et al., 2007, 2008, 2009; Saito et al., 2010, 2011; Sun et al., 2007; Taylor et al., 2007; Uchida et al., 1999, 2002), including the morphology of hydrate film, the growth rate of hydrate film, the thickness of hydrate film, the mechanism of hydrate film growth, and so on. However, so far it is still a controversial topic on the growth mechanism of hydrate film. Recently, more attention has been paid to the mechanism of heat transfer on hydrate film growth at the guest/water interface than intrinsic kinetic and mass transfer mechanisms. In this part, different hydrate film growth models, especially heat transfer models that have been developed by various research groups are summarized.

Experimental and molecular dynamic simulation studies on the initial formation of hydrate at the guest/water interface suggest that the interface where there is a significant concentration gradient is the place to initiate and sustain hydrate formation (Moon, et al. 2003; Vysniauskas and Bishnoi, 1983). Englezos et al. (1987a, 1987b) studied the kinetics of formation of methane, ethane, and their mixture hydrates in a semi-batch stirred tank reactor. They presented an intrinsic kinetic model for the hydrate particle growth and the rate of growth per particle was given by:

$$\left(\frac{dn}{dt}\right)_P = K^* A_P \left(f - f_{eq}\right) \tag{1}$$

with

$$\frac{1}{K^*} = \frac{1}{K_r} + \frac{1}{K_d} \tag{2}$$

where n is the moles of gas consumed, t is the hydrate reaction time, K^* is the combined rate parameter, A_P is the surface area of particles, f is the gas fugacity, f_{eq} is the equilibrium fugacity, K_r and K_d are the reaction rate constant and mass transfer coefficient around the particle, respectively. Similarly, based on the assumption that the intrinsic kinetics is the control step of hydrate formation and growth, Ma et al. (2002) developed a model to correlate the lateral growth rate of hydrate film. The model was formulated as the following form:

$$r_f = A \times \left[e^{B(-\Delta g/(RT))} - 1 \right] \tag{3}$$

where r_f is the lateral growth of hydrate film. Parameters A and B are system composition dependent and were determined by fitting experimental data. The Gibbs free energy difference (Δg) was selected as the driving force to describe the hydrate growth process. The experimental results indicated that this model could correlate the lateral film growth rate perfectly (Ma et al., 2002; Sun et al., 2007).

Except for the models described above, some researchers suggested that the growth rate of hydrate film is controlled by heat diffusion and some models were developed correspondingly. For example, Uchida et al. (1999) presented a model analysis of the two-dimensional growth of a carbon dioxide hydrate film (Figure 1).

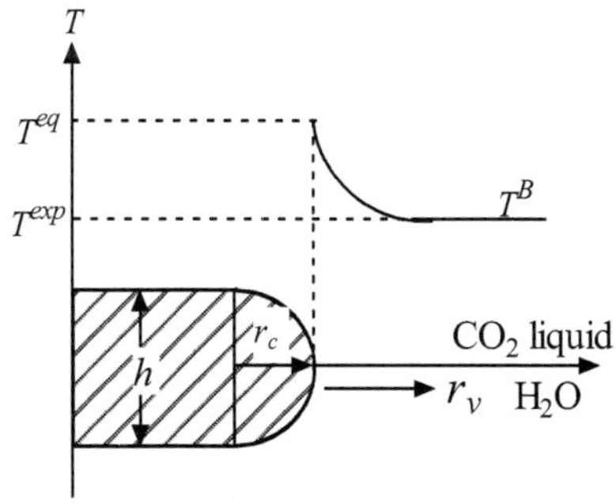

Fig. 1. Hydrate film model of Uchida et al (1999)

In this model, they assumed that one half of the film is in water phase, and the other half is in guest phase. The hydrate film has a semicircular front and is uniform in thickness. In addition, this model assumes that hydrate crystals successively form only at the front of the hydrate film and the front is maintained at the three-phase (water/guest-fluid/hydrate) equilibrium temperature. The heat released by the hydrate crystal formation is diffused away from the film front and into the water and guest-fluid phase. Based on these assumptions, they formulated the heat balance at the edge of the film as

$$v_f \rho_h \Delta h_h = \lambda_w \Delta T / r_c \tag{4}$$

where v_f is the rate of linear growth of the film, ρ_h is the mass density of the film, Δh_h is the heat of hydrate formation (per unit mass of hydrate), λ_w is the thermal conductivity of water, ΔT is the difference between the temperature at the film edge, T^{eq}, and the undisturbed temperature in the fluid phases, T^B, and r_c is the radius of curvature of the edge. Uchida et al. (1999) correlated their experimental data on v_f versus ΔT by means of a linear regression analysis as follows:

$$v_f = (1.73 \pm 0.16)\Delta T \tag{5}$$

In Uchida et al.'s model (Uchida et al., 1999), the conductive heat transfer from the film front was deduced from the temperature gradient, which was deemed as with little physical reasoning (Mochizuki and Mori, 2006; Mori, 2001). Mori (2001) presented an alternative model of hydrate film growth based on the idea that the front of hydrate film, which grew on the interface between stagnant water and guest fluid, could be viewed to be held in stratified flow of the two fluids with the velocity which was opposite in sign but equal in magnitude to the velocity of the hydrate film front. In his work, the heat removed from the film front to the liquid phases was treated as a steady convective heat transfer and other assumptions were same as those of Uchida et al. (1999). The heat balance over the hemicircular front of the film was formulated as follows:

$$v_f \delta \rho_g \Delta h_h = \frac{1}{4}\pi\delta\left(\overline{\alpha}_w + \overline{\alpha}_g\right)\Delta T \tag{6}$$

Mori (2001) assumed that the heat transfer coefficients, $\overline{\alpha}_w$ and $\overline{\alpha}_g$, could be given by the simplest type of convective heat transfer correlation in a dimensionless form and deduced a δ - v_f relation correspondingly:

$$v_f \delta = C\Delta T^{3/2} \tag{7}$$

where

$$C = \left[\frac{\pi A}{4}\frac{1}{\rho_h \Delta h_h}\left(\frac{\lambda_w}{\nu_w^{1/3}}\mathrm{Pr}_w^{1/3} + \frac{\lambda_g}{\nu_g^{1/3}}\mathrm{Pr}_g^{1/3}\right)\right]^{3/2} \tag{8}$$

In Equation 8, λ_w and λ_g are the thermal conductivity of water and the hydrate former, respectively, ν_w and ν_g are the kinematic viscosity of water and the hydrate former, respectively, and Pr_w and Pr_g are the Prandtl number of water and the hydrate former, respectively.

Mochizuki and Mori (2006) modified Mori's model and presented another model, as shown in Figure 2. They assumed that there is a transient two-dimensional conductive heat transfer from the film front to the water and guest-fluid phases plus the hydrate film itself. In this model, the hydrate film was assumed to exist on the water side of the water-guest fluid interface and the interface infinitely extend. No convection occurs in either of the water and guest-fluid and other assumptions were same as those of Uchida et al. (1999). The rate of heat removal from the front to the surroundings is balanced by the rate of heat generation of hydrate-crystal formation.

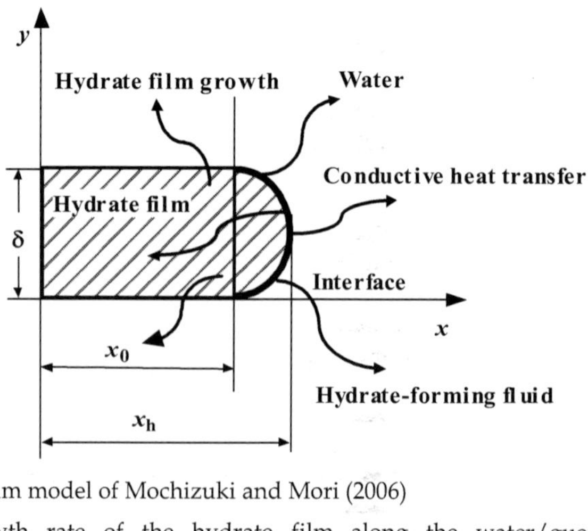

Fig. 2. Hydrate film model of Mochizuki and Mori (2006)

The linear growth rate of the hydrate film along the water/guest-fluid interface, $v_f = dx_h/dt$, was given in the following equation:

$$\rho_h \delta \Delta h_H v_f = \int_{-\pi/2}^{\pi/2} \left(\lambda_h \left.\frac{\partial T}{\partial r}\right|_{r=r_h-} - \lambda_w \left.\frac{\partial T}{\partial r}\right|_{r=r_h+} \right) r_h d\theta \tag{9}$$

where $\partial T / \partial r|_{r=r_h-}$ and $\partial T / \partial r|_{r=r_h+}$ are the radial temperature gradients on the hydrate side and the fluid side, respectively, at $r = r_h$ (i.e., the x position of the hydrate-film front). It should be pointed out that this model is computationally complicated and hence cumbersome to use. In addition, the assumptions adopted in this model, that is, the film front is in the water phase or one half in water phase and the other half in guest-fluid phase are too arbitrary. Therefore, Peng et al. (2007) proposed another hydrate film model based on Mori's model. In their model, they assumed part of thickness x of hydrate film is in guest phase and another part of thickness $\delta - x$ is in water phase, as shown in Figure 3. The value of x is guest composition dependent.

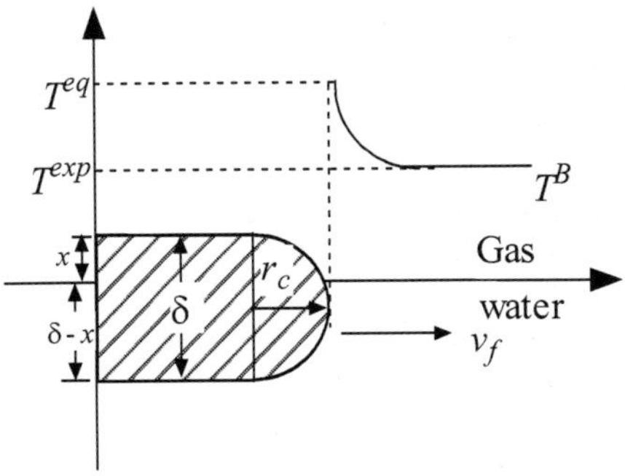

Fig. 3. Hydrate film model of Peng et al. (2007)

In Peng et al.'s model (Peng et al., 2007), the thickness of hydrate film was assumed to vary with driving force inversely, i.e.,

$$\delta = k / \Delta T \tag{10}$$

The lateral rate of hydrate film was then correlated by the following equation:

$$v_f = \frac{C}{k} \Delta T^{\frac{5}{2}} \tag{11}$$

where the constant C in Mori's model was reformulated by a generalized expression, as shown below:

$$C = \left[\frac{1}{6\rho_h \Delta h_h} \left(\lambda_w^{2/3} \rho_w^{1/3} c_{p,w}^{1/3} \left(\pi - \arccos(1 - \frac{2x}{\delta}) \right) + \lambda_g^{2/3} \rho_g^{1/3} c_{p,g}^{1/3} \arccos(1 - \frac{2x}{\delta}) \right) \right]^{3/2} \tag{12}$$

The modified convection heat transfer model presented by Peng et al.(2007), i.e., Equation 11, has been used to correlate the lateral growth rate of hydrate film of different crystal structures in wide temperature and pressure ranges (Peng et al., 2007, 2009). It can be concluded that validity of Equation 11 is independent of the composition of hydrate former and the structure type of hydrates (Peng et al., 2007).

For investigating the growth mechanism of the hydrate film, Freer et al. (2001) also studied methane hydrate film growth on the water/methane interface experimentally and proposed a model of lateral hydrate film growth. They calculated the v_f of methane hydrate film by assuming that one dimensional conductive heat transferred from the film front to water. As the calculated v_f was much lower than the experimentally measured v_f, they suggested that the hydrate film growth was controlled by both intrinsic kinetics and heat transfer. Their model was expressed as:

$$\lambda_h \rho_h v_f = K(T^{eq} - T^B) \tag{13}$$

$$\frac{1}{K} = \frac{1}{k} + \frac{1}{h} \tag{14}$$

where λ_h is the thermal conductivity of hydrate, K is the total resistance, h is the heat transfer coefficient, and k is the methane hydrate kinetic rate constant.

Additionally, for investigating which step is the main contribution to the hydrate film growth, Peng et al. (2008) also presented a model based on the assumption that hydrate lateral film growth is controlled by both intrinsic kinetics and heat transfer, as shown in Figure 4.

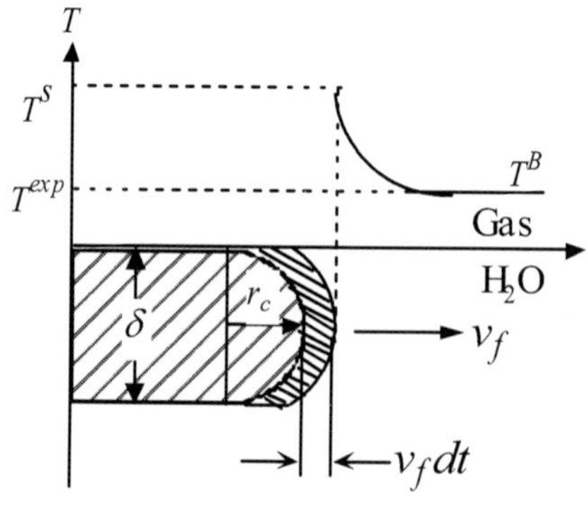

Fig. 4. Hydrate film model of Peng et al. (2008)

This model is similar to the model of Uchida et al. (1999). However, in this model Peng et al. kept the hydrate film at T^S rather than at the three-phase (water/guest/hydrate) equilibrium temperature and they proposed the intrinsic rate of hydrate film growth should not be of a linear relation with the driving force. Therefore, the balance of the heat removed from the film front with that generated by the hydrate formation was formulated by the equation proposed by Freer et al. (2001),

$$v_f \rho_h \Delta h_h = k_1 (T^S - T^B) \tag{15}$$

and the intrinsic kinetic equation was adopted as following:

$$v_f = k_2 (T^{eq} - T^S)^n \tag{16}$$

From Equations 15 and 16, the following equation can be obtained:

$$\frac{\rho_h \Delta h_h}{k_1} v_f + \left(\frac{v_f}{k_2}\right)^{1/n} = T^{eq} - T^B = \Delta T \tag{17}$$

Based on their experiment data, Peng et al. calculated the temperature differences between the hydrate film front and the bulk water at different driving forces, which were taken as an important factor on judging the dominating contribution for hydrate film growth at the gas/water interface. They found that the effect of heat transfer on hydrate film growth is much smaller than that of intrinsic kinetics, and suggested that the intrinsic kinetic is the main control step for hydrate film growth of methane and carbon dioxide hydrate.

It should be pointed out that for the models mentioned above, the parameters were obtained by correlating with different experimental data set. As the experimental data were obtained in different experiment apparatus and the stochastic induction time of hydrate nucleation may also affect the measurement of hydrate film growth rate for different experiment device, it is hard to draw a definite conclusion that which model is better. More efforts need to be made on hydrate film growth in the future.

4. Heat transfer in stirring or flowing hydrate system

Stirring is an important technique that can enhance heat and mass transfer, and thus accelerating the speed of hydrate formation/dissociation. The state of hydrates formed under stirring is usually in slurry, which is also the case when hydrates are formed in gas-oil-water multi-phase flowing systems containing hydrate anti-agglomerants (AA). As a result, the determination of heat transfer coefficient of hydrate slurry is crucial for investigating the heat transfer in hydrate forming/dissociating processes under stirring or flowing. Unfortunately, there are very few publications up to date, and thus only some results obtained by our group are introduced in this part.

4.1 Experimental apparatus

The experimental equipments adopted in our work are shown in Figure 5, which mainly contain the reactor with stirrer, constant temperature water bath, and temperature/pressure sensor.

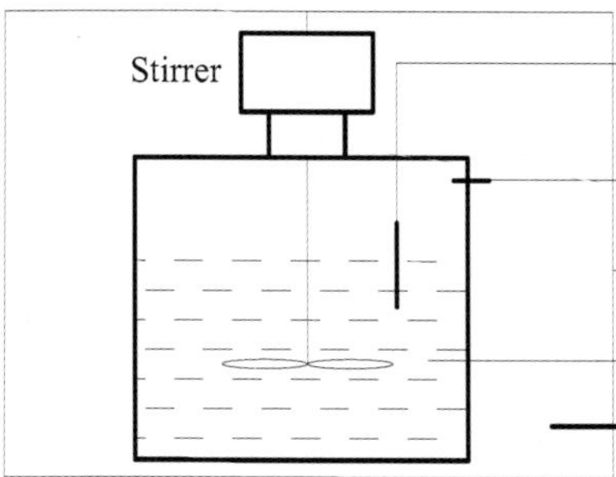

Fig. 5. The schematic outline of the experimental apparatus

4.2 Experimental principals and steps

In our work the measurements were performed via the following steps:

1. The equipments were washed three to four times using distilled water;
2. The reactor was wetted using the experimental liquid, then the experimental liquid was added to the reactor until the stirrer reaches half of the height of liquid added;
3. The experimental temperature and pressure was set. We started stirring until the hydrate slurry was formed completely, then we closed the stirrer;
4. When the hydrate slurry was formed completely and the temperature was constant, the temperature of the water bath was quickly decreased by about 10 K to make a difference in temperature between hydrate slurry and water bath;
5. The stirrer was open at a certain stirring speed and the temperatures of both water bath and hydrate slurry were recorded at a certain interval. The stirring was stopped when the change in temperature is very small in both the water bath and the reactor.

4.3 Experimental data analysis

(1) Calculation of the total heat transfer coefficient

If we assume the heat released by the slurry in the reactor is equal to that adsorbed by the water bath, that is, we neglect the heat loss during the measurement, the amount of heat transfer, Q, can be calculated with the following equation:

$$Q = cm(-\tfrac{dT}{dt}) \tag{18}$$

where m is the mass of the hydrate slurry, $\frac{dT}{dt}$ is the temperature increase/decrease per unit time, and c is the specific heat of the experimental liquid. The amount of heat transfer can also be calculated using the following equation:

$$Q = K_t A \Delta T_m \tag{19}$$

where K is the total heat transfer coefficient, A denotes the total area of heat transfer, and ΔT_m is the temperature difference between the water bath and the experimental liquid. The total heat transfer coefficient can be calculated from Equations 18 and 19:

$$K_t = \mathrm{cm}(-\frac{dT}{dt}) / A\Delta T_m \tag{20}$$

(2) Calculation of the heat transfer coefficient α_1 of hydrate slurry

The total heat transfer coefficient can be expressed by:

$$\frac{1}{K_t} = \frac{1}{\alpha_1}\frac{d_2}{d_1} + \frac{b}{\lambda}\frac{d_2}{d_m} + \frac{1}{\alpha_2} \tag{21}$$

where d_1, d_2, d_m represent the inner, outer, and the averaged radius of the reactor, respectively, and λ is the heat conduction coefficient of the reactor. Since in each run, $\frac{b}{\lambda}\frac{d_2}{d_m} + \frac{1}{\alpha_2}$ was kept constant, and thus this term can be eliminated by measuring the total heat transfer coefficient of pure water and hydrate slurry, and calculating the difference in the reciprocal total heat transfer coefficients of pure water and the hydrate slurry. The heat transfer coefficient of hydrate slurry can be calculated by Equation 22 then.

$$\frac{1}{K_{water}} - \frac{1}{K_{slurry}} = \frac{1}{\alpha_{water}}\frac{d_2}{d_1} - \frac{1}{\alpha_{slurry}}\frac{d_2}{d_1} \quad (22)$$

4.4 Experimental results

The heat transfer coefficients for both pure water and diesel oil/hydrate slurry systems with the volume fraction of hydrates of 5%, 10%, 15%, and 20% were measured at 270 K or so. The results are shown in Figure 6.

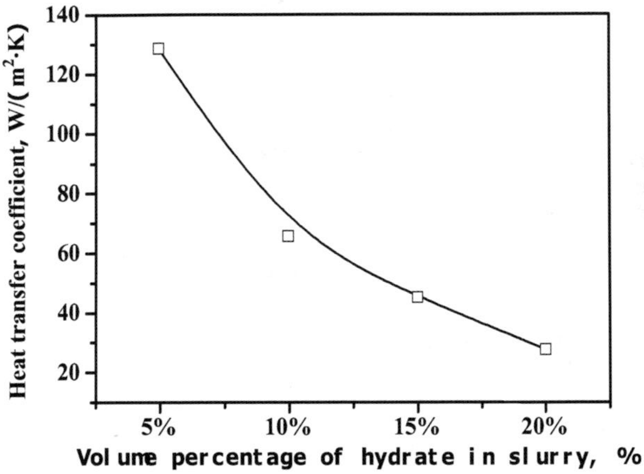

Fig. 6. Variation of heat transfer coefficient of hydrate slurry of different hydrate volume percentage

Figure 6 shows that the heat transfer coefficient decreases with increasing the content of hydrates in slurry, which can be attributed to the fact that the heat conductivity of hydrates is very small.

5. Heat transfer in quiescent hydrate formation/dissociation reactor

It has been well known that the hydrate formation rate can be increased drastically by adding low dose of suitable surfactants, such as sodium dodecyl sulfate (SDS). This kind of additives can enhance mass transfer involved in hydrate formation by decreasing gas/liquid interfacial tension and increasing the solubility of gas in liquid water. Then it is possible to produce gas hydrate rapidly without stirring (Lin et al., 2004; Xie et al., 2005; Zhong and Rogers, 2000). The advantage of quiescent formation of hydrate is that the cost on manufacture and maintenance of the reactor could be reduced largely. Although the mass transfer has been enhanced satisfyingly by adding SDS to water, the heat transfer becomes a serious limitation to the application of quiescent reactor as hydrate formation is an exothermic process. Rogers et al. (2005) designed a scaled-up quiescent process to store 5000 scf of natural gas in a vessel. They thought that the primary challenge of the scale-up design was to provide a surface area/volume ratio in the larger vessel. Therefore they devised an arrangement of finned-tube heat exchanger inside the hydrate formation vessel. Their elementary tests on this process indicated that the hydrate formation in the vessel lasted more than 9 hours, which is still too long for real applications. In order to suit the large scale industrial applications, we devised a multi-deck cell-type vessel as the internals of the reactor to reduce or eliminate the scale-up effect, which is schematically shown in Figure 7 (Pang et al., 2007). The vessel basically consists of a series of uniform boxes stacked up vertically and each box is divided into a series of uniform cells by metal plates. The metal plates were welded on the heat transfer tubes; therefore they also became the cool solid surface during the hydrate formation. The SDS aqueous solution was loaded in these cells with the same level. There are interspaces between two neighboring boxes such that the hydrate forming gas can flow into each deck of the vessel easily. The multi-deck cell-type vessel was placed in the high pressure reactor so that hydrate can form in each cell of the vessel uniformly and simultaneously. In this case, the reaction time depends mainly on the cell volume and the quantity of water loaded, and little on the total volume of the vessel and total quantity of water loaded. Thus the scale-up effect can be eliminated to a large extent as concluded by Pang et al. (2007). Since then, we carried out a systematical study on heat transfer in hydrate formation/dissociation process using this vessel. Experimental details and most recent results obtained by our group which have not been published are introduced in this part.

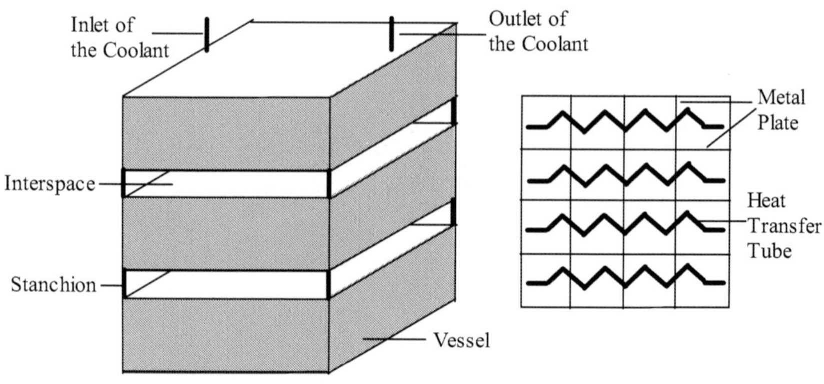

(a) Outline (b) Cell-type inner structure

Fig. 7. The schematic outline of multi-deck cell-type vessel

5.1 Experimental apparatus

In order to investigate the heat transfer performance of this kind of inner structure during hydrate formation/dissociation, a middle scale reactor of a volume of 10 liter as well as an inner multi-deck cell-type vessel suitable for this reactor were manufactured and an experimental set-up, as shown in Figure 8, was established correspondingly. The reactor is 200 mm in diameter, 320 mm in height, and has a volume of 10 L. It was sealed with a blank flange bolted to its top.

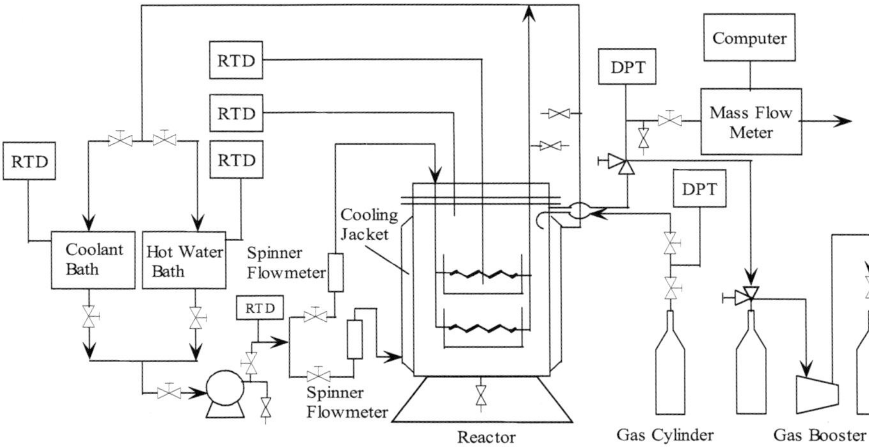

Fig. 8. Flow diagram of the experimental apparatus

Besides the stainless steel quiescent reactor, the primary components of the experimental apparatus consist of a refrigeration system and a coolant or hot water recycling system. A cooling/heating jacket was welded to the outside of the reactor and coiled copper tubes were placed inside the multi-deck cell-type vessel uniformly. Coolant or hot water was circulated through the copper tubes to cool or heat the reactant system for the hydrate formation or dissociation. The flow rate of the coolant or hot water was measured using a spinner flow meter. The coolant is a mixture of water and glycol and has a freezing point lower than 253 K. The coolant bath was controlled by the refrigeration system within a temperature range from 253 K to ambient with an uncertainty of 0.1 K. For the sake of accuracy, another platinum resistance thermometer of ±0.1 K accuracy was placed behind the outlet of the pump to measure the temperatures of the coolant. In addition, a gas recycle system was used to recover the residual gas after the completion of hydrate formation as well as the gas released during hydrate dissociation. Two platinum resistance thermometers of ±0.1 K accuracy were placed into the reactor from the flange to measure the temperatures of the vapor phase and the liquid (or hydrate) phase, respectively. The pressure in the reactor and the gas cylinder was measured using two calibrated Heise pressure gauges with a precision of 0.05 MPa. A mass flow meter and computer system were used to measure and record the flow rate of methane gas during the process of hydrates dissociation at atmospheric pressure. The accuracy of the flow meter is ±1% F.S and the interval of record time is 5s.

5.2 Experimental procedures

5.2.1 Hydrate formation

Prior to any experiments the reactor was washed with water, loaded with 1920 g aqueous solution of 2000 mg/L SDS, and evacuated to remove air. The refrigeration system and coolant pump were then turned on to set the temperature of the aqueous solution in the reactor to 276.15 K. After the desired temperature of the aqueous solution was achieved, methane gas was charged into the reactor until the desired pressure was achieved. During a hydrate formation process, the gas was charged into the reactor continuously to keep the pressure within the range of 6.4 to 6.8 MPa. When the drop in system pressure was less than 0.1 MPa over 20 minutes, the hydrate formation process was assumed to be complete. Subsequently, the hydrate was frozen for dissociation test. In the experiment, the temperature or flux of coolant was changed to determine the effect of heat transfer on the hydrate formation rate. In each experimental run, the change of the pressure and temperature of the gas in the reactor with the elapsing time was recorded, and the mole number, Δn_i, of methane consumed in the time period of Δt_i was then determined with the following equation:

$$\Delta n_i = \frac{P_t V_t}{Z_t R T_t} - \frac{P_{t+\Delta t_i} V_{t+\Delta t_i}}{Z_{t+\Delta t_i} R T_{t+\Delta t_i}} \tag{23}$$

where V_t, T_t, P_t, and Z_t are the volume, temperature, pressure, and compressibility of the gas phase in the reactor at time t, respectively. $V_{t+\Delta t}$, $T_{t+\Delta t}$, $P_{t+\Delta t}$, and $Z_{t+\Delta t}$ are the volume, temperature, pressure, and compressibility of the gas phase in the reactor at the time $t+\Delta t$, respectively. It should be noted that no gas was charged into the reactor from the gas cylinder during the time period of Δt_i. The cumulative moles of gas consumed could be calculated readily through

$$\Delta n = \sum_i \Delta n_i \tag{24}$$

5.2.2 Hydrate dissociation

After hydrate samples were formed with the procedure described above and cooled down to 268.15 K, the system was left for about 3 hours with a less than 0.1 K fluctuation in reactor temperature. Next, the vent was opened slowly to reduce the system pressure gradually to a bit above the equilibrium formation pressure of methane hydrate at the set temperature. Subsequently, the system was depressurized rapidly to near the atmospheric pressure of 0.1 MPa and the vent was then turned off. The methane hydrate samples were then heated with hot water at a fixed temperature to a complete dissociation. The history of pressure and temperature of the reactor, and the temperatures of the inlet and vent hot water were recorded with the elapsed time. During the dissociation process, the hot water was charged only into the coiled copper tubes and the coolant in the cooling jacket was expelled by air before the dissociation experiment.

The cumulative moles of methane dissociated at time t without considering the shrinking of methane hydrates volume during the dissociation can be calculated with the following equation (Pang et al., 2009)

$$\Delta n_t = \frac{P_t V_i}{Z_t R T_t} - \frac{P_i V_i}{Z_i R T_i} \tag{25}$$

where T_t, P_t, and Z_t are the temperature, pressure, and compressibility, respectively, of the gas phase in the reactor at time t. P_i, V_i, T_i, and Z_i are the initial dissociation pressure, initial volume, initial temperature, and initial compressibility, respectively, of the gas phase in the reactor.

5.3 Modeling of heat transfer dependence of hydrate formation

Englezos et al. (1987a) proposed that the hydrate formation rate was proportional to its driving force and reaction area. In our work we defined the supercooling, i.e., the difference between the equilibrium temperature and the actual reaction temperature, as the driving force, then the rate of gas consumed for hydrate growth could be formulated as

$$\frac{dn}{dt} = K_d A_{gl}(T_e - T_w) \tag{26}$$

where K_d is the intrinsic hydrate formation rate constant, A_{gl} is the formation area, and T_e and T_w are the equilibrium temperature and actual temperature of bulk water, respectively. As an exothermic reaction, the temperature of the water solution will increase with the formation of hydrate. Thus the metal plates were welded on the heat transfer tubes to remove the formation heat in the multi-deck cell-type vessel. Because of the temperature difference between the bulk water and the metal plates, the heat transfer happened and the reaction heat of hydrate formation could be expressed by

$$\frac{dn}{dt}\Delta H = K_h(T_w - T_l)A_m + \frac{d(mC_pT)}{dt} \tag{27}$$

where ΔH is the reaction heat of hydrate formation, K_h is the heat transfer coefficient, T_l is the temperature of coolant, A_m is the heat transfer area(it is assumed to be equal to the area of the metal plates), m is the total quantity of water and hydrate in the vessel, C_p is the heat capacity of mixture of water and hydrate in the vessel, and T is the temperature of water and hydrate in the vessel (here we assumed that the temperature of water and hydrate is same).

As the change of temperature is not obvious during the hydrate formation, it can be assumed that most of the formation heat of hydrate formation is removed by the coolant. So Equation 27 can be simplified as

$$\frac{dn}{dt}\Delta H = K_h(T_w - T_l)A_m \tag{28}$$

Combining Equations 26 and 28 yields

$$\frac{dn}{dt} = \left(\frac{1}{\frac{\Delta H}{K_h A_m} + \frac{1}{K_d A_{gl}}}\right)(T_e - T_l) \tag{29}$$

Considering that the gas/water interface area decreases with the proceeding of hydrate formation, the formation area A_{gl} in Equation 29 is empirically formulated as

$$A_{gl} = A_s\left(1 - \frac{n}{n_s}\right)^b \tag{30}$$

where A_s is the interface area of water and gas at the beginning of reaction, n_s is the total quantity of consumed gas at the end of the reaction, n is the quantity of consumed gas at time t, and b is the reduced exponent of reaction area. In our work, the shape of vessel is round, and the vessel was divided into a series of uniform cells by metal plates with a 25 mm span. So the shape of most of the cells is square except the cell at the edge of the vessel, which is not regular. In our model, those irregular cells were transformed into the square cell according to the volume ratio, and the reaction status of hydrate formation in every cell was assumed to be same.

The dependence of K_d on temperature is formulated with the following Arrhenius-type equation,

$$K_d = k_0 e^{-\frac{\Delta E}{RT}} \tag{31}$$

where k_0 is the pre-exponential coefficient and ΔE is the activation energy of intrinsic hydrate formation.

5.4 Heat transfer dependence in quiescent hydrate formation

Heat transfer is very important for the hydrate formation. The heat of hydrate formation must be removed in time; otherwise the hydrate formation could not proceed continuously. There are two factors significantly affecting the heat transfer rate: one is the structure of the reactor, i.e., total area for heat exchange and the other one is the coolant. The effect of the structure of the reactor has been investigated by Pang et al. (2007). Our present work focuses on the effect of the coolant on the hydrate formation, including the influence of the temperature and flux of the coolant. To study the effect of the coolant on the hydrate formation and ensure all of the hydrate formation experiments are performed on the same basis, the initial temperature of reaction system, the pressure, the mass of water solution, and the concentration of surfactant SDS were uniformly specified as 276.15 K, 6.4-6.8 MPa, 1920 g, and 2000 mg/L, respectively. By changing the temperature and flux of coolant, a series of experiments have been performed, and the results are shown in Figures 9 and 10, where the hydrate formation rate was manifested by the cumulative mole numbers of consumed gas vs. elapsed time.

5.4.1 Influence of coolant temperature

The effect of the temperature of coolant on the hydrate formation rate is shown in Figure 9. One can see that hydrate formation rate increases with the decreasing of the coolant temperature. This is easy to understand as when the temperature of coolant is lower, the formation heat can be more easily removed. As a result, the temperature of the water solution keeps lower and the driving force of hydrate formation keeps larger, resulting in the hydrate forms faster. However, when the temperature of coolant is low enough, the consumed rate of methane gas changed little with the decreasing of the coolant temperature. It means that the effect of coolant temperature on the hydrate formation rate is limited. When the temperature of coolant is low enough, the further decreasing of temperature have little effect on the hydrate formation rate. At that time, the hydrate formation rate is not controlled by the heat transfer, but controlled by the intrinsic kinetics of hydrate formation.

For comparison, the calculated results by Equation 29 are also presented in Figure 9. One can see that the agreement between experimental data and calculated results is satisfying. The parameters of model were correlated using the experimental data and are shown in Table 2, where the flux of the coolant was fixed at 5.5 L/min and the hydrate formation heat was set to 54.20 KJ/mol (Makogon,1997).

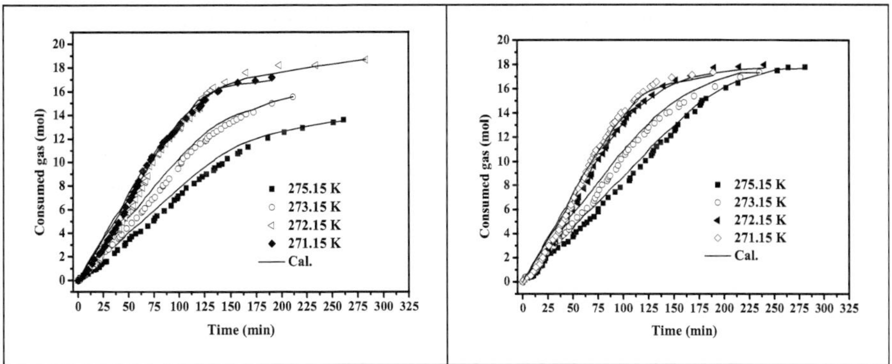

(a) (b)

Fig. 9. The effect of the coolant temperature on hydrate formation rate at different flux of coolant: (a) 2.0 L/min and (b) 3.5 L/min

Temperature of coolant K	System temperature K	Heat transfer coefficient KW $m^{-2} \cdot K^{-1}$	Reaction rate constant mol $m^{-2} \cdot K^{-1} \cdot s^{-1}$	b	ΔE KJ mol^{-1}	k_0 mol $m^{-2} \cdot K^{-1} \cdot s^{-1}$
275.15	281.4096	0.1662	0.0723	1.0030		
273.15	280.8802	0.1582	0.0688	0.8052		
272.15	280.1633	0.1808	0.0635	0.9670	66.467	1.5754×10^{11}
271.15	279.3663	0.1784	0.0589	1.0938		
270.15	279.3233	0.1987	0.0585	0.9145		

Table 2. Model parameters in Equation 29 and the correlated value of ΔE and k_0

5.4.2 Influence of coolant flux

Experimental results showing the influence of the coolant flux on the hydrate formation rate are plotted in Figure 10. It could be seen that the hydrate formation rate increases with the increasing of coolant flux when the temperature is fixed. The effect of coolant flux on the formation rate is more significant when the temperature of coolant is higher. When the temperature of coolant is low enough, the further increase of coolant flux has little effect on increasing hydrate formation rate. In this case, hydrate formation is controlled not by the heat transfer, but by the intrinsic kinetics of hydrate formation.

5.5 Heat transfer dependence in quiescent hydrate dissociation

We performed a series of experiments to reveal the effect of heating on methane hydrate dissociation in the quiescent reactor (Pang et al., 2009). Representative profiles of pressure, temperature, and cumulative mole number of evolved methane during hydrate dissociation are depicted in Figure 11, where hydrate was formed from 1920 g water of 2000 mg/L SDS, the initial temperature of hydrate was set to 268.15 K, the temperature of the input hot water was set to 298.15 K, and the flow rate was set to 2.0 liter per minute (LPM). As shown in Figure 11, at the beginning of hydrate dissociation, the temperature of hydrate decreased drastically and rapidly, which indicates a brief-but-rapid dissociation induced by the rapid depressurization in a short time period. During this initial degassing event, the temperature of sample typically dropped 3 to 7 K due to adiabatic cooling of methane gas expansion as well as heat absorption of hydrate dissociation (Stern et al., 2001).

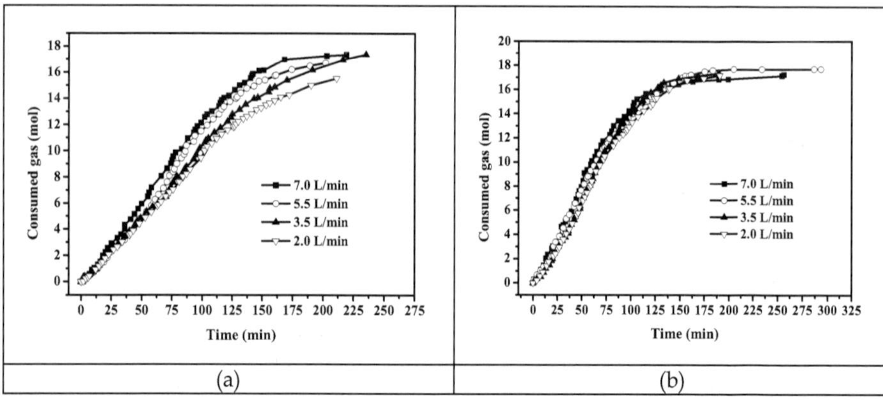

Fig. 10. The effect of the coolant flux on hydrate formation rate at different coolant temperatures: (a) 273.15 K and (b) 271.15 K

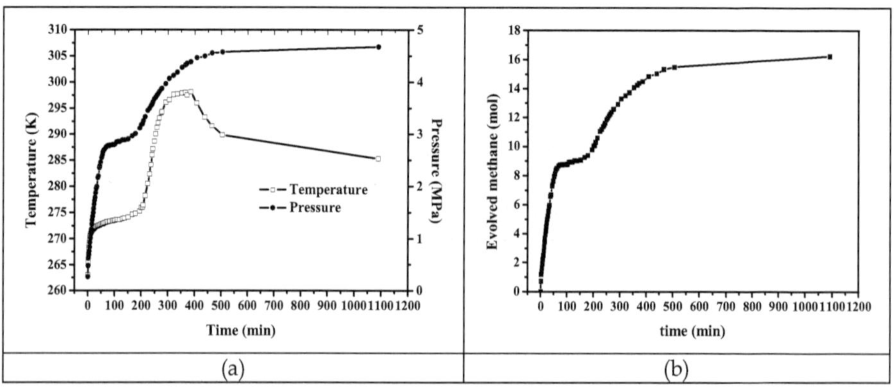

Fig. 11. Typical profiles of pressure, temperature and cumulative mole number of evolved methane during hydrate dissociation, heated by 298.15 K hot water of a flux of 2.0 LPM in a closed system

The temperature of hydrate and pressure of system increased sharply after the vent was turned off as the hydrates dissociated rapidly. This brief but rapid hydrate dissociation is driven mainly by the thermodynamic driving force, i.e., the difference between the equilibrium pressure and the present system pressure. Figure 12(a) shows the comparison of the experimental dissociation temperature/pressure locus with the equilibrium temperature/pressure curve of methane hydrate. It can be seen that the experimental pressure is obviously lower than the equilibrium pressure at this stage, which implies the driving force is large.

Then a stage of buffered dissociation occurred when the hydrates was heated to approach the melting point of ice. During this period, the sample temperature remains within a narrow range near the ice point, rises slowly with the proceeding of the hydrate dissociation, and the corresponding dissociation rate is very low. As shown in Figure 12(a), the temperature/pressure locus is very close to the equilibrium temperature/pressure curve, which implies that the hydrate is closely at equilibrium state at this buffering stage and the thermodynamic driving force for hydrate dissociation is very low. At this stage, the ice coming from the hydrate dissociation is heated to produce liquid water and much larger amount of heat is required for the thaw of ice. This phase transition process buffers the system temperature at nearly fixed temperature and pressure and the rate of heat transfer is the key rate-limiting factor for the hydrate dissociation. Circone et al. (2003, 2004) and Stern et al. (2003) also found this kind of thermal buffering during decomposing methane hydrates by heating them from low temperature through the melting point of ice. After the buffered dissociation stage, the rate of methane hydrate dissociation increases with the increasing of temperature of heating water. Therefore, the rate of heat transfer is an important factor that controls the rate of hydrate dissociation, especially during the buffering stage. In this buffering region, the hydrate dissociation rate is very low, which is obviously unfavorable to both the recovery of gas from hydrate in the NGH technology and the gas production from in situ natural gas hydrates with heating method. Measures should be taken to eliminate this effect.

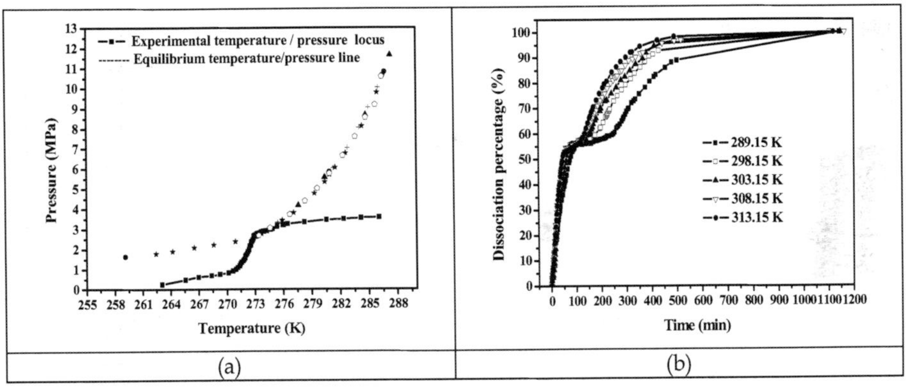

Fig. 12. (a) Comparison of the experimental dissociation temperature/pressure locus with the equilibrium temperature/pressure curve of methane hydrate, and (b) The influence of the temperature of heating water upon hydrate dissociation rate

Our previous work (Pang et al., 2009) demonstrated that the buffering effect can be eliminated and the dissociation rate can be improved by increasing the temperature of heating water or reducing the dissociation pressure. Figure 12(b) shows the influence of the temperature of heating water upon the dissociation rate of hydrate, where the dissociation rate is manifested by changing of cumulative dissociation percentage of hydrate with the elapsed time. It can be seen that the thermal buffering period can be shortened obviously by increasing heating water temperature, which suggests the important role of heat transfer in controlling the dissociation of hydrate.

6. Heat transfer in hydrate bearing sediment

6.1 Heat effect during hydrate formation in sediment

Hydrate formation process in sediment is an exothermic process. The variation of temperature is related to the hydrate formation rate and amount. For hydrate formation in large scale of sediment, the fluctuations of temperatures at different locations will be therefore different but will coincide with specified rules. A three-dimensional apparatus has been used by our group recently to study the heat transfer process during hydrate formation in sediment (Yang et al., 2010). Sixteen thermocouples were inserted into the sediment to detect the variation of temperature, which were placed in different radius and different depth of the reactor. Figures 13 and 14 show the pressure and temperature distribution during the hydrate formation using the three-dimensional apparatus. From point A to B in Figure 13, the top pressure decreases gradually, showing the process of hydrate formation. At the same time, the temperatures at different locations in the reactor first rise to a maximum value and then decreases, as shown in Figure 14. At the initial period, the hydrate formation rate is high and temperatures rise due to the exothermic reaction. With the decline of hydrate formation rate due to the decrease of the pressure driving force and pore space, the rising trend of temperature weakens and the temperatures even decrease gradually because the exothermic effect of hydrate formation can not make up the refrigeration effect of the water bath. From point B to C, the pores of the sediment are packed with formed hydrate, which resulting in that gas at the top of the reactor can hardly permeate into the inner of sediment. Therefore, little hydrate forms and pressure and temperature nearly keep constant in this stage. From point C to D, hydrate continues to form again, which can be implied from the magnitude of pressure decrease in Figure 13 and the temperatures increase in Figure 14. After point D, no hydrate formed and the top pressure and temperature kept constant gradually.

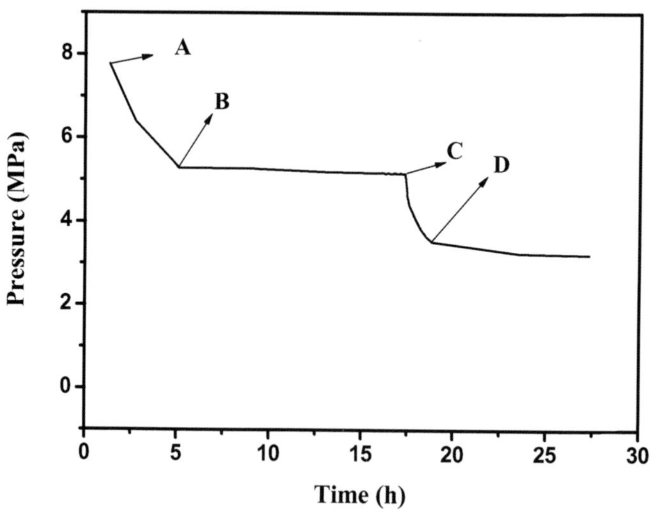

Fig. 13. Variation of pressure of the reactor versus time during hydrate formation

During the whole process of hydrate formation, there are two stages (from A to B and C to D) of rapid increase of temperature caused by the exothermic effect of hydrate formation. The heat transfer during hydrate formation in sediment can then be characterized from the appearance sequence of the temperature peak and its magnitude.

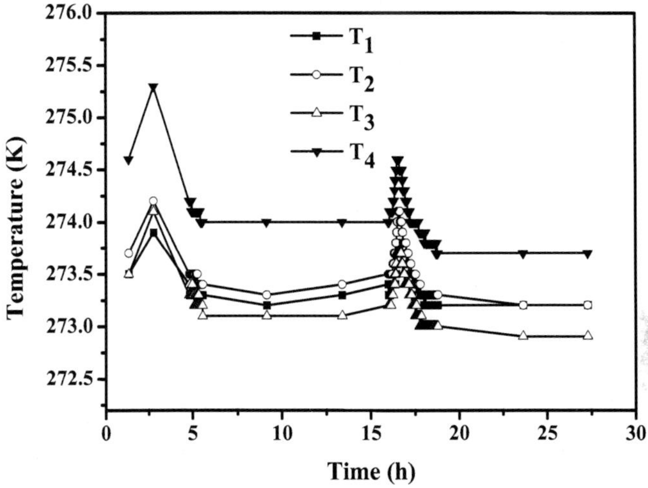

Fig. 14. Temperature distribution during hydrate formation with the same depth but different radius: T_1, T_2, T_3, T_4 were placed in different radius of 132 mm, 99 mm, 66 mm, and 33 mm, respectively

6.2 Heat effect during hydrate dissociation in sediment

Depressurization, to depressurize the gas hydrates reservoir below the equilibrium pressure at specified temperature, has been considered as one of conventional methods of producing gas from hydrate bearing sediment. Heat effect will arise from the Joule-Thomson effect and endothermic process of hydrate dissociation. For a typical hydrate depressurizing decomposition in a three-dimensional environment (Su et al., 2010), three steps were adopted. First, free gas is discharged until the pressure decreases to a specified value, which is close to the equilibrium pressure at the current temperature. Afterward, gas is discharged rapidly and the system pressure attains the experimental dissociation value quickly. During the rapid depressurizing stage, there exists Joule-Thomson effect, and remarkable temperature drop was observed. Hereafter, hydrate begins to decompose. The temperature begins to rise under the heat transfer of air bath and endothermic process of hydrate dissociation. The extent of temperature variation can be used to reflect the quantity of hydrate decomposed. The difference between temperature value before and after rapid depressurizing, ΔT, can be used to analyze the dissociation process. For the value of ΔT at different positions, both Joule-Thomson effect and thermal effect of hydrate decomposition contribute to ΔT. Since both are endothermic process, the quantity of hydrate decomposition can be indicated by ΔT value. If assuming that hydrate in the reactor decomposing at the same drive force and at the same apparent decomposition rate, the position sequence of hydrate decomposition can also be connected with ΔT value. That is, the larger the ΔT is, the earlier decomposition reaction begins. Hydrate at the top and bottom of the reactor decomposes earlier than that in the inner of the reactor.

Thermal stimulation, to produce gas from hydrates by increasing the temperature of the gas hydrates reservoir above the equilibrium temperature at specified pressure, is also an important method to produce gas from hydrate bearing sediment. Selim and Sloan (1990) found that the dissociation rate was a strong function of the thermal properties of the system and the porosity of the porous medium. Kamata et al. (2005) applied thermal recovery method to dissociate methane hydrate in hydrate sediment sample by hot-water injection from one side, gas production from another side. It was found that temperature and pressure in the sample fluctuated between stability region and decomposition region of methane hydrate sample when temperature of the hot water was high. We (Yang et al., 2010) studied the temperature distributions in the reactor during gas production by cyclic injecting hot water method. Figure 15 shows the variation of temperatures with time at different locations during the first cycle and the second cycle. It can be found that the overall temperature trend increases with hot-water injection and decreases with gas production. At initial stage of water injection, the temperatures rise slowly due to much heat is consumed to warm the flowing channel of hot-water. Afterward, the temperatures at different locations rise rapidly. It is noted that for the local positions of hydrate sample which are away from the injecting well and near the reactor wall, the temperature-jump phenomenon is not observed. It is known that the hot-water can hardly spread to those locations in a shorter time. Strong heat exchange with environment may also occur on those positions near the reactor wall with high conductivity. In contrast, at other locations, in view of low conductivity of hydrate sample, the fluctuation of temperature mainly depends on the seepage rate of hot-water in sediments instead of thermal diffusion. That is to say, the seepage flow controls the heat transport in this process at these locations. In addition, the temperature distribution also implied the injected hot-water distribution in the reactor to a certain extent. The location of well has an important effect on the temperature distribution in hydrate sediment, thereby affecting the gas production.

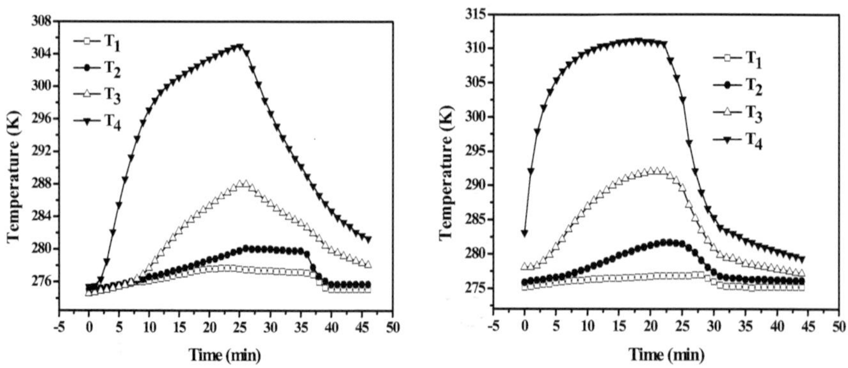

(a) first cycle (b) second cycle

Fig. 15.Variation of temperature with time at different locations with the same depth during the first cycle: T_1, T_2, T_3, T_4 were placed in different radius of 132 mm, 99 mm, 66 mm, and 33 mm, respectively

7. Conclusion

In this chapter we have discussed heat transfer involved in formation/dissociation of gas hydrates. The state of the art for the studies of the experimental measurement of thermal conductivities of different kinds of gas hydrates, the mechanism and modeling of heat transfer occurring in the growth of hydrate film at the guest/water interface, our experimental study on macroscopic heat transfer in stirring reactors or flowing pipes, the experimental and modeling studies on the heat transfer in quiescent reactors, and the mechanism of heat transfer in hydrate bearing-sediment are summarized. We believe this chapter will be helpful for readers to understand the critical position of heat transfer in the hydrate formation/dissociation kinetics, favor the ongoing hydrate formation/dissociation kinetic research, and promote the application of gas hydrate.

8. Acknowledgement

The financial supports of the NSFC (Nos. 20925623, 21006126), the Research Funds of China University of Petroleum, Beijing (BJBJRC-2010-01), and the National Basic Research Program of China (2009CB219504) are greatly appreciated.

9. References

Andersson P. & Ross R. G. (1983). Effect of guest molecule size on the thermal conductivity and heat capacity of clathrate hydrates. *J. Phys. C*, Vol. 16, pp. 1423-1432.

Circone S., Stern L. A., Kirby S. H., Durham W. B., Chakoumakos B. C., Rawn C. J., Rondinone A. J. & Ishii Y. (2003). CO_2 hydrate: synthesis, composition, structure, dissociation behavior, and a comparison to structure I CH_4 hydrate. *J. Phys. Chem. B*, Vol. 107, pp. 5529-5539.

Circone S., Stern L. A. & Kirby S. H. (2004). The role of water in gas hydrate dissociation. *J. Phys. Chem. B*, Vol. 108, pp. 5747-5755.

Cook J. G. & Laubitz M. J. (1983). In *Thermal Conductiuity*; Hust J. G., Ed.; Plenum: New York, Vol. 17, pp. 745.

Cortes D. D., Martin A. I., Yun T. S., Francisca F. M., Santamarina J. C. & Ruppel C. (2009). Thermal conductivity of hydrate-bearing sediments. *J. Geophys. Res.*, Vol. 114, pp. B11103.

deMartin B. J. (2001). *Laboratory measurements of the thermal conductivity and thermal diffusivity of methane hydrate at simulated in-situ conditions*. Georgia Institute of Technology: Atlanta, GA, USA.

Englezos P., Kalogerakis N. E., Dholabhai P. D. & Bishnoi P. R. (1987a). Kinetics of formation of methane and ethane gas hydrates. *Chem. Eng. Sci.*, Vol. 42, pp. 2647-2658.

Englezos P., Kalogerakis N. E., Dholabhai P. D. & Bishnoi P. R. (1987b). Kinetics of gas hydrate formation from mixtures of methane and ethane. *Chem. Eng. Sci.*, Vol. 42, pp. 2659-2666.

English, N. J. & Tse, J. S. (2010). Perspectives on hydrate thermal conductivity. *Energies*, Vol. 3, pp. 1934-1942.

Freer, E. M., Selim, M. S. & Sloan, E. D. (2001). Methane hydrate film growth kinetics. *Fluid Phase Equilib.*, Vol. 185, pp. 65-75.

Gustafsson S. E., Karawacki E. & Khan M. N. (1979). Transient hot-strip method for simultaneously measuring thermal conductivity and thermal diffusivity of solid sand fluids. *J. Phys. D: Appl. Phys.*, Vol. 12, pp.1411.

Gustafsson S. E., Karawacki E. & Chohan M. A. (1986). Thermal transport studies of electrically conducting materials using the transient hot-strip technique. *J. Phys. D: Appl. Phys.*, Vol. 19, pp. 727.

Handa Y. P. & Cook J. G. (1987). Thermal conductivity of xenon hydrate. *J. Phys. Chem.*, Vol. 91, pp. 6327-6328.

Huang D. Z. & Fan S. S. (2004). Thermal conductivity of methane hydrate formed from sodium dodecyl sulfate solution. *J. Chem. Eng. Data*, Vol. 49, pp. 1479–1482.

Huang D. Z., Fan S. S. & Shi L. (2004). Thermal conductivity of gas hydrate. *Huaxuetongbao*, Vol. 10, pp. 737-742. (in Chinese)

Kamata Y., Ebinuma T., Omura R., Minagawa H., Narita H., Masuda Y. & Konno Y. (2005). Decomposition experiment of methane hydrate sediment by thermal recovery method. *Proceedings of the 5th International Conference on Gas Hydrates*, Trondheim, Norway, pp. 81-85.

Krivchikov A. I., Gorodilov B. Y., Korolyuk O. A., Manzhelii V. G., Conrad H. & Press W. (2005). Thermal conductivity of methane-hydrate. *J. Low Temp. Phys.*, Vol. 139, pp. 693–702.

Krivchikov A. I., Gorodilov B. Y., Korolyuk O. A., Manzhelii V. G., Romantsova O. O., Conrad H., Press W., Tse J. S. & Klug D. D. (2006). Thermal conductivity of Xe clathrate hydrate at low temperatures. *Phys. Rev. B*, Vol. 73, pp. 064203.

Li D. L., Liang D. Q., Fan S. S. & Peng H. (2010). Estimation of ultra-stability of methane hydrate at 1 atm by thermal conductivity measurement. *J. Nat. Gas. Chem.*, Vol. 19, pp. 229-233.

Lin W., Chen G. J., Sun C. Y., Guo X. Q., Wu Z. K., Liang M. Y., Chen L. T. & Yang L. Y. (2004). Effect of surfactant on the formation and dissociation kinetic behavior of methane hydrate. *Chem. Eng. Sci.*, Vol. 59, pp. 4449-4455.

Lunden A., Trolle U., Azoulay M. & de Pablo J. (1986). Determination of kinetics and thermal conductivity of hydrates simultaneously by using thermogravimetry and transient hot strip method. *Thermochimica Acta*, Vol. 105, pp. 369-373.

Ma C. F., Chen G. J. & Guo T. M. (2002). Kinetics of hydrate formation using gas bubble suspended in water. *Science in China (Series B)*, Vol. 45, pp. 208-215.

Makogon Y. F. (1997). *Hydrates of Hydrocarbons.* Tulsa, Oaklahoma: Penn Well books.

Mochizuki T. & Mori Y. H. (2006). Clathrate-hydrate film growth along water/hydrate-former phase boundaries-numerical heat-transfer study. *J. Crystal. Growth*, Vol. 290, pp. 642-652.

Moon C., Taylor P. C. & Rodger P. M. (2003). Molecular dynamic study of gas hydrate formation. J. Am. Chem. Soc., Vol. 125, pp. 4706-4707.

Mori Y. H. (2001). Estimating the thickness of hydrate films from their lateral growth rates: application of a simplified heat transfer model. *J. Crystal. Growth*, Vol. 223, pp. 206-212.

Ohmura R., Kashiwazaki S. & Mori Y. H. (2000). Measurements of clathrate-hydrate film thickness using laser interferometry. *J. Crystal. Growth*, Vol. 218, pp. 372-380.

Ohmura R., Matsuda S., Uchida T., Ebinuma T. & Narita H. (2005). Clathrate hydrate crystal growth in liquid water saturated with a guest substance: observations in a methane + water system. *Cryst. Growth Des.*, Vol. 5, pp. 953-957.

Pang W. X., Chen G. J., Dandekar A., Sun C. Y. & Zhang C. L. (2007). Experimental study on the scale-up effect of gas storage in the form of hydrate in a quiescent reactor. *Chem. Eng. Sci.*, Vol. 62, pp. 2198-2208.

Pang W. X., Xu W. Y., Sun C. Y., Zhang C. L. & Chen G. J. (2009). Methane hydrate dissociation experiment in middle-sized quiescent reactor using thermal method. *Fuel*, Vol. 88, pp. 497-503.

Peng B. Z., Chen G. J., Sun C. Y., Yang L. Y. & Luo H. (2008). Is the gas hydrate film growth controlled by intrinsic kinetic or heat transfer? *Proceedings of the 6th international conference on gas hydrates*, Vancouver, British Columbia, Canada.

Peng B. Z., Dandekar A., Sun C. Y., Luo H., Ma Q. L., Pang W. X. & Chen G. J. (2007). Hydrate film growth on the surface of a gas bubble suspended in water. *J. Phys. Chem. B*, Vol. 111, pp. 12485-12493.

Peng B. Z., Sun C. Y., Chen G. J., Yang L. Y., Zhou W. & Pang W. X. (2009). Hydrate film growth at the interface between gaseous CO_2 and sodium chloride solution. *Science in China Series B: Chemistry*, Vol. 52, pp. 676-682.

Rogers R. E., Zhong Y., Etheridge J. A., Arunkumar R., Pearson L. E. & Hogancamp T. K. (2005). Micellar gas hydrates storage process. *Proceedings of the 5th international conference on gas hydrates*, Trondheim, Norway, pp. 1361-1365.

Rosenbaum E. J., English N. J., Johnson J. K., Shaw D. W. & Warzinski R. P. (2007). Thermal conductivity of methane hydrate from experiment and molecular simulation. *J. Phys. Chem. B*, Vol. 111, pp. 13194–13205.

Ross R. G. & Andersson P. (1982). Clathrate and other solid phases in the tetrahydrofuran–water system: thermal conductivity and heat capacity under pressure. *Can. J. Chem.*, Vol. 60, pp. 881-892.

Ross R. G., Andersson P. & Backstrom G. (1981). Unusual PT dependence of thermal-conductivity for a clathrate hydrate. *Nature*, Vol. 290, pp. 322.

Saito K., Kishimoto M., Tanaka R. & Ohmura R. (2011). Crystal growth of clathrate hydrate at the interface between hydrocarbon gas mixture and liquid water. *Crystal. Growth & Design*, Vol.11, pp. 295-301.

Saito K., Sum A. K. & Ohmura R. (2010). Correlation of hydrate-film growth rate at the guest/liquid-water interface to mass transfer resistance. *Ind. Eng. Chem. Res.*, Vol. 49, pp. 7102-7103.

Selim M. S. & Sloan E. D. (1990). Hydrate dissociation in sediments. *SPE Reserv. Eng.*, Vol. 5, pp. 245-251.

Sloan E. D. & Koh C. A. (2007). *Clathrate hydrate of natural gases*. 3rd ed, CRC Press: Boca Raton, FL.

Stern L. A., Circone S., Kirby S. H. & Durham W. B. (2001). Anomalous preservation of pure methane hydrate at 1 atm. *J. Phys. Chem. B*, Vol. 105, pp. 1756-1762.

Stern L. A., Circone S., Kirby S. H. & Durham W. B. (2003). Temperature, pressure, and compositional effects on anomalous or "self" preservation of gas hydrates. *Can. J. Phys.*, Vol. 81, pp. 271-283.

Stoll R. D. & Bryan, G. M. (1979). Physical properties of sediments containing gas hydrates. *J. Geophys. Res.*, Vol. 84, pp. 1629-1634.

Su K. H., Sun C. Y., Yang X., Chen G. J. & Fan S. S. (2010). Experimental investigation of methane hydrate decomposition by depressurizing in porous media with 3-Dimension device. *J. Nat. Gas Chem.*, Vol. 19, pp. 210–216.

Sun C. Y., Chen G. J., Ma C. F., Huang Q., Luo H. & Li Q. P. (2007). The growth kinetics of hydrate film on the surface of gas bubble suspended in water or aquesou surfactant solution. *J. Crystal. Growth*, Vol. 306, pp. 491-499.

Taylor C. J., Miller K. Y., Koh C. A. & Sloan E. D. (2007). Macroscopic investigation of hydrate film growth at the hydrocarbon/water interface. *Chem. Eng. Sci.*, Vol. 62, pp. 6524-6533.

Tse J. S. & White M. A. (1988). Origin of glassy crystalline behavior in the thermal properties of clathrate hydrates: A thermal conductivity study of tetrahydrofuran hydrate. *J. Phys. Chem.*, Vol. 92, pp. 5006-5011.

Tzirita A. (1992). *In situ detection of natural gas hydrates using electrical and thermal properties.* Ph.D. thesis, Texas A&M Univ., College Station.

Uchida T., Ebinuma T., Kawabata J. & Narita H. (1999). Microscopic observations of formation processes of clathrate-hydrate films at an interface between water and carbon dioxide. *J. Crystal. Growth*, Vol. 204, pp. 348-356.

Uchida T., Ikeda I. Y., Yakeya S., Ebinuma T., Nagao J. & Narita H. (2002). CO_2 hydrate film formation at the boundary between CO_2 and water: effects of temperature, pressure and additives on the formation rate. *J. Crystal. Growth*, Vol. 237, pp. 383-387.

Vysniauskas Y. & Bishnoi P. R. (1983). A kinetic study of methane hydrate formation. Chem. Eng. Sci., Vol. 38, pp. 1061-1072.

Waite W. F., Gilbert L. Y., Winters W. J. & Mason D. H. (2005). *Thermal property measurements in Tetrahydrofuran (THF) hydrate and hydrate bearing sediment between -25 and +4 0C, and their application to methane hydrate.* Paper presented at 5th International Conference on Gas Hydrates, Tapir Acad., Trondheim, Norway.

Waite W. F., Stern L. A., Kirby S. H., Winters W. J. & Mason D. H. (2007). Simultaneous determination of thermal conductivity, thermal diffusivity and specific heat in sI methane hydrate. *Geophys. J. Int.*, Vol. 169, pp. 767–774.

Xie Y. M., Guo K. H., Liang D. Q., Fan S. S. & Gu J.M. (2005). Steady gas hydrate growth along vertical heat transfer tube without stirring. *Chem. Eng. Sci.*, Vol. 60, pp. 777-786.

Yamamoto Y., Ohtake M., Kawamura T., Tsuji T. & Tsukada Y. (2008). Thermal conductivity measurement of methane hydrate with soil. *Proceedings of the 6th international conference on gas hydrates*, Vancouver, British Columbia, Canada.

Yang X., Sun C. Y., Yuan Q., Ma P. C. & Chen G. J. (2010). Experimental study on gas production from methane hydrate-bearing sand by hot-water cyclic injection. *Energy & Fuels*, Vol. 24, pp. 5912–5920.

Zhong Y. & Rogers R. E. (2000). Surfactant effects on gas hydrate formation. *Chem. Eng. Sci.*, Vol. 55, pp. 4175-4187.

Progress Works of High and Super High Temperature Heat Pipes

Wei Qu
Institute of Engineering Thermophysics, Chinese Academy of Sciences
North 4th Ring Road No.11, Beijing 100190
China

1. Introduction

Heat pipe is a high efficiency heat transfer element, depends on the evaporation, condensation and circulation of inside working fluid. The good performance of a heat pipe is due to that the working fluid evaporation of latent heat is generally large, so there needs no much working fluid and the circulation flow rate is usually small, and the flow resistance is small, when a heat pipe works. The temperature and pressure inside a heat pipe are nearly at uniform levels through the evaporator to the condenser, so a heat pipe has a high heat transfer ability at smaller temperature difference. The heat pipe has many other advantages, such as the heat fluxes of evaporator and condenser can be changed easily by adjusting the heat transfer lengths and areas etc..

The working fluid of a high temperature heat pipe is generally alkali metals, and their evaporations of latent heat are very large, and the saturation pressure is smaller at high temperature, then the case strength is not a problem. The high and super high temperature heat pipe can work over 500°C, 1000°C, 1500°C or much higher. Sodium heat pipe can operate during 500~1100°C, and lithium heat pipe can work over 1200°C (Ma et al, 1983). The high and super high temperature heat pipes have wide applications in space thermal control, solar storage power plant, nuclear thermal control, the constant industry stoves etc.. Usually, the heat pipes between 500~1100°C use the sodium as working fluid, they are called as the high temperature heat pipe; while the heat pipes over 1200°C, utilize lithium as the working fluid, they are named as the super high temperature heat pipes. Commonly, 1000°C can be supposed to divide the boundary of high and super high temperature heat pipes.

The first heat pipe in China was made in 1972, those stainless steel and sodium heat pipes realized the high temperature stove to make the semiconductor technology reliable. As shown in figure 1, the super high temperature heat pipe of W-Li by USA was tested in 1966 (Busse, 1992). The basic parameters and the startup process were given beside the figure, different temperature level tells the performances of the lithium heat pipe.

For the nuclear reactor cooled by the high temperature heat pipes, the fuel bars and the high temperature heat pipes are designed to be coupled together. By the high temperature heat pipes, the heat from the pile core is transferred to the heat exchanger, the technology of high temperature thermal radiation is realized (Zhang et al, 2009). The high temperature heat pipes connect the nuclear power suppliers to the outside and transfer the large amount of heat. When the space nuclear power suppliers work at full, the temperature of heat core can reach 3000~3300K, however, the maximum temperature to be controlled is 800K. So, the high temperature heat pipe is the key problem to be solved for the whole system. Such aspect relates also the miniaturization, and performance optimum of the high temperature heat pipe, and the shape adaptability and coupling with the high temperature radiator etc..

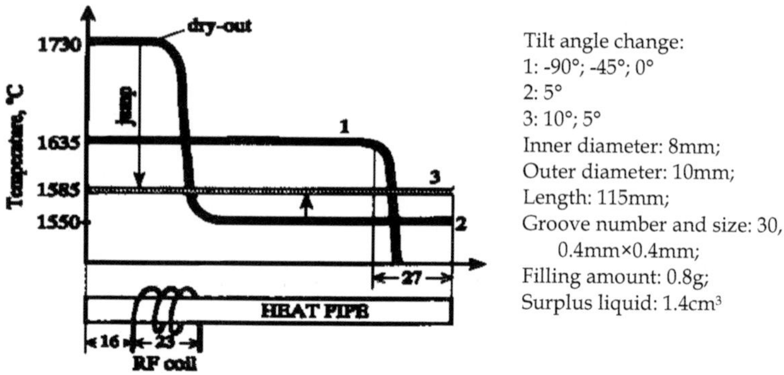

Fig. 1. Temperature levels of W-Li heat pipes when the heat flux is kept constant (Busse, 1992)

For the past few decades, the sodium and lithium heat pipes were developed a lot. There were different application in many fields as, the thermal control of super high Mach spacecraft, the heat storage and heat transfer of solar power plant, the heat recycling and application of high temperature gas discharge, the thermal control in nuclear energy, the constant temperature of industry stove, etc., in some situations, they are necessary and cannot be realized by other ways.

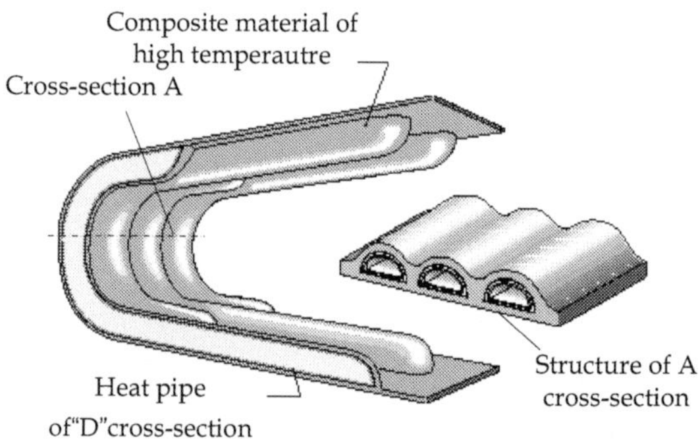

Fig. 2. Heat protection of high temperature heat pipe(by NASA) (David et al, 1998)

For the super high Mach spacecraft, the conduction heat protection is a kind of initiative method to realize thermal control, it is different from the heat protection of conventional surface ablation, or by the new material enduring much higher temperature level (Boman et al, 1990), as shown in figure 2. The "D" shape high temperature heat pipes were bended to couple the outside composite material of high temperature. During flight, the friction heat at cross-section A outside the cover will be conducted to the non-stagnant point of the cover by heat pipe. So the temperature of the stagnant points can be kept at lower level (Faghri et al, 1992, David et al, 1998).

There will be large aerodynamic heat at the leading edge of the flying supersonic aircraft and space shuttle due to the friction between the super high Mach flight and the ambience, especially for the stagnation point (Jiang et al, 2008).

In this chapter, the progress works of the high temperature and super high temperature heat pipes are summed up. Owing to the different levels of temperature and pressure, the strength of the heat pipe shell, the compatibility of material and wick should be considered and designed. Generally, the working fluid is solid, there will be startup problem, the startup is analyzed. The heat transfer limits are important to the heat pipe operation, the possible limitations of high and super high temperature heat pipes are calculated and discussed. The technology is the key problem to realize the high performance heat pipe, the distillation and non-distillation methods are introduced. At last, the technology development of heat pipe is introduced.

2. The compatibility of case material and working fluid

Because the material of high temperature is generally dense, and the fabricating of heat pipe will guarantee the welding connection to have enough thickness and strength, the eligible welding, it is not supposed to occur the reported "sodium and water reaction" (Zhuang et al, 1998,2000). For the failure of high temperature heat pipe, firstly, the purity of material and cleansing belong to the treatment method, no problem to be controlled, then, the real problem turns to the key cause of the compatibility (Jacobson et al, 1984).

When the metal contacts the ambient medium, the metal erosion may occur due to the chemical reaction and electrochemistry. The erosion mechanism of alkali metal heat pipe is important to take measures to keep the performance of high temperature heat pipes and reliability.

2.1 The high temperatue micro cell erosion

The erosions of oxidation and dissolution are introduced a lot in some reports or books (Ma et al, 1984; Zhuang et al, 1998; Zhang et al, 1987; Jacobson et al, 1984). Here, there are two points to be emphasized. (1) For the outside of the heat pipe, if the ambient oxygen concentration is significant, the heat pipe surface will be oxidized directly, the higher temperature is, the much more erodible is the surface. However, if a dense layer of oxides is formed, the existence of oxide will increase the anti-erosion ability of the outside surface. However, if there is a little alkali produced, they will dissolve the protection layer of the base metal, then the reaction will be increased. (2) For the inside of the heat pipe, some components of the composite metal may also dissolve in the liquid alkali metal. The inner surface will form rust speckles, so the effective thickness of the heat pipe wall is decreased. The alkali metal evaporates continuously, the dissolved speckles will deposit and jam up the liquid flow passages. The failure reason is liquid return resistance is increased a lot instead of wall metal consumption.

Here, another important erosion is put forward, it is the most possible in high and super high temperature heat pipes. The components and purity of metal are important for the

$$Y = \frac{1}{6}[2 + (\frac{D_n}{2h_n})^2] \tag{4}$$

In equation (4), h_n is the highness inside the heat pipe end cap.
For the material of GH128, the pull strength is 850MPa at room temperature. While at temperature 1000°C, the yielding strength is 350MPa; the pull strength is 140MPa, the extension rate is 85%. And at this temperature, the enduring life is 200 hours at 30MPa, the extension rate is 40%. For the operating temperature of 1000°C, the heat pipe shell and end cap have the lowest pull strength and yielding strength, the inside alkali metal vapor pressure is the highest. So, if the worst situation can satisfy the demand, then the lower temperature can also meet the strength demand. The explosion coefficient is set 2.0, the yielding coefficient is given as 1.5, the calculated results are shown in table 3. By the results, the design can meet the strength demands.

	Inner groove height (mm)	[σ] (MPa)	Enduring most pressure shell/end cap (MPa)	Enduring most pressure of end, shell/end cap (MPa)	Vapor pressure (MPa)
φ8×1	0.5	70	10/2.86	15.75/4.5	2.71
φ8×2	0.5	70	23/6.67	31.50/9.0	2.71

Table 3. The strength of high temperature heat pipe GH128-Na (1000°C)

2.3 The material selection of super high temperature heat pipe

For the situations of coupling the heat pipe and the outside material, the expanding characteristics of both them at operating higher temperatures can lead to the thermal stress due to the material distortion. If this problem is not considered, the crevice between the heat pipe and the outside material can worsen the heat transfer. The linear expanding coefficients of some high temperature materials are listed by table 4.

material	20~100°C	20~400°C	20~600°C	20~800°C	20~1000°C
GH3030	12.8	15	16.1	17.5	—
GH128	11.2	12.8	13.7	15.2	16.3
304	18.2	—	—	—	—
Inconel600	13.3	14.5	15.3	16.1	—

Table 4. Linear expanding coefficient (10^{-6}/°C)

In order to decrease the crevice between the heat pipe and the outside material, the material niobium is the best candidate, owing to its smallest linear expanding coefficient. Niobium belongs to the VB list in the chemistry element periodic table. Its melting point is 2468°C, the strength is high during 1093~1427°C. Its thermal expanding coefficient is 7.2×10^{-6}/°C, the density is similar to the steel, the strength can be kept at 1649.9°C, and it can endure a certain mechanical deformation. Niobium is stable in eroding medium, and has good thermal conductivity. Based on the good performance of physical and chemical properties, the niobium base alloys are best candidate for the super high temperature heat pipes. However, such alloy has lower anti-oxidization. As shown in figure 4, the niobium bars and strips are products by a certain company.

Fig. 4. Niobium sticks and straps, made by a certain company

For the heat leading problem of super Mach flight, the used heat pipes need to be coupled with the outside material. The thermal expanding coefficients of both them should be considered. The contact thermal resistance due to crevice by the different expanding can fail the heat leading design.

2.4 Alkali metal lithium

Alkali metals are commonly used to be the working fluid of high and super high temperature heat pipes, such as lithium, sodium, potassium, and their alloy. There are many advantages to select alkali metals. Alkali metals can work at high temperature, they have high latent heat of vaporization, they have stable characteristics and lower saturation vapor pressure and good thermal conductivity.

At room temperature, alkali metals are solid states. They will be oxidized in air. So the heat pipe filling process should keep them from contacting the air. Additionally, the purity of alkali metals should be guaranteed to fill the heat pipe, there is no other oxidization.

For the lithium, the obtained physical, thermodynamic properties, the erosion property etc. are extremely not consistent, especially for the high temperature situations. The experimental results are different for researchers, the tentative explanations can be owing to the different purity of lithium materials. The other reason may be due to the different accuracies used. The key reason may be that the lithium is not purified carefully before filling. Additionally, lithium is active to react with some impurities. The impurities have very effects on the properties.

Lithium is the lightest metal. Its density is from 0.531~0.534g/cm³, it can be believed to be 0.533g/cm³. The melting point is from 179~186 °C, it can be supposed to be 180.5±0.1°C. The saturation pressure of solid and liquid lithium (P_s and P_l, Pa) can be obtained by theory and experiments (Gelishen, 1966). Between the temperature T=298~453.7K, there is,

$$\lg(P_s/133.3) = 14.2121 - \frac{8551.18}{T} - 3.79295\lg T + 6.7167\times10^{-3}T \tag{5}$$

If the temperature T=453.7~1350K, then the relation is,

$$\lg(P_l/133.3) = 10.1184 - \frac{8442.53}{T} - 1.64098\lg T + 2.5968\times10^{-4}T \tag{6}$$

The simple equation of saturation pressure of liquid lithium is as,

$$\lg(P_l/133.3) = 8.0 - \frac{8143}{T} \tag{7}$$

When the temperature is between 700~1400 K, equation (7) coincides with equation (6).

2.5 The compatibility

Eliminating the fabrication factors, the compatibility of high and super high temperature heat pipes are due to the micro cell erosions, the shell and wick materials dissolve in the working fluids. In addition, the micro cell erosion can make the inner surface be granulating erosion and make the shell wall thinner. The temperature level will influence the compatibility essentially. Busse found that for tungsten and rhenium alloy-lithium heat pipe, the heat pipe longevity is several years, one year and one month respectively, corresponding to the temperatures 1600°C, 1700°C and 1800°C (Busse 1992). The effects of temperature level to the longevity are very obvious.

Table 5 shows the general results of compatibility, this is the basic principle to select the couple of shell material and alkali metal.

	SS	Ni	Ni alloy	W	Ta	Mo	Re	Ti	Nb
Lithium	×	×	×	√	√	√	√	×	√
Sodium	√	√	√	—	—	—	—	×	√

√: compatibility tested; ×: non-compatibility tested

Table 5. General results of compatibility

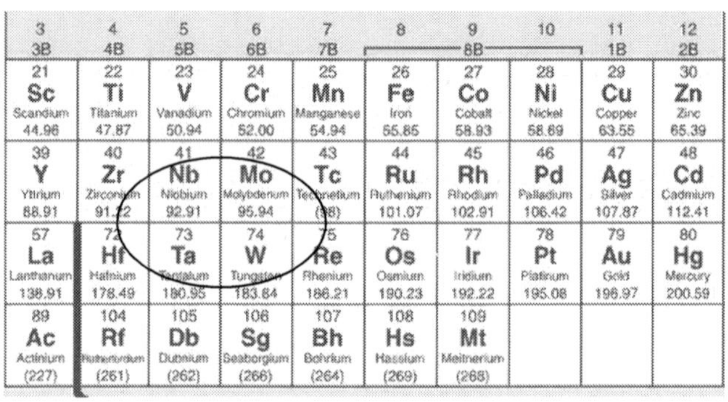

Fig. 5. Positions on element periodic table of materials suitable for super high temperature heat pipes

Considering all the important factors, the high temperature material, the above mentioned mciro cell erosion, high temperature alloy properties, the compatibility with the alkali metals, and the experimental results, figure 5 gives the selected heat pipe materials, signed as "ellipse" in the chemistry element periodic table. From 4B~7B rows, tungsten, tantalum, molybdenum, rhenium and niobium are good candidates to the high and super high temperature heat pipes.

In the mean time, the material selection should consider the material machining properties and availability, the price and other factors.

3. Startup analysis of alkali metal heat pipe

For alkali metal heat pipes, commonly, the working fluid in heat pipes is solid state. The heat transfer into the evaporator make the solid working fluid melt, then one equivalent heating section is formed. If the heat pipe is started up by heating one end, such as the heat pipe has only one end heating section for the high mach stagnant point. As shown in figure 6, when the solid working fluid is melted completely the temperature distribution along the heat pipe is given.

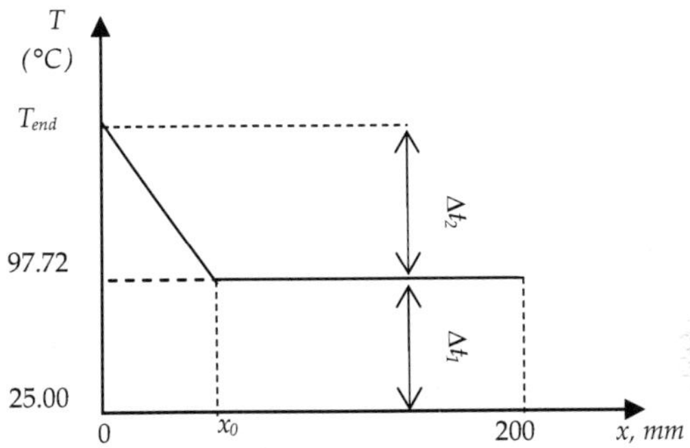

Fig. 6. Temperature distribution along the heat pipe when the solid working fluid is melted completely

For one concrete high temperature heat pipe, the cross-section area is 0.7536cm². At 700°C, the thermal conductivity of heat pipe shell material is 25W/(m.°C). If the startup power is 50W, the axial temperature difference will reach 26°C/mm by Fourier law. Obviously, there will be bigger temperature difference along the axial direction. When the heat pipe is started up, if the melted working fluid is not enough to make all the solid working fluid melt wholly, then the startup will fail. In order to control the highest temperature of the stagnant point, the heat applied should be lower than one level.

3.1 Analysis of startup time

For the horizontal sodium heat pipe, the ambience temperature is 25°C, then the sodium is solid before startup. Between the temperature range 25~97°C, the thermal conductivity and the heat capacity of sodium are considered as constants. And from 25°C to 800°C, the heat pipe shell is taken as constant. The axial conduction of the wick is ignored.

The sodium should be heated to melt completely, from room temperature 25°C to the melting point 97.72°C. The solid working fluid inside heat pipe is assumed to distribute uniformly. The thermal capacity of the heat pipe is C_{tube}=53J/K, the thermal capacity of the working fluid is C_{Na}=3.14 J/K. For the sodium, the melting latent heat is L_{melt}=113kJ/kg, the latent heat of evaporation is h_{fg}=4215kJ/kg. The heat to evaporate the working fluid should be large than that, the heat to increase the temperature to the melting point and the heat to melt the solid working fluid completely in the condenser, there is,

$$W_1 = h_{fg} m \frac{x}{200} = \left(1 - \frac{x}{200}\right)\left[\left(C_{tube} + C_{Na}\right)\Delta T_1 + mL_{melt}\right] \quad (8)$$

When the solid working fluid melt wholly, the temperature distribution is given in figure 4. ΔT_1 is the temperature difference between the melting point to the room temperature. x_0 is the needed length to evaporate the working fluid. For the given sodium heat pipe A without groove, x_0 = 58.6mm; while for the heat pipe B with groove, x_0 = 33.24mm by equation (8).
It is assumed that there is no heat transfer between the heat pipe condenser and the ambience. Then the startup heat transfer is estimated as,

$$W_2 = (C_{tube} + C_{Na})\Delta T_1 + mL_{melt} + \frac{x}{200} C_{tube} \frac{\Delta T_2}{2} \tag{9}$$

$\Delta T_2/2$ is the mean temperature difference between the evaporator temperature and the sodium melting point.
For the evaporator, there is only the axial conduction along the heat pipe shell. The heat transfer rate is given by Fourier law as

$$Q = -\lambda A \, dT/dx \tag{10}$$

Considering equations (8) and (10), the relation between the input power and the end temperature can be obtained. The startup time is given also by equation (8) and (10) as,

$$t = W_2/Q \tag{11}$$

3.2 Results of startup time

From the data of the sodium melting heat with pressure and temperature, it is known that the melting point changes little with the pressure. The melting point and melting heat are taken for one standard atmosphere pressure to calculate the startup time. The startup power should be controlled not to exceed one value, which can make the evaporator dryout before all the solid working fluid is melted completely.

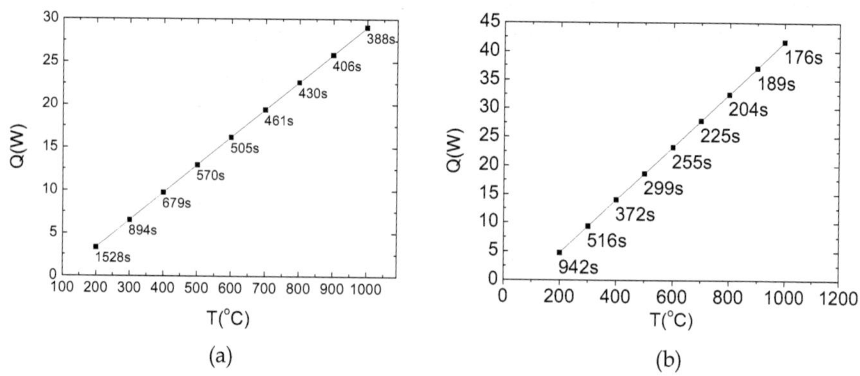

Fig. 7. Relations of heat transfer rate, startup time and temperature. (a) with screen wick, without groove (b) with screen wick and with grooves

For the sodium heat pipe A and B, the results are shown in figure 7. At the same startup heat transfer rate, 20W, the startup time of A and B is 450S and 290S respectively.

4. Technology control of alkali metal heat pipe

The performance of heat pipe depends on the fabrication technology. The charging process should guarantee the vacuum level before filling. The working fluid quantity charged should be controlled. The working fluid has enough purity, the oxidization and impurities are at the endurable levels, the seal is soldered and guaranteed etc..

4.1 The vacuum level control

Commonly, if the pressure is less one atmosphere, 1.01×10^5 Pa , the vacuum is divided by several regions, as shown in table 6 (Zhang et al,1987).

Region	Pressure (Pa)	Density of molecule number, n(cm^{-3})	Mean free journey, λ(cm)
Little vacuum	1.01×10^5~10^3	~10^{18}	10^{-4}
Low vacuum	10^3~10^{-1}	~10^{15}	10^{-1}
High vacuum (HV)	10^{-1}~10^{-6}	~10^{10}	10^3
Super high vacuum (UHV)	10^{-6}~10^{-12}	~10^6	10^9
Extreme high vacuum (XHV)	<10^{-12}	<10^2	>10^{12}

Table 6. The vacuum region partitions

For the high and super high temperature heat pipes, the vacuum region had better reach the levels of HV, if the vacuum is UHV or XHV, the technology will last long and the cost is increased a lot. If the vacuum is little or low, then the heat pipes have worse performances as shown in figure 8.

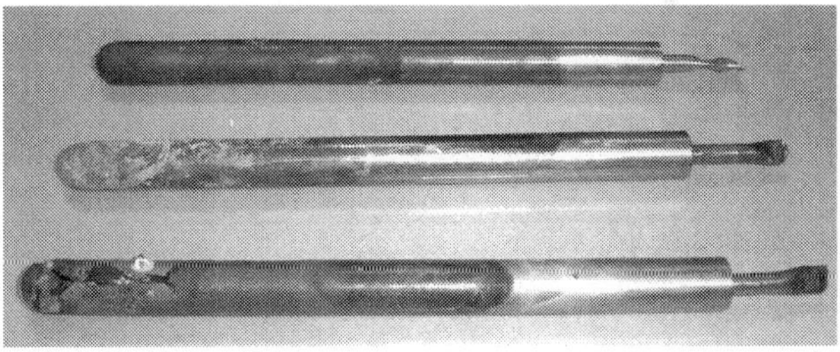

Fig. 8. The heat pipes have worse performances if the vacuums was low

4.2 The distillation technology

The distillation technology can make the alkali metal melt and evaporate. By controlling the temperature of distillation, the alkali metal is purified a lot, then the liquid alkali metal is charged into the candidate heat pipe. Such method keeps the system to be active vacuum, the vacuum equipment works continuously. This can keep the alkali metal purity, not to be oxidized by a little leakage air.

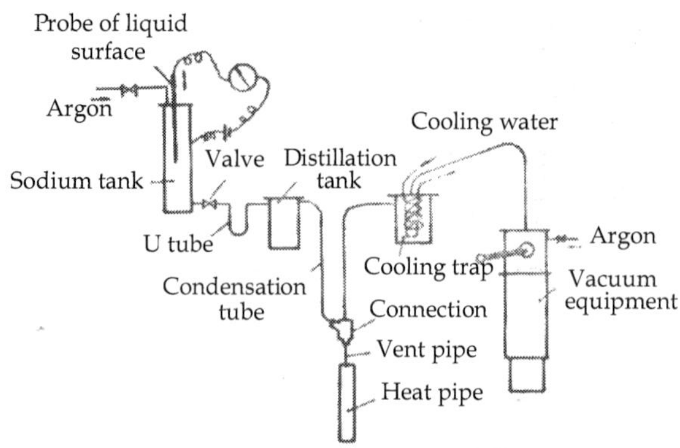

Fig. 9. Distillation technology of high temperature heat pipe

As shown in figure 9, the charging system consists of sodium tank, U tube, distillation tank, connection, cooling trap and vacuum equipment. The argon can protect the alkali metal in the sodium tank. The charging process includes two steps. Firstly, the set amount of alkali metal is filled from the sodium tank to the distillation tank. Secondly, the alkali metal is distilled and charged into the heat pipe.

For sodium, after pumping the system and the vacuum is permitted, the distillation tank is heated to a certain temperature, in the mean time the other part of the system has different temperature, such as, the outlet of the condensation tube should be controlled between 150~200°C. The much higher temperature can make sodium vapor be pumped into the vacuum equipment. The much lower temperature will lead to the higher viscosity of liquid sodium, then the small vent pipe can be jammed. Based on the same reasons, the connections, heat pipe, especially for the vent pipe also should be heated to about 200°C. After the other parts of the system reach the set temperature, the distillation tank is heated to a temperature between 480~500°C, and this temperature is kept constant to distill the sodium. The temperatures at every part are monitored. If the sodium is vaporized totally, the temperature of distillation tank will increase a lot, then stop heating the distillation tank.

Obviously, the distillation technology is complicated a little, and the consumptions of time, water and power are large. Once, only one heat pipe can be charged. And the after-treatment is also complicated, the sodium remains in the tubes are hard to be cleaned up.

4.3 The non-distillation technology

In order to make the charging process simple and several or many heat pipes can be filled simultaneously, the three-path-equipment of alkali metal charging was invented. As shown in figure 10.

The non-distillation charging system is composed of the vacuum equipment, the transparent glove chamber with argon protection, the valves of super high vacuum and tubes. There are three paths, can realize three alkali metal heat pipes charging simultaneously. For example, the flange of the first path is disconnected, the empty lower tank is put downward into the transparent glove chamber with argon protection. Also, the heat pipe end is inserted into one small tube, by which the air is replaced by argon. In this glove chamber, the alkali metal is cut, weighed and put into the tank, which outlet is set stainless steel screen. Then the tank with alkali metal is lifted to couple the flange and the system is closed by bolts and valves.

During this process, the main tube of the system is also blowed by argon. The system is pumped some time, and the argon in the system is evacuated as much as possible.
Here it is pointed out that the argon in the heat pipe is pumped out through the tank of alkali metal, the argon will cross the alkali metal by the aperture passage. By the bypass designed near the outlet of the alkali metal tank, the vacuum of heat pipe can be increased a lot.

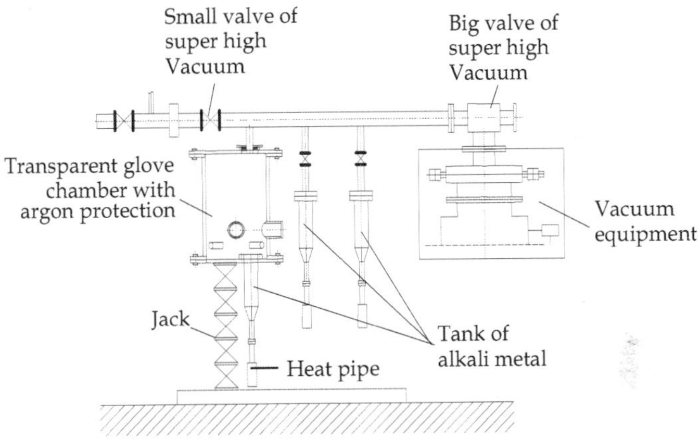

Fig. 10. Non-distillation technology of alkali metal heat pipe

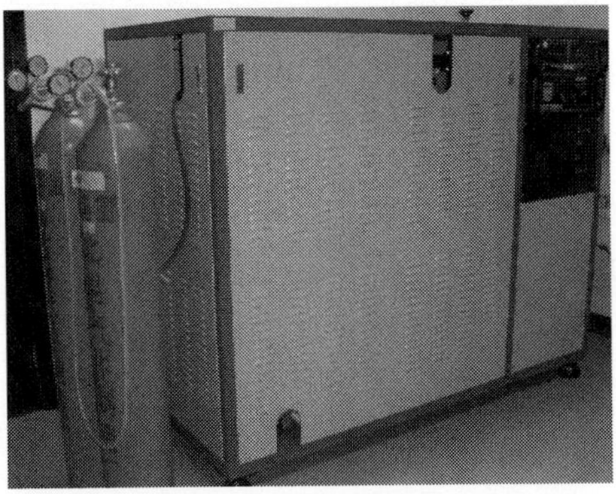

Fig. 11. Three-path-equipment of alkali metal charging (figure 10)

After that, the alkali metal tank, the lower connection and the heat pipe is heated by the outside heaters. The temperature can reach 150~160°C or so for sodium. If the alkali metal inside is melted completely, then the big valve of high vacuum is closed, the pump equipment is cut off. The small valve of high vacuum is opened and the argon will push the alkali metal into the heat pipe. Then the small valve is closed and the big valve is opened. The system is evacuated again to a high vacuum level. Finally, the heat pipe is sealed by a special plier, soldered by a welder. A heat pipe is charged successfully.
Figure 11 is the photo of three-path-equipment.

4.4 The technology monitoring

The monitor equipment of technology is shown in figure 12. The power increase can be set to heaters. The thermal couples and resistances are connected to the equipment. The computer and the inserted instruments are two-level system. The computer, digital instruments, controllable silicon, switches, contactors, buttons etc., are installed into the instrumental cabinet. By the computer, the technology process can be realized.

Fig. 12. Monitor equipment of heat pipe technology

The instruments and the sensors are connected to collect data and control the process. By the RS485 communication bus, the computer can display the process on time, the interface is displayed by Chinese. The data can be storaged in the computer.
The performance of heat pipe depends on the process technology essentially.

5. Experimental results of alkali metal heat pipes

5.1 Startup from ambience

The startup experiments can test the heat pipe performance before the heat pipe is applied. As shown in figure 11(a), the evaporator is heated by high frequency heater, the heat pipe is set inside the high frequency loops, and the thermocouples are set along the condenser, which is in the ambient air. By the high frequency heater, the heat flux can be very large, and the some dryout point may be displayed by thermocouples.

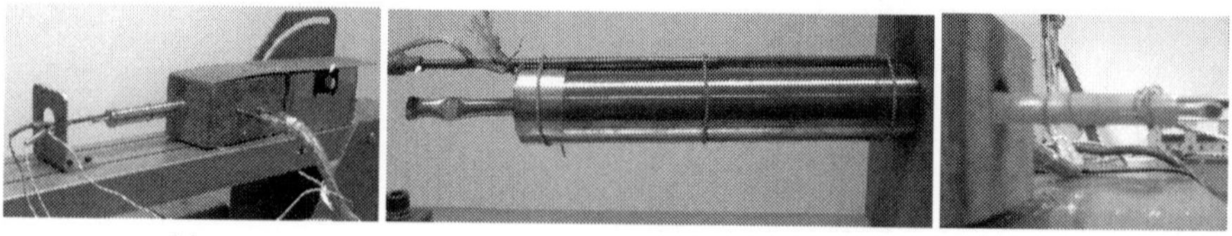

(a) (b) (c)

Fig. 13. Test rig of alkali metal heat pipe and startup photos of sodium heat pipes. (a) Test rig of heat pipe (b) Startup experiment, 5° anti-gravity (c) Startup, horizontal

The experimental photos are shown in figure 13 (b) and (c). By the color of the condenser, the sodium heat pipe can be started up successfully. The color is uniform along the condenser, the isothermal performance of the heat pipes are good. The purity of working

fluid and the high vacuum technology can guarantee that there is no noncondensible gas in the fabricated heat pipes.
As in figure 14, temperatures of three condenser points are demonstrated for horizontal position. For about 100S, the heat pipe can be started up. For 10 degree tilt angle, the evaporator is set lower, the results of sodium heat pipe startup is shown in figure 15. For another power step, the heat pipe performance is also satisfied.

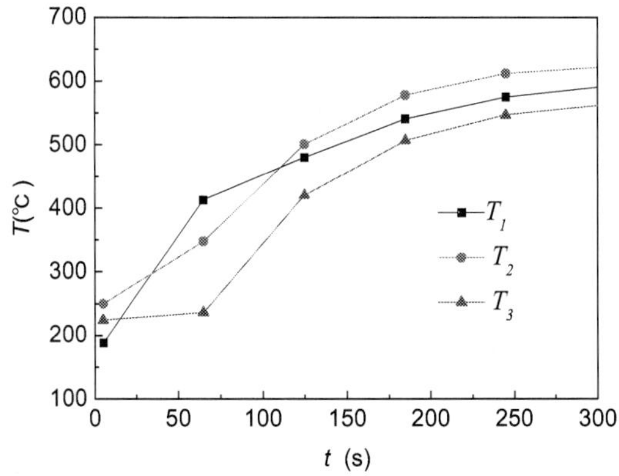

Fig. 14. Startup of horizontal sodium heat pipe

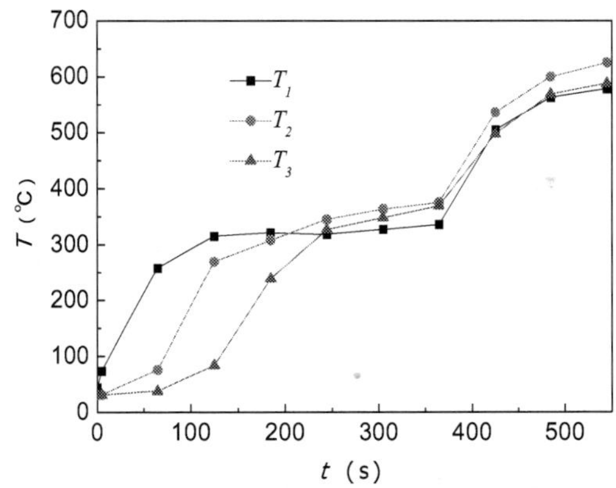

Fig. 15. Startup and power increased test of 10° degree tilt angle

5.2 Experiments in wind tunnel of electricity arc

By the high and super high temperature heat pipes, the local higher heat flux is moved to the lower heat flux region, the heat is moved from the "peak point" to the "valley region" by heat pipe, then the highest temperature is decreased a lot.
In a wind tunnel of electricity arc, three heat transfer elements as CC material, high conduction CC material and heat pipes are tested and compared, as shown in figure 16.

From figure 16, after 1200s, the heat pipe is started up successfully. The operation lasts nearly 5 minutes.
The upper three curves are the temperature histories of the stagnant points. The temperature of heat pipe stagnant point is lower than the other CC and high conduction CC elements, 120°Cand 50°C lower respectively. The heat pipe behaves good performance.

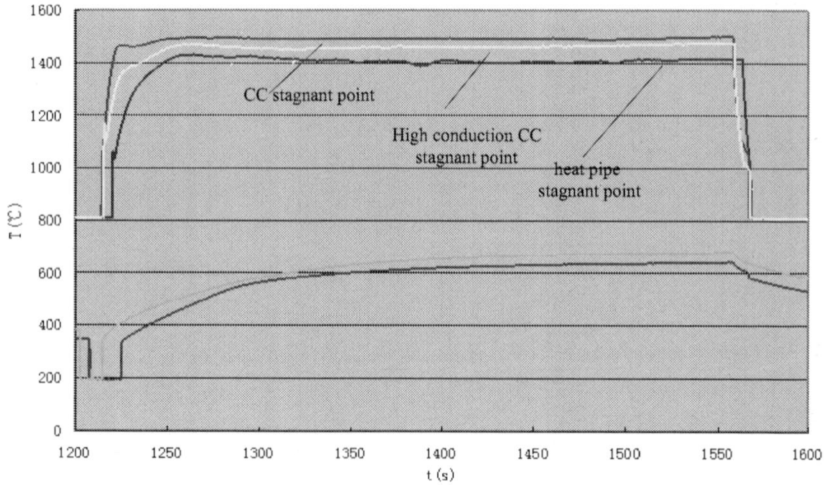

Fig. 16. Experimental temperatures of different method by arc tunnel heating (By China Academy 11)

6. Limits of alkali metal heat pipes

6.1 Continuum flow limit

With the decreasing of dynamic diameter, the heat pipe vapor flow may transit from the continuum flow to the free molecule flow. The continuum limit can be judged by the Knudsen number as,

$$Kn = \frac{\lambda}{D} \tag{12}$$

Here, λ is the free length of molecules, D is the minimum size of the vapor flow in heat pipe. If $Kn \le 0.01$, the flow is continuum; if $Kn > 0.01$, the flow belongs to the free molecules. For the latter situation, the heat pipe may lose its performance.
Cao and Fahgri derived the transition temperature as (Faghri, 1992),

$$T_{tr} = \frac{P_0}{\rho_{tr} R_g} \exp\left[-\frac{h_{fg}}{R_g}\left(\frac{1}{T_{tr}} - \frac{1}{T_0} \right) \right] \tag{13}$$

Substitute Kn=0.01 into equation (13), the transition temperature can be obtained.
As shown in figure 17, for sodium heat pipe, change of transition temperature with the dynamic diameter is given. If the dynamic diameter is decreased less than 1mm, the

continuum limit occurs. The transition temperature will increase with decreasing the diameter. When the dynamic diameter is 50μm, the transition temperature will be 830°C. This temperature is in the range of normal temperature 500~1100°C. Such results mean that the heat pipe will work at much higher temperature than the designed, in the mean time, the heat transfer rate decreases.

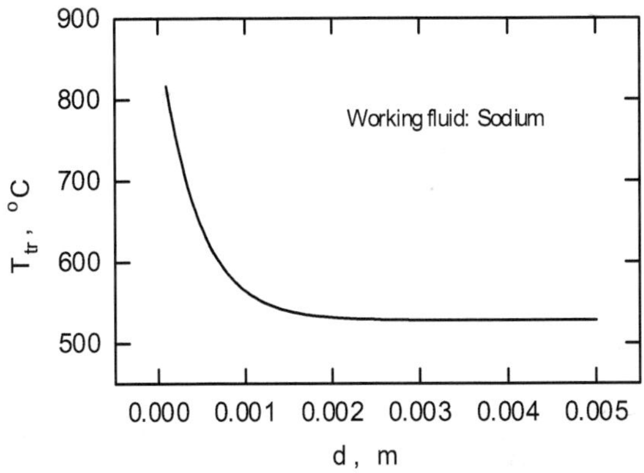

Fig. 17. For sodium heat pipes, change of transition temperature with the dynamic diameter

6.2 Other possible limits

When the alkali metal heat pipes are started up from low temperature, the vapor density is very small. The viscous resistance may dominate (Ma et al, 1983). At the end of the condenser, the vapor pressure decreases to extreme low, nearly zero. Then the viscous limit is reached as,

$$Q_{vis} = \frac{D_v{}^2 L}{64\mu_v l_{eff}} \rho_{v,o} P_{v,o} \tag{14}$$

If the alkali metal heat pipes operate at low vapor pressure, the vapor density is small and the velocity is big, then the sonic velocity may choke the heat transfer, the sonic limit is expressed as,

$$Q_s = 0.474 L A_v \sqrt{\rho_{v,o} P_{v,o}} \tag{15}$$

When the vapor flow can entrain the liquid, the inertial force is bigger enough, the entrainment limit is given as,

$$Q_E = A_v \sqrt{2\pi\sigma\rho_v L^2 / \lambda} \tag{16}$$

The capillary limit can be calculated by,

$$Q_{cap} = 2\left(\frac{\sigma\rho_l L}{\mu_l}\right)\left(\frac{K A_w}{r_e l_{eff}}\right) \tag{17}$$

The basic heat transfer limits are given in figure 18 for sodium heat pipe. If the temperature is lower than 500°C, the sonic limit should be paid attention. When the temperature is between 500°C and 900°C, the entrainment limit is easy to occur, the temperature 700°C corresponds to 756W, then the heat flux is 246W/cm².

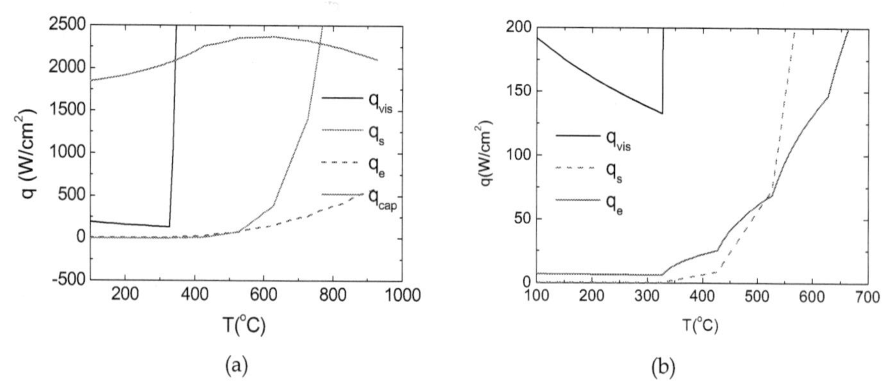

Fig. 18. Four limits of sodium heat pipes with temperature, (b) is the detail of local (a)

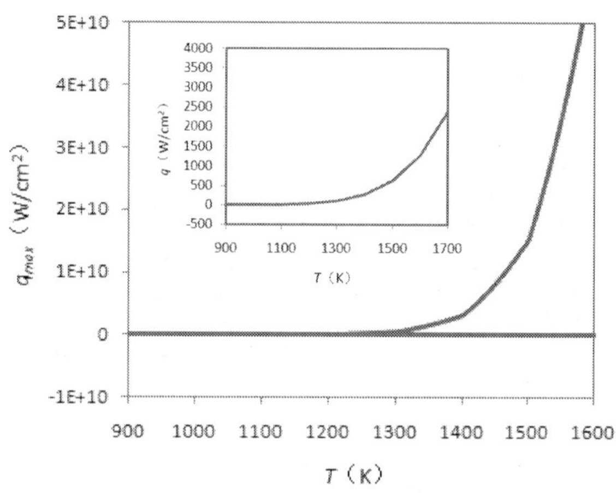

Fig. 19. Viscous and sonic limits of lithium heat pipe

The viscous and sonic limits of lithium heat pipes are illustrated in figure 19. The bottom line stands for the sonic limit. The details are also shown in the figure. From the results, the lithium should work at higher temperature and higher heat flux.

7. Chemical vapor deposit technology

The new material and new technology for high and super high temperature heat pipes are developed in recent years. There are some new technologies about alkali metal heat pipes.

Here the chemical vapor deposition (CVD) is introduced (Fortini, et al, 2010). The CVD method can be used to fabricate the heat pipes, the number of wicking grooves, their location, the cross-sectional shape, and the overall geometry of the heat pipe are easily varied. The integral grooves also eliminate the need for screens, thus allowing for greater design flexibility. Figure 20 shows the process schematically. The manufacturing process starts with a mandrel whose outer contour matches the desired inner contour of the finished product. For a heat pipe with a simple circular cross section, a tubular mandrel can be used to define the vapor channel, and smaller rods can be attached to the mandrel to define the liquid return arteries. After assembling the mandrel, the part is coated by CVD. The final step is etching away by mandrel with acid.

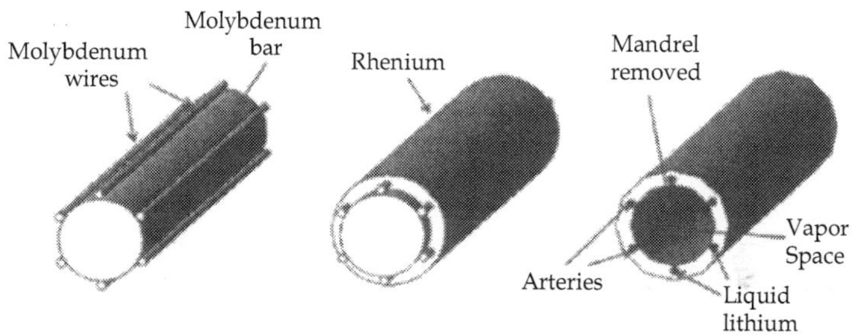

A: Mandrel B: CVD coating C: Heat pipe

Fig. 20. Schematic of heat pipe fabrication process by chemical vapor deposition (Fortini, et al., 2010)

The advantage of CVD is that the grooves are incorporated into the machining grooves into the pipe wall after the pipe itself is fabricated. It also provides excellent flexibility in that it enables any number of grooves to be incorporated into the heat pipe. Furthermore, the grooves don't have to be evenly spaced; they can be concentrated on one side or the other if desired. Screens can be used in conjunction with the arteries at the evaporator end to further delocalize the liquid. Also, because CVD is not a line-of-sight process, the cross section of the heat pipe can be of virtually any geometry. It is not limited to straight heat pipes, nor is it limited to pipes of circular symmetry. Finally, for heat pipes that are reasonably straight, multiple pipes can be fabricated simultaneously.

Another benefit of CVD manufacturing is that multiple materials can be used to fabricate the pipe. For example, rhenium was chosen because of its chemical compatibility with lithium, ductility, and strength at high temperature. Rhenium, however, is expensive. To reduce the amount of rhenium used, one could apply a thin film of rhenium to the mandrel and then switch to say tantalum. Tantalum is extremely ductile and is more than an order of magnitude less expensive than rhenium. Tantalum also has good strength at high temperature, though not as good as rhenium. So even though a somewhat thicker layer of tantalum would be needed over the initial thin film of rhenium, the material cost would be reduced by more than a factor of 10.

8. Summings-up

In this chapter, the progress works of high and super high temperature heat pipes are introduced. The micro cell erosion mechanism to the high temperature heat pipe case and the compatibility are given, the selections of the case material and the working fluid should be coupled to satisfy with the compatibility. The technology is the key problem to realize a high performance heat pipe, the alkali metal distillation and non-distillation technology are innovated, and the technology monitor is important for fabrication. Generally the working fluid is solid before high and super high temperature heat pipe startups, the startup possibility and time are analyzed and experimented. The heat transfer should be designed much smaller than those of the operating limits, the possible heat transfer limits of high and super high temperature heat pipes are calculated and discussed. The experimental and theoretical results show that the fabricated heat pipes have good performances. The new CVD methods can be used to fabricate the heat pipes, the integral grooves also eliminate the need for screens, thus allowing for greater design flexibility.

9. References

Ma T. Z., Hou Z.Q. and Wu G.W. (1983). *Heat Pipe* (in Chinese), Science Press, pp.277-282, ISBN 7-03-002011-1

Busse C.A. (1992). Heat Pipe Science, Advances in Heat Pipe Science and Technology, *Proc. of 8th Int. Heat Pipe Conference*, pp.3-8, Bejing, Int. Academic Publishers

Zhang J.X. and Ning X.W. (2009). Study of Thermal Control Technology on Space Reactor Power Supplier (in Chinese), *Proc. of 9th China Space Thermophysics Conference*, pp.1-5, Beijing

Boman B.L., et al. (1990). Heat Pipes for Wing Leading Edges of Hypersonic Vehicles, NASA CP-181922

Faghri A. and Cao Y. (1992). Numerical Analysis of Leading Edge and Nosecap Heat Pipes, Advances in Heat Pipe Science and Technology, Int. Academic Publishers, pp.303-308, *Proc. of 8th Int. Heat Pipe Conference*, Bejing

David E. Glas.(1998). Closed Form Equations for the Preliminary Design of a Heat Pipe-Cooled Leading Edge, NASA CR-1998-208962

Jiang G.Q., Ai B.C., Yu J. J., Chen L.Z. (2008). Application of high temperature heat pipe in the protection technology of heat conduction (in Chinese), pp.74-80, *Proc. of 11th China Heat Pipe Conference*, Weihai, China

Zhuang J. and Zhang H.(2000). *Heat Pipe Technology and Engineering Application*, pp.166-173, Chemical Industry Press

Zhuang J. and Zhang G.Y.(1998). Sodium and water reaction in sodium heat pipe, pp.151-155, *Proc. of 6th China Heat Pipe Conference,*Wuyishan, China

Jacobson D.L. and Wang J.H.(1984). Failure Analysis of a Sodium, Inconel 617 Heat Pipe, Preprints of Proceedings of 5th Heat Pipe Conference, Tsukuba, Japan, 121-125

Zhang S.H. and Shen H.J. (1987), *Molecule Physics and Thermodynamics*, Beijing Science and Technology Press, pp.75-76, ISBN7-5304-0023-1/Z

Jacobson D.L. and Soundararajan P.(1984). Failure Analysis of a Sodium Heat Pipe with Integral Lithium Fluoride Thermal Energy Storage, pp.115-120, Preprints of Proc. of 5th Heat Pipe Conference, Tsukuba, Japan

Li T.H. and Hua C.S. (1987). *Heat Pipe Design and Application* (in Chinese), pp.102-108, Chemical Industry Press

GB 9222-88 (1988). Strength Calculation of Pressure Parts for Water-tube Boilers, pp.1-72, National Standards Press of China

Gelishen, Gejizunuofu (Russian accent) et al. (1966). Properties of Lithium (Interpretation from Russian) , pp.10-12, China Industry Press

Fortini A.J., Arrieta V.M. (2010). Rhenium Heat Pipes for Hypersonic Leading Edges, Preprints of 15th Int. Heat Pipe Conference, pp.1-6, Clemson, USA

Design of the Heat Conduction Structure Based on the Topology Optimization

Yongcun Zhang, Shutian Liu and Heting Qiao
Dalian University of Technology
China

1. Introduction

The progress toward smaller scales in electronics makes the cooling of integrate circuits become an important issue. The conventional convective cooling method which is feasible and often used to control the temperature of a system becomes impractical because the channels of heat transfer take up too much space for high compacted integrate circuit. Hence, it is necessary to build heat conduct structures with high conductivity materials so that the heat can be collected, transferred and exchanged with external environment automatically and rapidly[1-2]. A key problem is how to design the structures with a rational distribution of high conductive materials, which not only benefits to the temperature control but also can reduce material and manufacturing costs and bring possibilities for further miniaturization.

Studies designing the optimal heat conduction structure have attracted much attention and many achievements have been obtained [1-24], including mathematical models and the corresponding solving methods. For example, Bejan and co-workers put forward a tree-like network construction method based on the constructal theory [1-8], Guo and co-workers proposed some practical design criteria and developed the corresponding optimization methods for the heat conduction structure based on the least dissipation principle of heat transport potential capacity [9-15]. The topology optimization method has also been applied for heat conduction structural optimization [16-22]. In all these cases, the nature of optimization design for heat conduction structures is to build a mathematical model that maximizes or minimizes an objective function (e.g. the thermal performance index) subjected to certain constrains. Thus, it is a key to define a suitable thermal performance index in such an optimization model.

Statistical data show that the failure of real devices with a fraction of 55% is caused by the high temperature and this fraction increases exponentially with increasing temperature [25, 26]. Thus, the highest temperature is a primary factor that induces the failure of practical cooling structure and should be well controlled. In practice, it is natural to define the highest temperature as an objective function of the optimization model. However, the location of the highest temperature usually changes with the change of material distribution in the topology optimization process and is a discontinuous function of design variables, which may introduce numerical difficulties in optimization. Therefore, instead of a directing optimization of the highest temperature, it is more convenient to define another proper

thermal performance index as the objective function in an optimization model to accomplish indirectly the goal of minimizing the highest temperature.
In the optimization model of heat conduction structure, the objective function can be selected as

$$f(\boldsymbol{X}) = \int_{\Omega} \frac{1}{2}(-\boldsymbol{q}(\boldsymbol{X})\nabla T(\boldsymbol{X}))\mathrm{d}\Omega \tag{1}$$

where $\boldsymbol{X}$ is the design variable used to describe the distribution of material, $\boldsymbol{q}(\boldsymbol{X})$ is the flux density and $\nabla T(\boldsymbol{X})$ is the temperature gradient. Using the finite element formulation, Eq. (1) can be also written as

$$f(\boldsymbol{X}) = \boldsymbol{T}^{\mathrm{T}}\boldsymbol{K}(\boldsymbol{X})\boldsymbol{T} \tag{2}$$

where $\boldsymbol{T}$ is the global temperature vector and $\boldsymbol{K}(\boldsymbol{X})$ is the thermal conductivity matrix. Generally, Eq. (1) is defined as the dissipation of heat transport potential capacity (DHTPC)[11], and the least dissipation principle of heat transport potential capacity is presented based on this definition; Eq. (2) is defined as the heat dissipation efficiency[17,18], which is the objective function of the heat conduction topology optimization.
Using the DHTPC (or the heat dissipation efficiency) as a thermal performance index, some good design results have been obtained. However, this index can only tell us the heat dissipative capability rather than the highest temperature. How much difference between DHTPC and the present design goal, that is, the control of the highest temperature? Is there any better thermal performance index? Answers to these questions are the motivation of this study.
Firstly, the difference between the DHTPC and the present design goal is evaluated by a one-dimensional heat conduction problem for a planar plate exchanger. Then, the geometric average temperature (GAT) is proposed as a new thermal performance index and the corresponding heat conduction optimization model is developed, the validity of optimization model is proved by two example. Finally, some useful conclusions are given.

2. Heat conduction optimization of the planner plate exchanger

In many practical cooling structures, a commonly used design criterion is that the highest temperature must not exceed a specified value. However, the optimization objective in many existing heat conduction optimization models is the DHTPC. To evaluate the quality of these exiting models, we compare their results with those obtained from an optimization model with the highest temperature as the objective function. For simplicity, the presented example is a one-dimensional heat conduction problem for a planar plate, which can be solved analytically.

2.1 Problem description

A rectangular planar plate exchanger, with length l, width W ($W \gg l$) and thickness t, is embedded in the heater. The heat generated by heater flows into the exchanger uniformly. The heat flowing into the exchanger is q'' per unit time and area. Only one side along the width direction of exchanger contacts with a thermostat with a constant temperature T_0 and others are adiabatic. This problem can be described as a planar heat conduction model with uniform heat source, as shown in Fig. 1. Furthermore, this model can be simplified into a one-dimensional heat transfer problem because the thickness t and the internal heat source q do not change along the width direction. The goal is to obtain the optimal heat conduction performance by designing the thickness t along the length direction of exchanger.

Since thermal conductivity is proportional to thickness t, the thickness design can be transformed into the conductivity field design. That is to say, the limitation of material, $\int_0^l t(x)W\mathrm{d}x = \mathrm{const}$, can be written as the capability of conductivity, $\int_0^l k(x)\mathrm{d}x = K_0$, where K_0 is a constant. The governing equation of heat conduction in the exchanger can be described as

$$\begin{aligned} &q(x) = -k(x)\frac{\mathrm{d}T}{\mathrm{d}x}, \quad \frac{\mathrm{d}q}{\mathrm{d}x} + q'' = 0, \quad 0 < x < l \\ &T(x=0) = T_0, \quad q(x=l) = 0 \end{aligned} \tag{3}$$

where, $k(x)$ is the thermal conductivity, $q(x)$ is the heat flux density and $T(x)$ is the temperature. In addition, the heat flux is assumed to be positive along the x direction. Solving equations (3), we can obtain

$$q = -q''(l-x), \quad T(x) = T_0 + \int_0^x \nabla T \mathrm{d}x = T_0 + q'' \int_0^x (l-x)/k(x)\mathrm{d}x. \tag{4}$$

Then, the optimization design for the exchanger is to determine the optimal heat conduction performance by designing the conductivity filed under a given integral of thermal conductivity (or material volume) over the design domain. Let $f(k)$ denotes a thermal performance index. The heat conduction optimization problem can be formulated as

$$\begin{aligned} &\text{Find}: \; k(x) \\ &\min: \; f(k) \\ &s.t.: \; \int_0^l k(x)\mathrm{d}x = K_0 \end{aligned} \tag{5}$$

Using the Lagrange multiplier method, the solution of the thermal conductivity field can be determined by

$$\delta_k(f(k)) + \lambda \int_0^l \delta k \mathrm{d}x = 0, \quad \delta\lambda\left(\int_0^l k\mathrm{d}x - K_0\right) = 0 \tag{6}$$

2.2 Minimization of the highest temperature

According to the heat conduction theory, the highest temperature is located on the boundary of $x = l$ and can be written as

$$T_{\max}(k) = T_0 + q'' \int_0^l (l-x)/k(x)\mathrm{d}x \tag{7}$$

Substituting Eq. (7) into Eq. (6), we have

$$\int_0^l \left[-q''(l-x)/k^2(x) + \lambda\right]\delta k(x)\mathrm{d}x = 0, \quad \int_0^l k\mathrm{d}x = K_0 \tag{8}$$

The optimal thermal conductivity field $k_{T_{\max}}(x)$ can be obtained by solving Eq. (8), which is

$$k_{T_{\max}}(x) = \frac{3K_0}{2l^{3/2}}(l-x)^{1/2} \tag{9}$$

and the corresponding temperature distribution is

$$T_{T_{\max}}(x) = T_0 + \frac{4q''l^{3/2}}{9K_0}(l^{3/2} - (l-x)^{3/2}) \tag{10}$$

Introducing a dimensionless parameter

$$\tilde{x} = x/l \tag{11}$$

the conductivity field and the temperature distribution can be expressed in the dimensionless space as

$$\tilde{k}_{T_{\max}}(\tilde{x}) = \frac{k(x)}{K_0/l} = 3(1-\tilde{x})^{1/2}/2 \tag{12}$$

and

$$\tilde{T}_{T_{\max}}(\tilde{x}) = \frac{T(x)-T_0}{q''l^3/K_0} = 4(1-(1-\tilde{x})^{3/2})/9 \tag{13}$$

where subscript $T_{\max}$ denotes that the optimization objective is to minimize the highest temperature.

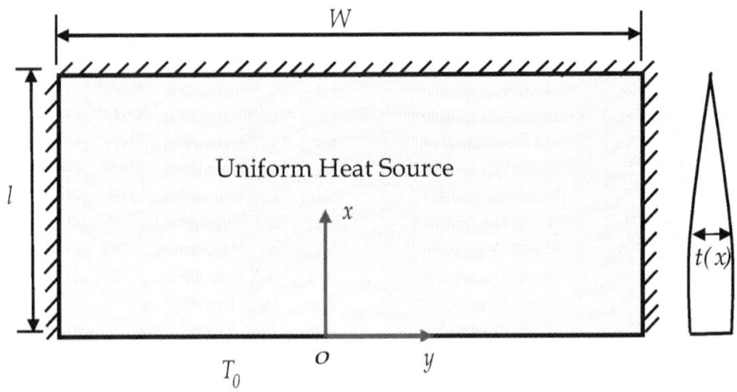

Fig. 1. A theoretical model of a planar plate exchanger

2.3 Minimization of the dissipation of heat transport potential capacity

For the planar plate exchanger, the DHTPC can be expressed as

$$f(k) = \frac{(q'')^2}{2}\int_0^l \frac{(l-x)^2}{k}\mathrm{d}x \tag{14}$$

When the DHTPC is considered as an optimization objective function, the optimal thermal conductivity field should obey the following necessary conditions

$$\int_0^l \left[-\frac{(q'')^2 (l-x)^2}{2k^2(x)} + \lambda \right] \delta k(x) \mathrm{d}x = 0, \quad \int_0^l k \mathrm{d}x = K_0 \tag{15}$$

The thermal conductivity field can be obtained by solving Eq. (15), which is

$$k_{\mathrm{dis}}(x) = \frac{2K_0}{l^2}(l-x) \tag{16}$$

and the corresponding temperature distribution is

$$T_{\mathrm{dis}}(x) = T_0 + \frac{q'' l^2}{2K_0} x \tag{17}$$

The dimensionless thermal conductivity field and temperature distribution are

$$\tilde{k}_{\mathrm{dis}}(\tilde{x}) = \frac{k(x)}{K_0 /(l)} = 2(1-\tilde{x}) \tag{18}$$

and

$$\tilde{T}_{\mathrm{dis}}(\tilde{x}) = \frac{T(x) - T_0}{q'' l^3 / K_0} = \tilde{x}/2 \tag{19}$$

where subscript dis denotes that the optimization objective is to minimize the DHTPC.

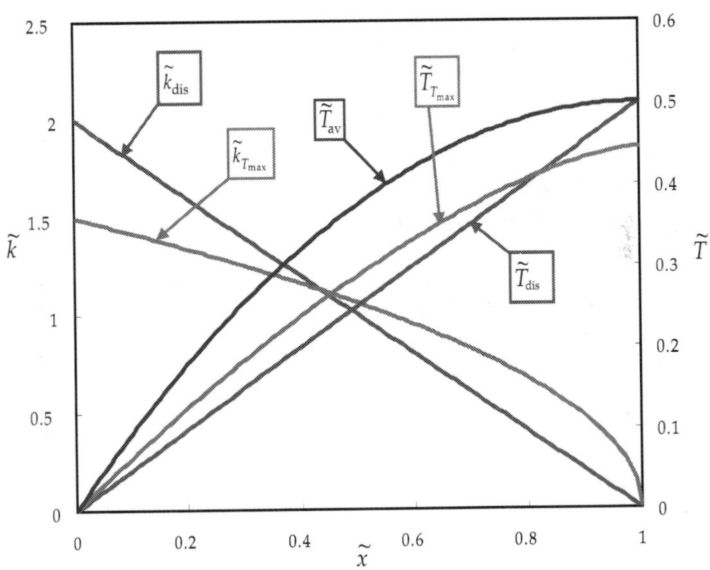

Fig. 2. Comparisons of thermal conductivity fields and temperature distributions from different optimization models. av: the uniform conductivity field; dis: the dissipation of heat transport potential capacity; $T_{\max}$: the highest temperature

2.4 Comparisons of two different optimization models

The dimensionless thermal conductivity fields and the corresponding dimensionless temperature distributions from the two different optimization models are shown in Fig. 2. To facilitate the comparisons, the temperature distribution with uniformly distributed thermal conductivity (denoted by 'av') is analyzed, which can be expressed as

$$T_{av} = T_0 + \int_0^x \nabla T \mathrm{d}x = T_0 + \frac{q''l(lx - x^2/2)}{K_0} \tag{20}$$

and the corresponding dimensionless temperature distribution is

$$\tilde{T}_{av} = \tilde{x} - \tilde{x}^2/2 \tag{21}$$

which is also plotted in Fig. 2. It can be found that the temperature distribution from the model with an objective function of the DHTPC has an obvious reduction in the internal exchanger when compared with the temperature field from the model with a uniform thermal conductivity field. However, these two models give the same highest temperature. In addition, when compared with the model with an objective function of the highest temperature, large differences in thermal conductivity field can be found and the highest temperature increases by 12.5%, which indicates that the optimization model with an objective function of the DHTPC sometimes cannot fulfill the present design goal. Thus, it is necessary to propose new thermal performance indexes for the optimization model.

3. Optimization model based on the geometric average temperature

3.1 Objective function and optimization model

As mentioned above, the optimal design by the optimization model with DHTPC as an objective function sometimes introduces large errors compared with the present design goal. Furthermore, since the highest temperature is a discontinuous function of design variables, direct optimization of it will bring numerical difficulties. To achieve a good tradeoff between the optimization performance and numerical cost, a new thermal performance index called the geometric average temperature T_{geoav} is proposed, which can be expressed as

$$T_{geoav} = \left(\frac{1}{|\Omega|} \int_\Omega (T(\mathbf{x}))^n \mathrm{d}\mathbf{x} \right)^{1/n}, x \in \Omega \tag{22}$$

Where $|\Omega|$ denotes the area (or volume) over the design region. Theoretically, the geometric average temperature is close to the highest temperature when n is infinitely large, i.e. $T_{geoav} \xrightarrow{n\to\infty} T_{max}$. Thus, the geometric average temperature is an appropriate approximation of the highest temperature. The new heat conduction optimization model can be written as

$$\begin{aligned} &\text{Find}: X = k(x), \quad x \in \Omega \\ &\min: \ T_{geoav}(X) = \left(\frac{1}{|\Omega|} \int_\Omega (T(\mathbf{x}))^n \mathrm{d}\mathbf{x} \right)^{1/n} \\ &s.t.: \quad \int_\Omega k \mathrm{d}\Omega = K_0, \quad K_0 = \text{const} \end{aligned} \tag{23}$$

Here, the finite element method is used to solve the optimization problem. Suppose that the material is uniformly distributed and has the same conductivity in each element. Then the distribution of material can be described by the different thermal conductivity in each element mesh, which can be expressed by the finite element method

$$k(x) = k_e, \quad x \in \Omega_e, \quad e = 1, 2, \cdots, Ne \tag{24}$$

where k_e $(e = 1, 2, \cdots, Ne)$ denotes the thermal conductivity of the e-th element, $\Omega_e \in \Omega$ the region of the e-th element and Ne the total number of elements. Then, the temperature and its n power in an element can be written as

$$T(x) = [N(x)]\{\overline{T}\}, \quad T^n(x) = [N(x)]\{\overline{T}^n\} \tag{25}$$

$$\{\overline{T}\} = (T_1, T_2, \cdots, T_{Np})^{\mathrm{T}}, \quad \{\overline{T}^n\} = [(T_1)^n, (T_2)^n, \cdots, (T_{Np})^n)^{\mathrm{T}} \tag{26}$$

where $\{\overline{T}\}$ and $\{\overline{T}^n\}$ denote the temperature vector of nodes and the corresponding n power, respectively. T_n $(n = 1, 2, \cdots N_{\mathrm{p}})$ is the temperature of the n-th node, N_{p} is the total number of nodes and $[N(x)]$ is the shape function matrix. The node temperature can be solved by the following governing equation

$$[K]\{\overline{T}\} = \{Q\} \tag{27}$$

where $\{Q\}$ is the thermal flux vector and $[K]$ is the thermal conductivity matrix which can be assembled by the element thermal conductivity matrix

$$[K] = \sum_{e=1}^{Ne} [K_e], \quad [K_e] = k_e [K_e^0] \tag{28}$$

where $[K_e^0]$ is the e-th element thermal conductivity matrix with a unit thermal conductivity. The geometric average temperature can be rewritten as

$$T_{\mathrm{geoav}} = \left([B]\{\overline{T}^n\}\right)^{1/n}, \quad [B] = \frac{1}{|\Omega|}\int_{\Omega}[N(x)]\mathrm{d}x \tag{29}$$

Therefore, the heat conduction optimization problem can be expressed as

$$\begin{aligned} &\text{Find}: \boldsymbol{X} = (k_1, k_2, \cdots k_{Ne})^{\mathrm{T}} \\ &\min: \ T_{geoav}(\boldsymbol{X}) = \left([B]\{\overline{T}^n\}\right)^{1/n} \\ &s.t.: \ \sum_{e=1}^{Ne} k_e V_e = K_0, \quad V_e = |\Omega_e| \end{aligned} \tag{30}$$

The feasible direction method was employed to find the optimal solution. The sensitivity of the objective function (the geometric average temperature) can be expressed as

$$\frac{\partial T_{geoav}}{\partial k_e} = \frac{1}{n}\left([B]\{\overline{T}^n\}\right)^{-(n-1)/n}[B]\frac{\partial}{\partial k_e}\{\overline{T}^n\} \tag{31}$$

where

$$\frac{\partial}{\partial k_e}\{\bar{T}^n\} = diag(nT_1^{(n-1)}, nT_2^{(n-1)}, \cdots, nT_{Np}^{(n-1)})\left\{\frac{\partial \bar{T}}{\partial k_e}\right\} \tag{32}$$

and

$$\left\{\frac{\partial \bar{T}}{\partial k_e}\right\} = -[K]^{-1}(\frac{\partial}{\partial k_e}[K])\{\bar{T}\} = -[K]^{-1}[K_e^0]\{\bar{T}\} \tag{33}$$

3.2 Example 1

The planar plate exchanger is analyzed again by the new optimization model (23), in which the one-dimensional heat conduction element with two nodes was used to mesh the design domain and the feasible direction method was employed to find the optimal solution. The obtained thermal conductivity field (material distribution) and the corresponding temperature distribution are shown in Fig. 3. To facilitate comparisons, the solutions of the optimization models with the DHPC and the highest temperature as objective functions are also shown in Fig. 3. It can be seen that the thermal conductivity field and the temperature distribution are close to the present design goal (the solution of minimizing the highest temperature) when the power index n is larger than 16. Thus, the geometric average temperature is an ideal thermal performance index. The new optimization model with the geometric average temperature as the objective function is a more accurate description for the design goal than that with the DHTPC as the objective function. The results obtained from the convectional optimization model with the DHTPC as the objective function is equal to that from the new optimization model when the power index n is 1. With the increasing

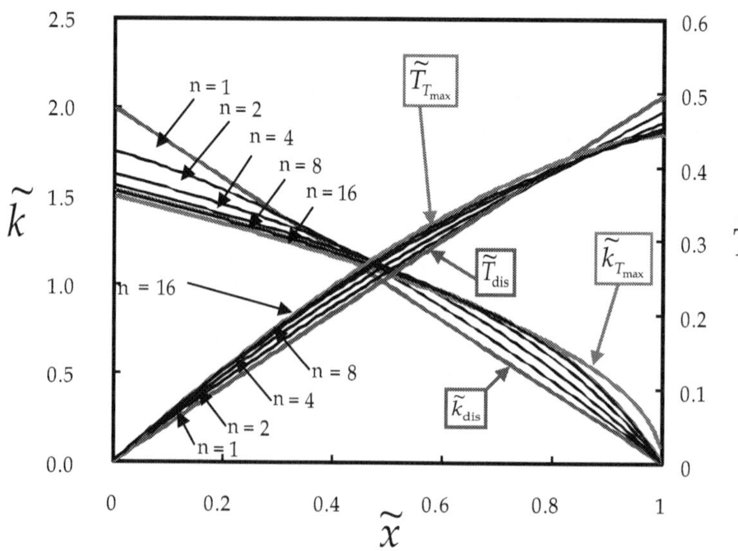

Fig. 3. The solutions of the optimization model based on the geometric average temperature with different power indexes. Results from other models are also shown for comparison

of power index n, the thermal conductivity field and temperature field obtained by the new model is rapidly close to the ideal design. The change of the corresponding highest temperature with the increasing power index n is shown in Fig. 4. Since the approximate level tends to stabilize with the increasing power index n, an appropriate value is required to select for the power index n in a practical optimization process.

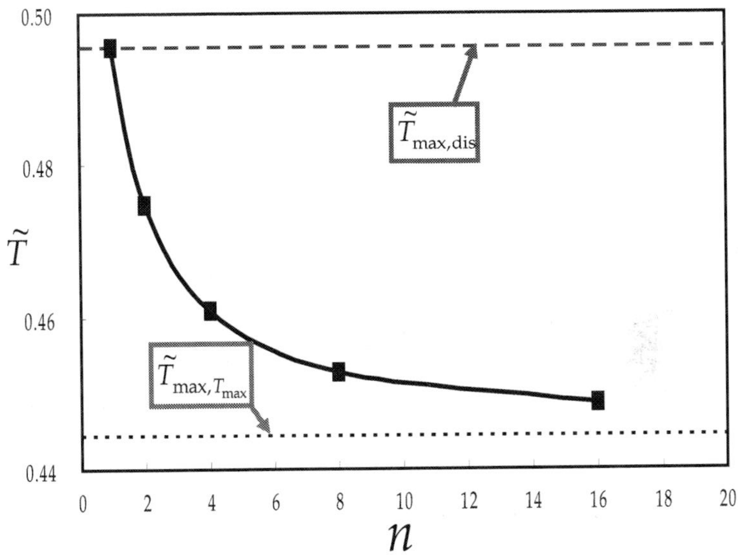

Fig. 4. The highest temperature obtained by the optimization model based on the geometric average temperature *versus* the power index n. $\tilde{T}_{max,dis}$ and $\tilde{T}_{max,T_{max}}$ denote the corresponding highest temperatures of the optimal solution of optimization models with the DHTPC and the highest temperature as the objective function, respectively

3.3 Example 2

To demonstrate the difference between these two objective functions used in the topology optimization method, a thermal structure with five heat sources is presented as an example in this section.

A square planar plate, with dimension 50mm×50mm, is meshed by 50×50 discrete rectangular elements. Temperature is 0 centigrade around the boundary, and five heat sources are set symmetrically in the centre and around the plate with heat flux 1kW/m^2. Material with thermal conductivity k_p = 200W/ (m·K) is used to filled this structure, and the volume fraction V_f of this high heat-conductivity material is given to be 0.35, as shown in fig. 5.

A model using minimum heat dissipation as its objective function is adopted to solve this problem, and its structural topology result is shown in fig. 6.(a), corresponding the contours of temperature distribution and temperature gradient are shown in fig. 6.(b) and fig. 6.(c) respectively. Likewise, the topology result of a model using geometric average temperature as its objective function is shown in fig. 7.

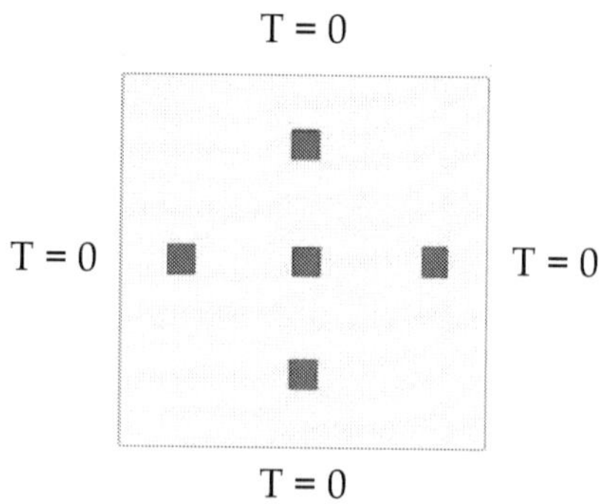

Fig. 5. Initial design domain with five heat sources

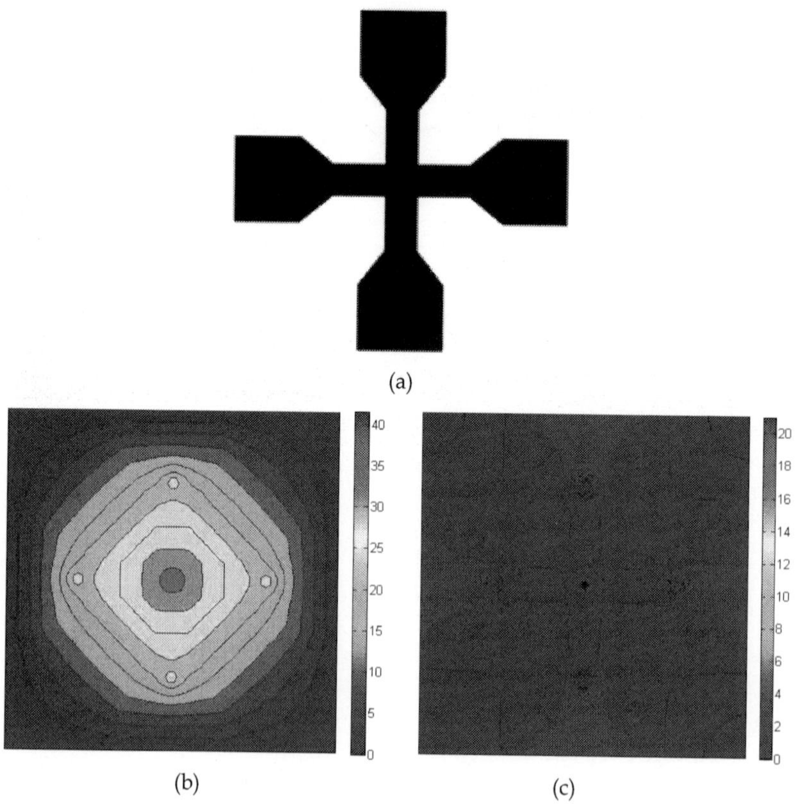

Fig. 6. (a) The optimal design generated from the optimization model 1 (b) the temperature distribution isoline map (c) the temperature gradient isoline map

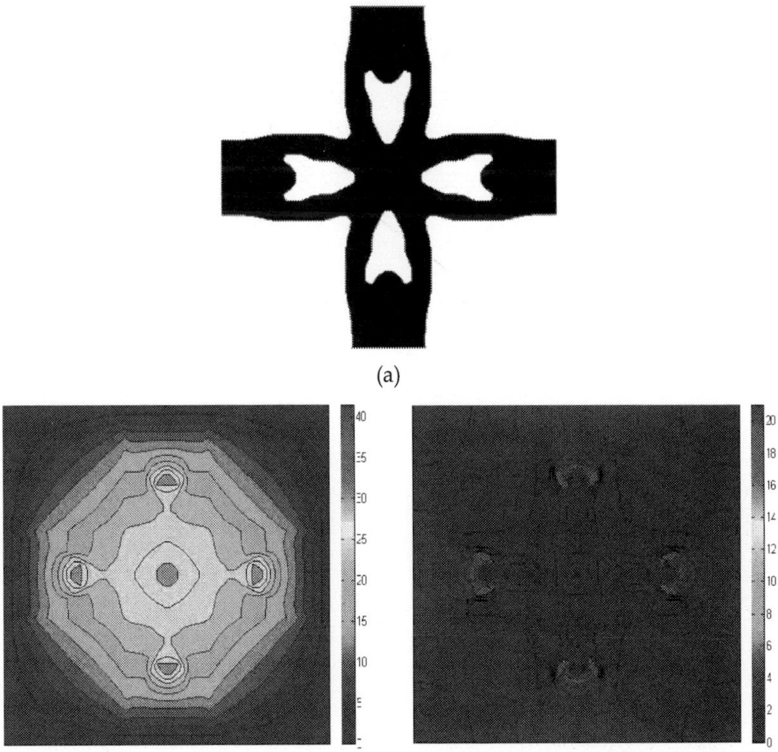

(a) (b) (c)

Fig. 7. (a) The optimal design generated from the optimization model 3 (b) the temperature distribution isoline map (c) the temperature gradient isoline map

Optimization model	Heat dissipation	Maximum temperature (°C)	Maximum temperature gradient
Model 1	5.43e5	37.81	4.24
Model 2	6.23e5	33.72	6.71

Table 1. Results of the different objective

As shown in Fig. 6-7, these two different topology optimization models obtain entirely different topology results. Using minimum heat dissipation as objective function, the result shows that the high-heat conductivity material connects the central heat source with surrounding heat sources directly and then extend to the outer thermal edge, as shown in fig. 4.(a). This kind of heat transfer path will cause the temperature of the central heat source much higher than the temperature of the surrounding heat sources. On the contrary, using geometric average temperature as objective function, the result shows that, instead of connecting the central heat source with surrounding heat sources directly, the high-heat conductivity material bypass the surrounding heat sources and then connect to the outer thermal edge, as shown in fig. 6.(a). This kind of heat transfer path will cause the temperature of the central heat source and the surrounding heat sources equal to the maximum temperature simultaneously. According to the topology results of these two different objectives as shown in table 1, heat dissipation and maximum temperature gradient in model 1 is smaller than model 2, and the maximum temperature in model 2 is smaller than model 1.

According to the analysis of this example, we can conclude that: the model using heat dissipation as its objective function can be used in problem which considers the homogenization of the temperature gradient; the mode using geometric average temperature as its objective function can be used in problem which considers oversize maximum temperature as its failure mode.

4. Conclusion

We have discussed how to minimize the highest temperature of a heat conduction structure by designing the material distribution with a specified material volume (conductivity ability). The large error sometimes occurs between the results by the usual optimization model with an objective function of the DHTPC and the theoretical optimal design. A geometric average temperature has been proposed, which is a better thermal performance index as the objective function. The solution of the new model with the geometric average temperature as the objective function is close to the theoretical optimal solution.

5. Acknowledgment

This work was supported by National Natural Science Foundation of China (10902019, 90816025), by the national key basic research program of china (Grant No.2011CB610304) and by the fundamental research funds for the central universities (DUT11LK07).

6. Referencess

Bejan A. Constructal theory network of conducting paths for cooling paths for cooling a heat generating volume. *International Journal of Heat Mass Transfer*1997; 40(4): 799-816.

Bejan A. Shape and Structure, from Engineering to Nature. Cambridge: *Cambridge University Press*, 2000.

Ledezma GA, Bejan A, Errera MR. Constructal tree network for heat transfer. *Journal of Applied Physics*1997; 82(1): 89-100.

Almogbel M, Bejan A. Conduction trees with spacing at tips. *International Journal of Heat Mass Transfer* 1999; 42 (20): 3739-56.

Almogbel M, Bejan A. Constructal optimization of nonuniformly distributed tree-shaped flow structures for conduction. *International Journal of Heat Mass Transfer* 2001; 44 (22): 4185-94.

Rocha LAO, Lorente S, Bejan A. Constructal design for cooling a disc-shaped area by conduction. *International Journal of Heat and Mass Transfer* 2002; 45(8): 1643-52.

Silva AKD, Vasile C, Bejan A. Disc cooling with high-conductivity inserts that extend inward from the perimeter. *International Journal of Heat and Mass Transfer* 2004; 47(19-20): 4257-63.

Rocha LAO, Lorente S, Bejan A. Conduction tree networks with loops for cooling a heat generating volume. *International Journal of Heat and Mass Transfer* 2006; 49(15-16): 2626-35.

Cheng XG, Xia ZZ, Li ZX, et al. Optimization of heat conduction: Thermal dissipation and optimal thermal conductivity distribution. *Journal of Engineering Thermophysics* 2002; 23(6): 715-17 [in Chinese].

Xia ZZ, Guo ZY. Heat conduction optimization simulated by life evolution. *Progress in Natural Science* 2001; 11(8): 845-52 [in Chinese].

Guo ZY, Cheng XG, Xia ZZ. Least dissipation principle of heat transport potential capacity and its application in heat conduction optimization. *Chinese Science Bulletin* 2003; 48(4): 406-10.

Cheng XG, Li ZX, Guo ZY. Variational principles in heat conduction. *Journal of Engineering Thermophysics 2004*; 25(3): 457-9 [in Chinese].

Guo ZY, Liang XG, Zhu HY. Entranspy - A new physical property of the overall capacity of heat transport. *Progress in Natural Science* 2006; 16(10): 1288-96 [in Chinese].

Cheng XG, Meng JA, Guo ZY. Potential capacity dissipation minimization and entropy generation minimization in heat conduction optimization. *Journal of Engineering Thermophysics* 2005; 6(6): 1034-6 [in Chinese].

Han GZ, Guo ZY. Two different thermal optimization objective functions: Dissipation of heat transport potential capacity and entropy production. *Journal of Engineering Thermophysics* 2006; 27(5): 811-3 [in Chinese].

Li Q, Steven GP, Xie YM, et al. Evolutionary topology optimization for temperature reduction of heat conducting fields. *International Journal of heat mass transfer 2004*; 47 (23): 5071-83.

Jia HP. Topology optimization of structure and compliant mechanism. *Doctor Dissertation, Dalian University of Technology*. 2004; 95-118 [in Chinese].

Zuo KT, Chen LP, Zhang YQ, et al. Structural optimal design of heat conductive body with topology optimization method. *Chinese Journal of Mechanical Engineering* 2005; 41(4):13-21 [in Chinese].

Zhang YC, Liu ST. Design of conducting paths based on topology optimization. *Heat and Mass Transfer, online*; DOI 10.1007/s00231-007-0365

Li JC, Ye BY, Tang Y, et al. Algorithm of topology optimization criteria for heat conduction structure based on density approach. *Journal of South China University of Technology (Natural Science Edition)* 2006; 34(2):27-32[in Chinese].

Xu XH, Liang XG, Xun RJ. Optimization of heat conduction using combinatorial optimization algorithms. *International Journal of Heat Mass Transfer* 2007; 50 (9-10): 1675-82.

Francois MP, Louis G. Optimal conduction pathways for a heat generating body: a comparison exercise. *International Journal of Heat Mass Transfer,* 2007; 50(15-16): 2996-3006.

Wu WJ, Chen LG, Sun FR. Improvement of tree-like network constructal method for heat conduction optimization, *Science in China: Series E Technological Sciences* 2006; 49(3): 1-10.

Zhou Shengbin, Chen lingen, Sun Fengrui. Optimization of constructual volume-point conduction with variable cross section conducting path. *Energy Conversion and Management* 2007; 48(1):106-11.

Liu MA, Dong QX, Chen S. Investigation on thermal design for opening cabinet of electronic equipment by numerical simulation. *Journal of air force engineering university (natural science edition)* 2005; 6(2):62-5 [in Chinese].

Qi YQ, He YL, Zhang W, et al. Thermal analysis and design of electronic equipments. *Modern Electronic Technology* 2003; 144:73-76 [in Chinese].

Thermal Modelling for Laser Treatment of Port Wine Stains

Li Dong[1], Wang Guo-Xiang[2] and He Ya-Ling[1]
[1]*State Key Laboratory of Multi-Phase Flow in Power Engineering Xi'an Jiaotong University; Xi'an,*
[2]*Department of Mechanical Engineering; The University of Akron; Akron,*
[1]*China*
[2]*USA*

1. Introduction

Port Wine Stains (PWS) are congenital vascular birthmarks (see Fig. 1) that occur in approximately 0.3% of children (Alper & Holmes, 1983). Clinically, PWS ranges in appearance from pale pink to red to purple. Most of the lesions are pale pink at birth and can progressively be darken and thicken with age. PWS can be associated with significant cosmetic disfigurement and psychologic distress. Histopathological analysis of PWS reveals a normal epidermis overlying an abnormal plexus of benign vascular malformations consist of ectatic capillaries of diameters varying from 10 to 300 μm. Laser treatment of PWS started in the late 1960s with continuous-wave lasers such as carbon-dioxide and argon lasers (Gemert et al., 1987; Dixon et al., 1984). During the laser treatment, light is absorbed by hemoglobin (Hb) and oxyhemoglobin (HbO_2) (the principle chromophores for light absorption in human tissue) within the blood vessels and then converted into heat which damages the endothelium and surrounding vessel walls (Alora & Anderson, 2000). Unacceptable side effects such as scarring and permanent dyspigmentation were the two major shortcomings of the early laser treatment of PWS. Those side effects result from unselective heating of both PWS and nearby healthy dermal tissues during laser irradiation.

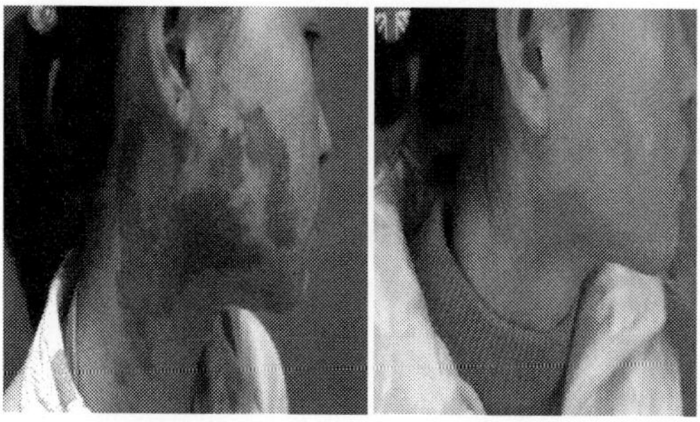

Fig. 1. PWS before and after laser treatment (Curtsey of Drs. Wang and Ying at Laser Cosmetic Centre of 2nd hospital of Xi'an Jiaotong University, Xi'an, China)

The principle of selective photothermolysis of Anderson and Parrish (Anderson & Parrish, 1983) revolutionized the laser treatment of vascular lesions such as PWS. Based on this principle, one can selectively destruct the PWS while keep the surrounding tissues untouched by properly choosing right laser wavelength and pulse duration. In the case of PWS, the primary absorption peaks of oxyhemoglobin and deoxyhemoglobin fall in the visible range (418, 542, or 577 nm) (see Fig. 2). For those wavelengths the water (another major chromophore for light absorption within tissue) has almost no absorption. Therefore, lasers with these wavelengths can effectively damage the PWS while does not cause any irreversible effect on surrounding tissues. As shown in Fig. 2, however, the melanin (another chromophore for light absorption) within epidermis also has strong absorption over the same wavelength range. Strong absorption of melanin and resulting epidermal heating not only lead to undesired skin injury but also reduce the heating effect of PWS which is buried deep underneath the epidermis. As a result, long wavelength lasers, such as the pulsed tunable dye lasers (PDL) with wavelengths of 595 - 600 nm, are now widely employed in laser treatment of PWS due to relatively weak absorption of melanin at the chosen wavelengths. In addition, the modern PDL lasers are often equipped with adjustable laser pulse width from 0.45 ms to 40 ms as well as surface cooling devices (Kelly et al., 2005).

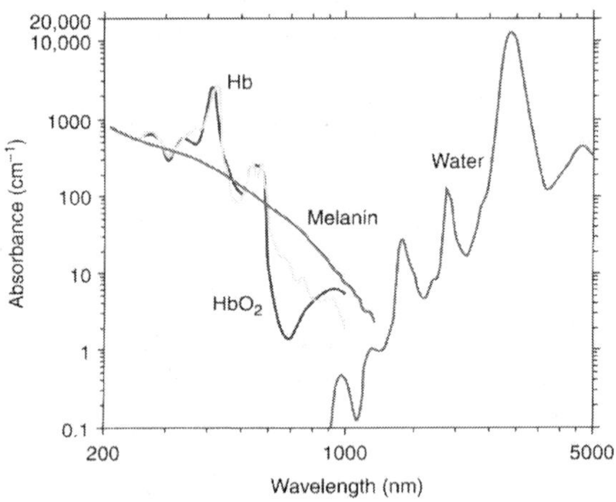

Fig. 2. Absorption spectra of major chromophores in skin (Anderson, 1997)

Surface cooling is critical in success of modern laser treatment of PWS (Kelly et al. 2005). The objective of surface cooling is to increase the laser fluence while prevent epidermal heating due to competitive absorption by epidermal melanin. Various cooling devices have been developed from early ice cubes (Gilchrest et al., 1982), chilled sapphire windows cooled by circulating refrigerant (Kelly et al., 2005), to present chilled air cooling (Hammes et al., 2005) and cryogen spray cooling (Nelson et al., 1995). The cryogen spray cooling technique provides spatially selective cooling of epidermis with pulse spray of volatile refrigerant R134a (tetrafluoroethane [C2H2F4]; boiling point at −26.2°C). Due to its short spurt, the PWS buried deep within dermis will be not affected. Consequently, PDL laser with cryogen spray cooling becomes standard protocol of PWS treatment (see Fig. 3).

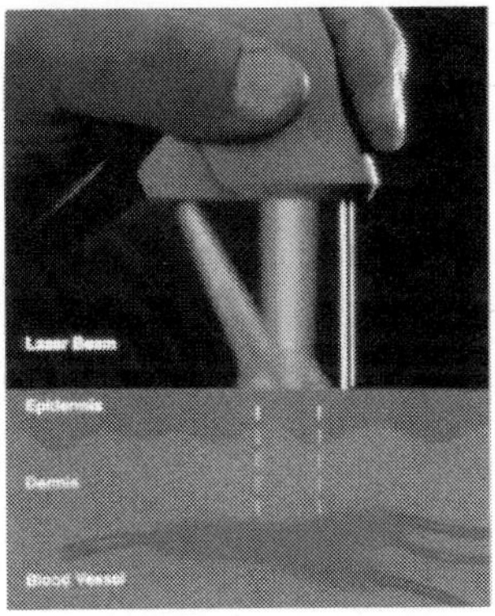

Fig. 3. Schematic of the laser treatment of PWS together with cryogen spray cooling (CSC). (Lanigan, 2000)

The laser surgery process of PWS (Laser PWS) has been studied extensively through mathematical modeling. Various models have been developed to better understand the laser-tissue interaction, in particular the thermal response of biological tissues and physiological mechanisms of thermal damage of PWS (see e.g., Pfefer et al., 2000; Pickering et al., 1989; Shafirstein et al., 2004; Tunnell et al., 2003). The models have also been used for optimization of the surgery parameters and for development of new treatment protocols (Jia et al., 2006, 2007). To model laser PWS, various simple skin models have been proposed to simplify the complex anatomic structure of the skin with PWS (Gemert et al., 1995). Mathematic techniques have been developed to calculate the light propagation and energy deposition within the PWS (Jacques & Wang, 1995). Energy absorbed by PWS can then be incorporated into the temperature equation as a heat source to determine the corresponding temperature change in the irradiated tissues. A thermal injury sub-model is also used to quantify the thermal damage to the PWS (Pearce & Thomsen, 1995).

The propagation of light in turbid media such as skin can be described by the radiative transport equation (Ishimaru, 1989) which cannot be solved analytically for tissue geometries. Techniques such as diffusion approximation (Gemert et al., 1997) and Beer's law (Verkruysse et al., 1993) have been used in the literature but they are limited to highly scattering or coherent radiance materials (Niemz, 1996). For skin tissue, the Monte-Carlo (MC) method, which offers a flexible approach to track the photon transportation in a biological tissue, is widely accepted now as an accurate method and has been used in modeling of laser surgery (Keijzer et al., 1991; Lucassen et al., 1996; Pfefer et al., 1996; Smith & Butler 1995; Wang et al., 1995 & Wilson & Adam, 1983).

The bio-heat transfer equation such as Penn equation based on the skin models, in consistent with that used for light simulation, can be solved for tissue temperatures in laser PWS. Due to limited irradiation time of pulsed laser in laser PWS, the perfusion effect in Penn equation is usually neglected so that the bio-heat equation becomes a heat conduction equation. The thermal properties of PWS skin are scarce and those for normal skin tissues and blood are usually employed as an approximation. For laser PWS, cryogen spray cooling is widely employed by spraying volatile refrigerant such as R134a on skin surface before laser irradiation. Therefore, the cooling of skins before laser irradiation should be part of the model of laser PWS (Jia et al., 2006; Li et al., 2007a,b)

In this chapter, we will first present a brief review of the development of various skin models proposed in the literature for thermal modeling of laser PWS. Next, we explain the basic principle of the multi-layer Monte-Carlo method which is widely used for simulation of light propagation in the skin with PWS. Then we introduce a general relation developed recently by the present authors to quantify the heat transfer of cryogen spray cooling in laser PWS. Finally, results from a multi-layer, homogeneous model of laser surgery of PWS are presented to illustrate typical thermal characteristics of laser surgery of PWS. The effect of laser wavelength and laser pulse duration on the heating of the PWS layer is also examined briefly. Future needs for modeling the laser surgery of PWS is then discussed in conclusion.

2. Skin models and numerical simulation of laser surgery of PWS

The human skin has three main layers, the epidermis, the dermis, and the subcutaneous layer (fat) (Brannon, 2007). The epidermis is the outer layer of skin with its thickness varying in different types of skin. It is made up of cells called keratinocytes, which are stacked on top of each other, forming different sub-layers. In the stratum basale sub-layer, there are cells called melanocytes that produce melanin which is one of three major chromophores for light absorption. Melanin is a pigment that is absorbed into the dividing skin cells to help protect them against damage from sunlight (ultraviolet light). The amount of melanin in skin is determined by genes and by how much exposure to sunlight. Below the epidermis is the layer called dermis, which is a thick layer of fibrous and elastic tissue (made mostly of collagen, elastin, and fibrillin) that gives the skin its flexibility and strength. The dermis contains nerve endings, sweat glands and oil glands, hair follicles, and blood vessels. The blood vessels of the dermis provide nutrients to the skin and help regulate body temperature. Port Wine Stains (PWS) are the result of malformation of significant amount blood vessels within dermis. The subcutaneous tissue is a layer of fat and connective tissue that houses larger blood vessels and nerves. This layer is important in the regulation of temperature of the skin itself and the body. The size of this layer varies throughout the body and from person to person.

Ideally, a real three-dimensional multi-component skin models with detailed anatomic structure of PWS vessels should be used to evaluate the light distribution and temperature variation during laser PWS (Pfefer et al., 1996). However, low resolution of noninvasive imaging technique and extremely complex human tissue structure makes this impossible at moment. Thus, simplified skin models have been developed to analyze the thermal response of complex human skin with PWS under laser irradiation (Gemert et al., 1995). A simplified skin model should represent important histological characters of target chromophores and physical events during and after laser surgery and should be easily handled mathematically. The simplest model for laser PWS is probably the two-layer homogeneous model proposed first by Gemert and Hulsbergen (Gemert & Hulsbergen, 1981) who simplified the human skin containing PWS blood vessels to two-layer structure that parallel to the skin surface. In their model, the upper epidermis layer includes uniformly distributed melanin and the lower dermis layer is mixed with extra blood representing PWS. Gemert and Hulsbergen were the first who attempted to calculate the light distribution within PWS theoretically by employing the Kubelka-Munk method which significantly simplifies the light transport equation. Refined skin models as shown in Figure 4 with multi-layers (Gemert et al., 1982, 1995; Miller & Veith, 1993) have then subsequently been developed, all including an extra blood layer buried between dermis to represent the PWS. The Kubelka-Munk method and later the Monte-Carlo method were used to predict the light deposition within the PWS.

The multi-layer model with homogeneously distributed chromophore (blood) in the PWS layer has been widely used in numerical simulation of laser treatment of skin lesions. For example, Pickering and Gemert (Pickering & Gemert, 1991) used this model to theoretically investigate the mechanism of superior treatment of PWS by the laser of 585nm over that of 577nm. They found that the 585nm laser produced a deeper depth of vascular injury. The effects of various laser parameters such as pulsed duration, the repetition rate of pulses, the beam spot size, and the radiant energy fluencies on the outcomes of the laser surgery in PWS were systematically investigated (Gemert et al., 1995, 1997; Kienle & Hibst, 1995, 1997; Verkruysse et al., 1993). In all these studies, the PWS layer is treated as homogeneous mixture of dermal tissue and blood of a given volumetric fraction. The average optical and thermal properties of the chromophores and non-chromophore tissues weighted by their volumetric fractions are employed in the analysis. It is found that the calculated temperature in the PWS layer seems to be lower than expected (Aguilar et al., 2002), probably due to the use of the averaging properties in the calculations.

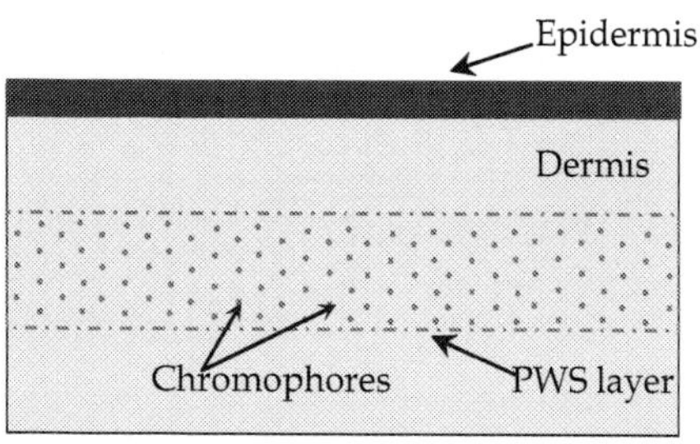

Fig. 4. Schematic of a multi-layer skin model including a PWS layer within dermis. The PWS layer is made of normal dermal tissue mixed homogeneously with blood

The multi-layer skin model simply treats the PWS as a homogeneous mixture of dermal tissue and blood, ignore the detailed structural characteristics of blood vessels in PWS. To represent more realistic anatomic features of PWS, skin models that contain individual blood vessels have been developed (Gemert et al., 1986; Lahaye & Gemert, 1985; Lucassen et al., 1995; Pickering et al., 1989), as shown in Figure 5. For example, Anderson and Parrish (Anderson & Parrish, 1983) performed a thermal analysis directly on an isolated blood vessel that was buried in dermal tissue and under laser irradiation. This analysis led to their famous "selective photothermolysis" theory that provides theoretical foundation for laser design and selection and for development of clinic protocol of laser treatment of PWS and other cutaneous diseases (Lanigan, 2000). With the rapid growth of the computing power, skin models with discretely distributed individual blood vessels became the favored among researchers. Skin models with the blood vessels that are parallel to the skin surface and regularly (aligned or staggered) buried in the dermis underneath the epidermal layer have been widely employed. Recently, models with randomly distributed blood vessels with varying sizes have also been utilized. Occasionally, the model with a single blood vessel was also used to test the clinic outcome of new treating protocols of laser PWS (Jia et al., 2006, 2007). Concerns have been raised for possible lack of the scattering and shielding effects of numerous blood vessels in real PWS for those discrete blood vessel models (Tan et al., 1990; Verkruysse et al., 1993).

The skin model with discrete blood vessels can provide direct visualization of heating of the blood vessels in laser PWS (Baumler et al., 2005; Shafirstein et al., 2004, 2007; Tunnell et al., 2003). For example, Smith and Butler (Smith & Butler, 1995) have shown that the blood vessels are heated up first during laser irradiation, and the dermis surrounding the vessel are then heated conductively by the high temperature vessels. The effect of local heating within the vessel and the shadowing effects of front vessels over the bottom vessels have also been demonstrated.

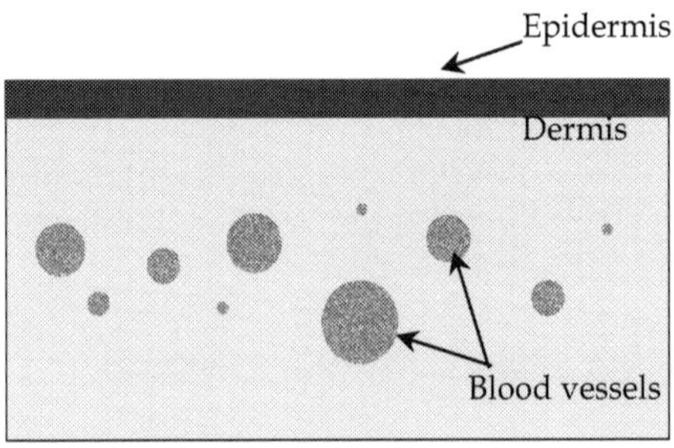

Fig. 5. Schematic of a skin model with discrete blood vessels buried in dermis

3. Multi-layer Monte-Carlo (MC) method for light propagation in skin tissue

All above skin models can be treated as a multi-layered structure, even for those with discretely distributed blood vessels (see Figs. 4 & 5). The light propagation within such multi-layered structure can be accurately simulated by the multi-layered Monte-Carlo (MLMC) method developed by Wang et al., (Wang et al., 1995). Here a brief introduction of the MLMC method is presented and the details can be found elsewhere (Jacques & Wang, 1995; Wang et al., 1995).

The MLMC method deals with the transport of a laser beam within a medium made of layered tissues that are parallel to each other. In MLMC, a photon packet is first injected perpendicular to the tissue surface with a unit weight (W = 1) with the position of the photon packet on the surface (position O in Figure 6) is determined from the shape of the beam profile (e.g., Gaussian). The propagation or movement of the photon packet in the tissue is then tracked step by step. At each step, the step size between two collisions within the tissue is determined from a logarithmic distribution function with a random number generated by the computer. Since the photon packet may hit a boundary of the current layer, before each photon movement, the distance between the current photon location and the boundary of the current layer in the direction of the photon propagation needs to be computed. If the step size is smaller than the calculated distance, the step will move within the current layer (segment OA, see Figure 6). Otherwise, if the step size is greater than this distance (segment AB), the photon packet will hit the boundary and the photon packet can be either internally reflected or transmitted across the boundary. If the photon is internally reflected, the photon packet continues propagation with an updated step size and changes its direction as a mirror reflection (segment BC). If the photon is transmitted, it continues its propagation with an updated step size but with the direction newly calculated according to the Snell's law (segment BD). At the end of each step, the photon packet will interact with the tissue. A fraction of the photon weight ($\Delta W = \mu_a/\mu_t$) is first absorbed at the interaction site, where μ_a and μ_t are the absorption coefficient and the attenuation coefficient, respectively, of the medium. The absorbed weight is scored into the absorption array $A[i,j]$ at the local volume element (i,j) in a two-dimensional grid system, where i and j, are the indices for volume elements. After absorption, the photon packet updates its weight to $W-\Delta W$ and is then scattered at the interaction site (segment DE) by choosing a new direction of propagation according to a given phase function and another random number generated by the computer.

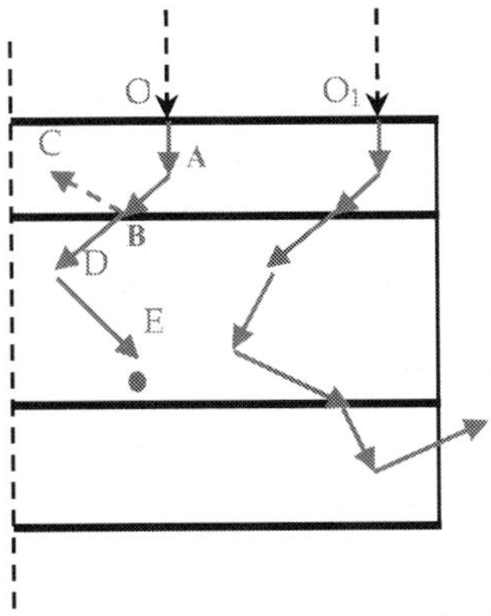

Fig. 6. A schematic showing movement of photon packets in multi-layer Monte Carlo simulation

The weight of a launched photon packet is continuously decreased due to absorption by the medium as the photon packet moves within the tissue step by step. The photon packet is terminated naturally if it moves out of the tissue domain due to either reflection or transmission. For a photon packet still propagating inside the medium, the Russian roulette technique is used to terminate the photon packet when the photon weight W is reduced to be lower than a given threshold value. Once the photon packet is terminated, a new photon packet enters the skin domain, probably at a new position (for example O_1 in Figure 6), and the above process is then repeated.

After all photon packets terminate, the MLMC simulation completes and the final absorption array $A[i,j]$ is obtained for the entire domain. Then, the energy density deposited in each grid element can be calculated as follows:

$$Q[i,j] = E \cdot \pi r^2 \cdot \frac{A[i,j]}{N \cdot dv} \tag{1}$$

where the array $Q[i, j]$ is the rate of energy storage in the volume element (i,j), E is the incident energy of the laser, r is the spot radius of the incident laser, N is the total number of photo packets used in the simulation, and dv is the volume of the element. Although the light transport in the layered-skin structure is gridless, the selection of the size of the volume element in the grid affects the simulation through Equation (1). The size of the volume element should be selected appropriately for proper scoring of the absorbed energy within the element by satisfying the statistical requirement of the MC simulation. There should have enough photons that have arrived at any given element in order to rightly reflect the photon absorption in the medium. If the size of the volume element becomes too small and the total number of photons used in the simulation is limited, one may find the case that no photons at all arrive at certain elements and no energy deposition is registered in those elements. Recently, the present authors have developed criteria that can be used to choose right element size and the total number of photon packets needed in the MC simulation (Li et al., 2011).

4. Cryogen spray cooling (CSC)

During the laser treatment of PWS, a significant amount of laser energy is absorbed by melanin within the basal layer of the epidermis (Tunnell et al., 2000). To allow higher laser energy doses used in clinic, one needs to cool the epidermis prior to laser irradiation. Cryogen spray cooling (CSC) is one of the techniques developed to protect the epidermis from non-specific heating by pre-cooling the skin prior to laser irradiation (Nelson et al., 1995). The cryogen utilized is 1,1,1,2 tetrafluoroethane, also known as R134a, with boiling temperature of T_b = - 26.2 °C at atmospheric pressure. Cryogen is usually kept in a container at saturation pressure, which is approximately 660 kPa at 25 °C (95.7 psi), and delivered through a standard high pressure hose to an electronically controlled fuel injector, to which a straight-tube nozzle is attached. As the cryogen reaches the tissue surface, tissue experiences a quick cooling process to reach low temperatures (Kelly et al., 2005).

To model laser PWS with CSC, a convective thermal boundary condition is usually used (Aguilar et al., 2002; Jia et al., 2007; Li et al., 2007a,b; Majaron et al., 2001; Pfefer et al., 2000). Due to lack of experimental data, the early models of laser PWS have employed a constant heat transfer coefficient to quantify the short-pulsed cooling process (Aguilar et al., 2002; Majaron et al., 2001; Pfefer et al., 2000). Recently, extensive experimental and numerical investigations have been conducted to characterize the cryogen spray (Pikkula et al., 2001; Zhou et al., 2008a,b) and to quantify the convection heat transfer during spray (Aguilar et al., 2003a,b; Franco et al., 2004, 2005; Jia et al., 2004, 2007). Based on these experimental data, the present authors have developed a quantitative relation that can be used to estimate the convection heat transfer coefficient, $h(r,t)$, as a function of both the space (r on the skin surface) and time (t) during CSC in laser PWS (Li et al., 2007a,b). The non-dimensional form of the relation is given as follows:

$$\begin{cases} h^*(r^*,\tau) = \begin{cases} h_o^*(\tau) & 0 \le r^* \le 0.4 \\ [5(1-r^*)]/[3h_o^*(\tau)] & 0.4 < r^* \le 1.0 \end{cases} & \tau \le 1.0 \\ h^*(r^*,\tau) = \begin{cases} h_o^*(\tau) & 0 \le r^* \le 0.2(\tau+1) \\ h_o^*(\tau)] + \dfrac{(5r^*-\tau-1)}{(4-\tau)} \times [0.09(\tau-1) - h_o^*(\tau)] & 0.2(\tau+1) < r^* \le 1.0 \end{cases} & 1.0 < \tau < 4 \\ h^*(r^*,\tau) = h_o^*(\tau) \quad 0 \le r^* \le 1.0 & \tau \ge 4.0 \end{cases} \tag{2}$$

with $h^*_o(\tau)$ as the non-dimensional heat transfer coefficient at the center of spray:

$$h_o^*(\tau) = \frac{h_o(t)}{h_{o,\max}} = \begin{cases} \tau & \tau \le 1.0 \\ 1.0 - 0.35(\tau - 1) & 1.0 < \tau \le 3.0 \\ 0.3 - 0.02(\tau - 3) & 3.0 < \tau \le 8.0 \\ 0.2 - 0.0125(\tau - 8) & 8.0 < \tau \end{cases} \quad (3)$$

where $h_o(t)$ is the dimensional heat transfer coefficient at the center of spray and $h_{o,\max}$ is the maximum value of h_o at time $t_{\max}$, both estimated from experimental data (Li et al., 2007a). In Eqs. (2) and (3), the non-dimensional heat transfer coefficient, h^*, the non-dimensional radius coordinate, r^*, and the non-dimensional time,τ, are defined as follows:

$$h^*(r^*,\tau) = h(r,t) / h_{o,\max}, \; r^* = r / r_{spray}, \text{ and } \; \tau = t / t_{\max}$$

with r_{spray} the radius of spray spot on the skin surface.
Equations (2) and (3) can be employed in any thermal models for laser PWS before laser irradiation. Figure 7 shows typical dynamic variations of the temperature in the skin during CSC calculated by our own model (to be discussed in detail below). In Figure 7, the temperature distributions within the skin are plotted at two time instants during spray. The corresponding spray distance is 30mm and the spray spurt is 100 ms. One point to be noticed is that the skin surface temperature drops to about -20 °C at 30 ms after the spray starts (see Figure 7a). At this time, however, the temperature within the skin including majority of the epidermal layer still remains the same initial value. After 60 ms of CSC (Fig. 7b), the entire epidermal layer is cooled down below -10 °C, but the temperature of the PWS layer still remains almost the same initial temperature and is not affected by the spray. Such an effect of the cryogen spray is desirable for laser PWS since the only purpose of cooling is to prevent the skin overheating. Since the skin tissue is a very poor thermal conductor, the cooling effect at the surface is hardly felt by the PWS buried deep in the dermis if the spray spurt is kept short.

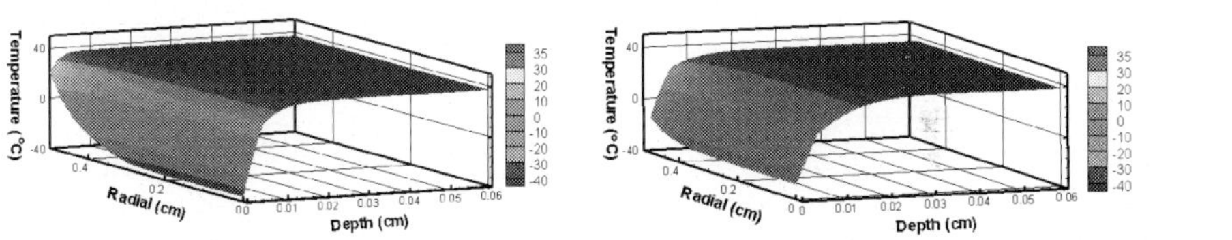

(a) t = 30 ms (b) t = 60 ms

Fig. 7. Calculated temperature distributions within skin during cryogen spray cooling at times after 30ms (a) and 60ms (b). (Spray distance: 30 mm)

5. A model for laser treatment of PWS with CSC

To illustrate the thermal characteristics of the laser surgery process of PWS, we present here a multi-layer thermal model of laser PWS with CSC. The skin model chosen consists of four layers: an epidermal layer containing melanin, two dermal layers with a PWS layer sandwiched in between. The PWS layer is assumed to be a homogeneous mixture of dermis and blood of a given volumetric fraction. The bulk optical and thermal properties of the epidermal layer and the PWS layer are determined based on the corresponding volume contents of melanin and hemoglobin (Verkruysse et al., 1993), respectively. The laser beam has a Gauss profile and the MLMC method discussed in Section 3 is used to quantify the energy deposition in various layers. The resulting rate of energy deposition in the tissue, given in Eq. (1), is included as the source term, Q, in the following heat conduction equation:

$$\rho_i c_{p,i}\frac{\partial T}{\partial t}=\frac{k_i}{r}\frac{\partial}{\partial r}\left(r\frac{\partial T}{\partial r}\right)+k_i\frac{\partial^2 T}{\partial z^2}+Q_i \tag{4}$$

where the subscript i (= e, d, p) represents, respectively, the epidermal layer, the dermal layer, and the PWS layer; ρ is the density, c_p the specific heat, k the thermal conductivity, T the temperature, and t the time. A cylindrical coordinate system as shown in Figure 8 is used with the origin located at the center of the laser beam on the skin surface.

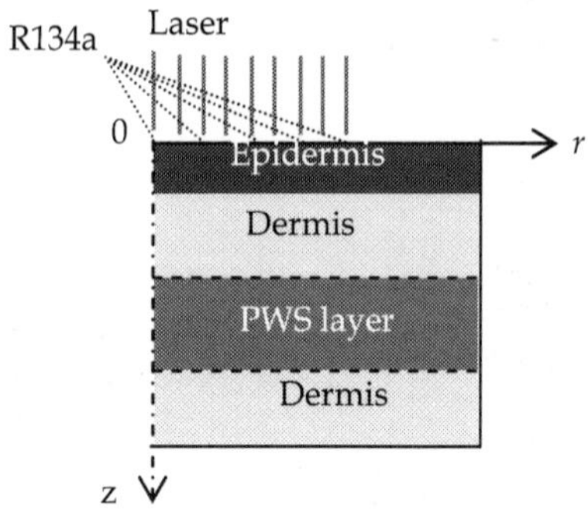

Fig. 8. Schematic of the multi-layer skin model and the corresponding two-dimensional cylindrical coordinate system

On the skin surface (z=0), a convection heat transfer condition is used to describe the cooling effect of CSC and the heating effect of the environment afterward:

$$\begin{aligned} k_e\left.\frac{\partial T}{\partial z}\right|_{z=0} &= h_c(r,t)[T(r,0,t)-T_c] && \textit{during CSC} \\ k_e\left.\frac{\partial T}{\partial z}\right|_{z=0} &= h_a[T(r,0,t)-T_a] && \textit{after CSC or without CSC} \end{aligned} \tag{5}$$

where T_c and T_a are the temperatures of liquid cryogen on the skin surface and the environment, respectively. h_a is the convective heat transfer coefficient between the cold surface and air, and is treated as a constant. $h_c(r, t)$ is the heat transfer coefficient between the surface and the liquid cryogen on the surface during short-pulsed cryogen spray and is determined from Eqs. (2) and (3).

The above model is use to simulate the thermal process in the laser treatment of PWS with CSC. The laser spot diameter is chosen as 5 mm while the CSC spray spot size is fixed at 10 mm. For the skin model, the thickness of the epidermal layer is 50 μm with melanin of five volumetric percent (5%); the thickness of the PWS layer is 200 μm with the volumetric fraction of hemoglobin of 30%. Here a high volumetric fraction of hemoglobin is used to simulate the mature PWS lesion (dark red or purple). The initial skin temperature is assumed to be at 37 °C and the surrounding temperature T_a is 25 °C. Two typical wavelengths (585 nm and 595 nm) of the pulsed-dye lasers are examined with laser pulse duration varying from 1.5 ms to 40 ms, corresponding to the working range of the clinically-used PDL lasers (Kelly et al., 2005).

5.1 Light and temperature distribution without CSC

A typical result from the multi-layer Monte-Carlo simulation is given in Figure 9a which shows the distribution of the photon weight $A(r, z)$ within the skin after irradiation of a PDL beam of 585 nm. As expected, a strong absorption occurs within both the epidermal layer ($0 < z < 50$ μm) and the PWS layer ($250 < z < 450$ μm) due to the existence of melanin and hemoglobin within the two layers, respectively. Less absorption in the PWS layer than that in the epidermal layer is due to the screening effect of the epidermal layer. It is such absorption by the melanin in the epidermal layer that prohibits the application of a high energy dose in laser PWS. Comparing to these two layers, the dermal layer shows very little absorption of light because of a low absorption coefficient of dermis at the given wavelength (585 nm). Figure 9a also shows a decrease of the photon absorption along the radial direction, due to the nature of the incoming laser beam which has a Gaussian profile.
The photon absorption given in Figure 9a can be converted into the rate of heat generation within the skin in the energy equation (4). Solving the energy equation gives the temperature distribution within the skin, as shown in Figure 9b. Figure 9b plots the temperature distribution within the skin at the end of 1.5 ms laser pulse. In this case, no cryogen spray cooling is used. The incident laser fluence was 4 J/cm^2. It is found that both the epidermal and the PWS layers experience high temperatures due to strong photon absorption of the two layers. At the present level of laser fluence, the temperatures of both the epidermal and PWS layers exceed the critical coagulation temperature of about 70 °C (Pearce & Thomsen, 1995). Although a high PWS temperature is desired, a high epidermal temperature causes unspecified skin injury and should be avoided. The cryogen spray cooling technique can be used to protect the skin from such overheating as will be shown below.

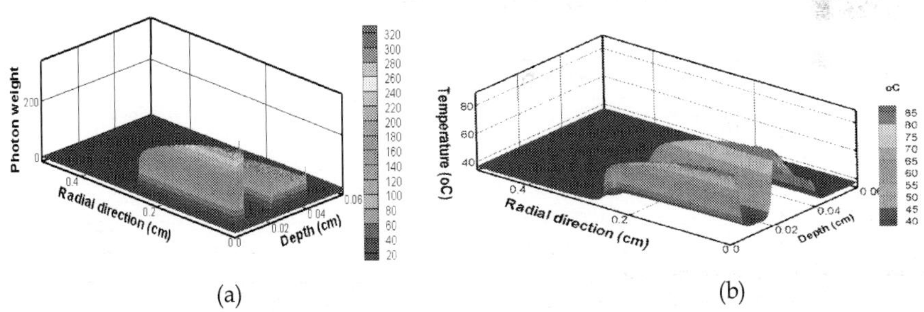

Fig. 9. Calculated distributions of the photon absorption (a) and temperature (b) within the skin at the end of 1.5 ms laser pulse of 585 nm. (Laser fluence E = 4 J/cm^2, without CSC)

5.2 Thermal characteristics of laser treatment of PWS with CSC

The thermal characteristics of laser PWS with CSC are illustrated in Figure 10 which presents the temperature distributions within the skin at four time instants: 100ms (a),

101.5ms (b), 121.5ms (c) and 301.5ms (d) after the CSC starts. Laser was fired after 100 ms spurt cryogen spray. The laser fluence is 6 J/cm^2 and the laser pulse duration is 1.5ms. At the end of the CSC (Fig. 10a), the skin surface temperature decreases to below -10 °C while the temperature of the PWS layer remains almost the same as the initial value. At the end of laser irradiation (Fig. 10b at 101.5 ms), the temperatures of both the epidermal and the PWS layers rise as a result of energy absorption, while the temperature of the dermal layer keeps almost unchanged due to a low absorption of the dermis. After laser irradiation, the temperatures in both the epidermal and the PWS layers decrease as a result of heat conduction. As we can see from Fig. 10c (20 ms after laser irradiation), the peak temperatures of both the epidermal and the PWS layers lower down while the temperature of the dermal layer rises. As heat conduction continues, the entire skin layers approach almost uniform temperature, see Fig. 10d at 200 ms after laser irradiation.

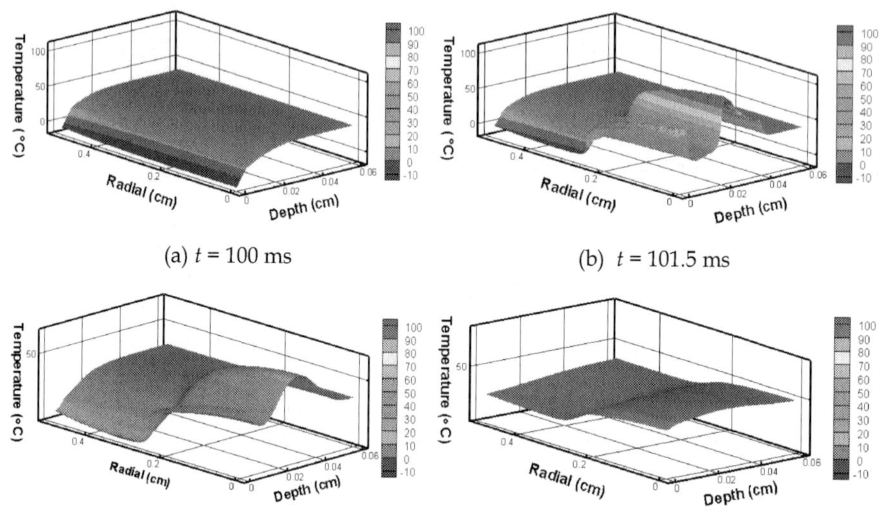

(a) t = 100 ms (b) t = 101.5 ms

(c) t = 121.5 ms (d) t = 301.5 ms

Fig. 10. Calculated temperature distributions within the skin during laser surgery of PWS with CSC at times after 100ms (a), 101.5ms (b), 121.5ms (c) and 301.5ms. (CSC spurt duration: 100ms, Spray distance: 30 mm, laser fluence: 6 J/cm^2, and laser pulse duration: 1.5ms)

5.3 Effect of laser wavelength

Appropriate selection of the wavelength of the laser beam is a critical issue in laser PWS due to the sensitivity of the light absorption of PWS to the wavelength. Figure 11 shows the temperature distributions of the skin at the end of 1.5 ms pulsed laser irradiation under two wavelengths, 585 (a) and 595nm (b), respectively, with all other conditions remain the same. A comparison of the temperature profiles along the central axial direction corresponding to Figure 11 is also given in Fig. 12a. Figure 12b plots similar temperature distributions for a thicker PWS layer of 300 μm. Inspecting these figures, one finds that the laser wavelength affects the highest temperature possible in the PWS layer and the evenness of the temperature distribution over the PWS layer. A much higher possible temperature in PWS is achieved for the shorter 585 nm laser than that for the longer 595 nm laser. The longer 595 nm laser, however, produces a much more even heating over the PWS layer. This can be more clearly demonstrated in the case of a thicker PWS layer as shown in Figure 12b.

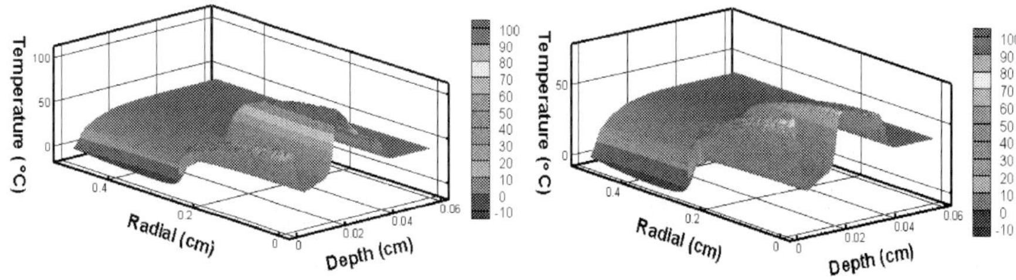

(a) Wavelength: 585nm (b) Wavelength: 595nm

Fig. 11. Temperature distributions within the skin at the end of 1.5 ms laser irradiation. (PWS layer thickness: 200 μm, laser fluence: 6 J/cm², with CSC)

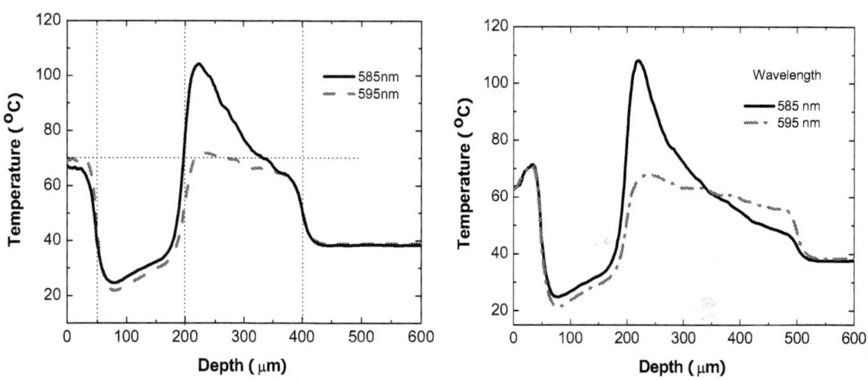

(a) 200 μm thick PWS layer (b) 300 μm thick PWS layer

Fig. 12. Temperature distribution along the tissue depth direction at the spray center for two wavelengths (585 and 595nm): PWS layer thickness (a) 200 μm and (b) 300 μm. (Laser fluence: 6 J/cm², pulse duration: 1.5 ms, with CSC)

5.4 Effect of laser pulse duration

The pulse duration of the laser beam is another important parameter that needs to be carefully chosen in clinic practice. Figure 13 shows the calculated temperature distributions within the skin at the end of laser irradiation for three pulse durations: 1.5 ms (a), 10 ms (b) and 40ms (c), respectively. The laser fluence is 6 J/cm². The comparison of the central temperature profile within skin for three cases is given correspondingly in Figure 13d.

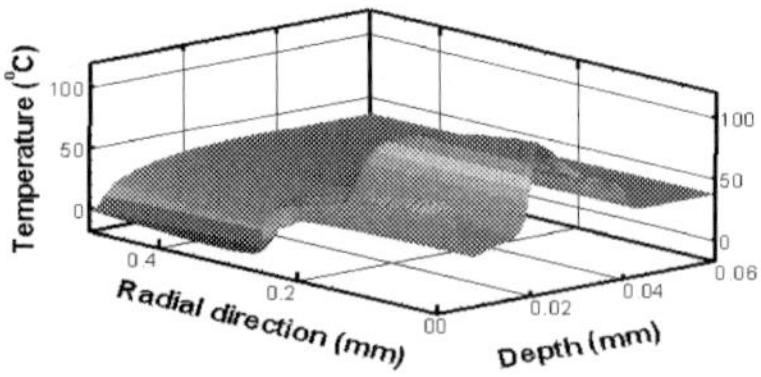

(a) Pulse duration: 1.5 ms

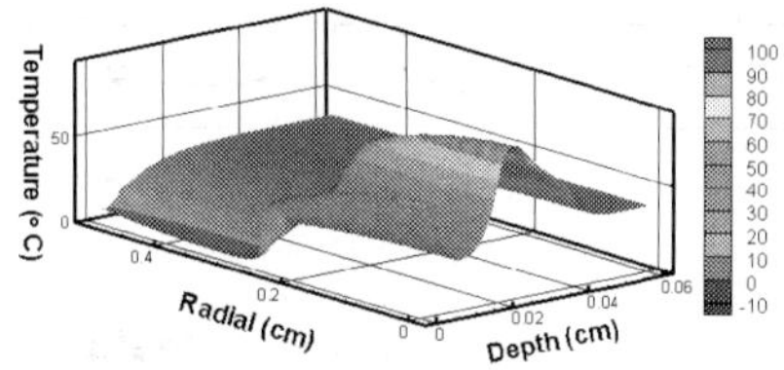

(b) Pulse duration: 10 ms

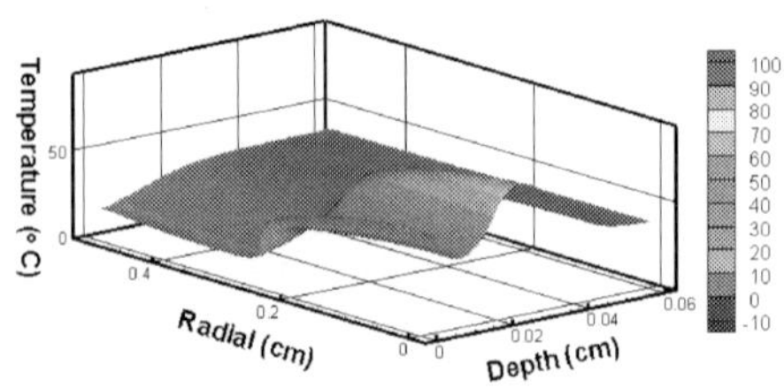

(c) Pulse duration: 40 ms

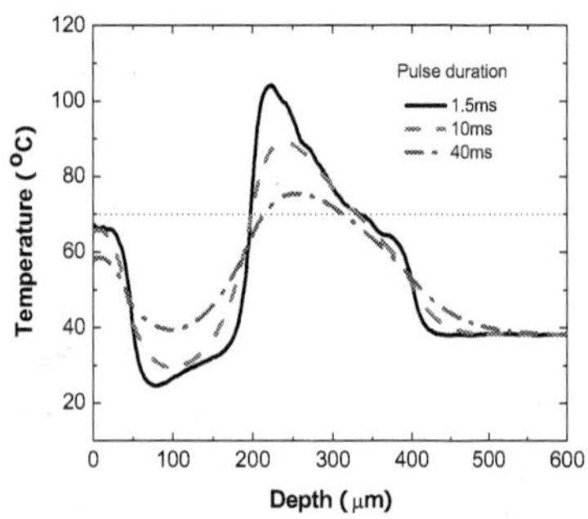

(d) Central temperature profile

Fig. 13. Calculated temperature distributions within the skin at the end of laser irradiation for three pulse durations: (a) 1.5 ms, (b) 10 ms and (c) 40 ms; (d) Comparison of the central temperature profiles for three pulse durations. (Laser fluence: 6 J/cm^2 with CSC)

Inspecting these plots finds immediately that the peak temperature of the PWS layer at the end of laser irradiation shows a continuous reduction, from 105 °C to 73 °C as the pulse duration increases from 1.5 ms to 40 ms. In the case of a short pulse duration (e.g., 1.5 ms), the PWS layer is heated up quickly at the end of laser irradiation with little heating of the neighbor dermal tissue. A significant portion of the PWS layer is heated up over the critical coagulation temperature of 70 °C. When the pulse duration increases to 40 ms, not only the peak temperature of the PWS layer reduces to a lower value, the percentage of the PWS layer that is above the critical coagulation temperature is also significantly reduced to a smaller portion. Meanwhile, the neighbor dermal tissue is significantly heated up to close to the coagulation temperature, which may lead to the damage of the healthy tissues. Such a variation in the peak temperature of the PWS with pulse duration can be understood by considering the combined effect of the laser heating and heat conduction of the heated PWS to the neighbor colder dermis. As the laser heating prolongs over a long pulse, the heat conduction from the heated PWS layer to the surrounding colder dermal tissues prevents a further increase in the PWS temperature and thus reduces the peak PWS temperature. A longer laser heating time is also associated with more energy through conduction into the surrounding tissues, leading to continuous increase in the temperature of the dermal tissues. In clinic practice of laser PWS, a short laser pulse is usually preferred, except for the cases with extremely large blood vessels. For that case, however, a better model is needed to provide more quantitative description of the laser surgery process of PWS.

6. Conclusion and future work

In this chapter, we present a brief review of thermal modelling of the treatment of port wine stains with the pulsed dye laser. We show that laser treatment of port wine stains is primarily a thermal issue involving both radiative energy transport within the tissue during laser irradiation and tissue heat conduction during and after laser irradiation. Based on simplified skin models that reduce the complex anatomic structure of skins to simple layer structures, the process can be successfully simulated by solving the corresponding radiative energy transport with the multi-layer Monte-Carlo method and the heat conduction equation with traditional numerical methods. We have used a simple multi-layer homogeneous model to illustrate the basic thermal characteristics of laser treatment of PWS. We also demonstrated that the model can be used to make selections of the laser parameters such as wavelength and pulse width in clinical practice. Quantitative information for critical surface cooling technique, CSC, is also presented and included in our model.

Although great progresses have been achieved in both clinic practice and physical understanding of laser PWS after four decades' efforts, many issues remain. Clinically, the present protocol of PDL-based lasers could significantly eliminate the PWS vessels, but only less than 20% of complete clearance of the PWS has been achieved (Kelly et al., 2005). Recurrence has been observed with a rate up to 50% after five years (Orten et al., 1996). All these suggest a lack of fundamental understanding of the PWS destruction mechanisms in the present laser PWS process. From the modeling point of view, neither the multi-layer homogeneous model nor the discrete blood vessel model provides accurate representation of the real and complex anatomic configuration of the PWS vessels. Attempts to construct realistic PWS structure based on computer-reconstructed biopsy from PWS patients had only limited success (Pfefer et al., 1996). New models are desired that should combine the simplicity of the multi-layer homogeneous model while take into account the detailed effect of complex PWS configurations. In addition, quantitative predictions of the temperature change of the PWS in the laser treatment require accurate optical and thermal properties of PWS, which are scarce at the moment.

The ultimate objective of any model for laser PWS is to accurately predict the thermal damage after the laser irradiation. The existing PWS damage model is a pure thermal model based on simple Arrhenius rate process integral (Pearce & Thomsen, 1995). The model does not take into account the photochemical and photomechanical effect of laser on skin tissues and blood vessels. Recent experimental evidence suggests that the vessel damage in laser PWS is a multi-time scale phenomenon. The collateral damage of blood vessels in laser PWS is due to accumulative result of early photothermal effect and later photochemical and photomechanical effect. The recurrence of PWS involves a time scale that may last to more than five years. Active researches are being conducted to understand these long term phenomena in laser PWS.

7. Acknowledgments

We like to acknowledge valuable discussions with Drs. Y.X. Wang and Z.Y. Ying at Laser Cosmetic Centre of 2nd Hospital of Xi'an Jiaotong University. Special thanks to Prof. Guo Lie-jin, Prof. Chen Bin, Prof. Wang Yue-she, Dr. Zhou Zhi-fu and Dr. Wu Wen-juan for their help to the project. G.-X. Wang thanks the support of "Changjiang Scholar" program of Education Ministry of China. The work is supported in part by the special fund from the State Key Laboratory of Multiphase Flow in Power Engineering at Xi'an Jiaotong University.

8. References

Aguilar, G.; Diaz, S.H.; Lavernia, E.J. & Nelson, J.S. (2002). Cryogen spray cooling efficiency: improvement of port wine stain laser therapy through multiple-intermittent cryogen spurts and laser pulses. *Lasers in Surgery and Medicine*, Vol.31, No.4, (July 2002), pp. 27-35, ISSN 0196-8092

Aguilar, G.; Wang, G.-X. & Nelson, J.S. (2003a). Effect of spurt duration on the heat transfer dynamics during cryogen spray cooling. *Phys Med Biol*, Vol.48, No.14, (July 2003), pp. 2169-2181, ISSN 0031-9155

Aguilar, G.; Wang, G.-X. & Nelson, J.S. (2003b). Dynamic Cooling Behavior during Cryogen Spray Cooling: Effects of Spurt Duration and Spray Distance. *Lasers in Surgery and Medicine*, Vol. 32, No.2, (February 2003), pp. 152-159, ISSN 0196-8092

Anderson, R.R. (1997). Laser-tissue interaction in dermatology, In: *Lasers in cutaneous and aesthetic surgery*, K.A. Arndt (Ed.), pp. (26-51), Lippincott Williams and Wilkins, ISBN 0316051772, Boston, Massachusetts, USA

Anderson, R.R. & Parrish, J.A. (1983). Selective photothermolysis: precise microsurgery by selective absorption of pulsed radiation. *Science*, Vol.220, No.4596, (April 1983), pp. 524–527, ISSN 0036-8075 (Print), 1095-9203 (Online)

Alora, M. B. T. & Anderson, R. R. (2000). Recent developments in cutaneous lasers. *Lasers in Surgery and Medicine*. Vol.26, No.2, (April 2000), pp. 108–118, ISSN 0196-8092

Alper, J.C. & Holmes, L.B. (1983). The incidence and significance of birthmarks in a cohort of 4641 newborns. *Pediatr Dermatol.*, Vol. 1, No.1, (July 1983), pp. 58-68, ISSN 1525-1470 (Online)

Baumler, W.; Vogl, A.; Landthaler, M.; Waner, M. & Shafirstein, G. (2005). Port wine stain laser therapy and the computer-assisted modeling of vessel coagulation using the finite elements method. *Medical Laser Application*, Vol.20, No.4, (December 2005), pp. 247-254, ISSN 1615-1615

Brannon, H. (April 2007). Skin Anatomy, In: *About.com*, 10.04.2011, Available from: http://dermatology.about.com/cs/skinanatomy/a/anatomy.htm

Dixon, J.A.; Huether, S.; & Rotering, R. (1984). Hypertrophic scarring in argon laser treatment of port wine stains. *Plast Reconstruct Surg*, Vol.73, No.5, (May 1984), pp. 771-780, ISSN 0032-1052 (Print), 1529-4242 (Online)

Franco W.; Wang, G.-X.; Nelson, J.S. & Aguilar, G. (2004). Radial Heat Transfer Dynamics during Cryogen Spray Cooling. *Proceedings of IMECE 2004*, IMECE2004-59609, ASME, ISBN 0-7918-4711-X

Franco, W.; Liu, J.; Wang, G.-X.; Nelson, J.S. & Aguilar, G. (2005). Radial and temporal variations in surface heat transfer during cryogen spray cooling. *Phys Med Biol*, Vol.50, No.2, (January 2005), pp. 387-397, ISSN 0031-9155

Gemert, M.J.C. van & Hulsbergen, J.P.H. (1981). A model approach to laser coagulation of dermal vascular lesions. *Arch Dermatol Res*, Vol.270, No.4, (February 1981), pp. 429-439, ISSN: 0340-3696 (Printed) 1432-069X (Online)

Gemert, M.J.C. van; de Kleijn, W.J.A. & Hulsbergen J.P.H. (1982). Temperature behaviour of a model port wine stain during argon laser coagulation. *Phys Med Biol*, Vol.27, No.9, (September 1982), pp. 1089–1104, ISSN 0031-9155

Gemert, M.J.C. van; Welch, A.J.; & Amin, A.P. (1986). Is there an optimal laser treatment for port wine stains? *Lasers in Surgery and Medicine,* Vol.6, No.1, (January 1986), pp. 76-83, ISSN 0196-8092

Gemert, M.J.C. van; Welch, A.J.; Tan, O.T.; & Parrish, J.A. (1987). Limitations of carbon dioxide lasers for treatment of port-wine stains. *Arch Dermatol,* Vol.123, No.1, (January 1987), pp. 71-73, ISSN (Printed): 0340-3696 (Online): 1432-069X

Gemert, M.J.C. van; Welch, A.J.; Pickering, J.W. & Tan, O.T. (1995). Laser Treatment of Port Wine Stains, In: *Optical-Thermal Response of laser-Irradiated Tissue,* A.J. Welch & M.J.C. van Gemert, (Ed.), 789-829, Plenum Press, ISBN 0-306-44926-9, New York

Gemert, M.J.C. van; Smithies, D.J.; Verkruysse, W.; Milner, T.E. and Nelson, J.S. (1997). Wavelengths for port wine stain laser treatment: influence of vessel radius and skin anatomy. *Phys Med Biol,* Vol.42, No.1, (January 1997), pp. 41–50, ISSN 0031-9155

Gilchrest, B.A.; Rosen, S. & Noe, J.M. (1982). Chilling port wine stains improves the response to argon laser therapy. *J Plast Reconstr Surg,* Vol.69, No.2, (February 1982), pp. 278-283, ISSN 0032-1052 (Print), 1529-4242 (Online)

Hammes, S. & Raulin, C. (2005). Evaluation of different temperatures in cold air cooling with pulsed-dye laser treatment of facial telangiectasia. *Lasers in Surgery and Medicine,* Vol.36, No.2, (February 2005), pp. 136–140, ISSN 0196-8092

Ishimaru, A. (1989). Diffusion of light in turbid media. *Appl Opt,* Vol.28, No.12, (June 1989), pp. 2210-2215, ISSN 1559-128X (Print) 2155-3165 (Online)

Jacques, S.L. & Wang, L.H. (1995). Monte Carlo Modeling of Light Transport in Tissues, In: *Optical-Thermal Response of laser-Irradiated Tissue,* A.J. Welch & M.J.C. van Gemert, (Ed.), 73-100, Plenum Press, ISBN 0-306-44926-9, New York

Jia, W.C.; Aguilar, G.; Wang, G.-X. & Nelson, J.S. (2004). Heat transfer dynamics during cryogen spray cooling of substrate at different initial temperatures. *Phys Med Biol,* Vol. 49, No.23, (December 2004), pp. 5295-5308. ISSN 0031-9155

Jia, W.; Aguilar, G.; Verkruysse, W.; Franco, W. & Nelson, J.S. (2006). Improvement of port wine stain laser therapy by skin preheating prior to cryogen spray cooling: a numerical simulation. *Lasers in Surgery and Medicine,* Vol.38, No.2, (February 2006), pp. 155-162. ISSN 0196-8092

Jia, W.C.; Choi, B.; Franco, W.; Lotfi, J.; Majaron, B.; Aguilar, G.; & Nelson, J.S. (2007). Treatment of cutaneous vascular lesions using multiple-intermittent cryogen spurts and two-wavelength laser pulses: numerical and animal studies. *Lasers in Surgery and Medicine,* Vol.39, No.6, (July 2007), pp. 494-503, ISSN 0196-8092

Keijzer, M.; Pickering, J.W. & van Gemert, M.J.C. (1991). Laser beam diameter for port wine stain treatment. *Lasers in Surgery and Medicine,* Vol.11, No.6, (October 1991), pp. 601-605, ISSN 0196-8092

Kelly, K.M.; Choi B.; McFarlane, S; *et al.* (2005). Description and analysis of treatments for port-wine stain birthmarks. *Arch Facial Plast Surg,* Vol.7, No.5, (October 2005), pp. 287–94, ISSN 1521-2491

Kienle, A. & Hibst, R. (1995). A new optimal wavelength for treatment of port wine stains? *Phys Med Biol,* Vol.40, No.10, (October 1995), pp. 1559-76, ISSN 0031-9155

Kienle, A. & Hibst, R. (1997). Optimal parameters for laser treatment of leg telangiectasia. *Lasers in Surgery and Medicine,* Vol.20, No.3, (December 1998), pp. 346–353, ISSN 0196-8092

Lahaye C.T.W. & van Gemert, M.J.C. (1985). Optimal laser parameters for port wine stain therapy: a theoretical approach. *Phys Med Biol*, Vol.30, No.6, (June 1985), pp.573-87, ISSN 0031-9155

Lanigan, S. W. (2000). *Lasers in Dermatology*, Springer, London, ISBN 1852332778

Li, D.; He, Y.L.; Liu, Y.W. & Wang, G.-X. (2007a), Numerical analysis of cryogen spray cooling of skin in dermatologic laser surgery using realistic boundary conditions. *Proceedings of 22th Int. Congress Refrigeration conference*, (July) 2007, Beijing, China

Li, D.; He, Y.L.; Wang, G.-X.; Xiao, J. & Liu, Y.W. (2007b), Numerical analysis of cold injury of skin in cryogen spray cooling for dermatologic laser surgery. *Proceedings of 2007 ASME International Mechanical Engineering Congress*, IMECE2007-43876, Nov. 2007, Seattle, Washington, ISBN 0-7918-4302-5

Li, D.; Wang, G.-X. & He, Y.L. (2011), Criteria for section of voxel size and photon number in Monte Carlo Simulation, to be submitted

Lucassen, G.W.; Svaasand, L.O.; Verkruysse, W. & van Gemert, M.J.C. (1995). Laser energy threshold for thermal vascular injury in a port wine stain skin model. *Laser Med Sci*, Vol.10, No.4, (December 1995), pp. 231-234, 0268-8921 (Print), 1435-604X (Online)

Lucassen, G.W.; Verkruysse, W.; Keijzer, M. & van Gemer, M.J.C. (1996). Light distribution in a port wine stain model containing multiple cylindrical and curved blood vessels. *Lasers in Surgery and Medicine*, Vol.18, No.4, (April 1996), pp. 345-357, ISSN 0196-8092

Majaron, B.; Verkruysse, W.; Kelly, K.M.; & Nelson, J.S. (2001). Er:YAG laser skin resurfacing using repetitive long-pulse exposure and cryogen spray cooling: II. theoretical analysis. *Lasers in Surgery and Medicine*, Vol.28, No.2, (February 2001), pp. 131-137, ISSN 0196-8092

Miller, I.D. & Veith, A. R. (1993). Optical modelling of light distributions in skin tissue following laser irradiation. *Lasers in Surgery and Medicine*, Vol.13, No.5, (May 1993), pp. 565–571, ISSN 0196-8092

Nelson, J.S.; Milner, T.E.; Anvari, B.; Tanenbaum, B.S.; Kimel, S.; Svaasand, L.O.; & Jacques, S.L. (1995). Dynamic epidermal cooling during pulsed laser treatment of port-wine stain: A new methodology with preliminary clinical evaluation. *Arch Dermatol*, Vol.131, No.6, (June 1995), pp. 695–700, ISSN 1538-3652 (Print), 0003-987X (Online)

Niemz, M.H.; (1996). *Laser-Tissue Interactions: Fundamentals and Applications*, Springer Berlin Heidelberg, New York. Hardcover, ISBN 978-3-540-40553-5 Softcover, ISBN 978-3-540-72191-8

Orten S.S.; Waner M.; Flock S.; Roberson P.K.; Kincannon J. (1996). Port wine stains: an assessment of five years of treatment. *Arch Otolaryngol Head Neck Surg*, Vol.122, No.11, (November 1996), pp. 1174-9. ISSN 0886-4470 (Print) 1538-361X (Online)

Pearce, J. & Thomsen, S. (1995). Rate Process Analysis of Thermal Damage, In: *Optical-Thermal Response of laser-Irradiated Tissue*, A.J. Welch & M.J.C. van Gemert, (Ed.), 561-606, Plenum Press, ISBN 0-306-44926-9, New York

Pfefer, T.J.; Barton, J.K.; Chan, E.K.; Ducros, M.G.; Sorg, B.S.; Milner, T.E.; Nelson, J.S. & Welch, A.J. (1996). A three dimensional modular adaptable grid numerical model for light propagation during laser irradiation of skin tissue. *IEEE Journal of Selected in Quantum Electronics*, Vol.4, No.4, (December 1996), pp. 934–942, ISSN 1077-260X

Pfefer, T.J.; Smithies, D.J.; Milner, T.E.; van Gemert, M.J.C.; Nelson, J.S. & Welch, A.J. (2000). Bioheat transfer analysis of cryogen spray cooling during laser treatment of Port Wine Stains. *Lasers in Surgery and Medicine*, Vol.26, No.2, (February 2000), pp. 145-157, ISSN 0196-8092

Pickering, J.W.; Butler, P.H.; Ring, B.J. & Walker, E.P. (1989). Computed temperature distributions around ectatic capillaries exposed to yellow (578 nm) laser light. *Phys Med Biol*, Vol.34, pp. 1247-1258, ISSN 0031-9155

Pickering, J.W. & van Gemert, M.J.C. (1991). 585 nm for the laser treatment of port wine stains: A possible mechanism. *Lasers in Surgery and Medicine*, Vol.11, No.6, (June 1991), pp. 616-618, ISSN 0196-8092

Pikkula, B.M.; Torres, J. H.; Tunnell, J.W. & Anvari, B. (2001). Cryogen spray cooling: effects of droplet size and spray density on heat removal. *Lasers in Surgery and Medicine,* Vol. 28, No.2, (February 2001), pp. 103-112, ISSN 0196-8092

Shafirstein, G.; Bäumler, W.; Lapidoth, M.; Ferguson, S.; North, P.E. & Waner, M. (2004). A New Mathematical Approach to the Diffusion Approximation Theory for Selective Photothermolysis Modeling and Its Implication In Laser Treatment of Port-Wine Stains. *Lasers in Surgery and Medicine,* Vol. 34, No.4, (April 2004), pp. 335–347, ISSN 196-8092

Shafirstein, G. & Buckmiller, L.M.; Waner, M. & Bäumler, W. (2007). Mathematical modeling of selective photothermolysis to aid the treatment of vascular malformations and hemangioma with pulsed dye laser. *Lasers Med Sci*, Vol.22, No.2, (June 2007), pp. 111-118, ISSN 0268-8921(Print), 1435-604X (Online)

Smithies, D.J. & Butler, P.H. (1995). Modelling the distribution of laser light in port-wine stains with the Monte Carlo method. *Phys Med Biol*, Vol.40, No.5, (May 1995), pp. 701-33, ISSN 0031-9155

Tan, O.T.; Morrison, P. & Kurban, A.K. (1990). 585 nm for the treatment of port-wine stains. *Plast Reconstr Surg Med*, Vol.86, No.6, (December 1990), pp. 1112-1117, ISSN 0032-1052 (Print), 1529-4242 (Online)

Tunnell, J.W.; Nelson, J.S.; Torres, J.H.; & Anvari, B. (2000). Epidermal protection with cryogen spray cooling during high fluence pulsed dye laser irradiation: an ex vivo study. *Lasers in Surgery and Medicine,* Vol.27, No.4, (April 2000), pp. 373-383, ISSN 0196-8092

Tunnell, J.W.; Wang, L.V. & Anvari, B. (2003). Optimum pulse duration and radiant exposure for vascular laser therapy of dark portwine skin: A theoretical study. *Appl Opt*, Vol.42, No.7, (July 2003), pp. 1367–1378, ISSN 1559-128X (Print), 2155-3165 (Online)

Verkruysse, W.; Pickering, J.W.; Beek, J.F.; Keijzer, M.; van Gemert, M.J.C. (1993). Modeling the effect of wavelength on the pulsed dye laser treatment of port wine stains. *Appl Opt,* Vol.32, No.4, (February 1993), pp. 393–8, ISSN 1559-128X (Print), 2155-3165 (Online)

Wang, L.H.; Jacques, S.L. & Zheng, L.Q. (1995). MCML-Monte Carlo modeling of light transport in multi-layered tissues. *Comp Meth Prog Biol*, Vol.47, No.2, (July 1995), pp. 131-146, ISSN 0169-2607

Wilson, B.C. & Adam, G. (1983). A Monte-Carlo model for the absorption and flux distribution of light in tissue. *Med Phys*, Vol.10, No.6, (December 1983), pp. 824-830, ISSN 0094-2405

Zhou, Z.; Xin, H.; Chen, B. & Wang, G.-X. (2008a). Theoretical Evaporation Model of a Single Droplet in Laser Treatment of PWS in Conjunction with Cryogen Spray Cooling. *Proceedings of 2008 ASME Summer Heat Transfer Conference*, HT2008-56063, Aug. 10-14, 2008, Hyatt Regency Riverfront, Jackonsville, Florida, ISBN: 978-0-7918-4849-4

Zhou, Z.; Xin, H.; Chen, B. & Wang, G.-X. (2008b). Single Droplet Evaporation Model in Laser Treatment of PWS in Conjunction with Cryogen Spray Cooling. *Proceedings of 2008 International Conference on Bio-Medical Engineering and Informatics*, Vol. 1, pp. 551-556, May 28-30, Sanya, Hanan, China, ISBN: 978-0-7695-3118-2

28

Study of the Heat Transfer Effect in Moxibustion Practice

Chinlong Huang and Tony W. H. Sheu
Dept of Engineering Sciences and Ocean Engineering, National Taiwan University, No. 1, Sec. 4, Roosevelt Road, Taipei, 10617 Taiwan

1. Introduction

"First you use the needle (acupuncture), then the fire (moxibustion), and finally the herbs" (Tsuei, 1996) has been well known in traditional Chinese medicine (TCM). In fact, moxibustion has played an important role in Asia for many years (Zhang, 1993). In Huang Di Nei Jing (Maoshing (translator), 1995), we can find that when needle can't do a job, moxa is a better choice. Moxibustion rather than acupuncture was commonly known to be able to alleviate pains due to some severe diseases, manifested by vacuity cold and Yang deficiency. In clinical studies, many experiments have confirmed that moxibustion is capable of enhancing immunity, improving circulation, accommodating nerve, elevating internal secretion and adjusting respiration, digestion and procreation et al. (Wu et al., 2001; Liu, 1999). However, moxibustion has not been accepted as the modern therapy because of the lack of standard practice procedures. In addition, moxibustion is subject to the danger of scalding patients. More effort needs therefore to be made so as to increase our knowledge about the moxibustion and, hopefully, these research endeavors can be useful for the future instrumentation and standardization of the moxibustion by some emerging modern scientific techniques.

The existing moxibustion techniques can be separated into the direct and indirect moxibustion therapies. In direct moxibustion, the ignited cone-shaped moxa is normally placed on the skin surface near acupoints (Fig. 1(a)). Direct moxibustion can be further categorized into the scarring and non-scarring two types. During the scarring moxibustion, the ignited moxa is placed on the top of an acupoint till a time it burns out completely. This moxibustion type may lead to a localized scarring or blister. In non-scarring moxibustion, moxa cone is also burned directly on the skin. Such an ignited moxa will be removed when it may cause an intense pain (moxa temperature should be under 60 °C). Usually, this treatment will result in a small red circular mark on the local area of the skin surface.

Indirect moxibustion becomes more popular currently because of its lower risk of leading to pain or burning. A common way of administering the therapeutic properties of moxibustion is to place, for example, a piece of ginger, garlic, salt or pepper in between the burning moxa and the skin. One can also ignite a moxa stick, which is placed at a location that is closed to but not in contact with the proper acupoint (about 2cm to skin surface normally), for several minutes until the color of the skin surface near this acupoint turns red.

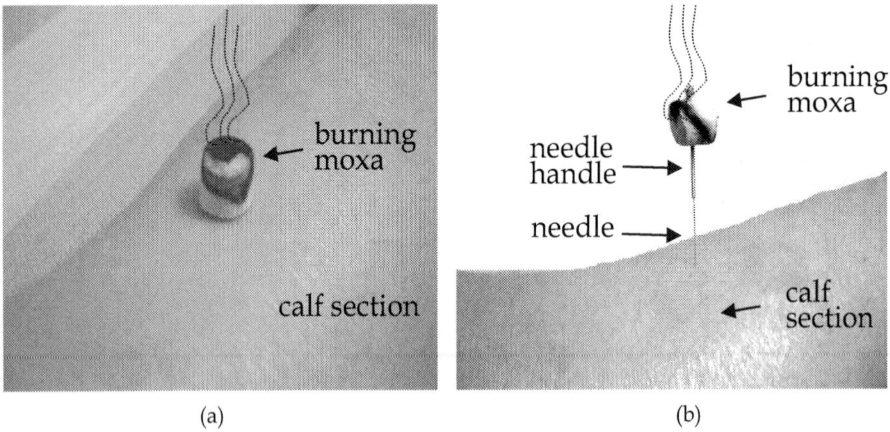

Fig. 1. Schematic of the moxibustions. (a) Direct moxibustion therapy; (b) Warm needle moxibustion therapy

Another indirect moxibustion involves the use of needles and moxa. One needle, along which there is a moxa, is inserted into the skin near an acupoint. The moxa cone placed on the inserted needle is then ignited. The heat generated from the burning moxa will propagate through the needle and transfers to the acupoint by heat conduction. During the treatment, a dried moxa is in contact with the handle of the acupuncture needle after the needle being inserted into the acupuncture point. This is followed by igniting the moxa and keeps it burning (Fig. 1(b)). Typically, the distance between the skin surface and the burning moxa stick is about 2 cm. Heat will be conducted from the needle handle to the needle itself and, finally, to the surrounding tissues. This acupuncture design with a burning moxa can result in a certain temperature gradient across the needle and enhances thus the Seebeck effect (Cohen, 1997). In Chinese medicine theory, this method is highly recommended for use to the patients with vacuity cold and wind damp (Wiseman, 1998) because of its functions of warming the meridians and promoting the qi- and blood-flow. This therapy is also applicable to release the cold-damp syndrome for the patients with rheumatoid arthritis (Li, 1999). The other technique is called as the fire needle, which involves holding the needle in a lamp flame until it becomes very hot. Afterwards, the needle is inserted to the appropriate depth in the body quickly and it will be removed later on (Unschuld, 1988). In comparison with the fire needle, the warming needle permits a longer retention and a gentler heating.

In the present study, our aim is to study two types of the moxibustion effect, which are the direct moxibustion and the warm needle moxibustion therapy. The acupoint GB 38 shown in Fig. 2 is one of the acupoints in gall bladder (GB) meridian, which has an association with the hemicrania and joint ache. Figure 3 shows the axial image of the right leg for the GB38 acupoint (Courtesy of Yang (1997)). At the calf section, the number of capillaries near the GB acupoints is greater than those at the other parts of the body (Fei, 2000). Also, the distances between the three acupoints GB37-GB38 and GB38-GB39 have about an inch. Therefore, the acupoint GB38 was also the focus of other investigations (Sheu and Huang, 2008; Huang and Sheu, 2008; Huang and Sheu, 2009).

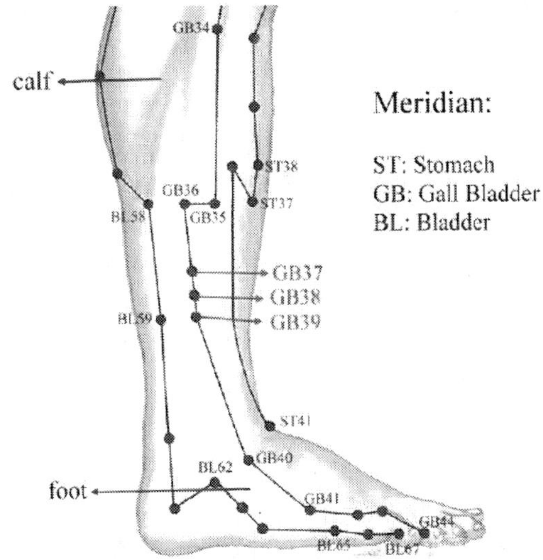

Fig. 2. Schematic of the stomach, gall bladder and bladder meridians. Note that GB38 is the acupoint under current investigation (Chang, 1999)

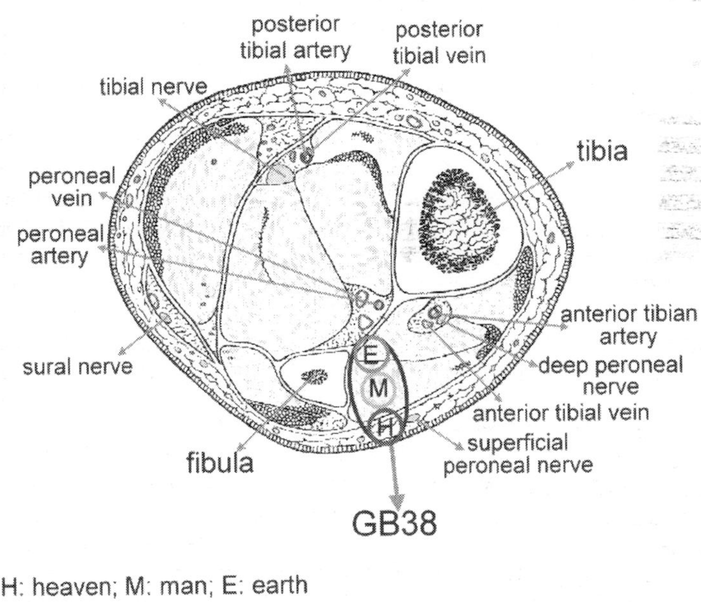

Fig. 3. One axial image of the right leg that contains the GB38 acupoint (Courtesy of Yang 1997). Heaven, man and earth are three depths beneath of the skin surface

2. Materials and methods

Heat transfer process will be modelled by solving the energy conservation equation. In this study, the energy equation cast in the following form for the total enthalpy will be employed

$$\frac{\partial(\rho h_0)}{\partial t} = \nabla \cdot (k \nabla T) \tag{1}$$

In the above equation, h_0, k, ρ and T denote the internal energy, thermal conductivity, density, and temperature, respectively.

Simulation of equation (1) will be carried out by employing the commercially available finite volume package, namely, CFDRC (CFD-ACE-GUI, 2003). This software package provides the modules CFD-GEOM for grid generation, CFD-ACE$^+$ for solution solver, and CFD-VIEW for post-processing. A convenient graphical user interface (GUI) is also available for us to specify the physical properties of the medium under investigation, and the specification of the boundary and initial conditions. In CFD-ACE$^+$ solver, the finite volume method employed together with the algebraic multigrid method and the conjugate gradient squared solution solver accelerates calculation. In this study, the central difference scheme is chosen to approximate the parabolic type partial differential equation.

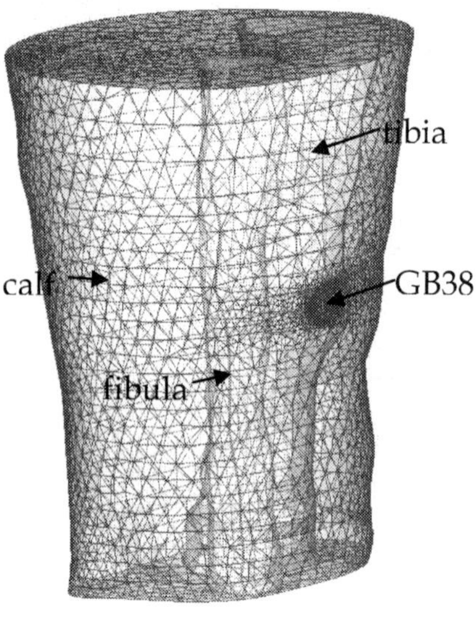

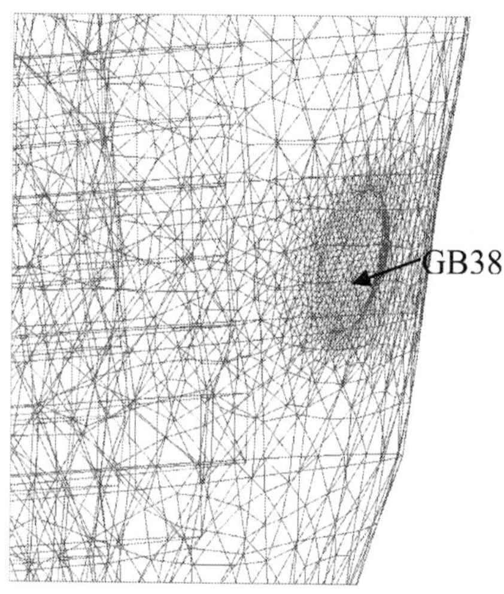

Fig. 4. (a) The generated surface meshes for the investigated calf section; (b) The zoom-in area around GB38

A hybrid mesh, containing the structured- and unstructured-type meshes, shown in Fig. 4, is generated from a total number of 17,000 nodal points. These mesh points have been properly distributed so that the predicted solutions can more accurately represent the physical phenomenon. In all the investigations, calculation of the enthalpy will be terminated when the residual norms fall below 10^{-15}. As the number of the employed nodal points (with 17,000 nodal points) is increased by 50% (25,500 nodal points), only a negligibly small difference is seen in the simulated results. On the contrary, when the number of the mesh points is decreased by 50% (or 8,500 nodal points) the computed difference shows an apparent difference. Hence the mesh generated by 17,000 nodal points will be employed in the present simulation.

In order to know the temperature distribution due to the ignited moxa, the IR images are collected under the almost dark condition. Both of the camera and the subject are kept apart from the external draft IR source. The room temperature and the relative humidity are kept at 22 ± 0.1 °C and 60 ± 7%, respectively. Thermography is taken using a calibrated IR camera (ThermaCAM® SC500 from FLIR Systems TM) (ThermaCAM SC500 Operator's Manual), which is equipped with a 45°close-up optic. Sensitivity, accuracy and resolution of the employed camera are kept at 0.07 °C, ± 2 °C and 320 × 240 pixels, respectively. The distance between the camera and the subject under current investigation is 0.1 m. The infrared images of the subject obtained at a sampling rate of 4 Hz will be directly recorded in the computer's hard disk.

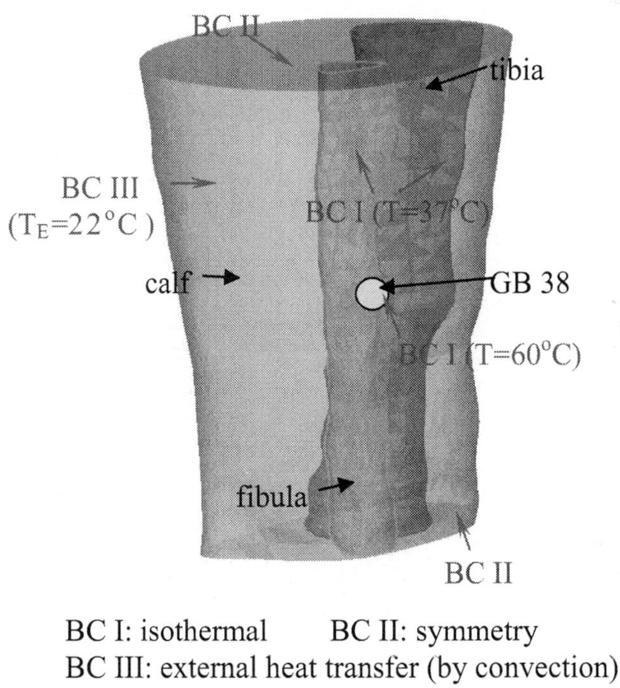

Fig. 5. Schematic of the calf section around the acupoint GB 38 and the specified boundary conditions

3. Results and discussion

In the present study, direct moxibustion and warm needle moxibustion treatments are described below:

(A) Direct moxibustion therapy

In direct moxibustion therapy, a burned moxa (moxa cone of 1 cm diameter and 1cm length, made by the dried Artemisia vulgaris leaves with a weight of 100 mg), was placed on acupoint GB38 shown in Fig. 4. The boundary conditions applied at the calf section with tibia and fibula are shown in Fig. 5, where BC I, II and III represent the isothermal, symmetric and heat transfer (by convection) boundary conditions, respectively. In BC III, the wall subtype is chosen to account for the transfer of heat to/from the external environment (i.e., the area outside of the computational system) by heat convection. On the boundary BC III, it is specified by the condition

$$q_w = h_c(T_E - T_w) \tag{2}$$

where h_c and T_E denote the heat transfer coefficient and the environment temperature. The wall temperature (T_w) is determined by balancing the heat fluxes between the environment and the skin surface.

The heat transfer coefficient of the skin surface, specific heat and the density of tissues in the investigated calf section are denoted as h_c, Cp_c and ρ_c, respectively. At the normal state, these coefficients are prescribed respectively with h_c = 3.7 W/m²°C (Nishi and Gagge, 1970), Cp_c = 3,594 J/kg°C (Blake et al., 2000) and ρ_c = 1,035 kg/m³. The thermal conductivity of the human tissues is assumed to change with the temperature (T) by the equation k_c = 0.840419+ 0.001403T W/m°C (Mura et al., 2006). The rest of the employed coefficients are tabulated in the Table 1.

k_s	16 W/m °C	Thermal conductivity of the stainless steel
k_c	0.840419+0.001403T W/m °C	Thermal conductivity of the calf
k_a	0.0299 W/m °C	Thermal conductivity of the air
Cp_s	460 J/kg °C	Specific heat of the stainless steel
Cp_c	3,594 J/kg °C	Specific heat of the calf
Cp_a	1,009 J/kg °C	Specific heat of the air
ρ_s	7,800 kg/m³	Density of the stainless steel
ρ_c	1,035 kg/m³	Density of the calf
ρ_a	1.0 kg/m³	Density of the air
h_s	7.9 W/m² °C	Heat transfer coefficient of the stainless steel
h_c	3.7 W/m² °C	Heat transfer coefficient of the calf
T_E	22 °C	Environment temperature

Table 1. Summary of the coefficients and the prescribed temperatures in the current simulation

The predicted temperature on the skin surface is plotted in Fig. 6. The moxa temperature on the skin surface is specified at 60 °C (non-scarring direct moxibustion), which is the highest temperature that our skin can possibly endure, to avoid scarring. On the tibia and fibula, their temperatures are given to be T= 37 °C, which is the same as the normal human body temperature. One can find from the upper and bottom planes of the simulated domain that the temperature near tibia and fibula has a larger value (Fig. 6). Since moxibustion takes place at a location near GB38, the area around this acupoint has a higher temperature. Figure 7 shows the predicted skin surface temperature that is distributed in a form similar to the experimentally measured temperature by IR image.

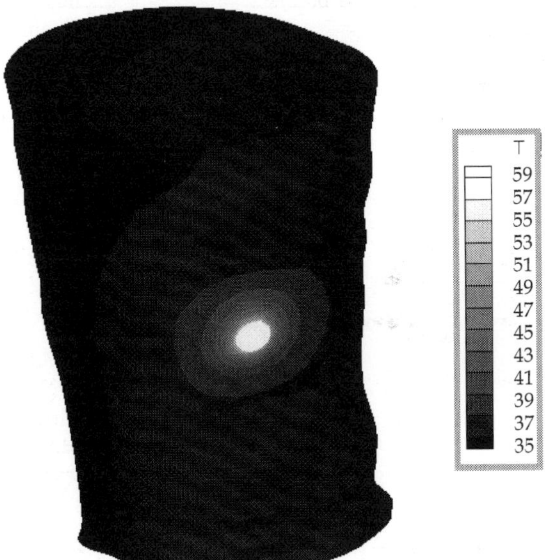

Fig. 6. The predicted skin surface temperature contours subject to the moxibustion

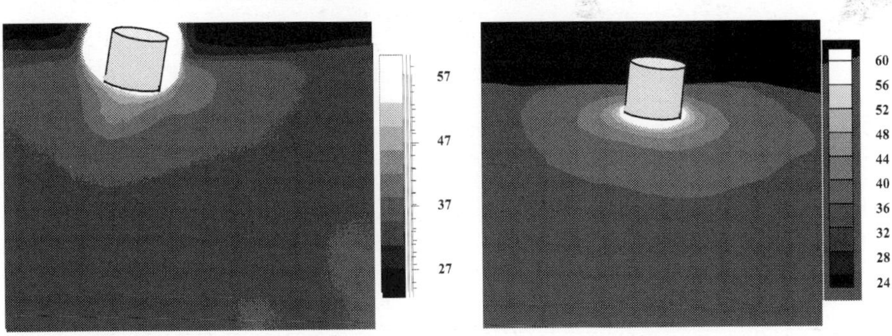

(a) (b)

Fig. 7. Comparison of the predicted and measured temperatures. (a) Measured by the IR image recording system; (b) Predicted by the numerical simulation

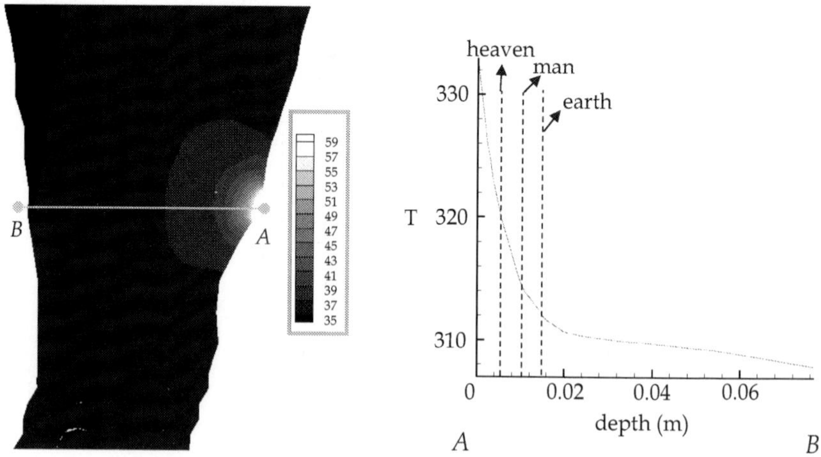

(a) (b)

Fig. 8. (a) The predicted temperature contours on the cutting plane that passes through the acupoint GB38 when the moxa temperature on the skin surface is 60 °C; (b) The predicted temperature profile along the line connected by two points *A* and *B* shown in Fig. 8(a). The "heaven", "man" and "earth" represent three depths of the investigated acupoints, respectively

In TCM, there is one concept, "When using acupuncture, shallow position treats diseases nearby; deep position treats diseases far away". Huang Di Nei Jing (Maoshing (translator), 1995) supported the similar concept that at the same acupoint shallow position treats mild diseases; deep position treats severe diseases. This enlightens that at the same acupoint different depths have different effects and even have an association with different diseases. For this reason, this study makes an effort to get the temperature distributions at different depths. Figure 8 (a) shows the temperature contours predicted at the plane of the acupoint GB38 for the case that the moxa temperature on the skin surface is 60 °C. From Fig. 8 (b), one can see the predicted temperature profile along a line that is connected by two nodes *A* and *B*. For the acupoints "heaven", "man" and "earth", they have three different depths at a location in between the skin surface and the associated connective tissue. When the moxa temperature is controlled at 60 °C, which is the temperature considered in the case of non-scarring moxibustion, the temperatures at the "heaven" (~ 0.5 cm beneath of the skin surface), "man" (~ 1.0 cm beneath of the skin surface) and "earth" (~ 1.5 cm beneath of the skin surface) are predicted as T^{60}_{heaven} = 47.8 °C, T^{60}_{man} = 41.7 °C and T^{60}_{earth} = 39.0 °C. From the skin surface to "heaven", we found that temperature decreases faster (12.2 °C in between) than those from the "heaven" to "man" (6.1 °C in between) and from the "man" to "earth" (2.7 °C in between) as well because our body has a bigger thermal capacity than that of the moxa. As a result, the temperature variation on the side of "heaven" is greater than that along the "earth". The relation between the predicted temperature (T) and the depth (x) can be expressed as

$$T = 60 - 5.31\times10^{-2}x + 5\times10^{-5}x^2 - 3\times10^{-8}x^3 + 7\times10^{-12}x^4 - 1\times10^{-15}x^5 + 6\times10^{-20}x^6 \quad (3)$$

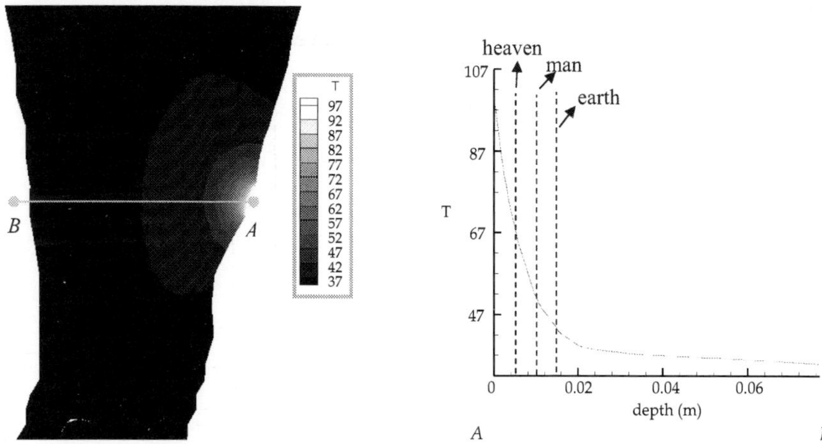

(a) (b)

Fig. 9. (a) The predicted temperature contours on the cutting plane that passes through the acupoint GB38 when the moxa temperature on the skin surface is 100 °C; (b) The predicted temperature profile along the line connected by two points *A* and *B* shown in Fig. 9(a). The "heaven", "man" and "earth" represent three depths of the investigated acupoints, respectively

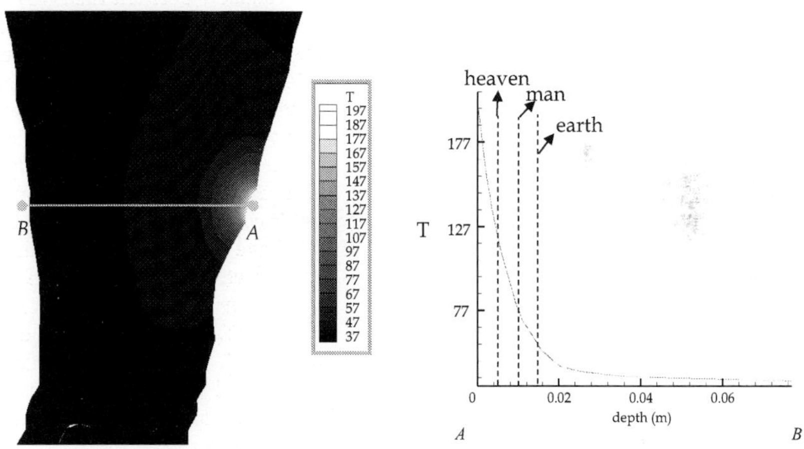

(a) (b)

Fig. 10. (a) The predicted temperature contours on the cutting plane that passes through the acupoint GB38 when the moxa temperature on the skin surface is 200 °C; (b) The predicted temperature profile along the line connected by two points *A* and *B* shown in Fig. 10(a). The "heaven", "man" and "earth" represent three depths of the investigated acupoints, respectively

We then consider the scarring moxibustion by specifying the moxa temperatures at two higher temperatures 100 and 200 °C. The temperatures at three locations are predicted to be T_{heaven}^{100} = 68.1°C, T_{man}^{100} = 51.3°C and T_{earth}^{100} = 43.5°C; T_{heaven}^{200} = 116.0°C, T_{man}^{200} = 72.8°C and T_{earth}^{200} = 53.4°, respectively. The predicted temperature contours on the plane of acupoint GB38 and the temperature profile along the line connected by nodes A and B are shown in Fig. 9 and Fig. 10. The relation of the predicted temperature (T) and the depth (x) is expressed by the equations given below

$$T = 100 - 1.374\times10^{-1}x + 1\times10^{-4}x^2 - 6\times10^{-8}x^3 + 2\times10^{-11}x^4 - 2\times10^{-15}x^5 + 1\times10^{-19}x^6 \quad (4)$$

$$T = 200 - 3.48\times10^{-1}x + 3\times10^{-4}x^2 - 2\times10^{-7}x^3 + 4\times10^{-11}x^4 - 6\times10^{-15}x^5 + 3\times10^{-19}x^6 \quad (5)$$

According to equations (3), (4) and (5) for the burning moxa cones with the temperatures at 60, 100 and 200 °C, the temperatures at the "heaven" are decreased to 47.8, 68.1 and 116.0 °C, respectively. In the scarring type (moxa temperatures are 100 and 200 °C), the temperature decrease from the skin surface to the "heaven" is faster (31.9 and 84 °C in between the skin surface and the "heaven" position) than that of the non-scarring type (12.2 °C in between the skin surface and the "heaven" position) since our human body has a larger thermal capacity than the moxa. The temperature, as a result, will quickly reach the human body's temperature (37 °C). In summary, the higher the moxa temperature, the more rapid temperature decrease will be.

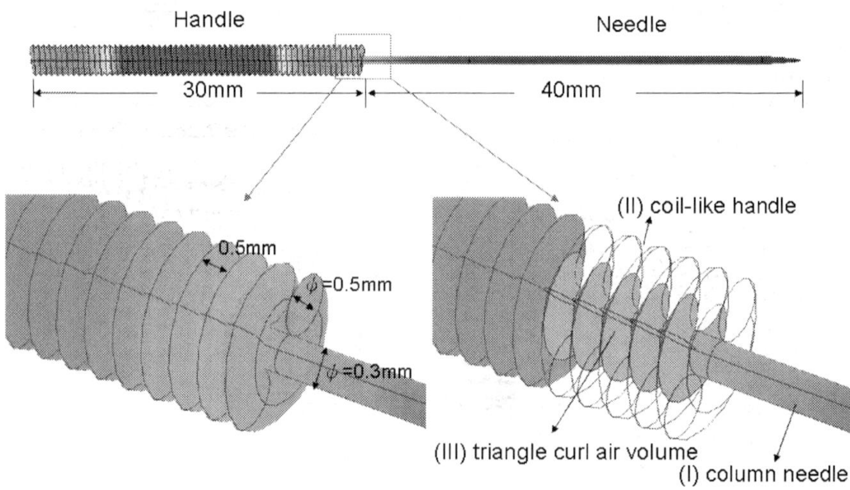

Fig. 11. Acupuncture needle is composed of one-column needle (portion I) and one coil-like handle (portion II), which covers the column needle

(B) Warm needle moxibustion therapy

In warm needle moxibustion, a needle assembly for the acupuncture includes a column needle (portion I) with its head having a sharp end and one coil-like handle (portion II), which covers the dull end of the column needle (Fig. 11). Between the column needle and the coiled handle, there is a triangle air volume (portion III). Figure 11 shows the investigated 1.5-inch acupuncture needle. A hybrid mesh system, containing both of the structured- and unstructured-type meshes shown in Fig. 12, is generated from a total number of 7,421 mesh points. The mesh density has been properly distributed so that the predicted solutions can accurately represent the physical phenomenon.

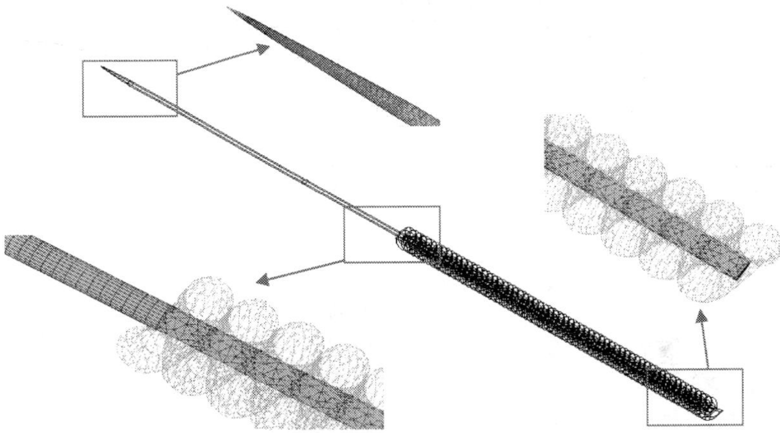

Fig. 12. The surface mesh points generated on the needle for acupuncture use

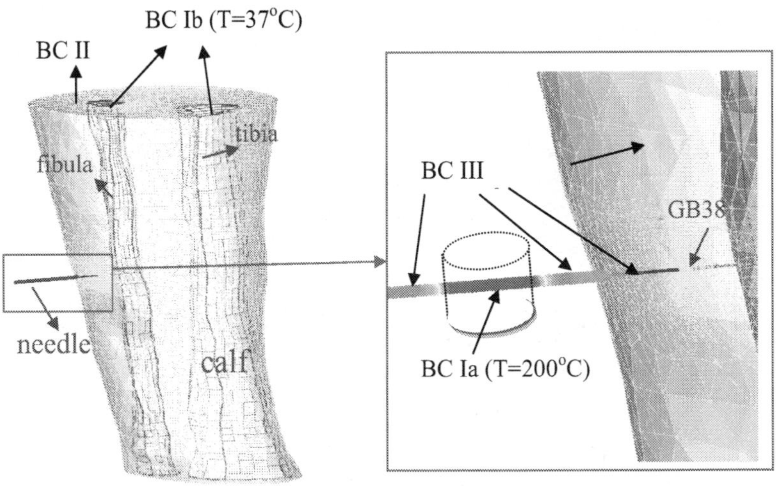

BC Ia: isothermal (T=200 °C) BC Ib: isothermal (T=37 °C)
BC II: symmetry BC III: external heat transfer (by convection)

Fig. 13. Schematic of the boundary condition types for the acupuncture needle inserted to the acupoint GB38

The boundary conditions are applied on the calf section with the needle shown in Fig. 13, where BC I, II and III represent the isothermal, symmetric and convection types of boundary conditions, respectively. BC Ia shows the boundary of needle handle which has a moxa. The temperatures in the bone zone (BC Ib), including both fibula and tibia, are specified as 37 °C to simulate the human body temperature. The symmetric conditions at the upper and bottom sides of the calf section (BC II), which connects knee and foot, are specified to account for the same structures of muscle tissues with knee and foot. BC III is normally chosen to simulate the transfer of heat to/from the surrounding (i.e., the area outside of the computational system) by convection. This subtype is used to fix either the wall temperature or the heat flux. On the wall of BC III, it is prescribed by equation (2).

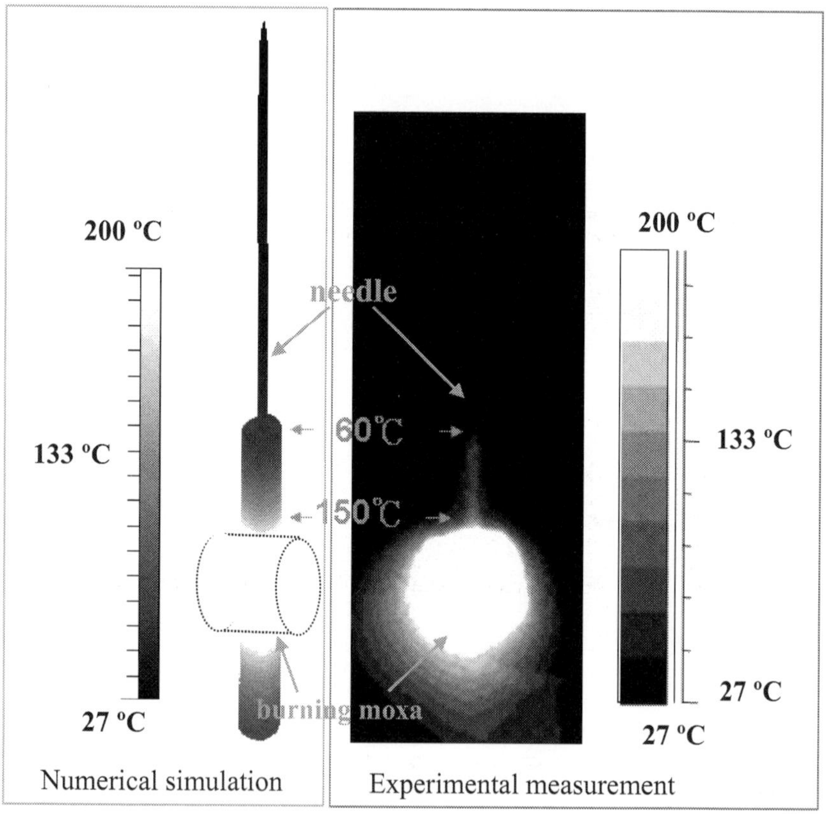

Fig. 14. Comparison of the numerically predicted and the experimentally measured temperatures for the burning moxa stick applied on the handle of the acupuncture needle. The temperature of burning moxa is 200 °C. Temperature gradient will be established across the needle, starting from the needle handle and ending at the needle head

In TCM, acupuncture needle can be made from stainless steel, iron, copper, silver and pottery etc. For the safety and cost effectiveness reasons, stainless steel is now more popular.

cause the same kind of electrophoresis, and lessen electrophoresis of the electrolyte. Electrophoresis of electrolyte is reduced by adding this in a large amount. Therefore, the potential gradient of the electrolyte becomes 0, and the electrophoresis term in equation (2) vanishes. As a result, only the diffusion term on the right side remains in equation (2).

$$J = -D\left(\frac{\partial C}{\partial y}\right) \tag{6}$$

J is proportional to the difference in concentration between the electrode surface and the fluid body, and is given by the following equation.

$$J = k_D\left(C_b - C_w\right) \tag{7}$$

where k_D is the mass transfer coefficient [cm/s], C_b and C_w are the fluid average concentration [mol/cm^3] and concentration of the electrode surface [mol/cm^3], respectively. The current obtained with diffusion control is defined as the limiting current I_d. The reaction in equation (4) is carried out extremely rapidly under diffusion control conditions, and thus concentration of Fe(CN) at the anode surface can be regarded as C_w=0. Thus equation (7) becomes:

$$J = k_D \cdot C_b \tag{8}$$

Substituting (3) into (8) and rearranging, the equation becomes:

$$k_D = \frac{I_d}{A \cdot Z \cdot F \cdot C_b} \tag{9}$$

Since A, Z and F are constants, the mass transfer coefficient k_D can be found by assaying fluid concentration C_b and measuring the limiting current. The Schmidt number (Sc) which corresponds to the Prandtl number (Pr) in heat transfer, and the Sherwood number (Sh) which corresponds to the Nusselt number (Nu), can be found as follows.

$$Sc = \frac{\nu}{D} \tag{10}$$

$$Sh = \frac{k_D \cdot d_i}{D} = \frac{I_d \cdot d_i}{A \cdot Z \cdot F \cdot C_b \cdot D} \tag{11}$$

Next, the wall shear stress is defined. Since the electrode surface is circular, the velocity gradient s_r of the wall surface can be calculated according to the following equation from the limiting current.

$$s_r = 3.55 \times 10^{-15} \frac{I_d^3}{D^2 \cdot C_b^3 \cdot d_E^5} \tag{12}$$

where d_E is the point electrode diameter.
From Newton's law of viscosity, shear stress is:

$$\tau = -\mu \frac{du}{dy} \tag{13}$$

where μ is the fluid viscosity [g/(cm s)].
Therefore, wall surface shear stress can be determined as follows.

$$\tau_w = \mu \cdot s_r = 3.55 \times 10^{-15} \frac{\mu \cdot I_d^3}{D^2 \cdot C_b^3 \cdot d_E^5} \tag{14}$$

It is also possible to predict the distribution of local mass transfer amounts from the wall surface shear stress when mass transfer occurs at the entire wall surface. If S is taken to be the distance in the flow direction from the stagnation point at the wall surface, and s_r (S) is taken to be the velocity gradient of the wall surface, then the local concentration gradient can be expressed as follows if separation does not occur in the flow.

$$\left(\frac{\partial C}{\partial y}\right)_s = \frac{C_b \left[s_r(S)\right]^{1/2}}{\Gamma\left(\frac{4}{3}\right)\left\{9D \int_0^s \left[s_r(x)\right]^{1/2} dx\right\}^{1/3}} \tag{15}$$

where $\Gamma\left(\frac{4}{3}\right)$ is the gamma function.
Substituting equation (6) and equation (8) into equation (15),

$$\frac{k_D \cdot C_b}{D} = \frac{C_b \left[s_r(S)\right]^{1/2}}{\Gamma\left(\frac{4}{3}\right)\left\{9D \int_0^s \left[s_r(x)\right]^{1/2} dx\right\}^{1/3}} \tag{16}$$

Thus, using the following equation, it is possible to calculate the flow direction distribution of local Sherwood numbers, which are the dimensionless mass transfer amounts.

$$Sh_L(S) = \frac{k_D \cdot d_i}{D} = \frac{d_i \left[s_r(S)\right]^{1/2}}{\Gamma\left(\frac{4}{3}\right)\left\{9D \int \left[s_r(x)\right]^{1/2} dx\right\}^{1/3}} \tag{17}$$

Rewriting this equation with dimensionless numbers, it becomes:

$$Sh_L(S) = \frac{Sc^{1/3} \cdot \mathrm{Re}_j^{2/3} \left[\frac{\tau_w}{\frac{1}{2}\rho \cdot u_i^2}\right]^{1/2}}{\Gamma\left(\frac{4}{3}\right) \cdot 18^{1/3} \left\{\int_0^s \left[\frac{\tau_w}{\frac{1}{2}\rho \cdot u_i^2}\right]^{1/2} dx \Big/ d_i\right\}^{1/3}} \tag{18}$$

where ρ is the fluid density [g/cm³], and u_i is the average velocity of the inner pipe nozzle exit.

2.3 Analogy of mass transfer and heat transfer

The amount of mass transfer is proportional to the concentration gradient, and is given by Fick's law.

$$J = -D\frac{dC}{dy} \tag{19}$$

The amount of heat transfer is proportional to the temperature gradient, and is given by Fourier's law.

$$q = -\lambda\frac{dT}{dy} \tag{20}$$

where λ is the thermal conductivity, and T is the fluid temperature.
Due to these relationships, there is a similarity in transport phenomena due to molecular motion between mass transfer and heat transfer, and an analogy holds between the two. In this experiment, since the electrode reaction is a diffusion controlled reaction and convection does not occur, the concentration boundary layer assumes a state very similar to a thermal boundary layer in the case of heat transfer. Therefore, mass transfer and heat transfer have the following sort of correspondence, and can be interchanged.

$$C \Leftrightarrow T,\ Sc \Leftrightarrow \mathrm{Pr},\ Sh \Leftrightarrow Nu$$

This experiment employed the Colburn j factor analogy, which is widely used in experiments and particularly effective for turbulence (Kobayashi, 1996). The J factors for mass transfer and heat transfer are, respectively:

$$j_D = \frac{Sh}{\mathrm{Re}_j \cdot Sc^{1/3}} \tag{21}$$

$$j_H = \frac{Nu}{\mathrm{Re}_j \cdot \mathrm{Pr}^{1/3}} \tag{22}$$

Here, the relationship $j_D = j_H$ holds, and thus the amount of heat transfer can be inferred by analogy as follows:

$$Nu = \frac{h \cdot d_i}{\lambda} = \frac{Sh}{Sc^{1/3}} \cdot \mathrm{Pr}^{1/3} \tag{23}$$

$$h = \frac{\lambda}{d_i}\frac{Sh \cdot \mathrm{Pr}^{1/3}}{Sc^{1/3}} \tag{24}$$

where h is the heat transfer coefficient.
Heat transfer coefficients in the jet-type cooling pipe of the die are predicted using values derived from an empirical equation (Mizushima et al., 1952) for the case where turbulence actually flows through the annular channel of a dual pipe, and calculation is done using the following equation.

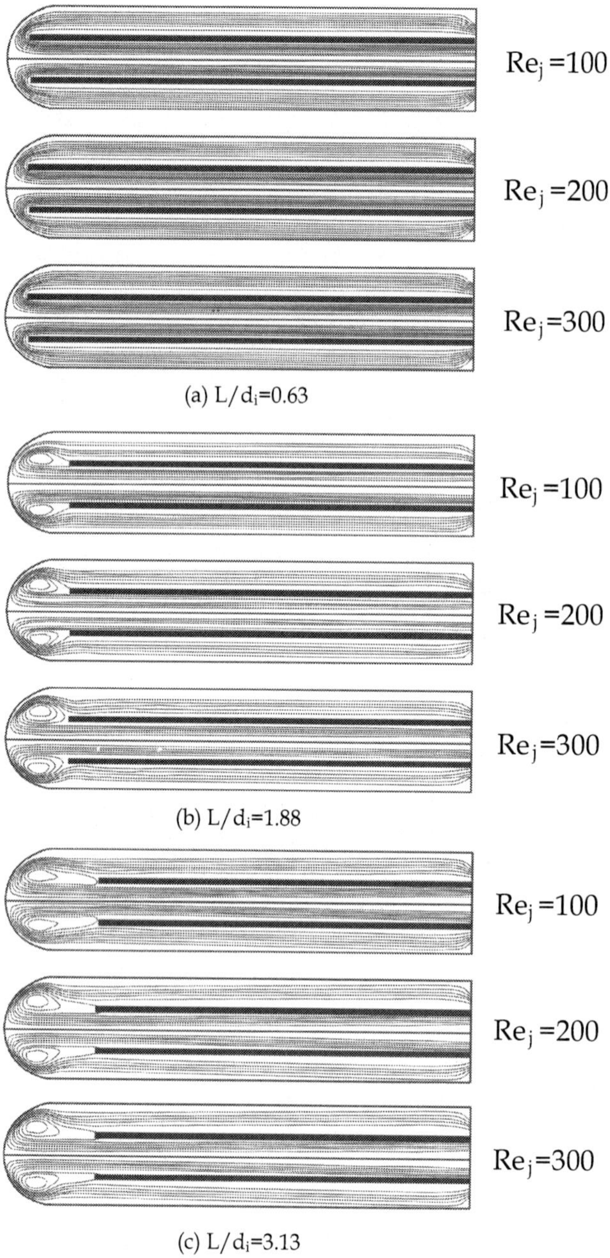

Fig. 3. Stream line distribution by numerical simulation

$$\frac{h \cdot d_e}{\lambda} = 0.02\left(\frac{d_{o2}}{d_i}\right)^{0.53} \mathrm{Re}^{*0.8} \cdot \mathrm{Pr}^{1/3} \tag{25}$$

Here, Re* are the Reynolds numbers for an annular channel, and are expressed as follows.

$$\mathrm{Re}^* = \frac{d_e \cdot u_o}{\nu} \tag{26}$$

where d_e is the equivalent diameter, and u_o is the cross section average velocity in the annular channel. The equivalent diameter used hydraulic diameter shown in the following.

$$d_e = \frac{4A_c}{l} \tag{27}$$

where A_c is the cross section of the annular channel, and l is the open verandah length.

3. Experimental results

3.1 Numerical simulation

In first, the numerical simulation before the experiment had forecast the flow in the jet flow type coolant pipe. Fig. 3 shows the result. The distance between the nozzle and the coolant pipe point has been changed with L/d_i=0.63, 1.88, and 3.13. The flow was assumed to be a stationary flow as calculation condition, in addition, ranges of the Rej number are from 100 to 300.The jet flow that occurred from the tube nozzle in L/d_i=0.63 of Fig. 3(a) flows along the wall, and forms the boundary layer flow, and the vortex is not generated, and even if the Rej number is raised, is not observed the vortex. The boundary layer flow is formed with the coolant pipe point for L/d_i=1.88 of Fig. 3(b) as well as Fig. 3(a), and the vortex occurs between the nozzle and the coolant pipe point. This vortex grows as Rej increases, and, the flow parts from the outside tube after it passes in this vortex, and bounds. This bounding flow becomes remarkable as Rej increases. And, because the flow is decelerated in the vicinity of the wall, the occurrence of the second vortex is forecast by flaking off. In the flow for L/d_i=3.13 of Fig. 3(c), the vortex becomes long and slender in the case of Fig. 3(b), and the bounding flow is unremarkable.

Therefore, in this experiment, the first vortex that occurs in the coolant pipe point is defined as Vortex A. And, the second vortex to which generation is forecast in the downstream is defined as Vortex B.

3.2 Flow visualization

Fig. 4 shows a flow visualization photograph for L/d_i=0.63. The white line in the figure indicates the position of the point electrodes. In this experiment, the vortex that arises near the cooling pipe tip is defined as a Vortex-A, and in addition, the vortex that arises in the downstream is defined as a Vortex-B. With Re_j=106, there is a flow along the wall surface, and Vortex-A does not occur. However, when the Re_j number is increased, Vortex-A occurs along the side surface of the inner pipe, and accompanying that, Vortex-B also appears. The flow for L/d_i=0.63 is steady, and Vortex-A and Vortex-B expand toward the downstream side up to around Re_j=1000. At Re_j numbers above that, the flow gradually becomes unsteady, but Vortex-A continues to exist. In the range $603<Re_j<856$, a vortex with rotation opposite to Vortex-A occurs on the side wall surface at the end of the inner pipe. This vortex is surrounded by Vortex-A, and here is called Vortex-C.

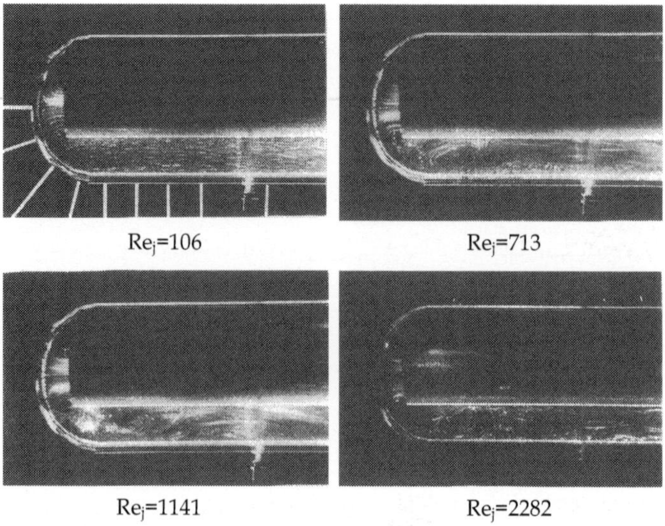

Fig. 4. Flow visualization photograph by the aluminium powder method for L/d_i=0.63

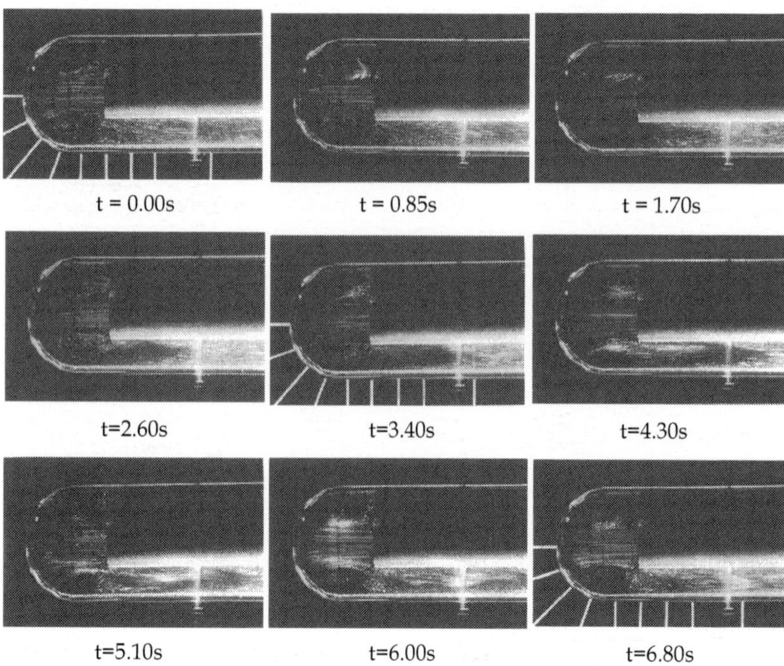

Fig. 5. Visualization photograph of the periodic fluctuation of the flow for L/d_i=1.88 and Re_j=302

Fig. 5 shows photographs of periodic fluctuation in the case where L/d_i=1.88 and Re_j=302. At t=0.0sec, Vortex-B is clearly visible on the downstream side of Vortex-A, but as time passes, Vortex-B disappears, as though pushed out to the downstream side, and the flow becomes parallel. However, Vortex-B appears again. Then this fluctuation is repeated with a period of 6-7 seconds. The instability of flow was evaluated based on these results, and Table 1 gives a summary. To the extent that the L spacing grows larger, the flow becomes unsteady more quickly, and large differences in flow were not observed in the range of high Re_j numbers.

Re_j	L/d_i=0.63	L/d_i=1.25	L/d_i=1.88	L/d_i=2.50	L/d_i=3.13
105	○	○	○	○	○
241	○	○	○	○	○
272	○	○	Δ	○	○
302	○	○	Δ	○	×
362	○	○	Δ	Δ	×
407	○	○	Δ	×	×
483	○	○	×	×	×
543	○	○	×	×	×
603	○	○	×	×	×
713	○	×	×	×	×
856	○	×	×	×	×
999	×	×	×	×	×
1141	×	×	×	×	×
1712	×	×	×	×	×
1997	×	×	×	×	×
2282	×	×	×	×	×
2853	×	×	×	×	×
5135	×	×	×	×	×

Table 1. Evaluation for the instability of the flow by the visualization

3.3 Wall surface stress

Fig. 6 shows the results of measuring wall surface shear stress. Here, I indicate the region where the flow is fluctuating periodically, and II indicates the transition zone from laminar flow to turbulence. At L/d_i=0.63, the shear stress values can be classified into No.2-5 and No. 6-9. For No. 2-5, the τ_w value is greatly increased from the low Re_j numbers, but the flow is constricted by Vortex-A at the position of these electrodes and the visualization photo shown in Fig. 4. Therefore, the rapid flow velocity is maintained. At Re_j=1000 or higher, the increase in τ_w slows down, and this may be due to the effects of changes in the flow velocity distribution at the nozzle outlet. There is also a tendency for No. 4 and 5 to match No. 6-9 at low Re_j numbers, and this shows that Vortex-A has still not developed. At No. 6-9 with low Re_j numbers, τ_w does not increase very much, and then it increases greatly after Re_j=1000 is exceeded. These electrodes are positioned at a location where the flow constricted at Vortex-A spreads out after passing Vortex-A, and since the cross-sectional area of the flow increases, the flow velocity near the wall surface slows down. Therefore the τ_w value is small. The fact that τ_w increases above Re_j=1000 may be because the flow becomes turbulent.

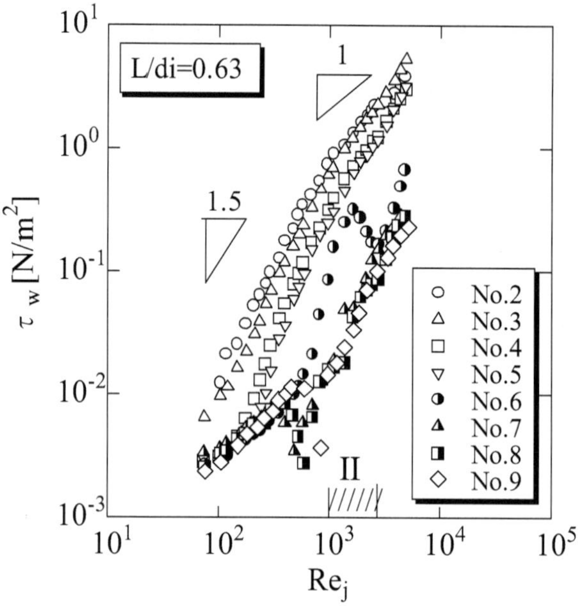
L/di=0.63
1
1.5
II
No.2
No.3
No.4
No.5
No.6
No.7
No.8
No.9
τ_w [N/m^2]
Re$_j$

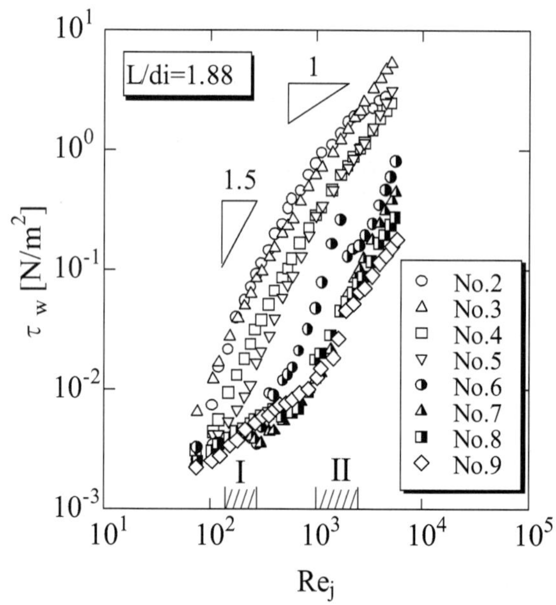
L/di=1.88
1
1.5
I
II
No.2
No.3
No.4
No.5
No.6
No.7
No.8
No.9
τ_w [N/m^2]
Re$_j$

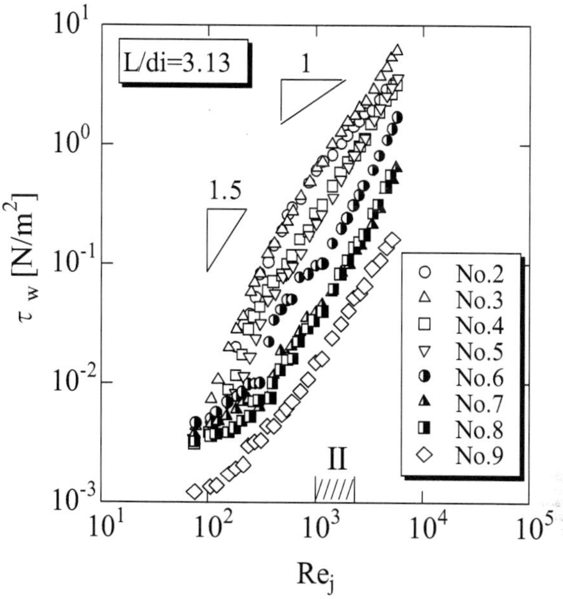

Fig. 6. Relationship between wall shear stress and Reynolds number

At L/d_i=0.63, Vortex-B grows large, but at No. 7-9, τ_w decreases rapidly in the range Re_j=500-900, and thus entry into the Vortex-B region can be confirmed. No. 6 exhibits a peculiar phenomenon where τ_w abruptly increases and decreases midway through. This may be because Vortex-A expanded to the No. 6 position as Re_j numbers increased, and then shrank again as the flow became turbulent. No. 6 is between Vortex-A and Vortex-B.

The results for L/d_i=1.88 are basically same as for L/d_i=0.63. The major difference is behavior relating to Vortex-B. Vortex B at L/d_i=1.88 fluctuates periodically, and thus does not remain steady and grow large. Therefore the reducing effect on τ_w due to Vortex-B does not appear to the same extent as at L/d_i=0.63. The phenomenon where τ_w decreases when the electrode enters the Vortex-B region is not observed at No. 9, and thus it can be seen that Vortex-B has not developed as far as No. 9. When L/d_i=3.13, the flow is more unstable than L/d_i=1.88, and thus the slope changes at an even lower Re_j number.

3.4 Mass transfer and heat transfer

Fig. 7 shows the results for average mass transfer measured using the overall electrodes No. (1)-(3). Here, the stagnation point is the value measured at point electrode No. 1. Average mass transfer measured at No. (1)-(3) is smaller than the value at the stagnation point. This is because the concentration boundary layer has developed taking the stagnation point as its base point, and its thickness is extremely small. Therefore, the boundary layer gradually develops and increases in thickness at No. (1)-(3) on the downstream side of that, and thus the Sh number becomes smaller. For the same reason, the Sh number gradually decreases moving from upstream to downstream at No. (1)-(3). With L/d_i=0.63, Vortex-B grows large at No. (2) and (3), and thus a large increase in the Sh number is not seen until close to Re_j=1000.

The heat transfer coefficients were calculated from these Sh numbers using the analogy equation for mass transfer and heat transfer in equation (24). Fig. 7 shows the average value of heat transfer coefficients for each electrode or No. (1)-(3). The solid line is the turbulence correlation equation (equation (25)) for an annular channel of a dual pipe. In order to compare with the annular channel of the dual pipe, the Re* numbers in the annular channel of the dual pipe were used as the Re numbers. All of the measured values are larger than the correlation equation for the annular channel of a dual pipe, and thus this correlation equation cannot be used for design of the equipment. Heat transfer increased locally, particularly in the No. (1) region. At No. (2) and (3), the slope of the heat transfer coefficient

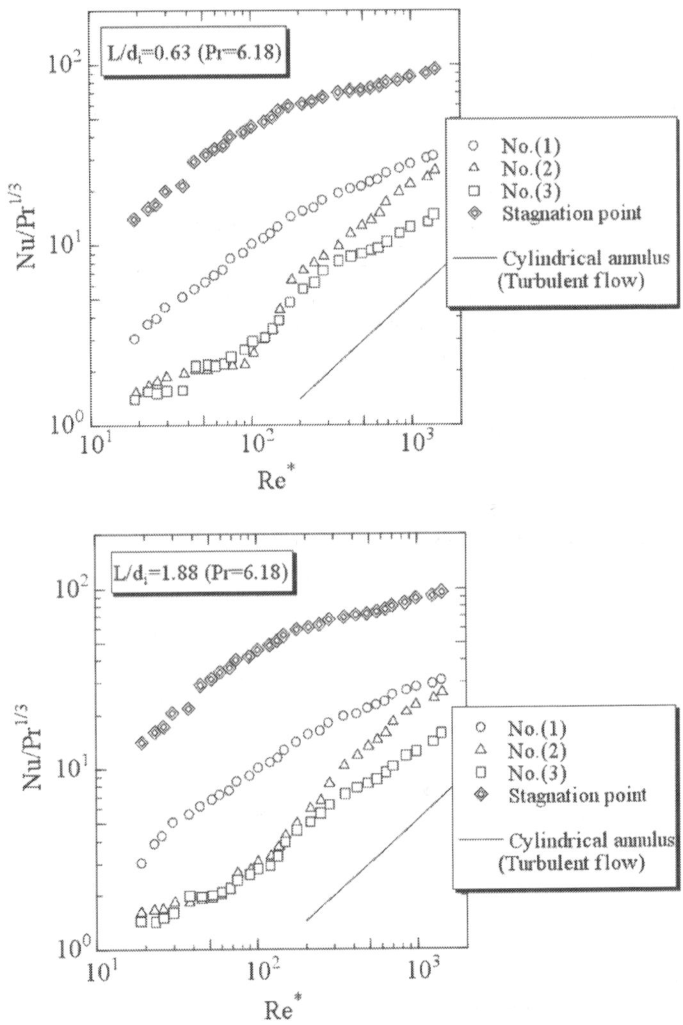

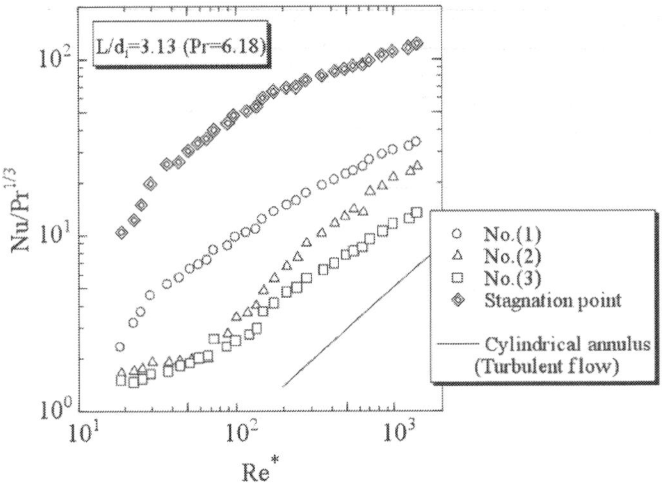

Fig. 7. Average mass transfer coefficient measured in the overall electrode

became equal to the slope of the correlation equation for the annular channel of a dual pipe, from the region where Re* numbers increased and the flow began to become unsteady. Fig. 8 shows the average values of heat transfer coefficients for No. (1)-(3). As a result, the average values of heat transfer coefficients form a straight line, as shown by the dotted line in the figure. The correlation equation is as given below.

$$Nu = 0.3 \cdot \mathrm{Re}^{*0.6} \cdot \mathrm{Pr}^{1/3} \tag{28}$$

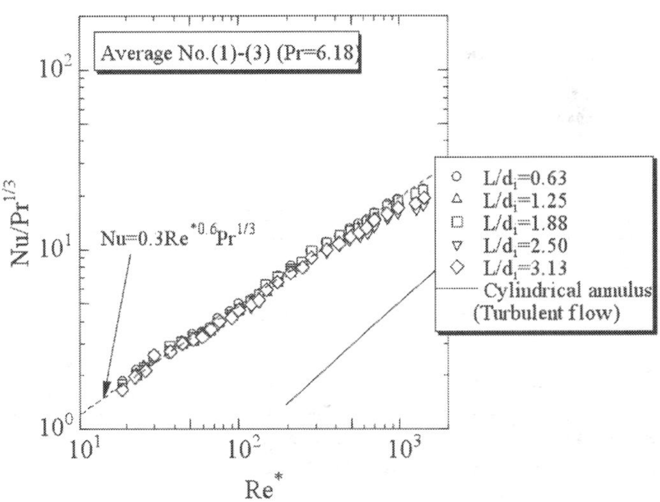

Fig. 8. Heat transfer coefficient in each electrode

4. Conclusions

In this research, the aim was to elucidate transfer phenomena in a jet-type cooling pipe. This was achieved by conducting flow visualization and measurement of mass transfer coefficients using a cooling pipe with the same shape as that actually used for die cooling, and predicting heat transfer based on analogy.
The findings of this research are indicated below.

1. Through visualization using aluminium powder, a primary vortex (Vortex-A) and secondary vortex (Vortex-B) were observed on the downstream side from the stagnation point of the impingement jet. These vortices vary greatly in size as the Re_j number increases. Also, the secondary vortex causes periodic fluctuations immediately before the flow starts to turn turbulent.
2. Point electrodes were embedded in the outer pipe of the cooling pipe, and wall surface shear stress was measured using the electrode reaction method. Those values varied in accordance with the expansion and shrinkage of the primary and secondary vortices observed through visualization. Thus, it was found that the primary vortex has the effect of accelerating flow near the wall surface, and the secondary vortex has the effect of decelerating flow near the wall surface.
3. The heat transfer coefficients were inferred from Sh numbers by using the analogy between mass transfer and heat transfer. The heat transfer coefficients showed almost the same behavior as the flow pattern, and tendencies similar to previous research were exhibited at the stagnation point. Also, a correlation equation for finding heat transfer coefficients was obtained from the average value for No. (1)-(3).

5. References

Miranda, J.M. & Campos, J.B.L.M. (1999), Impinging jets conical wall: laminar flow predictions, *AIChE Journal*, 45(11), pp.2273-2285, ISSN 1547-5905

Miranda, J.M. & Campos, J.B.L.M.(2001), Impinging jets confined by a conical wall-High Schmidt mass transfer in laminar flow, *International Journal of Heat and Mass Transfer*, 44, pp.1269-1284, ISSN 0017-9310

Kayansayan, N. & Kucuka, S. (2001), Impingement cooling of a semi-cylindrical concave channel by confined slot-air-jet, *Experimental Thermal and Fluid Science*, 25, pp.383-396, ISSN 0894-1777

Hrycak, P. (1981), Heat transfer from a row of impinging jets to concave cylindrical surface, *International Journal of Heat and Mass Transfer*, 24, pp.407-419, ISSN 0017-9310

Gau, C. & Chung, C.M. (1991), Surface curvature effect on slot-air-jet impinging cooling flow and heat transfer process, *Transaction of the ASME, Journal of Heat Transfer*, 113, pp.858-864, ISSN 0022-1481

Lee, D.H., Chung, Y.S. & Kim, D.S. (1997), Turbulent flow and heat transfer measurement on a curved surface with a fully developed round impinging jet, *International Journal of Heat and Fluid Flow*, 18, pp.160-169, ISSN 0142-727X

Kobayashi, S. & Iida (1996), K., *Idou-ron*, ISBN 978-4-254-13518-3 Asakura-shoten

Mizushima, T., Hara, K. & Kyuno, T., Heat and mass transfer coefficients in double tube cooler condenser, *Kagaku Kikai*, 16, pp.338-344, 1952.

30

Thermal State and Human Comfort in Underground Mining

Vidal F. Navarro Torres[1] and Raghu N. Singh[2]
[1]*Centre for Natural Resources and Environment of Technical University of Lisbon*
[2]*Nottingham Centre for Geomechanics, of University of Nottingham*
[1]*Portugal*
[2]*United Kingston*

1. Introduction

The human metabolism is accompanied by heat generation, with the body temperature remaining constant near 36.9°C and in contact with surrounding atmospheric temperature; people have cooler or hotter sensations.
When people are exposed to a temperature greater than the threshold limits, it causes physiological effects expressed as follows: loss of interest in people's activities, taking frequent rests or breaks, a desire to quickly complete the task, irritability, reduced concentration and reduction in sensitivity.
A prolonged exposure of people to unfavourable thermal conditions inevitably leads to increase in body temperature and consequently producing physiological effects that affect the work efficiency. Figure 1 shows a relationship between work efficiency and effective temperature and air, wet temperature and air velocity. It may be noted that the prolonged exposure of a worker to temperature exceeding 42°C may even cause death.

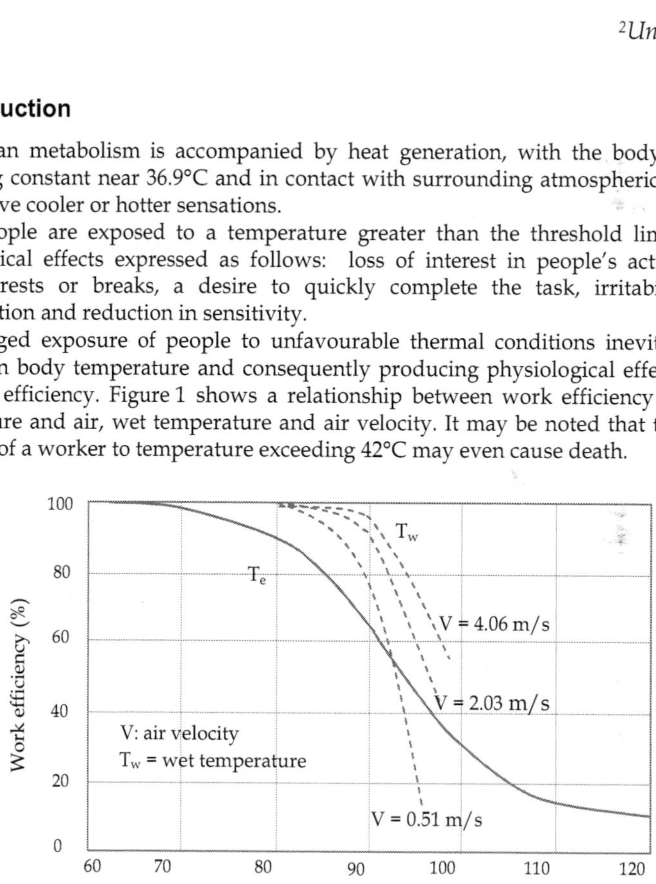

Fig. 1. Effect of temperature on work efficiency (Navarro, 2003, Ramani, 1992)

The temperature of intake air due to its passage through an underground opening gradually increases due to depth and the length of air travel through underground opening. The main cause of heat transfer to air flow in underground atmosphere is due to thermal properties of virgin rock, known as geothermal gradient. Other sources of heat to the air in underground atmosphere are air auto-compression, diesel emission, explosive detonation, human metabolism and influx of thermal water.

2. Mathematical model of heat transfer

The total variation of temperature in an underground environment Δt_{total} can be calculated by including the variation of temperature from air auto-compression Δt_a, thermal properties of rock Δt_r, heat emission from diesel equipments Δt_d, heat due to breaking of rocks with the use of explosives Δt_e, human metabolism Δt_h and thermal water Δt_w as outlined in equation (1):

$$\Delta t_{total} = \Delta t_a + \Delta t_r + \Delta t_d + \Delta t_e + \Delta t_h + \Delta t_w \quad (1)$$

With increasing mining depths, the influence of the thermal properties of the rock mass becomes more important (Navarro et al, 2008). Based on equation (1), the total underground atmosphere temperature T_2, will be expressed by equation (2), as a function of surface temperature t_s or underground opening initial temperature T_1.

$$T_2 = t_s + \Delta t_{total} \text{ or } T_2 = T_1 + \Delta t_{total} \quad (2)$$

2.1 Surface air temperature

It is well known that the surface air temperature varies with the seasons and is subjected to regional variations according to local weather conditions, so that the temperature variation is influenced by the ventilation current temperature in underground openings.

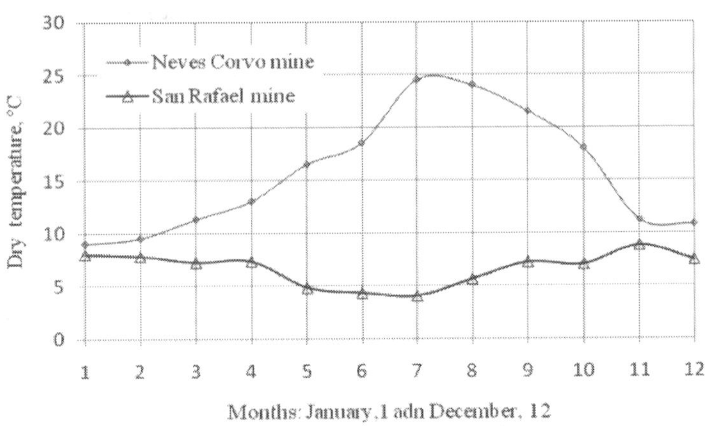

Fig. 2. Typical surface air dry temperature in Neves Corvo and San Rafael mines

Figure 2 indicates that average monthly surface temperature in Neves Corvo mine was maximum 24.5°C in July, minimum 9.0°C in January and mean being 15.6°C and in San Rafel mines was maximum temperature was 8.8°C in November, minimum 4°C in July and mean being 6.61°C. Surface air temperature variation throughout the year can be better illustrated with monthly average temperature measured in Neves Corvo and San Rafael mines (Figure 2). Neves Corvo mine is located in Portugal in North Hemisphere at 20° latitude and at altitude of 800m, but San Rafael mine located in Perú in South latitude at the altitude of about 4500 m. Temperature trend in Neves Corvo mine is similar to the metalliferous mines in Portugal with maximum variation range of 15 °C and temperature tendency in San Rafael mine is typical of the South American Andes with maximum variation range of 4.0 °C. Average dry bulb temperature measured in Neves Corvo mine stopes located at depth between 750 m to 770m, compared with variation of external dry bulb temperature, observed clearly the influence of outside temperature in underground openings (Figure 3). Therefore, the maximum temperature on the hottest month will be critical.

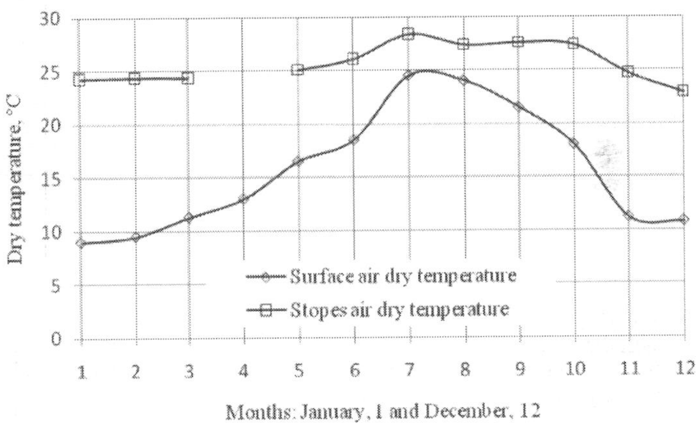

Fig. 3. Surface air temperature influencing underground openings air temperature

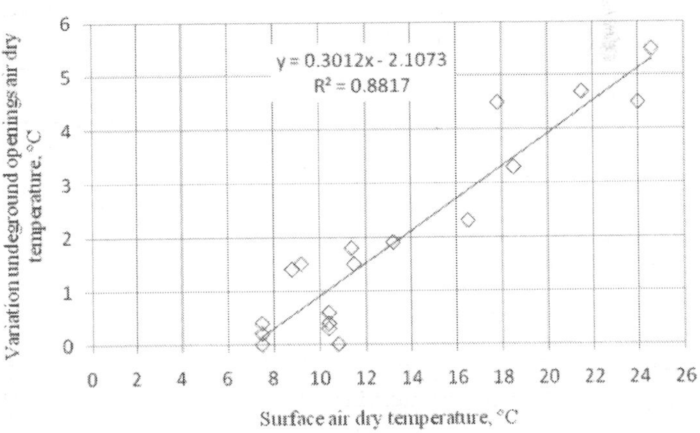

Fig. 4. Underground temperature variations as a function of surface air temperature

The surface air temperature can influence the temperature of air flow in the atmosphere of underground openings, since these are more than 7 °C, as in the Neves Corvo mine (Figure 4). This result indicates that during winter times or in mines located at large altitudes, such as the South America Andes, the outside temperature has little or no influence on the temperature of underground openings.

Moreover, as a part of an environmental thermal comfort assessment in deep underground mines, it is necessary to consider the surface temperature, because this is the initial temperature, t_1, of intake air to underground openings.

For similar conditions of Neves Corvo mine and at 750 m depth, variation in underground openings temperature, Δt_s, will be calculate by equation (3), based on surface air temperature, t_e.

$$\Delta t_s = 0.301 t_e - 2.107 \tag{3}$$

2.2 Heat transfer due to air auto-compression in vertical underground openings

Auto-compression process occurs during the air descent through the underground openings and due to its own compression. The mathematical model is deduced considering the equilibrium condition, air properties and the influenced of by vertical forces (Figure 5) expressed by air equilibrium condition as follows:

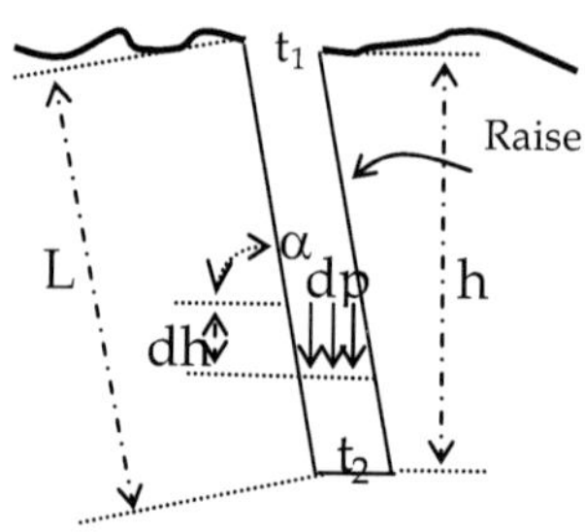

Fig. 5. Air auto-compression in inclined raise layout

$$g.dh - dp / \rho_a = 0 \tag{4}$$

Where, g is gravity, dh is depth differential, dp is pressure differential, ρ_a is air density. By substituting specific gravity γ and specific volume v *in* equation (4) the following expression is obtained:

$$dh = dp / \gamma = v dp \tag{5}$$

In adiabatic process $p.v^k$ = constant, when k is air adiabatic coefficient and differentiating results in equation (6) as follows:

$$v.dp + k.p.dp = 0 \tag{6}$$

Clapeyron equation $p.v = R.t_2$, where R is universal gases constant and t_2 is compressed air temperature, the following differential equation results:

$$p.dv = R.dt_2 - v.dp \tag{7}$$

Using equations (5), (6) and (7) equation (8) is obtained as follows:

$$dh + k(R.dt_2 - dh) = 0 \tag{8}$$

Integrating the equation (8) obtains the following expression (9) where C is constant:

$$(1-k)\int dh + k.R\int dt_2 = (1-k)h + k.R.t_2 + C = 0 \tag{9}$$

Rearranging equation (9), the temperature t_2 is obtained as follows:

$$t_2 = \frac{(k-1)h}{k.R} - C \tag{10}$$

For initial values *h*=0 and *t*=*t1*, the constant C=0, then C=-t_1, the adiabatic equation 9 result in equation 11 as follows:

$$\Delta t_a = t_2 - t_1 = \frac{(k-1)h}{k.R} \tag{11}$$

With numerical values of constant of perfect gases (R=29.27 kgf-m/kg-°K) and average air adiabatic index (1.302) the final equation is obtained as follows:

$$t_2 - t_1 = 0.0098h \tag{12}$$

In general condition depth h will be expressed as a function of underground opening length L (m) and inclination α (°), as $h = LSin\alpha$ and finally, the temperature increase due air auto-compression Δt_a (°C) results in following equation (Navarro Torres, 2003):

$$\Delta t_a = t_2 - t_1 = 0.0098L\sin\alpha \tag{13}$$

That means, when α=90° (vertical raise) for each 100 m air temperature increases by in 0.98°C, for 200 m 1.96°C, for 300 m 2.94°C, for 400 m 3.92°C, for 500 m 4.9°C, for 650 m 6.37°C, for 800 m 7.84 °C and for 1000 m 9.8°C. Therefore when α=0° (horizontal underground opening) auto-compression temperature is zero (Figure 6)

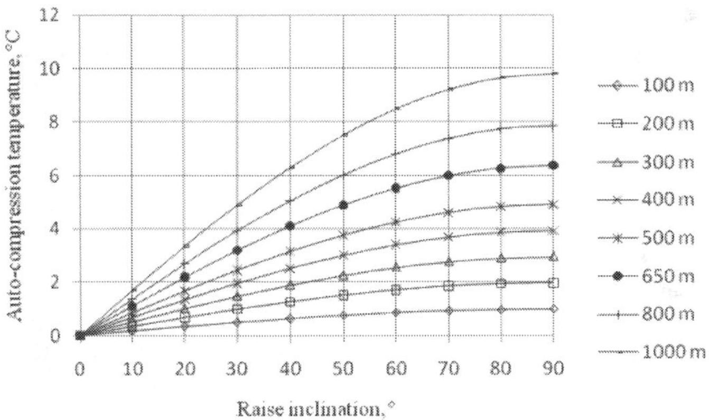

Fig. 6. Variation of auto-compression temperature with raise inclination

2.3 Heat transfer of thermal properties of rock mass to underground atmosphere

At a certain depth h_n from the surface defined as the thermal neutral zone (15 m according to Ramani 1992; 20 to 40 m indicated by Vutukuri & Lama, 1986) the temperature of rock masses varies during the year as a function of the changes of surface air temperature. The temperature of any rock mass at depth t_{hr} and underground atmosphere air temperature variation Δt_r can be calculated by the following equations:

$$t_{hr} = t_n + \frac{(h - h_n)}{g_g} \tag{14}$$

$$\Delta t_g = \frac{h_1 - h_n \pm L \sin\alpha}{g_g} \tag{15}$$

where; t_{hr} is the rock temperature at depth h (°C), t_n is the temperature of the rock mass above the thermal neutral zone (°C), h depth of mining excavation below the surface, h_n is the depth of the thermal neutral zone (m) and g_g geothermal gradient of the rock mass (m/°C).

In order to obtain the mathematical model for the calculation of heat transfer of thermal properties of rock mass, use of the heat transfer formulation of gas flow in pipes can be applied to underground openings.

Heat spreads from one point to another one in three distinct ways: conduction, radiation and convection. In most cases, the three processes occur simultaneously and therefore the amount of heat "q" supplied to a body of mass "m" and specific heat C_e, when the temperature increases from t_1 to t_2 is given by the general equation (16):

$$q = m.C_e(t_2 - t_1) = m.C_e.\Delta t \tag{16}$$

For the air flowing in the underground openings this equation can be expressed in function of the circulating air volume Q through:

$$q_r = 1000\rho_a.C_{e.}Q.\Delta t_r = 1000.\rho_a.C_e.Q.(t_2 - t_1) \tag{17}$$

Where q_r is the heat received by the air from the rock mass (W), ρ_a the air density (kg/m^3), C_e the specific heat of air (kJ/m.°C), Q the flow of air (m^3/s) and Δt_r the variation of temperature from t_1 to t_2 (Fig. 7). The heat coming out of the rock mass and received by the ventilation air in the underground environment can also be expressed in terms of coefficient of heat transfer λ (Holman, 1983) according to the equation (17):

$$dq = \lambda.P.dx(P_p - T_m) \tag{18}$$

Where T_p and T_m are the temperatures of rock wall and air mixture in the particular position x (°C), λ is the coefficient of heat transfer between the rock mass and the air mixture (W/m^2.°C) and P is the perimeter of the section of the underground opening (m). The total heat q_r transferred (W) can be calculated by using equation (19) as follows:

$$q_r = \lambda.P.L.(T_p - T_m)_{average} \tag{19}$$

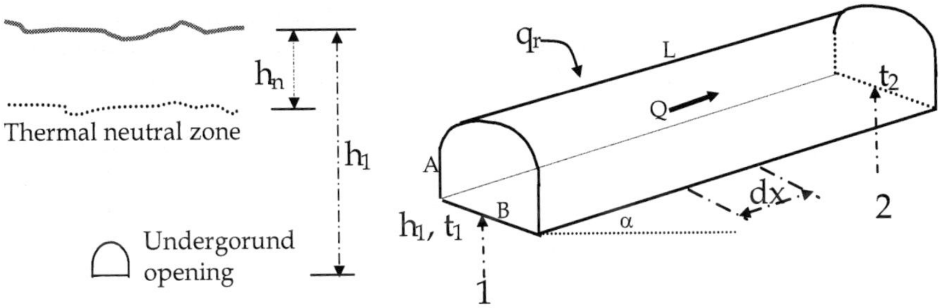

Fig. 7. Layer of rock influenced by external temperature and elementary parameters of an underground opening

Using equation (19) the average temperature of the rock mass may be given by equation (20) and (21):

$$T_p = \frac{1}{2}\left\{ t_1 + \left(t_1 + \frac{h_1 - h_n \pm L.\sin\alpha}{g_g} \right) \right\} \tag{20}$$

$$T_m = \frac{t_1 + t_2}{2} \tag{21}$$

By substituting equations (20) and (21) in equations (17) and (19) the following expression is obtained:

$$\left[\frac{\lambda.P.L}{2}\right]\left[\frac{h_1 - h_n \pm L.\sin\alpha}{g_g} + t_1 - t_2\right] = 1000.\rho_a.C_e.Q.(t_2 - t_1)$$

Finally the variation of temperature from t_1 to t_2 (Δt_r) may be expressed as follows:

$$\Delta t_r = t_2 - t_1 = \frac{\lambda.P.L.(h_1 - h_n \pm L.\sin\alpha)}{g_g(\lambda.P.L + 2000.\rho_a.C_e.Q)} \tag{22}$$

Resulting equation (22) is an innovative mathematical model developed for heat transfer of thermal properties of rock mass to underground openings (Navarro, 2003).
In raises or in any vertical underground openings, $h_1 = 0$, and the length which influences the temperature due to geothermal gradient is $L\ Sina - h_n$ and a =90°, thus, resulting in the following equation:

$$\Delta t_r = t_2 - t_1 = \frac{\lambda.P.(L - h_n)^2}{g_g\left[\lambda.P.(L - h_n) + 2000.\rho_a.C_e.Q\right]} \tag{23}$$

The coefficient of heat transfer λ is calculated as a function of the thermal conductivity K (W/m°C) which is the non-dimensional coefficient of Dittus and Boelter N_{db} and the diameter of section d (m); for horizontal and inclined underground openings $d = (B + A)/2$, where B is the width of the section (m) and A its height (m):

$$\lambda = \frac{k.N_{db}}{d} \tag{24}$$

The relation of Dittus and Boelter co-efficient N_{db}. (Holman, 1983) was studied in detail by Petukohov for gases (air) that derived the following equation:

$$N_{db} = \frac{\frac{f}{8}\text{Re}_d.\text{Pr}}{1.07 + 12.7(\frac{f}{8})^{0.5}(\text{Pr}^{0.67} - 1)} \tag{25}$$

Where R_{ed} is the Reynolds number (non-dimensional), given by:

$$R_{ed} = \frac{V.d}{\mu} \tag{26}$$

in which V is the average velocity of air (m/s), d the underground opening diameter (m) and μ the kinematic viscosity of air (Kg/m.s). In addition, f is the friction coefficient of the underground opening walls (Kg/m^3), P_r is the Prandtl number (non-dimensional) calculated by:

$$P_r = \frac{\rho_a.C_e.\mu}{K} \tag{27}$$

Air properties at atmospheric pressure will be determined based in temperatures (Table 1).

2.4 Heat transfer from diesel equipment

The equipments used in underground work generate the heat transfer to the ventilation current in underground atmosphere as follows:

1. Mobile diesel and electrical equipments, such as jumbo drills, trucks, LHDs, pumps, locomotives, etc.
2. Electrical and non-mobile equipments (fans, lighting, pumps, hoists, stations or transformer substations, etc.).

For the mobile and non-mobile equipments used in underground work, diesel equipments contributes significantly to heat transfer to the air flow in underground atmosphere. Diesel engines fuel consumption for mining equipment is 0.24 kg/kWh, with a calorific value of 44 MJ/kg (Vutukuri & Lama, 1986), so the total energy released is 0.24x 44x10^3 KJ/kWh = 10560 kJ/kWh = 176 kJ/mink = 2.9 kJ/s.KW = 2.9 kW/kW. Of the total 1kW energy release, (34%) is converted into mechanical energy and 1.9 kW (66%) is exhausted to air flow of underground atmosphere. This energy is not totally transferred to the air flow, because it depends to the effective time for which the equipment used, so it is different for each condition of underground work and the value is around 0.9 kW (31%).

Diesel equipment heat exhaust q_{ed} (KW) can be expressed by equation (28) as follows:

$$q_{ed} = f_m.f_t.q_d.P_d \tag{28}$$

Where q_d is the equivalent energy released by diesel fuel (2.9 kW/kW),
P_d is the equipment engine (kW),
f_m is mechanical efficiency and
f_t is equipment utilization efficiency.

T (ºK)	ρ a(kg/m³)	Ce (KJ/kg.ºC)	ν(kg/m.s) x10-5	μ(m²/s)x 10-6	K(W/m.ºC)	Dif.Térm. (m²/s)x10-4	Pr
100	3.6010	1.0266	0.6924	1.923	0.009246	0.02501	0.770
150	2.3675	1.0099	1.0283	4.343	0.013735	0.05745	0.753
200	1.7684	1.0061	1.3289	7.490	0.01809	0.10165	0.739
250	1.4128	1.0053	1.488	9.49	0.02227	0.13161	0.722
300	1.1774	1.0057	1.983	16.84	0.02624	0.22160	0.708
350	0.9980	1.0090	2.075	20.76	0.03003	0.2983	0.697
400	0.8826	1.0140	2.286	25.90	0.03365	0.3760	0.689
450	0.7833	1.0207	2.484	31.71	0.03707	0.4222	0.683
500	0.7048	1.0295	2.671	37.90	0.04038	0.5564	0.680
550	0.6423	1.0392	2.848	44.34	0.04360	0.6532	0.680
600	0.5879	1.0551	3.018	51.34	0.04659	0.7512	0.680
650	0.5430	1.0635	3.177	58.51	0.04953	0.8578	0.682
700	0.5030	1.0752	3.332	66.25	0.05230	0.9672	0.684
750	0.4709	1.0856	3.481	73.91	0.05509	1.0774	0.686
800	0.4405	1.0978	3.625	82.29	0.05779	1.1951	0.689
850	0.4149	1.1095	3.765	90.75	0.06028	1.3097	0.692
900	0.3925	1.1212	3.899	99.30	0.06279	1.4271	0.696
950	0.3716	1.1321	4.023	108.2	0.06525	1.5510	0.699
1000	0.3524	1.1417	4.152	117.8	0.06752	1.6779	0.702
1100	0.3204	1.160	4.44	138.2	0.0732	1.969	0.704
1200	0.2947	1.179	4.69	159.1	0.0782	2.251	0.707
1300	0.2707	1.197	4.93	182.1	0.0837	2.583	0.705
1400	0.2515	1.214	5.17	205.5	0.0891	2.920	0.705
1500	0.2355	1.230	5.40	229.1	0.0946	3.262	0.705
1600	0.2211	1.248	5.63	254.5	0.1000	3.609	0.705
1700	0.2082	1.267	5.85	280.5	0.105	3.977	0.705
1800	0.1970	1.287	6.07	308.1	0.111	4.379	0.704
1900	0.1858	1.309	6.29	338.5	0.117	4.811	0.704
2000	0.1762	1.338	6.50	369.0	0.124	5.260	0.702
2100	0.1682	1.372	6.72	399.6	0.131	5.715	0.700
2200	0.1602	1.419	6.93	432.6	0.139	6.120	0.707
2300	0.1538	1.482	7.14	464.0	0.149	6.540	0.710
2400	0.1458	1.574	7.35	504.0	0.161	7.020	0.718
2500	0.1394	1.688	7.57	543.5	0.175	7.441	0.730

Table 1. Air properties at atmospheric pressure (Holman, 1983, Navarro, 2003)

Based on equation (28), the temperature variation of air due to exhaust from the diesel equipment Δt_d (ºC) can be quantified by the following equation:

$$\Delta t_d = \frac{f_m \cdot f_t \cdot q_d \cdot p_d}{\rho_a . C_e . Q} \qquad (29)$$

It may be noted that the exhaust heat from the diesel engines to the underground atmosphere is from the local equipment use only.

2.5 Heat transfer from explosive blasting

The blasting process of explosive in underground environment generates heat that is transferred to the surrounding rock mass and to the ventilation current of the underground atmosphere.
Heat released by blasting q_e (kW) can be calculated by equation (30), based on calorific energy of explosive E_e (kJ/kg), and explosive quantity daily used q_e (Kg/day). For example, the calorific energy of ANFO is 3900 kJ/kg and the dynamite 60% varying between 4030 to 4650 kJ/kg.

$$q_e = \frac{E_e.q_e}{86400} \quad (30)$$

The thermal influence due to blasting Δt_e (°C) can be quantified by equation (31) as follows:

$$\Delta t_e = \frac{E_e.q_e}{86400.\rho_a.C_e.Q} \quad (31)$$

Similar to diesel exhaust heat, the heat due to explosive detonations influences the local atmosphere only.

2.6 Heat transfer due to human metabolism

The heat transfer of human metabolism in not significant and can be ignored (Hartman et al., 1997), for example 800 workers in normal working conditions leads to a total release of 192 kW (65000 BTU/hr), energy corresponding to each worker being 0.25 kW.
Thus, when the number of people or workers in an underground environment is large, temperature increase by human metabolism Δt_h (°C) can be expressed by equation (32), where q_h is the human heat release and it is a function of effective temperature (kW/person) and, n is the total number of human involved.

$$\Delta t_h = \frac{q_h.n}{\rho_a.C_e.Q} \quad (32)$$

2.7 Heat transfer from underground water

Two sources of water are encountered in mining: Groundwater or Mine water. All ground water, especially from hot fissures and natural rock reservoirs, is a prolific source of heat in mine workings. Since water and heat are both derived from the surrounding rock or geothermal sources, the water temperature will approach or even exceed the rock temperature.
The water transfers its heat to the mine air, mainly by evaporation increasing the latent heat of the air.
The total heat gain from hot underground water in open channel flow q_w (kW) can be calculated from the equation (33):

$$q_w = F_{tw}.c_w(t_{tw} - t_a) \quad (33)$$

Where F_{tw} is weight flow rate of thermal water (kg/s)
C_w is specific heat of water (4.187 kJ/kg°C), and
t_{tw} and t_a are water temperatures at points of emission and exit from the mine airway in (°C), respectively.

The thermal influence of underground ventilation air flow can be calculated by equation (34) as follows:

$$\Delta t_w = 4.187 \frac{F_{tw} \cdot (t_{tw} - t_a)}{\rho_a . C_e . Q} \tag{34}$$

3. Case studies of heat transfer in underground mining

3.1 Case study in Portuguese Neves Corvo mine

Vertical underground opening

The Neves Corvo mine is an operating underground copper and zinc mine in the western part of the Iberian Pyrite Belt which stretches through southern Spain into Portugal. The mine uses both bench and fill and drift and fills underground stoping methods. The copper plant has treated a maximum of 2.0 mt per annum of ore and in 2007 it was upgraded to treat up to 2.2 mt of ore per annum.

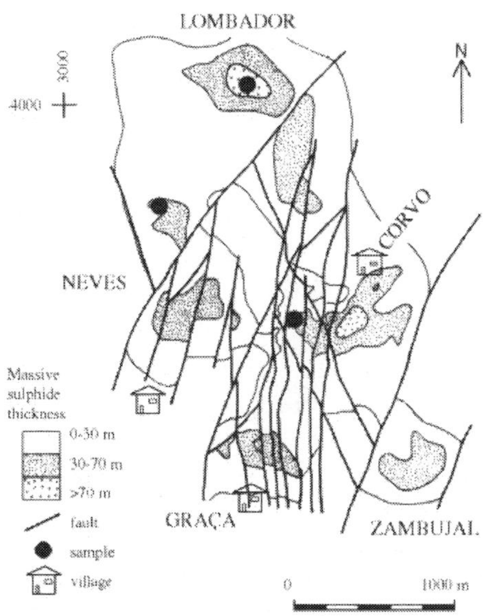

Fig. 8. Map of the Neves Corvo area showing the massive sulphide ore bodies Neves, Corvo, Gracia and Zambjal, main faults and the exploratory boreholes (Moura, 2005)

The ore bodies of the underground Neves Corvo copper mine (Fig. 8) were formed in a volcanic sedimentary submarine environment possibly linked with an intercontinental rift and, third order pull apart basins, not far from the collision zone and located in geological formations between Volcanic Lavas (V1) and Volcanic Sediments (V2). The V1 is composed of black shale/schist and has same silicfication but generally less than V2 volcanic. The V2 has a compact vitreous due the high quantity of silica (Riolitic) showing schistosity and alteration from Chlorite (Lobato, 2000).Mining areas are located between +200m and -450m, and they are referred to 0 level, equivalent to 0 m datum and transport level equivalent to -550m level (Fig. 9). The total length of underground vertical shafts, inclined and horizontal openings is about 80 kilometres.

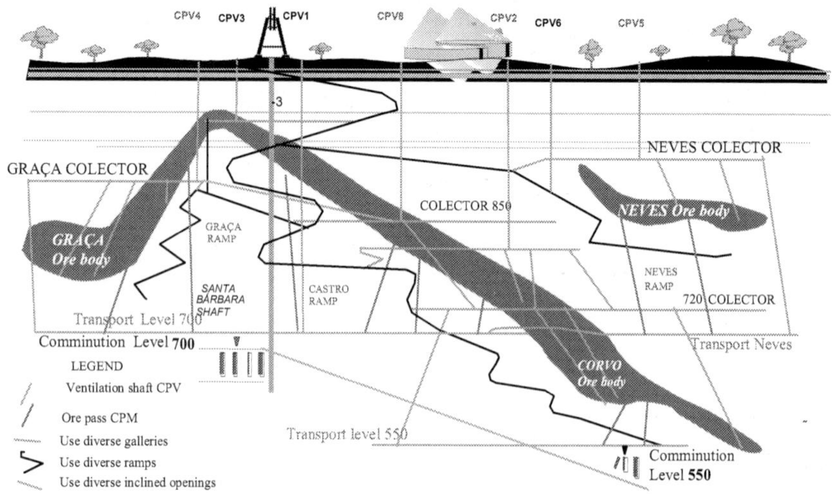

Fig. 9. Neves Corvo underground mine cross section (Navarro, 2003)

The air temperature in underground stopes of Neves Corvo mine is moderate averaging between 20°C to 33°C and in isolated areas in critical condition reaching 42°C

For applying the mathematical model a vertical underground opening the CPV3 shaft shown in Figure 10, was selected (Navarro Torres, et al, 2008). This shaft was constructed using a raise boring machine from the depth of 1222.40m level to 973.64m level with a length of 248.76m and a diameter of 4.2 m (perimeter 13.19 m and 13.85 m^2 in cross section area).

The wall friction factor corresponded to 0.0362 kg/m^3 with an average air velocity measured in July 2000 of 11.84 m/s and average exterior temperature of 24.61 °C (Figure 2 and Figure 3). Average airflow resulted in 164.03 m^3/s calculated based on air velocity measured, as indicated in Figure 11.

Fig. 10. Scheme and photograph of shaft CPV3

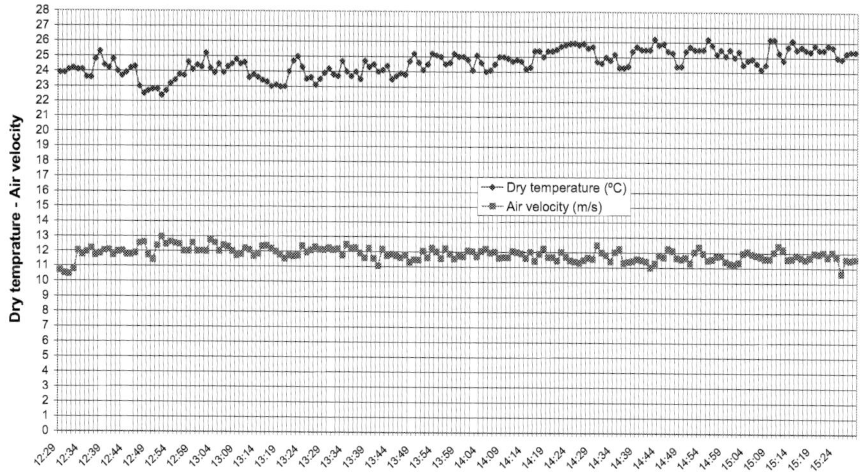

Fig. 11. Intake dry temperature and air velocity measured by Data LOGGER DL20K

The air temperature in CPV3 shaft is not influenced by the temperature rise due to diesel equipments (Δt_d), explosives (Δt_e), thermal water (Δt_w) and human metabolism (Δt_h). Therefore, only auto-compression (Δt_a) and geo-thermal properties of rock (Δt_r) were considered for the validation of the proposed model.
The physio-chemical properties of air shown in Table 2 extracted from Table 1 for 24.4°C enabled the calculation of the Prandtl number P_r, Reynolds number Re_d, related to Dittus and Boelter number N_{db} and coefficient of heat transfer λ, by applying equations (27), (26), (25) and (24), respectively as shown in Table 3.

ρ_a (kg/m^3)	C_e (kJ/kg.°C)	μ (m^2/s)	K(W/m.°C)
1.1888	1.0056	16.48 x10^{-6}	0.026

Table 2. Physio-chemical air properties at 24.4 °C (Navarro, 2008).

P_r	R_{ed}	N_{bd}	λ (W/ m^2.°C)
0.709	3.02 x 10^6	5162.02	76.106

Table 3. Coefficient of heat transfer and previous values calculated

Finally using the geothermal gradient as 30.3 m/°C for the rock mass (g_g) for the Neves Corvo mine (Fernández-Rubio. *et al.*, 1990) and using 30.0m as the depth of thermal neutral zone in the developed mathematical model in equation (23), the temperature rise of rock mass ($\Delta t_r = t_2 - t_1$) is calculated as 2.65°C. Applying these values to equation (13), the temperature increase due to air auto-compression can be obtained as 2.38°C. Then the total increase of the air temperature during the air flow in the shaft CVP3 results in 5.03°C (Fig. 12), calculated by following simplified equation:

$$\Delta t_a + \Delta t_r = 0.0098.L + \frac{0.033(L-30)^2}{2.38.Q + L - 30} \tag{34}$$

Obviously, when airflow is decreased, the total temperature increment (auto-compression + geothermal properties of rock + temperature due to depth increase) significantly raises the total temperature increment (Fig. 12).
By applying equation (2) to equation (34), the underground atmosphere's temperature of CPV3 shaft T_2 in Neves Corvo mine as a function of airflow quantity Q is represented by equation (35) given below and illustrated in Figure 13 for the average surface temperature 15.65°C (maximum 24.5°C, minimum 9°C and mean 15.65°C).

$$T_2 = t_s + 2.44 + \frac{1579.25}{2.38Q + 248.76} \tag{35}$$

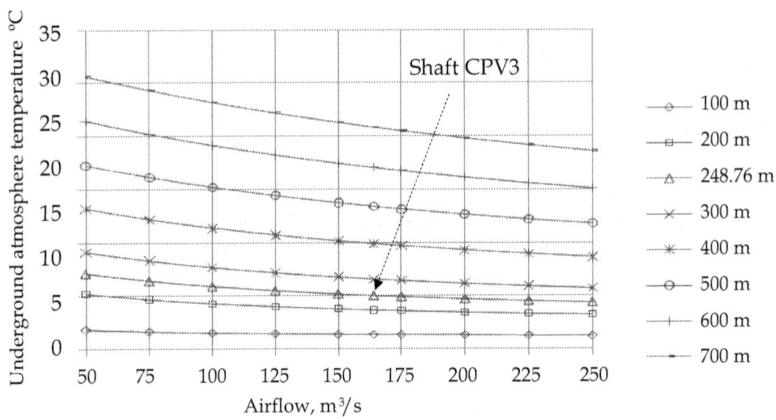

Fig. 12. Total increases in temperature due to auto-compression and geothermal properties of rock as a function of shaft depth and airflow in shaft CPV3

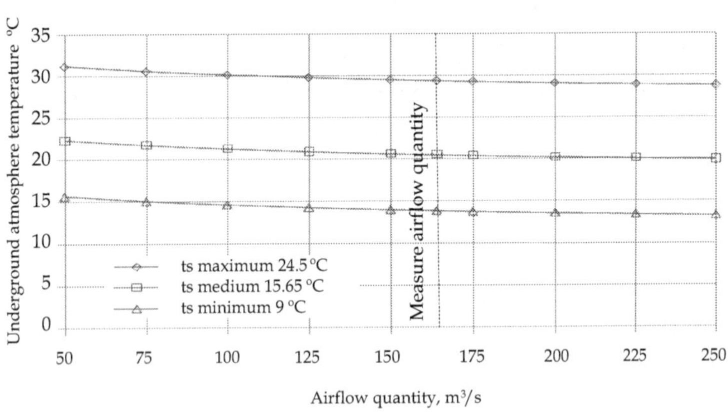

Fig. 13. Underground atmosphere temperature influenced by auto-compression and thermal properties of rock and airflow quantity in CPV3 shaft

It may be observed that a slight increase in the underground atmospheric temperature, results in slight decrease in the air flow. The average values measurement with Data LOGGER DL20K of ROTRONIC in the air shaft intake with a thermo/hygrometer Casella in the shaft (Fig. 14) was 29.52°C in the shaft bottom and 24.61°C in the intake, therefore the difference is 4.91°C.The comparison the results show a total variation of temperature (Δt_{total}) between the mathematical model and measured temperature in CPV3 shaft is only 0.12°C.

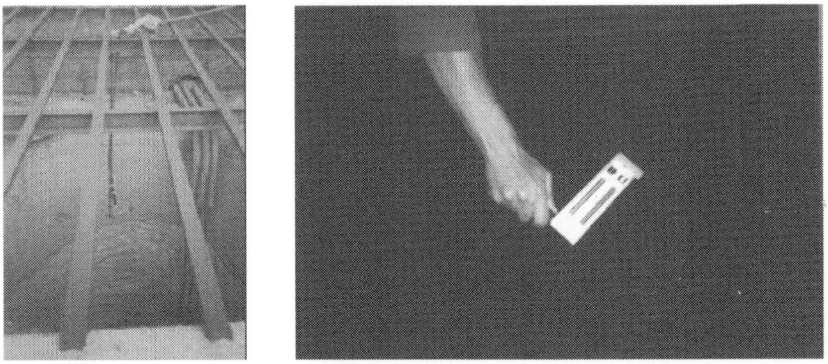

Fig. 14. Measurements with Data LOGGER DL20K and thermo/hygrometer Casel

3.2 Case study in Peruvian San Rafael tin mine

Sub-horizontal underground opening

The San Rafael mine belongs to the Peruvian company MINSUR S.A. and is located Southwest of the San Bartolomé de Quenamari mountain (altitude 5299 m), in the Department of Puno in the Eastern Mountains of Southern Peru. It is geographically located in the coordinates of 70°19' longitude West and 14°14' latitude south. This mine is the only producer of tin in Peru and ore production is 2500 tons per day, with 5.23% of tin (Sn). Geology of San Rafael mine involves silts and quartzite rocks of the tertiary Sandia formation with the intrusion of two granites. In the neighborhood there are rocks of the superior Paleozoic age. In the Sandia formation silts have dark gray colors with muscovite in the cleavage plans and the quartzites are intercalated with silts (Fig. 15).

The mineralized veins and ore bodies are located in the intrusive ore body of San Rafael along the NE - SW direction, with a length of 800 m to 1000 m, a width of 300 m and depth of up to 2000 m. These ore bodies have widths of 4 m to 30 m, lengths of 30 m to 180 m and heights of 60 m to 610 m, and in general have prismatic forms. The main existing minerals are cassiterite, stanhite and chalcopyrite.

The main access from surface is through 4523m ramp that communicates to 3825m level, constituting the principal infrastructure of underground transport, as well as the ventilation circuit (Fig. 16).

Figure 17 presents the clean air temperatures with normal trend until the level 3950 m (17°C), but in 3850m level, where a variation is only 100 m, temperature increases to 34°C, thus showing the effect of thermal water. A forecast for air temperature in the 3850m level without the influence of hot water leads to 20°C for the air flow of $8.11 m^3/s$.

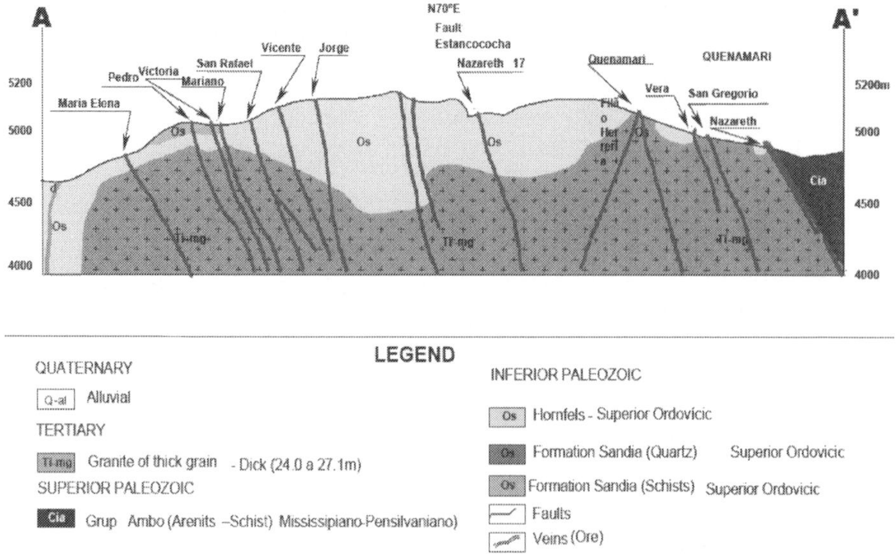

Fig. 15. Geological section with zones of mining in San Rafael along the direction N 70° E

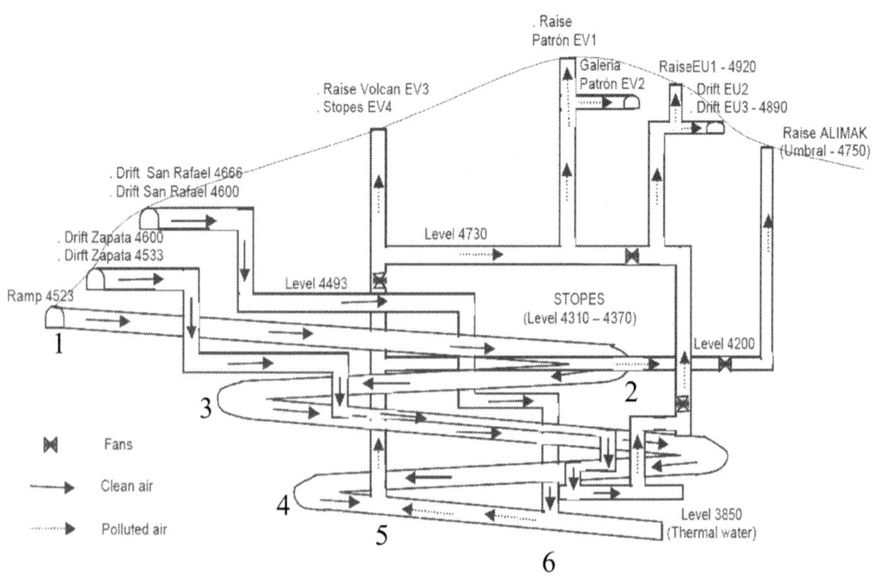

Fig. 16. General project of the underground environment of San Rafael mine (Navarro, 2001)

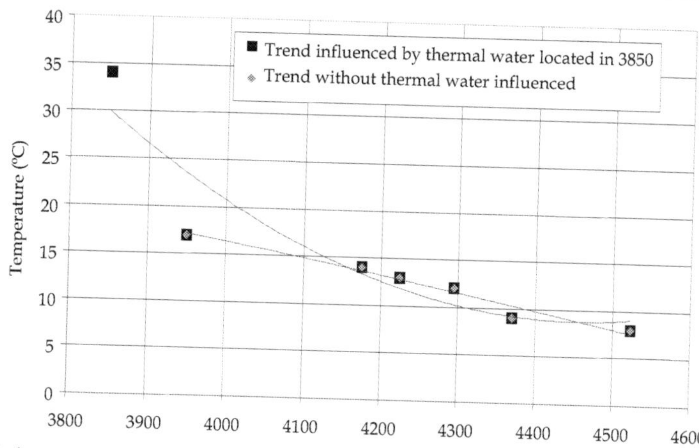

Fig. 17. Air temperature trends with and without the influence of thermal water (Navarro, 2003)

The thermal water temperature measured in the flowing water channel was 40°C and the influence of air temperature in underground opening was up to 34 °C (15 m^2 section). It may be observed that an increase in temperature by 12 °C is not a normal trend of air temperatures, because it represents an increase of about 60%.

In San Rafael mine case study, the calculated geothermal gradient g_g based on equations (1) and (22) and measured data and equation (22) needed recalculation of temperatures variation due to diesel exhaust, explosive detonations and thermal water, because in local level 3850, the temperature increment due to auto-compression and human metabolism are insignificant.

Diesel equipments heat transfer Δt_d, is calculated by applying equation (29) based on combined factors of mechanical efficiency and equipment utilization efficiency ($f_m.f_t$) of 0.005 for diesel equipments used in ramp 4523(level 3850) (Table 4), air density 1.2661 kg/m^3 and specific heat of air 1.0056 kJ/kg.°C (Table 1) and for depth local level 3850 of ramp 4523, temperature increase result is 0.85°C.

Heat transfer Δt_e due to explosive detonation is determined by applying equation (31), based on an average of 120 kg per day ANFO, air density and air specific heat (Table 1), resulting in temperature rise of 0.52°C.

The thermal water heat transfer calculated by applying equation (34), based on measured flow rate of 4.93 l/s of thermal water in channel F_{tw}, was 40°C water temperatures at points of emission t_{tw} and 34°C at the exit from the mine airway t_a; using air density and air specific heat (Table 1), result being 12°C.

Using these results and applying equation (1) based in measured 22°C (34°C-12°C) total temperature increase Δt_{total} the heat transfer of virgin rock Δt_r results in 8.63°C.

Finally, for the following conditions d = 4.5m, f = 0.0046kg/m^3, V = 0.39m/s, P = 18m, L = 7000m, h_1 = 30m, h_n = 30m, α = 7°, Q = 8.11m^3/s (branch 6-5 Figure 15) and for physical-chemical air conditions (Table 5) and calculated Prandtl number P_r, Reynolds number Re_d, relation of Dittus and Boelter N_{db} and coefficient of heat transfer λ, is calculated applying equations (27), (26), (25) and (24), respectively as shown in Table 6, and applying equation (22) the geothermal gradient result in 59.51m/°C or 1.68 °C for each 100 m.

methods that can accurately estimate temperatures. Fiber-optic distributed temperature sensing (DTS) is an approach available to provide coverage in both space and time that can be used continuously to monitor real-time data in different environments and at spatial scales that range from centimeters to kilometers. DTS was first developed in the mid 1980s (Dakin et al., 1985) and used in the oil and gas industry during the 1990s and early 2000s (Kersey, 2000), but only since the middle of this decade have DTS instruments achieved acceptable levels of spatial and temporal resolution, along with the high temperature accuracy and resolution needed to observe environmental processes. With their spatial and temporal coverage, DTS methods offer significant advantages over traditional measurement systems in the environment. For instance, the fiber-optic cables that serve as the temperature probe are low cost, with no issues of bias or fluid column disturbance, and variability due to different sensors and sensor measurement scale can be avoided.

The objective of this chapter is to provide the reader with an overview of the theory of fiber-optic DTS technology and a review of environmental applications to date, which will be used to investigate the main heat transfer mechanisms occurring in different environments. Important considerations, recent advances, and future trends are also discussed.

2. Fiber-optic distributed temperature sensing theory

Fiber-optic DTS technology uses Raman spectra scattering in an optical fiber to measure temperature along its length, i.e., the fiber-optic cable is the thermometer, achieving temperature resolutions as small as ±0.01 °C, and spatial and temporal resolutions of 1-2 m and 1-60 s, respectively, for cables up to 10 km (Selker et al., 2006a). Raman scattering can also be used to estimate temperatures in media other than optical fibers, such as atmospheric LIDAR (Eichinger et al., 1993), but can only be used to measure atmospheric temperatures and are far less widely used. To understand how fiber-optic DTS systems work, we first present the background of Raman scattering, describing how this scattering can be used to determine temperatures along the length of the optical fiber. Then, we present the governing equations that DTS systems use.

To measure the temperature along an optical fiber, the DTS instrument emits laser pulses at a known wavelength into an optical fiber. An optical fiber consists of a glass core surrounded by a glass cladding with a different refractive index than the core (Fig. 1). As light travels longitudinally along the fiber, a fraction of the incident light is scattered by interactions between the light and the crystalline structure, and vibration frequency (temperature) of the fiber itself (Hausner, 2010). Light scattering is classified as elastic or inelastic. Elastic (or Rayleigh) scattering occurs when the kinetic energy of the incident photons is conserved and thus, the frequency of the scattered photons is equal than that of the incident light. On the other hand, when the kinetic energy of the incident photons is not conserved, inelastic scattering occurs. As a result, the frequency of the incident and scattered photons differs. In optical fibers, the inelastic scattering typically has two components: Brillouin and Raman. The Brillouin scattering propagates as acoustic waves and is the result of density shifts caused by interaction between pulsed and continuous light waves counter-propagating in the optical fiber (Kurashima et al., 1990). Brillouin scattering occurs at a predictable amplitude but variable frequency. The Raman scattering is produced by interactions between the photons and vibrating molecules within the lattice of the glass fiber. This interaction results in a predictable frequency shift. The scattered light shifted to lower frequencies (longer wavelengths) than the incident light is termed Stokes, while the scattered light shifted to higher frequencies (shorter wavelengths) is called anti-Stokes. The backscattered Stokes signal is produced when a photon excites a molecule at a base vibrational state and the molecule returns to a slightly higher state (Fig. 2(b)). When the incident photon hits a previously-excited molecule and this molecule returns to the base state (Fig. 2(c)), the backscattered anti-Stokes signal is generated. The greater the temperature of the fiber, the more frequently these previously-excited molecules will be encountered (Smith & Dent, 2005).

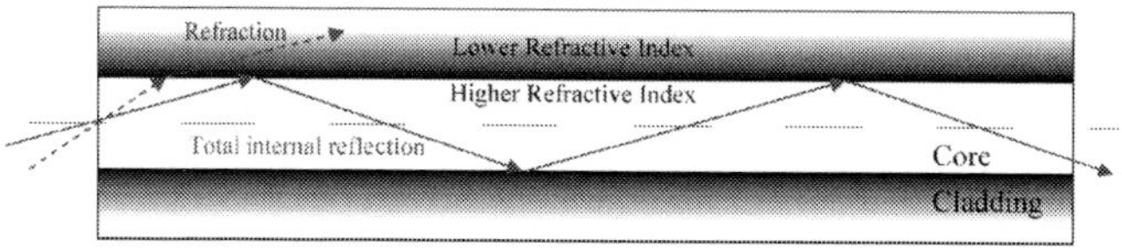

Fig. 1. Internal refractions of light in a cross section of an optical fiber

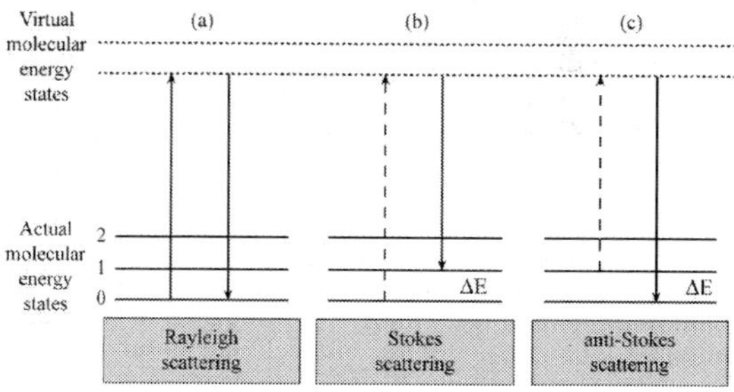

Fig. 2. Interactions between light and glass structure (Smith & Dent, 2005)

Although the frequencies of the Raman signals are predictable, their intensities are not. The intensity of the anti-Stokes scattering depends strongly on the temperature of the silica molecules of the fiber, while the intensity of the Stokes backscattering depends weakly on this temperature (Fig. 3). Because of this differential temperature dependence, the ratio of the anti-Stokes and Stokes signals can be used to determine the temperature of the fiber at the point of scattering (this is described below). The distance of the point of light scatter is calculated by time-domain reflectometry using the speed of light in the glass fiber, which is dependent on the frequency of the light and the index of refraction of the fiber itself. In standard optical fibers, the speed of light ranges between 1.7×10^8 and 2.0×10^8 m s^{-1} (Hausner, 2010). Commercial DTS instruments typically use a 10 or 20 ns laser pulse to illuminate the optical fiber. After emitting the laser pulse, the backscattered signals begin to return to the DTS instrument, where they are collected in discrete time periods by the detector unit. Because the light in the fiber travels approximately 2 m in 10 ns, the signals that return to the DTS instrument between the 0-10 ns following the injected pulse come from the first meter of fiber. If the backscattering detection unit is set to 10 ns, the DTS instrument will return temperature readings integrated over 1 m of fiber. The DTS instrument repeats the pulse and data collection continuously for temporal integration periods as specified by the user. A diagram of a typical DTS system is depicted in Fig. 4.

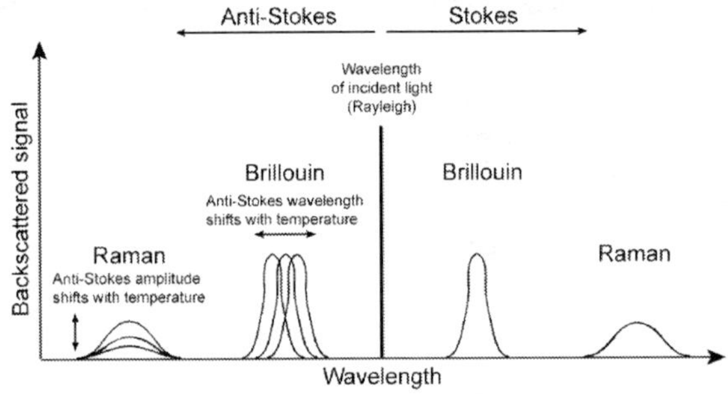

Fig. 3. Diagram of Rayleigh, Brillouin, and Raman backscattering in optical fibers

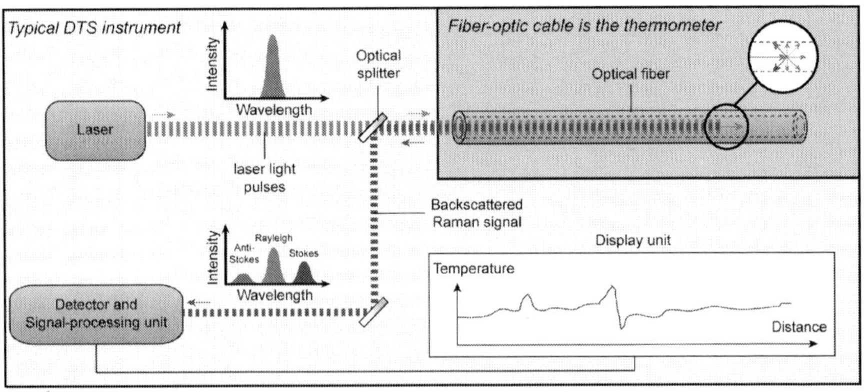

Fig. 4. Diagram of a typical distributed-temperature-sensing system

The intensities of the Raman anti-Stokes and Stokes signals collected at the DTS instrument can be expressed by (Rogers, 1999):

$$I_{aS}(z) = I_0 \exp(-\alpha_0 z) \exp(-\alpha_{aS} z) \wp_{aS} \Gamma_{aS} \quad (1)$$

$$I_S(z) = I_0 \exp(-\alpha_0 z) \exp(-\alpha_S z) \wp_S \Gamma_S \quad (2)$$

where $I_{aS}(z)$ and $I_S(z)$ are the Raman anti-Stokes and Stokes signals, respectively; I_0 is the intensity of the laser pulse emitted from the DTS instrument; α_0, α_{aS} and α_S are the attenuation coefficients of the emitted laser pulse, the backscattered anti-Stokes signal and the backscattered Stokes signal, respectively; $\wp_{aS}$ and $\wp_S$ are the Bose-Einstein probability distribution of phonons of the Raman anti-Stokes and Stokes signals; and Γ_{aS} and Γ_S are the capture coefficients of the Raman anti-Stokes and Stokes signals, which represent the fraction of the scattered light that is directed back towards the light source. The term "$\exp(\alpha_0)$" in equations (1) and (2) represents the attenuation of the emitted light as it travels away from the DTS instrument. The terms "$\exp(\alpha_{aS})$" and "$\exp(\alpha_S)$" represent the attenuation of the anti-Stokes and Stokes signals, respectively, as light travels back into the

DTS instrument. These attenuations are different because the Raman backscattered signals have different frequencies. The Bose-Einstein probability distribution of phonons can be described by (Farahani & Gogolla, 1999):

$$\wp_{aS} = \left[\exp\left(-\Delta E/_{kT}\right)\right]\left[1-\exp\left(-\Delta E/_{kT}\right)\right]^{-1} \tag{3}$$

$$\wp_{S} = \left[1-\exp\left(-\Delta E/_{kT}\right)\right]^{-1} \tag{4}$$

where ΔE represents the difference in molecular energy states that drive Raman scattering; k is the Boltzmann constant; and T is absolute temperature. Substituting equations (3) and (4) into equations (1) and (2), respectively, and taking the ratio between the anti-Stokes and Stokes intensities, $R(z)$, yields:

$$R(z) = \frac{I_{aS}(z)}{I_S(z)} = \frac{\Gamma_{aS}}{\Gamma_S}\exp\left(-\Delta E/_{kT}\right)\frac{\exp(-\alpha_{aS}z)}{\exp(-\alpha_S z)} \tag{5}$$

The ratio between the anti-Stokes and Stokes capture coefficients, Γ_{aS}/Γ_S, can be considered as a calibration parameter, C, that depends on the wavelength and frequency of the incident laser, the backscattered Raman signals, the instrument's photon detector, and the operating conditions of the DTS instrument. Defining the differential attenuation of the backscattered Stokes and anti-Stokes as $\Delta\alpha = \alpha_S - \alpha_{aS}$ and $\gamma = \Delta E/k$, equation (5) results in:

$$R(z) = C\exp\left(-\frac{\gamma}{T}\right)\exp(\Delta\alpha z) \tag{6}$$

and rearranging terms (Suárez et al., 2011):

$$T(z) = \frac{\gamma}{\ln[C] - \ln[R(z)] + \Delta\alpha z} \tag{7}$$

Equation (7) describes the temperature along the entire optical fiber. Here, for the sake of simplicity, the differential attenuation has been assumed to be constant along the fiber but in reality, this may change along the fiber due to differences in manufacturing, strain or other defects that scatter the two Stokes frequencies differently. From equation (7), it can be seen that integrated measurements of temperature, in both space and time, along the fiber can be estimated using the ratio of the anti-Stokes and the Stokes intensities. As explained before, when determining the temperature along the optical fiber, the location of the point of scatter must also be known. This location can be found by measuring the duration of the light reflection in the optical fiber. The distance, z, traveled by the light can be estimated by (Yilmaz & Karlik, 2006):

$$z = \frac{ct}{2n} \tag{8}$$

where c is the speed of light in vacuum, n is the refractive index of the fiber, and t is the propagation time of light in the forward and backward directions. More details about the theoretical basis of DTS system can be found in other investigations (Rogers, 1999; Selker et al., 2006a).

3. Use of fiber-optic distributed temperature sensing in the environment

The use of fiber-optic DTS in the environment began during the 1990s (Hurtig et al., 1994; Förster et al., 1997) when temperature in boreholes was monitored with resolutions of ±0.1 °C. In the middle of the 2000s, DTS systems achieved acceptable levels of spatial and temporal resolution, along with high temperature accuracy and resolution to monitor the environment. For example, when the systems are calibrated carefully, using integration times longer than 1 h and cables shorter than 5 km, a precision of ±0.01 °C can be achieved every 1 meter of cable (Selker et al. 2006a,b). This great ability to precisely observe temperatures at thousands of locations is the main thrust of DTS systems, which holds potential for transformative observation of many environmental processes. In this section, we present an overview of the DTS applications that have been performed in different environments. To show the promise of this technology, we present hydrological, ecological, engineering, and atmospheric DTS applications that have been studied.

3.1 Fiber-optic distributed temperature sensing in snow

Measurement of the thermal regime of snow has significant importance to a wide range of environmental processes, yet can be surprisingly challenging to measure. Snow depth and snow temperature strongly control the thermal balance and temperature profiles in the underlying soil (Lachenbruch, 1959) as well as heat transfer between the snow and atmosphere. The vertical distribution of temperature strongly controls metamorphism in the snowpack, important for both avalanche forecasting and snowmelt dynamics. In addition, the horizontal variation of snow temperatures can control the distribution of permafrost and frozen ground, both important factors for infiltration, runoff and many sub-snow biological processes. However, in almost all cases, snow temperature measurements are limited to one or at most a few points of measurement in space across the landscape.

While temperature sensing systems have been available for centuries, snow represents several unique challenges. In the past, measurements were often made by hand in snow pits, using simple thermometers but providing, at best, a very limited time series of the temperature evolution. Even with advances in continuous, low-cost, low-power thermal data collectors (Lundquist & Lott, 2008), measurements of the thermal regime of snow packs are still generally limited to a few points in the vertical dimension, and rarely more than a few different locales in any given watershed. Even continuous vertical measurements are challenging, as solar radiation and wind often cause melting, heat conduction and/or differential snow accumulation around measurement devices. The lack of ability to measure, continuously in time and space, the snow thermal environments has limited our understanding of the energy budgets of snow covered areas, as well as the spatial distribution of snow physical properties. Recently, Tyler et al. (2008) demonstrated the utility of DTS for measuring the thermal evolution of the snow/soil interface temperatures. In that work, standard telecommunication cables were deployed across several hundred meters at two experimental watersheds and monitored to determine the thermal differences between snow and bare ground, and also the effects of aspect on soil/snow interface temperatures. In both watersheds, deployed fibers withstood winter conditions and provided 300-400 individual measurements of snow/soil interface temperatures at temporal scales from minutes to hours (Tyler et al., 2008).

Since basal snow temperatures are now proven to be easily measured using DTS, the distribution of internal snowpack temperatures represents the next challenge in distributed sensing. While it has long been recognized that vertical gradients in the snowpack temperature lead to significant snow metamorphosis and instability, little study has been made on the horizontal gradients in temperature that may exist due to differences in shading and snow accumulation. Areas of steep horizontal gradient in snow temperatures may result in horizontal differences in vapor diffusion, snow morphology and strength, thereby producing local zones of weakness that may serve as avalanche trigger points.

Several methods have been proposed in the last few years to measure internal snowpack temperatures using fiber-optic sensing in harsh conditions during winter and snow loading. Researchers from École Polytechnique Fédérale de Lausanne and Oregon State University have suspended and stretched fiber-optic cables at fixed elevations above the soil prior to the onset of winter, and allowed the fiber-optic "fence" to be subsequently buried by snow. Such a design provides temperature data at fixed elevations above the land surface and precise knowledge of the cable location. Challenges to the fixed height method include cable sagging due to snow load, stress points at any cable attachment point and alteration of the snow deposition due to the presence of the pre-emplaced cable "fence". Alternatively, optical fiber can be deployed as snow accumulates during the winter season and allowed to follow both the snow surface, and any compaction/consolidation that occurs. Researchers at the University of Nevada, Reno and the University of California, Santa Barbara have used this deployment strategy in the Sierra Nevada of California, where storm accumulations are typical large and quickly bury the deployed fiber, and access to the site allows for easy deployment prior to major storms. Far less precise spatial location (both horizontal and depth) can be predicted for this method, as the optical fiber is deployed before the storm cycle begins, the optical fiber can move due to wind during the storm, and blowing snow may bury the cable differentially. Following the storm, the optical fiber should remain at the interface of the old and new snow, but its height above the ground surface will decrease as the snow consolidates through the season. This method of deployment therefore tracks snow layer temperatures, rather than at fixed depths in the snowpack. While there is potentially less strain on the fiber due to attachment points and less effect of the cable placement on snow accumulation, both methods are likely to alter the snow melting dynamics as the optical fiber approaches the snow surface during the melt season. While the "snow fence" method requires significant work at the beginning of the season, the "pre-storm" deployment requires significant work throughout the snow season and repeated access to the study site.

At the CRREL-UCSB Eastern Sierra Snow Study Site (CUES, Painter et al., 2000; Bales et al., 2006), fiber-optic DTS monitoring of basal and snowpack temperatures has been on-going since 2007 (Tyler et al., 2008). During the 2009/2010 and 2010/2011 winter seasons, fiber-optic cable (ADSS Flat Drop Cable, AFL Telecommunications, Duncan, SC) was deployed across the CUES field site to measure internal snowpack temperatures. Prior to the first snow, fiber-optic cable was deployed, including a 70m loop of cable south of the instrument shelter, and three coils of approximately 70m of fiber to be deployed during the winter season. Beginning on February 4, 2010, these coils were deployed on the snow surface over one another on successive storms (February 4, 21, and March 2, respectively). The snow depths at these times, as measured by an ultrasonic sounder, were 40 cm, 270 cm and approximately 350 cm, respectively. Throughout the winter season, DTS temperature measurements were made (Sensornet Halo, Sensornet Inc., Hertfordshire, UK) using 1 minute averaging intervals and post-processed using instrumented calibration baths (Tyler et al., 2008).

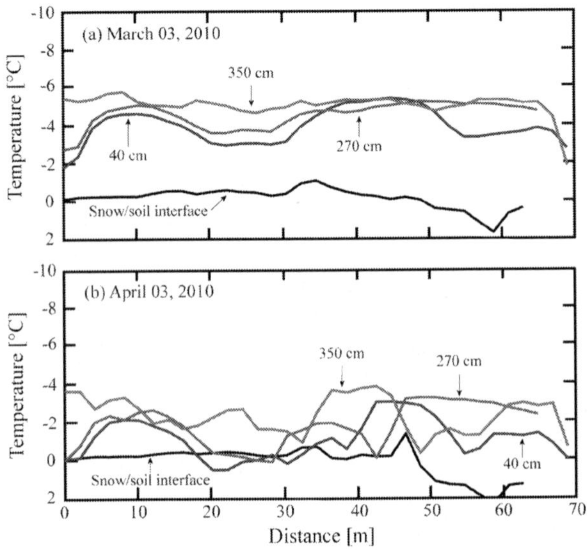

Fig. 5. (a) Horizontal distribution of snowpack temperatures from the CUES site on March 3, 2010. The elevations shown in the figures represent initial position of the fiber. Note that significant horizontal variation in temperature is visible in the 40 and 270 cm cables.
(b) Horizontal distribution of snowpack temperatures from the CUES site later in the season (April 3, 2010) shows much warmer snowpack as well as smaller vertical gradients in snow temperature. Horizontal differences continue to remain along each depth, and remain fixed in space when compared to the March 3, 2010 data

Fig. 5 (a) shows the temperature distribution from the soil surface up through the snowpack along the 70 meters of instrumented snow (Note that the y-axis is reversed to place the uppermost coil, which is also the coldest, at the highest elevation in the snowpack). Vertical thermal gradients, assuming the initial snow depths as representative lengths, range from 0.3 to approximately 8 °C m^{-1}. Horizontal gradients are also quite apparent (maximum ~0.25 °C m^{-1}), particularly in the 40 and 270 cm cables where two zones of colder conditions can be seen between 5 and 20 m and, to a lesser degree, between approximately 35 and 55 m. Both of these locations correspond with areas of snow ablation, with the zone between them characterized by drifted snow. Fig. 5 (b) shows the same transect data from one month later in the season (April 3, 2010). Snow temperatures are generally warmer, with much smaller differences with depth. However, horizontal differences have persisted, and remain stationary. At this time, vertical gradients have decreased to near zero in some areas, while the horizontal gradients reach a maximum of approximately 1 °C m^{-1}. Based upon visual observation of the transect late in the season, the zones of largest horizontal gradients correspond to rapid changes in snow depth, i.e., between drifts. In these zones, since the fiber-optic cable follows snow interfaces, the zones of coldest measurement correspond to areas of shallowest snow, where the fiber-optic cable is closer to the snow/air interface. While the horizontal gradients are small, their persistence and their gain in importance compared to the vertical gradient late in the snow season suggest that they may have significance in late season snow melt and snow stability. These transitions in temperature may lead to melt enhancement as well as serve as locations of weakness that could result in avalanche triggers on the slopes in the same area. Such horizontal variations in snow temperature are rarely recognized from individual snow pits or profiles and point to the utility of fiber-optic DTS approaches for snow thermal measurement.

The use of DTS for snow thermal process monitoring is relatively new and shows significant progress. Applications to snowmelt dynamics and frozen ground analysis are already underway and provide unique remotely sensed data that are important for hydrologic and environmental monitoring. Progress has been made in snowpack internal temperature measurement, and preliminary data suggest that horizontal gradients in snow temperature become significant late in the season as snowpacks approach melting conditions. Using fiber-optic DTS provides the first and only reliable method in which the spatial variability of snowpack temperatures can easily and remotely be measured. Measurement of both vertical and horizontal gradients and their spatial variability may provide important insights into snowpack dynamics, melting and avalanche susceptibility.

3.2 Fiber-optic distributed temperature sensing in ecohydrology

Devils Hole, a geothermally influenced groundwater-filled fracture in the carbonate aquifer of the southern Mojave Desert, represents a unique intersection of geology, hydrology, and ecology. The system offers a window into the carbonate aquifer of the Death Valley regional flow system (Riggs & Deacon, 2002), and is home to the world's only extant population of the endangered Devils Hole pupfish (*Cyprinodon diabolis*). Thought to comprise the smallest habitat containing the entire population of a vertebrate species (Moyle, 2002), Devils Hole was severely impacted by the development of nearby groundwater resources in the 1960s and early 1970s (Andersen & Deacon, 2001), leading to extensive litigation that to this day influences the United States' management of endangered species and water resources (Riggs & Deacon, 2002). The scientific footprint of Devils Hole is as significant as its legal impact; the system has been the site of pioneering work in palaeoclimate reconstructions (Winograd et al., 2006), as well as ichthyology (Minckley & Deacon, 1973) and evolution (Miller, 1950; Lema & Nevitt, 2006). In the late 1990's, the population of *C. diabolis* experienced an unexplained decline that has only recently stabilized, and a number of the hypotheses to explain this decline posit changes to the physical habitat of the system, especially changes in water temperatures.

Cyprinodon species tend to be very sensitive to water temperatures, and the 33-34 °C waters in Devils Hole are near the upper tolerance of most species (Brown & Feldmeth, 1971). The reproductive cycle of this fish is particularly temperature-dependent, and oogenesis in adult females, the viability of fertilized eggs, and the development of hatched larvae can all be severely retarded by exposure to higher temperatures (Shrode & Gerking, 1977). The thermal tolerances of *Cyprinodon* species also appear to be related to the variability of the water temperatures in which they live (Otto & Gerking, 1973) – the Devils Hole pupfish, which lives its entire life in water is seldom outside 33-35 °C, is especially susceptible to thermal stresses. Because the population of *C. diabolis* varies seasonally from highs of 200 to lows of fewer than 70 individuals, small variations in water temperature can have enormous impacts on the health of the population as a whole. Over the last three years, researchers have been using DTS instrumentation to characterize the thermal regime of Devils Hole.

Devils Hole comprises a 3 m x 8 m shallow shelf, which provides a spawning ground for the pupfish population, and a deep pool with a surface area of barely 30 m^2. The fracture that forms the deep pool strikes NE-SW and dips approximately 70°, with a uniform aperture of approximately 4 m. Between the surface and a depth of 30 m, the system is approximately 15 m wide along the NE-SW axis; below 30 m it opens into a cavern approximately 100 m wide, although the 4 m aperture remains. Devils Hole has been explored by divers to a depth of 130 m, and the clear waters allow visibility to almost 150 m (Riggs & Deacon, 2002). Below that point, the structure of the system is unknown. In January 2009, a fiber-optic cable was permanently installed in the deep pool to observe vertical temperature profiles.

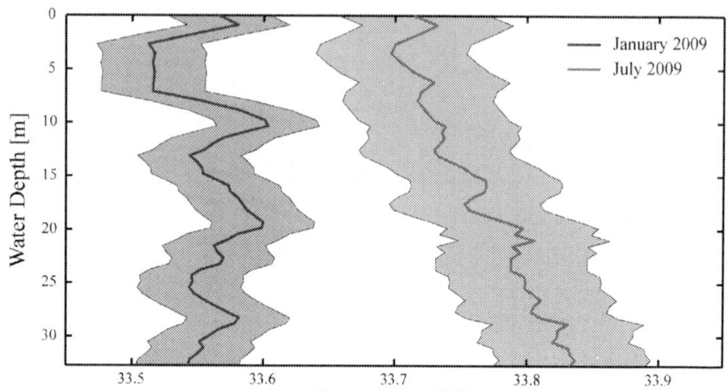

Fig. 6. Seasonal changes in the vertical temperature profiles of Devils Hole. The heavy lines indicate the mean temperature over 48 hours, and the shaded areas indicate the root mean square error of the calibrated DTS data (0.04° and 0.06° C in January and July, respectively)

Hausner et al. (2010) presented seasonal changes in the vertical temperature profiles that revealed a previously unidentified cycle of convective mixing. Fig. 6 shows the vertical temperature profiles observed in January and July, 2009. The near-uniform temperatures in the January profile indicate a system in which convective mixing is the dominant mode of heat transfer, while the constant temperature gradient observed in July fits a system dominated by conduction. In the summer, the water on the surface is warm, and tends to remain at the surface; the waters below the surface are warmed and stabilized by the natural geothermal gradient in the area, and the inverted temperature profile results. When water on the shallow shelf cools in the winter, the greater density causes it to plunge through the stratified water below, and the system mixes. The seasonal mixing affects the oxygen dynamics of the system, the winter distribution of allochthanous carbon (one of the pupfish's primary sources of food), and the availability of nutrients in the water column. However, these patterns would likely not have been noticed without the use of DTS in Devils Hole. This mixing model is based on observations of temperature gradients as small as 0.005 °C m^{-1}. Because a single common calibration is used to return DTS temperature observations along the entire length of the cable, fine gradients such as the one shown in Fig. 6 can be more easily observed and more confidently quantified with DTS instruments than with traditional sensors. Using a thermistor string, for example, the summer temperature gradient shown in Fig. 6 would likely have been lost in the noise generated by comparing multiple, separately calibrated instruments.

3.3 Fiber-optic distributed temperature sensing in salt-gradient solar ponds

A salt-gradient solar pond is a non-traditional solar collector that can provide long-term thermal storage and recovery for the collected energy. It is an artificially stratified water body that consists of three distinct zones (Suárez et al., 2010a): the upper convective zone, which is a thin layer of cooler and fresher water; the non-convective zone, comprised of a salt-gradient to suppress global circulation within the pond; and the lower-convective zone, in which salinity and temperature are the highest. The solar radiation that reaches the bottom of the pond is transformed into thermal energy and warms the brine in the lower convective zone. This warmer brine cannot rise beyond the lower convective zone because the effect of salinity on density is greater than the effect of temperature. The heat stored in the lower convective zone can only escape to the atmosphere by conduction, making the thickness of the non-convective zone a critical operating parameter for efficient solar pond operation (Suárez et al., 2010b). Because the brine has a low thermal conductivity, heat losses by conduction are relatively small. The hot brine in the lower convective zone may then be used directly for heating (Rabl and Nielsen, 1975), thermal desalination (Lu et al., 2001; Suárez et al., 2010c), or for other low-temperature thermal applications (Kumar and Kishore, 1999).

To investigate sustainable freshwater production using thermal desalination powered with solar energy, Suárez et al. (2010a, 2011) constructed a 1.0-m depth experimental salt-gradient solar pond. The pond was built inside a laboratory operated under controlled conditions, and was initially exposed to artificial lights 12 h per day. After reaching a thermal quasi steady-state, the pond was exposed continuously to the artificial lights until a new thermal steady-state was reached. Then, heat was extracted at approximately 0.5 m depth and was used to drive membrane distillation (Suárez, 2010). This pond was instrumented with a variety of sensors, including a vertical high-resolution DTS system to accurately monitor the thermal stratification and to investigate thermohaline circulation. This vertical high-resolution DTS system was constructed by wrapping a fiber-optic cable around a polyvinyl chloride pipe of approximately 1.3 m in length. This allowed temperature measurements every 1.1 cm in the vertical direction. A detailed explanation of the experimental salt-gradient solar pond and the characteristics of the vertical high-resolution DTS system are presented by Suárez et al. (2011).

Fig. 7 (a) presents the thermal evolution of the experimental salt-gradient solar pond as measured by the vertical high-resolution DTS system. Over the 15-days of the maturation period, the brine of the lower convective zone (0.65 m to 1.0 m) was warmed approximately 18 °C and showed strong internal convection when the lights were on. When the lights were off, the lower convective zone stratified due to the heat lost through the bottom and sides of the pond as well as through the non-convective zone. The upper convective zone (0.0 m to 0.1 m) also showed a diurnal pattern of strong stratification (due to radiation absorption) and subsequent mixing at night (due to penetrative convective mixing) during the first week of the experiment. After this, the upper convective zone showed strong internal convection during the day and night. This occurred because of the higher temperatures in the non-

usually shows a broader variation. For example, houses appliances and showers heat domestic waters, and temperatures in the range 30–90 °C could be expected in the drain. Even though the temperature drops as the domestic water travels before discharging into the sewer system, temperatures warmer than 20 °C could be found at the illicit connection (Hoes et al., 2009a).

Fig. 9 shows the DTS measurements performed by Hoes et al. (2009a) at an illicit connection that was located using this new technique. The temperatures in the sewer were relatively stable at approximately 12 °C, especially during the night. During the day and evening, anomalous peaks of temperatures were observed at this specific location. On-site verification by excavation and testing provided conclusive evidence that these temperature anomalies resulted from illicit connections in the storm water system.

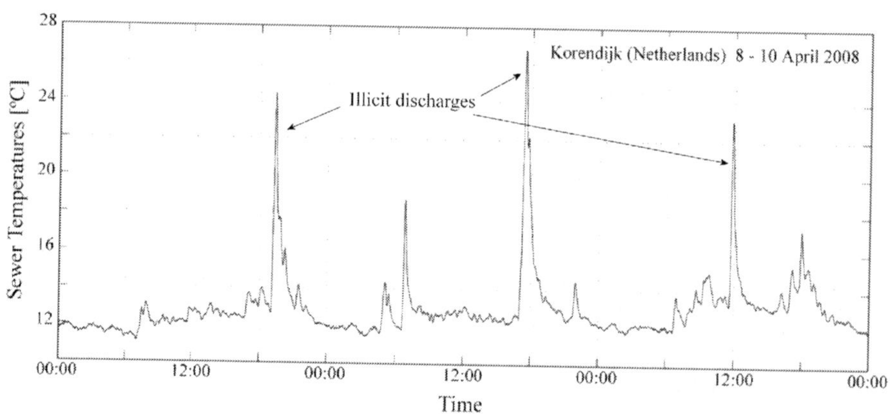

Fig. 9. DTS measurements performed in Korendijk (Netherlands) at the location of an illicit connection of a storm water system. Modified from Hoes et al. (2009a)

3.5 Distributed temperature sensing as an indirect tool for soil moisture estimation

Soil moisture is a key-state variable in water and energy balances at the land surface, and affects many different physical and environmental processes. The water content of a given soil is a controlling factor for agriculture and crop production, biological activity within the soil microbial community, and meteorological processes (Hausner, 2010). It also influences how efficiently water and solutes can move through the soil matrix. Despite its widespread influence and importance, measuring soil moisture is a difficult task, especially at field scale when soil heterogeneity could be important. The spatial scales of water content measurements present additional challenges. At small scales, single point measurements can be taken at specific intervals. However, these measurements are typically expensive, time consuming, and are difficult to extrapolate to field scale (e.g., over areas larger than 0.1 km^2). Remote sensing technologies can estimate near-surface soil moisture over larger areas, but there is a lack on methods to estimate in-situ water content at scales ranging from 0.1 to 80 km^2 (Robinson et al., 2008). By burying fiber-optic cables and recording the spatial distribution of soil temperatures over several days, Steele-Dunne et al. (2010) demonstrated that DTS can be used to estimate the distribution of soil moisture within these spatial scales. They first described heat transfer in a soil column using the diffusion equation:

$$\frac{\partial T}{\partial t} = D(\theta)\frac{\partial^2 T}{\partial z^2} = \frac{\kappa(\theta)}{C(\theta)}\frac{\partial^2 T}{\partial z^2} \tag{8}$$

where T is temperature; t is time; θ is volumetric water content (or soil moisture); $D(\theta)$, $\kappa(\theta)$ and $C(\theta)$ are the thermal diffusivity, thermal conductivity, and thermal heat capacity of the bulk soil, respectively, which are each functions of soil moisture; and z is depth. By measuring temperature at different depths, the soil thermal properties can be inferred by inversion, then the soil moisture can be estimated using a representative relationship between the soil thermal conductivity and soil moisture. Steele-Dunne et al. (2010) chose the Campbell exponential model (Campbell, 1985):

$$\kappa(\theta) = a + b\theta - (a-d)\exp\left[-(c\theta)^e\right] \tag{9}$$

where a, b, c, d and e are empirical parameters that depend on the volume fraction of soil minerals (e.g., quartz), organic matter, and on the clay fraction.

Steele-Dunne et al. (2010) buried two fiber-optic cables at approximately 8 and 10 cm below the surface of the ground to monitor the thermal response of the diurnal signal. A section of approximately 50 m was used to measure the transient temperature field at each depth and soil moisture was derived from these measurements. Even though they obtained reasonable results for both the soil thermal diffusivity and relative saturation (Fig. 10), it must be pointed out that their method relies on knowledge of the depth of the cable, and small variations were shown to cause large errors in the estimation of the thermal properties of the soil and thus, in the estimated values of soil moisture. The key of this method is the linkage between soil moisture and soil temperatures. Thus, the relationship between soil moisture and soil thermal conductivity must also be ascertained using in situ measurements over the entire range of saturation values (Steele-Dunne et al., 2010).

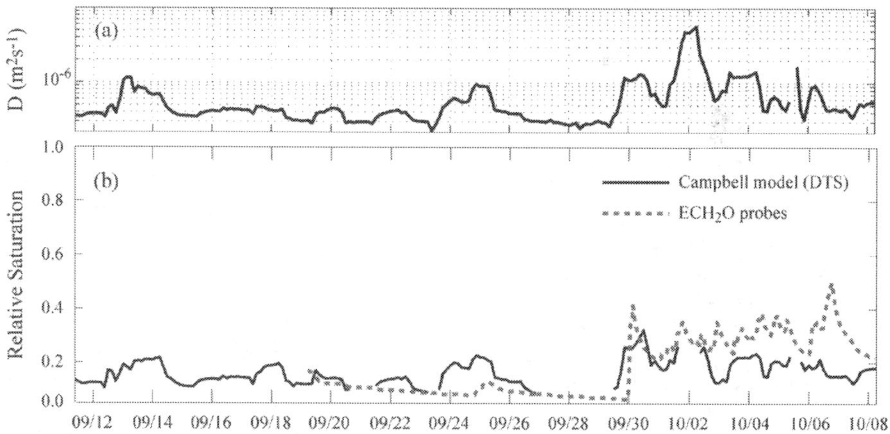

Fig. 10. (a) Estimated thermal diffusivity (D) at a selected location along the fiber-optic cable. (b) Mean value of relative saturation observed using ECH_2O probes and inferred relative saturation from the estimated thermal diffusivity using the Campbell model (1985). Modified from Steele-Dunne et al. (2010)

Sayde et al. (2010) also demonstrated the feasibility of using DTS to obtain accurate distributed measurements of soil water content. While Steele-Dunne et al. (2010) made passive observation of the natural diurnal cycles of heating and cooling, Sayde et al. (2010) actively heated a fiber-optic cable that was buried in variably saturated sand and then correlated the cumulative temperature increase and dissipation at different depths with the corresponding soil water content. Their correlations showed coefficients of determination (R^2) higher than 0.985, showing the promise of this method. Current efforts on spatially distributed soil moisture estimation are trying to combine both the passive and active DTS methods to obtain more precise measurements at field scale.

3.6 Atmospheric boundary layer estimation using distributed temperature sensing

The atmospheric boundary layer height, also called the mixed-layer or mixing height, is the lower part of the atmosphere in which the influence of heating and cooling, and surface friction is important (Brutsaert, 1982). The vertical mixing that occurs in this layer is driven by turbulence produced primarily by wind shear and buoyancy. The height of the atmospheric boundary layer is a key parameter for describing the physical state of the lower troposphere because it allows the prediction of air pollution concentrations or surface temperatures. For instance, air pollutants emitted at the ground surface will mix into this layer and, depending on meteorological conditions, can reach elevations from meters to a few kilometers (Brutsaert, 1982; Keller et al., 2011). Although it is one of the main parameters in atmospheric applications, the weakest point in meteorology is still the determination of the atmospheric boundary layer height (Builtjes, 2001).

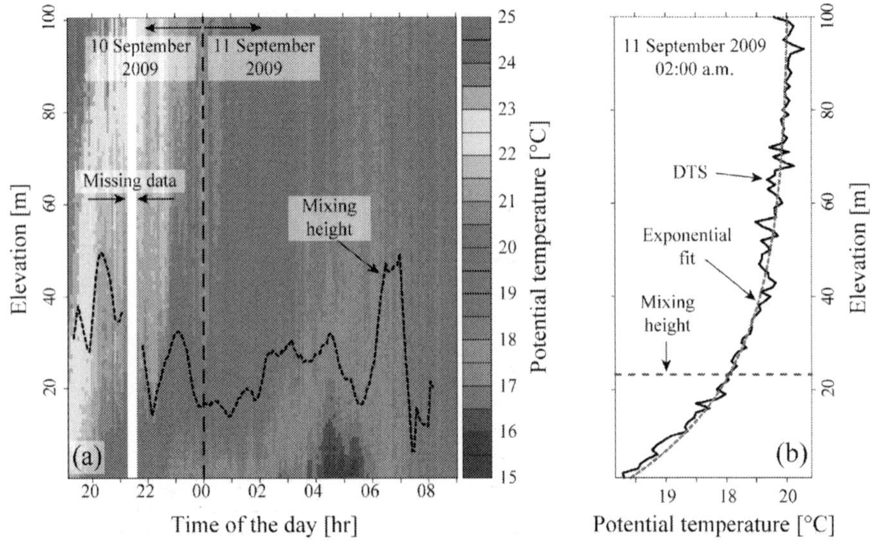

Fig. 11. (a) Evolution of potential temperature in the lowest 100 m of the atmosphere and mixing height as determined from exponential profile fits (bold-dashed line). (b) Example of a thermal profile measured by the DTS system with the corresponding best fit of an exponential curve. The mixing height is also shown. Modified from Keller et al. (2011)

Keller et al. (2011) presented a new method for measuring the air temperature profiles in the atmospheric boundary layer. A tethered balloon lifting system was used to suspend the optical fiber in the lowest 100 m of the atmosphere. As shown in Fig. 11, the temperature data were used to estimate the height of the stable boundary layer during the night, which varied between 5 and 50 m (September 11th data). They were also able to observe the erosion of the stable atmospheric stratification that occurred shortly after sunrise when the surface warmed. This resulted in an increase of turbulence, convection, and mixing due to latent and sensible heat being transferred to the atmosphere.

3.7 Other applications of fiber-optic distributed temperature sensing

Since it is difficult to describe all the work that has been carried out using this technology, in this section we present a brief summary of other environmental applications that were not described before.

Surface water-groundwater exchange has been widely studied using DTS methods in different environments such as estuaries (Henderson et al., 2009), rivers and streams (Lowry et al., 2007; Westhoff et al., 2007; Vogt et al., 2010; Slater et al., 2010), and ditches and canals (Hoes et al., 2009b). In general, these applications have used DTS technology to identify gaining sections of rivers, i.e., sections of the river where groundwater enters into it. The exception is the work of Vogt et al. (2010), where a vertical high-resolution DTS system was installed to measure both stream and streambed-sediment temperatures, which also allowed the estimation of the seepage rates from the river into the groundwater. Other environmental applications related with stream dynamics include the processes that controls thermal regime of saltmarsh channel beds (Moffett et al., 2008), development of distributed stream temperature models (Westhoff et al., 2007), response of stream temperatures in different riparian vegetation (Roth et al., 2010), and quantification of heat retardation along streams (Westhoff et al., 2010). The effects of radiative heating on fiber-optic cables used to monitor water temperatures have also been evaluated (Neilson et al., 2010; Suárez et al., 2011).

Early DTS work in hydrogeology was focused on thermal monitoring of geothermal wells and boreholes (Hurtig et al., 1994; Förster et al., 1997). Then, DTS systems were used to analyze the dynamic subsurface thermo-hydraulic conditions in aquifers (Macfarlane et al., 2002). Recently, Freifeld et al. (2008) developed a methodology to determine thermal conductivity in boreholes by combining a fiber-optic DTS system with a resistance heater, which created a controlled thermal perturbation in the borehole. The transient thermal data is inverted to estimate the thermal conductivity profile along the length of a wellbore with a spatial resolution equal to the spatial resolution of the DTS instrument. They also were able to determine the baseline geothermal profile and the ground surface temperature history in their study site (High Lake region, Nunavut, Canada).

4. Important considerations and future trends

Even though DTS systems have significant advantages over traditional measurement systems (as shown above), they also have limitations that must be assessed to maximize the potential of this technology. The performance of a DTS system is highly dependent upon the design of the experiment, the DTS instrument, the fiber-optic cables and connectors, the calibration, and the operating conditions. In this section we briefly discuss these factors in order to obtain thermal measurements with improved accuracy and precision.

4.1 Distributed-temperature-sensing instruments

There are a wide variety of commercial DTS instruments that use different methods for signal generation and data processing, which will affect the resolution and cost of the system. The cost of a DTS instrument ranges from approximately $15,000 to more than $150,000 (April 2011 U.S. Dollars). This cost depends on the features of each system (e.g., spatial and temporal resolution, ports to connect external thermocouples, number of channels). For temporary installations, there is also the option to lease DTS instruments for a lower cost (Tyler and Selker, 2009). The power requirements for DTS instruments range from approximately 20 to more than 100 W. To select the most cost-effective system, the required temporal and spatial repeatability must be evaluated by taking into account the goals and characteristics of each installation. The temporal and spatial averaging need to be long enough to provide suitable temperature resolutions, but short enough to observe the transient phenomena and the physics behind the system that is being measured.

4.2 Fiber-optic cables and connectors

As the cable acts as a thermometer, the selection of the cable is crucial in the experimental design. The cable typically includes a plastic jacket, tensile strength members, armoring to protect the optical fiber, a water-tight barrier, and the optical fiber(s). The cost of the cable can range from approximately $0.5 m^{-1} to more than $10 m^{-1} and cable weights can range from <1 kg km^{-1} to more than 30 kg km^{-1}. In general, the cost and weight of the cable are determined more by the armoring and construction than by the optical fiber itself. As cables get heavier, they are more difficult to handle and they can respond more slowly to temperature changes. Localized strains and stresses over the cable can result in greater signal losses, thus it is important to understand the environment of the installation before selecting the cable. Another important factor is the exposure of the cable to solar radiation, especially when used in the atmosphere or in shallow streams. Fiber-optic cables, as well as other thermal sensors, can absorb solar radiation and monitor higher temperatures than those of the surrounding ambient. This issue is more important when air or water velocities are small, and when the magnitude of solar radiation is large (Neilson et al., 2010). The effect of solar radiation absorption on cables can be reduced using reflective coatings or shielding the cable (Suárez et al., 2011). In addition, in a long-term installation with a cable exposed to sunlight, for example, it is important to select jackets that can withstand ultra-violet radiation and that can minimize radiative heating.

Connectors can also have an impact on signal strength. The standard connectors in the DTS industry typically produce a signal loss on the order of 0.1–0.2 dB (or even more if the connector is not clean or is incorrectly aligned). This loss is approximately the same loss that occurs in 300–700 m of fiber-optic cable, respectively (assuming an attenuation of ~0.3 dB km^{-1}). Ideally, the cable should only have connectors to physically connect the cable itself with the DTS instrument. If fibers need to be joined, fusion splices are recommended, since a properly fused fiber splice has a loss on the order of 0.01–0.03 dB.

4.3 Calibration and types of measurements

The calibration process is critical to achieve measurements with high resolution. An optimum deployment should have at least three sections of known temperature, each one with at least 10 sampling points within its length (e.g., 10 m when the spatial sampling resolution is 1 m). As suggested by Suárez et al. (2011), two of these sections can be at the same temperature and the other section needs to be at a different temperature. Ideally, the known-temperature sections should bracket the expected observations in the corresponding environment. If possible, the fiber-optic cable should have a loop to return the cable to the instrument (see Suárez et al. (2011) for more details about calibration procedures). This permits the DTS instrument to interrogate the fiber-optic from each end, i.e., allowing single- or double-ended measurements. Single-ended measurements refer to temperatures estimated from light transmission in only one direction along the optical fiber. These measurements assume a uniform rate of differential attenuation ($\Delta\alpha$) over the entire fiber, and provide greater precision near the instrument, degrading with distance because of the energy loss along the fiber length. Double-ended measurements refer to temperatures estimated from light transmission in both directions along the optical fiber. In these measurements, the temperature is estimated using single-ended measurements made from each end of the fiber, and can account for spatial variation in the differential attenuation of the anti-Stokes and Stokes backscattered signals, which typically occurs in strained fibers. Double-ended measurement results in a signal noise more evenly distributed across the entire length of the optical fiber, but uniformly greater than that obtained in a single-ended measurement (Tyler et al., 2009b; Suárez et al., 2011). Single-ended calibrations are encouraged for short cables (i.e., smaller than 1 or 2 km) since they provide more precision near the instrument. However, sometimes strains or sharp bends in the deployed fiber-optic cable yields large localized losses in the Stokes and anti-Stokes signals, which decrease the magnitude of the signals and add noise to the temperature data. Because these localized losses cannot be handled adequately by a single uniform value of the differential attenuation, further calibration is sometimes required to translate the scattered Raman signals into usable temperature data. In these cases, double-ended measurements are recommended because they allow the calculation of the differential attenuation along the entire length of the cable, and are much better able to handle the step losses introduced by strains and bends.

4.4 Operating conditions

An issue that has been observed in DTS installations is drift of the instrument. This drift typically occurs because of large variations in the instrument's temperature, particularly when the DTS instrument is subject to large daily temperature fluctuations in the field. The best solution to minimize this drift is to put the instrument in a controlled environment if possible. Other solution to minimize drift is to calibrate the DTS instrument at every measurement (sometimes referred to as dynamic calibration).

4.5 Current and future trends

As previously described, the ability to precisely measure temperature at thousands of locations is the main thrust of DTS systems. This capability has opened a new window for observation of environmental processes. Typical DTS instruments currently used in environmental applications can achieve temperature resolutions as small as ±0.01 °C, and spatial and temporal resolutions of 1-2 m and 10-60 s, respectively. At present, there are ongoing efforts to improve both spatial and temporal resolution of DTS systems. A high-resolution DTS instrument (Ultima, Silixa, Hertfordshire, UK) with temporal and spatial resolutions of 1 Hz and 12.5 cm, respectively, was recently commercialized and is under testing in environmental applications. This instrument simultaneously improved temporal precision by a factor of ten and spatial precision by a factor of four over previously available units. It was first deployed for observation of turbulent and stable atmospheric processes (http://oregonstate.edu/bmm/DONUTSS-2010/first-deployment-array), and it has also been utilized during a borehole heat tracer experiment designed to identify zones of high horizontal hydraulic conductivity and borehole through-flow. While this new DTS instrument has opened many possibilities, observation of atmospheric processes, for example, still needs improvement of temporal resolution to monitor turbulent processes. Instruments with this improved resolution are expected to be available in the near future and definitively will open new opportunities for observation of environmental processes.

5. Conclusion

In the environment, heat transfer mechanisms are combined in a variety of ways and span spatial scales that range from millimeters to kilometers. This extremely wide spatial scaling has been a barrier that limits observation, description, and modeling of environmental processes. The introduction of fiber-optic DTS has contributed to fill the gap between these two disparate scales. Fiber-optic DTS has proven effective to precisely observe temperatures at thousands of locations at the same time, with no issues of bias, and avoiding variability due to use of different sensors.

In this work, we have shown some of the environmental applications that have benefited from DTS methods. For instance, using fiber-optic DTS provides the first and only reliable method in which the spatial variability of snowpack temperatures can easily and remotely be measured. Measurement of both vertical and horizontal gradients and their spatial variability may provide important insights into snowpack dynamics, melting and avalanche susceptibility. DTS methods also have improved thermal measurements in natural and managed aquatic systems. For example, the hydrodynamic regimes in Devils Hole were observed at resolutions smaller than 0.1 °C, allowing observation of temperature gradients as small as 0.003 °C m^{-1}. This resolution allowed the examination of seasonal oxygen and nutrient distribution in the water column. In salt-gradient solar ponds, this temperature resolution allowed observation of both mixing and stratification, which is important for pond efficiency. In both Devils Hole and the solar pond, fiber-optic DTS provided high-resolution thermal measurements without disturbance of the water column. DTS methods also have been successfully utilized in other environments such as in atmosphere, streams, boreholes, and in many applications to understand the interdependence between groundwater and surface water. Novel extensions of DTS methods include spatially distributed soil moisture estimation, detection of illicit connections in storm water sewers, and there are many more to come in the near future, especially because the technology is growing and improving the spatial and temporal resolutions of DTS instruments, which will open new opportunities for environmental observations.

6. Acknowledgement

This work was funded by the National Science Foundation by Award NSF-EAR-0929638.

7. References

Alfe, D., Gillan, M.J., Vocadlo, L., Brodholt, J. & Price, G.D. (2002). The ab initio simulation of the earth's core. *Philosophical Transaction of the Royal Society of London A,* Vol.360, No.1795, (June 2002), pp. 1227-1244, ISSN 1364-503X

Alpers, M., Eixmann, R., Fricke-Begemann, C., Gerding, M. & Höffner, J. (2004). Temperature lidar measurements from 1 to 105 km altitude using resonance, Rayleigh, and Rotational Raman scattering. *Atmospheric Chemistry and Physics,* Vol.4, No.3, (2004), pp. 793-800, ISSN 1680-7316

Andersen, M.E & Deacon, J.E. (2001). Population size of Devils Hole pupfish (Cyprindon diabolis) correlates with water level. *Copeia,* Vol.2001, No.1, (February 2001), pp. 224-228, ISSN 0045-8511

Bales, R.C., Molotch, N.P, Painter, T.H, Dettinger, M.D., Rice, R. & Dozier, J. (2006). Mountain hydrology of the western United States. *Water Resources Research,* Vol.42, No.W08432, (2006), 13 pp., ISSN 0043-1397

Bense, V.F. & Kooi, H. (2004). Temporal and spatial variations of shallow subsurface temperature as a record of lateral variations in groundwater flow. *Journal of Geophysical Research,* Vol.109, No. B04103, (2004), 13 pp., ISSN 0148-0227

Branco, B.F. & Torgersen, T. (2009). Predicting the onset of thermal stratification in shallow inland waterbodies. *Aquatic Sciences,* Vol.71, No.1, (March 2009), pp. 65-79, ISSN 1015-1621

Brown, J.H. & Feldmeth, C.R. (1971). Evolution in constant and fluctuating environments: thermal tolerances of desert pupfish (Cyprinodon). *Evolution,* Vol.25, No.2, (June 1971), pp. 390-398, ISSN 0014-3820

Brutsaert, W. (1982). *Evaporation into the atmosphere: theory, history and applications* (1st Edition), Springer, ISBN 9789027712479, London, England

Builtjes, P.J.H. (2001). Major twentieth century milestones in air pollution modelling and its Applications, In: *Air Pollution Modeling and its Applications XIV*, Gryning, S.E. & Schiermeier (eds.), Springer, pp.3-16, Kluwer Academic/Plenum Publishers, ISBN 0306465345, New York

Campbell, G.S. (1985). *Soil physics with BASIC: transport models for soil-plant systems* (3rd Edition), Elsevier, ISBN 9780444425577, New York, USA

Dakin, J.P., Pratt, D.J., Bibby, G.W. & Ross, J.N. (1985). Distributed optical fiber Raman temperature sensor using a semiconductor light-source and detector. *Electronics Letters,* Vol.21, No.13, (1985), pp. 569-570, ISSN 0013-5194

Dürrenmatt, D.J. & Wanner, O. (2008). Simulation of the wastewater temperature in sewers with TEMPTEST. *Water Science and Technology,* Vol.57, No.11, (2008), pp.1809-1815, ISSN 0273-1223

Eichinger, W.E., Cooper, D.I., Parlange, M. & Katul, G. (1993). The Application of a Scanning, Water Raman-Lidar as a Probe of the Atmospheric Boundary Layer. *IEEE Transactions on Geoscience and Remote Sensing,* Vol.31, No.1, (January 1993), pp.70-79, ISSN 0196-2892

Farahani, M.A. & Gogolla, T. (1999). Spontaneous Raman Scattering in Optical Fibers with Modulated Probe Light for Distributed Temperature Raman Remote Sensing. *Journal of Lightwave Technology,* Vol.17, No.8, (August 1999), pp.1379-1391, ISSN 0733-8724

Förster, A., Schrötter, J., Merriam, D.F. & Blackwell, D. (1997). Application of optical-fiber temperature logging—An example in a sedimentary environment. *Geophysics,* Vol.62, No.4, (July-August 1997), pp.1107-1113, ISSN 0016-8033

Fridleifsson, I.B., Bertani, R., Huenges, E., Lund, J.W., Ragnarsson, A. & Rybach, L. (2008). The possible role and contribution of geothermal energy to the mitigation of climate change, *In: O. Hohmeyer and T. Trittin (Eds.) IPCC Scoping Meeting on Renewable Energy Sources, Proceedings*, pp. 20-25, Luebek, Germany, January 20-25, 2008

Freifeld, B.M., Finsterle, S., Onstott, T.C., Toole P. & Pratt, L.M. (2008). Ground surface temperature reconstructions: using in situ estimates for thermal conductivity acquired with a fiber-optic distributed thermal perturbation sensor. *Geophysical Research Letters*, Vol.35, No.L14309, (2008), 5 pp., ISSN 0094-8276

Hausner, M.B. (2010). *Estimating in situ integrated soil moisture content using fiber-optic distributed temperature sensing (DTS) measurements in the field.* M. Sc. Thesis, University of Nevada, Reno, 124 pp.

Hausner, M.B., Tyler, S.W., Wilson, K.P., Gaines, D.B. & Selker, J.S. (2010). Devils Hole: a window into the carbonate aquifer of the Death Valley regional flow system. *Abstract H31A-0977 presented at 2010 Fall Meeting, American Geophysical Union*, San Francisco, California, USA, December 2010

Henderson, R.D., Day-Lewis, F.D. & Harvey, C.F. (2009). Investigation of aquifer-estuary interaction using wavelet analysis of fiber-optic temperature data. *Geophysical Research Letters*, Vol.36, No.L06403, (2009), 6 pp., ISSN 0094-8276

Henderson-Sellers, B. (1984). *Engineering Limnology* (1st Edition), Pitman Published Limited, ISBN 0273085395, London, England

Hoes, O.A.C, Schilperoort, R.P.S., Luxemburg, W.M.J., Clemens, F.H.L.R. & van de Giesen, N.C. (2009a). Locating illicit connections in storm water sewers using fiber-optic distributed temperature sensing. *Water Research,* Vol.43, No.20, (December 2009), pp. 5187-5197, ISSN 0043-1354

Hoes, O.A.C., Luxemburg, W.M.J., Westhoff, M.C., van de Giesen, N.C. & Selker, J.S. (2009b). Identifying seepage in ditches and canals in polders in the Netherlands by distributed temperature sensing. *Lowland Technology International,* Vol.11, No.2, (December 2009), pp. 21-26, ISSN 1344-9656

Hurtig, E., S. Großwig, S., Jobmann, M., Kühn, K. & Marschall, P. (1994). Fibre-optic temperature measurements in shallow boreholes: experimental application for fluid logging. *Geotermics,* Vol.23, No.4, (August 1994), pp. 355-364, ISSN 0375-6505

Kelley, D.E., Fernando, H.J.S., Gargett, A.E., Tanny, J. & Özsoye, E. (2003). The diffusive regime of double-diffusive convection. *Progress in Oceanography,* Vol.86, No.3-4, (March 2003), pp. 461-481, ISSN 0079-6611

Keller, C.A., Huwald, H., Vollmer, M.K., Wenger, A., Hill, M., Parlange, M.B. & Reimann, S. (2011). Fiber optic distributed temperature sensing for the determination of the nocturnal atmospheric boundary layer height. *Atmospheric Measurement Techniques,* Vol.4, No.2, (2011), pp. 143-149, ISSN 1867-1381

Kersey, A.D. (2000). Optical fiber sensors for permanent downwell monitoring applications in the oil and gas industry. *IEICE Transactions on Electronics,* Vol.E83c, No.3, (March 2000), pp. 400-404, ISSN 0916-8524

Kumar, A. & Kishore, V. (1999). Construction and operational experience of a 6000 m2 solar pond at Kutch, India. *Solar Energy,* Vol.65, No.4, (March 1999), pp. 237-249, ISSN 0038-092X

Kurashima, T., Horiguchi, T. & Tateda, M. (1990). Distributed-temperature sensing using stimulated Brillouin scattering in optical silica fibers. *Optics Letters,* Vol.15, No.18, (1990), pp. 1038-1040, ISSN 0146-9592

Lachenbruch, A.H. (1959). Periodic heat flow in a stratified medium with applications to permafrost problems. U.S. Geological Survey Bulletin, 1083-A, 36 pp.

Lean, J. & Rind, D. (1998). Climate Forcing by Changing Solar Radiation. *Journal of Climate,* Vol.11, No.12, (December 1998), pp. 3069-3094, ISSN 0894-8755

Lee, K.K.M., Steinle-Neumann, G. & Akber-Knutson, S. (2009). Ab initio predictions of potassium partitioning between Fe and Al-bearing $MgSiO_3$ perovskite and post-perovskite. *Physics of the Earth and Planetary Interiors,* Vol.174, No.1-4, (May 2009), pp. 247-253, ISSN 0031-9201

Lema, S.C. & Nevitt, G.A. (2006). Testing an ecophysiological mechanism of morphological plasticity in pupfish and its relevance to conservation efforts for endangered Devils Hole pupfish. *The Journal of Experimental Biology,* Vol.209, No.18, (September 2006), pp. 3499-3509, ISSN 0022-0949

Lowry, C.S., Walker, J.F., Hunt, J.H. & Anderson, M.P. (2007). Identifying spatial variability of groundwater discharge in a wetland stream using a distributed temperature sensor. *Water Resources Research,* Vol.43, No.W10408, (2007), 9 pp., ISSN 0043-1397

Lu, H. , Walton, J. & Swift, A. (2001). Desalination coupled with salinity-gradient solar ponds. *Desalination,* Vol.136, No.1-3, (May 2001), pp. 13-23, ISSN 0011-9164

Lundquist, J.D. & Lott, F. (2008). Using inexpensive temperature sensors to monitor the duration and heterogeneity of snow-covered areas. *Water Resources Research,* Vol.44, No.W00D16, (2008), 6 pp., ISSN 0043-1397

Miller, R.R. (1950). Speciation in fishes of the genera Cyprinodon and Empetrichthys inhabiting the Death Valley region. *Evolution,* Vol.4, No.2, (June 1950), pp. 155-163, ISSN 0014-3820

Minckley, C.O & Deacon, J.E. (1973). Observations on the reproductive cycle of Cyprinodon diabolis. *Copeia,* Vol.1973, No.3, (August 1973), pp. 610-613, ISSN 0045-8511

Moffett, K.B., Tyler, S.W., Torgersen, T., Menon, M., Selker, J.S. & Gorelick, S.M. (2008). Processes controlling the thermal regime of saltmarsh channel beds. *Environmental Science and Technology,* Vol.42, No.3, (January 2008), pp. 671-676, ISSN 0013-936X

Moya-Laraño, J. (2010). Can Temperature and Water Availability Contribute to the Maintenance of Latitudinal Diversity by Increasing the Rate of Biotic Interactions? *The Open Ecology Journal,* Vol.3, No.1, (2010), pp. 1-13, ISSN 1874-2130

Moyle, P.B. (2002). *Inland Fishes of California: Revised and Expanded* (1st edition), University of California Press, ISBN 9780520227545, Berkeley, California, USA

Neilson, B.T., Hatch, C.E., Ban, H. & Tyler, S.W. (2010). Solar radiative heating of fiber-optic cables used to monitor temperatures in water. *Water Resources Research,* Vol.46, No.W08540, (2010), 17 pp., ISSN 0043-1397

Norand, C.W.B. (1920). Effect of High Temperature, Humidity, and Wind on the Human Body. *Quarterly Journal of the Royal Meteorological Society,* Vol.46, No.193, (January 1920), pp. 1-14, ISSN 0035-9009

Otto, R.G. & Gerking, S.D. (1973). Heat tolerance of a Death Valley pupfish (genus Cyprinodon). *Physiological Zoology,* Vol.46, No.1, (January 1973), pp. 43-49, ISSN 0031-935X

Painter, T.H., Donahue, D., Dozier, J., Li, W., Kattelmann, R., Dawson, D., Davis, R.E., Fiori, J., Harrington, B. & Pugner, P. (2000). The Mammoth Mountain cooperative snow study site: data acquisition, management, and dissemination. *Proceedings of the*

International Snow Science Workshop, Vol. ISSW2000, pp. 447-451, Big Sky, Montana, USA, October 2000

Rabl, A. & Nielsen, C. (1975). Solar ponds for space heating. *Solar Energy,* Vol.17, No.1, (April 1975), pp. 1-12, ISSN 0038-092X

Riggs, A. & Deacon, J.E. (2002). Connectivity in Desert Aquatic Ecosystems: The Devils Hole Story, *Proceedings of Spring-fed wetlands: important scientific and cultural resources of the intermountain region,* DHS Publication No. 41210, Las Vegas, Nevada, USA, May 2002, available from: http://www.dri.edu/spring-fed-wetlands

Robinson, D.A., Campbell, C.S., Hopmans, J.W., Hornbuckle, B.K., Jones, S.B., Knight, R., Ogden, F., Selker, J.S. & Wendroth, O. (2008). Soil moisture measurement for ecological and hydrological watershed-scale observatories: a review. *Vadose Zone Journal.* Vol.7, No.1, (February 2008), pp. 358-389, ISSN 1539-1663

Rogers, A. (1999). Distributed optical-fibre sensing. *Measurement Science and Technology,* Vol.10, No.8, (August 1999), pp. R75-R99, ISSN 0957-0233

Roth, T.R., Westhoff, M.C., Huwald, H., Huff, J.A., Rubin, J.F.,Barrenetxea, G., Vetterli, M., Parriaux, A., Selker, J.S. & Parlange, M.B. (2010). Stream Temperature Response to Three Riparian Vegetation Scenarios by Use of a Distributed Temperature Validated Model. *Environmental Science and Technology,* Vol.44, No.6, (February 2010), pp. 2072–2078, ISSN 0013-936X

Sayde, C., Gregory, C., Gil-Rodriguez, M., Tufillaro, N., Tyler, S.W., van de Diesen, N.C., English, M., Cuenca, R. & Selker, J.S. (2010). Feasibility of soil moisture monitoring with heated fiber optics. *Water Resources Research,* Vol.46, No.W06201, (2010), 8 pp., ISSN 0043-1397

Sclater, J.G, Parsons, B. & Jaupart, C. (1981). Oceans and Continents: Similarities and Differences in the Mechanisms of Heat Loss. *Journal of Geophysical Research,* Vol.86, No.B12, (1981), pp. 11535-11552, ISSN 0148-0227

Selker, J.S., Thevenaz, L., Huwald, H., Mallet, A., Luxemburg, W., van de Giesen, N.C., Stejskal, M., Zeman, J., Westhoff, M. & Parlange, M.B. (2006a). Distributed fiber-optic temperature sensing for hydrologic systems. *Water Resources Research,* Vol.42, No.W12202, (2006a), 8 pp., ISSN 0043-1397

Selker, J.S., van de Giesen, N.C., Westhoff, M., Luxemburg, W. & Parlange, M.B. (2006b). Fiber optics opens window on stream dynamics. *Geophysical Research Letters,* Vol.33, No.L24401, (2006), 4 pp., ISSN 0094-8276

Shrode, J.B. & Gerking, S.D. (1977). Effects of constant and fluctuating temperatures on reproductive performance of a desert pupfish, Cyprinodon n. nevadensis. *Physiological Zoology,* Vol.50, No.1, (January 1977), pp. 1-10, ISSN 0031-935X

Slater, L.D., Ntarlagiannis, D., Day-Lewis, F.D., Mwakanyamale, K., Versteeg, R.J., Ward, A., Strickland, C., Johnson, C.D. & Lane, J.W. (2010). Use of electrical imaging and distributed temperature sensing methods to characterize surface water-groundwater exchange regulating uranium transport at the Hanford 300 Area, Washington. *Water Resources Research,* Vol.46, No.W10533, (2010), 13 pp., ISSN 0043-1397

Smith, E. & Dent, G. (2005). *Modern Raman spectroscopy: a practical approach* (1st Edition), John Wiley and Sons, ISBN 978-0471497943, Sussex, England

Steele-Dunne, S.C., Rutten, M.M., Krzeminska, D.M., Hausner, M.B., Tyler, S.W., Selker, J.S., Bogaard, T.A. & van de Giesen, N.C. (2010). Feasibility of Soil Moisture Estimation using Passive Distributed Temperature Sensing. *Water Resources Research,* Vol.42, No.W03534, (2010), 12 pp., ISSN 0043-1397

Suárez, F. (2010). *Salt-gradient solar ponds for renewable energy, desalination and reclamation of terminal lakes.* Ph. D. Thesis, University of Nevada, Reno, 195 pp.

Suárez, F., Childress, A.E. & Tyler, S.W. (2010a). Temperature evolution of an experimental salt-gradient solar pond. *Journal of Water and Climate Change,* Vol.1, No.4, (2010), pp. 246-250, ISSN 2040-2244.

Suárez, F., Tyler, S.W. & Childress, A.E. (2010b). A fully coupled transient double-diffusive convective model for salt-gradient solar ponds. *International Journal of Heat and Mass Transfer,* Vol.53, No.9-10, (April 2010), pp. 1718-1730, ISSN 0017-9310.

Suárez, F., Tyler, S.W. & Childress, A.E. (2010c). A theoretical study of a direct contact membrane distillation system coupled to a salt-gradient solar pond for terminal lakes reclamation. *Water Research,* Vol.44, No.15, (August 2010), pp. 4601-4615, ISSN 0043-1354.

Suárez, F., Aravena, J.E., Hausner, M.B., Childress, A.E. & Tyler, S.W. (2011). Assessment of a vertical high-resolution distributed-temperature-sensing system in a shallow thermohaline environment. *Hydrology and Earth System Sciences,* Vol.15, No.3, (March 2011), pp. 1081-1093, ISSN 1027-5606

Tyler, S.W., Burak, S.A., Mcnamara, J.P., Lamontagne, A., Selker, J.S. & Dozier, J. (2008). Spatially distributed temperatures at the base of two mountain snowpacks measured with fiber-optic sensors. *Journal of Glaciology,* Vol.54, No.187, (December 2008), pp. 673-679, ISSN 0022-1430

Tyler, S.W. & Selker, J.S. (2009). New user facility for environmental sensing, *EOS, Transactions, American Geophysical Union,* Vol.90, No.50, (December 2009), pp. 483, ISSN 0096-3941

Tyler, S.W., Hausner, M.B., Suárez, F. & Selker, J.S. (2009a). Closing the energy budget: advances in assessing heat fluxes into shallow lakes and ponds, *EOS, Transactions, American Geophysical Union,* Vol.90, No.52 (Fall Meet. Suppl.), ISSN 0096-3941, San Francisco, California, USA, December 2009

Tyler, S.W., Selker, J.S., Hausner, M.B., Hatch, C.E., Torgersen, T., Thodal, C.E. & Schladow, S.G. (2009b). Environmental temperature sensing using Raman spectra DTS fiber-optic methods. *Water Resources Research,* Vol.45, No.W00D23, (2009), 11 pp., ISSN 0043-1397

Turner, J.S. (1974). Double-diffusive phenomena. *Annual Review of Fluid Mechanics,* Vol.6, (January 1974), pp. 37-54, ISSN 0066-4189

Uchida, Y., Sakura, Y. & Taniguchi, M. (2003). Shallow subsurface thermal regimes in major plains in Japan with reference to recent surface warming. *Physics and Chemistry of the Earth,* Vol.28, No.9-11, (2003), pp. 457-466, ISSN 1474-7065

Vogt, T., Schneider, P., Hahn-Woernle, L. & Cirpka, O.A. (2010). Estimation of seepage rates in a losing stream by means of fiber-optic high-resolution vertical temperature profiling. *Journal of Hydrology,* Vol.380, No.1-2, (January 2010), pp. 154-164, ISSN 0022-1694

Westhoff, M.C., Savenije, H.H.G., Luxemburg, W.M.J., Stelling, G.S., van de Giesen, N.C., Selker, J.S., Pfister, L., Uhlenbrook, S. (2007). A distributed stream temperature model using high resolution temperature observations. *Hydrology and Earth System Sciences,* Vol.11, No.4, (2007), pp. 1469-1480, ISSN 1027-5606

Westhoff, M.C., Bogaard, T.A. & Savenije, H.H.G. (2010). Quantifying the effect of in-stream rock clasts on the retardation of heat along a stream. *Advances in Water Resources*, Vol.33, No.11, (November 2010), pp. 1417-1425, ISSN 0309-1708

Winograd, I.J., Landwehr, J.M., Coplen, T.B., Sharp, W.D., Riggs, A.C., Ludwig, K.R. & Kolesar, P.T. (2006). Devils Hole, Nevada $\delta^{18}O$ record extended to the mid-Holocene. *Quaternary Research*, Vol.66, No.2, (September 2006), pp. 202-212, ISSN 0033-5894

Yilmaz, G. & Karlik, S.E. (2006). A distributed optical fiber sensor for temperature detection in power cables. *Sensors and Actuators A: Physical*, Vol.125, No.2, (January 2006), pp. 148-155, ISSN 0924-4247

Prandtl Number Effect on Heat Transfer Degradation in MHD Turbulent Shear Flows by Means of High-Resolution DNS

Yoshinobu Yamamoto and Tomoaki Kunugi
Department of Nuclear Engineering, Kyoto University
Japan

1. Introduction

Estimation of the heat transfer degradation effected by Magneto-Hydro-Dynamics (MHD) forces is one of the key issues of the fusion reactor designs utilized molten salt coolant. FLiBe which is the molten salt mixture of LiF and BeF, is one of the coolant candidates in the first wall and blanket of the fusion reactors, and has several advantages which are little MHD pressure loss, good chemical stability, less solubility of tritium and so on. In contrast, heat transfer degradation for the high Prandtl number, ($Pr=\nu/\alpha$, Prandtl number, ν is the kinetic viscosity, α is the thermal diffusivity) characteristics caused by the low thermal diffusivity and high viscosity (Sagara et al, 1995), was one of the issues of concern.

MHD turbulent wall-bounded flows have been investigated extensively by both experimental and numerical studies (Blum, 1967, Reed & Lykoudis, 1978, Simomura, 1991, Lee & Choi, 2001, Satake et al., 2006, Boeck et al, 2007, etc.) and much important information such as the drag reduction, the turbulent modulation, similarity of velocity profile, and heat transfer have been obtained.

On the other hands, MHD turbulent heat transfer in a high-Pr fluid has not been understood well. The previous experimental and direct numerical simulation (DNS) studies still have conducted for Prandtl number up to Pr=5.7. Therefore, the knowledge of the MHD heat transfer on higher-Pr fluids such as FLiBe (Pr=20–40), is highly demanded to verify and validate the MHD turbulent heat transfer models for the fusion reactor designs.

The objective of this study is to perform a direct numerical simulation of MHD turbulent channel flow for Prandtl number up to Pr=25, where all essential scales of turbulence are resolved. In this study, we report that the MHD turbulent heat transfer characteristics in Pr=25 for the first time and discuss that the MHD pressure loss and heat transfer degradation under the wide-range Pr conditions. The obtained database is of considerable value for the quantitative and qualitative studies of the MHD turbulent heat transfer models for the blanket design of a fusion reactor.

2. Target flow field and flow condition

The flow geometry and the coordinate system are shown in figure 1. The target flow fields are the 2-D fully-developed turbulent channel flows imposed wall-normal magnetic field and the streamwise and spanwise computational periods (L_x and L_z) are chosen to be $8h$ and $4h$, where h (=$L_y/2$) denotes channel half height.

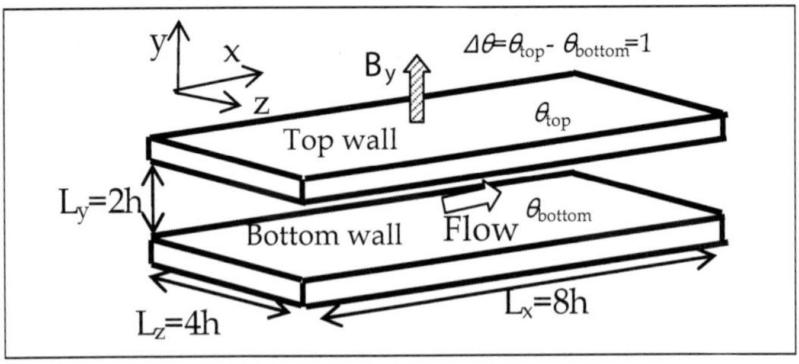

Fig. 1. Flow geometry and coordinate system

	Re_τ	Ha	Pr	Grid number N_x,N_y,N_z (M_x,N_y,M_z)	Resolution $\Delta x^+,\Delta y^+,\Delta z^+$ (temperature)
CASE1 (Lithium)	150	0,8,10,12	0.025	72,182,72	16,7,0.25-2.0,8.3
CASE2 (KOH)		0,6,8,10,12	5.7		
CASE2′ (fine grid)		0	5.0	432,182,216	2.8,0.25-2.0,2.8
CASE3 (FLiBe)		0,8,10,12	25.0	72,370,72 (320,370,160)	16.7,0.05-1.0,8.3 (3.8,0.05-1.0,3.8)
CASE3′ (fine grid)		0	25.0	648,370,324	1.9,0.05-1.0,1.9

Table 1. Numerical condition

Duo to the limitation of our utilizable computational resources, turbulent Reynolds number ($Re_\tau=u_\tau h/\nu$, u_τ: friction velocity) was limited to 150, and three thermal properties of the Lithium (Pr=0.025), KOH solution (Pr=5.7), and FLiBe (Pr=25) were covered. The KOH solution was used as the FLiBe simulant fluid in the previous experimental study (Yokomine et al., 2007) and the Lithium is a typical liquid metal coolant in a blanket of fusion reactors. To maintain the fully-developed turbulent status, Hartman number ($Ha=B_y 2h(\sigma/\rho\nu)^{1/2}$, B_y: wall-normal magnetic flux density, σ: electrical conductivity, ρ: density) was also limited around 12 in Re_τ=150 (Lee & Choi, 2001, Yamamoto et al., 2008). Numerical conditions are tableted in Table 1. Here, $N_x(\Delta x)$,$N_y(\Delta y)$, and $N_z(\Delta z)$ are the grid numbers (resolutions) in the streamwise, vertical, and spanwise directions, respectively. The super-script + denotes the nondimensional quantities normalized by the friction velocity, friction temperature and the kinematic viscosity. M_x and M_z are also the grid numbers in a horizontal direction temperature as mentioned 3.2, in case of adapting a different grid resolution for the flow and for the temperature field. In a wall-normal direction, the grid resolution resolved the Batchelor scale is ensured for all cases.

3. Numerical procedures

3.1 Governing equation and boundary condition

Governing equations of the present DNS are the continuity equation (1), the momentum equations (2) with the electric field described using the electrical potential approach (Simomura, 1991), Poisson equation (3) of the electrical potential, and the energy equation (4).

$$\frac{\partial u_i^*}{\partial x_i} = 0, \tag{1}$$

$$\frac{\partial u_i^*}{\partial t}+\frac{\partial u_i^* u_j^*}{\partial x_j}=\frac{F_i}{\rho}\delta_{i1}-\frac{\partial}{\partial x_i}\left(\frac{p^*}{\rho}\right)+\nu\frac{\partial^2 u_i^*}{\partial x_j \partial x_j}+\frac{\sigma}{\rho}B_{ijk}\left(-\frac{\partial \varphi^*}{\partial x_j}+\varepsilon_{jlm}u_l^* B_m\right)B_k, \tag{2}$$

$$\frac{\partial^2 \phi^*}{\partial x_i \partial x_i} = \frac{\partial}{\partial x_i}\left(\varepsilon_{ijk} u_j^* B_k\right), \tag{3}$$

$$\frac{\partial \theta^*}{\partial t}+\frac{\partial \theta^* u_j^*}{\partial x_j}=\alpha\frac{\partial^2 \theta^*}{\partial x_j \partial x_j}. \tag{4}$$

Here u_i and x_i are the streamwise (i=1), the vertical (i=2) and the spanwise (i=3) velocity and direction, respectively. t is time, F_i is the i-th competent mean pressure gradient, p is the pressure, ϕ is the electric potential, B_i=(0, B_y,0) is the Magnetic flux density, and θ is the temperature. Super script * denotes instantaneous value and δ_{ij}, ε_{ijk} (i,j,k=1-3) is the Kronecker delta and the Levi-Civita symbol, respectively.

Non-slip and periodic conditions are imposed for the boundary conditions of velocities and the constant temperature at top and bottom boundaries ($\theta_{top} > \theta_{bottom}$, θ_{top}: top wall temperature, θ_{bottom}: bottom wall temperature), and the periodic conditions are imposed for the temperature field. In this study, temperature transport is treated as a passive scalar.

The non-conducting conditions of the electric potential are applied to all walls and the periodic condition imposed on the horizontal directions. Total electric current in the spanwise flow domain is kept zero.

3.2 Numerical procedures

A hybrid Fourier spectral and the second-order central differencing method (Yamamoto et al, 2009) is used for the computations. The spectral method is used to compute the spatial discretization in the stream (x) and spanwise (z) directions. Nonlinear terms are computed with 1.5 times finer grids in horizontal (x and z) directions to remove the aliasing errors (Padding method). The derivative in the wall normal (y) direction is computed by a second-order finite difference scheme at the staggered grid arrangement (Satake et al, 2006). Time integration methods of the governing equations are the 3rd-order Runge-Kutta scheme for the convection terms, the Crank-Nicolson scheme for the viscous terms and the Euler Implicit scheme for the pressure terms, respectively. The Helmholtz equation for the viscous (diffusion) terms and the Poisson equations of the pressure and the electrical potential are solved by a Tri-Diagonal Matrix Algorithm, TDMA in Fourier space.

In DNS of the flow field, the Kolmogorov length scale has to be resolved. On the other hands, the length scales of the high-Pr temperature field are smaller than the smallest length scales of the velocity fields (Batchelor, 1959). To reduce the numerical costs in DNS of the high-Pr fluids, a different number of grid resolutions in the horizontal direction for velocity and temperature fields is adapted. In computing the temperature convection terms in (4) pseudo-spectrally, the grid points of velocities were expanded to the same grid points of the high-Pr temperature, as follow,

$$\Phi(k_x,y,k_z)=\begin{cases}\Phi(k_x,y,k_z)|k_x|\le \pi N_x/L_x,|k_z|\le \pi N_z/L_z \\ 0 \text{ otherwise}\end{cases},|k_x|\le \pi M_x/L_x,|k_z|\le \pi M_z/L_z. \quad (5)$$

Here, Φ denotes the velocities in Fourier space, k_x and k_z are the wavenumbers in the streamwise and spanwise directions, respectively. The phase-shift method (Patterson & Orszag, 1971) is used to remove the aliasing errors derived from the temperature convection terms. As a consequence, the grid size corresponded to the Batchelor length scale is retained for Pr=25.

Present DNS were calculated by using the T2K Open Supercomputer at Kyoto University. Elapsed time per one time step was about 1.2 [s] when using 8nodes (128cores) in CASE3.

4. Validation of present DNS

At the beginning of this study, we demonstrate the adequacy of the present DNS.

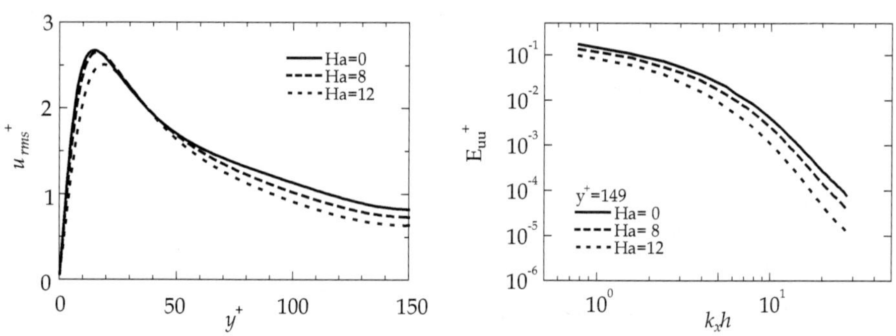

(a) Streamwise turbulent intensity, (b) Streamwise energy spectra,

Fig. 2. MHD suppression effects on turbulence

Figures 2 shows the turbulent intensities and the streamwise energy spectra at the channel center in Ha=0, 8, and 12. As well as the previous study (Lee & Choi, 2001), turbulent intensity was suppressed with increase of Ha as shown in Fig. 2-(a). Figure 2-(b) gives evidence that turbulent suppression effects can be remarkable in the high wave-numbers turbulence. It is clear that the effects of the grid dependency would be the biggest in Ha=0. Therefore, the convergences of the grid tendency were investigated in Ha=0, by using the DNS data fully-resolved the Batchelor length scale for Pr=5 or 25 in Ha=0 as tabled CASE2' and CASE3' in Table 1.

4.1 Medium high-Pr case

According to Na & Hanratty, 2000, the use of a higher resolution in horizontal direction does not produce significant changes to the first-order statistics from Pr=1 to 10. In this study, we investigated the grid dependency effects on the higher-order statistics such as the energy dissipation (=ε) and temperature energy dissipation (=ε_θ).

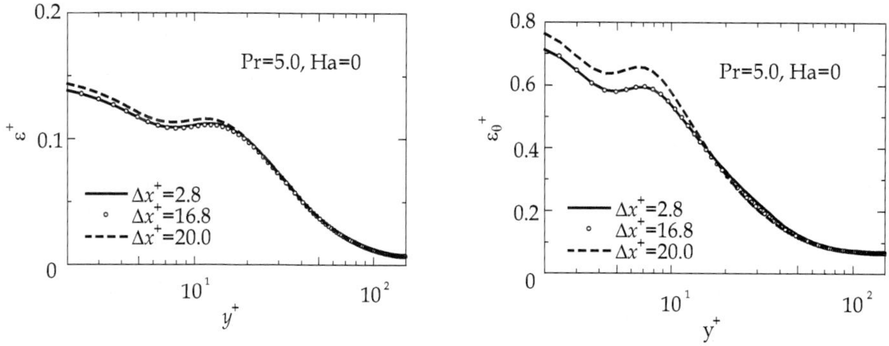

(a)energy dissipation, (b) temperature energy dissipation,

Fig. 3. Grid dependency on high-order statistics in medium high-Pr fluid

Figures 3 show the energy dissipation and temperature energy dissipation for Pr=5 with change of the horizontal resolutions. The required horizontal resolution for the reproductively of the energy dissipation and temperature dissipation, was estimated as Δx^+=16.7, and Δz^+=8.3 in this medium high-Pr fluid.

4.2 High-Pr case

For Pr=25, DNS in Re_τ=180 were conducted by means of a hierarchical algorithm in which only the scalar fields were solved on the grid dictated by the Batchelor scale (Schwertfirm &Manhart, 2007). However, the validation by using the different resolution for flow and high-Pr temperature field has not been reviewed yet. In this study, the adequacy of DNS by using a different resolution for flow and high-Pr temperature field is verified compared with DNS data fully-resolved the Batchelor length scale in the same grid size for flow and temperature.

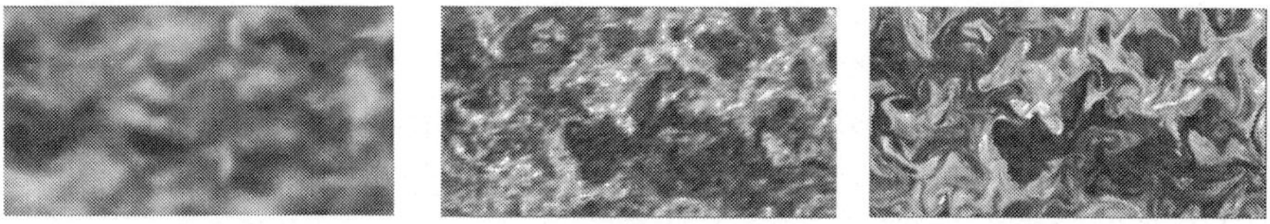

(a) (b) (c)

Fig. 4. Flow visualization, CASE3, Ha=0, Pr=25, y^+=149. (a) streamwise turbulent velocity, -2 (black) < u^+ < 2.0 (white), (b) turbulent temperature (coarse grid), -0.15 (black) < $\theta/\Delta\theta$ <0.15 (white) (c) turbulent temperature (fine grid), -0.15(black)<$\theta/\Delta\theta$<0.15 (white)

Figure 4 shows the flow visualization in results of Ha=0, Pr=25. In this case, 72x72 grids for flow (in Fig.4-(a)), 72x72 grids for the temperature filed (in Fig.4-(b)) and 320x160 grids for the temperature field (in Fig.4-(c)), were used in horizontal directions, respectively. Despites of the high wave-number flow fluctuations, the high wave- number temperature fluctuation can be computed as shown in Fig.4-(c).

Figure 5-(a) shows the temperature energy dissipation for Pr=25 with change of the horizontal resolutions. The required horizontal resolution for the reproductively of the temperature dissipation, was estimated as $\Delta x^+=8.3$, and $\Delta z^+=4.2$. This grid resolution is equivalent to twice as high for Pr=5; it is proportional to square root of the Pr ratio $(=(25/5)^{1/2})$. The effects of using the different resolution for flow and temperature cannot be found even in the temperature energy dissipation.

Figure 5-(b) shows the streamwise energy spectra near channel center for Pr=25. Compared with CASE3 and CASE3′, there is ninefold grid resolution in flow, but the variance of the spectra profile cannot be observed in this high-Pr temperature field. This indicates that the high wave-number velocity fluctuations less than the Kolmogorov scale can be ignored in a high-Pr passive scalar transport. As a consequence, we verify the adequacy of DNS by using the different resolution for flow and high-Pr temperature field and numerical cost in DNS of high-Pr fluids can be substantially reduced.

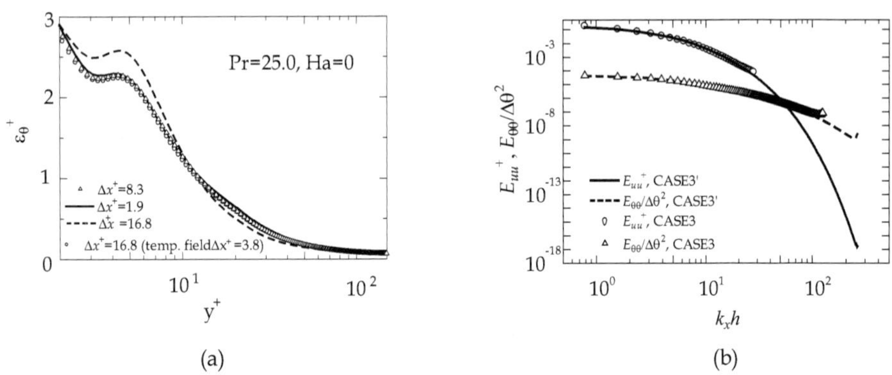

Fig. 5. Grid dependency and validation of different grid resolution for flow and high-Pr temperature field. (a) Temperature dissipation, (b) Streamwise energy spectra, streamwise velocity and temperature

5. MHD pressure loss and heat transfer

In this study, the friction drag confident (Cf) and Nusselt number (Nu) at the wall were expressed by

$$Cf=2u_\tau^2/U_b^2, \tag{6}$$

$$Nu=2h(d\Theta/dy)_{wall}/\Delta\theta. \tag{7}$$

Here, U_b and $(d\Theta/dy)_{wall}$ denotes the bulk mean velocity and mean temperature gradient at the wall.

Figure 6-(a) shows the friction drag coefficient as a function of the interaction parameter N $(=Ha^2/Re_b$, Re_b: Bulk Reynolds number$=U_b 2h/\nu)$, where the friction drag coefficients were normalized by that in Ha=0. The friction drag coefficients were monotonically decreased with increase of Ha; MHD pressure loss is less than the turbulent drag reduction effected by MHD. Therefore, all MHD cases of this study might be considered in a turbulent-laminar transition status. We need the DNS data in more higher Re to discuss the general relationships between MHD pressure loss and MHD turbulent drag reduction in turbulent condition.

Figure 6-(b) shows the Nusselt number as a function of N, where the Nusselt number were also normalized by that in Ha=0. Maximum heat transfer degradation in the low-Pr fluid was no more than 5% of the non-MHD condition. The usability of a low-Pr fluid was no doubt about heat transfer, however, Ha of Lithium was 700 times as large as one of FLiBe in the same Reynolds number (Re) and magnetic flux density (B_y) conditions.

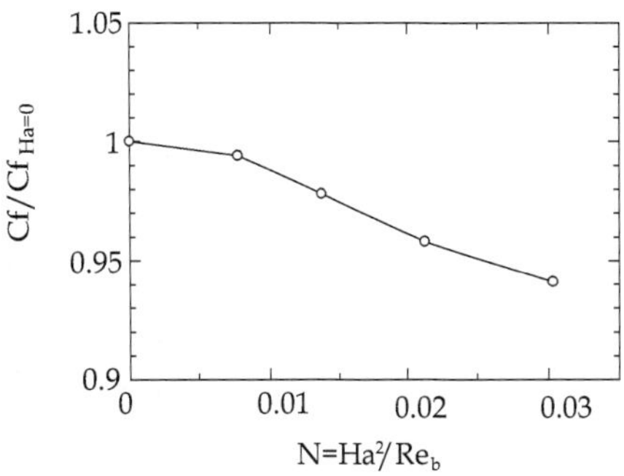

(a) friction drag coefficient,

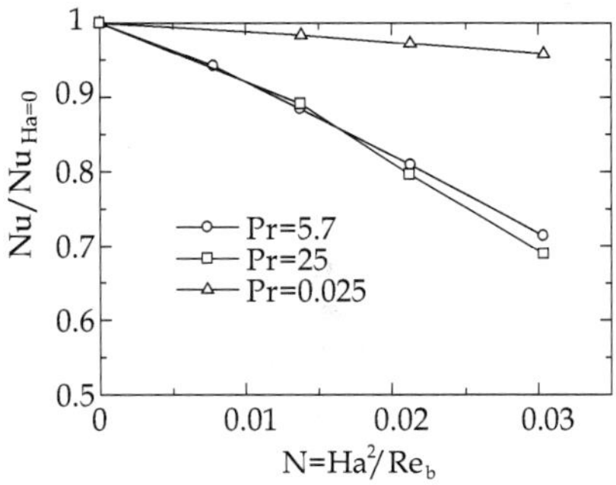

(b) Nusselt number,

Fig. 6. Friction drag coefficient and Nusselt number as a function N

On the other hands, heat transfer degradation in the high-Pr fluids (Pr=5.7 and 25) reached up to 30% without depending on Pr. This indicated that similarity of heat transfer degradation in high-Pr MHD flows might be existed.

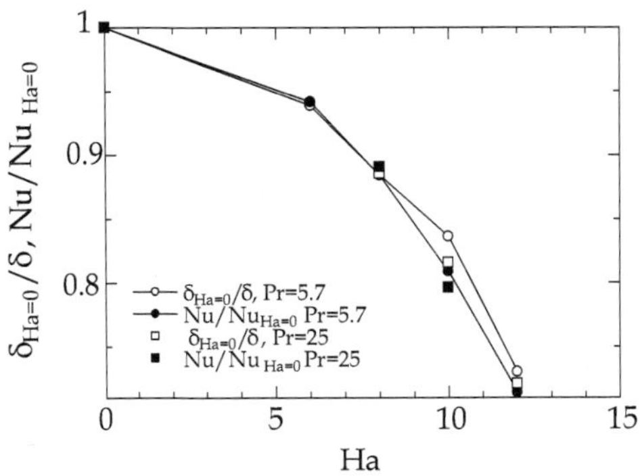

Fig. 7. Thermal viscosity thickness and Nusselt number as a function N

Figure 7 shows the thermal viscosity thickness (δ) and Nusselt number as a function of Ha in the high-Pr fluids, where thermal viscosity thickness was defined as

$$\delta = y^+ \text{ at } \Theta^+ = 0.99 \mathrm{Pr} y^+. \tag{8}$$

Thermal viscosity thickness was normalized by those in Ha=0. Heat transfer degradation was strongly correlated with change of the thermal viscosity thickness without depending on Pr.

6. Turbulence statistics

Figure 8 shows the profiles of temperature turbulent intensities for Pr=5.7 and 25. With increase of Ha, the peak position of turbulent intensity was shifted to the channel center side and the scale of it was decreased in both cases. In either case, the peak position was located below the wall-normal height y^+=15; thermal boundary layers for Pr=5.7 and 25 were thinner than the velocity boundary layer in the present MHD conditions.

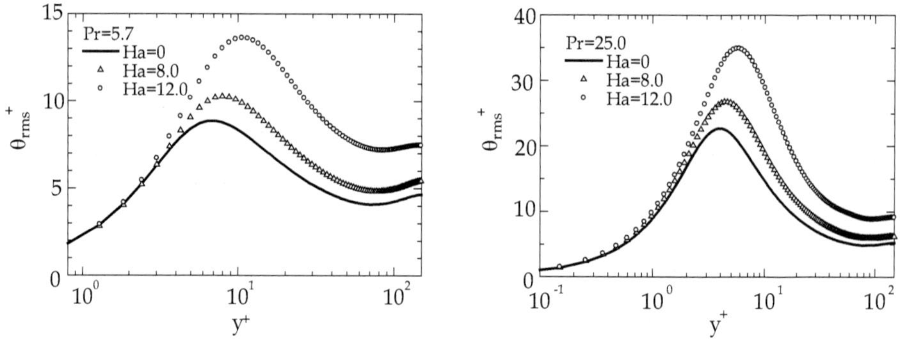

(a) Pr=5.7, (b) Pr=25,

Fig. 8. Turbulent temperature profiles

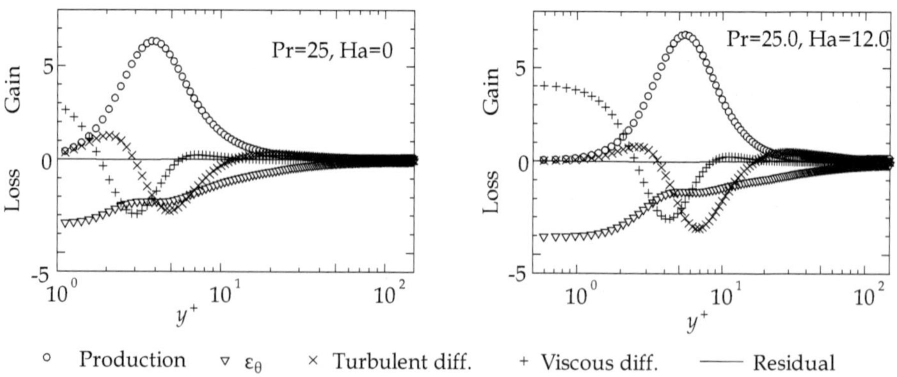

(a) Ha=0, Pr=25, (b) Ha=12, Pr=25,

Fig. 9. Budget of turbulent temperature energy

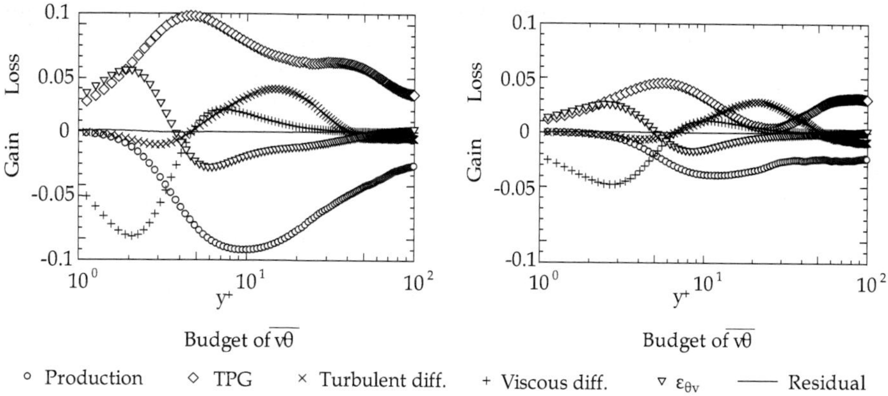

(a) Ha=0, Pr=25, (b) Ha=12, Pr=25,

Fig. 10. Budget of wall-normal turbulent heat flux

Figures 9 and 10 show the budget of turbulent temperature energy (K_θ) and wall-normal turbulent heat flux($v\theta$) for Pr=25, Ha=0 and 12. Transport equations (9) and (10) of turbulent temperature energy and wall-normal turbulent heat flux are expressed by

$$0=\underbrace{-\overline{v\theta}\frac{\partial\Theta}{\partial y}}_{\text{Production}}\underbrace{-\frac{1}{2}\frac{\partial\overline{\theta\theta v}}{\partial y}}_{\text{Turbulent diff.}}+\underbrace{\alpha\frac{\partial^2K_\theta}{\partial y^2}}_{\text{Viscous diff.}}-\underbrace{\alpha\overline{\left(\frac{\partial\theta}{\partial x_j}\right)^2}}_{\text{Dissipation:}\varepsilon_\theta}, \tag{9}$$

$$0=\underbrace{-\overline{vv}\frac{\partial\Theta}{\partial y}}_{\text{Production}}\underbrace{-\frac{\partial\overline{\theta vv}}{\partial y}}_{\text{Turbulent diff.}}+\underbrace{\frac{\partial}{\partial y}\left(\nu\overline{\theta\frac{\partial v}{\partial y}}+\alpha\overline{v\frac{\partial\theta}{\partial y}}\right)}_{\text{Viscous diff.}}+\underbrace{\overline{p\frac{\partial\theta}{\partial y}}-\frac{\partial\overline{p\theta}}{\partial y}}_{\text{Temp.Press-Grad.}}\underbrace{-(\nu+\alpha)\overline{\frac{\partial\theta}{\partial x_j}\frac{\partial v}{\partial x_j}}}_{\text{Dissipation}}. \tag{10}$$

Here, over bar denotes quantities estimated by ensemble average. In Fig. 9-(a), around the thermal buffer region (y^+=5), both diffusion terms of turbulent and viscous exceeded dissipation (ε_θ) term. Predominance of the diffusion terms in the high-Pr fluids (Pr>10) was confirmed in the previous DNS (Schwertfirm &Manhart, 2007). In Ha=12, predominance of diffusion terms was observed more clearly as shown in Fig. 9-(b). As well as turbulent temperature energy, turbulent diffusion term in Fig. 10-(b) was dominant at y^+=15-30 in Ha=12, however, the predominance of viscous diffusion term was indistinct. Compared with no-MHD case in Fig. 9-(a), the damping of turbulent diffusion term was small but the others were suppressed by the MHD effects; effects of turbulent diffusion on the MHD heat transfer were relatively larger with increase of Ha. These indicate that a sensitive model of the turbulent diffusion would be required in the prediction of MHD heat transfer in high-Pr fluids.

Figure 11 shows the turbulent Prandtl number (Pr_T) profiles for Pr=5.7 and 25. Turbulent Prandtl number was defined as

$$Pr_T=\overline{uv}\,/\,\overline{v\theta}\,. \tag{11}$$

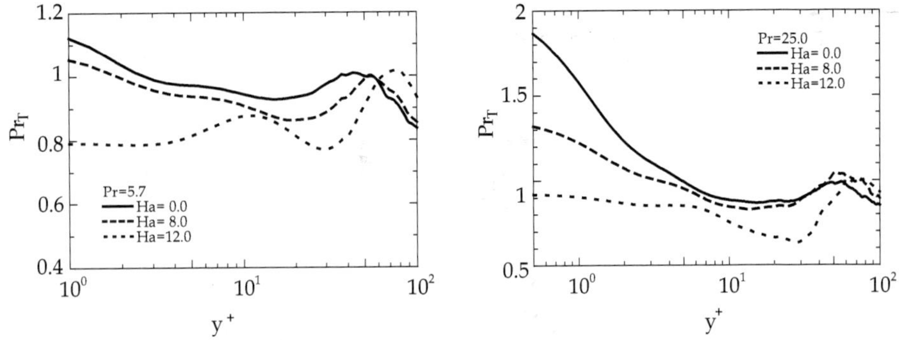

(a)Pr=5.7, (b) Pr=25,

Fig. 11. Turbulent Prandtl number profiles

Na & Hanratty, 2000 and Schwertfirm & Manhart, 2007 pointed out that turbulent Prandtl number close to the wall increases with increase of Pr. The turbulent Prandtl number profiles in the non-MHD case were good agreements with the results of Schwertfirm & Manhart, 2007, however, profiles in MHD case was decreased close to the wall for Pr=5.7 and 25 with increase of Ha. In Ha=12, the values of the turbulent Prandtl number in the vicinity of the wall fell into 1 for Pr=5.7 and 25. It was suggested that there was no MHD terms in balance of the heat transfer equation; turbulent effect on heat transfer might exceed that on momentum transfer as the limiting case of a turbulent-laminar transition status in Ha=12.

Figure 12 shows the time scale ratio for Pr=5.7 and 25. In non-MHD flow, time scale ratio had the weak peak at the buffer region for Pr=25 and 49 (Schwertfirm & Manhart, 2007 pointed out that). Time scale ratio profiles in MHD cases clearly had the peak in increase of Ha for Pr=5.7 and 25. At the buffer region, MHD effects on heat transfer might to be corresponded to the heat transfer in a higher-Pr fluid as shown in Figs. 9 and 12. However, these close to the wall might act on like a lower-Pr fluid as shown in Fig. 11.

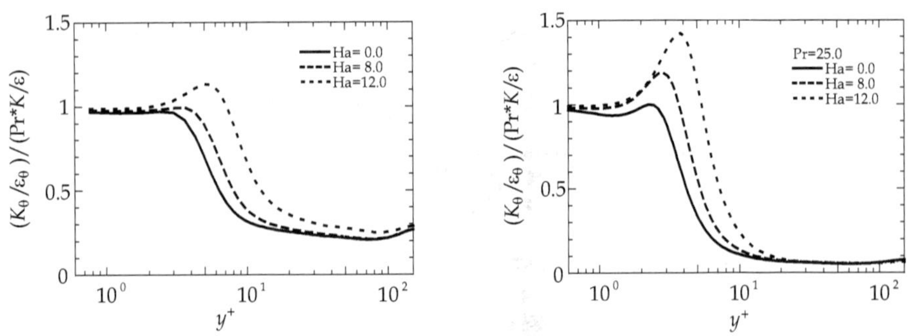

(a) Pr=5.7, (b) Pr=25,

Fig. 12. Time scale ratio profiles

Since both turbulent Prandtl number and time scale ratio were one of the dominant parameters in turbulent heat transfer modeling, change of profiles in increase of Ha might be caused the aggravation of the prediction accuracy.

7. Conclusion

In this study, direct numerical simulation of MHD turbulent channel flow for Prandtl number up to Pr=25 were performed. The adequacy of the present DNS data was verified by comparison with the DNS data fully-resolved the Batchelor length scale. As the results, the MHD turbulent heat transfer characteristics in Pr=25 were reported for the first time.
Maximum heat transfer degradation in the low-Pr fluid was no more than 5% of the non-MHD condition. On the other hands, heat transfer degradation in the high-Pr fluids (Pr=5.7 and 25) reached up to 30%. The similarity of heat transfer degradation in high-Pr MHD flows seemed be existed.
On the MHD heat transfer in high-Pr fluids, effects of turbulent diffusion were relatively larger. Turbulent Prandtl number and time scale ratio were considerably changed with increase of Ha.
The scaling of MHD heat transfer in high-Pr fluids was not understood yet. For the high-Ha and Re_τ condition (Ha>5, Re_τ>250), Boeck et al. 2007 reported the similarity of MHD mean velocity profiles on the parameter R(: Hartmann Reynolds number). To discuss the scaling of MHD heat transfer, we need DNS data of higher-Re and Ha conditions. In such cases, present DNS procedure by using a different resolution for flow and high-Pr temperature field will demonstrate a great advantage.

8. Acknowledgment

Present DNS were conducted by using the T2K open supercomputer at ACCMS and IIMC, Kyoto University. This study was supported by the Global COE program "Energy Science in the Age of Global Warming" and a Grant-in-aid for Young Scientists (B), KAKENHI (21760156) MEXT, Japan.

9. References

Batchelor, G.K., (1959), Small-scale variation of convected quantities like temperature in turbulent fluid. Part 1. General discussion and the case of small conductivity, *Journal of Fluid Mechanics,* Vol.5, pp.113–133.

Blum, E.YA., (1967), Effect of a magnetic field on heat transfer in the turbulent flow of conducting liquid, *High Temperature,*Vol. 5,pp. 68-74.

Boeck, T., Krasnov, D., and Zienicke, E., (2007), Numerical study of turbulent magnetohydrodynamic channel flow, *Journal of Fluid Mechanics,* Vol.572, pp.179-188.

Lee, D. & Choi, H., (2001), Magnetohydrodynamic turbulent flow in a channel at low magnetic Reynolds number, *Journal of Fluid Mechanics,* Vol.429, pp.367–394.

Na, Y. & T.J. Hanratty, T.J., (2000), Limiting behavior of turbulent scalar transport close to a wall, *International Journal of Heat and Mass Transfer,* Vol.43 , pp.1749-1758.

Patterson, S. & Orszag, S.A., (1971), Spectral calculations of isotropic turbulence: Efficient removal of aliasing interactions, *Physics of Fluids,* Vol.14, pp.2538-2541.

Reed, C.B. & Lykoudis, P.S., (1978), The effect of a transverse magnetic field on shear turbulence Journal of Fluid Mechanics, Vol.89, pp.147-171.

Sagara. A., Motojima, O., Watanabe, K., Imagawa, S., Yamanishi, H., Mitarai, O., Sato, T., Chikaraishi, H. and FFHR Group, Design and development of the Flibe blanket for helical-type fusion reactor FFHR, *Fusion Engineering and Design*, Vol.29, pp.51-56.

Satake, S., Kunugi, T., Takase, T., and Ose, Y., (2006), Direct numerical simulation of turbulent channel flow under a uniform magnetic field for large-scale structures at high Reynolds number, *Physics of Fluids*, Vol.18, 125106.

Schwertfirm, F. & Manhart, M., (2007), DNS of passive scalar transport in turbulent channel flow at high Schmidt numbers, *International Journal of Heat and Fluid Flow*, Vol.28, pp. 1204–1214.

Simomura Y., (1991), Large eddy simulation of magnetohydrodynamic turbulent channel flows under a uniform magnetic field, *Physics of Fluids* A 3, pp.3098-3106.

Yamamoto, Y.,Kunugi, T., Satake, S., and Smolentsev, S. (2008), DNS and k-ε model simulation of MHD turbulent channel flows with heat transfer, *Fusion Engineering and Design*, Vol.83, pp.1309-1312.

Yokomine, T., Takeuchi, J., Nakaharai, H., Satake, S., Kunugi, T., Morley, N.B., and M. A. Abdou, M.A., (2007), Experimental investigation of turbulent heat transfer of high Prandlt number fluid flow under strong magnetic field, *Fusion Science and Technology*, Vol.52, pp.625-629.

33

Effective Method of Microcapsules Production for Smart Fabrics

Luz Sánchez-Silva, Paula Sánchez and Juan F. Rodríguez
Department of Chemical Engineering/University of Castilla-La Mancha
Spain

1. Introduction

Nowadays, the attempts in the textile and clothing industry have moved towards more innovative and high quality products in order to differentiate themselves and be more competitive. The new demand for innovative textiles is increasingly oriented to match material innovation, new technologies and fashion. The new products are not only different for their lines, patterns and volumes but also for what they can do. Recently, microcapsules have been applied to many functional and technical textiles. Examples of this include fragrances, aromatic deodorants, cosmetics, insect repellents, antibiotics, polychromic, drug delivery for medical textiles and thermo-regulating systems. This kind of fabrics, that introduces new functionalities without affecting the look and feel of the textile, is commonly denominated as smart textiles.

Microencapsulated phase change materials (PCM) can be incorporated into textile structures to produce fabrics of enhanced thermal properties. A thermo-regulating fabric is an intelligent textile that has the property of offering suitable response to changes in external temperature changes or to external and environmental stimuli. The level of thermal comfort depends on the heat exchange between the human body and the environment that surrounds it.

Microcapsule production may be achieved by means of physical and chemical techniques. The use of some techniques has been limited to the high cost of processing, regulatory affairs, and the use of organic solvents, which are concern for health and the environment. In this way, a method based on a suspension like polymerization process for the encapsulation of phase change materials has been selected. This PCM encapsulation method is simply, inexpensive and technically easy.

Suspension like polymerization involves the dispersion of the monomer or monomers, mainly as a liquid in small droplets, into an agitated stabilizing medium consisting of water containing small amounts of suspension agents and without using aqueous phase inhibitors of secondary nucleation or modifiers. The initiator is dissolved in the monomer-PCM mixture and PCM material does not take part on the polymerization kinetic. In the proper conditions the polymer reacts mainly in the interface of the drop forming a shell around the PCM core since this interface is the only locus of polymerization. However, not all the recipes and conditions for suspension polymerization favour the formation of the polymer at the PCM/water interface all around the drop as desired and microparticles without complete phase separation into capsules are obtained.

Well-known PCM are linear chain hydrocarbons known as paraffin waxes (or n-alkanes), hydrated salts, polyethylene glycols (PEGs), fatty acids and mixture or eutectics of organic and non-organic compounds. PCM materials absorb energy during the heating process as phase change takes place and release energy to the environment in the phase change range during a reverse cooling process.

The required properties for a phase change materials depend on their specific application in textile fields. A wide spectrum of phase change materials are available with different heat storage capacity and phase change temperature. Different types of commercial PCMs can be encapsulated by means of suspension polymerization process. Rubitherm® RT20, Rubitherm® RT27, Rubitherm® RT31, Petrepar® n-C14 and Petrepar® n-C-13 have demonstrate their capability to be encapsulated and their thermal abilities to absorb and release energy. Their physical and chemical properties make them very attractive for thermal storage.

Thermal properties, air permeability, moisture vapour permeability and moisture regain of materials also influence the heat balance of the body and, consequently, affect clothing comfort (Ren & Ruckman, 2004). The incorporation of PCM microcapsules to textiles can affect other comfort-related properties and hand of the materials adversely, especially when the topical application of microcapsules results in drastic changes in the surface characteristics of materials. The extent of change in these properties depends on the loading amount of PCM microcapsules (Shin et al., 2005).

Several methods of incorporating PCM microcapsules into a fibrous structure have been developed. The microcapsules can be applied by stamping works, exhaustion dyeing, impregnation, spraying and coating or by direct incorporation in the fibre without highly modifying its touch and colour (Monllor et al., 2009; Dixit & Goal, 2007; Rodrigues et al., 2009). In previous applications of PCM technology in the textile industry, for garments and home furnishing products, microencapsulated PCM were incorporated into acrylic fibers (Bryant & Colvin, 1988) or polyurethane foams (Colvin & Bryant, 1996) or were embedded into a coating compound and topically applied to a fabric (Bryant & Colvin, 1994). Shin et al., (2005) incorporated melamine-formaldehyde microcapsules containing eicosane on polyester knit fabrics by means of a pad dry cure method with a polyurethane binder. Mengjin et al., (2008) developed a new kind of thermo-regulating fiber based on PVA and paraffin. Furthermore, Onder et al., (2008) studied the microencapsulation of three types of paraffin waxes by complex coacervation to improve thermal performances of woven fabrics. Recently, Koo et al. (2009) have attempted to demonstrate the application of PCM microcapsules on waterproof nylon fabrics and to enhance thermal insulation effect with ceramic materials (SiC) by using a dual coating method.

Binders play a crucial role in microcapsule coating formulation for various textile materials, as they are required to fix microcapsules on textile supports permanently. To a large extend, binders determine the quality, durability and washability of textile materials with microencapsulated ingredients. Some of the most frequently used binders in textile are water-soluble polymers, such as starch and modified starches, carboxymethyl cellulose; synthetic latexes, such as styrene-butadiene, polyvinylacetate or acrylate latexes; and aminoaldehyde resins (Boh & Knez, 2006).

In our previous work, the fixation of PCM microcapsules containing paraffin with a melting point around 40°C, into a cotton textile substrate by means of a coating technique were carried out. Furthermore, the influence of different coating formulations and mass ratio of microcapsules to coating formulation were evaluated in order to obtain an adequate textile with thermo-regulating properties (Sánchez et al., 2010). The coating fabric with 35 wt.% of microcapsules added related to commercial coating binder (WST SUPERMOR®) showed a energy storage capacity of 7.6 J g^{-1}, a high durability and an adequate stability after washing, rub fastness and ironing treatments. A difference of 8.8°C for 6 s was observed for textiles with thermo-regulating properties in comparison with a coated one without microcapsules. The different application areas of textiles with thermo-regulating properties imply the fixation to very different substrates. In this sense, there are few references in the literature studying the influence of the kind of textile on the fixation of microcapsules (Koo et al., 2009). In addition, the PCM microcapsules incorporation could degrade the original functionalities of the textile such as soft touch, vapor or moisture permeability and wearing comfort.

The aim of this work was to investigate the production of textiles with thermo-regulating properties by using PCM microcapsules and a coating technique. The influence of the type of used PCM on the heat capacity of microcapsules, the particle size distribution (PSD) and the microcapsules yield of each experiment was studied. On the other hand, different type of textile substrates depending on the field of their textile applications (apparel, blankets, insulation, protective clothing) were evaluated. Furthermore, a study of thermoregulatory effect of the coating fabrics produced was carried out using an infrared thermography camera. Thermal properties of textile samples were examined by Differential Scanning Calorimetry (DSC). Furthermore, Environmental Scanning Electron Microscopy (ESEM) and Optical Microscopy (OM) techniques were used to check the presence, surface distribution, preferred join position and to analyse the structure of microcapsules into the textile.

2. Experimental

2.1. Microcapsules synthesis

Styrene (99 wt.%) of reagent grade (Merck Chemical) previously purified by washing with sodium hydroxide and dried with calcium chloride was used as the monomer. Benzoyl peroxide (97 wt.%) was used as initiator (Fluka Chemical). PRS® paraffin wax, Rubitherm® RT20, Rubitherm® RT27, Rubitherm® RT31, Petrepar® n-C14 and Petrepar® n-C-13 were used as core materials. Polyvinylpyrrolidone (K30, Mw 40,000 gmol^{-1}) of reagent grade (Fluka Chemical) was used as stabilizer and methanol to pour the samples. All these reagents were used as received. Water was purified by distillation followed by deionization using ion-exchange resins. Nitrogen was of high-purity grade (99.999%).

A suspension like polymerization process was used for the microcapsules synthesis. A tubular type Shirasu porous glass membrane was used for a better control of microparticle size. Details of the synthesis process were previously described elsewhere (Sánchez et al., 2008b).

2.2 Preparation of textiles with thermo-regulating properties

Microcapsules were fixed into seven fabrics by means of a coating technique, using a motorized film applicator from Elcometer model 4340 according with ASTM D-823C (ASTM D-823-C, 1997). WST SUPERMOR® (supplied by Minerva Color Ltd.) were used as commercial coating binder. In a previous study (Sánchez et al., 2010), this binder was selected due to allow an efficient fixation of the PCM microcapsules on the fabrics. Every sample had 200 mm of wide and 290 mm of length due to requirements of the motorized film applicator.

The coating formulation consisted of WST SUPERMOR® commercial binder and Rubitherm® RT31 microcapsules (35 wt. % of the coating mixture).
The textile substrate was set on the motorized film applicator surface assuring the fabric with clips. In this study, the thickness selection of the coating layer was 0.1 mm to obtain a high thermal storage. The position of the motorized film applicator and the selection thickness was carried out manually. A dragging speed of 5 mm s^{-1} was chosen to allow a homogeneous coating along the film applicator.
Finally, the coated fabric was cured at 95 °C for 11 minutes.

2.3 Characterization

2.3.1 Differential Scanning Calorimetry (DSC)

Measurements of melting point and latent heat storage capacities of different materials were performed in a differential scanning calorimetry model DSC Q100 of TA Instruments equipped with a refrigerated cooling system and nitrogen as the purge gas. Measurements were carried out in the temperature range from -30°C to 80°C with heating and cooling rate of 10 °Cmin^{-1}.
Various samples of each experiment were analyzed at least three times and the average value was recorded. DSC analyses of coating textiles from random areas were done.
Furthermore, the encapsulation ratio of the different PCM in the microcapsule was calculated with the following equation based on enthalpy values:

$$\% \text{ PCM content by weight} = (\Delta Hm / \Delta Hpcm) \times 100\% \quad (1)$$

where ΔHm is the enthalpy for the analysed microcapsules (Jg^{-1}) and $\Delta Hpcm$ is the enthalpy of pure PCM.
In order to determine the thermal stability of the reversible phenomena of phase change, the coated textiles were subjected to repeated cycles of melting and crystallization.

2.3.2 Environmental scanning electron microscopy (ESEM)

ESEM was used to analyze the morphological structure of the microcapsules and the fixation and integrity of PCM microcapsules into the coating textile substrates. Textile samples were observed by using XL30 (LFD) ESEM with a wolfram filament operating at a working potential of 20 kV.

2.3.3 Calculation of number-average diameter and volume-average diameter

Particle size and particle size distribution (PSD) of microcapsules were determined on a Malvern Mastersizer Hydro 2000 SM light scattering apparatus with dilute dispersions of the particles in methanol.

2.3.4 Infrared thermography

The temperature distributions of the coated textiles with thermo-regulating properties were evaluated by means of an infrared and visible camera Fluke Ti25. This dispositive allows to obtain thermal and visual images in the range of temperatures from -20°C to 250°C with a precision of ±2 °C. The screen was observed from a distance of 30 cm at 24°C. Images were downloaded using Fluke SmartViewTM software for analysis. The coated fabrics were pre-heated at 60°C, time considered as zero, and then cooled to room temperature.

Thermal human comfort in summer conditions was tested, recording images from 25°C to outside temperature (35°C), comparing a reference textile with a prototype textile with thermo-regulating properties in contact with the body (shoulders in this specific case).

3. Results

3.1 Microencapsulation of different type of phase change materials (PCM)

The required properties for a phase change materials depend on their specific application in textile fields. A wide spectrum of phase change material is available with different heat storage capacity and phase change temperature. In this study different type of commercial PCM were assayed in order to know what PCM materials are suitable to be encapsulated by means of suspension polymerization. Thus, Rubitherm® RT20, Rubitherm® RT27, Rubitherm® RT31, Petrepar® n-C14 and Petrepar® n-C-13 were assayed due to their physical and chemical properties are very attractive for thermal storage (Table 1). All of them are saturated hydrocarbons

PCM	Molecular weight ($gmol^{-1}$)	Latent heat of fusion (Jg^{-1})	Melting temperature (°C)	Viscosity (mm^2s^{-1}) at 98°C
PRS® paraffin wax	168-240	206.8	40-45	2.43
Rubitherm® RT31	268	199.3	31	2.07
Rubitherm® RT27	258	214.6	28	1.64
Rubitherm® RT20	244	177.7	22	1.52
Petrepar® n-C14	198.4	225.0	3-7	0.99
Petrepar® n-C13	184.4	134.4	(-7)-(-5)	0.88

Table 1. Properties of different types of PCMs investigated

Figure 1 shows the particle size distributions (PSDs) in volume (Figure 1a) and in number (Figure 1b) of microcapsules obtained after the polymerization process using these PCM materials. It can be seen from Figure 1a that Petrepar® n-C14 and n-C13 exhibit bimodal PSDs with particles sizes smaller than 115 μm. However, PRS® paraffin wax, Rubitherm® RT31, RT27 and RT20 shows unimodal PSDs ranging in the interval between 149 to 251 μm. In all experiments a big difference between the average particle size in volume and in number was observed due to the heterogeneous sizes of obtained microcapsules. This behaviour was reported in previous works (Sánchez et al., 2007; Sánchez et al., 2008a; Sánchez et al., 2008b).

Table 2 reports average diameters ($dp_{0.5}$) in volume and in number, storage energy capacities and amount of PCM encapsulated of microcapsules produced using different PCMs. The mean diameter in number of the microcapsules increases as described: Rubitherm® RT27>PRS® paraffin wax>Rubitherm® RT31> Petrepar® C-14>Petrepar® C-13>Rubitherm® RT20. However, the average diameter in volume increases in the following way: Rubitherm® RT27>Rubitherm® RT31>Rubitherm® RT20>PRS® paraffin wax>Petrepar® n-C14>Petrepar® n-C13. Therefore, the average diameter of the microcapsules depends on phase change materials encapsulated. For PRS® paraffin wax, Petrepar® C-14 and Petrepar® C-13, the average diameter of microcapsules enlarged with increasing the viscosity. This fact could be due to the polymerization system is subject to nonnegligible effects of the viscosity ratio (dispersed phase/continuous phase viscosity). According to Hamielec & Tobita, (1992) in suspension polymerization an increase of the forming droplets viscosity and the density difference between the phases results in a tendency towards settling. However, this trend was not observed for Rubitherm materials.

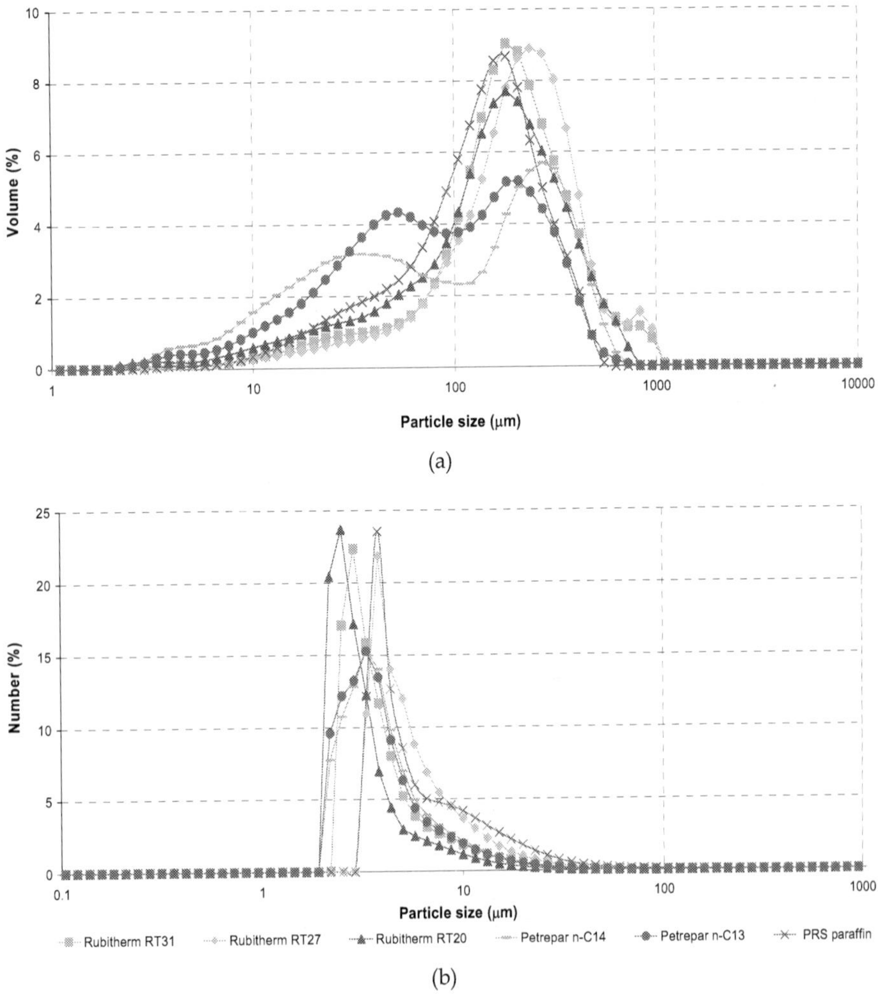

Fig. 1. Particle size distribution for microcapsules obtained using different PCM: (a) in volume and (b) in number

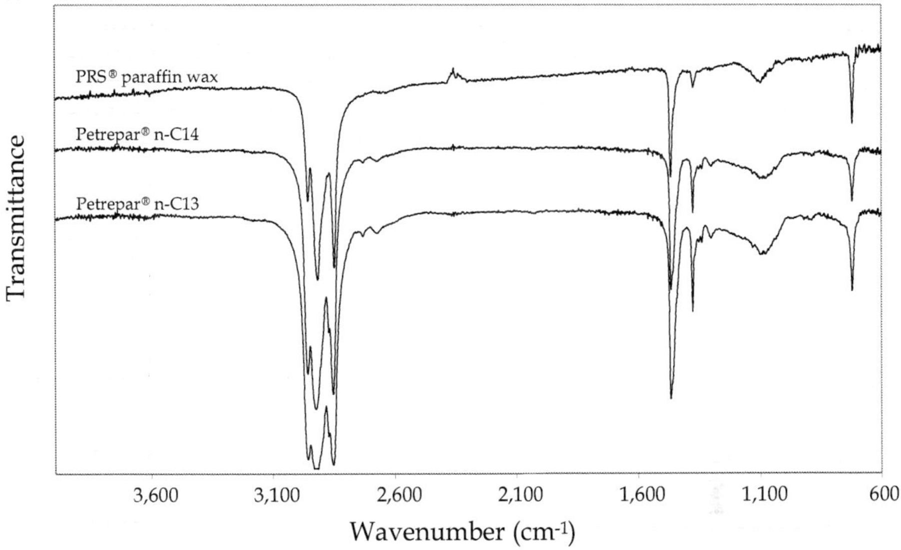

(a)

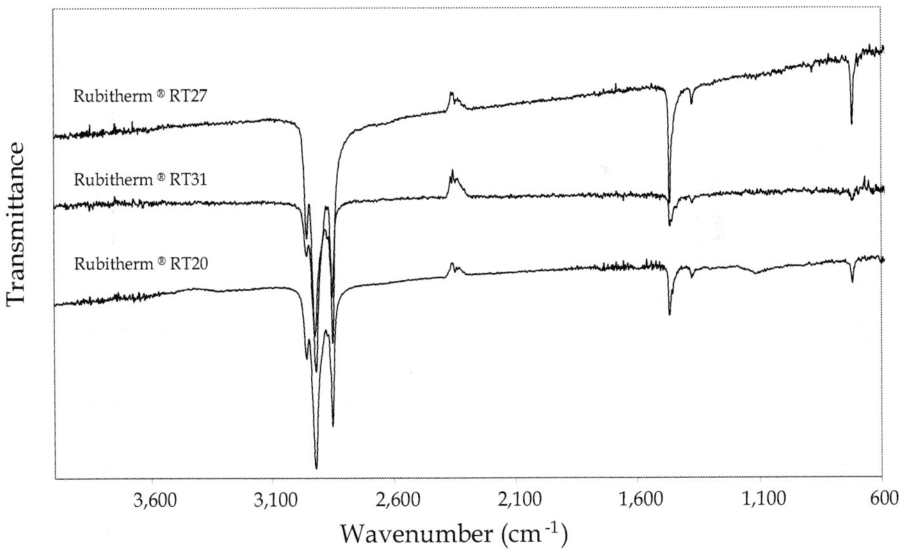

(b)

Fig. 2. FTIR spectra of: (a) PRS® paraffin wax, Petrepar® C-14 and Petrepar® C-13 and (b) Rubitherm® RT27, Rubitherm® RT31 and Rubitherm® RT20

This behaviour can be attributed to chemical nature of these commercial PCMs according to the FTIR spectra of the different PCMs obtained (Figure 2). In the spectrums of the PRS® paraffin wax, Petrepar® C-14 and Petrepar® C-13 materials (Figure 2b), in contrast to that of the Rubitherm products (Figure 2a), medium intensity bands in the 980-1,240 cm^{-1} were observed. These bands indicate the presence of ethyl and propyl groups branches. Therefore, PRS® paraffin wax, Petrepar® C-14 and Petrepar® C-13 are not straight chain alkanes.

PCM	$dpv_{0.5}$ (μm)[1]	$dpn_{0.5}$ (μm)[2]	Thermal storage energy (Jg^{-1})	PCM encapsulated (wt. %)
PRS® paraffin wax	149.0	4.53	104.7	50.1
Rubitherm® RT31	203.9	4.01	98.7	49.5
Rubitherm® RT27	250.9	6.02	100.2	46.7
Rubitherm® RT20	180.1	2.86	68.3	38.5
Petrepar® n-C14	110.7	3.97	79.0	35.1
Petrepar® n-C13	93.9	3.83	58.6	43.6

[1]$dpv_{0.5}$ represents 50% microcapsule particles whose mean volumetric diameter is less than this value.
[2]$dpn_{0.5}$ represents 50% microcapsule particles whose mean numeric diameter is less than this value.

Table 2. Average diameters and thermal properties of microcapsules containing different PCM

DSC measurements confirm that all PCMs studied have been successfully encapsulated inside the polystyrene microcapsules (Table 2). PRS® paraffin wax, Rubitherm® RT27 and Rubitherm® RT31 allow to obtain microcapsules with energy storage capacities higher than 98 Jg^{-1} and approximately 50 wt.% of encapsulation efficiency.

Figure 3 shows the environmental scanning electron microscopy (ESEM) micrographs of microcapsules prepared with different phase change materials. It can be seen from Figure 3 that microcapsules prepared with PRS® paraffin wax, Petrepar® C-14 and Petrepar® C-13 and Rubitherm® RT27 and Rubitherm® RT31 are regularly spherical shape and smooth surface. However, the particles obtained with Rubitherm® RT20 have are irregular with rough surface, which can be attribute to the polarity of this product (Sánchez et al., 2007).

Although microcapsules containing PRS® paraffin wax, Rubitherm® RT27 and Rubitherm® RT31 have similar average thermal storage energies and regular spherical microcapsules with smooth surface. The most convenient melting temperature for an effective utilization of this kind of materials in textiles field was obtained using Rubitherm® RT31. Obviously, its phase change temperature (31°C) is comfortable for the human body, and a higher efficiency of encapsulation (49.5%) than the others.

3.2 Influence of the kind of textile substrate on the production of smart fabric

Seven fabrics substrates for different textile applications using Rubitherm® RT31 as core material were used. Their description and properties are shown in Table 3. Samples were named A to G, according to the textile substrates employed.

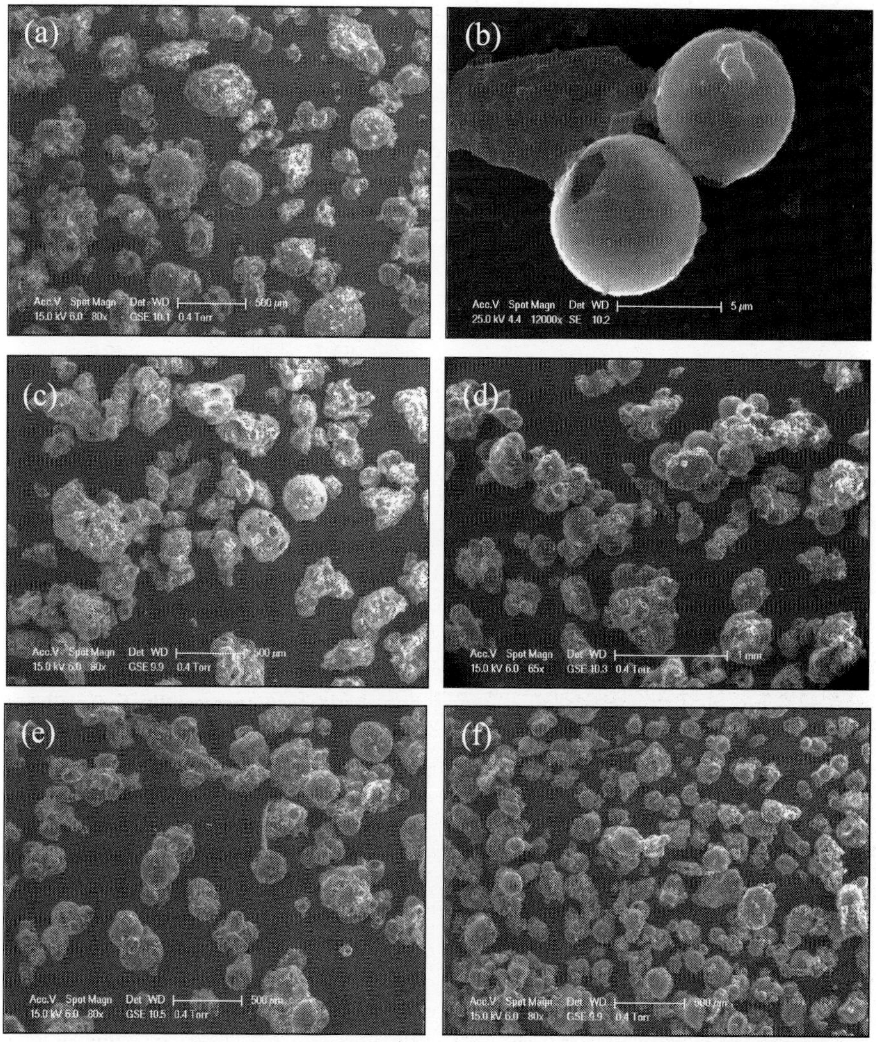

Fig. 3. ESEM micrographs of microcapsules containing: (a) Petrepar® C-13, (b) Petrepar® C-14, (c) Rubitherm® RT20, (d) Rubitherm® RT27, (e) Rubitherm® RT31 and (f) PRS® paraffin wax

Thermal performance of different coated textiles with thermo-regulating properties with 35 wt. % of PCM microcapsules as a function of the kind of textile substrate was evaluated by DSC analyses (Figure 4). It can be seen that all treated textile substrates allows to obtain thermo-regulating properties with acceptable latent heat storage capacities. The same result was observed by other authors (Salaün et al., 2010; Izzo Renzi et al., 2010). On the other hand, no significant differences on the phase change transition temperature of coated samples and the Rubitherm® RT31 microcapsules were observed. It was found that the melting transition points in the coated fabrics from A to G changed 0.72, 0.01, 0.13, 0.37, 0.13, -0.35 and 0.01 °C, respectively. This indicated that the kind of textile substrate have not significant effect on the microcapsules melting effect. In this sense, Koo et al., (2009) observed changes up to 0.58 °C by using a wet coating method.

Sample	Composition	Area Weight (gm^{-2})	Thickness (mm)	Uses
A	82% Polyester 18% Polyurethane	296	1.50	Soft-Shell fabric with an intermediate polyurethane membrane for cold protection
B	11% Elastane 35% Polyamide 54% Polyester	270	1.34	Soft-Shell fabric with an intermediate polyurethane membrane for cold protection
C	100% Polyamide	202	0.48	Green fabric for military uses
D	100% Polyester coating with 100% PVC	121	0.27	Yellow fabric for garments of high visibility
E	40% Polyester 60% Cotton	185	0.35	Openwork fabric in blue tone for medical garments uses
F	100% Polyamide	328	0.61	Military printed fabric for military uses
G	100% Cotton	158	0.30	Fabric used for upholstery, sheets, curtains and garments

Table 3. Textile substrates characterization and their textile uses

In our previous paper, textile substrate G (100% cotton) was used to obtain the textiles with thermo-regulating properties (Sánchez et al., 2010). In the present work the employed experimental conditions were the same except for the thickness of the coating layer. A thermal storage capacity of 14.4 Jg^{-1} was achieved using cotton textile substrate and a coating thickness of 0.1 mm. Comparing this result with the value of 7.6 Jg^{-1} obtained in the previous work with 0.01 mm of thickness, it is observed that the latent heat storage capacity of the treated fabric increased as the thickness of the coating layer increased.

Table 4 summarizes the latent heat storage capacity, the necessary time to decrease the temperature of coated textiles from 33 to 25 °C, the latent heat accumulated in 1 m^2 of fabric substrate associated with each sample and the amount of PCM microcapsules added on each textile substrate. There are not important differences in the latent heat storage capacity and the amount of retained PCM microcapsules depending on the kind of used substrate textile. Coated textiles A and B exhibit the highest latent heat and a long thermoregulatory effect, due to the soft shell characteristics and the large thickness of these fabrics that allow to accommodate a high amount of PCM microcapsules and improve the resistance of heat transfer, respectively. Sample D with 16.3 Jg^{-1} shows a short time of the heat release, this may be related to the small thickness of the used textile substrate (0.27 mm). For samples E and G the latent heat storage capacity, the amount of PCM microcapsules added on the textiles and the thickness of these textiles are similar, thus the duration of the thermal buffer effect was quite similar. Furthermore, the lowest latent heat storage capacity for samples having polyamide as textile substrate (samples F and C) suggests that this composition does not allow the incorporation of a large amount of PCM microcapsules into the fabric. Nevertheless, small differences of latent heat storage capacity were obtained. According to these results, the heat transfer through fabric depends on the quantity of PCM microcapsules added on the coating binder but also the textile composition and structure. Consequently, numerous factors must be considered in the evaluation of the heat transfer properties of fabrics (Koo et al., 2009).

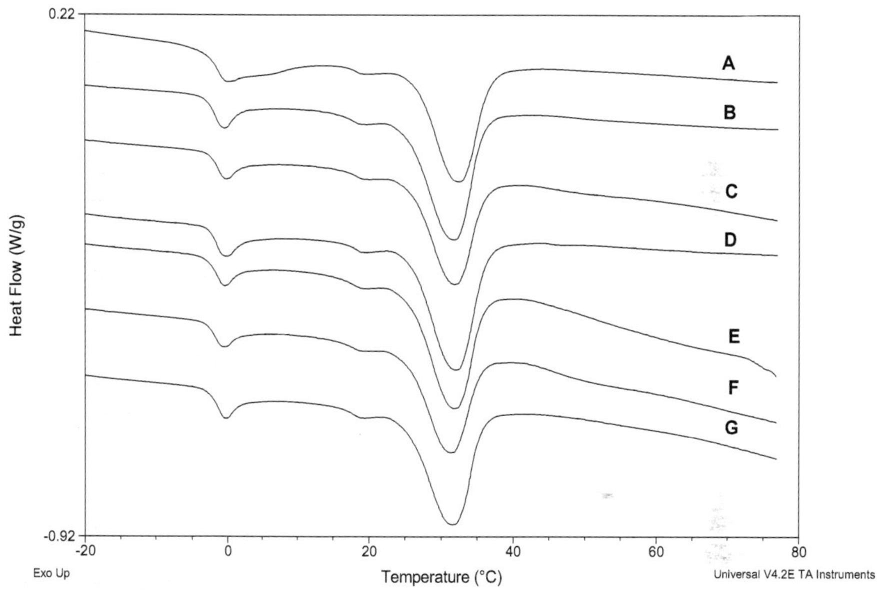

Fig. 4. DSC thermograms of the different coated textiles with 35 wt.% of microcapsules containing Rubitherm® RT31

Reliability tests to study the thermal performance of a textile with thermo-regulating properties (sample A) during thermal cycling were carried out using DSC analysis (Figure 5). It can be observed that the latent heat storage of the sample does not change when heating/cooling cycle is repeated (less than 2 % of latent heat variation). Furthermore, melting and freezing transition points of the coated fabrics shift to higher temperature than microcapsules containing Rubitherm® RT31. This fact could be attributed to the influence of the polymeric binder on the thermal response of textile (Salaün et al., 2010).

Therefore, these results demonstrated that textile substrates with different characteristics and applications are suitable for application of the microcapsules by means of a coating method. Similar results of latent heat storage capacity using natural leather and 40 wt.% of PCM microcapsules were obtained by Izzo Renzi et al., (2010).

Sample	ΔH (Jg^{-1})	Duration of the heat release from 33 to 25°C (s)	Latent heat accumulated in 1 m^2 of fabric substrate (kJm^{-2})	PCM microcapsules added on the textile (wt. %)
A	19.4	79	5.7	25.6
B	18.1	62	4.9	23.9
C	13.5	65	2.7	17.8
D	16.3	28	4.4	21.5
E	14.3	42	2.6	18.9
F	11.1	67	3.6	14.7
G	14.4	48	2.3	19.0

Table 4. Thermal properties of the thermo-regulating textiles

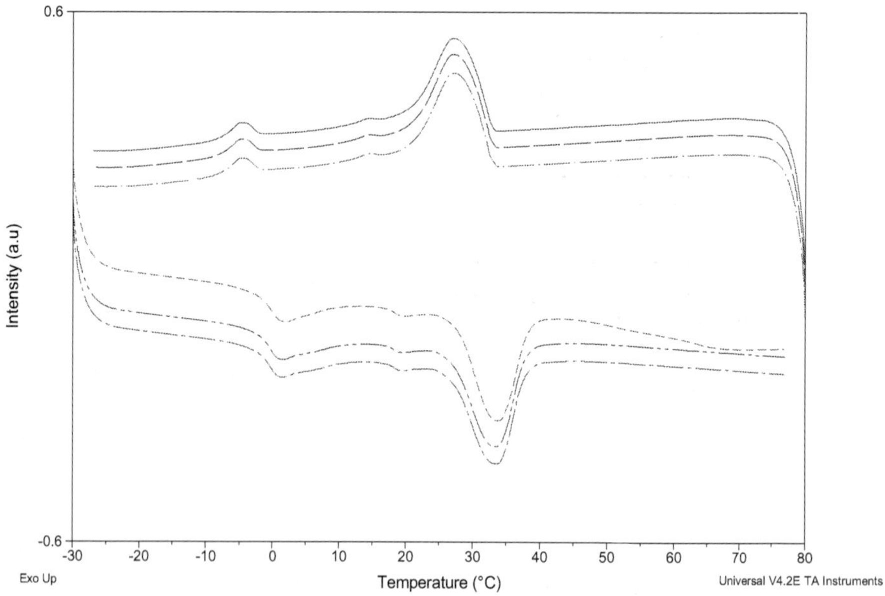

Fig. 5. DSC curves of a textile with thermo-regulating properties (sample A), triple scan

Figure 6 shows ESEM micrographs of the surface and the cross-sectional morphology of a representatively sample (coated textile A) in which PCM microcapsules were used as the thermal insulation material. It can be seen, the successful fixation between textile substrate and the microcapsules containing Rubitherm® RT31. The PCM microcapsules in the treated samples were mainly located at the spaces between fibers and the fiber surface. As shown in Figure 6b, the textile substrate A is remarkable thick leads to add a large amount of PCM microcapsules on the textile substrate according to DSC analysis obtained.

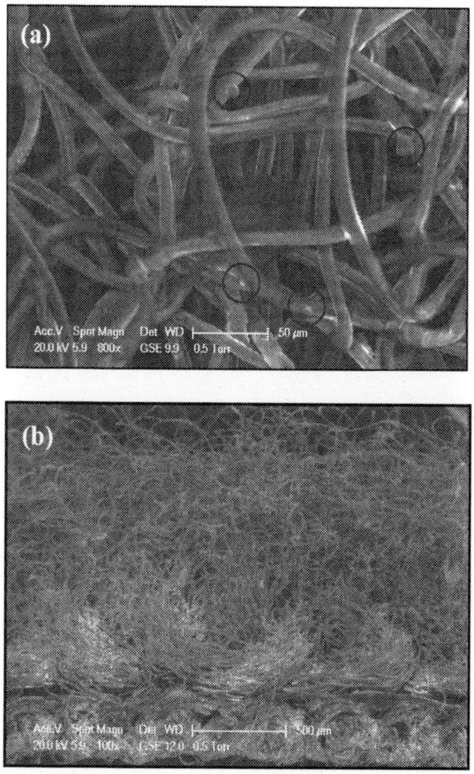

Fig. 6. ESEM micrographs of a thermo-regulating textile (sample A): (a) surface and (b) cross section

During the last years several testing methods have been developed for measuring the temperature-regulating ability of PCM in fabrics (Bryant & Colvin, 1994; Mengjin et al., 2008; Onder et al., 2008; Koo et al., 2009). In this work, the efficiency of the active thermal insulation effect of the textiles with incorporated microcapsules has been evaluated using thermal vision camera. Figure 7 shows a comparison of the maxima temperature reached on the surface temperature distribution for the seven samples with the time. In this study, the coated textiles with and without PCM microcapsules were cooled from 60°C to room temperature.

In all the cases, Rubitherm® RT31 buffer effect is observed. Therefore, during the cooling process there is a temperature range when the temperature of coated textile with PCM microcapsules is higher than the non-coated one. This indicates that the stored energy of PCM is transferred to the environment in the phase change range during the reverse cooling

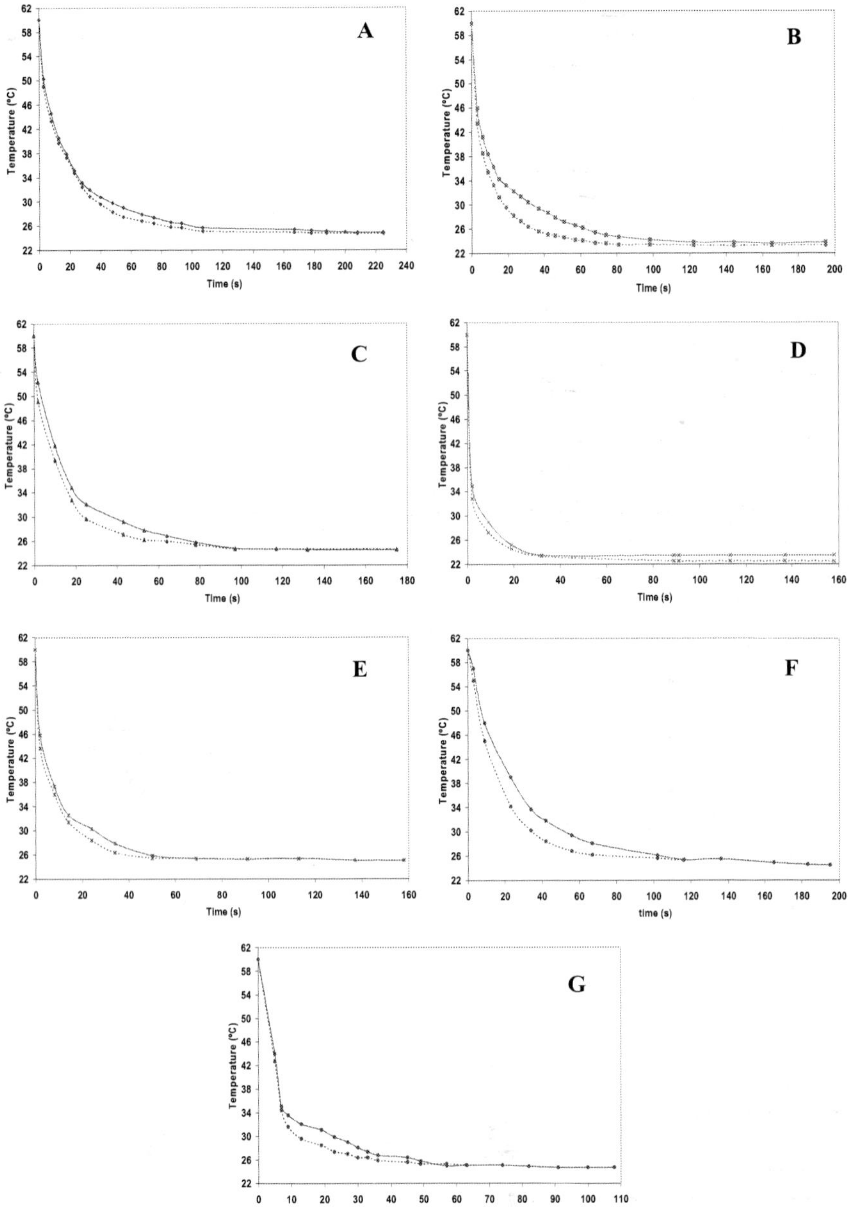

Fig. 7. Temperature distribution as a function of time for different coated textiles by thermal vision camera: ----- Textile without PCM microcapsules — Textile with PCM microcapsules

process (Salaün et al., 2010). As it was previously mentioned, coated textile D exhibits the highest rate of temperature decrease and samples A and B have a significant effect on the thermal insulation.

With the aim to test the thermal comfort in summer conditions, the effect of a reference textile and a prototype of a textile with thermo-regulating properties in contact with the body (shoulders in this specific case) was visualized by IR thermography images (Figure 8). The zero time was taken when the individual went out to the building, from 25 °C (inside temperature) to 35 °C (outside temperature).

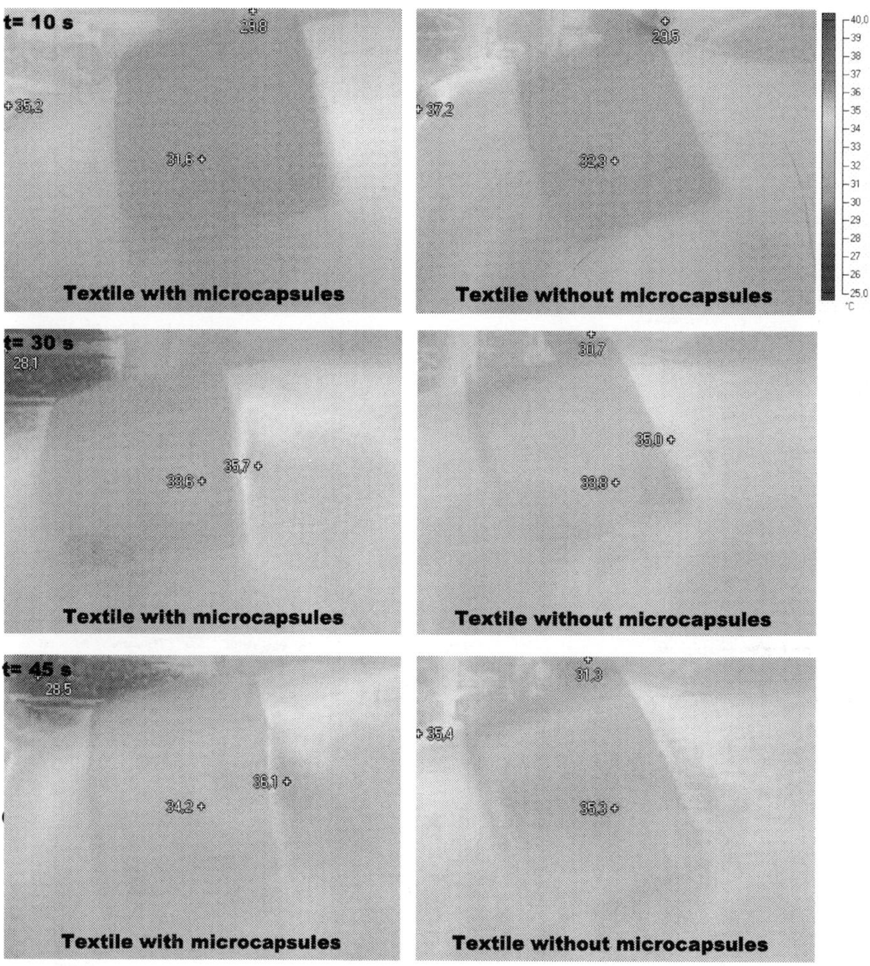

Fig. 8. Infrared camera images of coated textile A with and without PCM microcapsules at different times

Thermal imagines show that textiles with thermo-regulating properties undergoes a heating effect less marked than the textile without PCM microcapsules, as a consequence of the melting process of encapsulated Rubitherm® RT31. Thermoregulatory effect observed for textile with PCM microcapsules with respect to the normal fabric was 1, 0.9, 0.8 and 0.7 °C after 10, 30, 45 and 75 s.

Concluding, the thermal performance of textiles with thermo-regulating properties depends on the energy storage capacity, PCM microcapsules content, textile substrate structure and the correspondence between the phase change temperature and the application temperature range. Consequently, thermal effects can be improved according to the final application request using an appropriate combination of all these parameters.

4. Conclusion

A method based on a suspension free radical polymerization process has been used for the encapsulation. Different phase change materials (PRS® paraffin wax, Petrepar® C-14 and Petrepar® C-13 and Rubitherm® RT27 and Rubitherm® RT31) can be encapsulated by this method and form a core-shell structure. Although microcapsules containing PRS® paraffin wax, Rubitherm® RT27 and Rubitherm® RT31 have similar average thermal storage energies and regular spherical microcapsules with smooth surface. The most convenient melting temperature for an effective utilization of this kind of materials in textiles field was obtained using Rubitherm® RT31. Its phase change temperature (31°C) is comfortable for the human body and it has a higher efficiency of encapsulation (49.5%) than the others.

Fabrics with thermo-regulating properties had heat storage capacities of 11.1-19.4 Jg^{-1}, which depended on the textile substrate. The DSC and the thermal temperature distribution analyses indicated that all the coated textiles with 35 wt. % of microcapsules showed a high latent heat storage capacities and significant thermoregulatory effect. Furthermore, DSC curves demonstrated that the thermal energy storage of the textiles with thermo-regulating properties does not change when heating/cooling cycle was repeated. A thermoregulatory effect in contact with the body of 0.9 °C after 45 s was observed by means of IR thermography images.

5. Acknowledgment

Financial support from ASINTEC S.A. and Ref. PBC08-0243-1458 from Consejería de Ciencia y Tecnología (JCCM) are gratefully acknowledged.

6. References

ASTM D 823-C. (1997). Standard practices for producing films of uniform thickness of paint, varnish, and related products on test panels

Boh, B. & Knez, E. (2006). Microencapsulation of essential oils and phase change materials for applications in textile products, *Indian Journal of Fibre and Textile Research,* Vol.31, pp.72–82, ISSN 0971-0426

Bryant, Y.G. & Colvin, D.P. (1988). Fiber with reversible enhanced thermal storage properties and fabric made therefrom, US Patent 4,756,958

Bryant, Y.G. & Colvin, D.P. (1994). Fabric with reversible enhanced thermal properties, US Patent 5,366,807

Colvin, D.P. & Bryant, Y.G. (1996). Thermally enhanced foam insulation, US Patent 5,637,389

Dixit, S. & Goel, A. (2007). Microencapsulation in textile processing: an overview, *Asian Textile of Journal*, Vol.16, pp. 83–86, ISSN 1819-3358

Hamielec, A.E. & Tobita, H. (2003). Ullmann´s Encyclopedia of Industrial Chemistry (Weinheim: WILEY-VCH).

Izzo Renzi, A.; Carfagna, C. & Persico, P. (2010). Thermoregulated natural leather using phase change materials: An example of bioinsipiration, *Applied Thermal Engineering*, Vol.30, pp.1369-1376, ISSN 1359-4311

Koo, K.; Choe, J. & Park, Y. (2009). The application of PCMMcs and SiC by commercially direct dual-complex coating on textile polymer, *Applied Surface Science*, Vol.255, pp.8313-8318, ISSN 0169-4332

Monllor, P.; Sánchez, L.; Cases, F. & Bonet, M.A. (2009). Thermal behaviour of microencapsulated fragrances on cotton fabrics, *Textile Research Journal*, Vol. 79, pp. 365-380., ISSN 0040-5175

Mengjin, J.; Xiaoqing, S.; Jianjun, X. & Guangdou, Y. (2008). Preparation of a new thermal regulating fiber based on PVA and paraffin, *Solar Energy Materials & Solar Cells Solar*, Vol.92, pp. 1657-1660, ISSN 0927-0248

Onder, E.; Sarier, N. & Cimen, E. (2008). Encapsulation of phase change materials by complex coacervation to improve thermal performances of woven fabric, *Thermochimica Acta*, Vol.467, pp. 63-72, ISSN 0040-6031

Pause, B. (2003). Nonwoven protective garments with thermo-regulating properties, *Journal of Industrial Textiles*, Vol.33, No.2, pp. 93-99, ISSN 1528-0837

Ren, Y.J. & Ruckman, J.E. (2004). Condensation in Three-layer Waterproof Breathable Fabrics for Clothing, *International Journal of Clothing Science and Technology*, Vol.16, No.3, pp. 335-347, ISSN 0955-6222

Rodrigues, S.N.; Martins, I.M.; Fernandes, I.P.; Gomes, P.B.; Mata, V.G.; Barreiro, M.F. & Rodrigues, A.E. (2009). Microencapsulated perfumes for textile application, *Chemical Engineering Journal*, Vol.149, pp. 463-472, ISSN 1385-8947

Salaün, F.; Devaux, E.; Bourbigot, S. & Rumeau, P. (2010). Thermoregulating response of cotton fabric containing microencapsulated phase change materials, *Thermochimica Acta*. Vol. 506, No.1-2, pp. 82-93, ISSN 0040-6031

Sánchez, L.; Lacasa, E.; Carmona, M.; Rodríguez, J.F. & Sánchez, P. (2008b). Applying an experimental design to improve the characteristics of microcapsules containing PCMs for fabric use, *Industrial & Engineering Chemistry Research*, Vol.47, pp.9783-9790, ISSN 1520-5045

Sánchez, L.; Sánchez, P.; Carmona, M.; de Lucas, A. & Rodríguez, J.F. (2008a). Influence of operation conditions on the microencapsulation of PCMs by means of suspension-like polymerization. *Colloid and Polymer Science*, Vol.286, No.8-9, pp.1019-1027, ISSN 0303-402X

Sánchez, L.; Sánchez, P.; De Lucas, A.; Carmona, M. & Rodríguez, J.F. (2007). Microencapsulation of PCMs with a polystyrene shell. *Colloid and Polymer Science*, Vol.285, No.12, pp.1377-1385, ISSN 0303-402X

Sánchez, P.; Sánchez-Fernández, M.V.; Romero, A. ; Rodríguez, J.F. & Sánchez-Silva, L. (2010). Development of thermo-regulating textiles using paraffin wax microcapsules, *Thermochimica Acta*, Vol.498, pp.16-21, ISSN 0040-6031

Shin, Y. ; Yoo, D.I. & Son, K. (2005). Development of Thermoregulating Textile Materials with Microencapsulated Phase Change Materials (PCM). IV. Performance Properties and Hand of fabrics treated with PCM microcapsules, *Journal of Applied Polymer Science*, Vol.97, pp.910-915, ISSN 0021-8995

34

Heat Conduction in Nonlinear Media

Michael M. Tilleman
*Elbit Systems of America (Kollsman), LLC.,
220 Daniel Webster Highway, Merrimack, NH 03054
USA*

1. Introduction

The objective of this chapter is to demonstrate a closed form solution to a unique problem in heat transfer, that of heat transfer in nonlinear media. For this purpose identified are cases in which the nonlinear phenomenon dominates, where properties of a medium exhibit nonlinear response to heating, and the solution methodology is described. Further in this chapter presented is a set of analytical solutions applicable to various geometrical forms including the cases of: infinite and finite cylinders with axisymmetrical source, finite cylinder with axisymmetrical and axially varying sources, slender disk, infinite and finite parallelepiped with centrally symmetric source, finite parallelepiped with axially varying source and the resulting stress due to temperature distribution. In this abstract an example is shown of a solution for a particular case, that of an infinite cylinder.

In many cases equations governing the phenomenon of heat transfer are solved assuming constant physical properties of the media concerned. That is where a whole class of closed form solutions is found, if regular boundaries and boundary conditions are provided. However, at instances in which the coefficient of heat conduction, specific heat and thermal diffusivity, are functions of temperature, those solutions are no longer applicable. For those cases another family of closed form solutions is found and described herewith.

The analytical solution to the thermally nonlinear problem assumes a certain dependence of the coefficient of thermal conductivity, k, on temperature. This case is usually found in instances of large temperature gradients in media, for instance in active optical materials with intense electro-optical fields. They are realized in laser gain media, nonlinear optical crystals and saturable absorbers. One of the most frequently used host materials for lasers is Yttrium Aluminum Garnet, YAG, in which an inverse proportionality to temperature well approximates measured k values over a vast temperature range. What we applied for the solution of such a problem is the Kirchoff's transformation (Joyce, 1975), whereby the heat equation can be linearized and solved. However, the use of this method is limited to materials whose k is integrable in temperature, T. Only for these cases the linearization of the heat equation can be made. For instance in the case of $AgGaSe_2$, a nonlinear optical crystal useful for harmonic generation, the dependence is $k=A+B/T$ (Aggarwal & Fan, 2005) that is an integrable function in T, therefore solvable by the present method. Evidence has it, though, that most of the known optical crystals have similarly thermal coefficient enabling the use of the present solution. Another strategy used in many of the present solutions is the use of the Green's function as kernel in the integral expression. Owing to the strong

dependence of additional properties of these materials, for instance thermal expansion coefficient and thermal rate of refractive index (dn/dT) on temperature, the solution to the heat transfer problem presents an important tool to understand the reaction of these media to intense heating.
The analysis begins by defining the governing equation and the boundary conditions. Governing is the Poisson equation expressed as:

$$\nabla \cdot k(\nabla T)+Q=0 \tag{1}$$

where: k=k(T) is the thermal conductivity and Q is the heat source term defined as deposited power per unit volume. To solve for cylinders or rectangular parallelepipeds either a polar or Cartesian coordinate systems need be considered as necessary.

1.1 Coefficient of thermal conductivity

The coefficient of thermal conductivity is a product of the material density, thermal diffusivity and specific heat. Approximation to theory and the fitting to a broad body of measured values suggest an inverse linear approximation:

$$k(T)=k_0\frac{T_0}{T} \tag{2}$$

where k_0 is the coefficient of thermal conductivity at T_0 whose values for several materials are summarized.(Aggarwal et al., 2005) This simple function, similar to that suggested for Nd:YLF in (Pfistner et al., 1994) and (Hardman et al., 1999), closely fits theory and agrees very well with data in the range between cryogenic temperature and 770K. It is worth mentioning that though the specific approximation of Eq. 2 holds true for some materials, there are other materials for which alternative approximations may become a better fit.

1.2 The nonlinear poisson equation for a cylinder with axisymmetrical source

Let us consider the Poisson equation of Eq. 1 in cylindrical coordinates, using the expression for k of Eq. 2:

$$\nabla_{r,\varphi,z}\cdot\left[k_0\frac{T_0}{T}\left(\nabla_{r,\varphi,z}T\right)\right]+Q(r,\varphi,z)=0 \tag{3}$$

Consistent with Kirchoff's transformation this equation can be rewritten as:

$$k_0T_0\nabla_{r,\varphi,z}\cdot\frac{1}{T}\left(\nabla_{r,\varphi,z}T\right)+Q(r,\varphi,z)=k_0T_0\,\nabla^2_{r,\varphi,z}\ln\frac{T}{T_0}+Q(r,\varphi,z) \tag{4}$$

then one arrives at:

$$K\nabla^2_{r,\varphi,z}\theta(r,\varphi,z)+Q(r,\varphi,z)=0 \tag{5}$$

Being a linear equation in θ where:

$$\theta(r,\varphi,z)=\ln\frac{T(r,\varphi,z)}{T_0} \tag{6}$$

and K is a constant:

$$K = k_0 T_0 \tag{7}$$

The above notation is cylindrical. Notwithstanding, this linearization holds true for any orthogonal coordinate system. It should be emphasized that Eq. 5 represents any steady-state heat equation, where θ and K may be an arbitrary representation of any compound temperature-heat conduction function. Ultimately this case degenerates to the linear case where the heat conduction coefficient is independent of temperature.
Assuming a distributed heat source one has:

$$Q(r,\varphi,z) = \frac{P}{\pi r_P^2 L_{eff}} f_0(r,\varphi,z) \tag{8}$$

where P is the power deposited in the medium, r_P is the effective radius of the heating zone in the cylinder, L_{eff} is the effective length of the medium which may be, for instance, a cylindrical rod or parallelepiped and f_0 is an arbitrary spatial distribution function of the deposited power.
The boundary conditions for the problem are prescribed by a given cooling mechanism, which may consist of a coolant fluid or a solid heat sink, and physical surroundings of the device. Typically at least one side of the heated medium is held in good thermal contact with the cooling mechanism. Being either kept in vacuum or exposed to gas the other sides of the medium may be considered insulated due to lack of any considerable heat transfer. At the areas of the rod through which heat flow occurs, say at $r=r_W$, one may specify the boundary by a known temperature, i.e. a Dirichlet condition:

$$T(r_W,\varphi,z) = T_0 \tag{9}$$

Otherwise, in the case the heat flux out of the system is known, the boundary is specified by a Neumann condition:

$$k\frac{\partial T}{\partial \hat{n}} = q \tag{10}$$

and on all the insulated sides the heat flux vanishes, thus:

$$\frac{\partial T}{\partial \hat{n}} = 0 \tag{11}$$

2. Cylindrical rods

Consistent with Kirchoff's transformation (Joyce, 1975) Eq. 5 is expressed in cylindrical coordinates as:

$$\frac{K}{r}\frac{\partial}{\partial r}r\frac{\partial}{\partial r}\theta(r,\varphi,z) + \frac{K}{r^2}\frac{\partial^2}{\partial \varphi^2}\theta(r,\varphi,z) + K\frac{\partial^2}{\partial z^2}\theta(r\varphi,z) + Q(r,\varphi,z) = 0 \tag{12}$$

2.1 Rod of finite length

This case is illustrated in Figure 1(a). Sketched is a rod held by two conducting mounts at the ends, surrounded on its perimeter by a source emitting electromagnetic radiation. Such a source may be a black-body emitter, incandescent lamp, light emitting diode or laser, to name a few. The rod absorbs the radiation converting it to heat. In order to approximate cases of: 1) rod with evenly spaced side heating and varying axial heating rate or boundary condition, and 2) end heating with arbitrarily distributed source or boundary conditions, it is sufficient to assume an axisymmetrical case with a finite rod. Then the governing equation becomes:

$$\frac{K}{r}\frac{\partial}{\partial r}r\frac{\partial}{\partial r}\theta(r,z)+K\frac{\partial^2}{\partial z^2}\theta(r,z)+Q(r,z)=0 \tag{13}$$

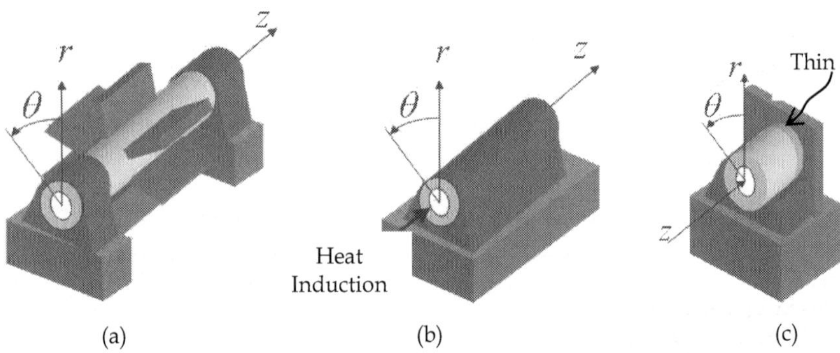

Fig. 1. Three cylindrical optical devices: a) rod held by heatsink mounts on both ends with radially symmetrical side induction heaters, b) rod held by a heatsink along its length with end induction heater and c) end induction heated thin disk (dark) with a cap

2.1.1 Side heating

The boundary conditions for a rod may model various cooling configurations including conductive and convective cooling means. A possible configuration is conduction cooling, where a heat sink with a mount holds the rod over part of its length. This configuration justifies the assumption of either Dirichlet or Neumann boundary conditions or their combination specified around the rod circumference:

$$T(r_W,z)=f_1(z) \quad or \quad k(T)\left.\frac{\partial T(r,z)}{\partial r}\right|_{r=r_W}=K\left.\frac{\partial\theta(r,z)}{\partial r}\right|_{r=r_W}=f_2(z) \tag{14}$$

The Dirichlet boundary condition is particularly suitable for cases where the boundary is held at a set temperature such as the case of heat sinking to Peltier junction or cryogenic cooling. In terms of the function θ it becomes:

$$\theta(r_W,z)=\ln\frac{f_1(z)}{T_0} \tag{15}$$

Then, the Neumann condition is suitable for cases where the boundary provides a certain heat flux across it, as specified in Eq. 14. At the rod ends the facets are assumed insulated, specified by Neumann condition, such that:

$$\left.\frac{\partial\theta}{\partial z}\right|_{z=0} = \left.\frac{\partial\theta}{\partial z}\right|_{z=L/2} = 0 \tag{16}$$

where the origin is at the rod center and z = L/2 is half the rod length. Modeling of the heat source assumes a region confined both radially and axially inside the rod:

$$Q(r,z) = \begin{cases} \dfrac{P}{\pi r_P^2 l/2} f_0(r) & ; \quad 0 \le z \le \dfrac{l}{2} \\ 0 & ; \quad \dfrac{l}{2} < z \end{cases} \tag{17}$$

where l is the length of the source region in the rod such that l≤L.

To solve the set of equations 13 – 16 it is convenient to employ the Green's function $G(r,z)$ in which case for the mixed boundary conditions the general solution becomes:

$$\theta(r,z) = r_W \int_0^{L/2} f_3(\zeta) V(r,z,r_W,\zeta)\, d\zeta + \frac{1}{K} \int_0^{r_W} \int_0^{L/2} Q(\xi,\zeta) G(r,z,\xi,\zeta)\, \xi\, d\xi\, d\zeta \tag{18}$$

where the function $f_3(\zeta)$ assumes either $f_1(\zeta)$ or $f_2(\zeta)$ defined in Eq. 14 and the function $V(r,z,r_W,\zeta)$ is either $\partial G(r,z,\xi,\zeta)/\partial\xi$ at $\xi = r_W$ or $G(r,z,r_W,\zeta)$, depending on the type of boundary condition around the rod circumference (Dirichlet or Neumann). Green's function is constructed by solving the homogeneous Eq. 13, or the Laplace equation, satisfying the specified boundary conditions of Eqs.14 and 16, and by the function holding throughout the domain.(Polianin, 2002) Thus G takes the form as demonstrated in (Polianin, 2002):

$$G_s(r,z,\xi,\zeta) = \frac{2}{r_W^2 L} \sum_{m=1}^{\infty} \sum_{n=1}^{\infty} \frac{J_0(\mu_{s,m} r) J_0(\mu_{s,m}\xi) \cos\left(2n\pi \frac{z}{L}\right) \cos\left(2n\pi \frac{\zeta}{L}\right)}{\left[\mu_{s,m}^2 + \left(\frac{2n\pi}{L}\right)^2\right] J_s^2(\mu_{s,m} r_W)} \tag{19}$$

where the subscript s assumes the value of either 0 or 1 corresponding to a Neumann or Dirichlet type boundary condition, respectively. The coefficients μ_m are the roots of the equation:

$$\begin{aligned} &\left.\frac{d}{dr} J_0(\mu_m r)\right|_{r_W} = 0 \quad ; \qquad s = 0 \\ &J_0(\mu_m r_W) = 0 \quad ; \qquad s = 1 \end{aligned} \tag{20}$$

To present a complete solution one still needs to define the functions $f_0(r)$ used in Eq. 17. Let two cases be considered:

1. Uniform heat source distribution where:

$$f_0(r) = \begin{cases} 1 & ; \quad 0 \le r \le r_P \\ 0 & ; \quad r_P < r \end{cases}$$

2. Gaussian heat source profile:

$$f_0(r) = \exp\left[-2\left(\frac{r}{r_P}\right)^2\right]$$

Considering a Dirichlet boundary condition for case 1 one may solve the problem for $\theta - \theta_W$, whereby the first term in Eq. 18 becomes θ_W and the solution is expressed as:

$$\begin{aligned}\theta(r,z) &= \theta_W + \frac{2P}{\pi r_P^2 l K}\int_0^{r_p}\int_0^{l/2} G_1(r,z,\xi,\zeta)\,d\zeta\,\xi\,d\xi \\ &= \theta_W + \frac{P}{\pi r_W^2 r_P L K}\sum_{m=1}^{\infty}\sum_{n=1}^{\infty}\operatorname{sinc}\left(n\pi\frac{l}{L}\right)\frac{J_1(\mu_{1,m}r_p)J_0(\mu_{1,m}r)\cos\left(2n\pi\frac{z}{L}\right)}{\mu_{1,m}\left[\mu_{1,m}^2+\left(\frac{2n\pi}{L}\right)^2\right]J_1^2(\mu_{1,m}r_W)}\end{aligned} \quad (21)$$

Considering a Neumann boundary condition for case (1), and assuming a case where the heat is relieved from the rod by conductive mounts holding the rod perimeter between ±S/2 and ±L/2 Eq. 18, the solution is:

$$\begin{aligned}\theta(r,z) &= \frac{q_B}{K}r_W\int_{S/2}^{L/2} G_0(r,z,r_W,\zeta)\,d\zeta + \frac{2\eta P_{abs}}{\pi r_P^2 l K}\int_0^{r_p}\int_0^{l/2} G_0(r,z,\xi,\zeta)\,d\zeta\,\xi\,d\xi \\ &= \frac{q_B}{K}\frac{S}{r_W L}\sum_{m=1}^{\infty}\sum_{n=1}^{\infty}\operatorname{sinc}\left(n\pi\frac{S}{L}\right)\frac{\mu_{1,m}J_0(\mu_{1,m}r)}{\left[\mu_{1,m}^2+\left(\frac{2n\pi}{L}\right)^2\right]J_1(\mu_{1,m}r_W)}\cos\left(2n\pi\frac{z}{L}\right) \\ &\quad + \frac{P}{\pi r_W^2 r_P L K}\sum_{m=1}^{\infty}\sum_{n=1}^{\infty}\operatorname{sinc}\left(n\pi\frac{l}{L}\right)\frac{J_1(\mu_{0,m}r_p)J_0(\mu_{0,m}r)\cos\left(2n\pi\frac{z}{L}\right)}{\mu_{0,m}\left[\mu_{0,m}^2+\left(\frac{2n\pi}{L}\right)^2\right]J_0^2(\mu_{0,m}r_W)}\end{aligned} \quad (22)$$

where q_B is the heat flux through the cooling mounts. For a very long rod the solutions in Eqs. 21 and 22 approach the asymptotic solution for a two dimensional, axisymmetrical geometry, which for the Dirichlet boundary condition becomes:

$$T(r) = \left(T_\infty + \frac{P}{2\pi L_{eff} h r_W}\right)\begin{cases}\left(\frac{r_W}{r_P}\right)^{\frac{P}{2\pi L_{eff}K}}\exp\left[\frac{P}{4\pi L_{eff}K}\left(1-\frac{r^2}{r_P^2}\right)\right] & ;\ 0 \le r \le r_P \\ \left(\frac{r_W}{r}\right)^{\frac{P}{2\pi L_{eff}K}} & ;\ r_P < r \le r_W\end{cases} \quad (23)$$

For the heat source of case 2 with Gaussian radial distribution the integrand becomes a product of Bessel function of the first kind of order zero and Gauss function. Then, the integral in the second term of Eq. 18 becomes:

$$\int_0^{r_W} J_0(\mu_m\xi)\exp\left[-2\left(\frac{\xi}{r_P}\right)\right]\xi\,d\xi \quad (24)$$

which has a closed form solution strictly for the case of $r_W\to\infty$, being (Abramowitz & Stegun, 1964):

$$\frac{r_P^2}{4}\exp\left[-\frac{\left(\mu_m r_P\right)^2}{8}\right] \tag{25}$$

It turns out that for r_W/r_P =1, 1.25, 1.5, 1.75 the above result has an error of 15%, 6%, 1.5% and less than 1% relative to a numerical approximation, respectively. Because in real scenarios r_W/r_P >1, this result is applicable to the present solution assuming a Dirichlet boundary condition, yielding:

$$\theta(r,z)=\theta_W+\frac{P}{4\pi r_W^2 LK}\sum_{m=1}^{\infty}\sum_{n=1}^{\infty}\exp\left[-\frac{\left(\mu_{1,m}r_P\right)^2}{8}\right]\operatorname{sinc}\left(n\pi\frac{l}{L}\right)\frac{J_0\left(\mu_{1,m}r\right)\cos\left(2n\pi\frac{z}{L}\right)}{\left[\mu_{1,m}^2+\left(\frac{2n\pi}{L}\right)^2\right]J_1^2\left(\mu_{1,m}r_W\right)} \tag{26}$$

The temperature in the cylinder for each case is derived via Eq. 6 using the values of θ in Eqs.21, 22 and 26.

2.1.2 End heating

This case is illustrated in Figure 1(b). Treating the problem of end heating is not much different than the above model for side heating. The main difference is in defining the heat source as an exponentially decaying function along the rod axis. Thus in end heating the temperature becomes maximized at the rod facet. The temperature surge is milder if an end-cap made up of a non-heating material, is bonded to the rod at the entrance. A non-heating material may be one which does not contain a heat source. The non-heating end-cap conducts the heat generated in the rod. Assumed in the present model is an identical coefficient of thermal conductivity in the heating and non- heating materials. For the sake of smooth transition from the previous section a notation is selected such where $L/2$ expresses the entire rod length. In sum the heat source in the rod becomes:

$$Q(r,z)=\begin{cases}0 & ; \quad \frac{L}{2}-z_C<z\\ \frac{P}{\pi r_P^2}\alpha\exp\left[-\alpha\left(\frac{L}{2}-z-z_C\right)\right]f_0(r) & ; \quad 0\le z\le\frac{L}{2}-z_C \cap 0\le r\le r_P\\ 0 & ; \quad r_P<r\end{cases} \tag{27}$$

where z_C is the thickness of the end cap, becoming zero for the case with no cap. As in the previous section also here considered are two cases of radial heat induction distribution: a flat-top beam and Gaussian. Also cooling of the rod is modeled by assuming either setting the radial rod wall at a constant temperature or by a conductive mount holding the rod between the lengths of *S/2 and L/2* on each side.

The solution for a rod with uniform heat source and Dirichlet conditions on the cylinder circumference becomes:

$$\theta(r,z)=\theta_W+\frac{2\alpha^2LP}{\pi r_P r_W^2 K}$$
$$\times\sum_{m=1}^{\infty}\sum_{n=1}^{\infty}\frac{\left\{(-1)^n\left[\cos\left(2n\pi\frac{z_C}{L}\right)+\frac{2n\pi}{\alpha L}\sin\left(2n\pi\frac{z_C}{L}\right)\right]-\exp\left[-\alpha\left(\frac{L}{2}-z_C\right)\right]\right\}}{\left[(2n\pi)^2+(\alpha L)^2\right]\left[\mu_{1,m}^2+\left(\frac{2n\pi}{L}\right)^2\right]}\frac{J_1(\mu_{1,m}r_P)J_0(\mu_{1,m}r)}{\mu_{1,m}J_1^2(\mu_{1,m}r_W)}\cos\left(2n\pi\frac{z}{L}\right) \tag{28}$$

Considering the Neumann boundary condition the solution is expressed as:

$$\theta(r,z)=\frac{q_B}{K}r_W\int_{S/2}^{L/2}G_0(r,z,r_W,\zeta)d\zeta+\frac{2P}{\pi r_P^2 lK}\int_0^{r_P}\int_0^{l/2}G_0(r,z,\xi,\zeta)d\zeta\xi d\xi$$
$$=\frac{q_B}{K}\frac{S}{r_W L}\sum_{m=1}^{\infty}\sum_{n=1}^{\infty}\operatorname{sinc}\left(n\pi\frac{S}{L}\right)\frac{\mu_{1,m}J_0(\mu_{1,m}r)}{\left[\mu_{1,m}^2+\left(\frac{2n\pi}{L}\right)^2\right]J_1(\mu_{1,m}r_W)}\cos\left(2n\pi\frac{z}{L}\right) \tag{29}$$
$$+\frac{P}{\pi r_W^2 r_P LK}\sum_{m=1}^{\infty}\sum_{n=1}^{\infty}\operatorname{sinc}\left(n\pi\frac{l}{L}\right)\frac{J_1(\mu_{0,m}r_p)J_0(\mu_{0,m}r)}{\mu_{0,m}\left[\mu_{0,m}^2+\left(\frac{2n\pi}{L}\right)^2\right]J_0^2(\mu_{0,m}r_W)}\cos\left(2n\pi\frac{z}{L}\right)$$

Finally, the solution for a rod with a Gaussian heat source and Dirichlet conditions on the cylinder circumference becomes:

$$\theta(r,z)=\theta_W+\frac{\alpha^2LP}{2\pi r_W^2K}$$
$$\times\sum_{m=1}^{\infty}\sum_{n=1}^{\infty}\frac{\exp\left[-\frac{(\mu_{1m}r_p)^2}{8}\right]\left\{(-1)^n\left[\cos\left(2n\pi\frac{z_C}{L}\right)+\frac{2n\pi}{\alpha L}\sin\left(2n\pi\frac{z_C}{L}\right)\right]-\exp\left[-\alpha\left(\frac{L}{2}-z_C\right)\right]\right\}}{\left[(2n\pi)^2+(\alpha L)^2\right]\left[\mu_{1,m}^2+\left(\frac{2n\pi}{L}\right)^2\right]}\frac{J_0(\mu_{1,m}r)}{J_1^2(\mu_{1,m}r_W)}\cos\left(2n\pi\frac{z}{L}\right) \tag{30}$$

2.2 Example for a cylindrical rod

Several cases are calculated showing the profiles of temperature in nonlinear materials. To allow a reasonable comparison between the various cases the following parameters are set: material Yb:YAG, rod diameter 5 mm, radiation absorbing zone diameter 2.5 mm and cryogenic cooling at 77K. Assumed is a circumferential, radially directed radiation forming a uniform heat zone.

To estimate the effect of the rod aspect ratio on the axial temperature distribution let a heat density of 410 W/cm³ be set while varying the rod length. Plotted in Figure 2(a) is the axial temperature for half a rod from center to facet with a varying length. Observe that for short rods the temperature is relatively small, growing with length to an asymptotic value. Further length increase causes the temperature profile to flatten out reaching an asymptotic value set by an infinitely long rod. A length of 50 mm may be considered as the value at which the rod is well approximated by an infinitely long rod. Then, the temperature profile is calculated assuming power magnitudes of P=240, 400 and 800 W and a rod length of 50 mm, corresponding to heat density rates of 245, 410 and 815 W/cm³, respectively. Shown in Figure 2(b) is the radial distribution through a rod center for using finite and infinite rod models yielding essentially identical results. Also in Figure 2(b) a comparison is made with a calculation based on the linear solution where k is assumed constant equal to that for the median temperature in the rod. In Figure 2(c) plotted are the temperature radial profiles for all three heat density levels, compared again to the linear approach, exhibiting a gradually growing discrepancy between the two for large power levels. It follows that the linear approach is unjustifiable for large heat loads, say above 200 W/cm³.

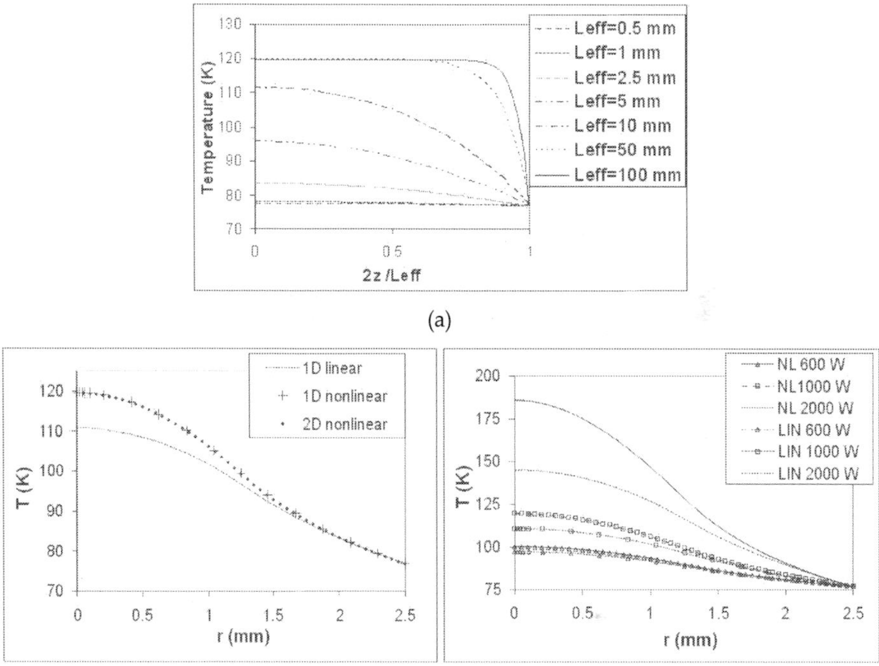

(b) (c)

Fig. 2. Temperature in a rod: a) axial distribution for heat density of 410 W/cm³, b) radial distribution for heat density of 410 W/cm³, comparing rod with finite and infinite length and with the linear solution, and c) radial distribution for heat density of 245, 410 and 815 W/cm³, comparing the nonlinear with linear solutions

2.3 Heated thin disk

Illustrated in Figure 1(c), the heated disk experiences greatly diminished radial temperature gradients due to its large area at an axial end being cooled. The thermal problem is solved as in the section above with the difference that here one of the axial facets is attached to a heat sink thus setting the temperature. The other facet is assumed either insulated or bonded to a non-heating layer, a cap. Specifying the boundary conditions for the disk radial wall is twofold: it is insulated for a non-heating disk whereas for a capped disk it has a set temperature, same as the heat-sunk end. Note that for the case of the uncapped end there is no substantial advantage to adding thermal contact to the radial wall due to the large aspect

ratio of the disk. On the other hand, for the capped-end case radial cooling must be assumed to render the cap useful.
Further assumed is a heat source with flat-top profile enveloped by a circular cylinder with a radius of r_P. Denoting the disk thickness as t the heat source density is expressed as:

$$Q(r)=\begin{cases} 0 & ; \quad t-z_C<z \\ \dfrac{P}{\pi r_p^2 t} & ; \quad 0\le r\le r_p \\ 0 & ; \quad r_p<r \end{cases} \tag{31}$$

Then, the boundary conditions are:

$$\left.\frac{\partial\theta(r,z)}{\partial z}\right|_{z=t}=0 \quad ; \quad \left.\frac{\partial\theta(r,z)}{\partial r}\right|_{r=r_W}=0 \quad \text{for uncapped disk}$$
$$\theta(r_W,z)=\theta_W \quad \text{for capped disk} \tag{32}$$

and the end cooled boundary condition becomes:

$$\theta(r,0)=\theta_W \tag{33}$$

Green's function for the Neumann and Dirichlet cases denoted by the subscript s (=0, 1) becomes:

$$G_s(r,z,\xi,\zeta)=\frac{2}{r_W^2 t}\sum_{m=1}^{\infty}\sum_{n=0}^{\infty}\frac{J_0(\mu_{sm}r)J_0(\mu_{sm}\xi)\sin\left[\left(n+\frac{1}{2}\right)\pi\frac{z}{t}\right]\sin\left[\left(n+\frac{1}{2}\right)\pi\frac{\zeta}{t}\right]}{\left\{\mu_{sm}^2+\left[\left(n+\frac{1}{2}\right)\frac{\pi}{t}\right]^2\right\}J_s^2(\mu_{sm}r_W)} \tag{34}$$

Thereby the solution takes the form:

$$\begin{aligned}\theta_s(r,z)&=\theta_W+\frac{1}{K}\int_0^{r_W}\int_0^t Q(\xi,\zeta)G_s(r,z,\xi,\zeta)d\zeta\xi d\xi \\ &=\theta_W+\frac{4P}{\pi^2 r_W^2 r_p t K}\sum_{m=1}^{\infty}\sum_{n=1}^{\infty}\frac{(-1)^n\sin\left[\left(n+\frac{1}{2}\right)\pi\frac{z_C}{t}\right]-1}{2n+1}\frac{J_1(\mu_{s,m}r_p)J_0(\mu_{s,m}r)\sin\left[\left(n+\frac{1}{2}\right)\pi\frac{z}{t}\right]}{\mu_{s,m}\left\{\mu_{s,m}^2+\left[\left(n+\frac{1}{2}\right)\frac{\pi}{t}\right]^2\right\}J_0^2(\mu_{s,m}r_W)}\end{aligned} \tag{35}$$

with the physical stipulation that for $s=0$ $z_C=0$. Thence the temperature is derived as:

$$T_s(r,z)=T_W\exp\left\{\frac{4P}{\pi^2 r_W^2 r_p t K}\sum_{m=1}^{\infty}\sum_{n=1}^{\infty}\frac{(-1)^n\sin\left[\left(n+\frac{1}{2}\right)\pi\frac{z_C}{t}\right]-1}{2n+1}\frac{J_1(\mu_{s,m}r_p)J_0(\mu_{s,m}r)\sin\left[\left(n+\frac{1}{2}\right)\pi\frac{z}{t}\right]}{\mu_{s,m}\left\{\mu_{s,m}^2+\left[\left(n+\frac{1}{2}\right)\frac{\pi}{t}\right]^2\right\}J_0^2(\mu_{s,m}r_W)}\right\} \tag{36}$$

Evidently the temperature grows exponentially with the dissipated heat density.

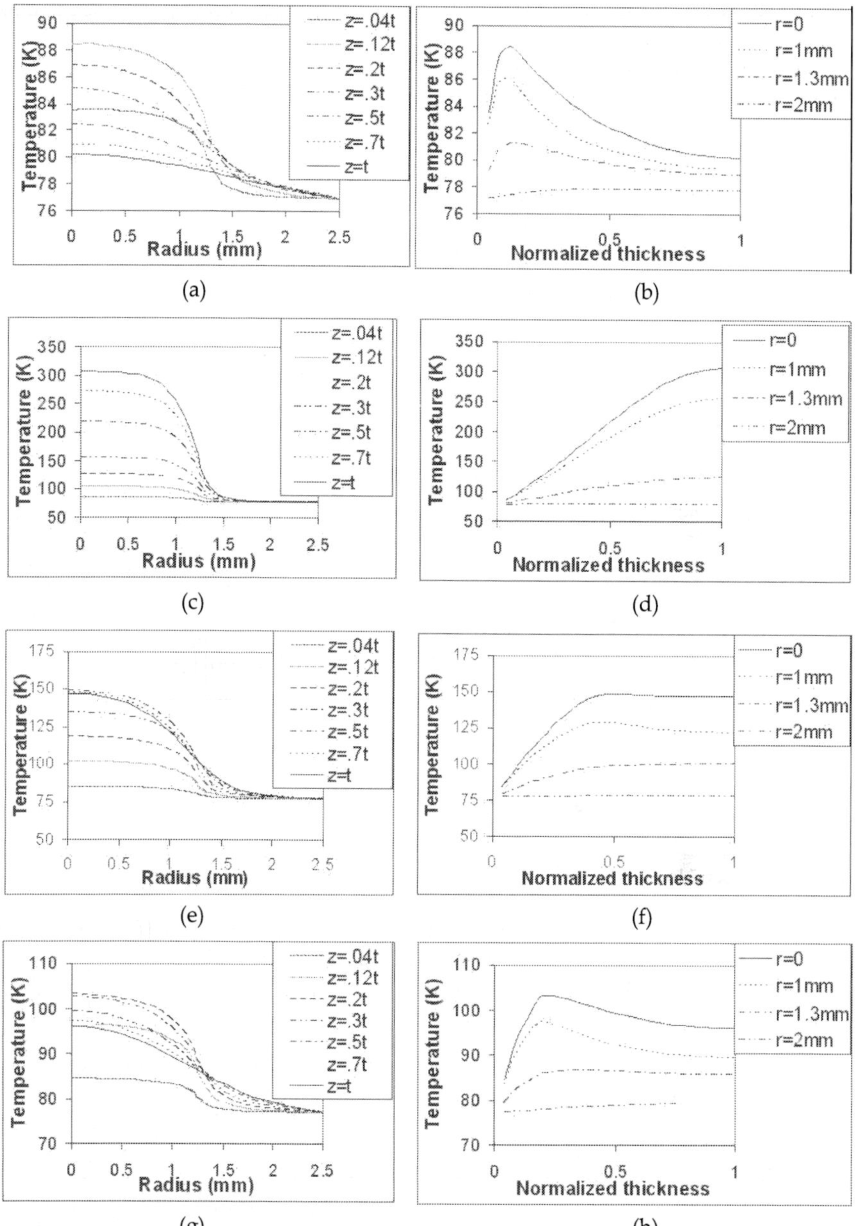

Fig. 3. Radial and axial temperature profiles in a thin disk assuming a cap length of 0 (a,b), 0.25 mm (c,d), 0.75 mm (e,f) and 1.25 mm (g,h)

2.4 Example for a thin disk

To calculate the temperature in the disk the following parameters are set: material Yb:YAG, disk diameter 5 mm, heat source diameter or thickness 2.5 mm, slab lengths 50 mm, disk thickness 0.25 mm, heat density rate 100 W/cm^3 and cryogenic cooling at 77K.

Consistent with most realistic scenarios consider end heating of the thin disk with uniform radial distribution within a radius of r_P of a disk attached to a heat sink. To enhance heat dissipation assumed is also a non-heating cap in good thermal contact with the disk and a heat sink around its perimeter. For simplicity the cap is considered possessing the identical values of coefficient of thermal conduction of the absorbing disk. Thus attached to the disk with the thickness of 0.25 mm is a cap with a varying thickness of 0, 0.25, 0.75 and 1.75 mm.

For a heat sink temperature of 77K, disk diameter of 5 mm, source diameter of 2.5 mm and power P=400 W the temperature axial and radial profiles are plotted in Figure 3. On inspecting the graphs (a) – (h) one finds a dramatic drop in the maximum temperature of above 300K to below 90K resulting from increasing the cap thickness from nil to 1.75 mm. Another interesting point is that the axial location of the hottest spot remains at roughly the doped disk facet. Next, the axial and radial profiles of Δn radial component are plotted in Figure 3. On inspecting the graphs (a) – (h) one finds a dramatic decrease in the Δn magnitude from 2.5×10^{-3} to below 4×10^{-5} with increasing the cap thickness. Also the axial location of the peak Δn remains at the doped disk facet.

3. Slab geometry

For active optical media having the shape of a parallelepiped it is most convenient to treat Eq. 5 in Cartesian coordinates, where consistent with Kirchoff's transformation (Joyce, 1975) it is expresses as:

$$K\left(\frac{\partial^2}{\partial x^2}+\frac{\partial^2}{\partial y^2}+\frac{\partial^2}{\partial z^2}\right)\theta(x,y,z)+Q(x,y,z)=0 \tag{37}$$

3.1 Slab of infinite length

The case of a very long slab relative to its lateral sides is well approximated by an infinitely long slab is schematically shown in Figure 4(a). Here the slab is held on one side in thermal contact with the heat sink, having its width in the z direction and thickness in x direction, both much shorter than its length in the y direction. Heating sources such as lasers, emit radiation which forms inside the slab a laser gain zone a fraction of which generates heat due to the quantum defect and other non-radiative relaxation processes.

Considered is a side-heated model having a slander spot shape on the slab side. It has predominantly a Gaussian intensity distribution in the fast-axis plane, i.e. along x, and a uniform distribution along y.

I choose to specify the boundary conditions as:

$$\left.\frac{\partial T(x,y)}{\partial y}\right|_{y=0}=\left.\frac{\partial T(x,y)}{\partial y}\right|_{y=b}=\left.\frac{\partial T(x,y)}{\partial x}\right|_{x=0}=0 \tag{38}$$

and:

$$T(a,y)=T_W \tag{39}$$

where T_W is the temperature of the heat-sink at the interface with the slab wall. In the notation of the non-dimensional temperature θ the two-dimensional Poisson equation becomes:

$$\left(\frac{\partial^2}{\partial x^2}+\frac{\partial^2}{\partial y^2}\right)\theta(x,y)+\frac{Q(x,y)}{K}=0 \tag{40}$$

with the boundary conditions thus converted to:

$$\left.\frac{\partial\theta(x,y)}{\partial y}\right|_{y=0}=\left.\frac{\partial\theta(x,y)}{\partial y}\right|_{y=b}=\left.\frac{\partial\theta(x,y)}{\partial x}\right|_{x=0}=0 \quad ; \quad \theta(a,y)=\theta_W \tag{41}$$

For a constant θ_W this allows a solution for the function $\theta(x,y)-\theta_W$ with a zero boundary value at x=a.
Assumed for this case is a heat source:

$$Q(x,y)=\frac{P\alpha}{2r_pL}\exp(-\alpha y)f_0(x) \tag{42}$$

where α is the absorption coefficient of the laser radiation, r_P is the small aperture of the radiation beam, *a*, *b* and *L* are the *x* , *y* and *z* dimensions of the slab ($a,b<<L$). If the slab side at $y=b$ reflects the inducing beam then the heat source becomes:

$$Q(x,y)=\frac{P\alpha\exp(-\alpha b)}{r_pL}\cosh\left[\alpha(b-y)\right]f_0(x) \tag{43}$$

The formal solution to the equation set 40 – 41 is:

$$\theta(x,y)=\theta_W+\frac{1}{K}\int_0^a\int_0^b Q(\xi,\eta)G(x,y,\xi,\eta)d\xi d\eta \tag{44}$$

where the Green's function construction is stipulated by solving the homogeneous Eq. 40, or the Laplace equation, by satisfying the boundary conditions and by its holding throughout the domain. It results in:

$$G(\xi,\eta,x,y)=\frac{4}{ab}\sum_{l=1}^{\infty}\sum_{m=0}^{\infty}\frac{\cos(p_ly)\cos(p_l\eta)\sin(q_mx)\sin(q_m\xi)}{p_l^2+q_m^2} \tag{45}$$

where:

$$p_l=\frac{\pi l}{b} \quad ; \quad q_m=\frac{\pi}{a}\left(m+\frac{1}{2}\right) \tag{46}$$

To present a complete solution one still needs to define the functions $f_0(x)$ used in Eq. 42. Let two cases be considered:

1. uniform heat source distribution in the *x* axis where:

$$f_0(x)=\begin{cases}1 & ;\ \frac{a}{2}-r_p\le x\le\frac{a}{2}+r_p\\ 0 & ;\ x<\frac{a}{2}-r_p\cap\ \frac{a}{2}+r_p<x\end{cases}$$

2. Gaussian heat source profile:

$$f_0(x) = \exp\left[-2\left(\frac{x - a/2}{r_P}\right)^2\right]$$

Considering case (1), Eq. 44 is integrated to the closed form solution:

$$\theta(x,y) = \theta_W + 4\frac{P\alpha^2}{r_P abLK}\sum_{l=0}^{\infty}\sum_{m=0}^{\infty}\frac{\left[1-(-1)^l e^{-\alpha b}\right]\sin\left(\frac{q_m a}{2}\right)\sin(q_m r_P)}{q_m\left(p_l^2+\alpha^2\right)\left(p_l^2+q_m^2\right)}\cos(p_l y)\sin(q_m x) \quad (47)$$

when no electromagnetic radiation is reflected from slab end, and to:

$$\theta(x,y) = \theta_W + 8\frac{P\alpha^2 \exp(-\alpha b)}{r_P abLK}\sum_{l=0}^{\infty}\sum_{m=0}^{\infty}\frac{\sinh(p_l b)\sin\left(\frac{q_m a}{2}\right)\sin(q_m r_P)}{q_m\left(p_l^2+\alpha^2\right)\left(p_l^2+q_m^2\right)}\cos(p_l y)\sin(q_m x) \quad (48)$$

when the electromagnetic radiation is fully reflected from slab end.
Then for case (2) the solution takes the form:

$$\begin{aligned}\theta(x,y) &= \theta_W - 4\sqrt{2\pi}\frac{P\alpha^2}{abLK}\sum_{l=1}^{\infty}\sum_{m=0}^{\infty}\frac{1-(-1)^l e^{-\alpha b}}{\left(p_m^2+\alpha^2\right)\left(p_l^2+q_m^2\right)}e^{-\frac{(q_m r_P)^2}{8}}\sin\left(\frac{q_m a}{2}\right)\\ &\times\left[\operatorname{erf}\left(\frac{iq_m r_P^2 - 2a}{2\sqrt{2}r_P}\right) - \operatorname{erf}\left(\frac{iq_m r_P^2 - 2a}{2\sqrt{2}r_P}\right)\right]\cos(p_l y)\sin(q_m x)\\ &\approx \theta_W - 8\sqrt{2\pi}\frac{P\alpha^2}{abLK}\sum_{l=1}^{\infty}\sum_{m=0}^{\infty}\frac{1-(-1)^l e^{-\alpha b}}{\left(p_m^2+\alpha^2\right)\left(p_l^2+q_m^2\right)}e^{-\frac{(q_m r_P)^2}{8}}\sin\left(\frac{q_m a}{2}\right)\cos(p_l y)\sin(q_m x)\\ &\times\left\{\operatorname{erf}\left(\frac{a}{r_P}\right) + \frac{e^{-\left(\frac{a}{r_P}\right)^2}}{2\pi\frac{a}{r_P}}\left(1-\cos(q_m a)\right) + \frac{2e^{-\left(\frac{a}{r_P}\right)^2}}{\pi}\sum_{n=1}^{\infty}\frac{e^{-\frac{1}{4}n^2}}{n^2+4\left(\frac{a}{r_P}\right)^2}\left[2\frac{a}{r_P}\left(1-\cosh\left(\frac{nq_m r_P}{2}\right)\cos(q_m a)\right)\right.\right.\\ &\qquad\left.\left. + n\sinh\left(\frac{nq_m r_P}{2}\right)\sin(q_m a)\right]\right\}\end{aligned} \quad (49)$$

where p_m is defined in Eq. 46. On inspecting Eqs. 47 and 49 one observes that for $a >> r_P$ the solution is insensitive to the magnitude of r_P.

3.2 Slab of finite length

The heat conduction problem in a slab of finite length is three dimensional, covering the cases illustrated by Figure 4(a) and (b), where a slab is held on one side in thermal contact with a heat sink and is either side heated (a), or end heated (b). The coordinate system is chosen to represent length in z axis and comparable width sizes in x and y axis. The governing equation is the Poisson equation in Cartesian coordinates, with respect to all three axes as given in Eq. 37. Several configurations of heat inductions are considered:

1) side heat induction in the y direction where the source has an oblong spot shape and a Gaussian intensity distribution in x, 2) side heat induction with alternating directions of $+y$ and $-y$, and 3) end heat source in the z direction having a circular spot shape and Gaussian intensity distribution. Also solved is a case for a heat source additional to that occurring in the heating path, which originates from fluorescence and photon trapping in the slab bulk and walls. Finally, several cases of slab cooling are considered: 1) contact with heat sink on one of the slab sides, 2) contact with heat sink on two of the slab sides, and 3) contact with heat sink on all four of the slab sides.

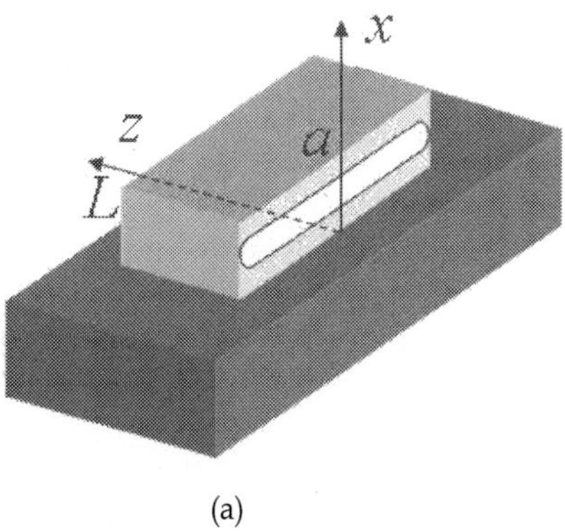

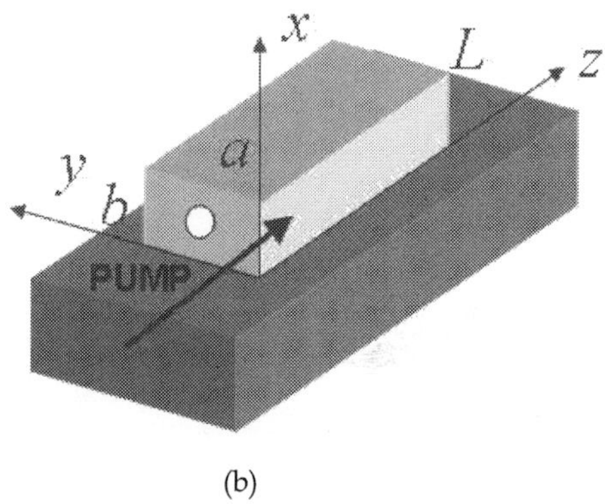

Fig. 4. Two parallelepipeds: a) slab attached to a heatsink on one side, with a slender side pump and, b) slab attached to heatsink on one side with an end pump

3.2.1 Side heating

In this case the crystal is considered three dimensional, in thermal contact with a heat sink on one side, thus there are five Neumann and a single Dirichlet type boundary conditions. They are expressed as:

$$\left.\frac{\partial\theta(x,y,z)}{\partial z}\right|_{z=0} = \left.\frac{\partial\theta(x,y,z)}{\partial z}\right|_{z=L} = \left.\frac{\partial\theta(x,y,z)}{\partial y}\right|_{y=0} = \left.\frac{\partial\theta(x,y,z)}{\partial y}\right|_{y=b} = \left.\frac{\partial\theta(x,y,z)}{\partial x}\right|_{x=0} = 0 \quad (50)$$

and: $\theta'(a,y,z) = \theta(a,y,z) - \theta_W = 0$

Consequently the Green's function becomes:

$$G(\xi,\eta,\zeta,x,y,z) = \frac{1}{ab}\sum_{n=0}^{\infty}\sum_{m=1}^{\infty}\frac{\cos(p_n x)\cos(p_n\xi)\cos(q_m y)\cos(q_m\eta)}{\beta_{nm}\cosh(\beta_{nm}L)}H_{nm}(z,\zeta) \quad (51)$$

where the function $H_{nm}(z,\zeta)$ is:

$$H_{nm}(z,\zeta) = \begin{cases}\cosh(\beta_{nm}\zeta)\cosh\left[\beta_{nm}(L-z)\right] & ; \quad L \geq z > \zeta \geq 0 \\ \cosh(\beta_{nm}z)\cosh\left[\beta_{nm}(L-\zeta)\right] & ; \quad L \geq \zeta > z \geq 0\end{cases} \quad (52)$$

and the coefficients p_n, q_m and β_{nm} are:

$$p_n = \frac{\left(n+\frac{1}{2}\right)\pi}{a} \quad ; \qquad q_m = \frac{m\pi}{b} \; ; \qquad \beta_{nm} = \sqrt{p_n^2 + q_m^2} \tag{53}$$

Further, allowing for a non-absorbing cap having thickness of b_C on the radiation side the heat source is defined as:

$$Q(x,y,z) = \begin{cases} 0 & ; \quad y < b_C \\ \dfrac{P\alpha}{2r_P L_P} \exp\left[-\alpha\left(y - b_C\right)\right] f_0(x,z) & ; \quad b_C \le y \le b \end{cases} \tag{54}$$

where L_P is the radiation length and the term f_0 defines the lateral profile of the source, assumed either as flat top or Gaussian:

1. flat top:

$$f_0(x,z) = \begin{cases} 1 \;\; ; \;\; 0 \le \left| x - \dfrac{a}{2} \right| \le r_P & \cap \qquad 0 \le z \le L_P \\ 0 \;\; ; \;\; \dfrac{a}{2} + r_P < x \;,\; x < \dfrac{a}{2} - r_P & \text{or} \qquad L_P < z \end{cases} \tag{55}$$

2. Gaussian:

$$f_0(x,z) = \begin{cases} 1 \;\; ; \;\; \exp\left[-2\left(\dfrac{x - a/2}{r_P} \right)^2 \right] & \cap \qquad 0 \le z \le L_P \\ 0 \;\; ; & L_P < z \end{cases} \tag{56}$$

The formal solution takes the shape:

$$\theta(x,y,z) = \theta_0 + \int_0^a \int_0^b \int_0^L Q(\xi,\eta,\zeta) G(x,y,z,\xi,\eta,\zeta)\, d\xi\, d\eta\, d\zeta \tag{57}$$

which after substituting the functions given in Eqs. 51 and 54 and carrying out the integration becomes:

$$\begin{aligned} \theta(x,y,z) = \theta_0 &+ \frac{P\alpha^2 e^{\alpha z_c}}{\pi r_P^2 abK\left\{1 - \exp\left[-\alpha\left(b - b_C\right)\right]\right\}} \\ &\times \sum_{n=0}^{\infty} \sum_{m=1}^{\infty} \frac{\left[1 - (-1)^m\right] \sin\left(p_n r_P\right) \cos\left(p_n \dfrac{a}{2}\right)}{p_n \left(\alpha^2 + q_m^2\right) \beta_{nm}^2} \cos\left(p_n x\right) \cos\left(q_m y\right) \\ &\times \left[\tanh \beta_{nm} L - \sinh \beta_{nm} \left(L - L_P\right) \frac{\cosh \beta_{nm} z}{\cosh \beta_{nm} L} \right] \end{aligned} \tag{58}$$

Observe that for r_P=$a/2$ the term $\sin(p_n r_P)\cos(p_n a/2)$ becomes $(-1)^n/2$, and for L=L_P the term in the square parenthesis becomes merely $\tanh \beta_{nm} L$ thus rendering the solution in Eq. 58 independent of z.

Often radiation sources are placed successively on both sides of a slab. In order to solve for this case one needs only to add a second source placed on the other side of the slab, whereby the governing equation and boundary conditions remain as stated above, however the heat source is defined as:

$$Q(x,y,z)=\begin{cases} 0 & ; \quad b-b_C<y<b_C \\ \dfrac{P}{2r_P L_P}\dfrac{\alpha\exp\left[-\alpha\left(b-b_C-y\right)\right]}{1-\exp\left[-\alpha\left(b-2b_C\right)\right]}f_0(x,z) & ; \quad b_C\le y\le b-b_C \end{cases} \tag{59}$$

where the term f_0 is assumed either flat top:

$$f_0(x,z)=\begin{cases} 1 \;;\; 0\le\left|x-\dfrac{a}{2}\right|\le r_P & \cap \quad L_P\le z\le 2L_P \\ 0 \;;\; \dfrac{a}{2}+r_P<x\;,\; x<\dfrac{a}{2}-r_P & \text{or} \quad z<L_P\;;\;2L_P<z \end{cases} \tag{60}$$

The solution to this problem becomes:

$$\begin{aligned}\theta(x,y,z)=\theta_0&+\frac{P\alpha^2e^{-\alpha B}}{2r_PL_PabK\left\{1-\exp\left[-\alpha\left(b-b_C\right)\right]\right\}}\\&\times\sum_{n=0}^{\infty}\sum_{m=1}^{\infty}\frac{\left[e^{\alpha B}\cos q_mB-e^{\alpha b_c}\cos q_mb_c\right]+\dfrac{q_m}{\alpha}\left[e^{\alpha B}\sin q_mB-e^{\alpha b_c}\sin q_mb_c\right]}{p_n\left(\alpha^2+q_m^2\right)\beta_{nm}^2}\\&\times\sin\left(p_nr_P\right)\cos\left(p_n\frac{a}{2}\right)\cos\left(p_nx\right)\cos\left(q_my\right)\\&\times\left[\tanh\beta_{nm}L-\sinh\beta_{nm}\left(L-2L_P\right)\frac{\cosh\beta_{nm}z}{\cosh\beta_{nm}L}-\sinh\beta_{nm}L_P\frac{\cosh\beta_{nm}\left(L-z\right)}{\cosh\beta_{nm}L}\right]\end{aligned} \tag{61}$$

The complete solution to the slab induced by two sources disposed successively on opposite sides of the slab is the combination of Eqs.58 and 61. Note that by this technique one can to arbitrarily add additional induction sources on the slab sides arriving at a closed form solution for each configuration. Finally, using this solution the temperature is again derived via Eq. 6.

3.2.2 End heating

Treating the problem of end heat induction is in fact exactly that as the above model for side heat induction however with changing the characteristics of the source. The heat source is now considered axisymmetrical with a Gaussian power distribution. In end heating the temperature is maximized at the slab facet. This surge may be relieved by bonding to the entrance facet of the slab a non-heating end-cap segment. Lacking a heat source yet being heat conductive contributes to removing heat from the slab facet. In the present model it is assumed that the coefficient of thermal conductivity in both materials is identical. For this case with the radiator aligned to the slab centerline the heat source term is rewritten as:

$$Q(x,y,z)=\begin{cases}0 & ; \quad z<z_C \\ \dfrac{P}{\pi r_P^2}\dfrac{\alpha\exp\left[-\alpha(z-z_C)\right]}{1-\exp\left[-\alpha(L-z_C)\right]}f_0(x,y) & ; \quad z_C\le z\le L\end{cases} \tag{62}$$

where f_0 is given by:

$$f_0(x,y)=\exp\left\{-2\left[\left(\frac{x-a/2}{r_P}\right)^2+\left(\frac{y-b/2}{r_P}\right)^2\right]\right\} \tag{63}$$

Again, the heat source in the slab can be expressed by Eq. 57, which after substituting the functions given in Eqs.51 and 54 and carrying out the integration becomes:

$$\begin{aligned}\theta(x,y,z)=\theta_0+&\frac{1}{4}\frac{P\alpha^2e^{\alpha z_c}}{abK\left\{1-\exp\left[-\alpha(b-b_C)\right]\right\}}\\ &\times\sum_{n=0}^{\infty}\sum_{m=1}^{\infty}F_n(p_n,a)F_m(q_m,b)\frac{\cos(p_nx)\cos(q_my)}{\beta_{nm}\cosh\beta_{nm}L}\\ &\times\frac{1}{\alpha^2-\beta_{nm}^2}\left[\cosh\beta_{nm}(L-z)-e^{-\alpha L}\cosh\beta_{nm}z-\frac{\beta_{nm}}{\alpha}e^{-\alpha z}\sinh\beta_{nm}L\right]\end{aligned} \tag{64}$$

where:

$$F_n(p_n,a)=\exp\left[-\frac{(p_nr_P)^2}{8}\right]\cos\left(\frac{p_na}{2}\right)\left\{\begin{aligned}&\operatorname{erf}\left(\frac{a}{\sqrt{2}r_0}\right)+\frac{2\sqrt{2}}{\pi}\frac{a}{r_0}e^{-\left(\frac{a}{2r_P}\right)^2}\\ &\times\sum_{k=0}^{\infty}\frac{e^{-\frac{k^2}{4}}}{k^2+2\left(\frac{a}{r_P}\right)^2}\left[1-\cos(p_na)\cosh\left(\frac{kp_nr_P}{2\sqrt{2}}\right)\right]\end{aligned}\right\} \tag{65}$$

Note that in the above equation the index n, coefficient p_n and length a can be replaced by m, coefficient q_m and length b, respectively. Next, the function F can be further simplified considering the values of the coefficients p_n and q_m as given in Eq. 53 provides the following values:

$$\begin{aligned}&\cos\left(\frac{p_na}{2}\right)=\frac{1}{\sqrt{2}},-\frac{1}{\sqrt{2}},-\frac{1}{\sqrt{2}},\frac{1}{\sqrt{2}},\ldots \quad \text{for } n=0,1,2,3,\ldots\\ &\cos(p_na)=0\\ &\cos\left(\frac{q_mb}{2}\right)=0,-1,0,1,\ldots \quad \text{for } m=1,2,3,4\ldots\\ &\cos(q_mb)=(-1)^m\end{aligned} \tag{66}$$

It should be mentioned that the F function is a close approximation to a combination of error functions of complex arguments as shown in (Abramowitz & Stegun, 1964).

3.2.3 Additional case – slab attached to a heat sink on two sides

Selecting next to carry out the solution for the case of a slab held in thermal contact with a heat sink on two adjacent walls, the boundary conditions are specified as:

$$\left.\frac{\partial T(x,y,z)}{\partial z}\right|_{z=0} = \left.\frac{\partial T(x,y,z)}{\partial z}\right|_{z=L} = \left.\frac{\partial T(x,y,z)}{\partial x}\right|_{x=0} = \left.\frac{\partial T(x,y,z)}{\partial y}\right|_{y=0} = 0 \qquad (67)$$

and:

$$T(a,y,z) = T(x,b,z) = T_W \qquad (68)$$

Relative to the dimensionless temperature θ the boundary conditions become:

$$\left.\frac{\partial \theta(x,y,z)}{\partial z}\right|_{z=0} = \left.\frac{\partial \theta(x,y,z)}{\partial z}\right|_{z=L} = \left.\frac{\partial \theta(x,y,z)}{\partial y}\right|_{y=0} = \left.\frac{\partial \theta(x,y,z)}{\partial x}\right|_{x=0} = 0 \qquad (69)$$

and: $\theta(a,y,z) = \theta(x,b,z) = 0$

The formal solution is given by Eq. 57 with the heat source Q given by Eq. 62. The Green's function is constructed according to that the rules explained in the previous section, thus becoming as in Eqs.51 and 52 however with the set of coefficients p_n, q_m and β_{nm} are

$$p_n = \left(n + \frac{1}{2}\right)\frac{\pi}{a} \quad : \quad q_m = \left(m + \frac{1}{2}\right)\frac{\pi}{b} \quad ; \quad \beta_{nm} = \sqrt{p_n^{\ 2} + q_m^{\ 2}} \qquad (70)$$

Because both walls attached to the heat sink are held at the uniform temperature of T_W, the solution is identical to that in Eqs.64 and 65 with the difference of the coefficient q_m and that the summation over m begins at 0.

3.2.4 Additional case – slab attached to a heat sink on four sides

In comparison with a slab in contact with a heat sink on two sides a surrounding heat sink on four sides is expected to further lower the temperature elevation due to heating. Unlike in the case where the slab is held by two sides, where the boundary conditions have four Neumann and two Dirichlet type conditions, for this case there are two Neumann and four Dirichlet type conditions expressed as:

$$\left.\frac{\partial \theta(x,y,z)}{\partial z}\right|_{z=0} = \left.\frac{\partial \theta(x,y,z)}{\partial z}\right|_{z=L} = 0 \qquad (71)$$

and: $\theta(0,y,z) = \theta(a,y,z) = \theta(x,0,z) = \theta(x,b,z) = 0$

One may approach a solution using symmetry considerations solving in the x and y coordinates instead of the domain $0 \rightarrow a$ and $0 \rightarrow b$, respectively, in the domain $0 \rightarrow a/2$ and $0 \rightarrow b/2$. Since in this case the heat source Q given by Eq. 62 is slightly modified to conform with the symmetry consideration such that the f_0 function is expressed as:

$$f_0(x,y) = \exp\left\{-2\left[\left(\frac{x}{r_p}\right)^2 + \left(\frac{y}{r_p}\right)^2\right]\right\} \qquad (72)$$

In this domain the solution is identical to the case of a slab held by a heat sink on two sides, for which the solution is given by Eq. 64 where the F_n function is expressed as:

$$F_n\left(p_n,a\right)=\exp\left[-\frac{\left(p_n r_P\right)^2}{8}\right]\left\{\begin{array}{l}\operatorname{erf}\left(\frac{a}{\sqrt{2}r_P}\right)+\frac{2\sqrt{2}}{\pi}\frac{a}{r_0}e^{-\left(\frac{a}{2r_P}\right)^2}\\ \times\sum_{k=0}^{\infty}\frac{e^{-\frac{k^2}{4}}}{k^2+2\left(\frac{a}{r_P}\right)^2}\left[1-\cos\left(p_n a\right)\cosh\left(\frac{k p_n r_P}{2\sqrt{2}}\right)\right]\end{array}\right\} \quad (73)$$

Further, the set of coefficients p_n, q_m and β_{nm} as expressed in Eq. 70.

3.3 Example for slab

To calculate the temperature in the slab the following parameters are set: material Yb:YAG, square slab side 5 mm, heat source diameter or thickness 2.5 mm, slab lengths 50 mm, heat density rate 100 W/cm^3 and cryogenic cooling at 77K.
Assumed in this case is a Yb:YAG slab side having heat induced by a flat top beam, with a reflecting opposite wall spaced by a varying width of 5, 10 and 20 mm from the first wall, where the slab height is 5 mm, its length is 50 mm and the radiation footprint is 2.5×50 mm. The assumed power is P=400 W and the slab is attached to a heat sink held at 77K by its wide side. For the three slab widths the vertical and horizontal temperature distributions are plotted in Figure 5(a) and (b). A considerable variation is found for the vertical temperature whereas horizontal the temperature remains nearly unchanged.

4. Concluding remarks

Closed form solutions to the nonlinear steady-state heat equation in a cylindrical rod, thin disk and parallelepiped are presented. Hinging on using Kirchoff's transformation, this solution is applicable to any material in which the coefficient of thermal conductivity is integrable in temperature. In turn, these solutions enable solving further the equation of elasticity and that of a propagating optical beam in an inhomogeneous medium where the refractive index is radially modulated.
Exemplary calculations are made for a Yb:YAG crystal with temperature spanning the range of 77K - 770K. It is shown that under the condition of large heating loads, say above 200 W/cm^3, predicting the temperature according to a linear approximation underestimates the temperature. Thus, in order to obtain accurate results one must use the nonlinear solution arrived at in this study. The predicted temperatures escalate rapidly with thermal loading reaching 120 K and almost 200 K on the optical path for a respective heating by 400 and 800 W, in a cryogenically cooled rod. If the cooling level is raised to room temperature, then by heating the rod at 400 W, the temperature at the center will exceed 700K.
Under the heat induction condition of heating at 500 W and cryogenic cooling the maximum temperature reached in a thin disk, having a thickness of ¼ mm, is less than 90K provided it has a cap at least seven times the disk thickness. Without the cap the maximum would exceed 300K. For a slab pumped with identical power and cryogenically cooled on one side the maximal temperature varies between 125K and 500K, depending on the slab width spanning the range from 20 to 5 mm, respectively.

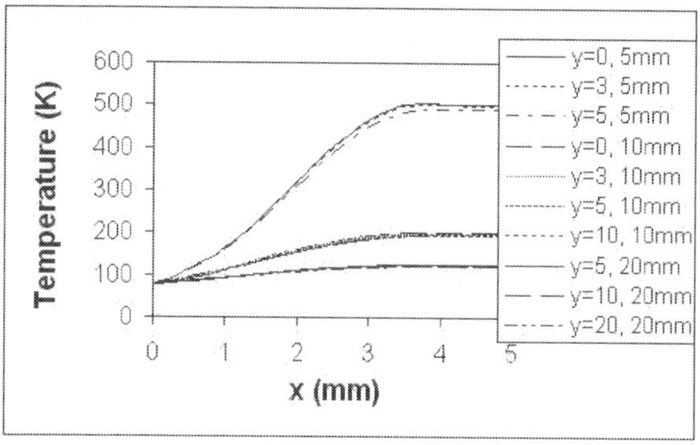

(a)

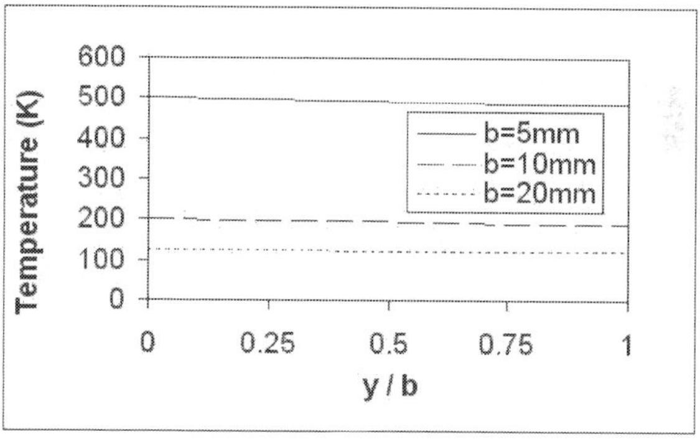

(b)

Fig. 5. Plot of characteristics in a slab: a) vertical temperature distribution, and b) horizontal temperature distribution

5. References

Abramowitz, M., and Stegun, I.A., (1964), *Handbook of Mathematical Functions with Formulas, Graphs, and Mathematical Tables*, p.299, Dover, New York.

Aggarwal, R.L. and Fan, T.Y., (2005), "Thermal diffusivity, specific heat, thermal conductivity, coefficient of thermal expansion, and refractive index change with temperature in $AgGaSe_2$ ", *Appl.Opt.*,44, (2005), 2673-2677.

Aggarwal, R.L., Ripin, D.J., Ochoa, J.R. and Fan, T.Y., (2005), "Measurement of Thermo-optic Properties of $Y_3Al_5O_{12}$, $Lu_3Al_5O_{12}$, $YAlO_3$, $LiYF_4$, $LiLuF_4$, BaY_2F_8, $KGd(WO_4)_2$, $KY(WO_4)_2$ Laser Crystals in the 80-300 K Temperature Range", *J.Appl.Phys.*98, (2005), 103514.

Hardman, P.J., Clarkson ,W.A., Friel, G.J., Pollnau, M. and Hanna, D.C., (1999) , "Energy-transfer upconversion and thermal lensing in high-power end-pumped Nd:YLF laser crystals", *IEEE J.Quant.Electron.*35, (1999), 647-655.

Joyce, W.B., (1975), "Thermal resistance of heat sinks with temperature-dependent conductivity", *Solid-State Electronics* 18, (1975), 321.

Pfistner, C., Weber, R., Weber, H.P., Merazzi, S., and Gruber, R., (1994) , "Thermal beam distortions in end-pumped Nd:YAG, Nd:GSGG and Nd:YLF rods", *IEEE J.Quantum Electron.*30, (1994), 1605-1615.

Polianin, A.D., (2002), *Handbook of linear partial differential equations for engineers and scientists,* Chapman & Hall/CRC.